SEA LICE BIOLOGY AND CONTROL

Edited by Jim Treasurer, Ian Bricknell and James Bron

5m Books

First published 2022

Published by
5M Books Ltd,
Lings, Great Easton,
Essex CM6 2HH, UK,
Tel: +44 (0)330 1333 580
www.5mbooks.com

A Catalogue record for this book is available from the British Library

ISBN 9781789181913
eISBN 9781789182194
DOI 10.52517/9781789182194

Book layout by Cheshire Typesetting Ltd, Cuddington, Cheshire
Printed and bound by CPI Group (UK) Ltd, Croydon, CR0 4YY
Photos by the authors unless otherwise indicated
Cover photos by James Bron, Ian Bricknell and by Argun Özak

Contents

Foreword

Sea lice emerged as a major health hazard for commercially farmed salmonids in Europe in the 1970s but their impact is now felt globally. Wherever marine finfish are taken into aquaculture, we now find that sea lice are becoming a significant problem. Although taxonomic information on sea lice had been slowly accumulating for over two centuries, and the salmon louse *Lepeophtheirus salmonis* had been described (as *Caligus salmonis*) by Henrik Nikolai Krøyer in 1837, little was known about sea lice biology in the 1970s and the need to devise methods for controlling infestation required a much better understanding of their biology. This stimulated a blossoming of research into all aspects of sea lice biology and in 1992 the time seemed ripe for an international workshop to bring researchers together to exchange their latest results and discuss novel approaches. Together with the late Andre Raibaut, I organised the first sea lice biology and control workshop in Paris, hosted during the First European Crustacean Conference. Over the course of two days we exchanged results and ideas about various aspects of sea lice biology and looked at methods for the control or management of sea lice outbreaks. A year later, and with financial support from the organisers of the host conference, the edited volume was published, containing 28 chapters, each looking at a different aspect of sea lice biology or control (Boxshall and Defaye (eds) (1993) *Pathogens of Wild and Farmed Fish: Sea Lice*). While the great majority of the papers focused on *Lepeophtheirus salmonis* (Krøyer, 1837) or used it as the test species, other sea lice species were considered including *Caligus elongatus* Nordmann, 1832, also found in salmonid farms around the North Atlantic. Interestingly, there were also two papers on *Caligus* species from marine and brackish finfish in South East Asia.

In the intervening three decades since that first international workshop, it has become clear that numerous species of sea lice are implicated. The major impact on commercial salmonid aquaculture still arises from infection by *Lepeophtheirus salmonis* and *Caligus elongatus* in the Northern Hemisphere, and by *C. rogercresseyi* Boxshall & Bravo, 2000 in the Southern Hemisphere. However, about 40 species of *Lepeophtheirus* and *Caligus* have been reported from finfish cultured in brackish and marine waters globally, and additional species are being found almost every year. While examining cultured rabbitfish (family Siganidae) and groupers (family Serranidae, subfamily Epinephelinae) at a parasitology workshop in Bali, Indonesia, I was struck by the massive numbers of sea lice of both of these genera infesting the fish. The live fish were from a recently established facility, and it was clear that problems with sea lice can emerge within a short time of any new species of finfish being taken into culture. Sea lice are truly a global problem in marine finfish aquaculture.

I always seek to remind sea lice specialists that sea lice are just one of more than 200 families of copepods, and that the majority of known copepods are free-living, found either up in the plankton or down in the sediments. Comparative biology has always been a hugely productive approach to biological research and our understanding of sea lice biology has benefitted enormously from rapid progress in research on various aspects of the biology of free-living copepods. For example, investigations into the reproductive behaviour of marine planktonic copepods have stimulated a productive line of research into the role of semiochemicals and pheromones in sea lice (see Chapter 6). Similarly, for nearly four decades the sea louse genus *Lepeophtheirus* was the odd one out, as the only copepod on Earth purported to have more than six post-naupliar stages in its life cycle. Comparative studies of development of the antennules across multiple orders of free-living and parasitic copepods finally provided a developmental criterion that allowed the errors in interpretation to be uncovered (see Chapter 2). All sea lice have only six post-naupliar stages.

Research on sea lice appeared to slow down from the late 1990s into the 2000s and, personally, I think this was due in part to complacency in the salmon farming industry: the use of in-feed medications as an anti-louse treatment had been so effective that the case for supporting continued research was harder to make. Subsequent emergence and rapid spread of drug resistance to these medications by sea lice populations shattered the illusion that sea lice were under control. Understanding the evolution of resistance in sea lice has required a new set of tools and the advent of molecular methods has brought about a step change in our ability to analyse and explore sea lice genetics, as well as many other aspects of their biology.

This volume covers an impressive array of topics, guiding the community on their journey towards awareness of the most recent advances in sea lice biology and through to the latest genomics studies. In parallel, it looks at the various methods employed to control sea lice and to manage sea lice outbreaks by the latest husbandry practices, as well as outlining the legislative framework as it relates to monitoring, regulation and accreditation. It will be an essential reference for anyone interested in sea lice and the problems they cause.

Prof. Geoff Boxshall FRS

PART I

Sea lice biology

CHAPTER 1

Introduction: a developing industry and the threat of sea lice

Jim Treasurer[1], Ian Bricknell[2] and James Bron[3]

[1]FAI Farms, UK
[2]Aquaculture Research Institute and School of Marine Sciences, University of Maine, USA
[3]Institute of Aquaculture, University of Stirling, UK

1.1 Introduction

This book seeks to provide an up-to-date overview of the biology and management of sea lice and of the approaches taken to regulate activities associated with their control. The book deals primarily with the species impacting Atlantic salmon *Salmo salar* L. 1758 and other salmonids, and largely focuses on the narrowly host-specific salmon louse *Lepeophtheirus salmonis* (Krøyer, 1837), the medium host-specificity Pacific species *Caligus rogercresseyi* (Boxshall & Bravo 2000) and the low host-specificity *Caligus elongatus* (von Nordmann 1832), which infects more than 100 host species (see Chapter 2, this volume). Following a period of intensive research on sea lice over the last three decades, aimed at protecting the health and welfare of both farmed and wild fish, our knowledge of these sea lice species has increased substantially, including our knowledge of host and environmental interactions.

During this period there has been industry, academic and government focus on the monitoring and management of this group of parasites. However, there has also been intense public awareness and interest, often associated with negative perceptions of the aquaculture industry. One issue of particular concern has been the suggestion that sea lice could be a contributing factor in declines of wild salmonids (Fig. 1.1), although it is clear that the timeline of decline often fails to match that of aquaculture growth and that the observed declines have been associated with a much wider range of anthropogenic, biological and environmental factors of which sea lice represent only one possible component. Additional threats include climate change and increasing water temperatures; diseases such as *Gyrodactlylus salaris* (Malmberg, 1957); invasive species, such as rainbow trout *Oncorhynchus mykiss* (Walbaum, 1792); genetic impacts from restocking and farm escapees; predators such as seals and cormorants; habitat degradation; barriers to migration; over-exploitation; and pollution *inter alia*.

Salmonids, and especially Atlantic salmon, are held in high regard in public consciousness both in terms of their role as a valued target for recreational fishing and in terms of their being a favoured food species whose consumption has a long history, going back to at least 11,500 YBP for chum salmon (Halffman et al., 2015). Salmon are further regarded as an iconic species, carrying important cultural significance and serving as symbols of the natural world and the health of the oceans. They

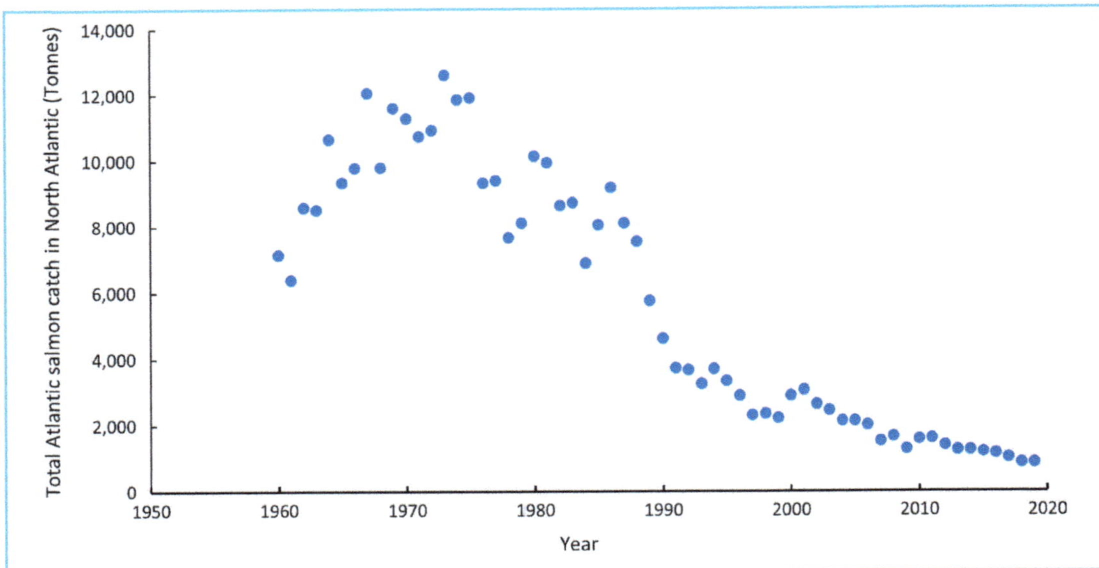

Fig. 1.1 The Atlantic salmon catch (tonnes) by all methods in the North Atlantic to 2019 (NASCO, 2020). These are fish captured in rivers, estuaries and within 12 nautical miles of the coasts. Credit: NASCO (2020).

also command admiration from the public because of their striking appearance, but also due to a life history that involves fighting up rivers to spawn, evading predators and swimming long distances to feeding grounds. The links, both real and perceived, between sea lice and the health and welfare of farmed and wild salmonids have driven a wealth of research, fostering the growth of many groups specialising in studies that serve to improve our understanding of their biology and control. The urgent need to provide solutions to the problems of sea lice management has also produced a surge of industrial innovation, particularly over the last decade.

The global salmon aquaculture industry has been estimated to generate $15.4 billion in global annual production in 2018 (Anon, 2018, 2021), not including job creation and wider economic contributions. By way of an example, farmed Atlantic salmon represents Scotland's leading food export, and generates over £2 billion of turnover for the Scottish economy (Marsh, 2018). In contrast, the overall economic cost of sea lice to the global Atlantic salmon aquaculture industry, including the value of direct impacts on production and the costs of sea lice control and other impact mitigation activities, has been estimated to amount to around $1 billion per annum globally (Brooker et al., 2018). For this reason, in addition to the considerable research spend from industry, large proportions of national fish and aquaculture research funding are also allocated to improving sea lice control and gaining a better understanding of their biology.

1.2 Rise of aquaculture

Salmon farming commenced in Norway in the 1960s with a harvest of a few hundred tonnes. The global harvest had increased to 2.5 million tonnes in 2019 (Fig. 1.2). In that year, the catch of wild salmonids, mainly the species chum, pink and sockeye salmon, was one third of aquaculture

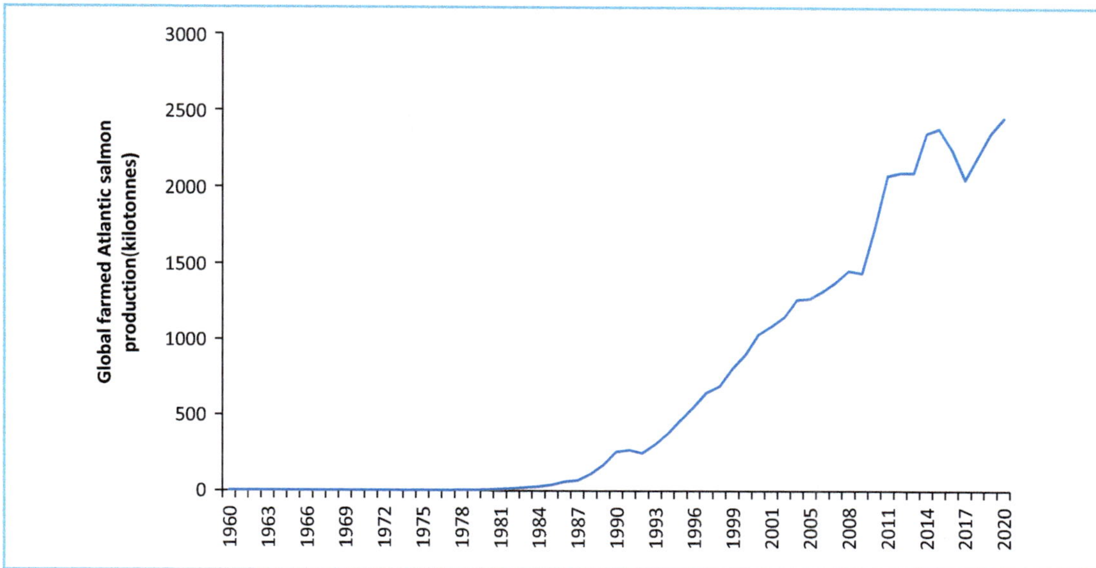

Fig. 1.2 Global farmed Atlantic salmon production (kilotonnes) to 2020. Figures as thousands of tonnes GWT. Data source Kontali Analyse, in MOWI (2021). Credit: Mowi Farming Industry Handbook (2021).

production (Mowi, 2021). Farming is confined to those areas where biological and temperature requirements are met and therefore farms for Atlantic salmon are located in Norway, Chile, the UK, USA, Canada, Iceland, the Faroe Islands, Ireland, Russia, New Zealand and Tasmania.

Prior to the commencement of salmon farming in Norway, sea lice received little public attention and, apart from isolated epizootics in river estuaries (for example Calderwood, 1905; White, 1940, 1942), were not perceived to be a threat to salmonids. Rather, sea lice have been seen as an indicator of a fresh run salmon. Nevertheless, an increase in sea lice infection of farmed salmon was seen from an early date and medicines such as Neguvon and dichlorvos were used to treat salmon infected with sea lice (Rae, 2002). High sea lice numbers have, however, been recorded even in those areas with low farmed salmon production (Costello, 2006). As the industry began to expand, fish had to be treated with medicines more regularly, and resistance to medicines, or more accurately reduced sensitivity, was soon observed (Pike and Wadsworth, 2000). As a consequence, low levels of sea lice clearance became more commonplace, with associated health issues and an increasing frequency of treatments following. The salmon farming sector, first in Scotland and then in Ireland and Norway in 1994, decided that alternative preventative and management measures were required. This included implementation of management measures such as single year classes and common stocking plans (Bron et al., 1993a). Fallowing and the instigation of management agreements were additional essential measures employed to break the cycle of sea lice infection and to improve sea lice control. Many novel medicines were explored (Roth et al., 1996), and the use of cleaner fish was also adopted (Bjordal, 1988, 1991).

In parallel with the growth of salmon farming it was suggested that there was evidence of related declines in salmon and sea trout stocks (Butler, 2002; Costello, 2009; Whelan, 2010; Krkošek et al., 2013), although limited evidence was presented of cause and effect. Scientific studies followed to verify this case and these became part of a political debate concerning the sustainability of salmon farming. Regardless of the sides of the debate there is no doubt that both aquacultured and wild salmonid populations often share the same coastal areas and river estuaries and hence the same parasite populations. Today, sea lice are the aquaculture industry's main infectious disease problem, affecting fish health and welfare, constraining productivity and capacity for expansion and impacting public perception of the industry. For all these reasons the industry has been the principal driver of research and innovation in the area of sea lice control.

Potential for impacts on wild salmonid stocks has led to criticism of the industry by an organised campaign involving activists, wild fisheries and environmentalists who have led very vocal campaigns. A contributing factor has been the established decline of Atlantic salmon and sea trout in certain areas on the west coasts of Ireland and Scotland (Butler, 2002). The factors affecting the decline in numbers of wild salmonids are complex and numerous. However, salmon farming has taken flak as a highly visible presence near the estuaries and approaches to salmon rivers in remote areas where wild fisheries are important to the tourist economy. Unravelling the respective importance and contribution of the various factors that impact salmonids through their life cycle is, however, difficult. Information is available in the case of sea trout, for example in Scotland (Fig. 1.3) (Marine Scotland/DAFS early records, 2021a, 2021b), that fish numbers were in decline from at least 1950. The 1950s and 1960s were times of major expansion in the net and line caught fishery for sea trout (Fig. 1.3(b)) and the numbers of fish caught increased by more than two and a half times the levels of 1950. The decline in catches occurs immediately after this period of increased fishing effort. Also, the sea trout is the same species as brown trout but with differing life histories and it may be that the behaviour and migratory patterns of the fish have changed in response to environmental changes, water chemistry, hydrology and food supply. In many ways the sea trout of the west coast of Scotland and Ireland may be more vulnerable than salmon as the fish migrating in spring enter inshore waters which they utilise as a nursery ground rather than heading for the open sea (Jackson, 2011; Moore et al., 2018). As the young sea trout inhabit shallow water this may make them more vulnerable to surface and wind driven concentrations of infective copepodids (McKibben and Hay, 2004; Murray and Gillibrand, 2006; Middlemas et al., 2009).

The adverse publicity derived from sea lice issues poses a major threat to the salmon farming industry, providing an excuse for objectors to block planning applications for new sea farm sites. Sea lice and impacts on wild fish are often, for instance, cited as a reason for restraint of the expansion of the salmon farming industry (Jansen et al., 2012), for example in British Columbia, Norway and Scotland. This restricts the growth of the industry and is a barrier to an uplift in the harvested tonnage. In Norway mitigating steps have to be taken if biomass is to be increased, or if planning applications are made for new farms. Examples of mitigating steps include the siting of shellfish farms nearby, the use of new technologies such as snorkel cages (Oppedal et al., 2019) or full containment systems, or the relocation of some farms offshore as in Scotland.

Industry concerns for farmed salmon welfare and protection and the potential for threats to wild fish stocks have led to a broad range of studies aimed at better understanding aspects of the biology, behaviour and host interactions of sea lice. There have been two principal avenues taken to improve sea lice control, the first comprising research into the biology of sea lice and the second looking to develop technological and farm management approaches to controlling lice, for example wellboat treatments (Jimenez et al., 2012; Shinn and Bron, 2012) and cage skirts (Oppedal et al., 2019). These research avenues are complementary and have led to huge advances in our understanding of sea lice and in the range of tools available to assist their management and control.

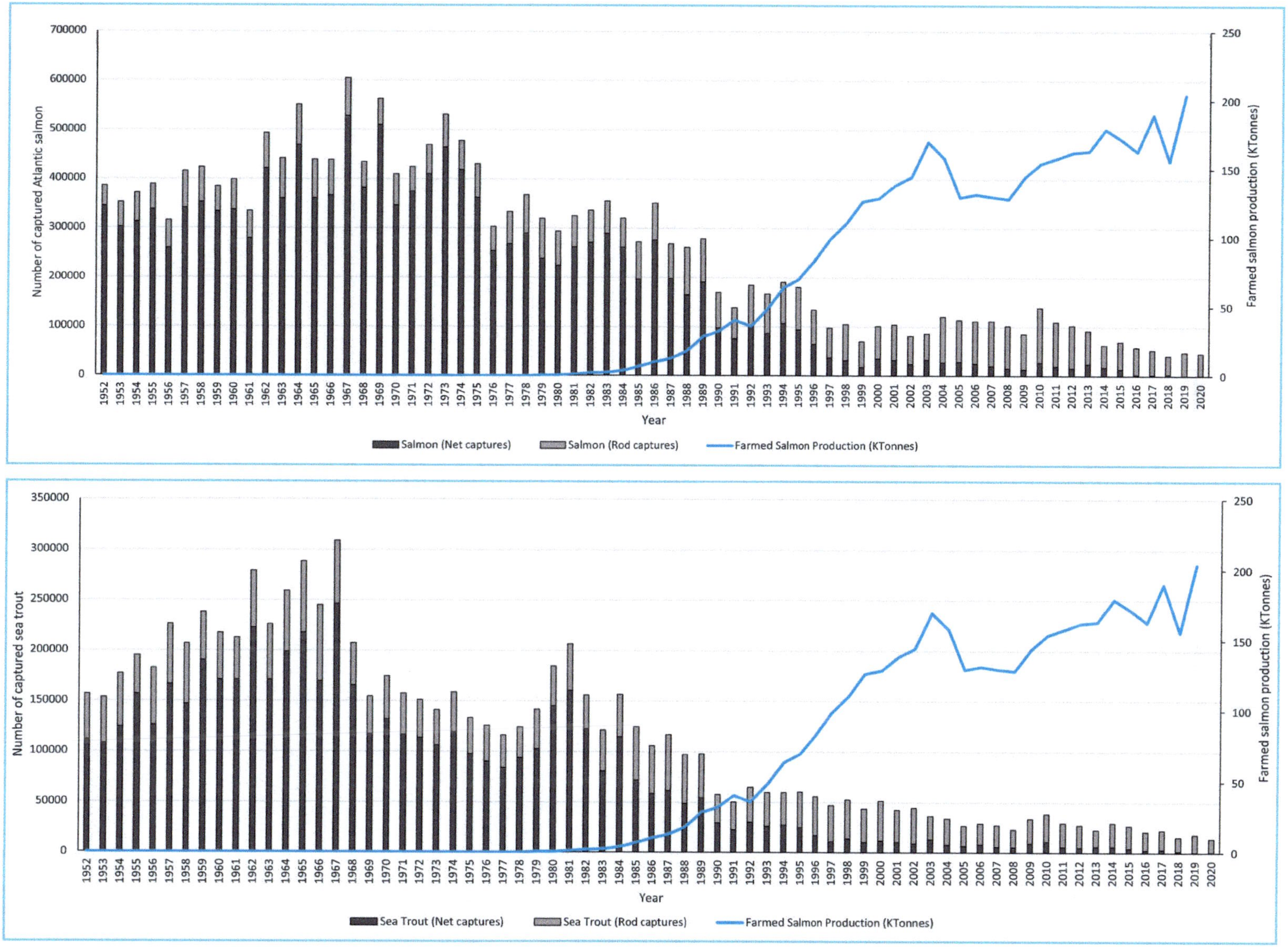

Fig. 1.3(a) and 1.3(b) Comparison of annual Scottish sea trout and Atlantic salmon capture fisheries catches with farmed salmon production 1952–2020. Data sourced from Marine Scotland (2021a, 2021b).

1.3 Statutory controls

Sea lice issues and treatments have involved many regulators including medicine directorates, environmental regulators, aquaculture and health inspectors, and latterly also fish quality and sustainable aquaculture labelling such as those associated with Freedom Foods and Aquaculture Sustainability Council (ASC). Therefore, to maintain good fish health and public and regulatory scrutiny, various controls have been placed on the industry. These are enforceable by government, for example by Marine Scotland. Most regions have defined upper thresholds for mean lice numbers on farmed fish, often differing inside and outside peak periods of wild salmonid migration, and with mandated responses to exceeded thresholds. To monitor numbers there is now frequently a requirement to carry out sea lice counts weekly and to report these within a week of the count. In several regions sea lice counts are undertaken by members of government-associated counting teams as in Ireland (Jackson, 2011), the Faroe Islands (Fiskaaling) and Chile (Sernapesca).

1.4 Management approaches and national treatment strategies

Increased collaboration between aquaculture companies and the instigation of area management agreements and national treatment plans have been a feature of more recent approaches to sea lice control. The first such approaches involved management agreements (Bron et al., 1993a; Grant and Treasurer, 1993). These involved all the farming companies in one area or loch/fjord system collaborating on sea lice control, with synchronisation of treatments, exchange of information, agreement on control strategies and use of single year class stocking and fallowing as part of the production cycle (Bron et al., 1993a; Grant and Treasurer, 1993). This strategy depended on a fallow period duration of at least six weeks to help break the infective cycle and also a separation of sea lice management areas based on a hydrographic model and certainly separation of at least 10 km between area management zones based on tidal flow and currents. This initiative was followed by further intercompany collaboration on a national basis, and these led to the formulation of national treatment strategies. It has often been more difficult to introduce measures such as fallowing over groups of farms in Chile and Canada due to the proximity of farms and lack of distinct loch/fjord areas, which makes it easier to fallow and stock farms in distinct fjords.

Now all countries have some form of national treatment plan. The main principle behind the national plan is to have synchronous treatments at certain times of the year and when sea lice numbers exceed a set amount, for example 0.5 adult females in the case of Norway, and also in Scotland from January to July. This approach leverages the general pattern of sea lice population dynamics, making use of a natural lull in sea lice infection due to reduced water temperature and development times of sea lice, which are temperature dependent (Johnson and Albright, 1991a,b). Lower temperatures and salinities in fjords serve to reduce sea lice infestation levels over the winter period.

1.5 Integrated pest management

Integrated pest management (IPM) can be defined as the use of diverse methods to control a pest species (Mordue and Pike, 2002) and such a strategy was first proposed for sea lice control by

Sommerville (1995). Following this approach, the industry evolved from a situation where sea lice control was largely reliant on sea lice medicines (Bron et al., 1993b), to one that relied on use of a broad selection of tools including new farm management approaches, functional feeds, cleaner fish, mechanical removal of lice and barrier methods for exclusion of infective copepodid stages. Preventative management approaches included fallowing, bay management, single year class stocking and coordinated treatments (Barrett et al., 2020). In later years, from 1989, the use of cleaner fish was introduced (Bjordal, 1988). Novel barrier methods now include the use of snorkel cages (Oppedal et al., 2019) to keep salmon at a lower level in the water column so as to reduce infection by copepodids that favour surface water layers. Skirts enclosing the upper collar of pens have also been used to exclude copepodids from entering pens (Grontvedt et al., 2018). Given the rise in resistance to medicines, physical methods of sea lice removal have been commonly introduced from 2015 involving thermolicers, hydrolicers and SkaMic licers to treat salmon with sea lice. This rise in physical control methods from 2015 to 2018 saw them increase from 5% to 80% of all treatments, leading to a large reduction in medicine use in Norway (see Chapter 15). This pattern has also been followed in Scotland. It is worth noting, however, that there have been welfare issues related to use of some physical methods as these may be of themselves stressful or injurious to fish. For example, temperatures may typically reach 28°C for the thermal treatments but also above 30°C on occasion (Gismervik et al., 2019). Physical abrasion, along with scale and mucus loss, may also occur with other physical methods or during crowding and pumping activities associated with treatments.

1.6 Research on sea lice in relation to salmon farming

Hastein and Bergsjo (1976) and Wootten et al. (1982) gave the first descriptions of pathology of farmed salmon caused by sea lice. Wootten et al. also examined the origin of the various medicinal control treatments. With sea lice control remaining a high priority for the industry, academics and universities in several countries have continued to work closely with industry and have applied their research to improving control of sea lice. Allied to this, companies also employ their own inhouse personnel, veterinarians and technical staff to support and develop sea lice treatment techniques, logistical approaches and equipment. Many specialist companies have developed novel technologies, including skirts, snorkel cages and lighting systems to drive down sea lice infection. The use of wellboats for treatment, while expensive, has been a major innovation, with specialist services allowing fish to be pumped on board wellboats to be treated with freshwater or hydrogen peroxide and other medicines at defined concentrations or to undergo physical treatments. Finally, after treatment, fish are transferred from the wellboat back to pens and the sea lice are filtered from the effluent water. In a recent system designed to remove treatment residue before water discharge, effluent from fish treatment in wellboats is processed and made safe prior to discharge.

Sea lice research, largely from the 1990s onwards and following the rapid growth of the salmon farming industry, has been aided by commercial support and the input of funding from national research bodies targeted at supporting the growth of aquaculture. Trade representative bodies have also played a key role in providing technical support to industry, by coordinating efforts to develop new treatment medicines and methods, and in providing and leveraging funding for sea lice research. For example Scottish Salmon Producers' Organisation (SSPO), the Scottish Aquaculture Research Forum and latterly the Sustainable Aquaculture Innovation Centre in the UK. International collaboration has been aided by a series of international conferences, the sea lice conferences, generally now held biannually. The first was initiated at the First Crustacean Conference in Paris in 1992 with the presentations at the sea lice conference being published in the book *Pathogens of Wild and Farmed Fish: Sea Lice*, edited by Boxshall and Defaye (1993).

The mix of academic research and applied research and development (R&D) has ensured that research has had commercial lead and encouragement. This has allowed researchers direct access to production salmon farms and the use of these for assessing new medicines and control measures and also, in some cases, for sourcing epidemiological data. In parallel, professional bodies such as the European Association of Fish Pathologists (EAFP) and the Fish Veterinary Society (FVS), for example, although both having a wider remit for fish health issues, have assisted greatly in the networking of sea lice researchers and the publishing of recent advances in sea lice biology and treatment. Now, much of the promotional dissemination work on technical advances and publications is included in the fish farming press, for example *The Fish Site*, *Fishfarmingexpert* and *Fish Farmer*.

Since the first text on this subject by Boxshall and Defaye (1993), there have been considerable advances in knowledge of sea lice biology and methods of control, including those achieved through new lines of science. Notable advances have been made in areas that include molecular biology of drug resistance, epidemiology and mathematical modelling of lice populations, sea lice genomics, wild/farmed fish interactions with sea lice, integrated pest management for sea lice, semiochemical interactions and the parasite's manipulation of the immune system. This new book summarises the current state of knowledge of sea lice across a number of key areas and is intended to follow the Boxshall and Defaye text as a further key volume in the field. It includes up-to-date chapters that present the latest developments across a wide scope of sea lice related knowledge and research. We hope that you enjoy the read!

Finally, we would like to thank over 60 authors who have contributed to this book. Thanks also to Prof Geoff Boxshall for the welcoming Foreword. We should like to thank Jeremy Toynbee and Alessandro Pasini at 5m Books and Sarah Binns, Jodie Marley and Joanne Phillips for their editing, proofreading and indexing expertise. Also, thanks to the colleagues, friends and families who have supported and encouraged us in our work on sea lice over many years. We also remember our absent friends at this time, including Peter Andreas Heuch, Thomas Schram, George Gettinby and Zbigniew 'Bob' Kabata, who together contributed so much to sea lice research and biological understanding and helped to build today's strong community of sea lice researchers and engaged stakeholders.

References

Anonymous (2018) International Salmon Farmers Association report (2018) Salmon farming sustaining communities and feeding the world. Available at: http://www.salmonfarming.org/cms/wp-content/uploads/2018/06/ISFA-JUNE-2018-REVISED-FOR-FOR-WEB1.pdf (accessed 8 November 2021).

Anonymous (2021) *Species Directory Atlantic Salmon (Protected)*. Washington DC: NOAA. Available at: https://www.fisheries.noaa.gov/species/atlantic-salmon-protected (accessed 1 September 2021).

Barrett, L., Oppedal, F., Robinson, N. and Dempster, T. (2020) Prevention not cure: a review of methods to avoid sea lice infestations in salmon aquaculture. *Reviews in Aquaculture* 1–17.

Bjordal, Å. (1988) Cleaning symbiosis between wrasse (Labridae) and lice infested salmon (*Salmo salar*) in mariculture. ICES Mariculture Committee 1988/F17.

Bjordal, Å. (1991) Wrasse as cleaner fish for farmed salmon. *Progress in Underwater Science* 16, 17–28.

Boxshall, G.A. and Defaye, D. (eds) (1993) *Pathogens of Wild and Farmed Fish: Sea Lice.* Ellis Horwood, Chichester, UK.

Bron, J.E., Sommerville, C., Wootten, R. and Rae, G.H. (1993a) Fallowing of marine Atlantic salmon, *Salmo salar* L, farms as a method for the control of sea lice, *Lepeophtheirus salmonis* (Krøyer, 1837). *Journal of Fish Diseases* 16, 487–493.

Bron, J.E., Sommerville, C., Wootten, R. and Rae, G.H. (1993b) Influence of treatment with dichlorvos on the epidemiology of *Lepeophtheirus salmonis* (Krøyer, 1837) and *Caligus elongatus* Nordmann, 1832 on Scottish salmon farms. In: Boxshall, G.A. and Defaye, D. (eds) *Pathogens of Wild and Farmed Fish: Sea Lice.* Ellis Horwood, Chichester, UK. pp. 263–275.

Brooker, A.J., Papadopoulou, A., Gutierrez, C., Rey, S., Davie, A. and Migaud, H. (2018) Sustainable production and use of cleaner fish for the biological control of sea lice: recent advances and current challenges. *Veterinary Record* 183, 383.

Butler, J. (2002) Wild salmonids and sea louse infestations on the west coast of Scotland: sources of infection and implications for the management of marine salmon farms. *Pest Management Science* 58, 595–608.

Calderwood, W.L. (1905) The "white spot" affecting salmon in the Island of Lewis. *Annual Report Fisheries Board of Scotland* 24, 77–79.

Costello, M.J. (2006) Ecology of sea lice parasitic on wild and farmed fish. *Trends in Parasitology* 22, 475–483.

Costello, M.J. (2009) The global economic cost of sea lice to the salmonid farming industry. *Journal of Fish Diseases* 32, 115–118.

Gismervik, K., Gasnes, S.K., Gu, J., Stien, L.H., Madaro, A. and Nilssen, J. (2019) Thermal injuries in Atlantic salmon in a pilot laboratory trial. *Veterinary and Animal Science* 8, 100081.

Grant, A. and Treasurer, J.W. (1993) The effects of fallowing on caligid infestation on farmed Atlantic salmon (*Salmo salar* L.) in Scotland. In: Boxshall, G.A. and Defaye, D. (eds) *Pathogens of Wild and Farmed Fish: Sea Lice.* Ellis Horwood, Chichester, UK. pp. 255–260.

Grontvedt, R.N., Kristoffersen, A.B. and Jansen, P.A. (2018) Reduced exposure of farmed salmon to salmon louse (*Lepeoptheirus salmonis* L.) infestation by use of plankton nets: estimating the shielding effect. *Aquaculture* 495, 865–872.

Halffman, C.M., Potter, B.A., McKinney, H.J., Finney, B.P., Rodrigues, A.T., Yang, D.Y. and Kemp, B.M. (2015) Early human use of anadromous salmon in North America at 11,500 y ago. Available at: https://www.pnas.org/content/112/40/12344 (accessed 8 November 2021).

Hastein, T. and Bergsjo, T. (1976) The salmon lice *Lepeophtheirus salmonis* as a cause of disease in farmed salmonids. *Rivista Italiana di Piscicoltura e Ittiopatologia* A11, 3–4.

Jackson, D. (2011) Ireland: the development of sea lice management methods. In: Jones, S. and Beamish, R. (eds) *Salmon Lice: an Integrated Approach to Understanding Parasite Abundance and Distribution.* Wiley, Chichester, UK. pp. 177–203.

Jansen, P.A., Kristoffersen, A.B., Viljugrein, H., Jimenez, D., Aldrin, M. and Stien, A. (2012) Sea lice as a density-dependent constraint to salmonid farming. *Proceedings Biological Sciences* 22, 279.

Jimenez, D.F., Heuch, P.A., Revie, C.W. and Gettinby, G. (2012) Confidence in assessing the effectiveness of bath treatments for the control of sea lice on Norwegian salmon farms. *Aquaculture* 344, 58–65.

Johnson, S.C. and Albright, L.J. (1991a) The developmental stages of *Lepeophtheirus salmonis* (Krøyer, 1837) (Copepoda: Caligidae). *Canadian Journal of Zoology* 69, 929–950.

Johnson, S.C. and Albright, L.J. (1991b) Development, growth and survival of *Lepeophtheirus salmonis* (Krøyer, 1837) (Copepoda: Caligidae) under laboratory conditions. *Journal of the Marine Biological Association UK* 71, 425–436.

Krkošek, M., Revie, C.W., Gargan, P.G., Skilbrei, O.T., Finstad, B. and Todd, C.D. (2013) Impact of parasites on salmon recruitment in the northeast Atlantic Ocean. *Proceedings of the Royal Society B: Biological Sciences* 280(1750), 20122359.

Marine Scotland (2021a) Salmon and Sea Trout fishery statistics: 1952–2020 season – reported catch and effort by method. Available at: www.gov.scot/publications/salmon-fishery-statistics-2020/ (accessed 8 November 2021).

Marine Scotland (2021b) Scottish Fish Farm Production Survey Data. Available at: www.gov.scot/publications/scottish-fish-farm-production-survey-2020/ (accessed 8 November 2021).

Marsh, R. (2018) Scottish salmon 2018 economic report. Available at: https://www.scottishsalmon.co.uk/sites/default/files/2020–01/Salmon-Impact.pdf (accessed 8 November 2021).

McKibben, M.A. and Hay, D.W. (2004) Distributions of planktonic sea lice larvae *Lepeophtheirus salmonis* in the inter-tidal zone of Loch Torridon, Western Scotland in relation to salmon farm production cycles. *Aquaculture Research* 35, 742–750.

Middlemas, S.J., Stewart, D.C., MacKay, S. and Armstrong, J.D. (2009) Habitat use and dispersal of post-smolt sea trout *Salmo trutta* in a Scottish sea loch system. *Journal of Fish Biology* 74, 639–651.

Moore, I., Dodd, J.A., Newton, M., Bean, C.W. and Lindsay, I. (2018) The influence of aquaculture unit proximity on the pattern of *Lepeophtheirus salmonis* infection of anadramous *Salmo trutta* populations on the Isle of Skye, Scotland. *Journal of Fish Biology* 92, 1849–1865.

Mordue (Luntz), A.J. and Pike, A.W. (2002) Special Issue: Integrated pest management in salmon farming. *Pest Management Science* 58, 513–635.

Murray, A.G. and Gillibrand, P.A. (2006) Modelling salmon lice dispersal in Loch Torridon, Scotland. *Marine Pollution Bulletin* 53, 128–135.

Mowi (2021) *Salmon Farming Industry Handbook*. Available at: https://corpsite.azureedge.net/corpsite/wp-content/uploads/2021/05/Salmon-Industry-Handbook-2021.pdf (accessed 4 July 2021).

NASCO (2020) The state of wild north Atlantic salmon. Available at: https://nasco.int/atlantic-salmon/state-of-salmon/ (accessed 2 July 2021).

Oppedal, F., Bui, S., Stien, L.H., Overton, K. and Dempster, T. (2019) Snorkel technology to reduce sea lice infestations: efficacy depends on salinity at the farm site, but snorkels have minimal effects on salmon production and welfare. *Aquaculture Environment Interactions* 11, 445–457.

Pike, A.W. and Wadsworth, S.L. (2000) Sealice on salmonids: their biology and control. In: Baker, J.R., Muller, R. and Rollinson, D. (eds) *Advances in Parasitology*. Vol 44. Elsevier Academic Press Inc, San Diego, California. pp. 233–337.

Rae, G.H. (2002) Sea louse control in Scotland, past and present. *Pest Management Science* 58, 515–520.

Roth, M., Richards, M.H., Dobson, D.P. and Rae, G.H. (1996) Field trial on the efficacy of the organophosphorus compound azamethiphos for the control of sea lice (Copepoda: Caligidae) infestations of farmed Atlantic salmon *Salmo salar*. *Aquaculture* 140, 217–239.

Shinn, A. and Bron, J.E. (2012) Chapter 8: Considerations in the use of anti-parasitic drugs in aquaculture. In: Austin, B.A. (ed.) *Infectious Disease in Aquaculture: Prevention and Control.* Woodhead Publishing Limited, Cambridge, UK. pp. 190–217.

Sommerville, C. (1995) Latest weapons in the war on lice. *Fish Farmer* March/April, 53–55.

Whelan, K. (2010) A review of the impacts of the salmon louse, *Lepeophtheirus salmonis* (Krøyer, 1837) on wild salmonids. Available at: https://atlanticsalmontrust.org/wp-content/uploads/2016/12/ast-sea-lice-impacts-review1.pdf (accessed 11 November 2021).

White, H.C. (1940) 'Sea-lice' (*Lepeophtheirus*) and death of salmon. *Journal Fisheries Research Board Canada* 5, 172–175.

White, H.C. (1942) Life history of *Lepeophtheirus salmonis*. *Journal Fisheries Research Board Canada* 5, 172–175.

Wootten, R., Smith, J.W. and Needham, E.A. (1982) Aspects of the biology of the parasitic copepods *Lepeophtheirus salmonis* and *Caligus elongatus* on farmed salmonids, and their treatment. *Proceedings Royal Society of Edinburgh* 81B, 185–197.

CHAPTER 2

Introduction to sea lice morphology and biology

Geoff A. Boxshall[1] and Argun A. Özak[2]

[1]Department of Life Sciences, Natural History Museum, London, UK
[2]Department of Fish Diseases, Faculty of Fisheries, University of Çukurova, Sarıçam, Adana, Turkey

2.1 Early growth of knowledge on sea lice (Caligidae)

Parasitic copepods have been known since classical times: both Aristotle and Pliny the Elder referred to large parasites on tuna in the Mediterranean, but these were probably *Brachiella thynni* Cuvier, 1830, a member of the family Lernaeopodidae (Damkaer, 2002). The first mentions of caligid sea lice came in the mid-18th century as naturalists began publishing brief accounts of caligids. They classified their species under a variety of generic names and often misunderstood their anatomy, with some mistakenly interpreting the egg strings as antennae (Strøm, 1762; Baster, 1765). However, some excellent illustrations of sea lice were appearing around this time (for example, Houttuyn, 1769) and the roots of the modern family can be traced back to Müller (1785) who established the type genus, *Caligus* Müller, 1785, and its type species *Caligus curtus* Müller, 1785.

Larval stages also caused considerable taxonomic confusion in the late 18th century. Müller (1785) established the genus *Nauplius* to accommodate small entomostracans that he assumed were adults but were in fact early developmental stages. The term nauplius is not valid as a genus name but is used for the first stage in the life cycle of almost all copepods and of many other crustacean groups including barnacles (Cirripedia) and krill (Euphausiacea). Similarly, *Chalimus* was first proposed as a generic name by Burmeister (1835) and was accepted as a valid genus by Milne Edwards (1840) and by Baird (1850), although Krøyer (1837) had already correctly recognised that these forms were developmental stages in the caligid life cycle.

In 1832, Nordmann established *Lepeophtheirus* Nordmann as the second genus of the Caligidae, which he distinguished from the type genus by the lack of lunules on the paired frontal plates. After that many new genera were gradually established: ten more in the 19th century, seventeen in the 20th and two more in the 21st. The vast majority of the early contributions to our knowledge were primarily taxonomic, resulting in a dramatic increase in the numbers of known genera and species and in an awareness of their host usage and geographical distributions. Knowledge of other aspects of the biology of sea lice, such as internal anatomy, accumulated only gradually (Scott, 1901), but in the latter half of the 20th century the focus expanded to include life cycles (Hwa, 1965; Izawa, 1969), locomotion (Kabata and Hewitt, 1971), feeding mechanisms (Kabata, 1974), population dynamics (Boxshall, 1974a, 1974b), histopathology (Boxshall, 1977), host location behaviour (Boxshall, 1976; Heuch et al., 1995; Heuch and Karlsen, 1997) and mating behaviour (Anstensrud, 1990, 1992). The increased research effort into the biology of sea lice at this time was, at least in part, a response to the expansion of salmon farming in the Northern Hemisphere and the emergence of sea lice as a major health hazard for this commercial enterprise.

The study of sea lice has now entered the era of molecular genetics. The power of these methods and their potential in developing targeted control strategies has had an enormous impact on the study of sea lice and will be explored elsewhere in this volume.

2.2 Phylogenetic position of the Caligidae

The family Caligidae belongs to a wholly parasitic order of copepods, the Siphonostomatoida, members of which are characterised by the possession of an oral cone (or mouth tube) enclosing stylet-like mandibles (Boxshall, 1990a). The Siphonostomatoida comprises over 2200 species, about 68% of them parasitic on fish hosts, and the Caligidae is the most species-rich family comprising over 500 valid species (Walter and Boxshall, 2022). There are currently ten orders recognised within the Copepoda, and the Siphonostomatoida is one of the eight orders classified within the superorder Podoplea, characterised by a body plan where the primary division between the anterior prosome and posterior urosome lies between the fourth and fifth pedigerous somites. The remaining two copepod orders, the Platycopioida (superorder Progymnoplea) and the Calanoida (superorder Gymnoplea), are both planktonic and share a different body plan.

The Copepoda has long been classified within the Crustacea, but phylogenetic studies based on the molecular sequence data that has become available on a vast scale have thrown the classification of the entire Arthropoda into a state of flux and it is now widely accepted that the traditional Crustacea does not exist as a monophyletic group. The traditional crustaceans represent a paraphyletic assemblage comprising three or four distinct lineages arising independently from the arthropod evolutionary tree somewhere between the Myriapoda and the Hexapoda (von Reumont et al., 2009; Regier et al., 2010; Edgecombe, 2010; Giribet and Edgecombe, 2013; Lozano-Fernandez et al., 2019). The robust molecular evidence indicates that the insects (Hexapoda) arose from within the 'Crustacea', although no consensus has yet been reached on the precise nearest relative within the crustaceans.

The phylogenetic analyses of the Arthropoda by Regier et al. (2010) and by Oakley et al. (2013) both placed the Copepoda in a large lineage together with the Malacostraca and the Thecostraca. This grouping is referred to as the Multicrustacea by Oakley et al. (2013) and was also recovered as a monophyletic grouping by Regier et al. (2010). Petrunina et al. (2014) subsequently confirmed the placement of the Tantulocarida as close relatives of the Thecostraca within the Multicrustacea. The Multicrustacea comprising the Copepoda, Thecostraca and Malacostraca was also recovered in the expanded analysis of Lozano-Fernandez et al. (2019) but these authors also noted that the position of the Copepoda in the phylogenetic tree was less stable than many of the other taxa.

These changes in our understanding of the higher level systematics of the arthropods have had relatively little impact on the integrity of the major constituent taxa of each lineage. Within the traditional 'Crustacea', for example, groups such as the Copepoda, Branchiura, Amphipoda and Isopoda are all still recognised as monophyletic taxa, as they were in overviews of crustacean classification before the molecular era, such as Bowman and Abele (1982). Indeed, the increasing flow of molecular sequence data has shown these taxa to be robust. So, the sea lice are still classified in the family Caligidae, the order Siphonostomatoida and the subclass Copepoda; it is simply the recognition that the 'Crustacea' is not a natural (that is, monophyletic) group that has changed.

2.3 Introduction to morphology

Body

The body of a caligid sea louse comprises two main functional regions: the anterior cephalothorax and the posterior genitoabdomen, comprising the fourth pedigerous somite, the genital complex plus the abdomen (Fig. 2.1). The cephalothorax includes the six somites of the cephalosome (Huys and Boxshall, 1991) plus the first, second and third pedigerous somites, and this precise composition helps to define the family Caligidae, distinguishing it from closely related families such as the Pandaridae, Trebiidae and Dissonidae which incorporate different numbers of pedigerous somites into the cephalothorax (Kabata, 1979; Hutson and Boxshall, 2016). It is covered by a flattened dorsal cephalothoracic shield. The posterior region is much more variable. Kabata (1979) recognised three different posterior tagmata but given that the fourth pedigerous somite is often fused to the genital complex and that the abdomen can be fully incorporated into the genital complex, this seems like an unnecessary subdivision of a single major genitoabdominal tagma.

Sea lice are ectoparasites and most species, including commercially important pest species such as *Lepeophtheirus salmonis* (Krøyer, 1837) and *Caligus elongatus* Nordmann, 1832 inhabit the outer body surface of the host as well as the internal surfaces of its buccal and branchial cavities. These sea lice display the typical dorsoventrally flattened body form that enables them to reduce resistance to water flow by adhering closely to the host's surface. On an effectively two-dimensional microhabitat, the flattened cephalothorax forms an efficient sucker that secures attachment to the surface of the host. The outer margins of this sucker are provided with membrane which extends along the paired frontal plates, the lateral zones of the dorsal cephalothoracic shield, and the free posterolateral and posterior margins of leg three. The fine structure of the marginal membrane was investigated by Ohtsuka et al. (2021). There are, however, a few genera that are secondarily adapted for attachment to the gill filaments of the host, or have invaded its nasal cavities. In these genera (for example, *Hermilius* Heller, 1865) the cephalothorax does not function as a sucker; instead the dorsal cephalothoracic shield is folded along the midline forming bilateral folds that facilitate clasping onto the gill filament. This longitudinal hinge is only weakly developed in the genus *Kabataella* Prabha and Pillai, 1983 which inhabits the nasal cavity of its pomadasyid host. The anterior margin of the dorsal cephalothoracic shield carries the paired frontal plates which are provided with marginal membrane that seals off the anterior rim of the cephalothoracic sucker between the paired antennules (Fig. 2.2A). In *Tuxophorus* Wilson, 1908, *Caligus* and related genera, each of these plates carries a lunule (Fig. 2.2A and B), an accessory sucker representing a modification of the marginal membrane (Kaji et al., 2012; Ohtsuka et al., 2021). The size of the lunules can vary between congeneric species, and within the genus *Abasia* Wilson, 1908, they can be secondarily reduced in size or even lost in some species as an adaptation to the use of the gill filament as a microhabitat. The genus *Alanlewisia* Boxshall, 2008 lacks lunules but also exhibits a modification of the frontal plate membrane; this genus is also unusual in having sexually dimorphic frontal plates.

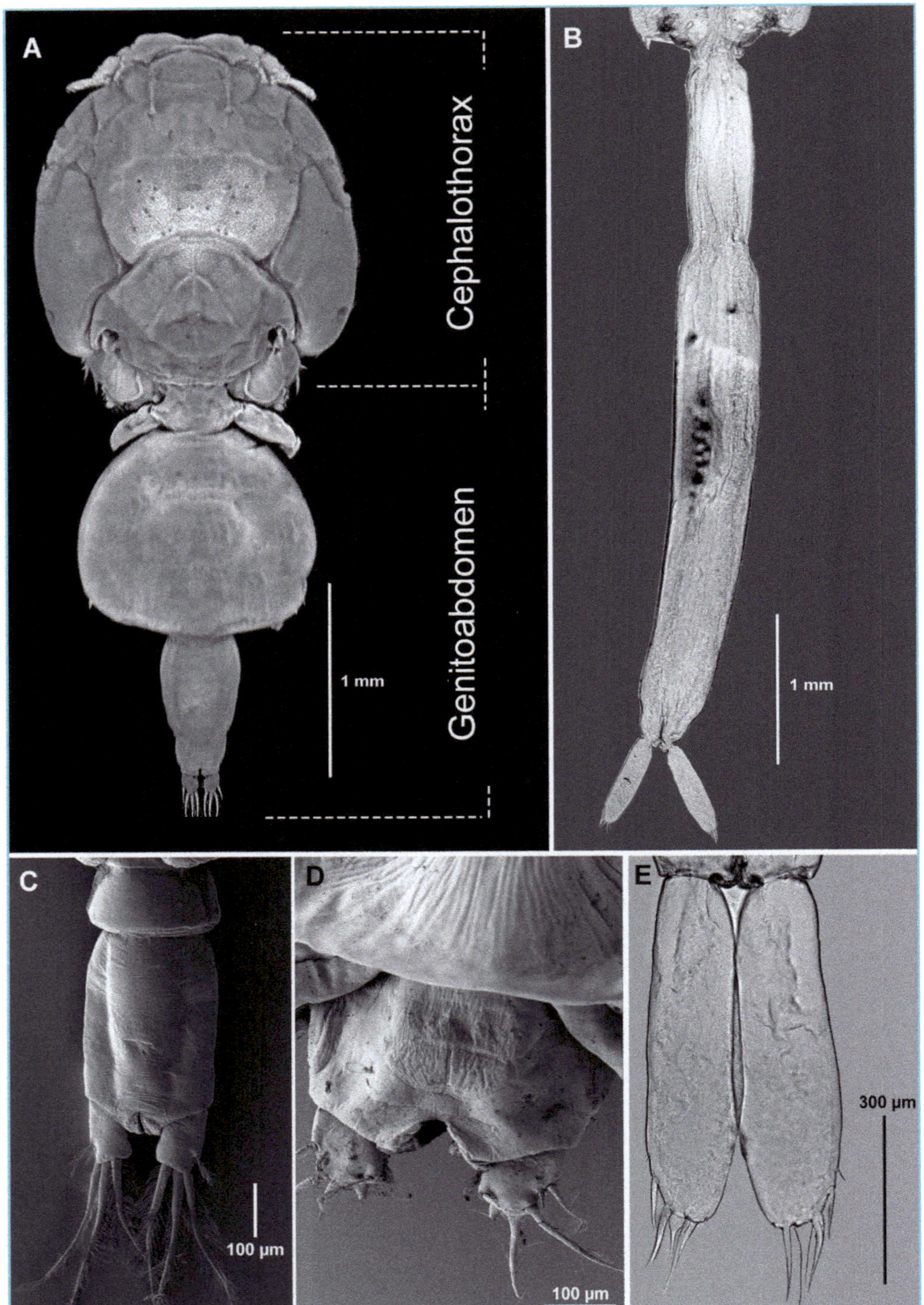

Fig. 2.1 A. Female *Caligus diaphanus* Nordmann, 1832 habitus, dorsal; B. Extremely long one-segmented abdomen of *Caligus macrurus* Heller, 1865; C. Two-segmented male abdomen and caudal rami of *Caligus solea* Demirkale, Özak, Yanar and Boxshall, 2014; D. Short one-segmented abdomen and caudal rami of *Caligus brevicaudatus* Scott, 1901; E. Extremely long caudal rami of *Caligus macrurus*.

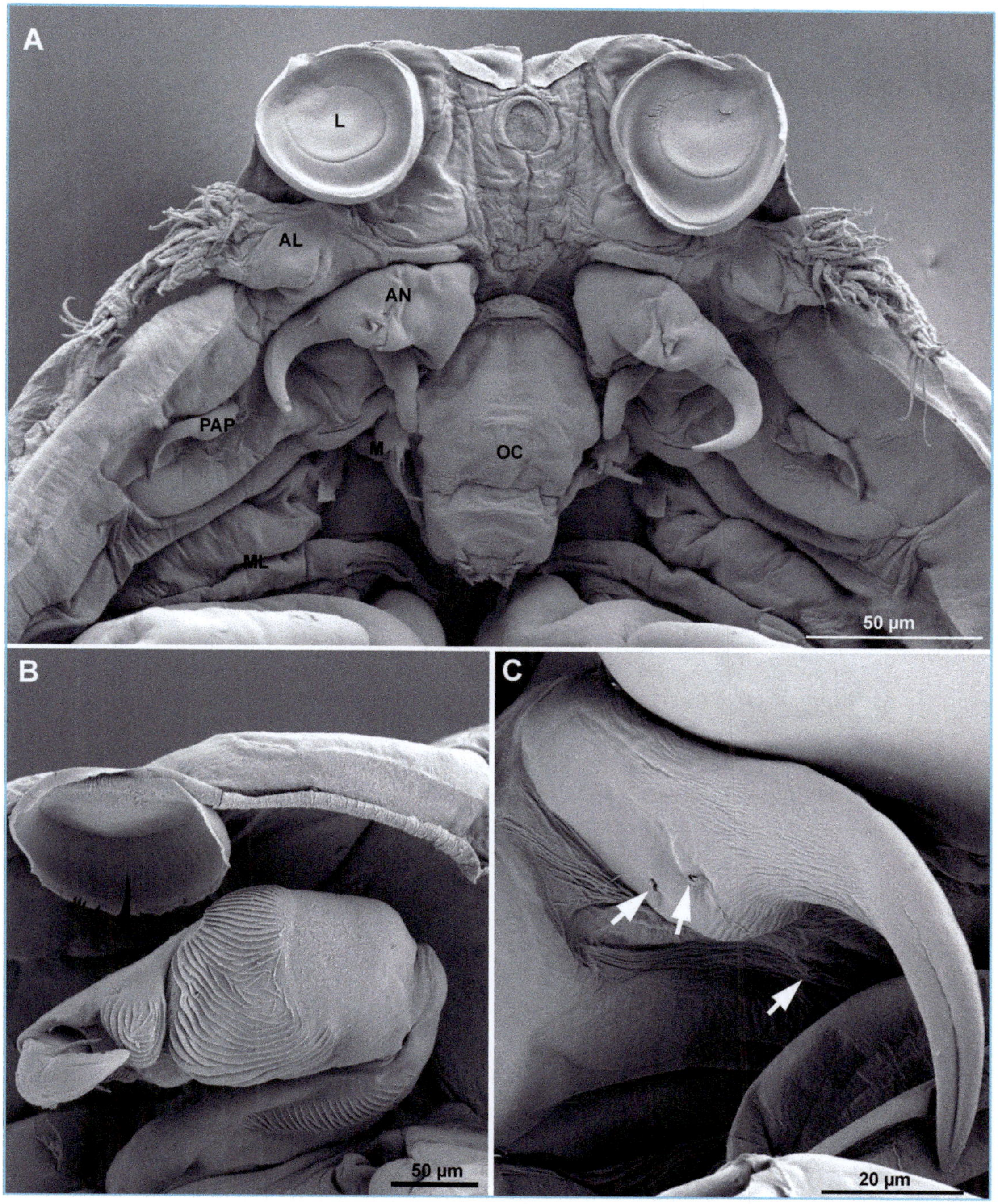

Fig. 2.2 A. Anteroventral part of *Caligus lagocephali* Pillai, 1961 showing paired lunules (L), antennule (AL), antenna (AN), postantennal process (PAP), oral cone (OC), maxillule (M) and maxilla (ML); B. Male antenna of *Caligus solea* Demirkale, Özak, Yanar and Boxshall, 2014 showing corrugated adhesion pads; C. Postantennal process of *Caligus solea*, with multi-sensillate papillae arrowed.

The genitoabdomen articulates with the cephalothorax between the third and fourth pedigerous somites and the region of the fourth somite often forms a 'waist' between the two tagmata. In the genus *Echetus* Krøyer, 1863 and its poorly-known relatives *Parechetus* Pillai, 1962 and *Pseudechetus* Prabha and Pillai, 1979, the fourth pedigerous somite is extremely elongate and is referred to as a 'neck'. The significance of the possession of a 'neck' has not yet been assessed in a comprehensive phylogenetic analysis and the genera based on this feature may not be robust. In four of the more basal genera in the family (that is *Euryphorus* Milne Edwards, 1840, *Alebion* Krøyer, 1863, *Gloiopotes* Steenstrup and Lütken, 1861 and *Tuxophorus*), the fourth pedigerous somite carries paired, flattened dorsal plates. In the past these dorsal plates were treated as a diagnostic feature to separate the family Euryphoridae (with plates) from the Caligidae (without plates) but these two families are now combined (Dojiri and Ho, 2013; Boxshall and Halsey, 2004).

According to Boxshall (1990a), the genital complex (Fig. 2.1A) of caligids comprises the fifth pedigerous somite, the genital somite and the first abdominal somite. It carries the paired genital openings in both sexes plus the paired copulatory pores in the female (into which the spermatophores discharge). Historically, the presence of paired posterolateral 'aliform expansions' (for example, *Parapetalus* Steenstrup and Lütken, 1861), posterolateral processes (for example, *Synestius* Steenstrup and Lütken, 1861) or posterolateral aliform processes (for example, *Caligodes* Heller, 1865) has been used to separate some small caligid genera from the large genus *Caligus*, but such distinctions have never been assessed in a comprehensive phylogenetic analysis. These genera all share the presence of lunules on the frontal plates with *Caligus*, and they appear to have been established primarily on the basis of differences, rather than on their shared derived characters. It seems likely that some, if not all, may eventually be subsumed within the genus *Caligus*. The posterolateral aliform processes of *Caligodes*, for example, appear to represent the fifth legs (Boxshall, 2018), and the validity of this genus needs to be assessed.

The free abdomen is the posterior-most division of a caligid and carries the paired caudal rami located either side of the median anal slit. It can be one- or two-segmented in both sexes and descriptions showing more subdivisions are based on misinterpretations. The abdomen varies in both size and morphology from taxon to taxon (Fig. 2.1B–D). Historically the possession of an extremely elongate abdomen, as found in *Sciaenophilus* Beneden, 1852, was often regarded as a generic discriminant, but Kabata (1979) noted that the males of *Sciaenophilus* and *Caligus* were identical and that the only difference was the length of the abdomen of the female. Various authors (for example, Capart, 1959) did not recognise *Sciaenophilus* as valid and Özak et al. (2017) formally treated *Sciaenophilus* as a junior synonym of *Caligus*. Similarly, several other genera have been characterised by the possession of lateral lobes/expansions/processes on the abdomen (Dojiri and Ho, 2013). Again, there is considerable doubt over the taxonomic validity of genera defined solely on the basis of such lobes. The monotypic genus *Midias* Wilson, 1911, which has paired lateral processes on both first and second abdominal somites, was synonymised with *Caligus* by Kabata (1979) but Dojiri and Ho (2013) continued to recognise it as valid, adding further character states to support their decision. Boxshall and Halsey (2004) recognised the synonymy and placed the type species *Midias lobodes* Wilson, 1911 within *Caligus* but these results have yet to be assessed in a comprehensive phylogenetic analysis.

Within the family a trend towards the reduction of the free abdomen can be recognised so that the genus *Anuretes* Heller, 1865 is distinguished from *Lepeophtheirus* primarily by the reduction or loss of the abdomen in the former (Dojiri and Ho, 2013). Again, however, there is continuous variation in the degree to which the abdomen is incorporated into the genital complex in species of *Anuretes*, and this character alone is unlikely to be a robust generic discriminant.

The paired caudal rami (Fig. 2.1A–E) vary in size and are fused to the abdomen in some taxa. The genus *Dartevellia* Brian, 1939 apparently lacks caudal rami but the presence of caudal setae on the paired abdominal aliform expansions (Dojiri and Ho, 2013) indicates that the caudal rami are incorporated into the abdomen and contribute to the formation of these posterolateral expansions on the abdomen.

Paired limbs and other appendages

The caligid antennule (Fig. 2.2A) is two-segmented and both segments carry an array of chemosensory aesthetascs and setae which may be mechanosensory or bimodal (Laverack and Hull, 1993). The proximal segment is usually more robust and typically carries up to 27 setae, 25 along the anteroventral margin and two set back on the dorsal surface. The setae on the anteroventral margin are often enlarged and hirsute; they come into close contact with the host's surface, helping to close off the anterior gaps between strips of marginal membrane around the rim of the cephalothoracic sucker. The distal segment carries a single seta on the posterior margin and an apical cluster of 13 or 14 setal elements, including at least one aesthetasc. Variation in the numbers of setae on both segments has been reported across the family but setae are often small and can be overlooked, so much of the reported variability needs confirmation.

The antenna (Fig. 2.2A and B) is an important grasping appendage, helping to secure attachment to the host in most caligids. Based on musculature patterns, Boxshall (1990a) concluded that the antenna of female *Lepeophtheirus* was three-segmented, comprising coxa, basis and a distal subchela consisting of the fused endopodal segments plus a terminal claw. The coxa is unarmed but often bears a posteriorly-directed spinous process. The basis is also unarmed but typically carries a corrugated adhesion pad on the dorsal surface. The subchela typically carries one or two vestigial setae derived from incorporated endopodal segments, and ends in a curved terminal claw. It can also carry an accessory claw in the female, which is especially prominent in *Hermilius*. The antenna is usually sexually dimorphic: the male antenna (Fig. 2.2B) typically has more extensive adhesion pads, the terminal claw is more compact, and is ornamented with adhesion pads in addition to two well developed setae, and it may have accessory claws or flattened apical plates. Males use their antennae to grasp onto females during the different stages of the mating process (Lewis, 1963; Anstensrud, 1990).

The postantennal process (Fig. 2.2C) has suffered varying interpretations over the centuries, having often been interpreted as one of the cephalic limbs. However, it has no extrinsic musculature and the consensus now considers it to be a modified sclerite of the ventral cephalic surface (Boxshall, 1990a). It typically appears hooked as it carries a single curved tine but in certain genera or species groups it has an accessory tine and appears bifid (for example, *Gloiopotes* and some *Caligus* species), or it forms a corrugated pad (for example, *Alebion*) or a sclerotised plate (for example, *Caligodes*), and it can be absent (for example, *Pseudanuretes* Yamaguti, 1936 and some species of *Caligus*). The ornamentation of the postantennal process consists of two sensillate papillae located proximally on the surface of the process and a third sensillate papilla positioned on the ventral cephalothoracic surface adjacent to the process. Even when the process itself is absent, the sensillate papillae are typically present on the cephalic surface. The numbers of sensillae on these papillae differ between taxa but are generally constant within species or species-groups.

The mandible of caligids is a stylet-like rod with a curved tip that bears teeth along its concave margin (Fig. 2.3B). The rod is subdivided along its length but the entire mandible is derived from the coxal gnathobase, a single-segment, so descriptions that refer to 'segments of the caligid mandible' are erroneous. The subdivisions allow some flexure of the mandible as it enters from the side at the base of the mouth tube (Fig. 2.3A) and passes along its length to the distal mouth opening. There is no trace of a mandibular palp in any adult caligid. There is remarkably little variation in the number of teeth on the mandible: it is 12 in almost every known species, but *Caligodes* and *Dartevellia* each have 13 teeth (Dojiri and Ho, 2013) and *Caligus madeirensis* Hamdi, Hermida, Kamanli, Benmansour, Özak and Boxshall, 2020 has 14 (Hamdi et al., 2021a: Fig. 2.3C). The mandible functions to scrape surface tissue from the host and help to sweep it up into the buccal cavity (Boxshall, 1990a; Ho and Lin, 2004). The mouth tube is formed by the anterior labrum and the posterior labium, itself derived from the fused paragnaths (Boxshall, 1990a). The muscular labrum and the labium interlock to create a tubular sucking cone that draws host fluids and tissue fragments up into the mouth.

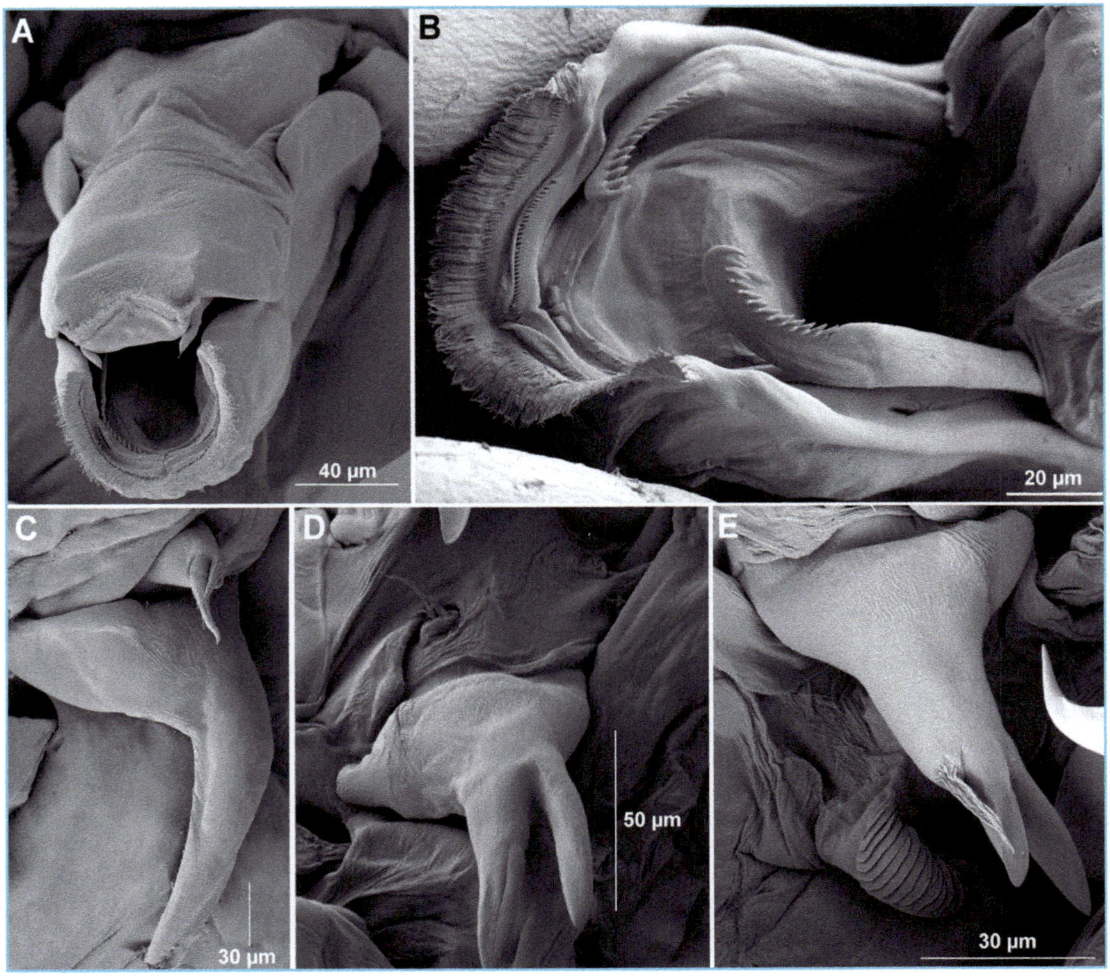

Fig. 2.3 A. Oral cone of *Lepeophtheirus acutus* Heegaard, 1943; B. Oral cone with labrum folded back to reveal mandibles lying in grooves in labium; C. Maxillule of *Caligus solea* Demirkale, Özak, Yanar and Boxshall, 2014 with simple dentiform posterior process; D. Maxillule of female *Lepeophtheirus acutus* showing bifid posterior process; E. Maxillule of male *Lepeophtheirus acutus* showing trifid posterior process and adjacent postoral adhesion pad.

The maxillule of caligids comprises a small anterior papilla carrying three setae and a posterior process (Fig. 2.3C). The posterior process is typically simple or bifid, but it can be sexually dimorphic and in many *Lepeophtheirus* species, for example, it is bifid in females (Fig. 2.3D) and trifid in males. In males of some *Caligus* species, the posterior process may carry a corrugated pad or there may be postoral adhesion pads on the adjacent body surface (Fig. 2.3E). The posterior process is typically conical but it can be reduced or even be absent, as in *Pseudanuretes*, or enlarged into a rounded process as in *Mappates*.

The maxilla is two-segmented, comprising a proximal segment, the basis or lacertus, which is unarmed, and the distal brachium (Fig. 2.4A). The brachium is ornamented subapically with a membranous structure, the flabellum (arrowed in Fig. 2.4A), and carries at its tip two curved claws, the calamus and canna. Both claws are typically ornamented with serrated membrane which often runs spirally across the surface towards the tip (Fig. 2.4B). At its base, the maxilla carries the opening of the excretory organ, the maxillary gland, and the reinforced duct leading to the opening is often visible through the cuticle. The maxilla varies little in structure across the family, although an additional lobate membranous structure is present on the outer margin of the brachium in *Euryphorus brachypterus* (Gerstaecker, 1853). The maxilla is multifunctional and has been observed assisting in settling movements on the host, in attachment, in manipulation of the frontal filament during the chalimus phase, and in grooming of other limbs.

The maxilliped (Fig. 2.4C) is an important limb involved in securing attachment to the host. It is subchelate, comprising a robust, muscular proximal segment (the corpus or protopod) and a claw-like distal subchela incorporating the endopodal segments and terminal claw. The claw can be inserted into the skin of the host for attachment. The medial or myxal margin of the corpus of the maxilliped can carry processes in both sexes or in males only, according to species (Fig. 2.4D). In some species, the proximal segment is ornamented and the claw can have longitudinal surface ridges (Fig. 2.4D). The myxal process may also be ornamented with surface corrugations. Such a sexually dimorphic structure is suggestive of a possible role in mating behaviour, and the male can use the maxillipeds to grasp the female during mating.

The sternal furca is located on the ventral cephalothoracic surface between the bases of the maxillipeds and the interpodal bar of the first swimming legs, and it consists of a proximal cylindrical part (the box) which carries paired processes (the tines) distally. The presence or absence of the sternal furca and the shape of its tines are useful for identification purposes (Fig. 2.5A–D). In addition, the shape, size and even the distance between the tines of the sternal furca can be sexually dimorphic, as in *Euryphorus brachypterus* (Fig. 2.5E and F), for example. However, drawings must be interpreted with caution as the furca can vary in appearance according to the angle of observation. The tines are typically simple but they can be bifurcate in some species of *Lepeophtheirus* and *Tuxophorus* and in all species of *Gloiopotes*. The homology of the sternal furca has been subject to many differing interpretations but the evidence from study of the skeletomusculature (Boxshall, 1990a) suggests that the muscles operating the sternal furca are derived as modifications of the ventral longitudinal trunk muscles and, therefore, that the furca is derived as an elaboration of the ventral wall of the cephalosome (strictly, of the median intersomitic sclerite). The sternal furca is supplied with muscles which can raise it from its posteriorly-directed resting position to perpendicular with the ventral body surface but its function is unclear.

Members of the first swimming leg pair (Fig. 2.6A) are laterally directed beneath the dorsal cephalothoracic shield and are connected medially by an intercoxal sclerite which ensures the legs beat together. Leg one has a proximal sympod consisting of the incompletely fused coxa and basis, and

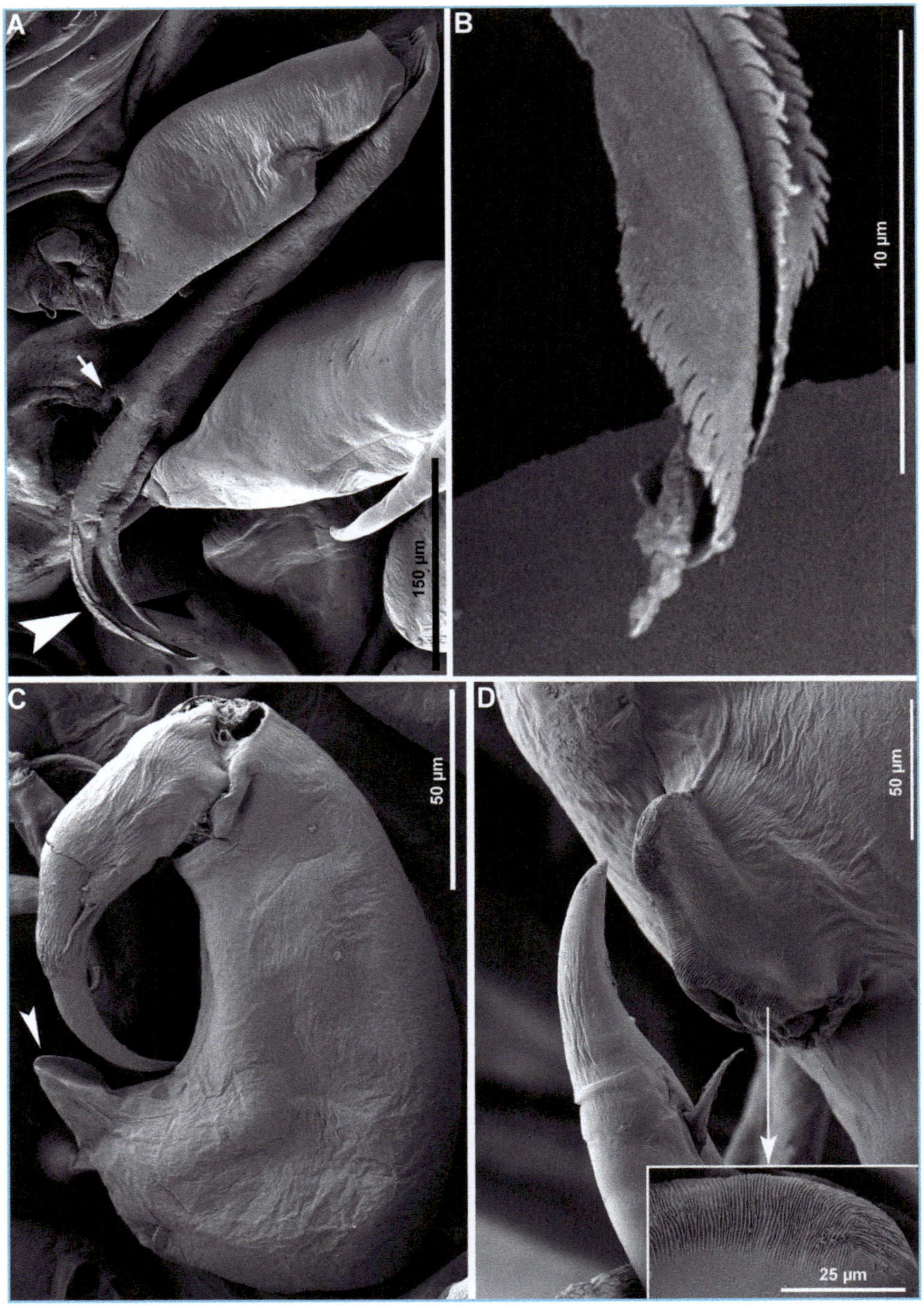

Fig. 2.4 *Caligus diaphanus* Nordmann, 1832. A. Female maxilla showing the flabellum (arrowed) plus calamus (white arrowhead) and canna (black arrowhead) at tip; B. Spirally twisted ornamentation at tip of calamus on maxilla; C. Maxilliped of male *Caligus madeirensis* Hamdi, Hermida, Kamanli, Benmansour, Özak & Boxshall, 2020 showing myxal process (arrowhead); D. Maxilliped showing corrugated outer margin of myxal process (*inset*) plus the small seta at the base of the terminal claw of the subchela.

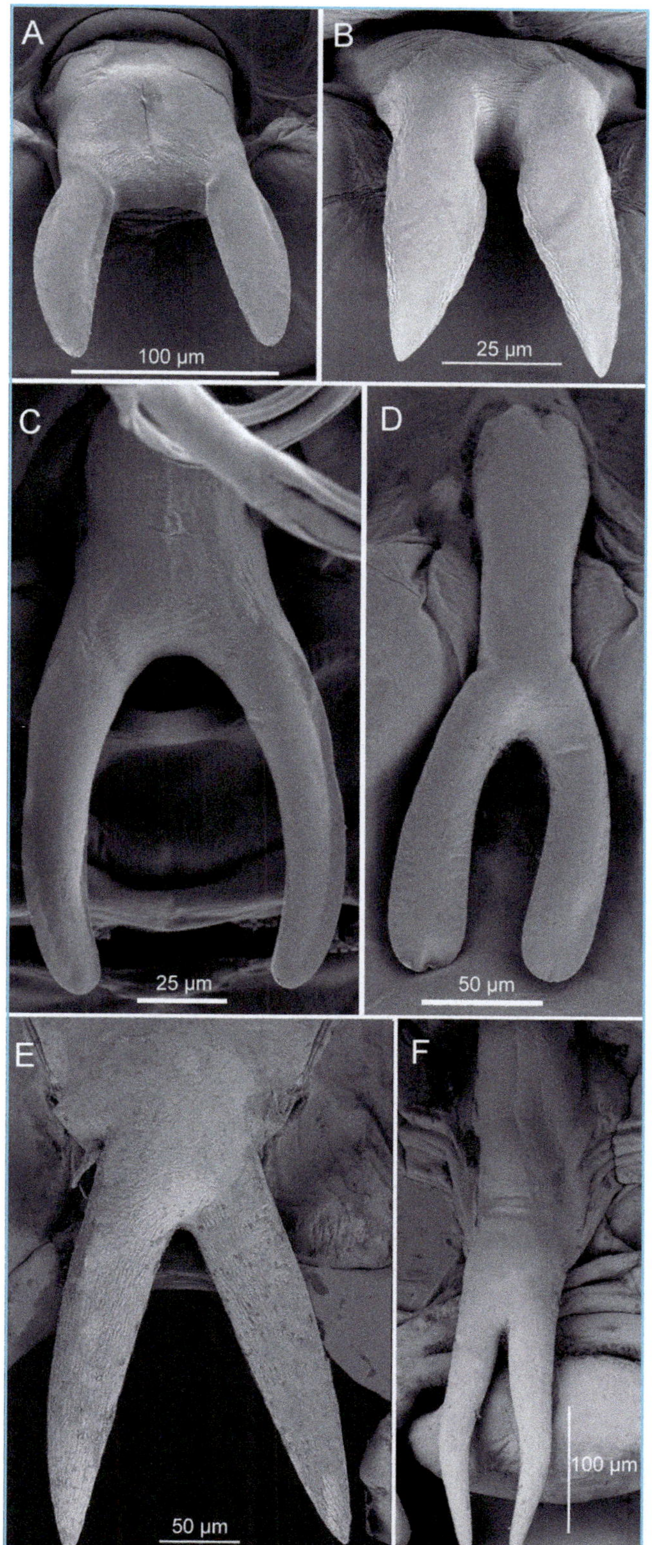

Fig. 2.5 Diversity in form of sternal furca. A. *Caligus solea* Demirkale, Özak, Yanar & Boxshall, 2014; B. *Lepeophtheirus azoricus* Özak, Rodrigues, Vieira, Rosa, Yanar, Koyuncu and Boxshall, 2017; C. *Caligus haemulonis* Krøyer, 1863; D. *Caligus apodus* (Brian, 1924); E. *Euryphorus brachypterus* (Gerstaecker, 1853) female; F. *Euryphorus brachypterus* male.

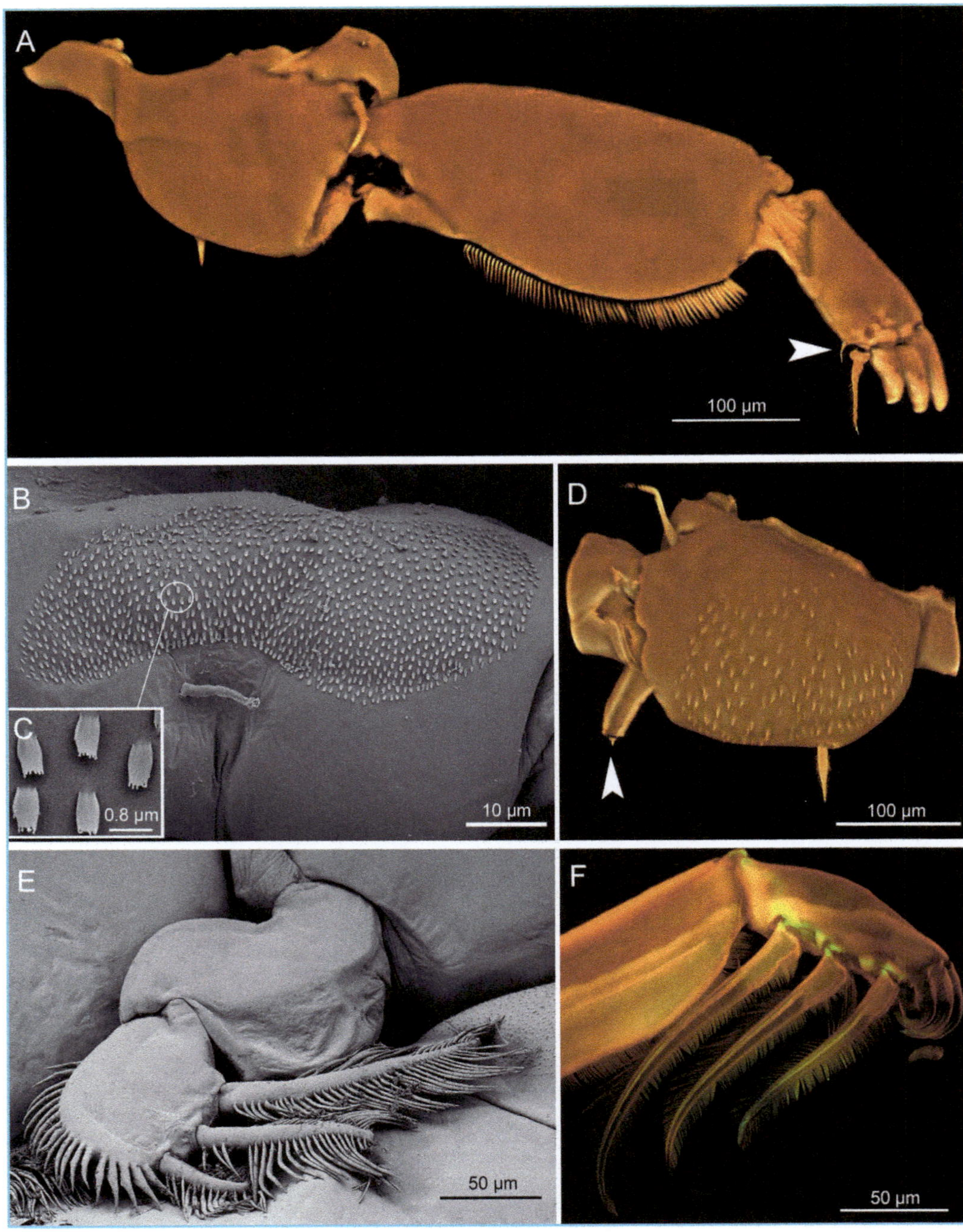

Fig. 2.6 A. *Caligus temnodontis* Brian, 1924, swimming leg one showing two-segmented exopod bearing four terminal elements (spines one to three and seta four) plus vestige of inner seta (arrowhead); B. *Caligus ligusticus* Brian, 1906, patch of scale-like ornamentation on ventral surface of leg one sympod; C. Detail of scale-like ornamentation; D. *Caligus tunisiensis* Hamdi, Benmansour, Zouari-Tlig, Kamanli, Özak, & Boxshall, 2020, showing vestigial endopod (arrowed) and spinular ornamentation on ventral surface of leg one sympod; E. *Euryphorus brachypterus* (Gerstaecker, 1853), two-segmented endopod of leg one; F. *Caligus minimus* Otto, 1821, distal exopodal segment of leg one with four terminal elements and three plumose setae on free posterior margin.

two rami. In most species, the sympod is ornamented with an irregular patch of cuticular scales (Fig. 2.6B and C) or spinules (Fig. 2.6D). In the vast majority of caligids the endopod is vestigial and carries only setal vestiges at its tip (arrowhead in Fig. 2.6D), but in five genera the endopod is two-segmented and bears three setae on the distal segment (Fig. 2.6E). *Alanlewisia* exhibits an intermediate state where the endopod is vestigial but carries two well-developed apical setae. The exopod is two-segmented and the precise setation of the distal segment (Fig. 2.6F) can provide useful characters for species-level identification, for example in the genus *Caligus.*

Leg two is biramous with a separate coxa and basis plus three-segmented rami (Fig. 2.7A). As with leg one, the second legs are connected medially by an intercoxal sclerite which ensures the legs beat together. Leg two is provided with extensive flaps of membrane along the free posterior margins of the intercoxal sclerite and basis, and there are flaps of membrane along the lateral margins of the coxa and first exopodal segment. These lateral flaps are reflexed dorsally and perform the important function of closing the gap between the limb and the ventral body surface during swimming motions (Ohtsuka et al., 2021). Leg two generates the major power stroke during swimming (Kabata and Hewitt, 1971) and varies little in structure between genera. The exceptions are the sessile genera that inhabit the gill filaments (for example, *Arrama* Dojiri & Cressey, 1991) and nasal cavities (*Kabataella*), in which leg two is reduced with only two-segmented rami and the marginal membrane flaps can be reduced or lost. The precise form of the setation of the exopod (Fig. 2.7B) and the nature and extent of the ornamentation on the endopod (Fig. 2.7C and D) can provide useful taxonomic characters at the species level.

Leg three (Fig. 2.8A) forms a large, flattened plate (the apron) consisting of the fused protopodal segments and intercoxal sclerite, and this carries the rami posterolaterally. The surface of the apron is often ornamented (Fig. 2.8B) and the precise form of the ornamentation can be a useful taxonomic character (for example, in the *Caligus confusus*-group, see Boxshall (2018)). The exopod is typically three-segmented but can be two-segmented in some taxa. The endopod is three-segmented in relatively primitive genera but is more usually two-segmented and can be further reduced or even absent, as in *Dartevellia*. The first endopodal segment can be transversely expanded to form a flattened plate, the velum (Fig. 2.8B), which closes off the space between the bases of the two rami preventing water flow. Legs one and two are involved in generating the swimming motions while leg three regulates water flow out under the posterior rim of the cephalothoracic suction cup (Ohtsuka et al., 2021). Leg three is reduced in the sessile genera which inhabit the gill filaments and nasal cavity, such as *Abasia* and *Kabataella*.

Leg four is biramous in *Euryphorus* and *Avitocaligus* Boxshall & Justine, 2005, but is uniramous (Fig. 2.9A), retaining only the exopod, in all other caligids. The endopod in both these basal genera is two-segmented. The exopod is primitively three-segmented but the segmentation and setation patterns vary and have proved extremely useful in species discrimination in all genera, but especially in *Caligus*. The whole limb may also be reduced (Fig. 2.9B), although Freeman et al. (2013) have demonstrated that such reduction has happened independently in different lineages and that higher taxa based on this character, for example *Pseudocaligus* Scott, 1901, are not valid.

Legs five (Fig. 2.9C) and six are represented by setiferous lobes carried posteriorly on the genital complex in both sexes. Leg five may be elongate and even spiniform, as in some species of *Lepeophtheirus* (Sakarya et al., 2019: Fig. 6C). Leg six (Fig. 2.9D) is typically part of the opercular apparatus closing off the paired genital apertures but may lack setation in females.

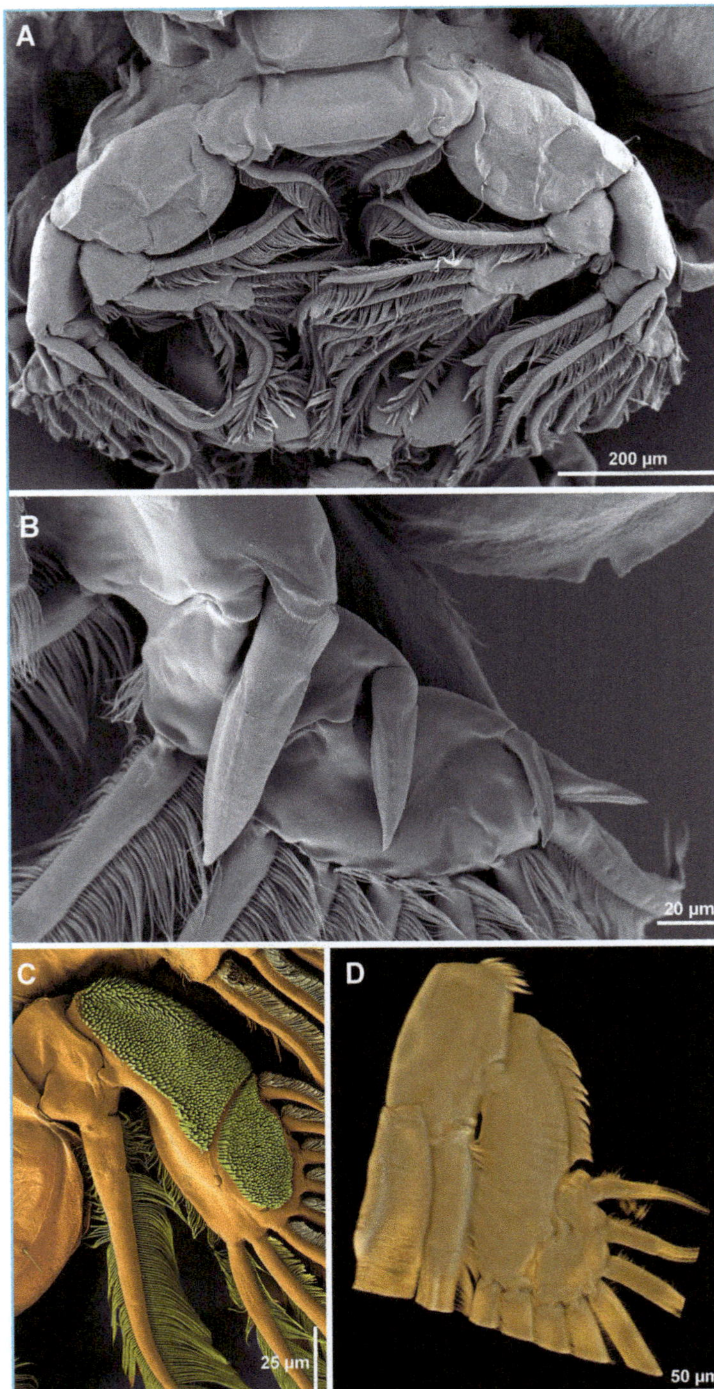

Fig. 2.7 *Caligus solea* Demirkale, Özak, Yanar & Boxshall, 2014. A. Leg two, ventral view; B. Exopodal segments of leg two showing outer spines lying obliquely across ramus; C. *Caligus ligusticus* Brian, 1906, surface ornamentation on leg two endopod (false coloured SEM image), D. *Caligus quadratus* Shiino, 1954, spinular ornamentation on outer margin of the first and second endopodal segments of leg two (Drishti applied CLSM image).

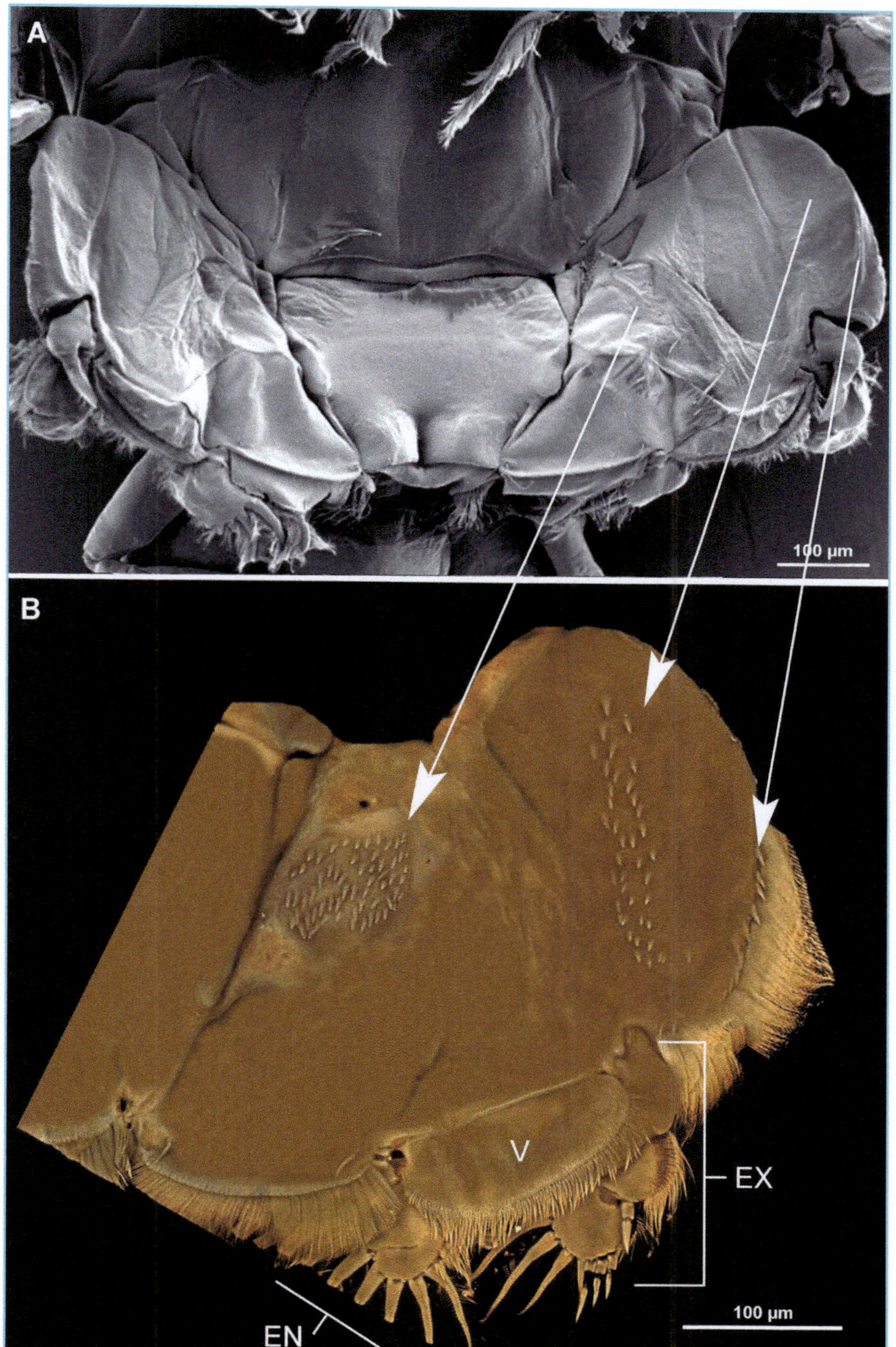

Fig. 2.8 *Caligus solea* Demirkale, Özak, Yanar & Boxshall, 2014. A. Leg three (SEM image); B. *Caligus tunisiensis* Hamdi, Benmansour, Zouari-Tlig, Kamanli, Özak & Boxshall, 2020, showing spinular ornamentation on ventral surface of apron of leg three and location of velum (V) closing gap between bases of rami.

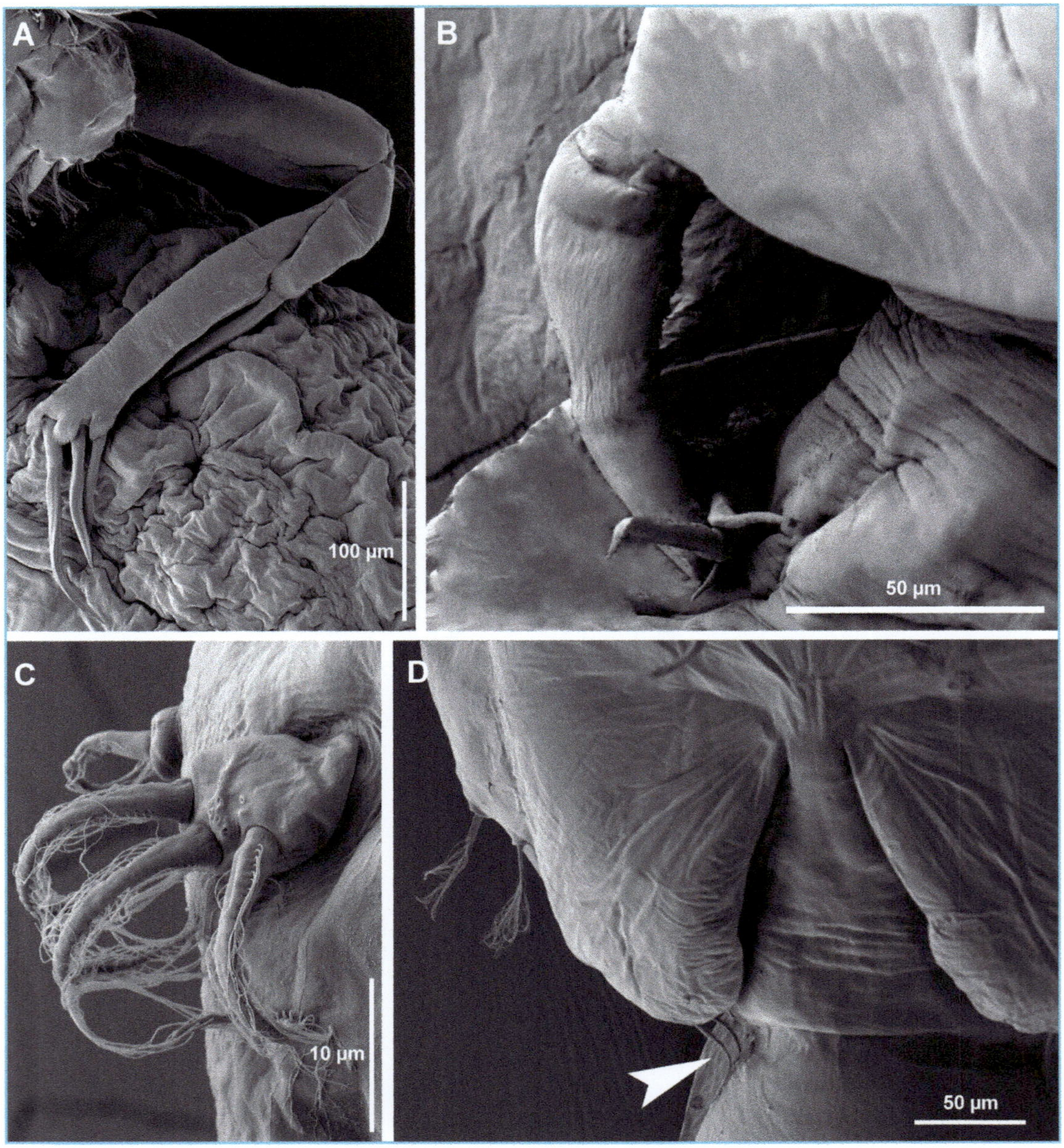

Fig. 2.9 A. Leg four of *Caligus solea* Demirkale, Özak, Yanar & Boxshall, 2014; B. Vestigial leg four of *Caligus apodus* (Brian, 1924); C. Leg five on surface of genital complex of *Caligus diaphanus* Nordmann, 1832; D. Genital openings of male *C. solea*, ventral view showing setae derived from leg six (arrowhead).

2.4 Composition of the Caligidae

The family Caligidae currently comprises 512 valid species classified in 31 genera (Walter and Boxshall, 2022), but only two genera are rich in species: *Caligus* has 273 species and *Lepeophtheirus* has 125. The remaining genera are much smaller, the third largest being

Anuretes with only 21 species, and 11 genera are monotypic (Table 2.1). The validity of many of the currently included genera has yet to be robustly tested using molecular data but on morphological grounds, several are of doubtful validity. A key to genera can be found in Dojiri and Ho (2013).

It is difficult to navigate around the genus *Caligus* because it contains over half of all caligid species. There are currently no recognised subgenera but work has started on recognising species groups within this large genus. Thus far seven species groups have been recognised accommodating about 140 species, but the remaining 130 species have not yet been assigned to a group (cf. Hamdi et al., 2021b). Each species group is based on the common possession of a suite of characters but the phylogenetic status of these groupings has not been tested (Boxshall, 2018).

Caligus bonito-group proposed by Boxshall (2018) is characterised by the following character set: three-segmented leg four with first and second exopodal segments bearing I and IV spines; three plumose setae present on posterior margin of distal exopodal segment of leg one; ornamentation of large denticles present along outer margin of second endopodal segment of leg two; antenna with process present on proximal segment but often small or weakly developed. This group currently contains 15 species.

Caligus diaphanus-group proposed by Boxshall (2018) is characterised by: four-segmented leg four with exopodal segments armed with I, I, III spines and ornamented with strip of membrane (modified pecten) associated with each spine; three plumose setae present on posterior margin of distal exopodal segment of leg one; spines two and three on distal exopodal segment of leg one apparently lacking accessory processes; ornamentation of fine setules extending onto surface of second and third endopodal segments of leg two (Fig. 2.7C); outer spines of first and second exopodal segments of leg two aligned close to longitudinal axis of ramus; antenna lacking posterior process on proximal segment; tine on postantennal process vestigial or weakly developed. This group currently contains 15 species.

Table 2.1 Species richness in genera of the family Caligidae (data from Walter and Boxshall, 2022).

Genus	No. species	Genus	No. species
Abasia Wilson, 1908	6	*Kabataella* Prabha & Pillai, 1983	1
Alanlewisia Boxshall, 2008	1	*Lepeophtheirus* von Nordmann, 1832	125
Alebion Krøyer, 1863	9	*Mappates* Rangnekar, 1958	2
Anchicaligus Stebbing, 1900	1	*Markevichus* Ödzikmen, 2008	1
Anuretes Heller, 1865	21	*Metacaligus* Thomsen, 1949	5
Arrama Dojiri & Cressey, 1991	2	*Paralebion* Wilson, 1911	4
Avitocaligus Boxshall & Justine, 2005	2	*Parapetalus* Steenstrup & Lütken, 1861	5
Belizia Cressey, 1990	1	*Parechetus* Pillai, 1962	1
Caligodes Heller, 1865	3	*Pseudanuretes* Yamaguti, 1936	10
Caligus Müller, 1785	273	*Pseudechetus* Prabha & Pillai, 1979	1
Caritus Cressey, 1967	2	*Pseudopetalus* Pillai, 1962	3
Dartevellia Brian, 1939	1	*Pupulina* van Beneden, 1892	7
Echetus Krøyer, 1863	1	*Sinocaligus* Shen, 1957	1
Euryphorus Milne Edwards, 1840	3	*Synestius* Steenstrup & Lütken, 1861	1
Gloiopotes Steenstrup & Lütken, 1861	5	*Tuxophorus* Wilson, 1908	6
Hermilius Heller, 1865	8	Total	512

Caligus confusus-group proposed by Boxshall (2018) is characterised by: four-segmented leg four armed with I, I, III spines on exopodal segments; raised cuticular rib (often with bifid tip) and circular array of large denticles present on apron of leg three; outer margin spine on first exopodal segment of leg three forming large recurved hook; antenna with posterior process on proximal segment (process typically spatulate); accessory tine present on postantennal process, and accessory tine present on posterior maxillulary process. This group currently contains 24 species.

Caligus macarovi-group proposed by Boxshall and Gurney (1980) is characterised by: three-segmented leg four with first and second exopodal segments bearing I and III spines; distal exopodal segment of leg one armed with three posterior margin plumose setae and with spines one, two and three all subequal in length, only spines two and three with accessory processes, and seta four markedly longer than spines; antenna with posterior process on proximal segment; distal margin of brachium of maxilla typically ornamented with marginal serrations; abdomen one-segmented in female. This group currently contains 44 species.

Caligus productus-group proposed by Boxshall and El-Rashidy (2009) is characterised by: posterior margin of distal exopodal segment of leg one lacking typical plumose setae, or retaining single vestigial seta; three-segmented leg four with first and second exopodal segments bearing I and IV spines; ornamentation of long, tapering spinules present along outer margin of second endopodal segment of leg two; antenna with well-developed posterior process on proximal segment. This group currently contains 17 species.

Caligus pseudorhombi-group proposed by Ohtsuka and Boxshall (2019) is characterised by: three-segmented leg four with first and second exopodal segments bearing I and IV spines; genital complex of female as long as wide, without posterolateral lobes, and about twice as long as abdomen; genital complex of male subquadrate with legs five and six located close together at posterolateral corner; abdomen of female about as long as wide; proximal spine on outer margin of third exopodal segment of leg two markedly reduced; myxal surface of maxilliped of male carrying up to three pointed or rounded processes. This group currently contains 18 species.

Caligus undulatus-group proposed by Ohtsuka et al. (2020) is characterised by: three-segmented leg four with first and second exopodal segments bearing I and IV spines; distal exopodal segment of leg one armed with three posterior margin plumose setae; outer spines on distal exopodal segment of leg two small or reduced; body with relatively narrow frontal plates; female genital complex longer than wide; male urosome slender, with two-segmented abdomen. This group currently contains six species.

2.5 Common pest species

Intensive farming of marine finfish has grown rapidly over the past half century and caligid sea lice are responsible for over 60% of the reported disease outbreaks according to Johnson et al. (2004). Species of *Caligus* and *Lepeophtheirus* have been responsible for infestations that have resulted in mortalities of cultured fish species and significant commercial losses (Costello, 2009). While much of the research on sea lice biology and control has focused on species that impact commercial salmonid aquaculture, that is, *Lepeophtheirus salmonis* and *Caligus elongatus* in the Northern Hemisphere and *Caligus rogercresseyi* Boxshall & Bravo 2000 in the Southern Hemisphere, it is clear that numerous species are now implicated. Johnson et al. (2004) listed over 30 named species of *Caligus* and *Lepeophtheirus*, plus one species of *Parapetalus* that had been reported from finfish cultured in brackish and marine waters. Since then numerous additional species have been reported (for example, Venmathi Maran et al., 2009, 2012a). It seems that problems with sea lice emerge within a short time of any new species of finfish being taken into culture. When an exotic fish species is taken into culture in an area, such as salmonids in Chile (where there are no natural salmonids), then sea lice infestation involves host switching by local caligid species: an example being *Caligus rogercresseyi* which switched onto farmed salmonids from local hosts in Chilean waters which included members of the families Nototheniidae, Atherinidae and Paralichthyidae (Carvajal et al., 1998). When the fish taken into culture is local then local sea lice species are already present in the environment and problems can emerge very quickly.

The best known sea louse species is *Lepeophtheirus salmonis*. The formal scientific name of this species is correctly attributed to Krøyer (1837) although the full description only appeared a year later with Krøyer (1838). Kabata (1979) considered that *Lepeophtheirus salmonis* had a circumpolar distribution in the Northern Hemisphere where it occurs primarily on salmonids, although he noted a few records from non-salmonid hosts. Slight morphological differences had been noted between Pacific and Atlantic *Lepeophtheirus salmonis*, but molecular genetic approaches have allowed the differences to be explored more fully. In a large-scale study, Yazawa et al. (2008) analysed about 15,000 expressed sequence tags (ESTs) and sequence data covering the full mitochondrial genome and found significant divergence between Pacific and Atlantic populations. The divergence was about 3.2% for nuclear genes and 7.1% for the mitochondrial genome (including 6.1% for Cytochrome Oxidase 1 (CO1) gene). Using molecular clock methodology, they inferred that *Lepeophtheirus salmonis* was an Atlantic species which had colonised the Pacific via the Bering Strait when it opened up about five million years ago (Yazawa et al., 2008).

The status of the Atlantic and Pacific forms was clarified by Skern-Mauritzen et al. (2014) who demonstrated reproductive compatibility between *Lepeophtheirus salmonis* from the two oceans and successfully established F2 hybrid strains using separate maternal lines from both oceans. They considered that Atlantic and Pacific forms should not be regarded as separate species based on reproductive segregation or sequence divergence levels; instead they formally established two subspecies: *Lepeophtheirus salmonis salmonis* for the Atlantic form and *Lepeophtheirus salmonis oncorhynchi* for the Pacific form. The recognition of formal status for the subspecies names brings them within the rules of nomenclature for zoology and, unfortunately, there are earlier names proposed for Pacific *Lepeophtheirus* on salmonid hosts which may have priority. This nomenclatural problem remains to be resolved.

Caligus elongatus is an extremely common species around the North Atlantic and is also a serious health hazard for farmed salmonids (Pike and Wadsworth, 1999). Kabata (1979) noted that it had been recorded from more than 80 different host species including both actinopterygian and elasmobranch fishes. Since Kabata's work many new hosts have been reported and this total now exceeds 100 species, by far the largest host range known for any caligid. Known hosts now include the common minke whale, *Balaenoptera acutorostrata* Lacepède, 1804, in Icelandic waters where the prevalence rate of 11.9% and mean intensity of 95.5 (Ólafsdóttir and Shinn, 2013) indicate that this is not an accidental host record. There is, however, an identity problem with *Caligus elongatus* because preliminary molecular data suggest that the morphological species comprises two distinct genotypes (Øines and Heuch, 2005).

Sea lice are a global problem in finfish aquaculture although the particular species varies with geographical region and with the host fish being cultured. In Chile, for example, where aquaculture is focused on salmonids, the dominant sea louse is *Caligus rogercresseyi*. In East and South-East Asia a much wider variety of fish species is cultured across the region and a wider range of caligids is implicated (Johnson et al., 2004). However, infestations by a few species appear to be spreading rapidly: *Caligus epidemicus* Hewitt, 1971, for example, was first described from Victoria in Australia in 1971 (Hewitt, 1971) but has since been reported across the Indo-Pacific region from Australia to Malaysia, the Philippines, Thailand and Taiwan on 39 different hosts (Boxshall, 2018). It has been reported from caged finfish from Australia to Malaysia and Taiwan, and has even been recorded on cultured decapod prawns (Ruangpan and Kabata, 1984). A similar situation involves the very distinctive species *Caligus sclerotinosus* Roubal, Armitage & Rohde, 1983. It was first described from Australia in 1983 (Roubal et al., 1983), but has since become a problem for cultured *Pagrus major* (Temminck & Schlegel, 1843) in Japan and Korea (Venmathi Maran et al., 2012a), and has recently been reported from Mexican Pacific waters. Currently, we don't know whether these pest species have always had a wide geographical range but have hitherto been overlooked, or are spreading naturally at a rapid rate, or are being inadvertently spread by translocation of hosts or equipment used in finfish farming.

2.6 Reproductive behaviour

Copepods have separate sexes and females become sexually receptive at the final moult to the adult stage. The adult female has paired ovaries located anteriorly in the cephalothorax (Fig. 2.10A and B) and the paired oviducts pass posteriorly into the genital complex. In mature females the oviducts are typically full of discoid eggs (Fig. 2.10C) ready for release as the next pair of egg strings is formed. The material comprising the egg strings themselves is produced by paired cement glands located posteriorly in the genital complex (Fig. 2.10D). Sperm are stored in the seminal receptacle (Fig. 2.10E) of the female, and are released into the distal part of the oviduct as the egg strings are extruded. The paired testes of the adult male are also located anteriorly in the cephalothorax. Paired vasa deferentia pass posteriorly into the genital complex where they expand, forming a chamber within which the mature sperm are packaged into spermatophores.

In sea lice there is typically a precopulatory mate guarding phase, in which adult males attach to immature females and remain attached until the female moults into the sexually receptive adult (Boxshall, 1990b; Anstensrud, 1990). In this precopulatory position males hold onto the dorsal side of the female, using their antennae to grasp the anterior end of the female's genital complex. Adult males have been reported guarding late chalimus stages in *Caligus* (Ho and Lin, 2004). In *Lepeophtheirus* the pre-adult stages are the most commonly guarded (Boxshall, 1990b; Anstensrud, 1992) but mate guarding has also been reported as early as a female chalimus II stage (Boxshall, 1990b, as chalimus IV). Mate guarding represents a significant investment by males although this would be reduced by selection of later female stages, and Boxshall (1990b) noted that factors determining the benefit to the male of mate guarding may include the probability of encountering a sexually receptive female, sex ratio and sperm competition.

Mating in caligids occurs on the host and multiple matings are typical (Anstensrud, 1992; Hull et al., 1998). In planktonic copepods, females release pheromones into the water column, effectively laying down a chemical trail that the adult males can use to locate them (Yen et al., 1998). Diffusable pheromones are also involved in mate location in sea lice. Female sea lice release pheromones which the male uses to locate the female. Ingvarsdóttir et al. (2002) showed that adult males of *Lepeophtheirus salmonis* were most strongly attracted towards a semiochemical produced by pre-adult II virgin females and preferred pre-adult virgins (or water preconditioned with pre-adult virgins)

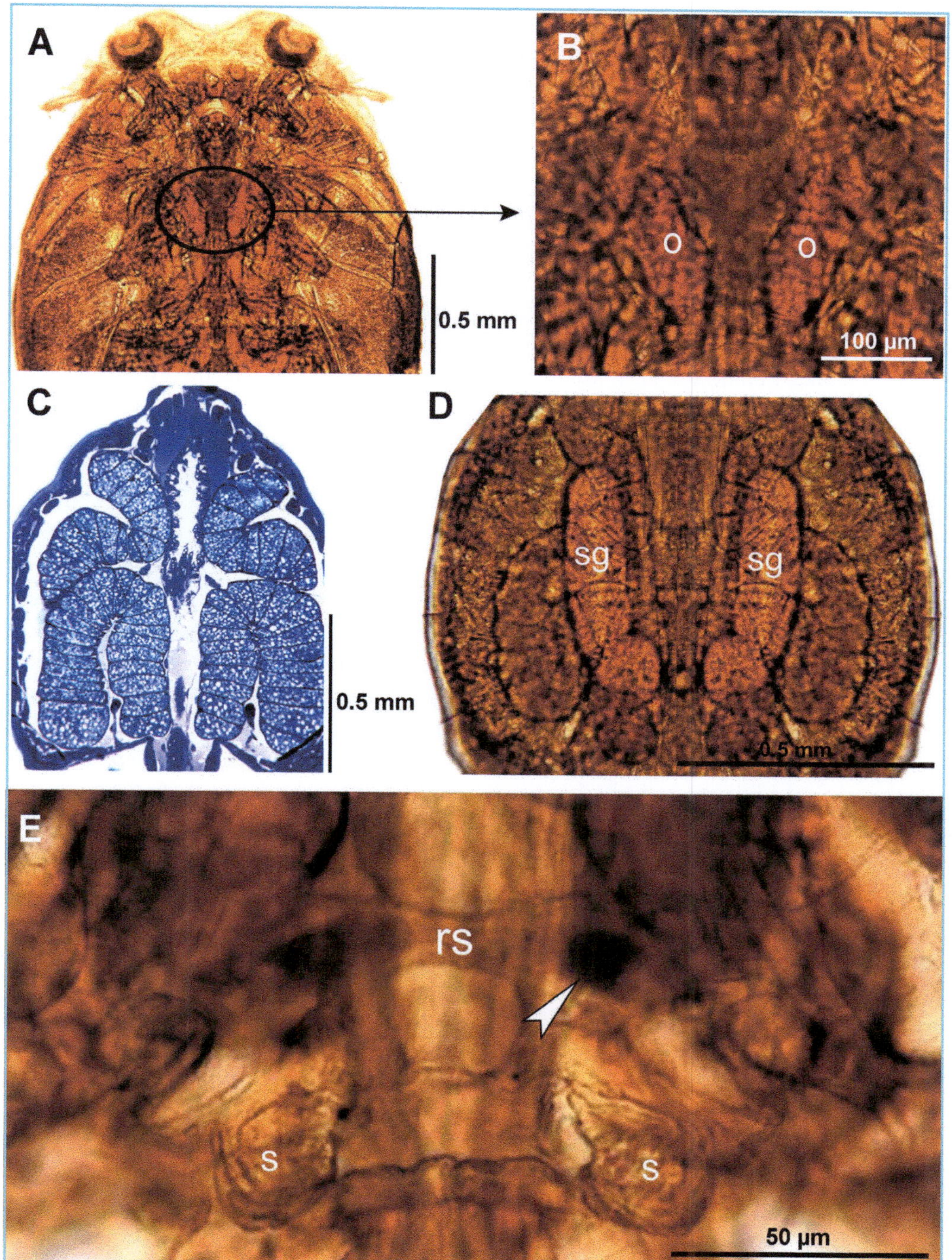

Fig. 2.10 *Caligus minimus* Otto, 1821 (female). A and B. Location of ovaries (o) in the cephalothorax; C. Horizontal section through genital complex showing coiled oviducts containing developing eggs; D. Paired cement glands (sg) in female genital complex; E. Posteroventral part of female genital complex showing attached spermatophores (s) and mass of spermatozoa (arrowhead) stored in the receptaculum seminis (rs). [A, B, D, E: Acetocarmine stained specimen; C: Toludine blue stained section (3 µm) of the genital complex.]

to mated females. It is not known whether males are able to distinguish the smell from virgins and mated females, or whether females stop emitting pheromones once they have mated. While the female only becomes sexually receptive and ready to receive spermatophores from the adult male after the final moult, it seems highly likely that the preceding pre-adult stages (and the eqivalent late chalimus stages in *Caligus*) also produce pheromones which have signalled their presence to mate-guarding males.

Mating, the transfer of paired spermatophores from the male to the female, was described in detail in *Lepeophtheirus pectoralis* by Anstensrud (1990). Prior to the final moult the second pre-adult female produces a temporary frontal filament securing it to the host during moulting. The exuvium splits and is shed posteriorly by contractions of the body (Anstensrud, 1990). After the new exoskeleton hardens, the sexually receptive adult female detaches itself from the frontal filament. During moulting the mate-guarding male releases its hold on the female and usually is positioned close by on the surface of the host. Remaining in close proximity is important for the male since Anstensrud (1990) found that during the moulting of the female, a new male may move in and take over the female. During copulation, the male holds onto the ventral side of the female's genital complex using its antennae and maxillipeds. The male then extrudes a pair of ovoid spermatophores which it deposits on the ventral surface of the genital complex of the female (Fig. 2.11A) using the second pair of swimming legs. Spermatophores are always transferred in pairs and are placed in the vicinity of the paired copulatory pores located either side of the ventral midline. Each spermatophore typically discharges into the copulatory pore on the opposite side of the body (Huys and Boxshall, 1991) resulting in a characteristic crossing of the spermatophore tubules (Fig. 2.11B and C). The sperm pass through paired copulatory ducts into the fused seminal receptacles where they are stored.

As eggs pass out of the oviduct opening they are fertilised by sperm stored in the seminal receptacles and mated females can produce multiple pairs of fertilised egg strings from a single mating. After the eggs in the egg strings hatch, the empty strings detach and after a short time a new pair of egg strings is extruded containing mature eggs which were stored in the paired oviducts within the genital complex. In *Caligus epidemicus* the delay between the spent egg string being detached and the production of a new clutch (that is, a new pair of egg strings) is in the order of 20 to 25 min (Lin and Ho, 1993). Few data are available on the number of pairs of egg strings produced, but it appears that females typically produce multiple clutches. Meeüs et al. (1993) noted that up to 16 clutches could be produced by a single female of *Lepeophtheirus europaensis* Zeddam, Berrebi, Renaud, Raibaut & Gabrion, 1988.

2.7 Life cycles

The life cycles of only 3.5% of caligid species are known (Venmathi Maran et al., 2016) and the only complete cycles that are known belong to the two large genera *Caligus* (13 species) and *Lepeophtheirus* (five species). The adult females of all caligids produce paired uniseriate egg strings (Fig. 2.12A), each string typically containing in the range of 5 to 50 flattened, discoid eggs although some of the larger species produce many more. The number of eggs per egg string in *Lepeophtheirus salmonis*, for example, varies from less than 100 to 700 (Wootten et al., 1982), and part of this variation may be due to the age of the female as egg strings tend to be longer in older females. The egg strings are typically linear and trail out behind the female but in *Avitocaligus* the egg strings are loosely coiled within a space protected by the lamellar plates on the genital complex. The eggs hatch into a free-swimming

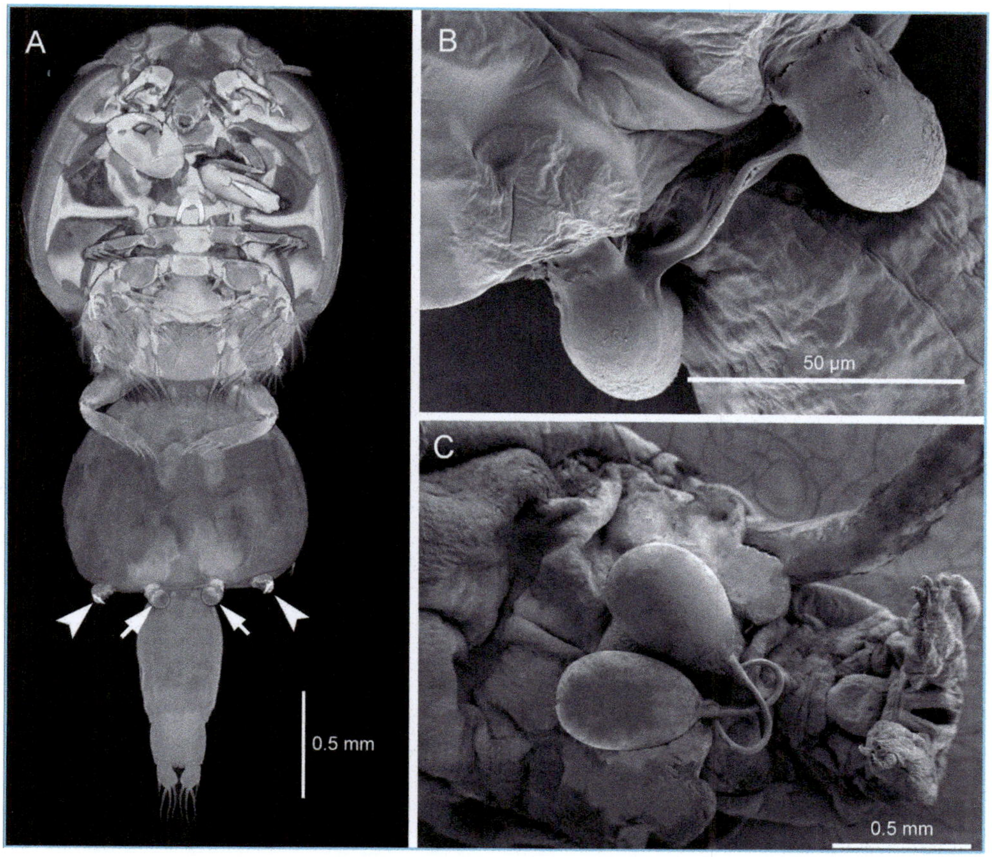

Fig. 2.11 *Caligus diaphanus* Nordmann, 1832 (female). A. Habitus, ventral view showing position of attached spermatophores (arrows) and oviduct openings where egg strings extruded (arrowheads); B. Attached spermatophores with crossed spermatophore tubules; C. Female *Euryphorus brachypterus* (Gerstaecker, 1853) with pear-shaped spermatophores attached.

and non-feeding larval stage, the nauplius, which is lecithotrophic (depending on the yolk it contains) and does not have a functional mouth.

The first nauplius stage (Fig. 2.12B and C) has a small ovoid body and is characterised by the possession of three pairs of cephalic limbs, antennules, antennae and mandibles. The antennule comprises just a single axis and is two-segmented. The antennae and mandibles are biramous; the rami are setose in both and these limbs function as swimming appendages. Both lack feeding processes on the coxa, as is typical of non-feeding lecithotrophic larvae. The caligid nauplius carries a pair of balancers at the rear end of the body. Despite their name, the function of these elements is unclear, although we know that they are homologous with caudal seta IV of other copepods (Huys et al., 2007). The nauplius phase in the life cycle of all caligids comprises two stages, the first and second nauplii, separated by a moult (Fig. 2.12D–E). The second nauplius (Fig. 2.12F) is typically a little longer than the first, and differs in limb setation patterns and in the presence of a rudimentary fourth cephalic limb in the midline on the ventral body surface. The second nauplius moults into the free-swimming, first copepodid stage (Fig. 2.12G–I). The nauplii are relatively short-lived and serve as a dispersal phase in the caligid life cycle.

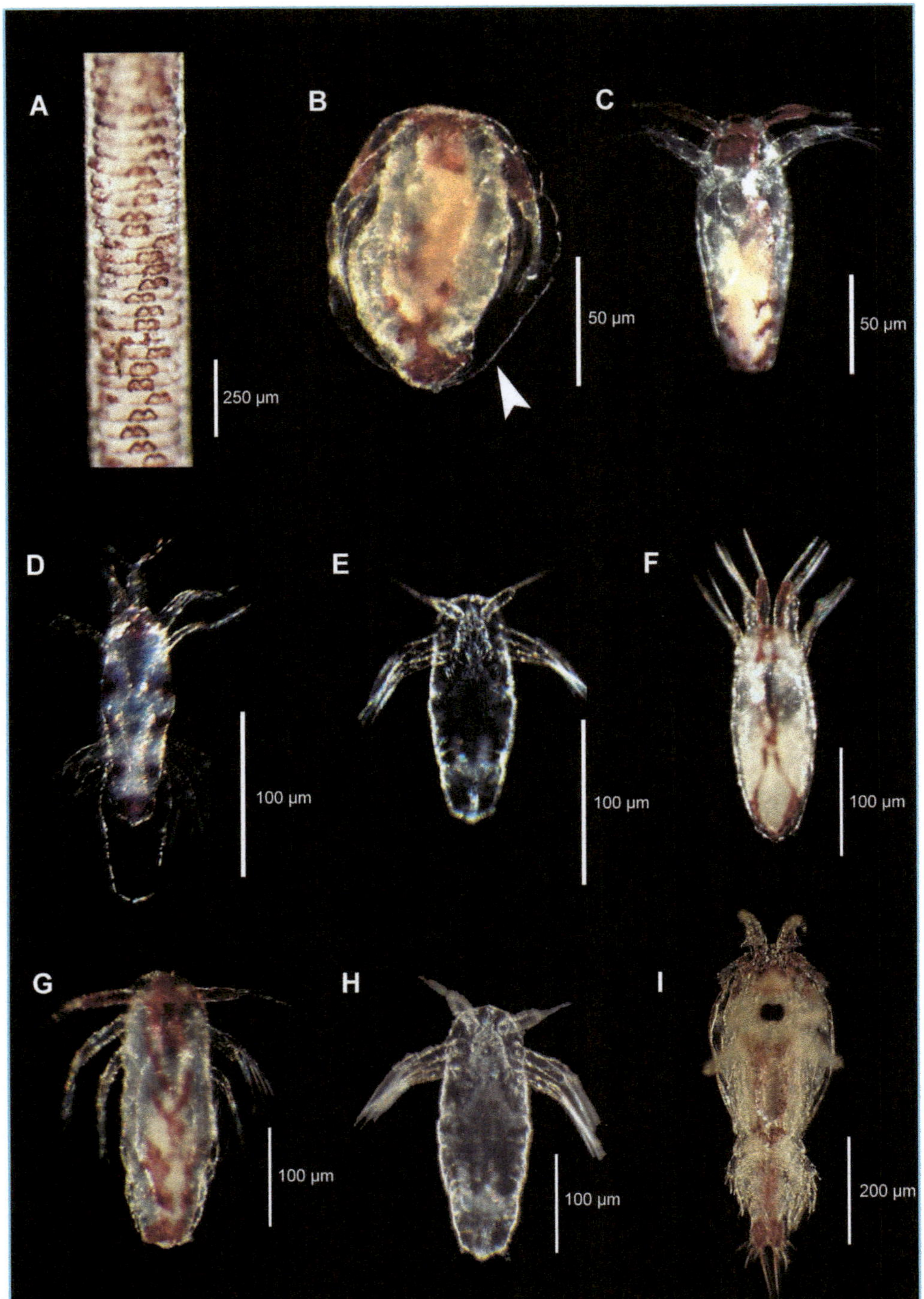

Fig. 2.12 Free-living stages of *Caligus minimus* Otto, 1821. A. Pigmented egg string close to hatching; B. Detached egg showing nauplius ready to hatch but still within egg membrane (arrowhead); C. Newly-hatched first nauplius; D. First nauplius undergoing moult; E. Empty exuvium of first nauplius; F. Second nauplius; G. Late second nauplius containing developing copepodid; H. Exuvium of second nauplius; I. Infective copepodid.

The first copepodid stage (Fig. 2.12I) is able to swim freely in the water column. Its body comprises an anterior cephalothorax consisting of the cephalosome plus the incorporated first pedigerous somite, and four posterior somites – the second to fourth pedigerous somites plus the anal somite. The cephalothorax bears the antennules, antennae, mandibles (contained within the mouth tube), maxillules, maxillae, maxillipeds and first swimming legs in anterior to posterior sequence. At this stage, the mandible retains a vestige of the palp in a few species of *Caligus* (Ohtsuka et al., 2009). In addition, the developing postantennal processes are already visible laterally. The cephalosomic limbs of the copepodid are clearly segmented and functional. Behind the cephalothorax, the free second pedigerous somite bears a pair of biramous swimming legs, similar to the first pair. The third pedigerous somite carries a rudiment of the developing leg three. The following somite lacks appendages and the final (anal) somite bears the paired caudal rami. The copepodid swims actively using the two pairs of biramous swimming legs.

The first copepodid stage is the infective stage, locating and securing attachment to the host. It is equipped with an array of mechano- and chemoreceptors on the paired antennules and has a nauplius eye with a cuticular lens in each of the paired dorsal ocelli. Different species probably detect and recognise their hosts using differing mixes of visual, mechanical and chemical cues (Boxshall, 1976; Novales Flamarique et al., 2000; Bailey et al., 2006; Fields et al., 2017). Having landed on a potential host, the copepodid attaches initially using its prehensile antennae and maxillipeds but later extrudes a frontal filament which securely attaches the developing larval stage to the host. In *Caligus* species, a pre-formed frontal filament is carried within the frontal region of the infective copepodid. This is everted and attached to the host before the next moult into the chalimus I stage (Bron et al., 1991; Ohtsuka et al., 2009).

Chalimus stages are characterised by being permanently attached via a frontal filament which is present throughout the entire stage. In *Caligus* species there are four chalimus stages (chalimus I – chalimus IV) and at the moult from copepodid to chalimus I an additional bulb of material (an ecdysial extension lobe) is secreted at the origin of the frontal filament, around its base. At each of the next three moults a further bulb of material is secreted at the origin of the filament, so chalimus II has two ecdysial extension lobes at the base of its filament, chalimus III has three lobes and the final chalimus IV has four (Fig. 2.13). In *Caligus* species the chalimus IV stage undergoes a final moult into the adult which is motile. In this pattern, first suggested for *Caligus centrodonti* Baird, 1850 by Gurney (1934), an extension lobe of filament secretion is released from a frontal gland in the larval cephalothorax at each moult and serves to attach the newly moulted parasite to its original filament, which has remained securely embedded in the host throughout the moulting process. This pattern is shown in several other *Caligus* species, including *Caligus elongatus* (Piasecki and MacKinnon, 1993), *Caligus punctatus* Shiino, 1955 (Kim, 1993), *Caligus orientalis* Gusev, 1951 (Hwa, 1965), *Caligus pageti* Russell, 1925 (Ben Hassine, 1983), *Caligus stromatei* Krøyer, 1863 (Lin et al., 1997 under the name of *Caligus multispinosus*) and *Caligus fugu* (Yamaguti, 1936) (Ohtsuka et al., 2009, under the name *Pseudocaligus fugu*). According to the reinterpretation of Ohtsuka et al. (2009), *Caligus epidemicus* probably also shares this pattern even though Ho and Lin (2004) originally inferred that a new frontal filament was formed during each moult during the chalimus phase. However, not all *Caligus* species conform to this pattern: species such as *Caligus clemensi* Parker & Margolis, 1964 have an enlarged frontal zone in the chalimus and appear to have a simple frontal filament lacking ecdysial extension lobes (Kabata, 1972).

It was long thought that the life cycle of *Lepeophtheirus* species differed from the *Caligus* pattern described above by the inclusion of two additional pre-adult stages (for example, Boxshall, 1974c;

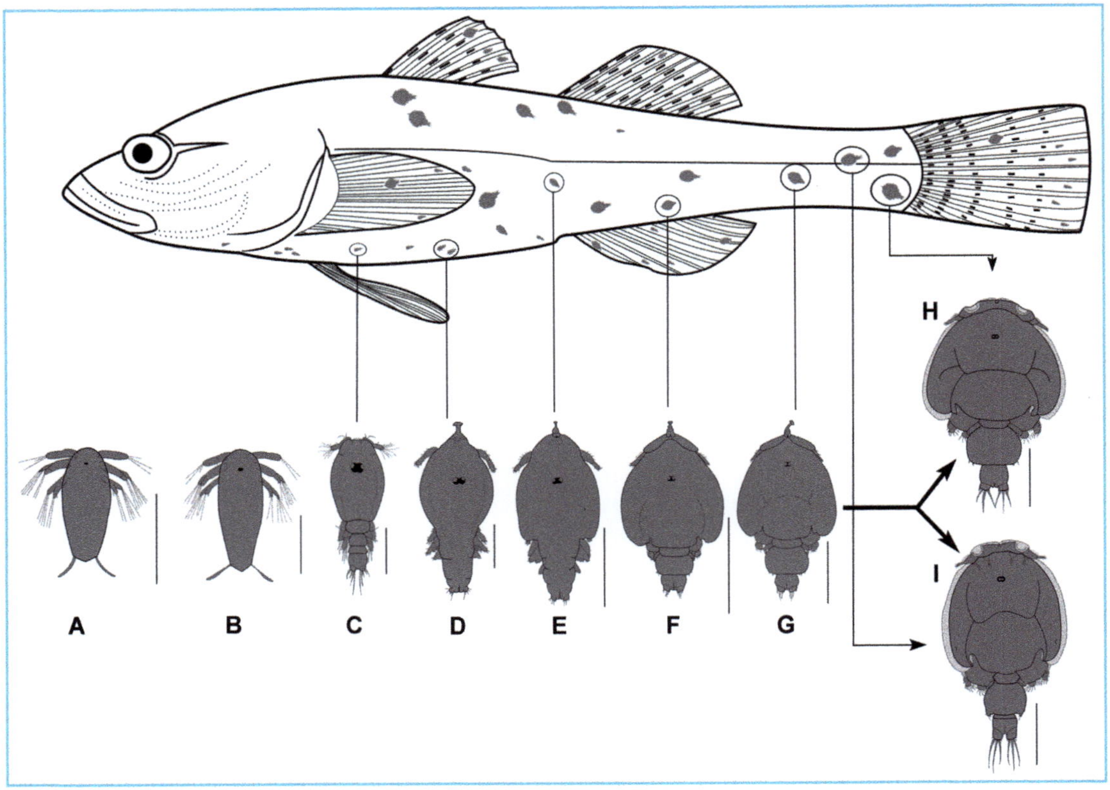

Fig. 2.13 The life cycle of *Caligus* (based on *Caligus punctatus*). A. First nauplius stage; B. Second nauplius; C. Infective copepodid stage; D. Chalimus I; E. Chalimus II; F. Chalimus III; G. Chalimus IV; H. Adult female; I. Adult male. Scale bars: A and B, 0.1 mm; C and D, 0.2 mm; E and F, 0.5 mm; G–I, 1.0 mm.

Johnson and Albright, 1991) but it has recently been shown that the life cycle of *Lepeophtheirus* species comprises the same total number of stages as found in *Caligus* (Venmathi Maran et al., 2013; Hamre et al., 2013). In *Lepeophtheirus salmonis*, for example, there are two naupliar stages, the copepodid, two chalimus stages, two pre-adult stages and the adult (Fig. 2.14). The third and fourth post-copepodid stages do not have permanent frontal filaments; instead they produce a temporary frontal filament which secures them only during the moult to the next stage but which is then detached leaving the newly emerged stage freely motile over the surface of the host. These motile stages are referred to as pre-adults. The second pre-adult undergoes a definitive (final) moult into the mature adult.

The frontal filament of a *Lepeophtheirus* chalimus differs in possessing a surface sheath which is continuous with the integument on the frontal region of the chalimus (Pike et al., 1993), and it does not add a freshly secreted bulb of material at each moult through the chalimus phase (and thus has no ecdysial extension lobes). There is some uncertainty whether the frontal filament in *Lepeophtheirus* is replaced at each moult or whether the moulting chalimus adheres to the original frontal filament during ecdysis. Given that the two mobile pre-adult stages in *Lepeophtheirus* life cycles secure themselves to the host during their final two moults (from pre-adult I to pre-adult II, and from pre-adult II to adult) by secreting a novel but temporary frontal filament (Anstensrud,

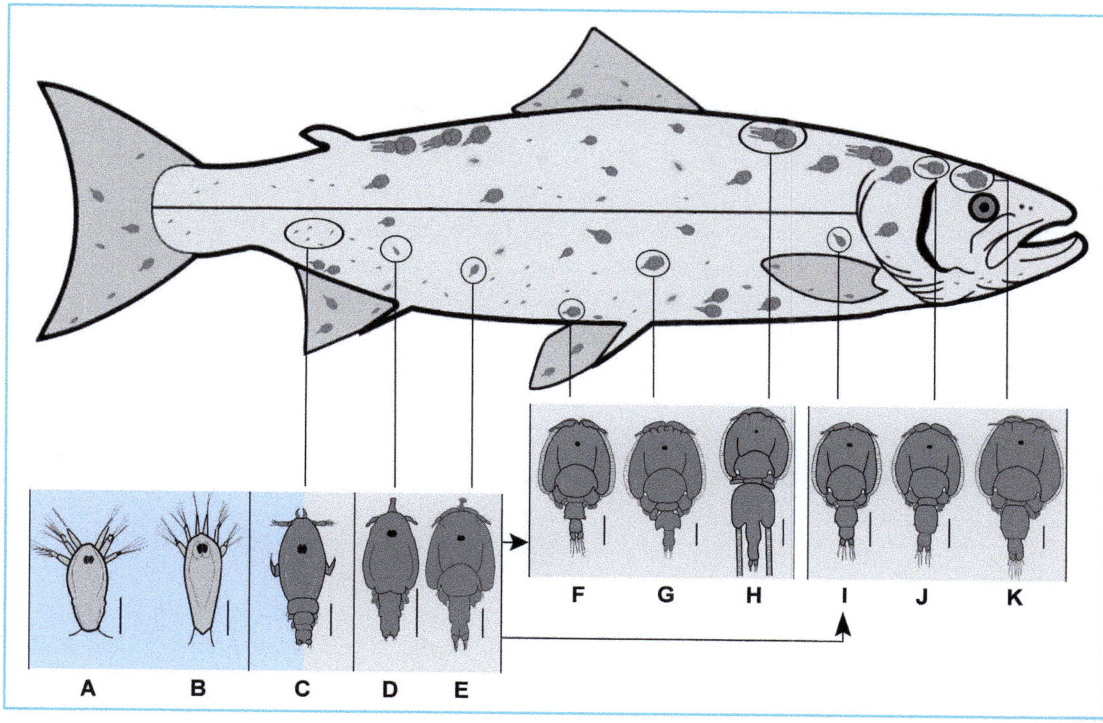

Fig. 2.14 Life cycle of *Lepeophtheirus* (based on *Lepeophtheirus salmonis*). A. First nauplius stage; B. Second nauplius; C. Infective copepodid stage; D. Chalimus I; E. Chalimus II; F. First pre-adult female; G. Second pre-adult female; H. Adult female; I. First pre-adult male; J. Second pre-adult male; K. Adult male. Scale bars: A–C, 0.2 mm; D and E, 0.5 mm; F, G, I–K, 1.0 mm; H, 2.0 mm.

1990), Ohtsuka et al. (2009) concluded it was likely that *Lepeophtheirus* species secrete a new filament at all moults throughout the chalimus phase.

The chalimus phase in the life cycle of sea lice can be viewed as a metamorphic phase. The body changes from the basically cylindrical form of the copepodid to the dorsoventrally flattened form of the adult, gradually attaining adult tagmosis by incorporating the first to third pedigerous somites into the cephalothorax and adding the fourth pedigerous somite, genital complex and free abdomen. This metamorphosis is largely achieved in only two moult stages in *Lepeophtheirus* but takes four moult stages in *Caligus*. As noted by Ohtsuka et al. (2009), this metamorphosis comprises an initial regression (that is, loss of, or weaker expression of, limb and body segmentation, and reduction in size of setal elements) followed by progressive development (that is, stronger expression of limb and body segmentation, increased setation) towards the adult condition. The limbs most affected by this metamorphosis are the antennae, maxillipeds and legs one and two.

Caligids have separate sexes and both females and males pass through the same number of moult stages during development. In *Caligus* species there are two naupliar stages, the infective copepodid, four chalimus stages and the adult whereas in *Lepeophtheirus* species there are two naupliar stages, the infective copepodid, two chalimus stages, two pre-adult stages and the adult. In older literature on sea lice life cycles, there was considerable disparity in the number of stages recognised between the copepodid and the adult, but recent studies (Ohtsuka et al., 2009; Venmathi Maran et al., 2013; Hamre et al., 2013) have largely resolved this confusion. In addition to the erroneous recognition of four chalimus stages in *Lepeophtheirus* cycles, another major source of confusion concerned the status of the so-called pre-adult stages described in certain *Caligus* cycles. Ohtsuka et al. (2009) concluded that there was no pre-adult in *Caligus* and that the pre-adult stages that had been described, for example in *Caligus clemensi*, *Caligus spinosus* and *Caligus epidemicus* (Kabata, 1972; Izawa, 1969; Ho and Lin, 2004), were merely young adults as they were not separated by a moult from the fully grown adults. Young females typically emerge from the final moult with a small genital complex which expands considerably after mating and as it fills with developing eggs. This post-mating metamorphosis can also be recognised in females of other genera and is probably common to all caligids. In *Lepeophtheirus pectoralis* (Müller, 1776), for example, the adult female immediately post-moulting has a mean length of 3.9 mm whereas the gravid female is 5.6 mm and this increase in length is due entirely to expansion of the genital complex (Boxshall, 1974c).

2.8 Alternative life cycles

In all caligid species where it has been fully elucidated, the life cycle involves a single host. The free-swimming phase (the nauplii and copepodid) alternates with an ectoparasitic phase (copepodid to adult), with the infective copepodid bridging the two phases. However, there is some fragmentary evidence that in a few caligids the life cycle routinely involves more than one host. Hayward et al. (2008) first noted that only adult sea lice (*Caligus chiastos* Lin and Ho, 2003) were present on ranched south bluefin tuna (*Thunnus maccoyii* (Castlenau, 1872)) in sea cages and later (Hayward et al., 2011) found developing chalimus stages on up to 93% of *Thamncornus degeni* (Regan, 1903) outside of the sea cages. The conspecificity of the chalimus stages on *Thamncornus degeni* and the adults on *Thunnus maccoyii* was confirmed by molecular studies (Hayward et al., 2011). This may indicate a change of host between chalimus and adult. Similarly, Cressey and Cressey (1980) considered that the immature stages of *Caligus biseriodentatus* Shen, 1957 were found on a different host (*Scomberomorus* spp.) from the adult (*Auxis thazard* Lacepède, 1800). The possibility of host switching during the life cycle requires further investigation as it may help to explain some of the reports of adult caligids caught in plankton nets while swimming freely in the water column (Venmathi Maran et al., 2016).

Reports of adult sea lice caught up in the water column are relatively frequent and most concern species of *Caligus* (Todd et al., 1996; Suàrez-Morales et al., 2012; Ho and Lin, 2004; Venmathi Maran et al., 2012b, 2012c, 2016; Kim et al., 2019) although adults of *Lepeophtheirus* and *Metacaligus* Thomsen, 1949 have also been found (Venmathi Maran et al., 2012c). Remarkably, 13 species of *Caligus* were originally established based on material collected in plankton samples. Various hypotheses explaining the presence of caligids in the water column for extended periods have been suggested (Venmathi Maran et al., 2016). The case of *Caligus undulatus* Shen and Li, 1959 is especially interesting: it was first discovered in plankton samples taken off China (Shen and Li, 1959) and has since been collected widely from Indo-West Pacific and Atlantic waters, but for more than 60 years it had not been found in association with any host. Ohtsuka et al. (2018) suggested that this species attaches temporarily to its host to feed and frequently shifts between host and water column. Similar behaviour is found in the branchiuran genus *Argulus* Müller, 1785, although the adaptive significance of this behaviour is more apparent in the Branchiura as the females leave their hosts after feeding in order to lay eggs on submerged hard substrates (Boxshall, 2005). The natural host of *Caligus undulatus* was finally identified by Ohtsuka et al. (2020) who found both sexes on *Sardinella zunasi* (Bleeker, 1854) caught in the Seto Inland Sea of Japan (East Sea). They also noted that ovigerous females in the water column carried fewer eggs per egg string than those found on the host, and explored the suggestion of an evolutionary trade-off between the various possible drivers of this behaviour. *Sardinella zunasi* is euryhaline and can be found in brackish as well as marine coastal waters, and Ohtsuka et al. (2020) inferred that *Caligus undulatus* probably detached from its host as it entered lower salinity waters.

Seven of the 13 species of *Caligus* collected in plankton samples were represented only by males (Venmathi Maran et al., 2016) and Hull et al. (1998) demonstrated that males of *Lepeophtheirus salmonis* are more mobile than females and switched between host individuals more frequently than females did (in aquaculture facilities). Given that a single unfertilised parasite alone on a host is effectively at a reproductive dead end and that in copepods it is the male that actively locates a mate, then the higher likelihood of finding males in the water column seems readily explicable.

2.9 Host location and recognition

Host location is a critical process in the caligid life cycle: the infective copepodid stage is less than a millimetre in length and it is a big ocean. The initial phase of host location involves the behaviour of the infective copepodid taking it into the part of the habitat that will optimise the probability of encountering a host. Boxshall (1976) noted that after an initial period of positively phototactic behaviour, the copepodid of *Lepeophtheirus pectoralis* became negatively phototactic. This change in behaviour would tend to take it down to the base of the water column to where the probability of encountering a suitable flatfish host is much greater. In contrast, the copepodid larvae of *Lepeophtheirus salmonis* exhibit diurnal vertical migration (Heuch et al., 1995), remain positively phototactic and swim upwards into surface waters until low salinities are encountered (Novales Flamarique et al., 2000). This behaviour results in the accumulation of infective copepodids in the subsurface halocline around the depth where they are most likely to encounter out-migrating salmon smolts and sea trout (see discussion in Crosbie et al., 2019).

Once in the optimum part of the water column, host detection and recognition by copepodids of *Lepeophtheirus salmonis* may involve various sensory modalities and behaviour patterns operating in sequence or simultaneously. Initial cues may be visual, such as fluctuating light intensity caused by light reflecting from the sides of a potential host or the shadows cast by fish swimming overhead (Fields et al., 2017). Infective copepodids strongly increased their swimming speed in response to changes in light intensity but such cues probably operate at relatively long range and are unlikely to be species specific (Novales Flamarique et al., 2000). Subsequent behavioural cues may be mechanosensory (Heuch et al., 2007) and olfactory (Fields et al., 2007): both probably operate at close range and are more likely to be species specific. Heuch et al. (2007) demonstrated that hydromechanical signals in the high velocity gradient around the head of a fish can trigger a rapid swimming response and subsequent attachment to the host by its flatfish infective copepodid. Similarly, Boxshall (1976) demonstrated that attaching *Lepeophtheirus pectoralis* copepodids showed a positive rheotactic response resulting in their concentration at sites on its flatish host associated with water flow, such as the branchial opening and the marginal fins. Mechanoreception acts by switching on specific behaviours that enable landing on a host fish.

According to Mordue and Birkett (2009), chemoreception plays a defining role in host location and recognition – and this role has been investigated experimentally using conditioned water (that is, water containing water-soluble chemicals released from the skin of the host). Bailey et al. (2006) demonstrated that *Lepeophtheirus salmonis* are attracted to salmon conditioned water and showed directional choice in response to extracts in y-tube choice experiments. However, Fields et al. (2017) were unable to clearly demonstrate whether olfactory cues provide directional information although they found that a diffuse host odour cue stimulated swimming by the infective copepodid. Mordue and Birkett (2009) concluded that the detection of host chemical cues switches on behaviour patterns associated with host searching and also induced landing responses, whereas non-host chemical cues failed to induce attraction or significant landing behaviour.

Núñez-Acuña et al. (2016) identified a small peptide, Cath-2, found in the mucus of salmon, which activates a chemosensory transduction pathway involved in host recognition by copepodids of *Caligus rogercresseyi*. This pathway is involved in the detection of host-specific chemical stimuli and recent work on the genetics of chemoreceptors in *Lepeophtheirus salmonis* has demonstrated the role of a specific receptor (Núñez-Acuña et al., 2019). These molecular genetics studies are beginning to generate insights into the precise mechanisms by which host specificity could be maintained.

2.10 Patterns of host usage

Caligids are primarily ectoparasites of marine vertebrates although there are a few records of sea lice on invertebrate hosts. With the exception of *Anchicaligus nautili* Stebbing, 1900 which is exclusively found on its cephalopod mollusc host, *Nautilus pompilus* Linnaeus, 1758, and possibly *Caligus lolliguncula*e Capart, 1941 which was described on the basis of a single adult of each sex collected from the squid *Afrololigo mercatoris* (Adam, 1941) (Capart, 1941), these extremely rare records appear to be either accidental, such as the isolated report of a *Lepeophtheirus* on the echinoderm *Asterias* Linnaeus, 1758 (Hansen, 1923), or refer to an artificial aquaculture facility infestation on decapod crustaceans (Ruangpan and Kabata, 1984). Although there are several reports of caligids from marine mammals and some document well-established infestations (Ólafsdóttir and Shinn, 2013), caligid sea lice overwhelmingly utilise fish as hosts. The genera *Alebion*, *Paralebion* Wilson, 1911 and *Pupulina* Beneden, 1892 are parasites of elasmobranch hosts while the remaining genera primarily use actinopterygian fishes as hosts.

The two most species-rich genera, *Caligus* and *Lepeophtheirus*, each contain a small number of species that occur on chondrichthyan (elasmobranch and holocephalan) fishes. Reports exist of 12 species of *Lepeophtheirus* parasitic on elasmobranch hosts, but eight of these species are known predominantly from actinopterygian hosts (Özak et al., 2018). One of the remaining four species has not been seen since its original description more than 150 years ago and was considered as *species inquirendum* by Özak et al. (2018), so effectively only 3 out of 125 species of *Lepeophtheirus* routinely occur on elasmobranchs. Similarly, only 6 of the 273 valid species of *Caligus* exclusively utilise elasmobranch fishes as hosts although another 13 species include one or more chondrichthyan species in their lists of reported hosts (which overwhelmingly comprise actinopterygian fishes).

The mean number of host species recorded per species in the Caligidae is 4.2 (based on host data collated for 470 species). A small number of species described from specimens caught in the plankton have no host data, and other species have no identified hosts, but about 45% of species are known from a single host species only (Fig. 2.15). At the opposite end of the host spectrum is *Caligus elongatus* with over 100 recorded host species, but as noted above there is uncertainty over the identity

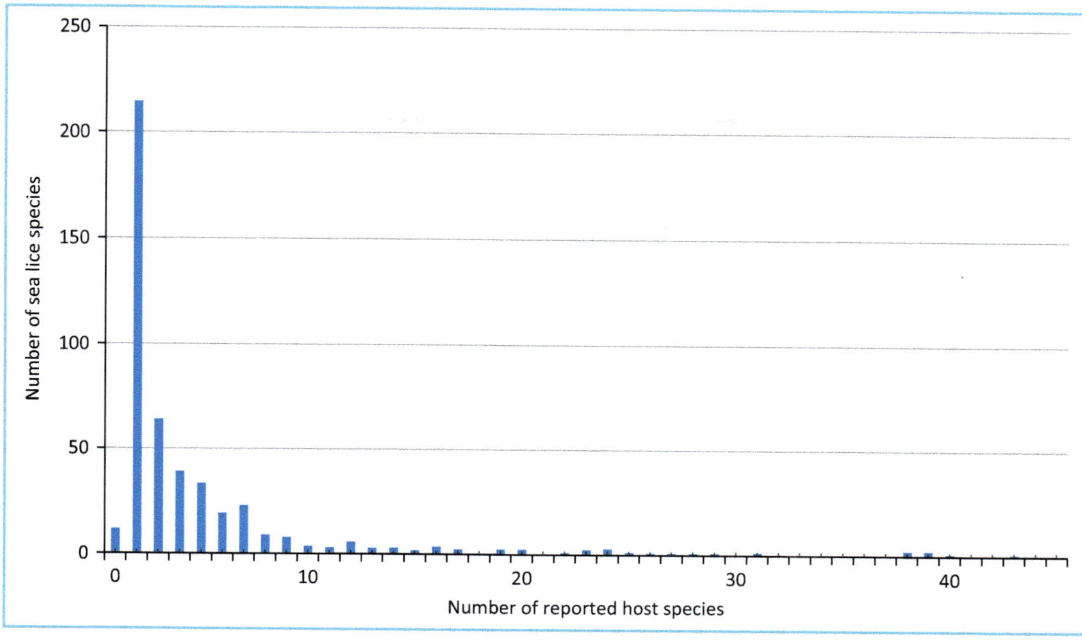

Fig. 2.15 Host specificity of sea lice species: number of hosts recorded per species of Caligidae (excluding *Caligus elongatus* which has been recorded from more than 100 host species but is probably a complex of two sibling species).

of this parasite. Host data are often fragmentary or even unreliable due to misidentifications of either host or parasite, and the absolute number of host species utilised by a single caligid species may vary as a function of sampling intensity and/or of the length of time since a species was described. However, the breadth of the host spectrum may be an indicator of the propensity for any given sea lice species to switch hosts and potentially become problematic in finfish farming facilities.

Acknowledgements

We would like to thank Dr Seyit Ali Kamanli (Burdur Mehmet Akif Ersoy University, Turkey) for help with confocal microscopy and Yetkin Sakarya (Cukurova University, Turkey) for his technical assistance. In addition, we are grateful to Prof. Selma Erat, Prof. Suphan Karaytug and Seher Kuru (Mersin University, Turkey) for their help and technical support during our SEM and CLSM studies conducted in MEITAM (Mersin University Advanced Technology Education Research and Application Center). Finally, we acknowledge the funding support from Cukurova University Academic Research Projects Unit (BAP): Project No. FBA-2017-7490.

References

Anstensrud, M. (1990) Moulting and mating in *Lepeophtheirus pectoralis* (Copepoda: Caligidae). *Journal of the Marine Biological Association of the UK* 70, 269–281.

Anstensrud, M. (1992) Mate guarding and mate choice in two copepods, *Lernaeocera branchialis* (L.) (Pennellidae) and *Lepeophtheirus pectoralis* (Müller) (Caligidae), parasitic on flounder. *Journal of Crustacean Biology* 12, 31–40.

Bailey, R.J., Birkett, M.A., Ingvarsdóttir, A., Mordue (Luntz), A.J., Mordue, W., O'Shea, B., Pickett, J.A. and Wadhams, L.J. (2006) The role of semiochemicals in host location and non-host avoidance by salmon louse (*Lepeophtheirus salmonis*) copepodids. *Canadian Journal of Fisheries and Aquatic Science* 63, 448–456.

Baird, W. (1850) *The Natural History of the British Entomostraca*. The Ray Society, London, UK.

Baster, J. (1765) Van de Luizen der Vissen, en de Zee-Luizen, In: *Natuurkundige Uitspenningen, behelzende eenige Waarneemingen, over sommige Zee-Planten en Zee-Insecten, benevens derzelver Zaadhuisjes en Eijernesten*. 2(2–3), 106–108, 150–167, pl. 8. J. Bosch, Haarlem.

Ben Hassine, O.K. (1983) Les Copépodes parasites de poissons Mugilidae en Méditerranée occidentale (côtes françaises et tunisiennes). Morphologie, bio-écologie, cycles évolutifs. [Thèse de Doctorat d'Etat]. Université des Sciences et Techniques du Languedoc, Montpellier, France.

Bowman, T.E. and Abele, L.G. (1982) Classification of the recent Crustacea. In: Bliss, D.E. (ed.) *The Biology of Crustacea* Vol 1. Academic Press, New York. pp. 1–27.

Boxshall, G.A. (1974a) The population dynamics of *Lepeophtheirus pectoralis* (Müller, 1776): seasonal variation in abundance and age structure. *Parasitology* 69, 361–371.

Boxshall, G.A. (1974b) The population dynamics of *Lepeophtheirus pectoralis* (Müller): dispersion pattern. *Parasitology* 69, 373–390.

Boxshall, G.A. (1974c) The developmental stages of *Lepeophtheirus pectoralis* (Müller, 1776) (Copepoda: Caligidae). *Journal of Natural History* 8, 681–700.

Boxshall, G.A. (1976) The host specificity of *Lepeophtheirus pectoralis* (Müller) (Copepoda: Caligidae). *Journal of Fish Biology* 8, 255–264.

Boxshall, G.A. (1977) The histopathology of infection by *Lepeophtheirus pectoralis* (Müller, 1776) (Copepoda: Caligidae). *Journal of Fish Biology* 10, 411–415.

Boxshall, G.A. (1990a) The skeletomusculature of siphonostomatoid copepods, with an analysis of adaptive radiation in structure of the oral cone. *Philosophical Transactions of the Royal Society of London* 328, 167–212.

Boxshall, G.A. (1990b) Precopulatory mate guarding in copepods. *Bijdragen tot de Dierkunde* 60, 209–213.

Boxshall, G.A. (2005) Branchiura. In: Rohde, K. (ed.) *Marine Parasitology*, CSIRO Publishing, Collingwood & CABI Publishing, Wallingford, UK. pp. 145–147.

Boxshall, G.A. (2018) The sea lice (Copepoda: Caligidae) of Moreton Bay (Queensland, Australia), with descriptions of thirteen new species. *Zootaxa* 4398, 1–172.

Boxshall, G.A. and El-Rashidy, H.H. (2009) A review of the *Caligus productus* species group, with the description of a new species, new synonymies and supplementary descriptions. *Zootaxa* 2271, 1–26.

Boxshall, G.A. and Gurney, A.R. (1980) Descriptions of two new and one poorly known species of the genus *Caligus* Müller, 1785 (Copepoda: Siphonostomatoida). *Bulletin of the British Museum (Natural History) Zoology Series* 39, 161–178.

Boxshall, G.A. and Halsey, S.H. (2004) *An Introduction to Copepod Diversity*. The Ray Society, London, UK.

Bron, J.E., Sommerville, C., Jones. M. and Rae, G.H. (1991) The settlement and attachment of early stages of the salmon louse, *Lepeophtheirus salmonis* (Copepoda: Caligidae) on the salmon host, *Salmo salar*. *Journal of Zoology* 224, 201–212.

Burmeister, H. (1835) Beschreibung einiger neuen oder weniger bekannten Schmarotzer-krebse, nebst allgemeinen Betrachtungen uber die Grubben, welcher sie angehoren. *Acta Academiae Caesareae Leopoldina-Carolina Naturae Curiosorum*, 17(1), 269–336.

Capart, A. (1941) Copepoda parasitica. In: Résultats scientifiques des croisières du Navire-école belge 'Mercator', 3(5). *Mémoires du Musée Royal d'Histoire Naturelle de Belgique* 21(2), 171–197, Fig. 1–9.

Carvajal, J., González, L. and George-Nascimento, M. (1998) Native sea lice (Copepoda: Caligidae) infestation of salmonids reared in netpen systems in southern Chile. *Aquaculture* 66, 241–246.

Costello, M.J. (2009) The global economic cost of sea lice to the salmonid farming industry. *Journal of Fish Diseases* 32, 115–118.

Cressey, R.F. and Cressey, H.B. (1980) Parasitic copepods of mackerel- and tuna-like fishes (Scombridae) of the world. *Smithsonian Contributions to Zoology* 311, 1–86.

Crosbie, T., Wright, D.W., Oppedal, F., Johnsen, I.A., Samsing, F. and Dempster, T. (2019) Effects of step salinity gradients on salmon lice larvae behaviour and dispersal. *Aquaculture Environment Interactions* 11, 181–190.

Damkaer, D.M. (2002) *The Copepodologist's Cabinet. A Biographical and Bibliographical History*. American Philosophical Society, Philadelphia, Pennsylvania.

Dojiri, M. and Ho, J.-S. (2013) *Systematics of the Caligidae, Copepods Parasitic on Marine Fishes*. Crustaceana Monographs 18, Brill, Leiden, Boston.

Edgecombe, G.D. (2010) Arthropod phylogeny: an overview from the perspectives of morphology, molecular data and the fossil record. *Arthropod Structure and Development* 39, 74–87.

Fields, D.M., Weissburg, M.J., and Browman, H.I. (2007) Chemoreception in the salmon louse *Lepeophtheirus salmonis*: an electrophysiology approach. *Diseases of Aquatic Organisms* 78, 161–168.

Fields, D.M., Skiftesvik, A.B. and Browman, H.I. (2017) Behavioural responses of infective-stage copepodids of the salmon louse (*Lepeophtheirus salmonis*, Copepoda: Caligidae) to host-related sensory cues. *Journal of Fish Diseases* 2017, 1–10.

Freeman, M.A., Anshary, H. and Ogawa, K. (2013) Multiple gene analyses of caligid copepods indicate that the reduction of a thoracic appendage in *Pseudocaligus* represents convergent evolution. *Parasites & Vectors* 6, 336.

Giribet, G. and Edgecombe, G.D. (2013) The Arthropoda: a phylogenetic framework. In: Minelli, A., Boxshall, G.A. and Fusco, G. (eds) *Arthropod Biology and Evolution*. Springer Verlag, Berlin, Heidelberg. pp. 17–40.

Gurney, R. (1934) The development of certain parasitic Copepoda of the families Caligidae and Clavellidae. *Proceedings of the Zoological Society of London* 1934, 177–217.

Hamdi, I., Benmansour, B., Hermida, M., Kamanli, S.A., Özak, A.A. and Boxshall, G.A. (2021a) *Caligus maderiensis* sp. nov. (Copepoda: Caligidae) parasitic on pompano, *Trachinotus ovatus* (Linnaeus, 1758), from eastern Atlantic waters, surrounding the Madeira archipelago, Portugal. *Acta Parasitologica* 65 (4), 1–16.

Hamdi, I., Benmansour, B., Zouari-Tlig, S., Kamanli, S.A., Özak, A.A. and Boxshall, G.A. (2021b) *Caligus tunisiensis* n. sp. (Copepoda: Caligidae) parasitic on the painted comber *Serranus scriba* (L.) (Perciformes: Serranidae) from the Mediterranean Sea, off the Tunisian coast. *Systematic Parasitology* 98, 57–71.

Hamre, L.A., Eichner, C., Caipang, C.A.M., Dalvin, S.T., Bron, J.E., Nilsen, F., Boxshall, G.A. and Skern-Mauritzen, R. (2013) The parasitic copepod *Lepeophtheirus salmonis* (Krøyer, 1838) has two chalimus stages. *PLoS ONE* 8(9), e73539.

Hansen, H.J. (1923) Crustacea Copepoda. II. Copepoda parasita and hemiparasita. *Danish Ingolf Expedition* 3(7), 1–92, pls 1–5.

Hayward, C.J., Aiken, H.M. and Nowak, B.F. (2008) An epizootic of *Caligus chiastos* on farmed southern Bluefin tuna *Thunnus maccoyii* off South Australia. *Diseases of Aquatic Organisms* 79, 57–63.

Hayward, C.J., Svane, I., Lachimpadi, S.K., Itoh, N., Bott, N.J. and Nowak, B.F. (2011) Sea lice infections of wild fishes near ranched southern bluefin tuna (*Thunnus maccoyii*) in South Australia. *Aquaculture* 320, 178–182.

Heuch, P.A. and Karlsen, H.E. (1997) Detection of infrasonic water oscillations by copepodids of *Lepeophtheirus salmonis* (Copepoda: Caligidae). *Journal of Plankton Research* 19, 735–747.

Heuch, P.A., Parsons, A. and Boxaspen, K. (1995) Diel vertical migration: a possible host-finding mechanism in salmon louse (*Lepeophtheirus salmonis*) copepodids? *Canadian Journal of Fisheries and Aquatic Sciences* 52, 681–689.

Heuch, P.A., Doall, M. and Yen, J. (2007) Water flow around a fish mimic attracts a parasitic and deters a planktonic copepod. *Journal of Plankton Research* 29, Suppl. 1: i3–i16.

Hewitt, G.C. (1971) Two species of *Caligus* (Copepoda: Caligidae) from Australian waters, with a description of some developmental stages. *Pacific Science* 25, 145–164.

Ho, J.-S. and Lin, C.-L. (2004) *Sea Lice of Taiwan (Copepoda: Siphonostomatoida: Caligidae)*. The Sueichan Press, Keelung.

Houttuyn, M. (1769) *Natuurlyke Historie of Uitvoerige Beschrijving der Dieren, Planten en Mineraalen, volgens het Samenstel van der Heer Linnaeus*.13(1), 460–464, 476–477, pl. 106.

Hull, M.Q., Mordue (Luntz), A.J., Pike, A. and Rae, G. (1998) Patterns of pair formation and mating in an ectoparasitic caligid copepod *Lepeophtheirus salmonis* (Krøyer, 1837); implications of its sensory and mating biology. *Philosophical Transactions of the Royal Society B* 353, 753–764.

Hutson, K. and Boxshall, G.A. (2016) Discovery of the male of the rare caligiform copepod *Kabataia* Kazachenko, Korotaeva & Kurochkin, 1972 (Copepoda: Siphonostomatoida), with a reconsideration of its phylogenetic affinities. *Zootaxa* 4174, 122–136.

Huys, R. and Boxshall, G.A. (1991) *Copepod Evolution*. The Ray Society, London, UK.

Huys, R., Llewellyn-Hughes, J., Conroy-Dalton, S., Olson, P.D., Spinks, J.N. and Johnston, D.A. (2007) Extraordinary host switching in siphonostomatoid copepods and the demise of the Monstrilloida: integrating molecular data, ontogeny and antennulary morphology. *Molecular Phylogenetics and Evolution* 43, 368–378.

Hwa, T.-K. (1965) Studies on the life history of a fish louse (*Caligus orientalis* Gusev). *Acta Zoologica Sinica* 17, 48–57.

Ingvarsdóttir, A., Birkett, M.A., Duce, I., Mordue, W., Pickett, J.A., Wadhams, L.J. and Mordue (Luntz), A.J. (2002) Role of semiochemicals in mate location by parasitic sea louse, *Lepeophtheirus salmonis*. *Journal of Chemical Ecology* 28, 2107–2117.

Izawa, K. (1969) Life history of *Caligus spinosus* Yamaguti, 1939 obtained from cultured yellowtail, *Seriola quinqueradiata* T. & S. *Report of the Faculty of Fisheries, Prefectural University of Mie* 6, 127–157.

Johnson, S.C. and Albright, L.J. (1991) The developmental stages of *Lepeophtheirus salmonis* (Krøyer, 1837) (Copepoda: Caligidae). *Canadian Journal of Zoology* 69, 929–950.

Johnson, S.C., Treasurer, J.W., Bravo, S., Nagasawa, K. and Kabata, Z. (2004) A review of the impact of parasitic copepods on marine aquaculture. *Zoological Studies* 43, 229–243.

Kabata, Z. (1972) Developmental stages of *Caligus clemensi* (Copepoda: Caligidae). *Journal of the Fisheries Research Board of Canada* 29, 1571–1593.

Kabata, Z. (1974) Mouth and mode of feeding of Caligidae (Copepoda), parasites of fishes, as determined by light and scanning microscopy. *Journal of the Fisheries Research Board of Canada* 31, 1583–1588.

Kabata, Z. (1979) *Parasitic Copepoda of British Fishes*. The Ray Society, London, UK.

Kabata, Z. and Hewitt, G.C. (1971) Locomotory mechanisms in Caligidae (Crustacea: Copepoda). *Journal of the Fisheries Research Board of Canada* 28(8), 1143–1151.

Kaji, T., Venmathi Maran, B.A., Kondoh, Y., Ohtsuka, S., Boxshall, G.A. and Tsukagoshi, A. (2012) The lunule of caligid copepods: an evolutionarily novel structure. *Evolution & Development* 14, 465–475.

Kim, I.-H. (1993) Developmental stages of *Caligus punctatus* Shiino, 1955 (Copepoda: Caligidae). In: Boxshall, G.A. and Defaye, D. (eds) *Pathogens of Wild and Farmed Fish: Sea Lice*. Ellis Horwood Ltd, West Sussex, UK. pp. 16–29.

Kim, I., Suárez-Morales, E. and Márquez-Rojas, B. (2019) Caligid copepods (Copepoda: Siphonostomatoida: Caligidae) as zooplankters off the Venezuelan coast, western Caribbean Sea. *Thalassas: An International Journal of Marine Sciences* 35(2), 607–618.

Krøyer, H. (1837) Om Snyltekrebsene, isaer med Hensyn til den danske Fauna. *Naturhistoriske Tiddskrift* 1(6), 605–628, pl. 6.

Krøyer, H. (1838) Om Snyltekrebsene, isaer med Hensyn til den danske Fauna. *Naturhistoriske Tiddskrift* 2(1), 8–52, pl. 7.

Laverack, M.S. and Hull, M.Q. (1993) Sensory innervation of the antennule of the pre-adult male of *Caligus elongatus*. In: Boxshall, G.A. and Defaye, D. (eds), *Pathogens of Wild and Farmed Fish: Sea Lice*. Ellis Horwood Ltd, West Sussex, UK. pp. 114–122.

Lewis, A.G. (1963) Life history of the caligid copepod *Lepeophtheirus dissimulatus* Wilson, 1905 (Crustacea: Caligoida). *Pacific Science* 17, 195–242.

Lin, C.-L. and Ho, J.-S. (1993) Developmental stages of *Caligus epidemicus* Hewitt, a copepod parasite of tilapia cultured in brackish water. In: Boxshall, G.A. and Defaye, D. (eds) *Pathogens of Wild and Farmed Fish: Sea Lice*. Ellis Horwood Ltd, West Sussex, UK. pp. 5–15.

Lin, C.-L., Ho, J.-S. and Chen, S.N. (1997) Development of *Caligus multispinosus* Shen, a caligid copepod parasitic on the black sea bream (*Acanthopagrus schlegeli*) cultured in Taiwan. *Journal of Natural History* 31, 1483–1500.

Lozano-Fernandez, J., Giacomelli, M., Fleming, J.F., Chen, A., Vinther, J., Thomsen, P.F., Glenner, H., Palero, F., Legg, D.A., Iliffe, T.M., Pisani, D. and Olesen, J. (2019) Pancrustacean evolution illuminated by taxon-rich genomic-scale data sets with an expanded remipede sampling. *Genome Biology and Evolution* 11(8), 2055–2070.

Meeüs, T. De, Raibaut, A. and Renaud, F. (1993) Comparative life history of two species of sea lice. In: Boxshall, G.A. and Defaye, D. (eds) *Pathogens of Wild and Farmed Fish: Sea Lice*. Ellis Horwood Ltd, West Sussex, UK. pp. 61–67.

Milne Edwards, H. (1840) *Histoire Naturelle des Crustacés, Comprenant l'Anatomie, la Physiologie et la Classification de ces Animaux*. Encyclopédique Roret, Paris. Vol. III, pls 1–42.

Mordue (Luntz), A.J. and Birkett, M.A. (2009) A review of host finding behaviour in the parasitic sea louse, *Lepeophtheirus salmonis* (Caligidae: Copepoda). *Journal of Fish Diseases* 32 (1), 3–13.

Müller, O.F. (1785) *Entomostraca seu Insecta Testacea, quae in aquis Daniae et Norvegiae reperit, descripsit et iconibus illustravit Otho Fridericus Müller*. Thiele, Lipsiae & Havniae, pls 1–21.

Nordmann, A. von. (1832) *Mikrographische Beiträge zur Naturgeschichte der Wirbellosen Thiere*. G. Reimer, Berlin.

Novales Flamarique, I., Browman, H.I., Bélanger, M. and Boxaspen, K. (2000) Ontogenetic changes in visual sensitivity of the parasitic salmon louse *Lepeophtheirus salmonis*. *Journal of Experimental Biology* 203, 1649-1657.

Núñez-Acuña, G., Marambio, J.P., Valenzuela, T., Wadsworth, S. and Gallardo-Escárate, C. (2016) Antimicrobial peptides from *Salmon salar* skin induce frontal filament development and olfactory/cuticle-related genes in the sea louse *Caligus rogercresseyi*. *Aquaculture* 464, 171–177.

Núñez-Acuña, G., Gallardo-Escárate, C., Berit Skiftesvik, A., Fields, D.M. and Browman, H.I. (2019) Silencing of ionotropic receptor 25a decreases chemosensory activity in the salmon louse *Lepeophtheirus salmonis* during the infective stage. *Gene* 697, 35–39.

Oakley, T.H., Wolfe, J.M., Lindgren, A.R. and Zaharoff, A.K. (2013) Phylotranscriptomics to bring the understudied into the fold: monophyletic Ostracoda, fossil placement, and pancrustacean phylogeny. *Molecular Biology and Evolution* 30, 215–233.

Ohtsuka, S. and Boxshall, G.A. (2019) Two new species of the genus *Caligus* (Copepoda: Siphonostomatoida) from the Sea of Japan (East Sea), with a note on the establishment of a new species group. *Zookeys* 893, 91–113.

Ohtsuka, S., Takami, I., Venmathi Maran, B.A., Ogawa, K., Shimono, T., Fujita, Y., Asakawa, M. and Boxshall, G.A. (2009) Developmental stages and growth of *Pseudocaligus fugu* Yamaguti, 1936 (Copepoda: Siphonostomatoida: Caligidae) host-specific to puffer. *Journal of Natural History* 43, 1779–1804.

Ohtsuka, S., Madinabeitia, I., Yamashita, H., Maran, B.A.V., Suárez-Morales, E. and Ho, J,-S. (2018) Planktonic phases in symbiotic copepods: a review. *Bulletin, Southern California Academy of Sciences* 117(2), 104–119.

Ohtsuka, S., Nawata, M., Nishida, Y., Nitta, M., Hirano, K., Adachi, K., Kondo, Y., Venmathi Maran, B.A. and Suárez Morales, E. (2020) Discovery of the fish host of the 'planktonic' caligid *Caligus undulatus* Shen & Li, 1959 (Crustacea: Copepoda: Siphonostomatoida). *Biodiversity Data Journal* 8, e52271.

Ohtsuka, S., Nishida, Y., Hirano, K., Fuji, T., Kaji, T., Kondo, Y., Komeda, S., Tasumi, S., Koike, K. and Boxshall, G.A. (2021) The cephalothoracic sucker of sea lice (Crustacea: Copepoda: Caligidae): the functional importance of cuticular membrane ultrastructure. *Arthropod Structure and Development* 62, 101046.

Øines, Ø. and Heuch, P.A. (2005) Identification of sea louse species of the genus *Caligus* using mtDNA. *Journal of the Marine Biological Association of the United Kingdom* 85, 73–79.

Ólafsdóttir, D. and Shinn, A.P. (2013) Epibiotic macrofauna on common minke whales, *Balaenoptera acutorostrata* Lacépède, 1804, in Icelandic waters. *Parasites & Vectors* 6, 105.

Özak, A.A., Yanar, A. and Boxshall, G.A. (2017) The discovery of *Caligus macrurus* Heller, 1865 (Copepoda: Caligidae) in the Mediterranean Sea, and the recognition of *Sciaenophilus* van Beneden, 1852 as a junior synonym of *Caligus* Müller, 1785. *Systematic Parasitology* 94, 97–109.

Özak, A.A., Yanar, A., Sakarya, Y. and Boxshall, G.A. (2018) The discovery of *Lepeophtheirus acutus* Heegaard, 1943 (Copepoda: Caligidae) from two new elasmobranch hosts in the Mediterranean Sea, and a comparative redescription of *Lepeophtheirus rhinobati* Luque, Chaves et Cezar, 1998. *Acta Parasitologica* 63(3), 454–473.

Petrunina, A.S., Neretina, T.V., Mugue, N.S. and Kolbasov, G.A. (2014) Tantulocarida versus Thecostraca: inside or outside? First attempts to resolve phylogenetic position of Tantulocarida using gene sequences. *Journal of Zoological Systematics and Evolutionary Research* 52, 100–108.

Piasecki. W. and MacKinnon, B.M. (1993) Changes in structure of the frontal filament in sequential developmental stages of *Caligus elongatus* von Nordmann, 1832 (Crustacea: Copepoda: Siphonostomatoida). *Canadian Journal of Zoology* 71, 889–895.

Pike, A.W. and Wadsworth, S.L. (1999) Sea lice on salmonids; their biology and control. *Advances in Parasitology* 44, 233–337.

Pike, A.W., MacKenzie, K. and Rowand, A. (1993) Ultrastructure of the frontal filament in chalimus larvae of *Caligus elongatus* and *Lepeophtheirus salmonis* from Atlantic salmon, *Salmo salar*. In: Boxshall, G.A. and Defaye, D. (eds) *Pathogens of Wild and Farmed Fish: Sea Lice*. Ellis Horwood Ltd, West Sussex, UK. pp. 99–113.

Regier, J.C., Shultz, J.W., Zwick, A., Hussey, A., Ball, B., Wetzer, R., Martin, J.W. and Cunningham, C.W. (2010) Arthropod relationships revealed by phylogenomic analysis of nuclear protein-coding sequences. *Nature* 463, 1079–1083.

von Reumont, B.M., Meusemann, K., Szucsich, N.U., Dell'Ampio, E., Gowri-Shankar, V., Bartel, D., Simon, S., Letsch, H.O., Stocsits, R., Luan, Y.X., Wägele, J.W., Pass, G., Hadrys, H. and Misof, B. (2009) Can comprehensive background knowledge be incorporated into substitution models to improve phylogenetic analyses? A case study on major arthropod relationships. *BMC Evolutionary Biology* 9, 119.

Roubal, F.R., Armitage, R. and Rohde, K. (1983) Taxonomy of metazoan ectoparasites of snapper, *Chrysophrys auratus* (family Sparidae), from southern Australia and New Zealand. *Australian Journal of Zoology* Supplement 94, 1–68.

Ruangpan, L. and Kabata, Z. (1984) An invertebrate host for *Caligus* (Copepoda: Caligidae). *Crustaceana* 47, 219–220.

Sakarya, Y., Özak, A.A. and Boxshall, G.A. (2019) The discovery of *Lepeophtheirus lichiae* Barnard, 1948 (Copepoda: Caligidae) parasitic on leerfish, *Lichia amia* (Linnaeus) in the Mediterranean Sea. *Systematic Parasitology* 96, 603–616.

Scott, A. (1901) *Lepeophtheirus and Lernaea. L.M.B.C. Memoirs of Typical British Marine Plants and Animals* No. 6.

Shen, C.J. and Li, H.L. (1959) Parasitic copepods from fishes of China, IV. Caligoida. Caligidae (3). *Acta Zoologica Sinica* 11, 12–23.

Skern-Mauritzen, R., Torrisen, O. and Glover, K.A. (2014) Pacific and Atlantic *Lepeophtheirus salmonis* (Krøyer, 1838) are allopatric subspecies: *Lepeophtheirus salmonis* and *L. salmonis oncorhynchi* subspecies novo. *BMC Genetics* 15, 32.

Strøm, H. (1762) Physiske og Oeconomisk Beskrivelske over Fogderiet Söndmör. I, Deel.

Suárez-Morales, E., Camisotti, H. and Martín, A. (2012) A new species of *Caligus* (Copepoda: Siphonostomatoida) from the plankton of the Caribbean coast of Venezuela with a key to species. *Zookeys* 201, 59–71.

Todd, C.D., Laverack, M.S. and Boxshall, G.A. (1996) *Coastal Marine Zooplankton: A Practical Manual for Students*. 2nd edn. Cambridge University Press, Cambridge, UK.

Venmathi Maran, B.A., Seng, L.T., Ohtsuka, S. and Nagasawa, K. (2009) Records of *Caligus* (Crustacea: Copepoda: Caligidae) from marine fish cultured in floating cages in Malaysia with a redescription of the male of *Caligus longipedis* Bassett-Smith, 1898. *Zoological Studies* 48, 797–807.

Venmathi Maran, B.A., Oh, S.-Y., Soh, H.Y., Choi, H.J. and Myoung, J.-G. (2012a) *Caligus sclerotinosus* (Copepoda: Caligidae), a serious pest of cultured red seabream *Pagrus major* (Sparidae) in Korea. *Veterinary Parasitology* 188, 355–361.

Venmathi Maran, B.A., Ohtsuka, S. and Jitchum, P. (2012b) Occurrences of caligid copepods (Crustacea) in plankton samples collected from Japan and Thailand, with the description of a new species. *Species Diversity* 17, 87–95.

Venmathi Maran, B.A., Ohtsuka, S. and Shang, X. (2012c) Records of adult caligiform copepods (Crustacea: Copepoda: Siphonostomatoida) in marine plankton from East Asia, including descriptions of two new species of *Caligus* (Caligidae). *Species Diversity* 17, 201–219.

Venmathi Maran, B.A., Moon, S.Y., Ohtsuka, S., Oh, S.-Y., Soh, H Y., Myoung, J.-G., Iglikowska, A. and Boxshall, G.A. (2013) The caligid life cycle: new evidence from *Lepeophtheirus elegans* Gusev, 1951 reconciles the cycles of *Caligus* and *Lepeophtheirus* (Copepoda: Caligidae). *Parasite* 20, 15.

Venmathi Maran, B.A., Suárez-Morales, E., Ohtsuka, S., Soh, H.Y. and Hwang U.W. (2016) On the occurrence of caligids (Copepoda: Siphonostomatoida) in the marine plankton: a review and checklist. *Zootaxa* 4174(1), 437–447.

Walter, T.C. and Boxshall, G.A. (2022) World of Copepods database. Caligidae Burmeister, 1835. Accessed through: World Register of Marine Species available at: https://www.marinespecies.org/aphia.php?p=taxdetails&id=135513 (accessed 7 January 2022).

Wootten, R., Smith, J.W. and Needham, E.A. (1982) Aspects of the biology of the parasitic copepods *Lepeophtheirus salmonis* and *Caligus elongatus* on farmed salmonids and their treatment. *Proceedings of the Royal Society of Edinburgh* 81B, 185–197.

Yazawa, R., Yasuike, M., Leong, J., von Schalburg, K.R., Cooper, G.A., Beetz-Sargent, M., Robb, A., Davidson, W.S., Jones, S.R.M. and Koop, B.F. (2008) EST and mitochondrial DNA sequences support a distinct Pacific form of salmon louse, *Lepeophtheirus salmonis*. *Marine Biotechnology* 10, 741–749.

Yen, J., Weissberg, M.J. and Doall, M.H. (1998) The fluid physics of signal perception by mate-tracking copepods. *Philosophical Transactions of the Royal Society – London B* 353, 787–804.

CHAPTER 3

Sea lice development

Christiane Eichner

Department of Biological Sciences, University of Bergen, Norway

3.1 Life cycle

Caligidae are arthropods and, as in all other members of this phylum, the body is covered by a protective chitinous cuticle. In order to grow and develop, this rigid exoskeleton has to be replaced in a process called moulting. The timespan between two moults, or between hatching and the first moult or after the last moult is defined as a stage. The life cycle of Caligidae L. has eight stages and is described for *Caligus* species, for example *C. rogercresseyi* (González and Carvajal, 2003), *C. minimus* (Khoa et al., 2019), *C. elongatus* (Piasecki and MacKinnon, 1995), *C. latigenitalis* (Madinabeitia and Nagasawa, 2011); for *Lepeophtheirus* species, for example *L. salmonis* (Hamre et al., 2013; Johnson and Albright, 1991a), *L. elegans* (Maran et al., 2013; Lewis, 1963), *L. hospitalis* (Voth, 1971), *L. pectoralis* (Boxshall, 1974), *L. simplex* (Neptali Morales-Serna et al., 2015); and for *Pseudocaligus* species, for example *P. fugu* (Ohtsuka et al., 2009). In some cases, more than eight stages have been described (for example *C. minimus* (Khoa et al., 2019), *C. epidemicus* (Lin et al., 1996), *L. salmonis* (Johnson and Albright, 1991a) and *L. pectoralis* (Boxshall, 1974)). However, most descriptions are based on studies of lice successively sampled from host populations and the determination of the stage is not always straightforward. Interstage size discrepancy has contributed to this confusion and shall be clarified on the description of growth of *L. salmonis* below. Subsequent reinterpretation revealed only eight stages for *Caligus* (Ohtsuka et al., 2009) and the use of the numbers of setae was suggested for *Lepeophtheirus* as a stage marker in *L. elegans* (Maran et al., 2013). Experimental investigation confirmed that the life cycle of *Lepeophtheirus salmonis* also consists of only eight stages (Hamre et al., 2013), the same number as found in the described free-living copepods.

The life cycle of all Caligidae comprises three planktonic stages, two nauplius and one copepodid stage, the infective stage. On the host various stages can be found: chalimus stages, characterized by a frontal filament, by which the louse is firmly attached to its host, and pre-adult and adult stages, which can freely move on the surface of the host. After attachment to the host, the copepodid moults into the first chalimus stage. In *Caligus* and *Pseudocaligus* species the following three stages are also chalimus, before moulting to the final adult stage. *Lepeophtheirus* comprises only two chalimus stages followed by two pre-adult stages before the definitive adult stage. Visual discrimination between chalimus stages can be observed in *Caligus* spp., in addition to size differences, based

Fig. 3.1 Frontal filament development in *Lepeophtheirus salmonis*. The anterior region of the cephalothorax of chalimus 2 larvae is shown. Lice are sorted after increasing instar age, separated by using the cephalothorax length (CL) to total length (TL) ratio (Eichner et al., 2015). CL to TL ratios are descending (from left to right: CL/TL= 0.692, CL/TL=0.626, CL/TL=0.611, CL/TL=0.597). A structure resembling a developing frontal filament can be seen in the two oldest instars.

on characteristic structures of their frontal filament (Kim, 1993; Piasecki and MacKinnon, 1993; González and Carvajal, 2003), which is originating from production of a new lobe to the existing permanently attached filament at each moult (Piasecki and MacKinnon, 1993). In *Lepeophtheirus*, on the other hand, the frontal filament is most likely produced for each moult (Bron et al., 1991; Eichner et al., 2015b; Gonzalez-Alanis et al., 2001; Johnson and Albright, 1992) (Fig. 3.1) and does not show visible stage-specific structural characteristics. Visual distinction between the two chalimus stages can be made by size differences (Hamre et al., 2013).

Sexual dimorphism can be distinguished in *Caligus* spp. by morphology of the antenna from the third chalimus stage, and in addition, by sexual dimorphism of the genital segment in the fourth chalimus stage (Kim, 1993; González and Carvajal, 2003; Madinabeitia and Nagasawa, 2011). In *Lepeophtheirus salmonis* sex differences in the morphology of appendages and genital segment can be seen from the first pre-adult stage (Johnson and Albright, 1991a; Schram, 1993a). Distinction by minor size differences of the cephalothorax is possible from the second chalimus stage (Eichner et al., 2015b).

3.2 Moulting

The protective exoskeleton of sea lice consists, as for other arthropods, mainly of the polysaccharide chitin and different structural proteins. Growth and development depend on moulting, or ecdysis – the shedding of this rigid exoskeleton and its replacement by a new and larger one. As moulting is restricted to invertebrates, it is a process suitable for pest control methods in sea lice. In the search for new drug targets, investigation of ecdysis on a molecular basis has been addressed in several studies (referred to later in this chapter). The moulting process comprises separate phases: the actual ecdysis, post-moult, intermoult, and pre-moult. During ecdysis the old cuticle is shed and the new cuticle stretches by water absorption; during post-moult, the new cuticle hardens. The intermoult is the longest phase where the animal gains weight and accumulates energy reserves. During pre-moult the new and larger cuticle is built underneath the old one and resorption of substances from the old exoskeleton takes place (Skinner, 1985). Precise timing of both physiological and behavioral processes is decisive for successful moulting. In arthropods the moulting process is regulated hormonally with ecdysteroids as moulting hormones (Chang and Mykles, 2011). Internal and external cues regulate the hormone secretion. The steroid hormone signals are mediated through nuclear receptors

(Nakagawa and Henrich, 2009; Bonneton and Laudet, 2012) – transcription factors which modulate the rate of transcription of specific target genes. These nuclear receptors are organized in a specific structure and are conserved between arthropods (Bonneton and Laudet, 2012). In insects the two nuclear receptors ecdysonreceptor (EcR) and ultraspiracle/retinoid X receptor (USP/RXR) form the functional ecdysteroid receptor, being the base of a signaling cascade regulating gene expression of early and late response genes, and are by that responsible for the correct transition through the moult cycle. This regulation is well studied in insects (Ashburner and Richards, 1976; Thummel, 2001; King-Jones and Thummel, 2005). In addition to their function in moulting, nuclear receptors play a role in development, reproduction, and limb regeneration (Durica et al., 2002). Different isoforms of the nuclear receptors with distinct spatial and temporal expression can result in specific responses (Davis et al., 2005). Nuclear receptors have been used as targets in agricultural pest control (Fahrbach et al., 2012). In the genome of *L. salmonis*, homologous genes for most of the nuclear receptors conserved in arthropods (Bonneton and Laudet, 2012) can be found. In *L. salmonis* some nuclear receptors are characterized (ecdysteroidreceptor (EcR) (Sandlund et al., 2015, 2016), retinoid X receptor (RXR) (Eichner et al., 2015a), fushi tarazu factor-1 (FTZ-F1) (Brunet et al., 2021). Functional studies of these revealed their importance for moulting and reproduction in *L. salmonis* (Eichner et al., 2015a; Sandlund et al., 2016; Brunet et al., 2021). Moulting arrest could be shown for combined knock down of *EcR* and *RXR* (Sandlund et al., 2016) and for knock down of one of the isoforms of *FTZ-F1* (Brunet et al., 2021). Mostly 20-hydroxyecdysone is the active moulting hormone, but in some organisms its precursor ecdysone or structurally related compounds, such as ponasterone A, can also act as moulting hormones. In *L. salmonis*, ecdysone, 20-hydroxyecdysone, and ponasterone A were characterized (Sandlund et al., 2018a). In the parasitic copepodids, ponasterone A was the predominant active ecdysteroid. In adult female lice changing levels of ecdysone, 20-hydroxyecdysone, and ponasterone A were found during oocyte maturation. Steroid hormones are synthesized from cholesterol through several reactions catalyzed by cytochrome P450 enzymes encoded by the so called Halloween genes (Mykles, 2011). These genes, except phantom, are also found in *L. salmonis* (Sandlund et al., 2018a; Humble et al., 2019) and in *Caligus rogercresseyi* (Gonçalves et al., 2014).

In crustaceans ecdysteroid levels rise at the onset of the pre-moult triggering the release of hydrolytic enzymes digesting the inner layer of the old exoskeleton, while materials for the new exoskeleton are also secreted. Ecdysteroid levels fall again directly before ecdysis (Borst, 2003). Ecdysteroid levels at different stages of the moult cycle have not been measured in sea lice, but gene expression of the Halloween genes during the moult cycle were measured for *L. salmonis* (Eichner et al., 2018) and showed a distinct pattern for the cytochrome P450 mono-oxygenase spook, which was upregulated in the middle of the stage. Expression patterns for the other genes were less clear. A gene homologous to a known *Drosophila* steroid inactivation gene, CYP18A1, was upregulated directly before the moult.

An essential process during the moult cycle is the synthesis and degradation of chitin. Different enzymes of the chitin synthesis pathway have been investigated in *L. salmonis* (Braden et al., 2020; Harðardóttir et al., 2019, 2021). Chitin synthase is the final enzyme that synthesizes chitin from the nucleotide sugar, uridine diphosphate N-acetylglucosamine. Two chitin synthases with different functions are described for *L. salmonis* (Harðardóttir et al., 2019, 2021). Downregulation of the expression of these genes in pre-adult male and female lice had severe implications on survival and reproduction (Harðardóttir et al., 2021). Chitin is found in several tissues of *L. salmonis* (Harðardóttir et al., 2021) and chitin synthase 1 was found to be essential for exoskeleton formation. Downregulation of the expression of this gene in *L. salmonis* larvae also resulted in deformed copepodids (Braden et al., 2020). Other enzymes of the chitin synthesis pathway were also experimentally downregulated in *L. salmonis* larvae and two other enzymes critical for chitin synthesis were found. Downregulation of fructose-6-phosphate aminotransferase (GFAT), one of the early enzymes in the chitin synthesis pathway, and of chitin deacetylase LsCDA5956 in nauplii resulted in copepodids with abolished swimming capability and inability to infect fish (Braden et al., 2020). Downregulation of these enzymes also showed an effect on other enzymes involved in chitin metabolism. Degradation of chitin is catalyzed by chitinases. Three chitinases were described in *L. salmonis* (Eichner et al., 2015c). Temporal (Eichner et al., 2018) and spatial expression of chitinase 2 suggests its role in moulting and knock down of this gene in larvae showed deformation of copepodids with abnormal locomotive behavior, unable to infect fish (Eichner et al., 2015c).

3.3 Growth

Due to the rigidity of the exoskeleton, growth in arthropods is considered to be restricted to moulting. The new and larger exoskeleton produced beneath the old one is folded and can expand after the larvae leaves the old exoskeleton. However, in *L. salmonis* intra- and post-moult growth also takes place. Post-moult growth in adult female lice can be easily observed. After the moult to the adult stage the genital segment and abdomen show substantial growth while the cephalothorax remains about the same size (Ritchie et al., 1996; Eichner et al., 2008) (Fig. 3.2d). Growth of the attached larvae is more difficult to observe. A crucial factor for this, in addition to its small size, is that the development rate of lice is not uniform. There is individual variability as well as sex-biased differences, with male lice entering the pre-adult stage before females (Johnson and Albright, 1991b; Hamre et al., 2013; Eichner et al., 2015b). This leads to lice of different instar ages, or even different stages, at any given sampling time point after infection. Even a tight sampling regime will therefore fail to show successive development of larvae. The elucidation of the growth pattern in *L. salmonis* chalimus larvae can explain why there has been a confusion in the number of stages for *Lepeophtheirus*. In an experimental study significant instar growth of the chalimus larvae was found (Hamre et al., 2013; Eichner et al., 2015b) leading to a strong variability in size of larvae within the stage. The size difference between lice newly moulted into the stage compared to lice that are in the state shortly before moulting to the next stage can easily be wrongly interpreted as different stages when applying conventional growth patterns of arthropods. The investigation by Eichner et al. (2015b) revealed growth patterns of the lice and showed that instar growth in the chalimus larvae constituted about half of the total growth. Further, growth of the chalimus larvae was not equal in the different body parts. Growth within the stage was mainly due to a size increase of the posterior segments of the larvae and to a lesser degree of the cephalothorax. The size increase of the cephalothorax occurred mainly at moulting. In the study the posterior segments of chalimus 1 larvae increased their length by 78% due to instar growth, while the increase of the cephalothorax within the stage was only 13%. On the other hand, size increase by moulting to the chalimus 2 stage was dominated by the increase of the cephalothorax with an average increase of 64%, while increase of the posterior segments was on average 10%. The dominant increase of the posterior segments during instar growth led to a change in the ratio of cephalothorax length (CL) to total length (TL). For the youngest chalimus 1 larvae this ratio was 0.69 and for the oldest 0.59. By comparing length measurements of individual chalimus 2 larvae before and after the moult to pre-adult lice where sex differences are visible, a sex bias in the length of the cephalothorax could be shown. This allowed the possibility of dividing chalimus 2 larvae into lice of different sex and, by that, the investigation of sexual size dimorphism in this stage. However, this sexual size

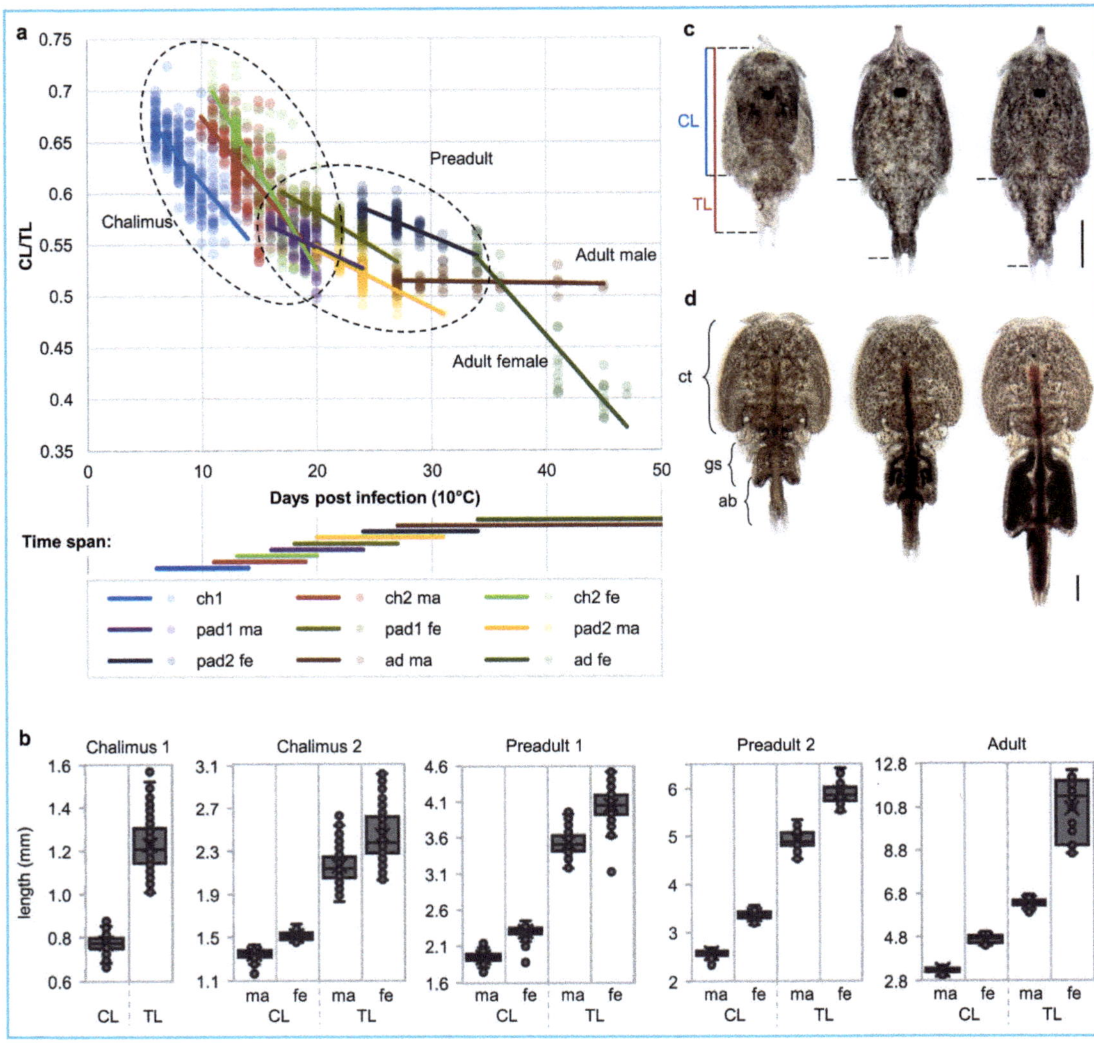

Fig. 3.2 Length measurements of *Lepeophtheirus salmonis* in a time series. a) Cephalothorax length (CL) to total length (TL) ratio of attached *Lepeophtheirus salmonis* sampled between 5 and 47 days post-infection from Atlantic salmon reared at 10°C. The trend from a higher to lower CL to TL ratio with time within the stage can be seen for chalimus and pre-adult lice. A strong trend can also be seen for adult female lice until a fully stretched genital segment. The average CL/TL of fully stretched adult female lice for lice sampled after 60 days was 0.38 (not shown) (data for chalimus and pre-adult measurements are taken from Eichner et al., 2018). The time span in which lice of the different stages were found is indicated by lines underneath for better visualization. b) Boxplots showing the CL and TL of the same lice. c) Photographs of chalimus 2 male larvae are shown as examples for visual changes of the CL to TL ratio within the stage, from left to right: directly after moulting to the stage, in the middle of the stage, and before moulting to pre-adult 1. d) Photographs of adult female lice with differently developed genital segment illustrating post moulting growth. On the left: lice shortly after moulting to the adult stage with a very short genital segment where no oocytes are visible yet; in the middle: louse with more developed genital segment where oocytes are visible; on the right: louse with fully developed genital segment. While the cephalothorax (ct) stays about the same length, the genital segment (gs) and abdomen (ab) are growing. ch=chalimus, pad=pre-adult, ad=adult, ma=male, fe=female; bars are 0.5 mm.

dimorphism accounted for only about 10% of the size variation in chalimus 2. Instar growth, and to a lesser degree sexual size dimorphism, can be the main reason for the previous misinterpretation of the number of chalimus stages of *L. salmonis*. (Hamre et al., 2013; Eichner et al., 2015b). The CL to TL ratio was unaffected by sex; the mean ratio for the youngest chalimus 2 was 0.68 and for the oldest 0.56. The change of the CL to TL ratio with instar age allowed for the determination of the relative age within the stage of chalimus larvae and could also be confirmed for pre-adult lice (Eichner et al., 2018). This method of distinguishing lice of different instar age is especially of interest when investigating moulting-related processes (for example, as seen in Harðardóttir et al. (2019) and Brunet et al. (2021)) but is also essential when comparative lice are necessary for functional studies or gene expression comparisons. Changes of the overall gene expression within the stage will be detailed later. In Fig. 3.2a the CL to TL ratios of *L. salmonis* chalimus, pre-adult, and adult lice sampled on consecutive days (data from Eichner et al. (2018)) are shown. Changes in the CL to TL ratio of pre-adult lice show a more gradual gradient with stage progression than chalimus.

The size to which the louse is growing is, as for other aquatic ectotherms, temperature-dependent with smaller lice at higher temperatures in a temperature range from 5°C to 20°C (Tully, 1989; Ritchie et al., 1993; Nordhagen et al., 2000; Costello, 2006; Samsing et al., 2016), but not at very low temperature (Samsing et al., 2016). The correlation between temperature and adult body size of *L. salmonis* for a temperature range from 5°C to 20°C followed universal models of temperature dependence, whereas at 3°C body size was decreased compared to body size at 5°C (Samsing et al., 2016).

3.4 Sexual maturation

The reproductive system of the adult salmon louse consisting of ovaries, oviducts, cement glands, and a receptaculum seminis in female lice; and testis, vasa deferentia, spermatophore sacs, and cement gland in male lice is described for *Lepeophtheirus salmonis* (Ritchie et al., 1996; Schram, 2000). Reproductive structures are already visible at the pre-adult 1 stage but are first mature in the adult stages (Ritchie et al., 1996). Dalvin et al. (2013) investigated a molecular germ cell marker in *Lepeophtheirus salmonis*, a Vasa homologue, generally having a crucial role in the formation and maintenance of germ cells. *LsVasa* transcript was found in all analyzed stages with a modest increase during the planktonic and attached stages and highest expression in reproductive adults, probably earlier in male than female lice. RNA sequencing data (Eichner et al., 2018; Skern-Mauritzen et al., 2021) with higher sampling resolution for attached stages until pre-adult 1 showed an increase in the transcript levels from chalimus 2 for male lice and pre-adult 1 for female lice (Fig. 3.3), also suggesting an earlier timepoint for male gonad formation.

After the final moult, a substantial growth of the genital segment in adult female lice takes place (see also Fig. 3.1d). During this process, in addition to genes related to growth, genes involved in reproduction are upregulated indicating that sexual maturation occurs (Eichner et al., 2008). Oocytes develop in both fertilized and unfertilized adult females. First occurrence of immature oocytes can be found in the genital segment when 32% of the size increase has taken place (Eichner et al., 2008). Upregulation of transcripts for the vitellogenin starts at the same time point (Eichner et al., 2008). Also *LsVasa* transcript becomes upregulated with maturation in the same way as vitellogenin in that data set (data not shown). Growth and maturation of oogonia takes place in the oviduct, a process divided into pre-vitellogenic and vitellogenic development, where in the latter yolk proteins are

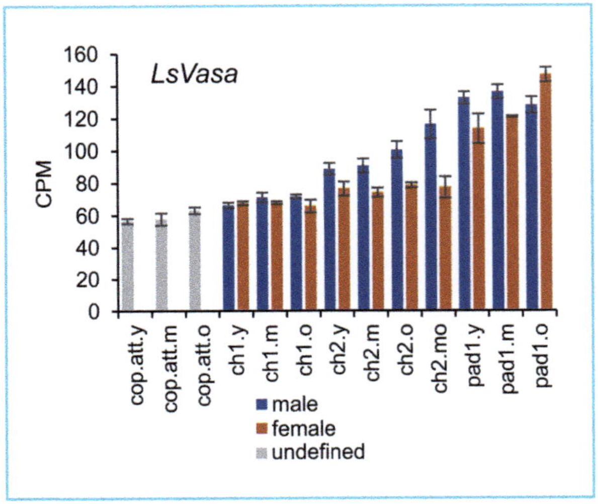

Fig. 3.3 Expression of *LsVasa* (Dalvin et al., 2013) transcript in parasitic *Lepeophtheirus salmonis* larvae measured by RNA sequencing. (Data from Eichner et al. (2018) and from Skern-Mauritzen et al. (2021)). Lice are divided into groups of different instar age within each stage: shortly after moulting (young (y)), in the middle of the stage (middle(m)), before moulting to next stage (old(o)). For chalimus 2 expression in moulting lice moult (mo) is also shown. CPM: counts per million.

incorporated. In *Lepeophtheirus salmonis* the main yolk proteins are vitellogenin 1, vitellogenin 2, and the maternal yolk-associated protein (LsYAP) (Dalvin et al., 2009, 2011). These are produced in the subcuticular tissue and transported by the hemolymph to the maturing oocytes. The process is usually controlled by ecdysteroids. In *L. salmonis* an increase in ecdysteroid level during oogenesis was measured indicating its role in oocyte maturation in salmon lice (Sandlund et al., 2018a). Lipids are also deposited maternally into the oocytes (Khan et al., 2017). Another protein maternally deposited is the LsTryp10 protein, which is probably involved in regulation of the yolk degradome (Skern-Mauritzen et al., 2009). A sufficient amount of energy has to be available to ensure development until food can be consumed, which in this case is until the attachment of the copepodid to the host. Egg development and maturation was found to be regulated by the TOR pathway (Sandlund et al., 2018b). The upstream regulator Ras homolog enriched in brain (Rheb) especially seems to play a crucial role in oocyte maturation in the genital segment, and knock down of Rheb halted egg maturation. The ecdysone receptor also mediates the egg maturation (Sandlund et al., 2015). The progression of oocyte maturation becomes visible in the genital segment of the adult female louse (Fig. 3.4). Eggs are produced in batches and are embedded into a cement, produced by the cement glands (Schram, 2000), where they form paired egg strings. The cement consists of six main proteins (Borchel et al., 2019). The first pair of egg strings emerging is shorter and contains fewer eggs than the following ones (Schram, 1993b; Heuch et al., 2000). After starting egg production salmon lice continuously produce eggs for life (Heuch et al., 2000). The number and size of eggs produced is dependent on temperature (Ritchie et al., 1993; Heuch et al., 2000; Nordhagen et al., 2000; Bravo et al., 2009; Samsing et al., 2016) and can be influenced by other factors such as louse size or population (Jackson and Minchin, 1992; Tully and Whelan, 1993). The frequency of egg batch production is also temperature-dependent and was more than four times faster at 18°C than at 6°C (Hamre et al., 2019).

3.5 Early development

Not much is known about embryogenesis in sea lice. Embryogenesis takes place within the egg strings which are attached to the adult female louse by a hook structure (Schram, 2000). Eggs

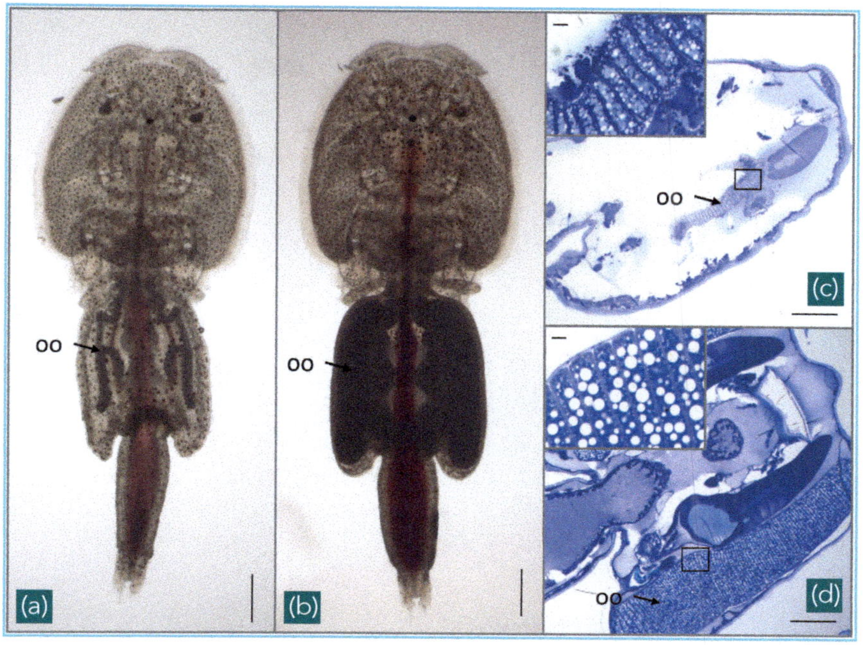

Fig. 3.4 Oocyte maturation in *Lepeophtheirus salmonis*. Photographs (a and b) and histological sections (c and d) of adult female lice. Lice with differently developed oocytes (oo) are shown. a) Louse with oocytes just deposited into the genital segment. b) Louse with oocytes ready to extrude. c) Histological sections of the genital segment of the same louse shown in a). d) Histological sections of the genital segment of the same louse shown in b). Inserts are enlargements of the oocytes as marked with a square. Bars are 0.5 mm (a and b), 500 µm (c and d), 10 µm (insert).

are fertilized while extruded through the copulatory duct by sperm stored in a spermatheca in the female upon mating (Ritchie et al., 1996). The embryos in the egg string lie flattened on top of each other and are oriented laterally with the front pointing to the outer side of the egg string, as can be seen by the pigmentation of the eye and the antennules in further developed embryos (Schram, 1993b, 2000; Skern-Mauritzen et al., 2009). Increasing pigmentation in the fertilized eggs can be observed in the last part of development before hatching. Hatching of nauplii larvae begins at the distal end of the egg string. The embryos begin to absorb water by a pulsing movement and become more rounded (Fig. 3.5a). The hatching itself proceeds quickly with the larvae bouncing out of the shell. Hatching of *C. elongatus* is described in a similar way (Piasecki and MacKinnon, 1995). Development of the planktonic stages is relatively synchronized. The duration of the nauplius 1 stage is much shorter than that of the nauplius 2 stage: about three to five times for *L. salmonis* (Johnson and Albright, 1991b; Eichner et al., 2014) and two to three times for *C. rogercresseyi* (Montory et al., 2018). The development rate is temperature-dependent and has been investigated in several studies on *L. salmonis* (Johnson and Albright, 1991b; Boxaspen and Naess, 2000; Samsing et al., 2016) and on *Caligus rogercresseyi* (González and Carvajal, 2003). The two nauplius stages of *L. salmonis* are very similar. The second nauplius is slightly longer (Johnson and Albright, 1991a; Schram, 1993a). Thinning of the developing urosome becomes visible in older nauplius 2 larvae (Schram, 1993a) (Fig. 3.6). Also in *Caligus clemensi* the two nauplius stages are described as very similar, with the second being a bit longer and having relatively shorter balancers, and the outline of the copepodid can be found in the second nauplius (Kabata, 1972). Moulting of nauplius larvae

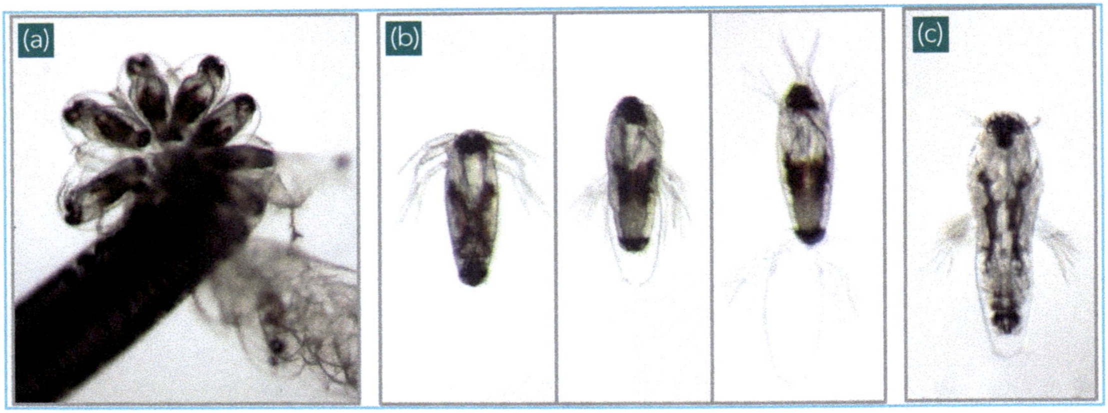

Fig. 3.5 Hatching and moulting of planktonic *L. salmonis* larvae.
a) Hatching starts at the distal end of the egg string. b) Moulting of nauplius 1 larva. The moulting starts with folding down the antennae as shown in the left photograph. With a pulsing movement the nauplius 2 is leaving the exuvia through an opening in the cephalic site and unfolds its antennae shortly after. c) Moulting of nauplius 2 larvae.

begins in *L. salmonis* with folding down of the antennae (Fig. 3.5b). With a pulsing movement the larva slowly leaves the exuvia through an opening at the cephalic site. The moult is accomplished by unfolding of the antenna shortly after leaving the exuvia. Moulting to the copepodid stage takes place in a similar manner. The planktonic stages are lecithotrophic, and are subsisting on maternally derived nutrients, mainly lipid reserves. The lifespan of the copepodid is therefore limited by energy reserves (Tucker et al., 2000b).

3.6 Development rate

The development rate of egg, planktonic, and parasitic stages of sea lice has been investigated in several studies. The majority of these are performed on *L. salmonis* (Bjørn and Finstad, 1998; Boxaspen and Naess, 2000; Grimnes and Jakobsen, 1996; Hamre et al., 2019; Johnson and Albright, 1991b; Johnson, 1993; Ritchie et al., 1993; Samsing et al., 2016; Tucker et al., 2000a). Other studies of development rates are accomplished on *C. rogercresseyi* (Bravo, 2010; González and Carvajal, 2003; Montory et al., 2018, 2020), *C. elongatus* (Piasecki and MacKinnon, 1995), *Pseudocaligus fugu* (Ohtsuka et al., 2009), and *L. simplex* (Neptali Morales-Serna et al., 2015). The development rate in sea lice is, as in other poikilotherms, temperature-dependent with a negative correlation of water temperature to development time in a given temperature range (Bjørn and Finstad, 1998; González and Carvajal, 2003; Hamre et al., 2019; Heuch et al., 2000; Johnson and Albright, 1991b; Samsing et al., 2016; Tucker et al., 2000a).

Different studies of *L. salmonis* showed deficient development at very low temperatures. Boxaspen and Naess (2000) found that even though hatching was possible at 2°C and 3°C, the development of nauplii to copepodids was strongly reduced at these temperatures but not at 4°C. Samsing et al. (2016) even observed reduced hatching at 3°C and development to the copepodid stage was disabled, while successful development to the copepodid stage was possible at 5°C. However, the infection success was reduced at 5°C. Development of the parasitic stages of *L. salmonis* was also found to be

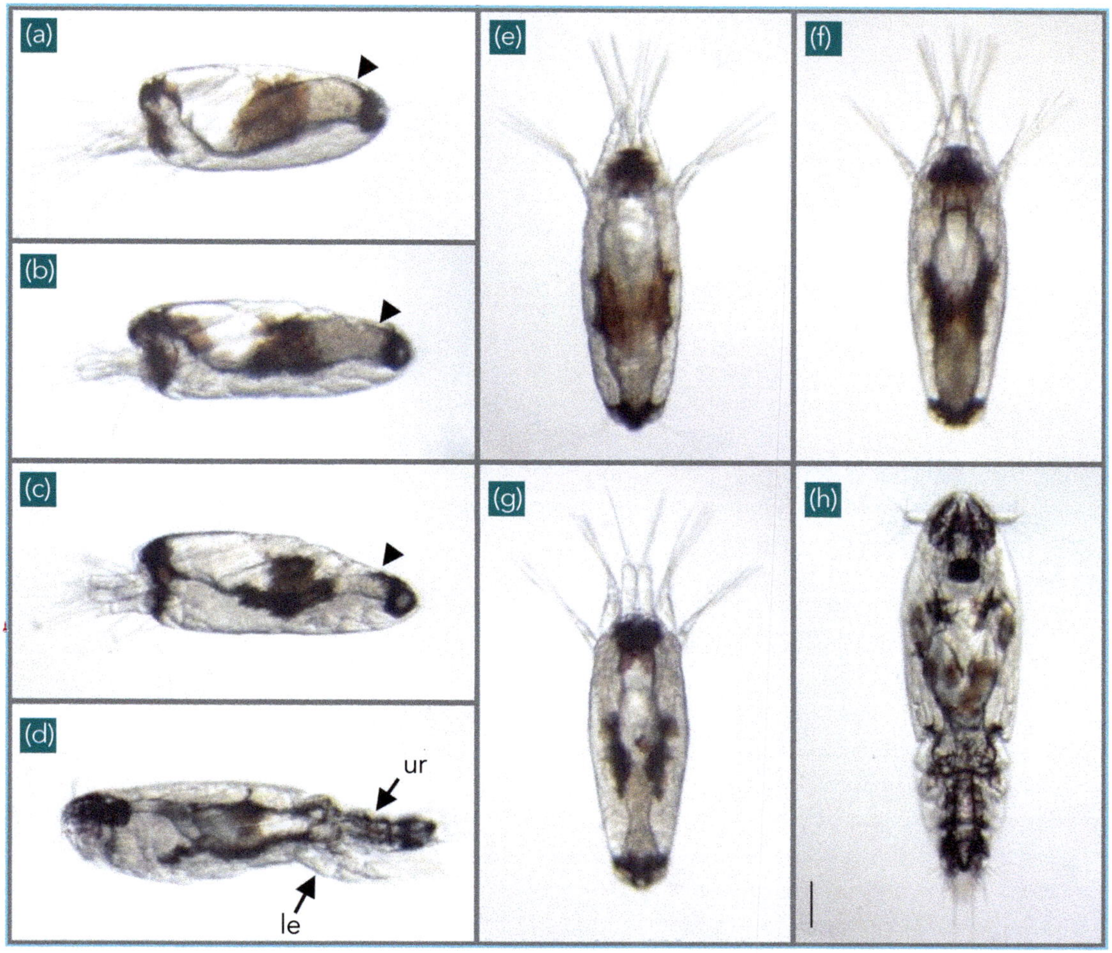

Fig. 3.6 Photographs of planktonic stages, lateral view (a–d), dorsal view (e–h). Newly hatched nauplius 1 (a and e), nauplius 2 short after the moult (b and f), nauplius 2 shortly before moulting to the copepodid stage (c and g), copepodid (d and h). Urusome (ur) and legs (le) are marked in (d). The part developing to the urusome (marked with triangle) is distinguished in the developing nauplius. Developing legs can be seen in nauplius 2 (b and c).

compromised at 3°C by Hamre et al. (2019). In the same study, Hamre et al. (2019) showed that a temperature of 24°C also led to severely impaired development of the parasitic stages, while this was not the case at 21°C.

Strong variations in the developmental time with different temperatures were found for all developmental stages. The time for hatching of eggs was investigated at temperatures between 2°C and 10°C in *L. salmonis* (Boxaspen and Naess, 2000; Johnson and Albright, 1991b) and varied between 8.6 and 45.1 days at the highest and lowest temperatures, respectively. Also, development time from hatching to the infective copepodid showed high variation in *L. salmonis* with 1.7 days at 20°C (Samsing et al., 2016) and 68.5 days at 2°C (Boxaspen and Naess, 2000). In *C. rogercresseyi* the development of nauplii from hatching to the copepodid stage varied between around 1.5 days at 18°C to around 6 days at 6°C (González and Carvajal, 2003). The development time of the parasitic stages from copepodid to adult differs between sexes in *L. salmonis* (Johnson and Albright, 1991b; Tucker et al., 2002; Eichner et al., 2015b; Hamre et al., 2013, 2019) and slightly in *C. rogercresseyi* (González and Carvajal, 2003), with male lice developing quicker than females. This was not found in *C. elongatus* (Piasecki and MacKinnon, 1995). A large variability in the development time between individual lice was found for *L. salmonis* (Johnson and Albright, 1991b; Hamre et al., 2013, 2019; Eichner et al., 2015b, 2018). Development times for the parasitic stages of *L. salmonis* from infection to the adult stage varied strongly with temperature and the fastest development was measured at 21°C with 11 and 13 days for male and female lice respectively, and the slowest at 3°C with 110 and 135 days for male and female lice respectively (Hamre et al., 2019). For *C. rogercresseyi* development times between 15 and 16 days on average at 16.7°C, to 31 and 33 days on average at 10.3°C for males and females respectively, were observed (González and Carvajal, 2003). In most studies the development rate of *L. salmonis* was investigated on one or few given temperatures only. Stien et al. (2005) implemented these data for egg, planktonic, and parasitic development rates into a model and found equiproportional development, which means that each stage uses a constant proportion of the time that it takes to become adult at all temperatures, as described generally for copepods. The correlation between temperature and the minimum developmental time could be described by Belehrádek's function, a commonly used function in models of copepod development (Stien et al., 2005). However, for the parasitic development, only data for temperatures between 7°C and 15°C were available. In two later studies the development rate of *L. salmonis* was examined over a wider temperature range (Samsing et al., 2016; Hamre et al., 2019). Samsing et al. (2016) investigated the time for production of eggs, hatching of eggs, and the development times of planktonic larvae at six different temperatures between 3°C and 20°C. The results from temperatures between 5°C and 20°C could be described with universal theoretical models of temperature dependence that give the possibility to predict the duration of larval development and egg production in this temperature range. At 3°C, the production of eggs was also decreased in addition to reduced hatching and disabled development (as described above), indicating that the minimum temperature for development is expected around this temperature. Hamre et al. (2019) investigated the parasitic development of *L. salmonis* at eight different temperatures between 3°C and 24°C by repeated sampling throughout the life cycle. Lice developed normally at temperatures between 6°C and 21°C while development at 3°C and 24°C was severely impaired with scarcely any lice, and in the case of 24°C, no females, reaching the adult stage. A maximum temperature limit was therefore expected somewhere between 21°C and 24°C, and a minimum around 3°C whereby the authors pointed out that no absolute limit is likely, but rather a range with increasing deteriorations of development. The moult rate increased with temperature and was higher for males. The duration of all developmental stages was about equal, males and females viewed separately. As a measure of lice age, Hamre et al. (2019) introduced the 'StageAge,' representing the number of moults since infection plus the elapsed proportion of the present stage. The mean daily growth rate expressed in stages per day could be described by a second-order polynomial function of temperature for each sex. Relative age was defined as the age of a louse expressed as the percentage of the development time that the majority of the population needs to become adult. This enabled implementation of a temperature-independent *x*-axis to compare developmental events across temperature and sex. With the StageAge growth rate model the timing of developmental events can be calculated as a function of temperature. Also, temperature-independent descriptions of life cycle patterns are possible. In *C. rogercresseyi* four experiments with different temperature ranges between 9°C and 19.5°C were implemented (González and Carvajal, 2003). In these experiments growth rates were not investigated at stable temperatures but at a wider range of temperatures during development. The time to complete a full life cycle was between 45 days at an average 10.3°C and 26 days at an average 15.2°C. The authors proposed a simple degree-day model with the development rate being a linear function of the average water temperature for a temperature range from 4°C to 17°C. From their model the authors also calculated the lower threshold for development as 4.2°C with a maximum–minimum range of 3.2–4.9°C.

3.7 Genes regulated in growth and development of *L. salmonis*

Due to its characteristic arthropod development involving moulting, development in the louse cannot be seen as a progression only, but recurring patterns in each moulting cycle will also cause a cyclic component. Growth and development in sea lice comprises a number of different processes. In addition to accumulation of body mass and the moulting process itself, restructuring of the body and appendages characterizes the transition between different larval stages. Sexual maturation also accompanies the development. All these processes will be reflected in the overall gene expression. Therefore, it can be expected that differences in gene expression will be found not only between stages, but also within a stage depending on the status of the individual regarding its moult cycle.

Investigation of specific gene expression is a common approach in functional studies. Overall gene expression measurement by mRNA sequencing or microarray analysis is a prevalent method to find unknown genes affected by disease or special treatments. It has also been used in sea lice research to find effects of drugs on the parasite (Chávez-Mardones and Gallardo-Escárate, 2015; Sutherland et al., 2015; Valenzuela-Muñoz et al., 2015; Valenzuela-Miranda and Gallardo-Escárate, 2016; Agusti-Ridaura et al., 2020; Núñez-Acuña et al., 2020), or reactions to different stressors (Sutherland et al., 2012; Agusti-Ridaura et al., 2020; Borchel et al., 2021) or feed (Sutherland et al., 2017). The genomes of both *C. rogercresseyi* (Gallardo-Escárate et al., 2021) and *L. salmonis* (Skern-Mauritzen et al., 2021) are sequenced and expressed sequence tags (ESTs) are available also for *C. clemensi* and *Lernaeocera branchialis* (Yasuike et al., 2012). However, the non-uniform development rate of lice caused by individual variability as well as sex-biased differences (Johnson and Albright, 1991b; Tucker et al., 2002; Hamre et al., 2013, 2019; Eichner et al., 2015b) can lead to difficulties when investigating gene expression in parasitic larval stages as samples taken at the same time point after infection can consist of lice of different stages and instar ages.

Gene expression changes during development of chalimus and pre-adult 1 larvae of different instar age independent of sampling time point were investigated in a time series study (Eichner et al., 2018). Individuals were separated by using the cephalothorax to total length ratio (Eichner et al., 2015b) as a proxy for instar age. For each stage and sex, groups of individuals shortly after the moult to the respective stage (post-moult, young), in the middle of the stage (intermoult, middle), and shortly before moulting to the next stage (pre-moult, old), and also for chalimus larvae that had started the moult (ecdysis), were investigated. Thousands of differentially expressed genes were found from the around 13,000 annotated genes of *L. salmonis* (Skern-Mauritzen et al., 2021) between lice from different stage groups but also between lice from the same stage but different instar age. The overall gene expression pattern was more similar between lice from different stages of the same instar age group than within stages (Eichner et al., 2018). In Table 3.1 the number of differentially expressed genes between different conditions (stage, instar age, and sex) is shown. Even when combining samples

Table 3.1 Number of differentially expressed (DE) genes between two conditions.

Condition	Number of genes DE
Stage:	
Chalimus1 vs chalimus2[#]	5039
Chalimus2 vs pre-adult1[#]	3415
Chalimus1 vs pre-adult1[#]	7610
Instar age all stages:	
Young vs middle*	3979
Middle vs old*	4112
Young vs old*	5487
Instar age of each stage:	
Chalimus1 young vs middle	1463
Chalimus1 middle vs old	2561
Chalimus1 young vs old	3537
Chalimus2 young vs middle	4612
Chalimus2 middle vs old	6374
Chalimus2 young vs old	6522
Pre-adult1 young vs middle	4542
Pre-adult1 middle vs old	2518
Pre-adult1 young vs old	4472
Sex:	
Chalimus2[§]	162
Pre-adult1[§]	1346

[#] samples of all instar ages and both sexes combined, *samples of all stages and both sexes combined, [§]samples of all instar ages combined. Data from Eichner et al. (2018).

from all stages of a given instar age several thousand differentially expressed genes can be found between lice from different age groups with respect to the moult, indicating quite a conserved moulting pattern in all investigated stages. In contrast, sex-biased gene expression was less dominant and was, as expected, more prevalent in pre-adult lice than in chalimus. More than 2000 genes were found to be cyclically expressed in all investigated stages. As expected, in the genes upregulated in all pre-moult groups there were many genes annotated as being involved in chitin metabolism or in the cellular response to ecdysone. The known nuclear receptors also showed clear expression patterns upregulated in lice from either the intermoult or pre-moult stage. In addition to the characterized genes many genes with unknown functions were also under the cyclically regulated genes and a role within the moulting cycle can be anticipated. In this group possible targets for new drugs could be found. The study shows that the gene expression of developing larvae reflects the cyclic component of moulting. The rate is individual for each louse.

After the final moult male lice are fully grown and little change in gene expression is expected. In contrast the female louse is not fully grown after the final moult (as described previously) (Fig. 3.2d) and gene expression changes due to maturation of the louse are found (Eichner et al., 2008). In addition, it can be expected that genes involved in oocyte maturation could have an expression pattern in coordination with the egg laying cycle. Varying amounts of ecdysteroid were measured during oocyte maturation (Sandlund et al., 2018a) and indicate that ecdysteroid target genes could also be regulated differently. However, for genes encoding for the main components of the cement, which could also be expected to be regulated this could not been shown (Borchel et al., 2019).

These results mean for gene expression studies that the non-uniform development rate of lice will result in a bias, caused by underlying gene expression patterns due to moulting and development, when relying on the sampling time point only. This has to be considered in gene expression studies involving larvae or immature female lice, and careful planning of experiments is needed.

References

Agusti-Ridaura, C., Bakke, M.J., Helgesen, K.O., Sundaram, A.Y.M., Bakke, S.J., Kaur, K. and Horsberg, T.E. (2020) Candidate genes for monitoring hydrogen peroxide resistance in the salmon louse, *Lepeophtheirus salmonis*. *Parasites & Vectors* 13, 344.

Ashburner, M. and Richards, G. (1976) Sequential gene activation by ecdysone in polytene chromosomes of *Drosophila melanogaster*. 3. Consequences of ecdysone withdrawal. *Developmental Biology* 54, 241–255.

Bjørn, P.A. and Finstad, B. (1998) The development of salmon lice (*Lepeophtheirus salmonis*) on artificially infected post smolts of sea trout (*Salmo trutta*). *Canadian Journal of Zoology* 76, 970–977.

Bonneton, F. and Laudet, V. (2012) 6 – Evolution of nuclear receptors in insects. In: Gilbert, L.I. (ed.) *Insect Endocrinology*. Academic Press, San Diego, California. pp. 219–252.

Borchel, A., Kongshaug, H. and Nilsen, F. (2019) Identification and description of the key molecular components of the egg strings of the salmon louse (*Lepeophtheirus salmonis*). *Genes* 10, 1004.

Borchel, A., Heggland, E.I. and Nilsen, F. (2021) The transcriptomic response of adult salmon lice (*Lepeophtheirus salmonis*) to reduced salinity. *Comparative Biochemistry and Physiology Part D: Genomics and Proteomics* 37, 100778.

Borst, D. (2003) Crustacean endocrine systems. In: Henry, H.L. and Norman, A.W. (eds) *Encyclopedia of Hormones*. Academic Press, New York. pp. 340–351.

Boxaspen, K. and Naess, T. (2000) Development of eggs and the planktonic stages of salmon lice (*Lepeophtheirus salmonis*) at low temperatures. *Contributions to Zoology* 69, 51–55.

Boxshall, G.A. (1974) Developmental stages of *Lepeophtheirus pectoralis* (Müller, 1776) (Copepoda: Caligidae). *Journal of Natural History* 8, 681–700.

Braden, L., Michaud, D., Igboeli, O.O., Dondrup, M., Hamre, L., Dalvin, S., Purcell, S.L., Kongshaug, H., Eichner, C., Nilsen, F. and Fast, M.D. (2020) Identification of critical enzymes in the salmon louse chitin synthesis pathway as revealed by RNA interference-mediated abrogation of infectivity. *International Journal for Parasitology* 50, 873–889.

Bravo, S., Erranz, F. and Lagos, C. (2009) A comparison of sea lice, *Caligus rogercresseyi*, fecundity in four areas in southern Chile. *Journal of Fish Diseases* 32, 107–113.

Bravo, S. (2010) The reproductive output of sea lice *Caligus rogercresseyi* under controlled conditions. *Experimental Parasitology* 125, 51–54.

Bron, J.E., Sommerville, C., Jones, M. and Rae, G.H. (1991) The settlement and attachment of early stages of the salmon louse, *Lepeophtheirus salmonis* (Copepoda: Caligidae) on the salmon host, *Salmo salar*. *Journal of Zoology* 224, 201–212.

Brunet, J., Eichner, C. and Male, R. (2021) The FTZ-F1 gene encodes two functionally distinct nuclear receptor isoforms in the ectoparasitic copepod salmon louse (*Lepeophtheirus salmonis*). *PLoS ONE* 16(5), e0251575.

Chang, E.S. and Mykles, D.L. (2011) Regulation of crustacean molting: a review and our perspectives. *General and Comparative Endocrinology* 172, 323–330.

Chávez-Mardones, J. and Gallardo-Escárate, C. (2015) Next-generation transcriptome profiling of the salmon louse *Caligus rogercresseyi* exposed to deltamethrin (AlphaMax (TM)): discovery of relevant genes and sex-related differences. *Marine Biotechnology* 17, 793–810.

Costello, M.J. (2006) Ecology of sea lice parasitic on farmed and wild fish. *Trends in Parasitology* 22, 475–483.

Dalvin, S., Frost, P., Biering, E., Hamre, L.A., Eichner, C., Krossoy, B. and Nilsen, F. (2009) Functional characterisation of the maternal yolk-associated protein (LsYAP) utilising systemic RNA interference in the salmon louse (*Lepeophtheirus salmonis*) (Crustacea: Copepoda). *International Journal for Parasitology* 39, 1407–1415.

Dalvin, S., Frost, P., Loeffen, P., Skern-Mauritzen, R., Baban, J., Ronnestad, I. and Nilsen, F. (2011) Characterisation of two vitellogenins in the salmon louse *Lepeophtheirus salmonis*: molecular, functional and evolutional analysis. *Diseases of Aquatic Organisms* 94, 211–224.

Dalvin, S., Nilsen, F. and Skern-Mauritzen, R. (2013) Localization and transcription patterns of LsVasa, a molecular marker of germ cells in *Lepeophtheirus salmonis* (Krøyer). *Journal of Natural History* 47, 889–900.

Davis, M.B., Carney, G.E., Robertson, A.E. and Bender, M. (2005) Phenotypic analysis of EcR-A mutants suggests that EcR isoforms have unique functions during *Drosophila* development. *Developmental Biology* 282, 385–396.

Durica, D.S., Wu, X.H., Anilkumar, G., Hopkins, P.M. and Chung, A.C.K. (2002) Characterization of crab EcR and RXR homologs and expression during limb regeneration and oocyte maturation. *Molecular and Cellular Endocrinology* 189, 59–76.

Eichner, C., Frost, P., Dysvik, B., Jonassen, I., Kristiansen, B. and Nilsen, F. (2008) Salmon louse (*Lepeophtheirus salmonis*) transcriptomes during post molting maturation and egg production, revealed using EST-sequencing and microarray analysis. *BMC Genomics* 126.

Eichner, C., Nilsen, F., Grotmol, S. and Dalvin, S. (2014) A method for stable gene knock-down by RNA interference in larvae of the salmon louse (*Lepeophtheirus salmonis*). *Experimental Parasitology* 140, 44–51.

Eichner, C., Dalvin, S., Skern-Mauritzen, R., Malde, K., Kongshaug, H. and Nilsen, F. (2015a) Characterization of a novel RXR receptor in the salmon louse (*Lepeophtheirus salmonis*, Copepoda) regulating growth and female reproduction. *BMC Genomics* 16, 81.

Eichner, C., Hamre, L.A. and Nilsen, F. (2015b) Instar growth and molt increments in *Lepeophtheirus salmonis* (Copepoda: Caligidae) chalimus larvae. *Parasitology International* 64, 86–96.

Eichner, C., Harasimczuk, E., Nilsen, F., Grotmol, S. and Dalvin, S. (2015c) Molecular characterisation and functional analysis of LsChi2, a chitinase found in the salmon louse (*Lepeophtheirus salmonis salmonis*, Krøyer 1838). *Experimental Parasitology* 151–152c, 39–48.

Eichner, C., Dondrup, M. and Nilsen, F. (2018) RNA sequencing reveals distinct gene expression patterns during the development of parasitic larval stages of the salmon louse (*Lepeophtheirus salmonis*). *Journal of Fish Diseases* 41, 1005–1029.

Fahrbach, S.E., Smagghe, G. and Velarde, R.A. (2012) Insect nuclear receptors. *Annual Review of Entomology* 57, 83–106.

Gallardo-Escárate, C., Valenzuela-Muñoz, V., Nuñez-Acuña, G., Valenzuela-Miranda, D., Gonçalves, A.T., Escobar-Sepulveda, H., Liachko, I., Nelson, B., Roberts, S. and Warren, W. (2021) Chromosome-scale genome assembly of the sea louse *Caligus rogercresseyi* by SMRT sequencing and Hi-C analysis. *Scientific Data* 8, 60.

Gonçalves, A.T., Farlora, R. and Gallardo-Escárate, C. (2014) Transcriptome survey of the lipid metabolic pathways involved in energy production and ecdysteroid synthesis in the salmon louse *Caligus rogercresseyi* (Crustacea: Copepoda). *Comparative Biochemistry and Physiology Part B: Biochemistry and Molecular Biology* 176, 9–17.

Gonzalez-Alanis, P., Wright, G.M., Johnson, S.C. and Burka, J.F. (2001) Frontal filament morphogenesis in the salmon louse *Lepeophtheirus salmonis*. *Journal of Parasitology* 87, 561–574.

González, L. and Carvajal, J. (2003) Life cycle of *Caligus rogercresseyi*, (Copepoda: Caligidae) parasite of Chilean reared salmonids. *Aquaculture* 220, 101–117.

Grimnes, A. and Jakobsen, P.J. (1996) The physiological effects of salmon lice infection on post-smolt of Atlantic salmon. *Journal of Fish Biology* 48, 1179–1194.

Hamre, L.A., Eichner, C., Caipang, C.M.A., Dalvin, S.T., Bron, J.E., Nilsen, F., Boxshall, G. and Skern-Mauritzen, R. (2013) The salmon louse *Lepeophtheirus salmonis* (Copepoda: Caligidae) life cycle has only two chalimus stages. *PLoS ONE* 8, e73539.

Hamre, L., Bui, S., Oppedal, F., Skern-Mauritzen, R. and Dalvin, S. (2019) Development of the salmon louse *Lepeophtheirus salmonis* parasitic stages in temperatures ranging from 3 to 24 °C. *Aquaculture Environment Interactions* 429–443.

Harðardóttir, H.M., Male, R., Nilsen, F., Eichner, C., Dondrup, M. and Dalvin, S. (2019) Chitin synthesis and degradation in *Lepeophtheirus salmonis*: molecular characterization and gene expression profile during synthesis of a new exoskeleton. *Comparative Biochemistry and Physiology Part A: Molecular and Integrative Physiology* 227, 123–133.

Harðardóttir, H.M., Male, R., Nilsen, F. and Dalvin, S. (2021) Chitin synthases are critical for reproduction, molting, and digestion in the salmon louse (*Lepeophtheirus salmonis*). *Life* 11(1), 47.

Heuch, P.A., Nordhagen, J.R. and Schram, T.A. (2000) Egg production in the salmon louse [*Lepeophtheirus salmonis* (Krøyer)] in relation to origin and water temperature. *Aquaculture Research* 31, 805–814.

Humble, J.L., Carmona-Antonanzas, G., McNair, C.M., Nelson, D.R., Bassett, D.I., Egholm, I., Bron, J.E., Bekaert, M. and Sturm, A. (2019) Genome-wide survey of cytochrome P450 genes in the salmon louse *Lepeophtheirus salmonis* (Krøyer, 1837). *Parasites & Vectors* 12, 563.

Jackson, D. and Minchin, D. (1992) Aspects of the reproductive output of two caligid copepod species parasitic on cultivated salmon. *Invertebrate Reproduction and Development* 22, 87–90.

Johnson, S.C. (1993) A comparison of development and growth rates of *Lepeophtheirus salmonis* (Copepoda: Caligidae) on naive Atlantic (*Salmo salar*) and chinook (*Oncorhynchus tshawytscha*) salmon. In: Defaye, D. and Boxshall, G. A. (eds) *Pathogens of Wild and Farmed Fish: Sea Lice*. Ellis Horwood, Chichester, UK. pp. 68–80.

Johnson, S.C. and Albright, L.J. (1991a) The developmental stages of *Lepeophtheirus salmonis* (Krøyer, 1837) (Copepoda: Caligidae). *Canadian Journal of Zoology* 69, 929–950.

Johnson, S.C. and Albright, L.J. (1991b) Development, growth, and survival of *Lepeophtheirus salmonis* (Copepoda: Caligidae) under laboratory conditions. *Journal of the Marine Biological Association of the United Kingdom* 71, 425–436.

Johnson, S.C. and Albright, L.J. (1992) Comparative susceptibility and histopathology of the response of naive Atlantic, chinook and coho salmon to experimental infection with *Lepeophtheirus salmonis* (Copepoda: Caligidae). *Diseases of Aquatic Organisms* 14(3), 179–193.

Kabata, Z. (1972) Developmental stages of *Caligus clemensi* (Copepoda: Caligidae). *Journal of the Fisheries Research Board of Canada* 29, 1571–1593.

Khan, M.T., Dalvin, S., Nilsen, F. and Male, R. (2017) Microsomal triglyceride transfer protein in the ectoparasitic crustacean salmon louse (*Lepeophtheirus salmonis*). *Journal of Lipid Research* 58, 1613–1623.

Khoa, T.N.D., Mazelan, S., Muda, S. and Shaharom-Harrison, F. (2019) The life cycle of *Caligus minimus* on seabass (*Lates calcarifer*) from floating cage culture. *Thalassas* 35, 7785.

Kim, I.-H. (1993) Developmental stages of *Caligus punctatus* Shiino, 1955 (Copepoda: Caligidae). In: Boxshall, G.A. and Defaye, D. (eds) *Pathogens of Wild and Farmed Fish: Sea Lice*. Ellis Horwood Ltd, West Sussex, UK. pp. 16–29.

King-Jones, K. and Thummel, C.S. (2005) Nuclear receptors – a perspective from *Drosophila*. *Nature Reviews Genetics* 6, 311–323.

Lewis, A.G. (1963) Life history of the caligid copepod *Lepeophtheirus dissimulatus* Wilson, 1905 (Crustacea: Caligoida). *Science* 17, 195–242.

Lin, C.L., Ho, J.-S. and Chen, S.N. (1996) Developmental stages of *Caligus epidemicus* Hewitt, a copepod parasite of tilapia cultured in brackish water. *Journal of Natural History* 30, 661–684.

Madinabeitia, I. and Nagasawa, K. (2011) Chalimus stages of *Caligus latigenitalis* (Copepoda: Caligidae) parasitic on blackhead seabream from waters, with discussion of terminology used for developmental stages of caligids. *Journal of Parasitology* 97, 221–236.

Maran, B.A.V., Moon, S.Y., Ohtsuka, S., Oh, S.Y., Soh, H.Y., Myoung, J.G., Iglikowska, A. and Boxshall, G.A. (2013) The caligid life cycle: new evidence from *Lepeophtheirus elegans* reconciles the cycles of *Caligus* and *Lepeophtheirus* (Copepoda: Caligidae). *Parasite* 20(15).

Montory, J.A., Cumillaf, J.P., Cubillos, V.M., Paschke, K., Urbina, M.A. and Gebauer, P. (2018) Early development of the ectoparasite *Caligus rogercresseyi* under combined salinity and temperature gradients. *Aquaculture* 486, 68–74.

Montory, J.A., Cumillaf, J.P., Gebauer, P., Urbina, M., Cubillos, V.M., Navarro, J.M., Marín, S.L. and Cruces, E. (2020) Early development and metabolic rate of the sea louse *Caligus rogercresseyi* under different scenarios of temperature and pCO_2. *Marine Environmental Research* 162, 105154.

Mykles, D.L. (2011) Ecdysteroid metabolism in crustaceans. *Journal of Steroid Biochemistry and Molecular Biology* 127, 196–203.

Nakagawa, Y. and Henrich, V.C. (2009) Arthropod nuclear receptors and their role in molting. *FEBS Journal* 276, 6128–6157.

Neptali Morales-Serna, F., Rivas-Salas, A.I., Gomez, S. and Josefina Fajer-Avila, E. (2015) Developmental stages and fecundity of *Lepeophtheirus simplex* (Copepoda: Caligidae) parasitic on bullseye puffer fish (*Sphoeroides annulatus*). *Folia Parasitologica (Ceske Budejovice)* 62, 004.

Nordhagen, J.R., Heuch, P.A. and Schram, T.A. (2000) Size as indicator of origin of salmon lice *Lepeophtheirus salmonis* (Copepoda: Caligidae). *Contributions to Zoology* 69, 99–108.

Núñez-Acuña, G., Sáez-Vera, C., Valenzuela-Muñoz, V., Valenzuela-Miranda, D., Arriagada, G. and Gallardo-Escárate, C. (2020) Tackling the molecular drug sensitivity in the sea louse *Caligus rogercresseyi* based on mRNA and lncRNA interactions. *Genes* 11, 857.

Ohtsuka, S., Takami, I., Maran, B.A.V., Ogawa, K., Shimono, T., Fujita, Y., Asakawa, M. and Boxshall, G.A. (2009) Developmental stages and growth of *Pseudocaligus fugu* Yamaguti, 1936 (Copepoda: Siphonostomatoida: Caligidae) host-specific to puffer. *Journal of Natural History* 43, 1779–1804.

Piasecki, W. and MacKinnon, B.M. (1993) Changes in structure of the frontal filament in sequential developmental stages of *Caligus elongatus* von Nordmann, 1832 (Crustacea: Copepoda: Siphonostomatoida). *Canadian Journal of Zoology* 71, 889–895.

Piasecki, W. and MacKinnon, B.M. (1995) Life-cycle of a sea louse, *Caligus elongatus* von Nordmann, 1832 (Copepoda: Siphonostomatoida: Caligidae). *Canadian Journal of Zoology* 73, 74–82.

Ritchie, G., Mordue (Luntz), A.J., Pike, A.W., Rae, G.H., Boxshall, G.A. and Defaye, D. (1993) The reproductive output of *Lepeophtheirus salmonis* adult females in relation to seasonal variability of temperature and photoperiod. In: Boxshall, G.A. and Defaye, D. (eds) *Pathogens of Wild and Farmed Fish: Sea Lice.* Ellis Horwood, West Sussex, UK. pp. 153–165.

Ritchie, G., Mordue, A.J., Pike, A.W. and Rae, G.H. (1996) Morphology and ultrastructure of the reproductive system of *Lepeophtheirus salmonis* (Krøyer, 1837) (Copepoda: Caligidae). *Journal of Crustacean Biology* 16, 330–346.

Samsing, F., Oppedal, F., Dalvin, S., Johnsen, I., Vagseth, T. and Dempster, T. (2016) Salmon lice (*Lepeophtheirus salmonis*) development times, body size, and reproductive outputs follow universal models of temperature dependence. *Canadian Journal of Fisheries and Aquatic Sciences* 73, 1841–1851.

Sandlund, L., Nilsen, F., Male, R., Grotmol, S., Kongshaug, H. and Dalvin, S. (2015) Molecular characterisation of the salmon louse, *Lepeophtheirus salmonis salmonis* (Krøyer, 1837), ecdysone receptor with emphasis on functional studies of female reproduction. *International Journal for Parasitology* 45, 175–185.

Sandlund, L., Nilsen, F., Male, R. and Dalvin, S. (2016) The ecdysone receptor (EcR) is a major regulator of tissue development and growth in the marine salmonid ectoparasite, *Lepeophtheirus salmonis* (Copepoda: Caligidae). *Molecular and Biochemical Parasitology* 208, 65–73.

Sandlund, L., Kongshaug, H., Horsberg, T.E., Male, R., Nilsen, F. and Dalvin, S. (2018a) Identification and characterisation of the ecdysone biosynthetic genes neverland, disembodied and shade in the salmon louse *Lepeophtheirus salmonis* (Copepoda: Caligidae). *PLoS ONE* 13(2), e0191995.

Sandlund, L., Kongshaug, H.D., Nilsen, F. and Dalvin, S. (2018b) Molecular characterization and functional analysis of components of the TOR pathway of the salmon louse, *Lepeophtheirus salmonis* (Krøyer, 1838). *Experimental Parasitology* 188, 83–92.

Schram, T.A. (1993a) Supplementary description of the developmental stages of *Lepeophtheirus salmonis* (Krøyer, 1837) (Copepoda: Caligidae). In: Boxshall, G.A. and Defaye, D. (eds) *Pathogens Of Wild And Farmed Fish: Sea Lice.* Ellis Horwood, West Sussex, UK. pp. 30–47.

Schram, T.A. (1993b) Life cycle of *Lepeophtheirus salmonis* (Krøyer, 1837): the morphology and biology of the developmental stages. *Norsk Veterinaertidsskrift* 105, 1207–1215.

Schram, T.A. (2000) The egg string attachment mechanism in salmon lice *Lepeophtheirus salmonis* (Copepoda: Caligidae). *Contributions to Zoology* 69, 21–29.

Skern-Mauritzen, R., Frost, P., Dalvin, S., Kvamme, B.O., Sommerset, I. and Nilsen, F. (2009) A trypsin-like protease with apparent dual function in early *Lepeophtheirus salmonis* (Krøyer) development. *Molecular Biology* 10(44).

Skern-Mauritzen, R., Malde, K., Eichner, C., Dondrup, M., Furmanek, T., Besnier, F., Komisarczuk, A.Z., Nuhn, M., Dalvin, S., Edvardsen, R.B., Grotmol, S., Karlsbakk, E., Kersey, P., Leong, J.S., Glover, K.A., Lien, S., Jonassen, I., Koop, B.F. and Nilsen, F. (2021) The salmon louse genome: copepod features and parasitic adaptations. *Genomics* 113(6), 3666–3680.

Skinner, D.M. (1985) Moulting and regeneration. In: Bliss, D.E. (ed.) *The Biology of Crustacea – Integument, Pigments, and Hormonal Processes.* Academic Press, Inc, New York.

Stien, A., Bjorn, P.A., Heuch, P.A. and Elston, D.A. (2005) Population dynamics of salmon lice *Lepeophtheirus salmonis* on Atlantic salmon and sea trout. *Marine Ecology-Progress Series* 290, 263–275.

Sutherland, B.G., Covello, J.M., Friend, S.E., Poley, J.D., Koczka, K.W., Purcell, S.L., MacLeod, T.L., Donovan, B.R., Pino, J., Gonzalez-Vecino, J.L., Gonzalez, J., Troncoso, J., Koop, B.F., Wadsworth, S.L. and Fast, M.D. (2017) Host-parasite transcriptomics during immunostimulant-enhanced rejection of salmon lice (*Lepeophtheirus salmonis*) by Atlantic salmon (*Salmo salar*). *Facets* 2, 477–495.

Sutherland, B.J., Poley, J.D., Igboeli, O.O., Jantzen, J.R., Fast, M.D., Koop, B.F. and Jones, S.R. (2015) Transcriptomic responses to emamectin benzoate in Pacific and Atlantic Canada salmon lice *Lepeophtheirus salmonis* with differing levels of drug resistance. *Evolutionary Applications* 8, 133–148.

Sutherland, B.J.G., Jantzen, S.G., Yasuike, M., Sanderson, D.S., Koop, B.F. and Jones, S.R.M. (2012) Transcriptomics of coping strategies in free-swimming *Lepeophtheirus salmonis* (Copepoda) larvae responding to abiotic stress. *Molecular Ecology* 21, 6000–6014.

Thummel, C.S. (2001) Molecular mechanisms of developmental timing in *C. elegans* and *Drosophila*. *Developmental Cell* 1, 453–465.

Tucker, C.S., Sommerville, C. and Wootten, R. (2000a) The effect of temperature and salinity on the settlement and survival of copepodids of *Lepeophtheirus salmonis* (Krøyer, 1837) on Atlantic salmon, *Salmo salar* L. *Journal of Fish Diseases* 23, 309–320.

Tucker, C.S., Sommerville, C. and Wootten, R. (2000b) An investigation into the larval energetics and settlement of the sea louse, *Lepeophtheirus salmonis*, an ectoparasitic copepod of Atlantic salmon, *Salmo salar*. *Fish Pathology* 35, 137–143.

Tucker, C.S., Norman, R., Shinn, A.P., Bron, J.E., Sommerville, C. and Wootten, R. (2002) A single cohort time delay model of the life-cycle of the salmon louse *Lepeophtheirus salmonis* on Atlantic salmon *Salmo salar*. *Fish Pathology* 37, 107–118.

Tully, O. (1989) The succession of generations and growth of the caligid copepods *Caligus-elongatus* and *Lepeophtheirus salmonis* parasitizing farmed Atlantic salmon smolts (*Salmo salar* L). *Journal of the Marine Biological Association of the United Kingdom* 69, 279–287.

Tully, O. and Whelan, K.F. (1993) Production of nauplii of *Lepeophtheirus salmonis* (Krøyer) (Copepoda: Caligidae) from farmed and wild salmon and its relation to the infestation of wild sea trout (*Salmo trutta* L.) off the west coast of Ireland in 1991. *Fisheries Research* 17, 187–200.

Valenzuela-Miranda, D. and Gallardo-Escárate, C. (2016) *Caligus rogercresseyi* serine proteases: transcriptomic analysis in response to delousing drugs treatments. *Aquaculture* 465 65–77.

Valenzuela-Muñoz, V., Chavez-Mardones, J. and Gallardo-Escárate, C. (2015) RNA-seq analysis evidences multiple gene responses in *Caligus rogercresseyi* exposed to the anti-salmon lice drug azamethiphos. *Aquaculture* 446, 156–166.

Voth, D.R. (1971) Life history of the caligid copepod *Lepeophtheirus hospitalis* Fraser, 1920 (Crustacea:Caligoida). PhD thesis, Oregon State University.

Yasuike, M., Leong, J., Jantzen, S.G., von Schalburg, K.R., Nilsen, F., Jones, S.R.M. and Koop, B.F. (2012) Genomic resources for sea lice: analysis of ESTs and mitochondrial genomes. *Marine Biotechnology* 14, 155–166.

CHAPTER 4

Caligids in global aquaculture

Ken MacKenzie

School of Biological Sciences (Zoology), University of Aberdeen, Scotland

4.1 Introduction

Parasitic copepods display an impressive variety of body forms, many quite bizarre, reflecting their adaptations to parasitic lifestyles. Most of these forms would be completely unrecognisable as copepods to those familiar with free-living species only. Members of the family Caligidae, however, are more readily recognisable as copepods than most other parasitic families. Kabata (1979) traced the progressive adaptation of copepods from free-living to parasitic forms. He described changes in morphology associated with this adaptation and showed that the Caligidae show less extreme changes than most families of parasitic copepods, suggesting that they represent a relatively early stage in the evolution of parasitism in the Copepoda. This is supported by the fact that, although apparently obligate parasites that depend on their hosts for nutrition, many species of caligids are capable of surviving for extended periods as free-living organisms. This is evidenced by the many records of adult forms of caligid species, mostly of the genus *Caligus*, found in plankton samples (Venmathi Maran et al., 2016), with 13 species recorded exclusively from plankton samples with no known hosts. Such versatility in lifestyle is further enhanced by the low host specificity of many caligids, especially those of the genus *Caligus*. Caligids are thus supremely well adapted to take advantage of the opportunities presented to them by dense populations of farmed fish. In addition to being pathogens in their own right, some species are also suspected of serving as vectors for viruses and bacteria that cause disease in farmed fish (Nylund et al., 1993; Oelckers et al., 2014).

The family Caligidae comprises more than 30 genera (Kabata, 1979) and more than 500 valid species (Walter and Boxshall, 2020). It has been estimated that in marine and brackish water fish culture, 61% of copepod infestations are caused by members of the family Caligidae, 40% of which are caused by species of *Caligus* and 14% by species of *Lepeophtheirus* (Johnson et al., 2004). The aims of this review are to list those caligid species that are known or potential parasites of cultured fish in global aquaculture and to provide information on their geographical distributions, host specificity (or lack of it) and the cultured species they infect. Finally, the potential effects of predicted climate change on the distributions and biology of caligids in relation to aquaculture are briefly discussed.

4.2 Which caligid species represent the most serious threats to global aquaculture?

Thirty-three species of *Caligus*, six of *Lepeophtheirus* and one each of *Euryphorus* and *Parapetalus* have been reported from cultured fish to date (Table 4.1), while several other species not yet found on cultured fish are considered potential threats.

Those species considered to be the most serious pests in global aquaculture are discussed individually below, including those highlighted by Ho (2000) and Venmathi Maran et al. (2009) as the most problematic in Asian aquaculture.

4.2.1. Genus Caligus Müller, 1785

Caligus acanthopagri Lin, Ho & Chen, 1994

This species is distributed throughout the South China Sea, Indonesia and South Africa, where it appears to be mainly a parasite of fishes of the family Sparidae. It infects the gills and gill cavities, causing congestion, mucous proliferation and damage, leading to mortalities of cultured black porgy (Lin et al., 1994; Lin, 1996 cited in Ho, 2000).

Caligus chiastos Lin & Ho, 2003

This species is found throughout the Eastern Pacific from Taiwan to South Australia where, despite being considered to be a rare species (Hayward et al., 2008; Venmathi Maran et al., 2009), it has been reported from at least 11 families of actinopterygian fishes and one elasmobranch (Boxshall, 2018). To date it has also been reported from six species of farmed fish in Malaysia and Australia (Hayward et al., 2007, 2008, 2010; Venmathi Maran et al., 2009; Nowak et al., 2011; Muhd-Faizul et al., 2012). Given its low host specificity, future reports from other species of farmed fish in East Asia seem very likely. The most severe effects of *C. chiastos* infections on farmed fish have been on southern bluefin tuna *Thunnus maccoyi* in Australia, where it has been associated with loss of condition and blindness (Hayward et al., 2008).

Caligus clemensi Parker & Margolis, 1964

This species is native to the Northeast Pacific where it has been reported from a wide range of mainly pelagic fishes (Parker and Margolis, 1964). Jones and Johnson (2014) listed 13 fish species as reported hosts, including Atlantic salmon and six species of the genus *Oncorhynchus*. *Caligus clemensi* is a common parasite of farmed Atlantic salmon in British Columbia, Canada, although not found at such high abundances as *Lepeophtheirus salmonis*. The results of a study carried out by Saksida et al. (2007) suggested that pre-adult and adult *C. clemensi* are highly mobile and likely to move freely on and off the farmed salmon, making it difficult to obtain accurate estimates of their abundance levels.

Caligus elongatus von Nordmann, 1832

This species appears to be most abundant in the North Atlantic and may be restricted to this region, as Hayward et al. (2008) considered that earlier records from Australia and New Zealand were probably of *Caligus chiastos*. It has a very low host specificity and has been reported from more than 80 wild fish species and one cetacean. Several publications have identified lumpfish *Cyclopterus lumpus* as

Table 4.1 List of caligid species infecting cultured fish globally, with their geographical distributions, natural and cultured hosts and key references.

Caligid species	Known distribution	Natural hosts	Cultured hosts	Key references
Genus *Caligus* Müller, 1785				
C. acanthopagri Lin, Ho & Chen, 1994	South China Sea, Indonesia, South Africa	Goldsilk seabream *Acanthopagrus berda*, Cape stumpnose *Rhabdosargus holubi*, spotted cat *Scatophagus argus*	Black sea bream *Acanthopagrus schlegeli*, Malabar rock cod *Epinephalus malabaricus*, Mozambique tilapia *Oreochromis mossambicus*, spotted cat	Lin et al. (1994); Lin (1996) cited in Ho (2000); Yuniar et al. (2007)
C. antennatus Boxshall & Gurney, 1980	Persian Gulf	Breams (Sparidae)	Yellow-fin seabream *Acanthopagrus latus*, Sobaity sea bream *Sparidentex hasta*	Boxshall and Gurney (1980); Tareen (1986); Al-Marzouk (2002)
C. apodus (Brian, 1924)	Mediterranean, west coast of Africa	Mullets (Mullidae)	Flathead grey mullet *Mugil cephalus*, sea bass *Dicentrarchus labrax*	Paperna (1975); Ragias et al. (2004); Özak et al. (2013)
C. brevicaudatus Scott, 1901	Northeast Atlantic, Mediterranean, South Korea	Gurnards (Triglidae), sand sole *Pegusa lascaris*, common sole *Solea solea*, Senegalese sole *Solea senegalensis*	Bastard halibut *Paralichthys olivaceus*	Choi et al. (1995); Mortelmans et al. (2017)
C. chiastos Lin & Ho, 2003	Taiwan, Malaysia, Australia	Wide host range, reported from at least 11 families of teleosts and one elasmobranch	Southern bluefin tuna *Thunnus maccoyi*, Japanese meagre *Argyrosomus japonicus*, yellowtail amberjack *Seriola lalandi*, Asian seabass *Lates calcalifer*, John's snapper *Lutjanus johnsi*, rainbow trout	Hayward et al. (2007, 2011); Muhd-Faizul et al. (2012)
C. clemensi Parker & Margolis, 1964	Northeast Pacific	Wide range of mainly pelagic fishes	Pacific salmon *Oncorhynchus* spp., Atlantic salmon *Salmo salar*	Parker and Margolis (1964); Kabata (1972); Margolis and Kabata (1988); Jones and Johnson (2014)
C. curtus Müller, 1785	Arctic-Boreal Atlantic and contiguous waters	Predominantly gadid fishes, but also reported from a variety of other fish, including elasmobranchs	Atlantic salmon, Atlantic cod *Gadus morhua*	Parker et al. (1968); Hogans and Trudeau (1989); Heuch et al. (2011)
C. dakari van Beneden, 1892	Mediterranean, Southeast Atlantic	Wide host range	Common dentex *Dentex dentex*	Boxshall and El-Rashidy (2009); Öktener (2009)

Table 4.1 (Continued)

Caligid species	Known distribution	Natural hosts	Cultured hosts	Key references
C. elongatus von Nordmann, 1832	North and South Atlantic	Very wide host range (>80 species of teleosts)	Atlantic salmon, brook trout *Salvelinus fontinalis*, rainbow trout, arctic charr *Salvelinus alpinus*, Atlantic cod	Parker (1969); Sutterlin et al. (1976); Hogans and Trudeau (1989); Mustafa et al. (2005); Øines et al. (2006); Heuch et al. (2011); Hemmingsen et al. (2020)
C. epidemicus Hewitt, 1971	Malaysia, Thailand, India, Vietnam, Philippines, Australia	Very wide host range (at least 13 families of teleosts)	Elongate surgeonfish *Acanthurus mata*, yellowfin bream *Acanthopagrus australis*, *A. schlegeli*, *Lates calcalifer*, orange-spotted grouper *Epinephelus coioides*, *E. bleekeri*, *E. malabaricus*, blue tilapia *Oreochromis aurea*, Mozambique tilapia, Wami tilapia *O. urolepis hornorum*, spotted cat, grey mullet *Mugil cephalus*, large scale mullet *Liza macrolepis*, broad-mouthed mullet *Paramugil parmatus*, milk fish *Chanos chanos*, snubnose pompano *Trachinotus blochii*, three-striped tiger fish *Terapon jarbua*, Celebes goby *Glossogobius celebius*, mangrove red snapper *Lutjanus argentimaculatus*, silver moony *Monodactylus argenteus*, cobia *Rachycentron canadum*, orange-spotted spinefoot *Siganus guttatus*. Crustacean *Penaeus monodon*	Ruangpan and Kabata (1984); Regidor and Arthur (1986); Lin and Ho (1993); Roubal (1994, 1995, 1997); Lin (1996) cited in Ho (2000); Lin et al. (1996a); Ho et al. (2004a); Vo et al. (2008); Venmathi Maran et al. (2009); Muhd-Faizul et al. (2012)
C. fugu Yamaguti & Yamasu, 1959	Japan, India, Eastern Mediterranean	Puffer fishes (Tetraodontidae)	Tiger puffer *Takifugu rubripes*	Ogawa and Inouye (1997); Ogawa and Yokoyama (1998); Özak et al. (2012); Nagasawa (2015)
C. lalandei Barnard, 1948	Widely distributed in the Pacific and reported from South Africa	Amberjacks (Carangidae)	Yellowtail amberjack *Seriola lalandi*, Japanese amberjack *Seriola quinqueradiata*	Ho et al. (2001); Nagasawa and Fukuda (2011); Nagasawa (2015); Bravo et al. (2017)
C. latigenitalis Shiino, 1954	Japan, Taiwan	Black sea bream *Acanthopagrus schlegeli*	Black sea bream	Lin et al. (1994); Izawa and Choi (2000)

C. longipedis Bassett-Smith, 1898	Malaysia, Philippines, India, Japan, Taiwan	Very wide range, but reported most frequently from Carangidae	Golden trevally *Gnathonodon specious*, striped jack *Pseudocaranx dentex*, coho salmon *Oncorhynchus kisutch*, Atlantic salmon, rainbow trout *Oncorhynchus mykiss*	Ogawa (1992); Madinabeitia et al. (2009); Venmathi Maran et al. (2009)
C. longirostris Heegaard, 1962	Australia	Wide range	Pacific bluefin tuna *Thunnus orientalis*	Heegard (1962); Nowak et al. (2011)
C. macarovi Gusev, 1951	Northern Pacific and adjacent seas	Wide range, but appears to favour Pacific saury *Cololabis saira*	Seabass *Dicentrarchus labrax*, Asian sea bass	Nagasawa (2011, 2015)
C. minimus Otto, 1821	Northeast Atlantic and adjacent seas, Malaysia	Wide range, but mainly perciform fishes	Yellow grouper *Epinephelus awoara*	Ragias et al. (2004); Khoa et al. (2019)
C. nanhaiensis Wu & Pan, 1997	China	Groupers *Epinephelus* spp.	*Epinephelus awoara*	Wu and Pan (1997); Wu et al. (1997)
C. nuenonnae Andrews, Bott, Battaglene & Nowak, 2009	Australia	Striped trumpeter *Latris lineata*	Striped trumpeter	Andrews et al. (2009); Nowak et al. (2011)
C. orientalis Gusev, 1951	Northwest Pacific Ocean: Russia, Japan, Korea and China	Wide range, reported from at least 22 species of fish belonging to eight orders in marine and freshwater	Pacific salmon *Oncorhynchus* spp., milkfish, Asian sea bass, Malabar rock cod, yellow-fin sea bream, black sea bream, Mozambique tilapia, largescale mullet, grey mullet, common carp *Cyprinus carpio*	Urawa and Kato (1991); Lin and Ho (1998); Nagasawa (2004, 2015)
C. oviceps Shiino, 1952	Japan, Taiwan, Australia	Wide host range	Rabbit fish *Siganus fuscescens*	Lin et al. (1996b); Boxshall (2018)
C. pageti Russell, 1925	Mediterranean	Mullets (Mugilidae)	Grey mullet, thin-lip grey mullet *Liza ramada*	Paperna (1975); Raibaut et al. (1980); Ragias et al. (2004)
C. patulus Wilson, 1937	Philippines, Florida, Panama, possibly cosmopolitan	Milkfish *Chanos* sp., dolphinfish *Coryphaena hippurus*	Milkfish	Burnett-Herkes (1974); Jones (1980)
C. pelamydis Krøyer, 1863	Cosmopolitan	Wide range, most frequently reported from Scombridae	Japanese sea perch *Lateolabrax japonicus*	Choi et al. (1995); Catalano and Hutson (2010)
C. punctatus Shiino, 1955	Malaysia, Japan, Taiwan, Australia	Wide range	Black sea bream, Asian sea bass, blue tilapia, Mozambique tilapia, grey mullet, thin-lipped grey mullet, Japanese sea bass, Malabar rock cod, milkfish	Lin (1996) cited in Ho (2000); Ho and Lin (2004); Venmathi Maran et al. (2009); Muhd-Faizul et al. (2012)

Table 4.1 (Continued)

Caligid species	Known distribution	Natural hosts	Cultured hosts	Key references
C. rogercresseyi (Boxshall & Bravo, 2000)	Southeast Pacific, southern Argentina	Wide range, but most favoured host appears to be the rock cod *Eleginus maclovinus.*	Coho salmon, rainbow trout and Atlantic salmon, Mozambique tilapia	Boxshall and Bravo (2000); Bravo et al. (2006); Bravo et al. (2011)
C. rotundigenitalis Yü, 1933	Cosmopolitan	Wide range	Crimson snapper *Lutjanus erythropterus*, spotted grouper, brown-marbled grouper *Epinephelus fuscoguttatus*, golden trevally *Gnathodon speciosus*, pearlspot *Etroplus suratensis*	Venmathi Maran et al. (2009); Leaw et al. (2012); Muhd-Faizul et al. (2012); Solanki et al. (2016)
C. sclerotinosus Roubal, Armitage & Rohde, 1983	Japan, Korea, Mexico	Sparidae and Lutjanidae	Red seabream *Pagrus major*, spotted rose snapper *Lutjanus guttatus*, Pacific red snapper *L. peru*	Ho et al. (2004b); Venmathi Maran et al. (2012b); Morales-Serna et al. (2014), Nagasawa (2015)
C. serratus Shiino, 1965	Eastern Pacific	Wide range	Coho salmon	Reyes (1985); Morales-Serna et al. (2013)
C. spinosus Yamaguti, 1939	Japan, Korea	Carangidae	Japanese amberjack, yellowtail amberjack	Ogawa and Yokoyama (1998); Nagasawa and Fukuda (2011); Cruz-Lacierda et al. (2011); Nagasawa (2015)
C. stromatei Krøyer, 1863	China, India	*Stromateus* sp., butterfish *Pampus argenteus*	Black sea bream	Lin et al. (1994); Hayes et al. (2012)
C. teres Wilson, 1905	Southern Chile	Wide range	Coho salmon, rainbow trout	Reyes and Bravo (1983); Carvajal et al. (1998)
C. vexator Heller, 1865	Mediterranean, Atlantic coast of Africa	Sparidae	Common dentex *Dentex dentex*	Alaş and Öktener (2018)
Genus *Euryphorus* Milne Edwards, 1840				
E. brachypterus (Gerstaecker, 1853)	Cosmopolitan	Large Scombridae	Southern bluefin tuna, Pacific bluefin tuna	Deveney et al. (2005); Nagasawa and Shirakashi (2018)

Genus *Lepeophtheirus* Nordmann, 1832				
L. atypicus Lin, Ho & Chen, 1996	Taiwan	Siganidae	Rabbit fish	Lin et al. (1996b)
L. cuneifer Kabata, 1974	Northeast Pacific	Wide range, including one elasmobranch	Rainbow trout, Atlantic salmon	Kabata (1974); Johnson and Albright (1991)
L. hippoglossi (Krøyer, 1837)	North Atlantic	Atlantic halibut *Hippoglossus hippoglossus*	Atlantic halibut	Haug (1988); Armstrong et al. (1999); Bergh et al. (2001)
L. longiventralis Yü & Wu, 1932	Northwest Pacific	Spotted halibut *Verasper variegatus*	Spotted halibut	Ho et al. (2004b); Nagasawa (2015)
L. paralichthydis Yamaguti & Yamasu, 1960	Japan	Bastard halibut *Paralichthys olivaceus*	Bastard halibut	Ho (2000); Nagasawa (2015)
L. salmonis (Krøyer, 1837)	North Atlantic, North Pacific	Salmonidae, but also two cyprinids	*Salmo* spp., *Oncorhynchus* spp.	Nagasawa (2004); Costello (2006); Jones and Johnson (2014)
Genus *Parapetalus* Steenstrup & Lütken, 1861				
P. occidentalis Wilson, C.B., 1908	Taiwan, India	Cobia *Rachycentron canadum*, *Johnius* sp.	Cobia	Ho and Lin (2001); Ku and Lu (2009)

a favoured host (Hemmingsen et al., 2020). *Caligus elongatus* is a persistent pest in salmonid farming in the North Atlantic, although not as pathogenic as *Lepeophtheirus salmonis*. The only regions where *C. elongatus* has been reported as being more abundant than *L. salmonis* on farmed Atlantic salmon are the Bay of Fundy in the Northwest Atlantic (Hogans and Trudeau, 1989) and possibly the coast of Finnmark in North Norway (Hemmingsen et al., 2020).

Caligus epidemicus Hewitt, 1971

This species is considered to be the predominant threat to aquaculture in Australian and Asian waters due to its abundance and low host specificity (Venmathi Maran et al., 2009). It is tolerant of a wide range of salinities and has been reported from fish belonging to at least 13 families of teleosts (Boxshall, 2018), and from at least 22 species of cultured fish in Malaysia (Venmathi Maran et al., 2009; Muhd-Faizul et al., 2012), Taiwan (Lin and Ho, 1993), Vietnam (Vo et al., 2008), the Philippines (Regidor and Arthur, 1986; Ho et al., 2004a; Cruz-Lacierda et al., 2011) and Australia (Roubal, 1994), and from a cultured crustacean in Thailand (Ruangpan and Kabata, 1984). *Caligus epidemicus* is also unique among those caligid species whose life cycles have been described to date in that it has two chalimus stages more than any of the others (Lin and Ho, 1993).

Caligus orientalis Gusev, 1951

This species appears to be confined to coastal temperate regions of the Northwest Pacific, where it has been reported from at least 22 species of fish belonging to 13 families (Nagasawa, 2004). Its regular occurrence in plankton samples suggests that it readily detaches from its hosts and actively swims in the water column (Venmathi Maran et al., 2012a). Another noteworthy feature of *C. orientalis* is its ability to thrive in a wide range of salinities. In Japan, mortalities in common carp *Cyprinus carpio* and rainbow trout *Oncorhynchus mykiss* reared in brackish water were described by Matumoto (1980) and Urawa and Kato (1991). Nagasawa (2015) reported that *C. orientalis* was not important in rainbow trout culture in Japan. It is, however, a serious problem in Taiwan and Korea, where it has been reported from eight species of farmed fish (Lin and Ho, 1998).

Caligus patulus Wilson, 1937

This species was listed by Venmathi Maran et al. (2009) as one of the ten species of sea lice considered to be killer species on Asian fish farms, but this seems to be limited to the culture of milkfish *Chanos chanos* in the Philippines (Jones, 1980). Originally described from milkfish caught in Panama Bay, Panama, it was later reported from dolphinfish *Coryphaena hippurus*, an oceanic species with a worldwide distribution in tropical and subtropical waters, caught in the Straits of Florida (Burnett-Herkes, 1974). The latter report prompted Jones (1980) to comment that *C. patulus* may have a cosmopolitan trans-oceanic distribution.

Caligus punctatus Shiino, 1955

This is another species with low host specificity and tolerant of a wide range of salinities. It is regarded as a pest with potential threats to aquaculture similar to those posed by *C. epidemicus* (Venmathi Maran et al., 2009). It is distributed throughout the West Pacific from Japan to Australia, where it has been found on at least 21 fish species belonging to 14 families (Catalano and Hutson, 2010). Most reports of *C. punctatus* on farmed fish have been from Taiwan, where it has been found on at least 12 species (Lin, 1996 cited in Ho, 2000). It has also been found on farmed Asian sea bass *Lates calcarifer* in Malaysia (Muhd-Faizul et al., (2012).

Caligus rogercresseyi (Boxshall & Bravo 2000)

This species is native to the Southeast Pacific and is the dominant and most pathogenic sea lice species affecting salmonid farming in Chile (Carvajal et al., 1998; González and Carvajal, 2003). Atlantic salmon and rainbow trout are particularly susceptible, with coho salmon *Oncorhynchus kisutch* less so (Bravo, 2003). *Caligus rogercresseyi* has been reported from a wide range of wild fish species, but its most favoured host appears to be the rock cod or robalo *Eleginus maclovinus*, which occurs commonly around salmonid cages, along with the Chilean silverside *Odontesthes regia*, which has also been reported as a host (Carvajal et al., 1998). The economic impact of *C. rogercresseyi* on the salmonid farming industry in Chile has been estimated at US$650 million/year (Arriagada et al., 2019).

Caligus rotundigenitalis Yü, 1933

This species is distributed worldwide with low host specificity, having been reported from at least 32 host species belonging to 19 families (Ho and Lin, 2004). It is one of the most common species in cultured and wild fish in East Asia (Venmathi Maran et al., 2009), and has been reported from at least five species of farmed fish in India, Malaysia and Taiwan (Venmathi Maran et al., 2009; Leaw et al., 2012; Muhd-Faizul et al., 2012; Solanki et al., 2016).

Caligus spinosus Yamaguti, 1939

This species has been reported only from Japan and Korea, but probably also occurs in other areas (Choe and Kim, 2010). It parasitises the oral cavity, especially gill arches and rakers, of farmed Japanese and yellowtail amberjacks, and has been reported to cause mass mortalities of the former species (Fujita et al., 1968). *Caligus spinosus* was confused with the very similar species *C. aesopus* by various authors (Fernandez and Villalba, 1986; Lin and Ho, 2007; Hutson et al., 2007) until Choe and Kim (2010) confirmed them to be separate species.

Caligus teres Wilson, 1905

Like *C. rogercresseyi*, this species is native to the Southeast Pacific, where it has been reported from a wide range of fish of various taxonomic groups. It was the first native caligid to transfer to farmed salmonids in Chile in the early 1980s, when it was found infesting coho salmon (Reyes and Bravo, 1983). When the culture of rainbow trout in Chile began in 1987, they were found to be highly susceptible to *C. teres* (Bravo, 2003). This species is not considered to be as great a threat as *C. rogercresseyi* to Chilean salmonid farming.

4.2.2 Genus Lepeophtheirus *Nordmann, 1832*

Lepeophtheirus longiventralis Yü & Wu, 1932

This species has only been reported from the spotted halibut *Verasper variegatus*, which has a distribution limited to the Northwest Pacific from Japan to the East China Sea (Froese and Pauly, 2019). According to Ho et al. (2004b), *L. longiventralis* can be considered a rare species, but one that is well capable of reproducing and proliferating on farmed spotted halibut. Adult females infect the branchial cavity where they feed on blood. Heavily infected fish produce excessive mucous at the site of infection and become anaemic and weak (Nagasawa, 2015).

Lepeophtheirus paralichthydis Yamaguti & Yamasu, 1960

This is one of the species cited by Venmathi Maran et al. (2009) as one of the ten species of sea lice considered killer species on fish farms in Asia. It appears to be specific to the bastard halibut *Paralichthys olivaceus*, which is distributed in the Western Pacific from Japan to the South China Sea (Froese and Pauly, 2019). The only report of its occurrence on farmed bastard halibut is that by Ho (2000), who reported mortalities attributed to the infections. Despite the importance of bastard halibut as an aquaculture species in Japan, no other report has been published on its occurrence on farmed fish (Nagasawa, 2015).

Lepeophtheirus salmonis (Krøyer 1837)

Commonly known as the 'salmon louse', this is undoubtedly one of the most important caligid species in terms of economic and financial costs to aquaculture. Although its geographical distribution is confined to the northern hemisphere, the fact that it affects the lucrative salmonid farming industry makes it such an important pest. Costello (2009) estimated that in 2006 the worldwide salmonid farming industry had a total loss of US$480 million due to salmon lice infestations. Abolofia and Wilen (2017) estimated that lice parasitism caused US$436 million damage, equivalent to 8.70% of total production value, to the Norwegian salmon farming industry alone in 2011. In northern Japan, however, where coho salmon is the main cultured salmonid species, *L. salmonis* is not a serious pest because coho are more resistant to it than most other salmonids (Wagner et al., 2008; Nagasawa, 2015). Nor was *L. salmonis* considered a serious health problem for farmed salmon in British Columbia, Canada (Saksida et al., 2015). In wild salmon in the Canadian Pacific, pink salmon *Oncorhynchus gorbuscha* are often the most heavily infected with *L. salmonis* (Saksida et al., 2015), but they mount a strong tissue response which leads to rejection of the parasite within the first week of infection (Wagner et al., 2008).

4.2.3 Genus *Parapetalus* Steenstrup & Lütken, 1861

Parapetalus occidentalis Wilson, C.B., 1908

This species appears to be specific to cobia *Rachycentron canadum*, which has a worldwide distribution in tropical and subtropical waters, although absent from the Eastern Pacific (Froese and Pauly, 2019). It was reported from farmed cobia in Taiwan by Ho and Lin (2001) and Ku and Lu (2009). The latter authors stated that although *P. occidentalis* had not yet caused a serious problem, its high abundance makes it a potentially serious risk.

4.3 Relationships with wild fish populations

Many of the *Caligus* species listed in the previous section have very low host specificity. This allows them to persist in the vicinity of fish farms on a variety of wild fish species which serve as reservoir hosts, thereby making control within the farm system itself extremely difficult. Despite their low host specificity, however, these copepods usually show a preference for particular host species. Identifying their favoured host species and controlling the numbers of these around farms is one potential method of reducing risk to the farmed fish. For example, the rock cod or robalo *E. maclovinus* appears to be the most favoured host for *C. rogercresseyi* (Carvajal et al., 1998). Physical methods of controlling or preventing transmission between wild and farmed fish populations include the use of plankton nets and 'snorkel' sea cages. These 'barrier technologies' were recently reviewed by Barrett et al. (2020).

4.4 Caligids and climate change

There are two main ways in which an ectoparasite with a direct single-host life cycle, such as a caligid, is affected by changes in its environment: through direct effects on the parasite itself, and through indirect effects on its host. Effects on the parasite itself are likely to be greater with caligids than with most ectoparasites because they are able to spend a considerable amount of time as free-living organisms in the plankton as early larval stages and adults. Effects on the host are arguably less important because of the ability of adult caligids to abandon a host when conditions are unfavourable. The probable effects of climate change on aquatic parasites in general have been reviewed by Marcogliese (2001, 2008), Löhmus and Björklund (2015) and Byers (2020). Among the many effects of current climate change, those likely to have the greatest effects on caligids are increasing sea temperatures and increasing acidification. Copepods and other crustaceans are particularly sensitive to increasing acidification because of its adverse effects on the maintenance of their calcium carbonate exoskeletons (Orr et al., 2005). There have been few studies on the effects of acidification on marine parasites, and only one to date on its effects on caligids – that of Thompson et al. (2019). In this study the early free-living stages (nauplii and copepodites) of *L. salmonis* were exposed to different pCO_2 levels, including a 'High' level representing the worst-case scenario end-of-century projection of the Intergovernmental Panel on Climate Change. The High treatment group showed significantly higher oxygen consumption, but there was no significant effect on carbon, nitrogen, lipid volume or fatty acid content. The authors concluded that *L. salmonis* has mechanisms to compensate for increasing acidification and that populations will be tolerant of projected future acidification scenarios. It is important to note, however, that *L. salmonis* and its Atlantic salmon host have evolved to handle the diverse environments of their predominantly coastal habitat. Other caligid species may show very different responses. There have been several studies on the effects of acidification and temperature on free-living marine copepods. These have revealed a complex and sometimes antagonistic relationship, with acidification partially counteracting some effects of increased temperature while adding to others (Garzke et al., 2016; Pedersen and Hanssen, 2017). Runge et al. (2016) reviewed the research on the response of copepods to increased acidification and found that, with only a few exceptions, the effects of increased acidification are small at the levels predicted to occur during the next century. Wang et al. (2018) reviewed the effects of ocean acidification on copepods and pointed out that the idea that copepods are resilient to increased acidification is primarily based on short-term (less than one generation) exposure of adult female copepods, whereas short-term and long-term (multigenerational) exposures showed significant negative impacts on growth and reproduction. One of the results of current climate warming is expansion of host geographical ranges, which can result in species that have evolved in isolation being brought into close contact. For caligids with low host specificity, such as many species of *Caligus*, this provides opportunities to infect new hosts, including cultured species. The predicted poleward shifts have serious implications for aquaculture in higher latitude regions with extensive fish-farming industries. For example, one of the most pathogenic species for aquaculture, *C. epidemicus*, has been reported from all mainland states of Australia, but not from Tasmania (Boxshall, 2018). It would take only a small poleward extension for this to change, with potentially serious consequences for Tasmanian aquaculture. Another effect of increasing temperature on caligids is a shortening of the developmental time, resulting in earlier seasonal peaks and more annual generations (Hogans and Trudeau, 1989; Groner et al., 2016; Hamre et al., 2019). This effect, too, is likely to be more problematic for aquaculture at higher latitudes.

4.5 Conclusion

There is no doubt that caligid infestations will continue to pose serious problems for global aquaculture. Of the 500-plus caligid species known, many not yet reported from cultured fish have the potential to become pests, and the predicted expansion of geographical ranges will provide new opportunities for both these and known problem species to colonise new hosts. To control these pests, much more research is needed into the biology of caligids and particularly on their responses to climate change.

References

Abolofia, J. and Wilen, J.E. (2017) The cost of lice: quantifying the impacts of parasitic sea lice on farmed salmon. *Marine Resource Economics* 32, 329–349.

Alaş, A. and Öktener, A. (2018) *Caligus vexator* Heller, 1865 including female and male descriptions from the Aegean Sea in Turkey. *Scientia Parasitologica* 18, 16–23.

Al-Marzouk, A. (2002) Preliminary study of direct freshwater bath treatment for a parasitic copepod, *Caligus antennatus* sp. nov. Boxshall and Gurney infesting cultured yellow-fin porgy, *Acanthopagrus latus* (Houttuyn) in Kuwait. *Journal of Aquaculture in the Tropics* 17, 1–6.

Andrews, M., Bott, N., Battaglene, S. and Nowak, B. (2009) A new species of copepod (Siphonostomatoida: Caligidae) parasite on the striped trumpeter, *Latris lineata* (Forster), in Tasmania. *Zootaxa* 1971, 59–68.

Armstrong, P.J., Sommerville, C. and Wootten, R. (1999) The life-cycle of the caligid copepod *Lepeophtheirus hippoglossi* (Krøyer,1837) and its pathology on the cultured Atlantic halibut (*Hippoglossus hippoglossus* L.). In: *Book of Abstracts. 5th International Symposium on Fish Parasites*. České Budějovice, Czech Republic, 9–13 August 1999, 6.

Arriagada, G., Valenzuela-Muñoz, V., Arriagada, A.M., Núñez-Acuña, P., Brossard, M., Montecino, K., Lara, M., Gallardo, A. and Gallardo-Escárate, C. (2019) First report of the sea louse *Caligus rogercresseyi* found in farmed Atlantic salmon in the Magellanes region, Chile. *Aquaculture* 512, 734386.

Barrett, L.T., Oppedal, F., Robinson, N. and Dempster, T. (2020) Prevention not cure: a review of methods to avoid sea lice infestations in salmon aquaculture. *Reviews in Aquaculture* 1–17.

Bergh, Ø., Nilsen, F. and Samuelsen, O.B. (2001) Diseases, prophylaxis and treatment of the Atlantic halibut *Hippoglossus hippoglossus:* a review. *Diseases of Aquatic Organisms* 48, 57–74.

Boxshall, G.A. and Gurney, A.R. (1980) Descriptions of two new and one poorly known species of the genus *Caligus* Müller, 1785 (Copepoda: Siphonostomatoida). *Bulletin of the British Museum (Natural History) Zoology* 39, 161–178.

Boxshall, G.A. and Bravo, S. (2000) On the identity of the common *Caligus* (Copepoda: Siphonostomatoida: Caligidae) from salmonid netpen systems in southern Chile. *Contributions to Zoology* 69, 137–146.

Boxshall, G.A. and El-Rashidy, H.H. (2009) A review of the *Caligus productus* species group, with a new species, new synonymies and supplementary descriptions. *Zootaxa* 2271, 1–26.

Boxshall, G.A. (2018) The sea lice (Copepoda: Caligidae) of Moreton Bay (Queensland, Australia), with descriptions of thirteen new species. *Zootaxa* 4398 (1), 1–172.

Bravo, S. (2003) Sea lice in Chilean salmon farms. *Bulletin of the European Association of Fish Pathologists* 23, 197–200.

Bravo, S., Perroni, E., Torres, E. and Silva, M.T. (2006) Report of *Caligus rogercresseyi* in the anadromous brown trout (*Salmo trutta*) in the Rio Gallegos Estuary, Argentina. *Bulletin of the European Association of Fish Pathologists* 26, 186–191.

Bravo, S., Boxshall, G.A. and Conroy, G. (2011) New cultured host and a significant expansion of the known geographical range of the sea louse *Caligus rogercresseyii. Bulletin of the European Association of Fish Pathologists* 31, 156–160.

Bravo, S. Hurtado, C.F. and Silva, M.T. (2017) Coinfection of *Caligus lalandei* and *Benedenia seriolae* on the yellowtail kingfish *Seriola lalandi* farmed in a net cage in northern Chile. *Latin American Journal of Aquatic Research* 45, 852–857.

Burnett-Herkes, J. (1974) Parasites of the gills and buccal cavity of the dolphin, *Coryphanea hippurus*, from the Straits of Florida. *Transactions of the American Fisheries Society* 103, 101–106.

Byers, J.E. (2020) Marine parasites and disease in the era of climate change. *Annual Review of Marine Science* 13, 3.1–3.24.

Carvajal, J, González, L. and George-Nascimento, M. (1998) Native sea lice (Copepoda: Caligidae) infestation of salmonids reared in netpen systems in southern Chile. *Aquaculture* 166, 241–246.

Catalano, S.R. and Hutson, K.S. (2010) Harmful parasitic crustaceans infecting wild arripids: a potential threat to southern Australian finfish aquaculture. *Aquaculture* 303, 101–104.

Choe, M.-K. and Kim, I.-H. (2010) Redescriptions of two morphologically confusing sea lice *Caligus aesopus* Wilson, 1921 and *C. spinosus* Yamaguti, 1939 (Copepoda: Siphonostomatoida: Caligidae) parasitic on amberjacks (*Seriola* spp.) from Korea. *Zootaxa* 2483, 23–34.

Choi, H.S., Hong, S.Y. and Lee, J.M. (1995) Two species of *Caligus* (Siphonostomatoida, Copepoda) parasitic on marine cultured fishes from Kamak Bay in Korea. *Bulletin of the National Fisheries Research and Development Institute (Korea)* 49, 157–165.

Costello, M.J. (2006) Ecology of sea lice parasitic on farmed and wild fish. *Trends in Parasitology* 22, 475–483.

Costello, M.J. (2009) The global economic cost of sea lice to the salmonid farming industry. *Journal of Fish Diseases* 32, 115–118.

Cruz-Lacierda, E.R., Pagador, G.E., Yamamoto, A. and Nagasawa, K. (2011) Parasitic caligid copepods of farmed marine fishes in the Philippines. In: Bondad-Reantaso M.G., Jones, J.B., Corsin, F. and Aoki, T. (eds) *Diseases in Asian Aquaculture* VII. Fish Health Section, Asian Fisheries Society, Selangor, Malaysia. pp. 53–62.

Deveney, M.R., Bayly, T.J., Johnston, C.J. and Nowak, B.F. (2005) A parasite survey of farmed southern bluefin tuna, *Thunnus maccoyii* (Castelnau). *Journal of Fish Diseases* 28, 279–284.

Fernandez, J. and Villalba, C. (1986) Contribucion al conocimiento del genero *Caligus* Müller, 1785 (Copepoda: Siphonostomatoida) en Chile. *Gayana, Zoology* 50, 37–62.

Froese, R. and Pauly, D. (2019) FishBase. World Wide Web electronic publication. Available at: www.fishbase.org.(accessed January 2022).

Fujita, S., Yoda, M. and Ugajin, I. (1968) Control of an ectoparasitic copepod, *Caligus spinosus* Yamaguti, on the cultured adult yellowtail. *Fish Pathology* 2, 122–127.

Garzke, J., Hansen, T., Ismar, S.M.H. and Sommer, U. (2016) Combined effects of ocean warming and acidification on copepod abundance, body size and fatty acid content. *PLoS ONE* 11(5), e015952.

González, L. and Carvajal, J. (2003) Life cycle of *Caligus rogercresseyi* (Copepoda: Caligidae) parasite of Chilean reared salmonids. *Aquaculture* 220, 101–117.

Groner, M.L., McEwan, G.F., Rees, Gettinby, G. and Revie, C.W. (2016) Quantifying the influence of salinity and temperature on the population dynamics of a marine ectoparasite. *Canadian Journal of Fisheries and Aquatic Sciences* 73, 1281–1291.

Hamre, L.A., Bui, S., Oppedal, F., Skern-Maritzen, R. and Dalvin, S. (2019) Development of the salmon louse *Lepeophtheirus salmonis* parasitic stages in temperatures ranging from 3 to 24°C. *Aquaculture Environment Interactions* 11, 429–443.

Hayes, P., Justine, J.-L. and Boxshall, G.A. (2012) The genus *Caligus* Müller, 1785 (Copepoda: Siphonostomatoida): two new species from reef associated fishes in New Caledonia, and some nomenclatural problems resolved. *Zootaxa* 3534, 21–39.

Hayward, C.J., Bott, N.J., Itoh, N., Iwashita, M., Okihiro, M. and Nowak, B.F. (2007) Three species of parasites emerging on the gills of mulloway, *Argyrosomus japonicus* (Temminck and Schlegel, 1843) cultured in Australia. *Aquaculture* 265, 27–40.

Hayward, C.J., Aiken, H.M. and Nowak, B.F. (2008) An epizootic of *Caligus chiastos* on farmed southern bluefin tuna *Thunnus maccoyii* off South Australia. *Diseases of Aquatic Organisms* 79, 57–63.

Hayward, C.J., Ellis, D., Foote, D., Wilkinson, R.J., Crosbie, P.B.B., Bott, N.J. and Nowak, B.F. (2010) Concurrent epizootic hyperinfections of sea lice (predominantly *Caligus chiastos*) and blood flukes (*Cardicola forsteri*) in ranched southern bluefin tuna. *Veterinary Parasitology* 173, 107–115.

Hayward, C.J., Svane, I., Lachimpadi, S.K., Itoh, N., Bott, N.J. and Nowak, B.F. (2011) Sea lice infections of wild fishes near ranched southern bluefin tuna (*Thunnus maccoyi*) in South Australia. *Aquaculture* 320, 178–182.

Heegard, P.E. (1962) Parasitic copepoda from Australian waters. *Records of the Australian Museum* 40, 149–234.

Hemmingsen, W., MacKenzie, K., Sagerup, K., Remen, N., Bloch-Hansen, K. and Dagbjartarson Imsland, A.K. (2020) *Caligus elongatus* and other sea lice of the genus *Caligus* as parasites of farmed salmonids: a review. *Aquaculture* 522, 735160.

Heuch, P.A., Jansen, P.A., Hansen, H., Sterud, E., MacKenzie, K., Haugen, P. and Hemmingsen, W. (2011) Parasite faunas of farmed cod and adjacent wild cod populations in Norway: a comparison. *Aquaculture Environment Interactions* 2, 1–13.

Ho, J.-S. (2000) The major problem of cage aquaculture in Asia relating to sea lice. In: Liao, I.C. and Lin, C.K. (eds) *Proceedings of the First International Symposium on Cage Aquaculture in Asia, 2–6 November 1999, Tungkang, Manila*. Asian Fisheries Society and World Aquaculture Society-Southeast Asian Chapter. pp. 13–19.

Ho, J.-S. and Lin, C.-L. (2001) *Parapetalus occidentalis* Wilson (Copepoda: Caligidae) parasitic on both wild and farmed cobia (*Rachycentron canadum*) in Taiwan. *Journal of the Fisheries Society of Taiwan* 28, 305–316.

Ho, J.-S. and Lin, C.-L. (2004) *Sea Lice of Taiwan (Copepoda: Siphonostomatoida: Caligidae)*. Sueichan Press, Taiwan, p. 388.

Ho, J.-S., Nagasawa, K., Kim, I.H. and Ogawa, K. (2001) Occurrence of *Caligus lalandei* Barnard, 1948 (Copepoda: Siphonostomatoida) on amberjacks (*Seriola* spp.) in the western North Pacific. *Zoological Science* 18, 423–431.

Ho, J.-S., Kim, I.-H., Cruz-Lacierda, E.R. and Nagasawa, K. (2004a) Sea lice (Copepoda: Caligidae) parasitic on marine cultured and wild fishes of the Philippines. *Journal of the Fisheries Society of Taiwan* 31, 235–249.

Ho, J.-S., Gómez, S., Ogawa, K. and Aritaki, M. (2004b) Two species of parasitic copepods (Caligidae) new to Japan. *Systematic Parasitology* 57, 19–34.

Hogans, W.E. and Trudeau, D.J. (1989) *Caligus elongatus* (Copepoda: Caligoida) from Atlantic salmon (*Salmo salar*) cultured in marine waters of the Lower Bay of Fundy. *Canadian Journal of Zoology* 67, 1080–1082.

Hutson, K.S., Ernst, I., Mooney, A.J. and Whittington, I.D. (2007) Metazoan parasite assemblages of wild *Seriola lalandi* (Carangidae) from eastern and southern Australia. *Parasitology International* 56, 95–105.

Izawa, K. and Choi, K.H. (2000) Redescription of *Caligus latigenitalis* Shiino, 1954 (Copepoda: Siphonostomatoida: Caligidae), parasitic on Japanese black sea bream *Acanthopagrus schlegeli* (Bleeker, 1854). *Crustaceana* 73, 995–1005.

Johnson, S.C. and Albright, L.J. (1991) *Lepeophtheirus cuneifer* Kabata, 1974 (Copepoda: Caligidae) from sea-water reared rainbow trout, *Oncorhynchus mykiss*, and Atlantic salmon, *Salmo salar*, in the Strait of Georgia, British Columbia, Canada. *Canadian Journal of Zoology* 69, 1414–1416.

Johnson, S.C., Treasurer, J.W., Bravo, S., Nagasawa, K. and Kabata, Z. (2004) A review of the impact of parasitic copepods on marine aquaculture. *Zoological Studies* 43, 229–243.

Jones, J.B. (1980) A redescription of *Caligus patulus* Wilson 1937 (Copepoda: Caligidae) from a fish farm in the Philippines. *Systematic Parasitology* 2, 103–116.

Jones, S. and Johnson, S. (2014) Biology of sea lice, *L. salmonis* and *Caligus* spp., in western and eastern Canada. *Canadian Science Advisory Secretariat Research Document* 19, I–V, 1–18.

Kabata, Z. (1972) Developmental stages of *Caligus clemensi* (Copepoda: Caligidae). *Journal of the Fisheries Research Board of Canada* 29, 1571–1593.

Kabata, Z. (1974) *Lepeophtheirus cuneifer* sp. nov. (Copepoda: Caligidae), a parasite of fishes from the Pacific Coast of North America. *Journal of the Fisheries Research Board of Canada* 31, 43–47.

Kabata, Z. (1979) *Parasitic Copepoda of British Fishes.* The Ray Society, London, UK. p. 468.

Khoa, T.N.D., Mazelan, S., Muda, S. and Shaharom-Harrison, F. (2019) The life cycle of *Caligus minimus* on seabass (*Lates calcarifer*) from floating cage culture. *Thalassas: An International Journal of Marine Sciences* 35, 77–85.

Ku, C.-C. and Lu, C.-H. (2009) Investigation and treatment of *Parapetalus occidentalis* Wilson (Copepoda: Caligidae) infestation in sea-cage cultured cobia (*Rachycentron canadum*) at Penghu Islands, Taiwan. *Journal of the Fisheries Society of Taiwan* 36, 161–168.

Leaw, Y.Y., Faizah, S., Anil, C. and Kua, B.C. (2012) Prevalence, mean intensity and site preference of *Caligus rotundigenitalis* Yü, 1933 (Copepoda: Caligidae) on cage-cultured crimson snapper (*Lutjanus erythropterus* Bloch, 1790) from Bukit Tambun, Penang, Malaysia. *Veterinary Parasitology* 187, 505–510.

Lin, C.-L. and Ho, J.-S. (1993) Life history of *Caligus epidemicus* Hewitt, parasitic on the tilapia (*Oreochromis mossambica*) cultured in salt water. In: Boxshall, G.A. and Defaye, D. (eds) *Pathogens of Wild and Farmed Fish: Sea Lice.* Ellis Horwood, Chichester, UK. pp. 5–15.

Lin, C.-L. and Ho, J.-S. (1998) On a species of copepod, *Caligus orientalis* Gusev, 1951, parasitic on the fishes cultured in saltwater. *Journal of the Fisheries Society of Taiwan* 25, 199–208.

Lin, C.-L. and Ho, J.-S. (2007) Six species of sea lice (Copepoda: Caligidae) new to Taiwan. *Journal of the Fisheries Society of Taiwan* 34, 41–67.

Lin, C.-L., Ho, J.-S. and Chen, S.-N. (1994) Two species of *Caligus* (Copepoda: Caligidae) parasitic on black sea bream (*Acanthopagrus schlegeli*) cultured in Taiwan. *Fish Pathology* 29, 253–264.

Lin, C.-L., Ho, J.-S. and Chen, S.-N. (1996a) Developmental stages of *Caligus epidemicus* (Hewitt), a copepod parasite of tilapia cultured in brackish water. *Journal of Natural History* 30, 661–684.

Lin, C.-L., Ho, J.-S. and Chen, S.-N. (1996b) Two species of Caligidae (Copepoda) parasitic on cultured rabbit fish (*Siganus fuscescens*) in Taiwan. *Fish Pathology* 31, 129–139.

Lõhmus, M. and Björklund, M. (2015) Climate change: what will it do to fish-parasite interactions? *Biological Journal of the Linnaean Society* 116, 397–411.

Madinabeitia, I., Ohtsuka, S., Okuda, J., Iwamoto, E., Yoshida, T., Furukawa, M., Nakaoka, N. and Nakai, T. (2009) Homogeneity among *Lactococcus garveiae* isolates from striped jack, *Pseudocaranx dentex* (Bloch & Schneider), and its ectoparasites. *Journal of Fish Diseases* 32, 901–905.

Marcogliese, D.J. (2001) Implications of climate change for parasitism of animals in the aquatic environment. *Canadian Journal of Zoology* 79, 1331–1352.

Marcogliese, D.J. (2008) The impact of climate change on the parasites and infectious diseases of aquatic animals. *Revue Scientifique et Technique – Office International des Epizooties* 27, 467–484.

Margolis, L. and Kabata, Z. (eds) (1988) Guide to the Parasites of Fishes of Canada. Part II – Crustacea. *Canadian Special Publication of Fisheries and Aquatic Sciences* 101, p. 184.

Matumoto, T. (1980) *Caligus orientalis* parasitic on cultured carp. *Fish Pathology* 14, 143–144.

Morales-Serna, F.N., Hernandéz-Inda, Z.L., Gómez, S. and Pérez-Ponce de León, G. (2013) Redescription of *Caligus serratus* Shiino, 1965 (Copepoda: Caligidae) parasitic on eleven fish species from Chamela Bay in the Mexican Pacific. *Acta Parasitologica* 58, 367–375.

Morales-Serna, F.N., Pinacho-Pinacho, C.D., Gómez, S. and Pérez-Ponce de León, G. (2014) Diversity of sea lice (Copepoda: Caligidae) parasitic on marine fishes with commercial and aquaculture importance in Chamela Bay, Pacific coast of Mexico by using morphology and DNA barcoding, with description of a new species of *Caligus*. *Parasitology International* 63, 69–79.

Mortelmans, J., Debusschere, E., Vranken, S., Deneudt, K., Hernandez, F. and Mees, J. (2017) Two marine parasitic crustaceans new to the Belgian fauna. *Belgian Journal of Zoology* 147, 143–153.

Muhd-Faizul H.A.H., Kua, B.C. and Leaw, Y.Y. (2012) Caligidae infestation in Asian seabass, *Lates calcalifer* Bloch, 1790 cultured at different salinity in Malaysia. *Veterinary Parasitology* 184, 68–72.

Mustafa, A., MacKinnon, B.M. and Piasecki, W. (2005) Interspecific differences between Atlantic salmon and Arctic charr in susceptibility to infection with larval and adult *Caligus elongatus*: effect of skin mucus protein profiles and epidermal histological differences. *Acta Ichthyologica et Piscatoria* 35, 7–13.

Nagasawa, K. (2004) Sea lice, *Lepeophtheirus salmonis* and *Caligus orientalis* (Copepoda: Caligidae), of wild and farmed fish in sea and brackish waters of Japan and adjacent regions: a review. *Zoological Studies* 43, 173–178.

Nagasawa, K. (2011) *Caligus macarovi* (Copepoda: Caligidae) from Pacific bluefin tuna, *Thunnus orientalis*, cultured in Japan. *Crustaceana* 84, 1145–1147.

Nagasawa, K. (2015) Parasitic copepods of marine fish cultured in Japan: a review. *Journal of Natural History* 49, 2891–2903.

Nagasawa, K. and Fukuda, Y. (2011) Two species of *Caligus* (Copepoda: Caligidae) from amberjacks (*Seriola* spp.) cultured in Oita Prefecture, Kyushu, western Japan. *Biosphere Science* 50, 49–53.

Nagasawa, K. and Shirakashi, S. (2018) *Euryphorus brachypterus* (Gerstaecker, 1853), a caligid copepod infecting Pacific bluefin tuna, *Thunnus orientalis* (Temminck & Schlegel, 1844) cultured in Japan. *Crustaceana* 91, 251–255.

Nowak, B.F., Hayward, C.J., González, L., Bott, N.J. and Lester, R.J.G. (2011) Sea lice infections of salmonids farmed in Australia. *Aquaculture* 320, 171–177.

Nylund, A., Wallace, C. and Hovland, T. (1993) The possible role of *Lepeophtheirus salmonis* (Krøyer) in the transmission of infectious salmon anaemia. In: Boxshall, G.A. and Defaye, D. (eds) *Pathogens of Wild and Farmed Fish: Sea Lice*. Ellis Horwood, Chichester, UK. pp. 367–373.

Oelckers, K., Vike, S., Duesund, H., Gonzalez, J., Wadsworth, S. and Nylund, A. (2014) *Caligus rogercresseyi* as a potential vector for transmission of infectious salmon anaemia (ISA) virus in Chile. *Aquaculture* 420–421, 126–132.

Ogawa, K. (1992) *Caligus longipedis* infection of cultured striped jack, *Pseudocaranx dentex* (Teleostei: Carangidae) in Japan. *Gyobyo Kenkyu* 27, 197–205.

Ogawa, K. and Inouye, K. (1997) Parasites of cultured tiger puffer (*Takifugu rubripes*) and their seasonal occurrences, with descriptions of two new species of *Gyrodactylus*. *Fish Pathology* 32, 7–14.

Ogawa, K. and Yokoyama, H. (1998) Parasitic diseases of cultured marine fish in Japan. *Fish Pathology* 33, 303–309.

Øines, Ø., Simonsen, J.H., Knutsen, J.A. and Heuch, P.A. (2006) Host preference of adult *Caligus elongatus* Nordmann in the laboratory and its implications for Atlantic cod aquaculture. *Journal of Fish Diseases* 29,167–174.

Öktener, A. (2009) *Caligus mauritanicus* Brian, 1924 a parasite of common cultured dentex, *Dentex dentex* L. from Aegean Sea and parasites of cultured fishes of Turkey. *Pakistan Journal of Zoology* 41, 417–420.

Orr, J.C., Fabry, V.J., Aumont, O., Bopp, L., Doney, S.C., Feely, R.A., Gnanadesikan, A., Gruber, N., Ishida, A., Joos, F., Key, R.M., Lindsay, K., Maier-Reimer, E., Matear, R., Monfray, P., Mouchet, A., Najjar, R.G., Plattner, G.K., Rodgers, K.B., Sabine, C.L., Sarmiento, J.L., Schlitzer, R., Slater, R.D., Totterdell, I.J.,

Weirig, M.F., Yamanaka, Y. and Yool, A. (2005) Anthropogenic ocean acidification over the twenty-first century and its impact on calcifying organisms. *Nature* 437(7059), 681–686.

Özak, A.A., Dermikale, I. and Yanar, A. (2012) First record of two species of parasitic copepods on immigrant pufferfishes (Tetraodontiformes: Tetraodontidae) caught in the Eastern Mediterranean Sea. *Turkish Journal of Fisheries and Aquatic Sciences* 12, 675–681.

Özak, A.A., Dermikale, I., Boxshall, G.A. and Etyemez, M. (2013) Parasitic copepods of the common sole, *Solea solea* (L.) from the Eastern Mediterranean coast of Turkey. *Systematic Parasitology* 86, 173–185.

Paperna, I. (1975) Parasites and disease of the grey mullet (Mugilidae) with special reference to the sea of the near east. *Aquaculture* 5, 65–80.

Parker, R.R. (1969) Validity of the binomen *Caligus elongatus* for a common parasitic copepod formerly misidentified with *Caligus rapax*. *Journal of the Fisheries Research Board of Canada* 26, 1013–1035.

Parker, R.R. and Margolis, L. (1964) A new species of parasitic copepod, *Caligus clemensi* sp. nov. (Caligoida: Caligidae), from pelagic fishes in the coastal waters of British Columbia. *Journal of the Fisheries Research Board of Canada* 21, 873–889.

Parker, R.R., Kabata, Z., Margolis, L. and Dean, M.D. (1968) A review and description of *Caligus curtus* Müller, 1785 (Caligidae: Copepoda), type species of its genus. *Journal of the Fisheries Research Board of Canada* 25, 1923–1969.

Pedersen, A.P. and Hanssen, A.E. (2017) Ocean acidification ameliorates harmful effects of warming in primary consumer. *Ecology and Evolution* 8, 396–404.

Ragias, V., Tontis, D. and Athanassopoulou, F. (2004) Incidence of an intense *Caligus minimus* Otto 1821, *C. pageti* Russel, 1925, *C. mugilis* Brian, 1935 and *C. apodus* Brian, 1924 infection in lagoon cultured sea bass (*Dicentrarchus labrax* L.) in Greece. *Aquaculture* 242, 27–733.

Raibaut, A., Divanach, P., Coste, F. and Maillard, C. (1980) Copepodose larvaire en ecloserie de poissons marins. *Pisciculture Francaise* Supplement 61–62, 49–51.

Regidor, S.E. and Arthur, J.R. (1986) Parasites of juvenile milkfish, *Chanos chanos*. In: *Proceedings of the First Asian Fisheries Forum, 26–31 May 1986. Manila, Philippines.* Asian Fisheries Society, Manila, Philippines. pp. 261–264.

Reyes, X. (1985) Reserche sur les maladies des salmonidés au Chili: Proposition d'une réglementation saniaire. *These pour obtenir le grade de Docteur de Troisieme Cycle.* L'Institut National Polytechnique de Toulouse, France. p 126.

Reyes, X. and Bravo, S. (1983) Salmón coho *Oncorhynchus kisutch*, cultivado en Puerto Montt, Chile, nuevo huésped para el copépodo *Caligus teres* (Caligidae). *Investigaciones Marinas, Valparaiso* 11, 55–57.

Roubal, F.R. (1994) Histopathology caused by *Caligus epidemicus* Hewitt (Copepoda: Caligidae) on captive *Acanthopagrus australis* (Günther) (Pisces: Sparidae). *Journal of Fish Diseases* 17, 631–640.

Roubal, F.R. (1995) Changes in monogenean and copepod infestation on captive *Acanthopagrus australis* (Sparidae). *Journal of Fish Biology* 46, 423–431.

Roubal, F.R. (1997) Survival and development of *Caligus epidemicus* Hewitt in sea water of different salinity. *Bulletin of the European Association of Fish Pathologists* 17, 78–80.

Ruangpan, L. and Kabata, Z. (1984) An invertebrate host for *Caligus* (Copepoda: Caligidae). *Crustaceana* 47, 219–220.

Runge, J.A., Fields, D.M., Thompson, C.R.S., Shema, S.D., Bjelland, R.M., Durif, C.M.F., Skiftesvik, A.B. and Browman, H.I. (2016) End of century CO_2 concentrations do not have a negative effect on vital rates of *Calanus finmarchicus*, an ecologically critical planktonic species in North Atlantic ecosystems. *ICES Journal of Marine Science* 73, 937–950.

Saksida, S., Constantine, J. Karreman, G.A. and Donald, A. (2007) Evaluation of sea lice abundance levels on farmed Atlantic salmon (*Salmo salar* L.) located in the Broughton Archipelago of British Columbia from 2003 to 2005. *Aquaculture Research* 38, 219–231.

Saksida, S., Bricknell, I., Robinson, S. and Jones, S. (2015) Population ecology and epidemiology of sea lice in Canadian waters. *Canadian Science Advisory Secretariat Research Document* 2015/004, V, 3–4.

Solanki, H.G., Patil, P.K., Vanza, J.G., Patel, P. Sethi, S. and Gopal, C. (2016) Sea lice, *Caligus rotundigenitalis* infestations and its management in pond cultured pearlspot *Etropus suratensis* in Gujarat: a case study. *Journal of Parasitic Diseases* 40, 565–567.

Sutterlin, A.M., Harmon, P. and Blanchard, H. (1976) The culture of brook trout in salt water. *Fisheries and Marine Services Research and Development Directorate Technical Report* No. 636.

Tareen, I.U. (1986) Parasitic infestations on cultured marine fish *Acanthopagrus cuvieri* (Sparidae) incidence and control. In: Vivares, C.P., Bonami, J.R. and Jaspers, E. (eds) *Pathology in Marine Aquaculture*. European Aquaculture Society, Bredene, Belgium. pp. 85–90.

Thompson, C.R.S., Fields, D.M., Bjelland, R.M., Chan, V.B.S., Durif, C.M.F., Mount, A., Runge, J.A., Shema, S.D., Skiftesvik, A.B. and Browman, H.I. (2019) The planktonic stages of the sea louse (*Lepeophtheirus salmonis*) are tolerant of end-of-century pCO_2 concentrations. *Peerj* 7, e7810 Available at: http://doi.org/10.7717peerj.7810 (accessed 18 March 2021).

Urawa, S. and Kato, T. (1991) Heavy infections of *Caligus orientalis* (Copepoda: Caligidae) on caged rainbow trout *Oncorhynchus mykiss* in brackish water. *Gyobyu Kenkyu* 26, 161–162.

Venmathi Maran, B.A., Seng, L.K., Ohtsuka, S. and Nagasawa, K. (2009) Records of *Caligus* (Crustacea: Copepoda: Caligidae) from marine fish cultured in floating cages in Malaysia with a redescription of the male of *Caligus longipedis* Barrett-Smith, 1898. *Zoological Studies* 48, 797–807.

Venmathi Maran, B.A., Ohtsuka, S. and Shang, X. (2012a) Records of adult caligiform copepods (Crustacea: Copepoda: Siphonostomatoida) in marine plankton from East Asia, including descriptions of two new species of *Caligus*. *Species Diversity* 17, 201–219.

Venmathi Maran, B.A., Oh, S.-Y., Soh, H.Y., Choi, H.J. and Myoung, J.-G. (2012b) *Caligus sclerotinosus* (Copepoda: Caligidae), a serious pest of cultured seabream *Pagrus major* (Sparidae) in Korea. *Veterinary Parasitology* 188, 355–361.

Venmathi Maran, B.A., Suárez-Morales, E., Ohtsuka, S., Soh, H.Y. and Hwang, U.W. (2016) On the occurrence of caligids (Copepoda: Siphonostomatoida) in the marine plankton: a review and checklist. *Zootaxa* 4174(1), 437–447.

Vo, D.T., Bristow, G.A., Nguyen, D.H. and Vo, D.T. (2008) Parasitism of two species of *Caligus* (Copepoda: Caligidae) on wild and cultured grouper in Viet Nam. *Journal of the Fisheries Society of Taiwan* 35, 35–43.

Wagner, G.N., Fast, M.D. and Johnson, S.C. (2008) Physiology and immunology of *Lepeophtheirus salmonis* infections of salmonids. *Trends in Parasitology* 24, 176–183.

Walter, T.C. and Boxshall, G.A. (2020) World of Copepods database. Caligidae Burmeister, 1835. Accessed through World Register of Marine Species available at: https://www.marinespecies.org/aphia.php?p=taxdetails&id=135513 (accessed 1 November 2020).

Wang, M., Jeong, C.-B., Lee, Y.H. and Lee, J.-S. (2018) Effects of ocean acidification on copepods. *Aquatic Toxicology* 196, 17–24.

Wu, Z. and Pan, J. (1997) A study on lice disease on cultured grouper *Epinephelus* sp. 1. Description on the pathogen *Caligus nanhaiensis* n. sp. *Tropic Oceanology* 16, 60–66.

Wu, Z., Pan, J. and Qin, Q. (1997) The lice disease in cultured banded grouper *Epinephelus awoara*. Pathology. *Acta Hydrobiologica Sinica* 21, 207–212.

Yuniar, A.T., Palm, H.W. and Walter, T. (2007) Crustacean fish parasites from Segara Anakan Lagoon, Java, Indonesia. *Parasitology Research* 100, 1193–1204.

CHAPTER 5

Sea lice behaviour

Samantha Bui[1], Lars A. Hamre[2], Rasmus Skern-Mauritzen[1], Sussie Dalvin[2] and James E. Bron[3]

[1]Institute of Marine Research, Bergen, Norway
[2]Sea Lice Research Centre, University of Bergen, Bergen, Norway
[3]Institute of Aquaculture, University of Stirling, UK

The life cycle of *Lepeophtheirus salmonis* (Krøyer, 1837) is divided into three free-living planktonic and five parasitic stages. Naturally, the behaviour of lice in the planktonic phase differs significantly from that of parasitic lice attached to the surface of the fish. Generally, the behaviour of planktonic lice is governed by their need to avoid unfavourable water conditions and to maintain a position in the upper part of the water column at the lowest energetic cost. Although behaviour in larvae may encompass a range of actions, the difficulty of observing their behaviour *in situ*, due to their small size and the usual difficulties associated with making observations in the marine environment, has been a limiting factor in characterising behaviour. Currently, behavioural studies of planktonic larvae have hence focused on detection of movement including direction. Larvae move by short intense bursts of swimming, which can bring them closer to the surface, whereas if they do not swim they will slowly sink through the water column. For the infective copepodid, the key to infection of a host and thus survival is to position itself in the part of the water column where the host fish is most likely to be present and, furthermore, to be able to detect, intercept, and settle on approaching fish. Once on the fish, correct identification of host species and settlement on a suitable part of the host body is necessary to continue the life cycle. For parasitic lice, behavioural responses encompass the need for the louse to remain on the skin of an actively swimming host, including during the moulting process. For adults, males need to locate and mate with females, and finally, both sexes need to respond to unfavourable water conditions such as brackish water in estuaries, where the salmonid host may spend a considerable amount of time according to species.

5.1 Behaviour in the planktonic phase

5.1.1 Response to salinity

The free-living salmon louse stages comprise two naupliar stages and the infective copepodid stage. These are planktonic and have limited ability to navigate, leaving them largely at the mercy of hydrodynamic forces. Since the majority of the sea louse population is found in coastal areas affected by freshwater runoff, the lice larvae are exposed to a dynamic environment varying from full strength seawater (34 ppt) to freshwater (0 ppt). Adult salmon lice are isosmotic at full seawater salinity, but exhibit hyperosmotic regulation in less saline environments, that is at lower salinities the intracellular osmolality decreases but exceeds that of the external environment (Hahnenkamp and Fyhn, 1985). Although these findings are based on studies on adult salmon lice, it is reasonable to expect similar physiological responses in larvae where a number of genes, including chaperones, ion pumps, and transporters, are upregulated in response to low salinity exposures (Sutherland et al., 2012; Borchel et al., 2018). These gene expression changes become increasingly pronounced as salinities decline, with a pronounced threshold at 27 ppt (Sutherland et al., 2012). The associated costs unsurprisingly have a number of consequences for the lecithotrophic planktonic stages that have to divert energy to physiological low salinity responses, rather than to merely staying alive while waiting for a host.

A significant effect of lowered salinities is reduced hatchability and hence reduced recruitment to the plankton component. Hatching appears unaffected by reduced salinity down to 25 ppt and to be increasingly compromised at lower salinities, and completely absent at 10 ppt (Johnson and Albright, 1991a). There are, however, significant inheritable differences in salmon louse low salinity tolerance (Ljungfeldt et al., 2017; Andrews and Horsberg, 2020), which may or may not be genetically defined (Andrews and Horsberg, 2020; Skern-Mauritzen and Mikkelsen, 2021), and the threshold salinities may thus vary between lineages. After hatching, lower salinities reduce naupliar activity progressively, with no active nauplii found below 15–20 ppt (Johnson and Albright 1991a; Gravil, 1996). While low salinity evidently affects naupliar behaviour, the reduced salinity effects are better described in the infective copepodid stage. Infective copepodids can survive exposure to salinities down to approximately 5 ppt for several hours (Johnson and Albright 1991a; Skern-Mauritzen, 2021, unpublished data) with progressively longer survival at less reduced salinities (Gravil, 1996; Bricknell et al., 2006). Bricknell et al. (2006) found that copepodids exposed to salinities of 16 ppt survived short-term exposure (less than four hours) without long-term impact on the surviving copepodids. Exposure of copepodids to less than 10 ppt for two hours, however, affected activity over at least one to two days (R. Skern-Mauritzen, 2021, unpublished data). While the ability to infect is heavily influenced by temperature and senescence (Brooker et al., 2018; Skern-Mauritzen et al., 2020), it is also governed by salinity. Hence, reduced salinities are associated with altered location of host attachment and increasingly reduced infectivity at declining salinities with close to no infection found at 12 ppt (Tucker et al., 2000; Bricknell et al., 2006). The ultimate goal of the larvae is to survive the predation-exposed planktonic stages (Brooker et al., 2018) while simultaneously positioning themselves to increase the chance of encountering a potential host. Since residing in reduced salinity environments is energetically costly, lice larvae generally avoided the low salinities commonly found in surface waters by passive sinking or active downward swimming. This behaviour has been observed at salinities of 27 ppt or lower (Bricknell et al., 2006), which corresponds to the salinity where an increasingly clear gene expression response to reduced salinity is found (Sutherland et al., 2012). However, lower salinities are not fully avoided, as copepodids aggregate around haloclines (Heuch, 1995). This is probably in order to increase the likelihood of host encounters, as salmonids have been observed to follow salinity gradients (Heuch, 1995; Lyse et al., 1998). Despite the general tendency of salmon louse copepodids to aggregate at haloclines in the upper water column, copepodids are nevertheless regularly found at depths exceeding 12 m (S. Bui, 2019, unpublished results).

5.1.2 Light sensors and behavioural responses to light intensity and wavelength

Optic receptors

Lepeophtheirus salmonis copepodids possess a relatively well-developed nauplius eye that reflects the importance of the optic receptors in this stage, and this eye structure is maintained in later developmental stages. The nauplius eye is medially placed and comprises three ocelli: two antero lateral facing, lensed dorsal ocelli and a downward directed ventral ocellus, which lacks a lens (Bron and Sommerville, 1998). The dorsal ocelli are screened by pigment cups with a narrow aperture through which the lens protrudes. The ventral ocellus is also screened by pigment cells. All the ocelli have a reflective tapetum underlying the sensory retinular cells, which maximises the light retained in the eyes. This structure suggests good light sensitivity and highly directional vision although it is unlikely to be image-forming. In addition to the nauplius eye, copepodids possess a small extra-optic photoreceptor, likely bilaterally paired and lacking a lens, tapetum, or pigment cell cup (Bron and Sommerville, 1998).

Response to light intensity and wavelengths

Light induces and influences the behavioural responses of nauplii and copepodids in the water column slightly differently. Light is also considered to affect the infection process and subsequent success of infection. This section addresses the sensitivity of larval stages, in terms of phototactic responses, to light intensity and wavelength, and suggests how the light environment may shape infection success.

The described visual system of copepodids led early investigators to conclude that light or shadows were likely to be important for detection of hosts (Boxshall, 1992) and subsequent studies have indeed shown copepodids to be positively phototactic (Bron, 1993; Fields et al., 2018; Novales Flamarique et al., 2000; Nordtug et al., 2021; Szetey et al., 2021).

The planktonic stages of *L. salmonis* can detect and respond to light cues even at relatively low intensities, with nauplius I and II stages reported to demonstrate responses to light at 200 and 85 lux (around 4.17×10^{18} and 1.77×10^{18} photons m^{-2} s^{-1}, calculated for halogen white light) respectively (Gravil, 1996). Novales Flamarique et al. (2000) demonstrated that the lowest tested intensity eliciting a response in nauplii was 1.22×10^{17} photons m^{-2} s^{-1}, with responses shown only to a change in lighting from on to off, and showing no response at 4.24×10^{16} photons m^{-2} s^{-1}. The same authors indicated that copepodids showed responses down to the lowest tested intensity of around 3×10^{13} photons m^{-2} s^{-1} (estimated from Graph 4C page 1653, Novales Flamarique et al., 2000) but no response at 8.07×10^{12} photons m^{-2} s^{-1}, with no response to switched-off light but strong responses to switched-on lights. In a laboratory setting, the behavioural response of copepodids is positively correlated with light intensity, with swimming under low intensities showing little distance gain and low directional movement, while at high intensities copepodids were observed to swim directly towards the light source (Bron, 1993). Copepodids will choose higher intensity light over medium or low intensity (Nordtug et al., 2021). Notably, exposure to extreme light intensity (20,000 lux) resulted in termination of any behaviour and sinking to the bottom (Bron, 1993). Novales Flamarique (2000) proposed that these results indicated that copepodids could use visual cues and stimuli to drive behaviour during crepuscular periods, which is discussed later in this chapter (also see Chapter 22), this being the period when chance of host encounter is higher due to the crossing of paths during vertical migration of host and parasite (Heuch et al., 1995).

Investigations of sensitivity of larval stages to light have often been conducted using white light. However, behavioural responses have been recorded for copepodids at 400–700 nm (Bron, 1993; Gravil, 1996) and for both nauplii and copepodids at 353–652 nm (Novales Flamarique et al., 2000). The peak response in nauplius II occurs at the 500 nm wavelength, and similarly at 500–550 nm for copepodids (cyan and green light according to Bron, 1993; Gravil, 1996), although Novales Flamarique et al. (2000) showed a consistently high response from 550–650 nm in copepodids and no response to any wavelength in nauplii. This supports the theory of behavioural adaption to the crepuscular periods, with the wavelengths of highest sensitivity corresponding to the maximum transmitted light during twilight hours (Bron, 1993). In preference tests of phototactic responses to simultaneously presented wavelengths, Nordtug et al. (2021) found that group movement was distinct and different in the presence of light stimuli, while scattered and random when in darkness. This study demonstrated that the collective preference of copepodids was for white and blue light, with the next preference for green, then UV or red light. These authors also reported that copepodids could detect and swim quickly towards UV light, similar to the response observed for white light. Aarseth and Schram (1999) found that copepodids swam deeper in the presence of UV light having a maximum wavelength of 313 nm when delivered with white light.

In reporting responses of sea lice to light it is, however, important to note that none of these studies employed natural lighting, with many of them employing light sources having a narrow directed beam. It is well established (Forward, 1988) that many zooplankton species show an artefactual positive phototactic response to such light sources, so that the reported behaviours may not therefore accurately reflect those occurring under natural conditions. All of the experiments were also conducted in very shallow depths of water, such that real-world responses might be different in terms of animals perceiving light at greater depths.

Response to varying intensity, polarisation, and shadow

Investigations of light responses in copepodids have focused not only on constant light, but also on their response to light polarisation, changing intensities, and shadows (scototaxis). The *L. salmonis* eye has retinular cell structuring and orientation that could potentially allow it to act as a polarisation (POL) detector, like other species with similar eye structures (Bron and Sommerville, 1998). For many animals, the ability to detect orientation of polarised light can facilitate object recognition, environmental orientation, detection of visual stimuli, and discrimination of cues through enhancing contrast. Theoretically, the highly reflective scales of a salmonid host could also generate a distinct POL signal that could be used for host recognition by *L. salmonis* copepodids; however, research by Novales Flamarique et al. (2000) demonstrated a lack of differential response both to linearly polarised light in comparison to diffuse light, and to rotation of the plane of polarisation. Fields et al. (2018) reported strong behavioural responses towards flickering light, which were intended to simulate the reflections off a salmon host. However, when presented with the choice of a constant light source or a flickering (on/off) stimulus in a horizontal plane, copepodids preferred and swam towards the constant light (Nordtug et al., 2021). *L. salmonis* copepodids, unlike those of *Caligus elongatus* Nordmann 1832 (Hogans and Trudeau, 1989), do not exhibit responses to shadows (Bron, 1993; Novales Flamarique et al., 2000). This has been interpreted in terms of them lacking/suppressing the more usual predator shadow avoidance response of free-living copepods, to allow them to make contact with hosts (Bron, 1993). In contrast, both nauplius and adult *L. salmonis* have been reported to show distinct shadow responses (Bron, 1993; Novales Flamarique et al., 2000), the nauplii possibly employing it for predator avoidance while adults use a strong shadow-directed (positively scototactic) response to re-establish contact with a host.

Response to the light environment and infection success

Partly because of positively phototactic behaviour, planktonic sea lice copepodids are likely to be distributed towards the surface during the day and sink down to deeper layers at night, this being the opposite direction of vertical migration to that observed in most zooplankton species (Longhurst, 1976). This reverse diurnal migration behaviour has been observed in the field for copepodids (Heuch et al., 1995); however, the opposite pattern has been reported for nauplii (Nelson et al., 2018). This likely arises from their differential response to light, comprising a potential strategy to avoid predation during the non-infective nauplius stages, something commonly seen in other zooplankton species (Bollens et al., 1992). Because of the challenges of conducting plankton surveys, capturing sea lice copepodids in large water masses, and processing samples (Nelson et al., 2018; Bui et al., 2021), few extensive field studies have proven the theoretical shallow distribution of copepodids. However, laboratory studies combined with a considerable amount of indirect evidence points towards this positively phototactic behaviour, with higher infections being recorded in fish kept between 0 m and 4 m when no halocline is present (Hevrøy et al., 2003) and lower infections being recorded when fish are prevented from accessing these depths (Oppedal et al., 2019). This phototactic behaviour and avoidance of low salinity waters (see section 5.1.1) is the basis of a recent paradigm shift in prevention strategies for salmon aquaculture, using sea cage modifications to create depth-based mismatch between hosts and parasites (see Chapter 22, this volume).

Perception of visual cues associated with distribution of potential hosts is undoubtedly important for locating a fish. However, this acts in synchrony with mechano- and chemosensory cues in order for successful settlement to occur. Despite the apparently strong photopositive behaviours in copepodids, infection success is similar if occurring in darkness or illuminated environments (Johnson and Albright, 1991a; Bron, 1993; Genna et al., 2005; Hamoutene et al., 2016). In fact, the frequency of 'attack' attempts on a fish mimic were similar in dark and light settings (Heuch et al., 2007). Browman et al. (2004) similarly observed no difference in infection success with darkness (or polarised or UV-A light) compared to light and suggested that light cues were likely to facilitate host detection at a distance (within metres) but that the primary mode for host detection at smaller distances (cm to mm) is probably through olfactory and mechano sensory cues. Thus, in the absence of explicit evidence that copepodids use light to locate individual hosts, it seems likely that light is employed more as a cue to bring copepodids into that part of the water column that overlaps with the highest host distribution. Genna et al. (2005) tested infection success under different light conditions and found that greater attachment occurred at low to medium light intensities, suggesting that there may have been increased copepodid swimming activity higher in the water column (thus reducing encounter rate) or overstimulation of the visual system. Either scenario could interrupt the other sensory responses recognised by Browman et al. (2004) to occur at the smaller spatial scales.

While not considered in depth here, it is worth noting that light parameters will also affect the behaviour of the host, for example swimming speed of fish hosts and therefore the encounter rate with respect to infective stages and potentially the success of attachment following initial contact. This relates not only to natural light experienced by hosts but also to the use of artificial lighting (Hevrøy et al., 2003).

5.1.3 Mechanoreceptors and response to mechanical stimuli

Although knowledge of the mechanosensory capabilities of nauplius stages remains poor, the sensory capabilities and behavioural responses of *Lepeophtheirus salmonis* copepodids, with respect to their ability to locate and attach to potential hosts, is well established. The antennules of copepodids have mechanosensory setae that are important for detecting mechanical signals (Johnson and Albright, 1991b; Gresty et al., 1993) and a large number of hair and peg sensilla are also distributed in symmetrically arranged pairs across the cuticular surface, providing additional mechanosensory capabilities (Bron, 1993). Copepodids show increasing levels of response to increasing intensities of infrasonic acceleration and are sensitive to low frequency water accelerations (Heuch and Karlsen, 1997). In response to any disturbances in the water, copepodids swim in a spiral movement that is characteristic of host searching and contact behaviour (Bron, 1993). This response is likely to be triggered by the anterolateral flow field created by a swimming fish, which results from the water being pushed in front of and along the sides of the potential host. This action would create near field movement and far field sound, the latter of which can produce low frequency signals below 10 Hz (Bleckmann et al., 1991). Under controlled conditions, *L. salmonis* copepodids exhibited considerable behavioural responses at 3 Hz stimulation. However, responses were much weaker at 10 Hz, indicating that they react to near field accelerations that are produced within centimetres of a host (Heuch and Karlsen, 1997). In reality, this means that behavioural responses can be induced when hosts create flow speeds (linear strain rates in the direction of flow) greater than 0.5 cm s^{-1} (Heuch et al., 2007). Notably, lice will exhibit these behaviours when presented with hydromechanical stimuli while chemosensory stimuli are absent. Specifically, copepodids will attack a mimic host fish even without the presence of host-derived compounds (Heuch et al., 2007). That study reported a circular 'attack' behaviour, where *L. salmonis* copepodids swam away then looped back towards the head, making contact with it or the body (Heuch et al., 2007). Thus, the vibrations created from the front of a moving fish will likely trigger host-attachment swimming behaviour in nearby copepodids, a distinct looping behaviour that would bring them into contact with the passing fish. If responses are initiated from water currents created by the fish's head, the louse would be at an advantage, having the maximal time available to infect as the host swims past. (See section 5.1.5 for the influence of host swimming speed on infection success.)

Response to pressure

Few studies have focused on the response of sea lice to hydrostatic pressure. In field studies of larval distribution, it is difficult to tease apart the influences of individual environmental factors in driving nauplius or copepodid behaviour. Pressure, light, temperature, and salinity, for instance, can all be intrinsically linked with increasing depth. In the laboratory setting, however, two studies have aimed to isolate the effect of pressure while keeping other variables constant. Bron (1993) found that copepodids gradually swam upwards when exposed to pressures corresponding to 3 m depth within five minutes, while higher pressures simulating 3–4 m and 4–5 m depth induced clear and immediate upward swimming in their characteristic pattern of short bursts. A more recent study used pressurised columns to simulate 0 m, 5 m, or 10 m depth and similarly recorded a strong response to hydrostatic pressure, with double the number of copepodids migrating upwards with the higher pressure (Coates et al., 2020). Both studies found that this response behaviour can be saturated, with a reduction in intensity of response over time and similar responses observed with small or large changes in pressure. Coates et al. (2020) also noted that it was likely the change in pressure that induced responses, rather than the absolute pressure, this also being observed in planktonic larval crustaceans (Lincoln, 1970). This overall negative barotaxic behaviour would likely keep copepodids distributed closer to the surface, with adjustments depending on light and salinity conditions, in order to maximise encounter chances with potential hosts.

Sampling of plankton in the vicinity of farms supports the conclusions drawn from these experiments, albeit that the copepodids are likely responding to a number of different environmental cues. Pumped plankton samples taken through a depth profile in daylight (500 litres at each depth) and obtained from three different farm sites (Frenzl, 2014) showed relatively consistent patterns of copepodid abundance. Maximum copepodid numbers were present at 2 m depth (Fig. 5.1), with copepodid abundance being low in surface water and at 8 m and below, and with numbers being intermediate at 4 m and 6 m sample depths. For all farms, numbers of copepodids at 0–6 m were significantly higher than at 8–12 m. It should, however, be re-emphasised that copepodids are also routinely found at depths exceeding 12 m (S. Bui, 2019, unpublished results).

5.1.4 Chemoreceptors and response to waterborne compounds

Chemoreceptors convey smell and taste in sea lice. For copepodids, olfactory cues are crucial to being able to detect and attach to suitable host species of fish. Chemoreceptors in arthropods are classified as gustatory, odorant, and ionotropic receptors, of which only the latter has been identified in the salmon louse genome (Skern-Mauritzen et al., 2021). The perception of taste affects behaviour

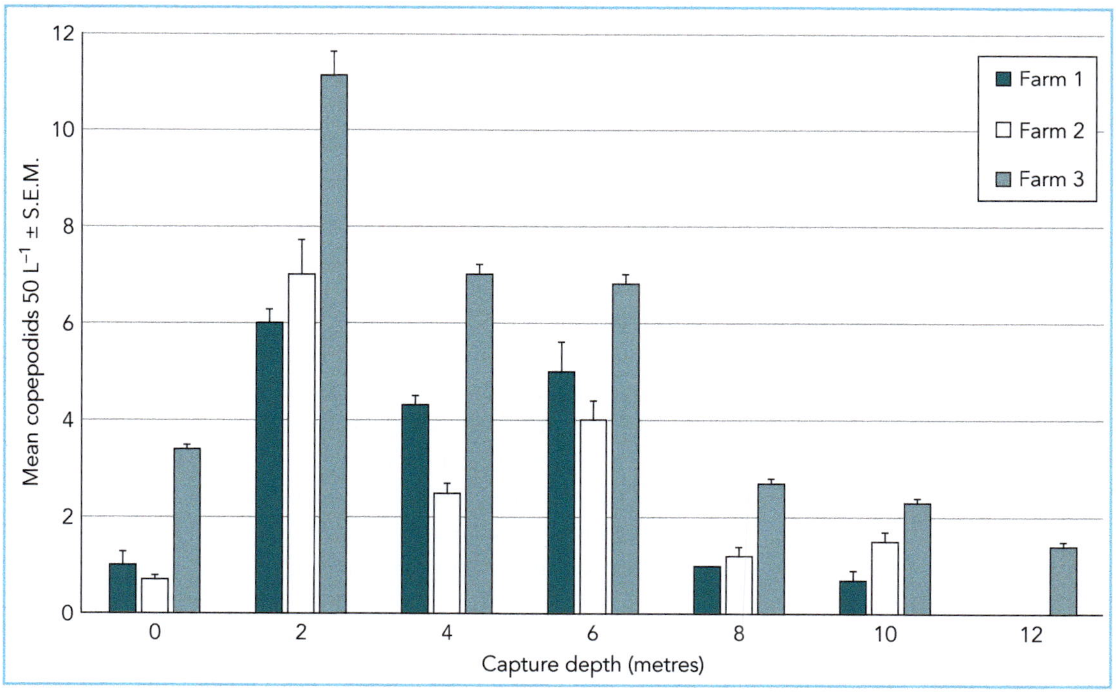

Fig. 5.1 Graph showing results of plankton samples taken at different depths at three Atlantic salmon farm sites. Each column represents the mean number of copepodids recovered from five replicate subsamples from an original 500 litre sample taken in daylight, with each subsample being representative of 50 litres. (Data replotted from Table 2.3, page 52, Frenzl, 2014.)

during the infection process. First, in terms of increased swimming activity of planktonic lice when exposed to semiochemicals from host fish in the surrounding environment. Second, in host recognition once the louse has made contact with a potential host fish. At this point the louse should settle on a suitable spot on the fish and begin to feed, in order to proceed with the life cycle by moulting into the sessile chalimus 1 stage. Alternatively, if the fish is not a correct host, the louse should leave the fish and continue searching for a host fish. Although the taste and confirmation process is not described in detail, salmon lice will only develop on host fish, that is salmonid fishes, with the exception of *Lepeophtheirus salmonis oncorhynchi* from the Pacific (Skern-Mauritzen et al., 2014), which will also develop on sticklebacks (*Gasterosteus aculeatus*) (Jones et al., 2006). Attraction to host fish is likely to be governed by a mixture of attractive and repellent signals. Furthermore, salmon lice with reduced expression of ionotropic receptors exhibit lowered species specificity and can settle on non-host fish (Komisarczuk et al., 2017).

The topic of semiochemicals and their effects on sea lice behaviour is thoroughly reviewed in Chapter 6 of this volume.

5.1.5 Infection success and attachment sites

Using the assortment of sensory modalities and behavioural tools available, the infective copepodids integrate their responses to optimise the chance of encountering and infecting an appropriate host. Host location, host recognition, and successful attachment to the external surface of a host are the crucial steps required for survival. The actual interaction between host and parasite at the point of exposure, and subsequent success or failure to attach, is dependent on favourable environmental and physical conditions. For example, host swimming speed (or conversely, current velocity) will influence the duration of the infection opportunity window for a copepodid and probability of attachment success. As host velocity increases, potential contact time decreases and the boundary layer at the fish's surface becomes thinner and less protective of settling copepodids. This boundary layer, representing the transitional layer within which the speed of water flow declines, through surface drag, from the full flow of the surrounding water to zero flow at the fish's surface, is critical to successful and secure attachment. As the attachment process requires the copepodid to grapple the fish and attach using their antennae to pierce the skin of the host (Bron et al., 1991), reduced contact time likely results in a lower probability of infection. Although not contributing equally to infection success compared to contact time, the increasing velocity of the host or currents will also positively influence encounter rate, with higher encounter rates if a host passes through a dense patch of infective lice. A laboratory trial using adapted swim flumes showed that optimal infection success occurred at moderate velocity (17 cm s^{-1}, or 9.7 body lengths s^{-1}), compared to low or high velocities; in fact, infection success was poorest on fish swimming at high speeds (Samsing et al., 2015).

The interaction of these factors will influence not only infection success, but the location of settled larvae on the host. Copepodids reportedly attach principally on the host's ventral surface along the body and on the fins, and on the dorsal fin (Bron et al., 1991; Bjørn and Finstad, 1998; Genna et al., 2005; Hamoutene et al., 2016; Bui et al., 2018b), although this observation could be confounded by the difficulty of finding copepodids on the dark dorsal side of salmon through visual inspection. In lower current velocities, fish flare out their fins to keep their stable position and therefore lice are more likely to be found on the fins (Genna et al., 2005; Samsing et al., 2015). The increased prevalence of attachment to fins is suggested to result from the protective effect of the fin rays being perpendicular to the current and thus having a deeper boundary layer on their downstream side that allows the copepodid to escape full flow, hence reducing frictional drag and tendency to be removed (Bron et al., 1991). Similarly, the thinner boundary layers and lack of eddies at the front of the fish leads to little settlement on the central lateral sides and head at higher speeds (Genna et al., 2005). An exception to the general lack of settlement on the flanks occurs when scale loss or other damage provides a more protected area for settlement. Under experimental challenge conditions, copepodids may also be frequently found in the gills. However, this is likely to be an artefact, attributable to lower swimming/current speeds in tanks and it is much more rarely observed in the field. While attachment location can be important for ectoparasites' success or their hosts' tolerance to infection, the initial settlement site where *Lepeophtheirus salmonis* copepodids remain until they moult through to mobile stages does not appear to affect development success or survival, at least under laboratory conditions (indirectly observed in Dalvin et al., 2020), although rates of development may vary according to site. Once the mobile stage is reached, the requirements determining their positioning on the host will shift with new goals to optimise feeding and mating opportunities.

5.2 Behaviour in the parasitic phase

5.2.1 *Attachment and movement on host during mobile stages*

Attachment mechanisms and behaviour

The mechanism and behaviours associated with attachment of mobile caligid stages (pre-adults and adults) to their hosts has long been of interest, with an early description of attachment by *Lepeophtheirus pectoralis* (Müller, 1776) being provided by Scott (1901) and a more detailed study of attachment of *Caligus clemensi* (Parker & Margolis, 1964) and *L. salmonis* being provided by Kabata and Hewitt (1971). Most recently, a highly detailed study of the morphological and behavioural aspects of attachment in *Caligus fugu* (Yamaguti, 1936) and *Lepeophtheirus sekii* (Yamaguti, 1936) has been authored by Ohtsuka et al. (2021). The principal means of attachment in pre-adult and adult caligids is the cephalothoracic sucker, whose anterior seal comprises the frontal plate and in some species the flattened first segment of the antennules, with additional purchase provided by the small anterior sucker-like lunules present in *Caligus* spp. The posterior seal is provided by the plate-like third swimming leg. The remaining edges of the cephalothoracic sucker are sealed by use of a flexible marginal membrane, whose surprisingly complex structure and ultrastructure have recently been described in detail by Ohtsuka et al. (2021). Conservation across caligid species of the form of the elements that comprise the cephalothoracic sucker (Ohtsuka et al., 2021) suggests that observations of attachment behaviour made on one species are relevant to others. In brief, from the work of Ohtsuka et al. (2021) on *C. fugu* and *L. sekii*, the initial phase of attachment involves driving of the anterior edge of the animal into contact with the host, including placement of the lunules in *Caligus*, to provide an initial point of contact/attachment. In both *Caligus* and *Lepeophtheirus* the activity of the antennae and maxillipeds was also considered to assist in gaining initial purchase, although epithelial penetration by the sharp tips of these appendages could not be confirmed with the experimental test substrate. With the frontal plate/lunules in contact with the host surface, an initial forward stroke of first and second swimming legs drives water from beneath the cephalothorax under the marginal membrane, the latter of which, in *Caligus*, is blown dorsally by the flow and then settles back to seal the edge of the cephalothorax. Further water is then forced from under the cephalothorax by the backstroke of these appendages, with most of it exiting under the lifted posterior margin of the third swimming leg, while the remainder is driven out through the valve-like dorso-posteriorly directed posterior sinus. Having largely forced the water out from under the cephalothorax, and with its edges sealed against the host surface, it was suggested that the strength of suction might be modulated through flexing of the 'muscular apron' of the third swimming leg, although this was not directly observed by Ohtsuka et al. (2021). In order to release the suction to move or leave the host, opening of any sealing structure would break the seal, for example by raising the margin of the third swimming leg or opening the posterior sinus. Ohtsuka et al. (2021) observed that the lateral movement of the maxillipeds might be employed to assist this action and, on one occasion, noted a strong contraction of the cephalothorax edge in *L. sekii* being used to help break the seal. While it is uncertain as to whether the antennae are constantly engaged in the skin of the host once suction has been established, they do provide an emergency attachment that prevents removal from the host under strong flow or other dislodging conditions, acting in the manner of staples to maintain contact (J. Bron, Stirling, Scotland, 2021, personal communication). A further mechanism of attachment, seen in mobile lice, involves the use of an alternative frontal filament, this being longer and more elastic than that of chalimus stages, which secures lice to the host while their cuticles harden post-moulting.

Preferences in attachment locations

Although initial settlement location depends on many physical influences (see section 5.1.5), sea lice in their pre-adult and adult stages are mobile ectoparasites, and can thus move around the surface of the host according to feeding behaviour and hydrodynamic advantage, and in response to conspecifics (see section 5.2.2 on mating behaviour).

Generally, adult male and female *L. salmonis* aggregate on the head, along the dorsal line, and on the peduncle of salmonid hosts swimming in the sea (Wootten et al., 1982; Jaworski and Holm, 1992; Bui et al., 2020) as well as in the protected area behind the anal fin. Adult females particularly are primarily found focused on the head area (Fig. 5.2) (Bui et al., 2020) or on the post-anal body region (Jaworski and Holm, 1992). Because of their larger body size, females are likely to position themselves to optimise feeding potential and take advantage of hydrodynamic streamlining (or shelter from mechanical disturbance) in these areas (Todd et al., 2000). Mobile lice may also insert the anterior edge of the cephalothorax under scales to gain additional protection from the current and may be positioned one behind the other, possibly through multiple individuals taking advantage of sheltered positions or as a consequence of individuals gaining from the additional protection from the current afforded by other lice. In *C. elongatus*, preferences differ slightly, with more adults recorded on the ventral surface (Treasurer and Bravo, 2011) but also the head (Hogans and Trudeau, 1989). There is evidence that preferences can be different in lice attached to wild versus farmed salmonids (Bui et al., 2020); however, the strongest driver of distribution preference is likely to be the short-term environmental conditions and intraspecific interactions. Even so, the experience of parasites on farmed salmon contrasts with that on wild salmonids, with novel pressures that may influence attachment locations specific to the farm environment. For example, farmed salmon are held in high densities, exhibit schooling behaviour in circular patterns within the sea cage, and may share the space with species of cleaner fish (see Chapter 19, this volume; Treasurer, 2002). These factors collectively may affect location preferences, with hydrodynamic implications or the threat of predation (by cleaner fish) affecting position.

Some studies have reported clear contrasts in the areas occupied by males compared to females, with a hypothesis that the larger females can compete aggressively with males for the more optimal positions, that is, those which are relatively flat, and protected from high current velocities or

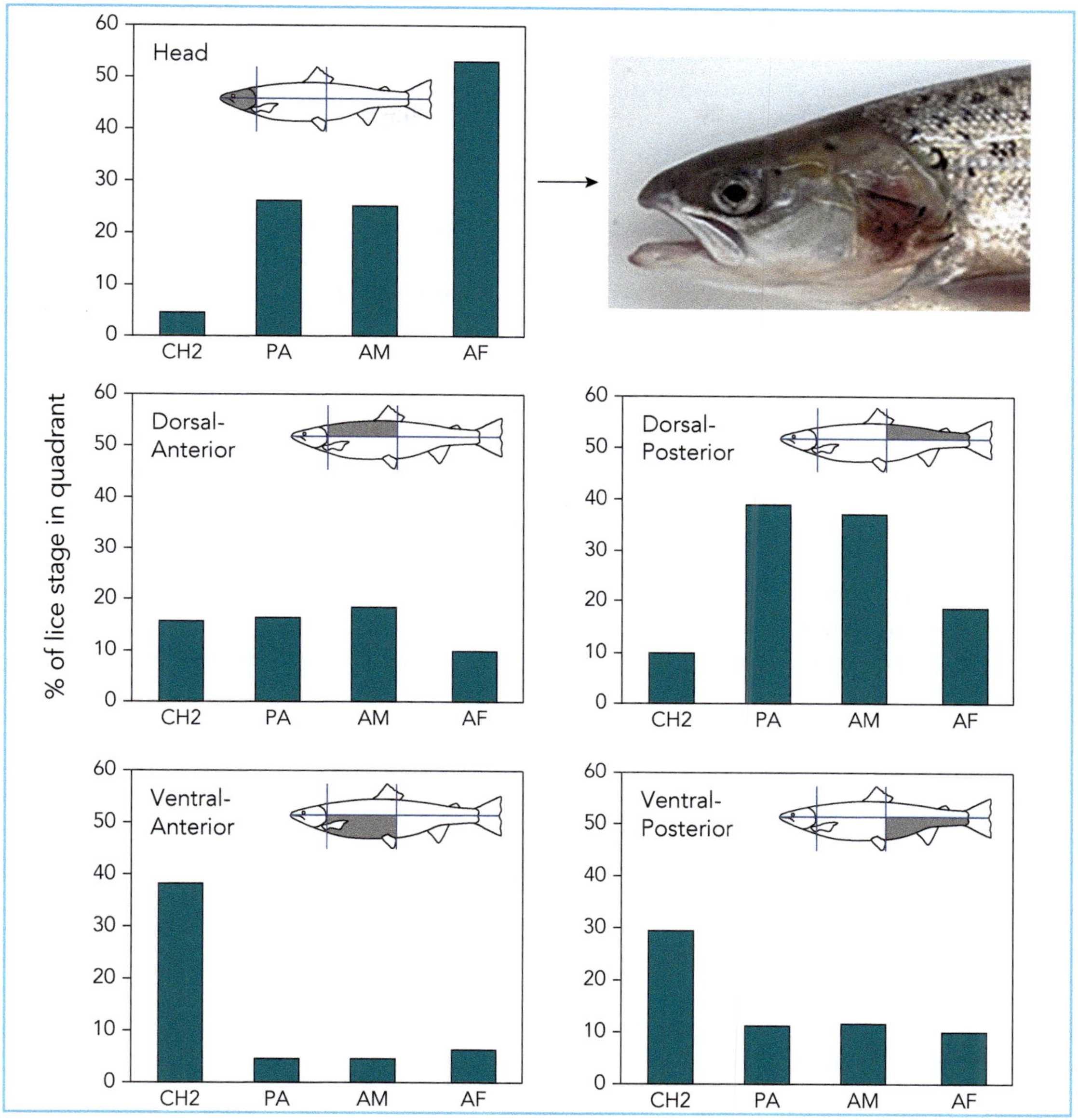

Fig. 5.2 An example of the distribution of *Lepeophtheirus salmonis* stages across the body (excluding fins) of Atlantic salmon hosts held in sea cages. Chalimus 2 (CH2) were recorded predominantly on the ventral sections, whereas pre-adults (PA) and adult males (AM) favoured the dorsal-posterior section, particularly the sheltered area behind the adipose fin. While PA and AM were also reasonably prevalent on the head, adult females (AF) by far dominated this section, and especially aggregated in the concave and protected area of the opercula (as seen in the photo, top right panel). (Adapted from Bui et al., 2020.)

mechanical disturbance (Todd et al., 2000). Because they are generally more sedentary and do not need to move around for mate searching, claiming the prime locations on the host is likely important for adult female survival and reproductive success. Alternatively, males might not be competitively displaced but instead are more mobile on the host surface, searching for potential presence of pre-adult females to form pre-copulatory pairs with. Nevertheless, there is likely a more complex amalgamation of physical, biological, social, and sensory drivers for the movement of sea lice on the host, which will facilitate survival, growth, and reproductive strategy in these ectoparasites (Ritchie, 1997; Pike and Wadsworth, 1999).

In terms of the probability of mobile lice leaving their hosts, adult female *L. salmonis* appear to be relatively sedentary, perhaps due to the risk of failing to find another host as *L. salmonis* has a relatively narrow host range. Male adults on the other hand appear to be much more likely to switch hosts, likely after mating, and reassortment among hosts is common, at least in challenge studies. *C. elongatus*, however, is far more mobile on the host and both sexes are far more likely to re-enter the plankton, this reflecting both their very broad host range and the fact that they have dedicated lipid stores that may allow them to survive longer in plankton.

5.2.2 Developmental loss of lice and density-dependent competition

Host fish are primarily infected by the planktonic copepodid stage and there is no reproduction on the fish host that causes direct re-infection. Thus, infection with one copepodid should lead to development of one adult, female or male. This does not, however, account for loss of lice from the surface of the fish caused either by developmental loss, encompassing processes such as failure to moult successfully, or through behavioural loss from the fish surface especially during moulting and mating events. Loss of lice from infected fish during development from the infective copepodid to the adult stage has for practical reasons only been studied in more detail in tank experiments, and thus extrapolation of these results to lice on wild fish is problematic. Survival numbers vary notoriously between tank experiments (Dalvin et al., 2020), even when performed in identical facilities and with similar fish stock. The question of survival is complicated by variation in infection success and the difficulty of counting all lice on the fish at the copepodid stage, meaning that in most experiments the precise initial load of lice is unknown. Nevertheless, significant loss of lice is often observed as lice develop from the sessile chalimus stages to the mobile pre-adult and the adult stage (Dawson et al., 1997; Bjørn and Finstad, 1998; Hamre et al., 2009; Hamre and Nilsen, 2011; Bui et al., 2016), coinciding with significantly different louse behaviour and localisation (see previous section 'Preferences in attachment locations'). Systematic recording of parasite success in many routine infections indicates, however, that host density and size is important (Hamre et al., 2009; Hamre and Nilsen, 2011), whereas louse behaviour has been studied to a lesser degree. In a natural setting, salmon louse infections are, as for many parasite infections, characterised by most hosts carrying none or very few parasites, whereas a few hosts will carry many parasites (Poulin, 2013). Fish can obtain high lice burdens by being exposed either to repeated infections or to large concentrations of infective copepodids, for example in coastal areas with a large biomass of farmed fish. Heavy infections expose the parasite to density-dependent competition from other lice from the same infection event but this can also result from a situation where there is repeated exposure to lice of various stages. Competition can be related both to simple physical features such as a limited suitable skin area on the fish (Fig. 5.3) and also to more systemic (general) effects such as immune response and changes to behaviour in the fish elicited or amplified by the presence of other lice.

Density-dependent loss is often observed as a higher reduction of lice from fish carrying heavy infections compared to fish with fewer lice as the mobile stages develop. Heavily infected fish kept at high densities can sometimes undergo a complete loss as lice develop to the mobile stages (S. Dalvin, Bergen, Norway, 2020, personal communication). Host jumping, that is lice moving between host

Fig. 5.3 Atlantic salmon carrying large numbers of pre-adult salmon lice (*Lepeophtheirus salmonis*) and illustrating the effects of competition on distribution. Mobile lice, especially adult females, are often found in hydrodynamically sheltered locations e.g. behind fins. On heavily infected fish, lice will experience competition for such advantageous sites and will therefore be displaced to other areas such as the flanks of the fish, as seen here. (Source: S. Bui, Institute of Marine Research, Norway.)

fish, can be significant in common garden tank experiments (Bui et al., 2018a) and may obscure results in, for example, vaccine trials. Host jumping is likely to also occur among farmed fish but the significance of this behaviour may be much less among wild fish, which typically experience a much more solitary lifestyle. Contrary to *L. salmonis*, host jumping from one species to another is probably much more common for *C. elongatus* (Øines et al., 2006). Once *L. salmonis* have reached the adult stage, loss typically ceases and a study of adult females kept on Atlantic salmon demonstrated that loss was not related to parasite load (Ugelvik et al., 2016). It should, however, be noted that the study was performed with relatively low parasite density.

Fish already infected with salmon lice may also be more or less exposed to new infections. In one experiment, salmon carrying lice from a previous infection acquired more new lice in a second round of infection than in non-infected control fish (Ugelvik et al., 2017; Overton et al., 2018). Other experiments with a different experimental setup do, however, show the opposite trend, with fewer new lice settling on previously infected fish (M. Ugelvik, Bergen, Norway, 2021, personal communication). The mechanisms supporting these observed patterns may depend on louse behaviour but are also likely to involve fish behaviour and response to the primary infection. Louse infections lead to a change in host locomotor activity, with infected fish moving more than non-infected (Øverli et al., 2014) but with a decrease in critical swimming speed (Bui et al., 2016).

5.2.3 Response to environmental factors

As noted in section 5.1.1, adult salmon lice are isosmotic at full seawater salinity but exhibit hyperosmotic regulation in less saline environments. Hence, at lower salinities the intracellular osmolality decreases but exceeds that of the external environment (Hahnenkamp and Fyhn, 1985). Decreasing osmolality of the haemolymph is reflected in a general decrease of chloride levels and is further mirrored in a general cellular decline in free amino acids, although certain amino acids, arginine and in particular lysine, increase in their relative abundance, with lysine also increasing in absolute concentration (Hahnenkamp and Fyhn, 1985). Notably, a recent gene expression study indicates proline biosynthesis to be upregulated by reduced salinity (Borchel et al., 2021). This, combined with the lack of increase in cellular proline concentrations (Hahnenkamp and Fyhn, 1985), indicates that proline may be secreted into the haemolymph (Borchel et al., 2021), in effect creating a haemolymph-mediated cuticle-bounded osmotic micro environment. Such a mechanism has previously been suggested also for barnacles (Sundell et al., 2019; Borchel et al., 2021). The capacity to sustain such an internal microenvironment does not appear to be entirely dependent on host-derived osmolytes as a hyperosmotic steady state condition is maintained when detached from a host (Hahnenkamp and Fyhn, 1985). This ability depends on the external environment. However, attached copepodids appear to tolerate reduced salinity better than free-living copepodids (Sievers et al., 2019) and host-bound adult lice may survive freshwater exposure on their host for weeks, although they detach egg strings after only a few hours (L. Hamre, 2005, unpublished data), while detached lice die within hours after transfer to freshwater (Hahnenkamp and Fyhn, 1985). The exposure to substantially reduced salinity seems to elicit a host-associated response whereby the salmon louse buries itself deeper into the host's mucus layers, as suggested by the decreasing rates of change in haemolymph osmolality in freshwater compared to those observed in diluted salt water (Hahnenkamp and Fyhn, 1985), which may explain why adult females exhibit no movement on their host in freshwater. Although host-attached mobile lice (pre-adults and adults) may survive extended periods of freshwater exposure (up to three weeks for adult females, less for adult males) (L. Hamre, Bergen, Norway, 2020, personal communication), there is a very pronounced freshwater induced loss of lice after exposure over a few hours to days depending on the stage (Wright et al., 2016). These effects are even more pronounced at earlier stages, with attached copepodids dying after minutes of freshwater exposure (Wright et al., 2016).

The responses of mobile lice stages to other natural environmental factors, such as light or temperature, have not been well documented. However, the advent of salmon aquaculture, and the fact that the largest proportion of *L. salmonis* populations now exists on farmed fish in some farming areas (Dempster et al., 2021), means that these parasites may adapt their behavioural responses to the novel conditions presented by farms. For example, drawing salmon deeper in the water column and using artificial lights (see Chapter 22) will result in modified light and temperature environments for lice, while other delousing approaches, for example using cleaner fish to predate them, might drive the adoption of new optimal positions on the host to promote survival.

5.2.4 Mating and moulting behaviour

The Caligidae family exhibits a diversity of strategies and behaviours related to mating, and the levels of sexual size dimorphism as well as the relative reproductive investment by males and females vary among genera and among species. Males can be either similar to females in size (*C. elongatus*), considerably smaller (*L. salmonis*, *L. pectoralis*), or sometimes larger (*C. curtus*, *C. pageti*). Sexual size dimorphism, spermatophore size, operational sex ratios, and risk taking are important factors in sexual behaviour and strategy but may also be indicative of the balance between male and female choice in the mating process (Emlen and Oring, 1977; Titelman et al., 2007).

One of the most studied caligids is *L. salmonis*. Adult *L. salmonis* males are close to half the size of adult females and engage in precopulatory mate guarding with pre-adult females (Fig. 5.4). In this process they attach to the dorsal side of the female genital segment and stay there in a guarding position until the females have finished the final moult to adult and mating commences. *L. salmonis* males develop faster than females and reach the adult stage at about the time females reach the pre-adult 2 stage (Johnson and Albright, 1991b; Hamre et al., 2019). In the period after the final moult, adult males go through a short 'puberty' phase before becoming sexually competent and engage in precopulatory mate guarding. The timing of these events is described in Hull et al. (1998). Here the first adult males and the first pre-adult 2 females were observed at day 15 post-infection in a cohort

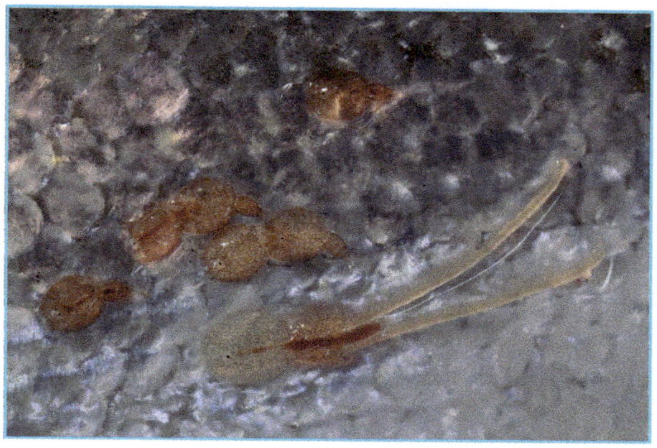

Fig. 5.4 Adult males and pre-adult 2 females in precopula on the skin of Atlantic salmon. Note the insertion of the leading edge of the pre-adult 2 female in the upper pair beneath a scale for more secure attachment. (Credit: L. Hamre, Sea Lice Research Centre.)

of developing lice. Unfortunately, the temperature at which the lice developed was not reported, but the first observation of adult males at day 15 suggests a temperature of around 15°C (Hamre et al., 2019). The first precopulatory pairs were observed 24 hours later, suggesting that the puberty phase lasted for at least 24 hours at around 15°C. This puberty phase is probably required for completing spermatophore development, and/or to obtain sufficient resources before mate guarding begins, since male feeding is restricted when in a guarding position.

Spermatophore development rate is probably temperature dependent, like other developmental processes in lice, and thus takes more time at lower temperatures. The duration of the male puberty phase reported in Hull et al. (1998) amounts to approximately 6.25% of the total development time to adult at 15°C; in other words, the phase assumedly lasts for at least 6.25 relative age units (RAU) (Hamre et al., 2019). At 10°C 6.25 RAU is equivalent to 43 hours. Ritchie et al. (1996a) also give details on development of male reproductive organs in pre-adults and adults.

A detailed description of mate searching behaviour of *L. salmonis* males was given by Ritchie et al. (1996b). Adult males situated posterior to females repeatedly moved from dorsal to ventral positions and back again while maintaining an anterior orientation, often stopping to raise the cephalothoraxes for several seconds. At a short distance from the females, movements were less frequent, and the males remained in the vicinity of the females. When a potential mate had been chosen, males approached the females from the rear and climbed onto the female genital segment to establish a firm grip, placing their antennae around the fourth pedigerous segment of the female. The maxillipeds were also often used to grip the female's abdomen. Males located anterior to the females were sometimes observed to move backwards, sliding over the cephalothorax before settling on the genital segment.

The majority of precopulatory pairs was observed on the dorsal side of the fish by Ritchie et al. (1996b). Mate testing behaviour was also often observed. Males established precopulatory positions but left after less than one minute, with this predominantly occurring during encounters with pre-adult 1 females (86%), and more rarely during encounters with pre-adult 2 females (18.5%). The order of preference for adult males was virgin adult females, pre-adult 2 females, then pre-adult 1 females.

Mating takes place shortly after the female has moulted to adult. The well-coordinated acrobatics carried out by the male when the female removes the old cuticle, and during the subsequent mating, is described in detail by Ritchie et al. (1996b). In brief, shortly before moulting the female attaches the frontal filament anchoring the precopulatory pair to the host. The old cuticle splits open anteriorly and is pulled backwards and off the female. During this process, the male moves from the genital segment to the emerging cephalothorax, keeping a grip with its antennae and maxillipeds. When the old cuticle is shed, the male resumes position over the genital segment. This process, from filament attachment until resuming position over the genital segment, lasts for about 66 minutes. During this phase other males aggregated around the pair and raised their cephalothoraxes for a few seconds each time. Attached to the filament, pairs were observed to drift off the fish surface for up to five minutes at a time. When the new cuticle had hardened the female actively pulled to break the filament. Upon release, the female with the attached male clamped down on the fish surface.

The males remained in precopula with the adult female for about 24 hours post-moulting before copulation began. In this period the males displayed two forms of behaviour. While still in position over the female genital segment, males raised their genital segments to a steep angle for up to six seconds at a time before pressing downwards again, curling the cephalothorax around the female genital segment while rapidly beating their swimming legs. In between, the males also raised their entire body for up to 30 seconds at a time, holding tight around the fourth pedigerous segment with their antennae.

Copulation starts when males slide round the female genital segment to obtain a ventral position. Once in position the male was active in manipulating the female genital orifices and other parts of the genital segment with his maxillipeds, but also, in between, rapidly beating his swimming legs against her genital segment. After a series of repetitions of these behaviours the male finally transferred the two spermatophores simultaneously to the female genital segment, close to the genital orifices, using the second swimming legs. This took about two to three seconds. The swimming legs pushed against the spermatophores several times and, after a while, the maxillipeds were also used to manipulate the spermatophores to a correct position. Following this, the female genital segment was raised again, and the male swimming legs beat rapidly over the spermatophore surface for up to four minutes before the male repositioned to the posterior side of the genital segment, identical to the precopula position. The whole process lasted for about seven minutes. Post-copula, the male held the spermatophores for a few minutes with the maxillipeds but remained on the female for about three hours before leaving her, moving backwards over the genital segment and onto the fish skin.

All the above observations on louse behaviour were obtained from passive hosts and may play out differently in a natural environment or in fish farms on hosts in motion, at least with respect to the timing of events. It would be reasonable to assume that some of the described male actions involve a high risk of detachment and are carried out in periods when the host is relatively calm (for example, at night), and not when the host is agitated or fast swimming. Furthermore, observations from wild salmon showed that the number of adult males almost always exceeded the number of reproductively available females on individual fish, yet a large proportion of the young adult females were unfertilised. This could indicate that fertilisation does not always occur as soon as the females undergo the final moult to adult (Todd et al., 2005). Also, females separated from males displayed a normal post-moult growth of the genital segment, and the virgin adult females produced normal but unfertilised egg strings in the absence of males (Hull et al., 1998; L. Hamre, Bergen, Norway, 2020, personal communication). Adult males introduced to old and fully grown virgin females successfully fertilised them, but they were also more rarely observed in a dorsal position on the females. This indicates that less time, or potentially no time, is spent in mate guarding adult females ready to be fertilised (Hull et al., 1998).

L. pectoralis displays a similar mating process to that of *L. salmonis*. However, when the female sheds the old cuticle during moulting, *L. salmonis* males remain attached to the female whereas most *L. pectoralis* males leave the female and are positioned next to her. Thus, while no *L. salmonis* males were observed to lose the female they guarded, there was a 40% chance that an *L. pectoralis* male lost the female to a competitor in this process (Anstensrud, 1990; Ritchie et al., 1996b). Furthermore, the male mating strategy in both *L. salmonis* and *L. pectoralis* males involves cementing relatively large spermatophores to the female and blocking the copulatory duct to hinder sperm competition from later males (Anstensrud, 1990; Ritchie et al., 1996a). Despite this strategy, polyandry is common in both species (Anstensrud, 1990; Todd et al., 2005). Little is known about how later mating events take place. Anstensrud (1990) observed that only half of non-virgin females without spermatophores were re-inseminated over a four week period and suggests that a female sexual signal may be required for males to fertilise non-virgins.

5.3 Summary

This chapter has discussed the sensory capabilities and behaviours that free-living planktonic stages of sea lice employ to avoid detrimental environments, bring them into the vicinity of hosts, detect and make contact with hosts, and finally infect them successfully. It has also examined aspects of the behaviour of host-attached parasitic stages, particularly pre-adult and adult lice, which allow them to maintain favourable positions on the host, locate mates, and survive in the face of competition from conspecifics and the movement of hosts into adverse environments. It should, however, be remembered that first, many of the reported behaviours have been established under experimental conditions that do not accurately reflect the real-world marine and host environments and hence experimentally determined behaviours may not be perfect reflections of actual behaviour under natural conditions. Second, behaviour, like any other property of an organism, is not fixed but is plastic and may therefore change through time in response to changing host and environmental pressures that impact 'success' of the parasite. This is particularly important to note in a time where climate change, pollution, loss of wild fish stocks, and the expansion of intensive aquaculture are all capable of affecting and therefore modifying established patterns of behaviour in sea lice.

References

Aarseth, K.A. and Schram, T.A. (1999) Wavelength-Specific Behaviour in *Lepeophtheirus salmonis* and *Calanus finmarchicus* to Ultraviolet and Visible Light in Laboratory Experiments (Crustacea: Copepoda). *Marine Ecology Progress Series* 186, 211–217.

Andrews, M. and Horsberg, T.E. (2020) Sensitivity Towards Low Salinity Determined by Bioassay in the Salmon Louse, *Lepeophtheirus salmonis* (Copepoda: Caligidae). *Aquaculture* 514, 8.

Anstensrud, M. (1990) Mating Strategies of Two Parasitic Copepods [(*Lernaeocera branchialis* (L.) (Pennellidae) and *Lepeophtheirus pectoralis* (Müller) (Caligidae)] on Flounder: Polygamy, Sex-Specific Age at Maturity and Sex Ratio. *Journal of Experimental Marine Biology and Ecology* 136, 141–158.

Bjørn, P.A. and Finstad, B. (1998) The Development of Salmon Lice (*Lepeophtheirus salmonis*) on Artificially Infected Post Smolts of Sea Trout (*Salmo trutta*). *Canadian Journal of Zoology* 76, 970–977.

Bleckmann, H., Breithaupt, T., Blickhan, R. and Tautz, J. (1991) The Time Course and Frequency Content of Hydrodynamic Events Caused by Moving Fish, Frogs, and Crustaceans. *Journal of Comparative Physiology A* 168, 749–757.

Bollens, S.M., Frost, B.W., Thoreson, D.S. and Watts, S.J. (1992) Diel Vertical Migration in Zooplankton: Field Evidence in Support of the Predator Avoidance Hypothesis. *Hydrobiologia* 234, 33–39.

Borchel, A., Komisarczuk, A.Z., Rebl, A., Goldammer, T. and Nilsen, F. (2018) Systematic Identification and Characterization of Stress-Inducible Heat Shock Proteins (HSPs) in the Salmon Louse (*Lepeophtheirus salmonis*). *Cell Stress and Chaperones* 23, 127–139.

Borchel, A., Heggland, E.I. and Nilsen, F. (2021) The Transcriptomic Response of Adult Salmon Lice (*Lepeophtheirus salmonis*) to Reduced Salinity. *Comparative Biochemistry and Physiology Part D: Genomics and Proteomics* 37, 100778.

Boxshall, G.A. (1992) Copepoda. In: Harrison, F.W. and Humes, A.G. (eds) *Microscopic Anatomy of Invertebrates Volume 9: Crustacea*. Wiley-Liss, New York. pp. 347–387.

Bricknell, I.R., Dalesman, S.J., O'Shea, B., Pert, C.C. and Mordue Luntz, A.J. (2006) Effect of Environmental Salinity on Sea Lice *Lepeophtheirus salmonis* Settlement Success. *Diseases of Aquatic Organisms* 71, 201–212.

Bron, J.E. (1993) A Study of the Biology and Behaviour of the Copepodid Larva of the Salmon Louse *Lepeophtheirus salmonis* (Krøyer, 1837) (Copepoda: Caligidae). PhD thesis, University of Stirling.

Bron, J.E. and Sommerville, C. (1998) The Functional and Comparative Morphology of the Photoreceptors of the Copepodid Larva of the Salmon Louse *Lepeophtheirus salmonis* (Krøyer, 1837) (Crustacea: Copepoda: Caligidae). *Zoologischer Anzeiger* 237, 113–126.

Bron, J.E., Sommerville, C., Jones, M. and Rae, G.H. (1991) The Settlement and Attachment of Early Stages of the Salmon Louse, *Lepeophtheirus salmonis* (Copepoda: Caligidae) on the Salmon Host, *Salmo salar*. *Journal of Zoology* 224, 201–212.

Brooker, A.J., Skern-Mauritzen, R. and Bron, J.E. (2018) Production, Mortality, and Infectivity of Planktonic Larval Sea Lice, *Lepeophtheirus salmonis* (Krøyer, 1837): Current Knowledge and Implications for Epidemiological Modelling. *ICES Journal of Marine Science* 75(4), 1214–1234.

Browman, H.I., Boxaspen, K. and Kuhn, P. (2004) The Effect of Light on the Settlement of the Salmon Louse, *Lepeophtheirus salmonis*, on Atlantic Salmon, *Salmo salar* L. *Journal of Fish Diseases* 27, 701–708.

Bui, S., Dempster, T., Remen, M. and Oppedal, F. (2016) Effect of Ectoparasite Infestation Density and Life-History Stages on the Swimming Performance of Atlantic Salmon *Salmo salar*. *Aquaculture Environment Interactions* 8, 387–395.

Bui, S., Halttunen, E., Mohn, A.M., Vagseth, T. and Oppedal, F. (2018a) Salmon Lice Evasion, Susceptibility, Retention, and Development Differ Amongst Host Salmonid Species. *ICES Journal of Marine Science* 75, 1071–1079.

Bui, S., Oppedal, F., Samsing, F. and Dempster, T. (2018b) Behaviour in Atlantic Salmon Confers Protection against an Ectoparasite. *Journal of Zoology* 304, 73–80.

Bui, S., Oppedal, F., Nola, V. and Barrett, L.T. (2020) Where Art Thou Louse? A Snapshot of Attachment Location Preferences in Salmon Lice on Atlantic Salmon Hosts in Sea Cages. *Journal of Fish Diseases* 43, 697–706.

Bui, S., Dalvin, S., Vågseth, T., Oppedal, F., Fossøy, F., Brandsegg, H., Jacobsen, Á., Á Norði, G., et al. (2021) Finding the Needle in the Haystack: Comparison of Methods for Salmon Louse Enumeration in Plankton Samples. *Aquaculture Research* 52, 3591–3604.

Coates, A., Phillips, B.L., Oppedal, F., Bui, S., Overton, K. and Dempster, T. (2020) Parasites under Pressure: Salmon Lice Have the Capacity to Adapt to Depth-Based Preventions in Aquaculture. *International Journal of Parasitology* 50, 865–872.

Dalvin, S., Hamre, L., Skern-Mauritzen, R., Vagseth, T., Stien, L., Oppedal, F. and Bui, S. (2020) The Effect of Temperature on Ability of *Lepeophtheirus salmonis* to Infect and Persist on Atlantic Salmon. *Journal of Fish Diseases* 43, 1519–1529.

Dawson, L.H.J., Pike, A.W., Houlihan, D.F. and Mcvicar, A.H. (1997) Comparison of the Susceptibility of Sea Trout (*Salmo trutta* L.) and Atlantic Salmon (*Salmo salar* L.) to Sea Lice (*Lepeophtheirus salmonis* (Krøyer, 1837)) Infections. *ICES Journal of Marine Science* 54, 1129–1139.

Dempster, T., Overton, K., Bui, S., Stien, L.H., Oppedal, F., Karlsen, Ø., Coates, A., Phillips, B.L. and Barrett, L.T. (2021) Farmed Salmonids Drive the Abundance, Ecology and Evolution of Parasitic Salmon Lice in Norway. *Aquaculture Environment Interactions* 13, 237–248.

Emlen S.T. and Oring L.W. (1977) Ecology, Sexual Selection, and the Evolution of Mating Systems. *Science* 197, 215–223.

Fields, D.M., Skiftesvik, A.B. and Browman, H.I. (2018) Behavioural Responses of Infective-Stage Copepodids of the Salmon Louse (*Lepeophtheirus salmonis*, Copepoda: Caligidae) to Host-Related Sensory Cues. *Journal of Fish Diseases* 41, 875–884.

Forward, R.B. (1988) Diel Vertical Migration: Zooplankton Photobiology and Behaviour. *Oceanography and Marine Biology* 26, 361–393.

Frenzl, B. (2014) Understanding Key Factors Associated with the Infection of Farmed Atlantic Salmon by the Salmon Louse *Lepeophtheirus salmonis*. PhD thesis, University of Stirling.

Genna, R.L., Mordue, W., Pike, A.W. and Mordue, A.J. (2005) Light Intensity, Salinity, and Host Velocity Influence Presettlement Intensity and Distribution on Hosts by Copepodids of Sea Lice, *Lepeophtheirus salmonis*. *Canadian Journal of Fisheries and Aquatic Sciences* 62, 2675–2682.

Gravil, H.R. (1996) Studies on the Biology and Ecology of the Free Swimming Larval Stages of *Lepeophtheirus salmonis* (Krøyer, 1838) and *Caligus elongatus* Nordmann, 1832 (Copepoda: Caligidae). PhD thesis, University of Stirling.

Gresty, A., Boxshall, G.A. and Nagasawa, K. (1993) Antennulary Sensors of the Infective Copepod Larva of the Salmon Louse *Lepeophtheirus salmonis* (Copepoda: Caligidae). In: Boxshall, G. A. and Defaye, D. (eds) *Pathogens of Wild and Farmed Fish: Sea Lice*. Ellis and Horwood Ltd, Chichester, UK. pp. 83–98.

Hahnenkamp, L. and Fyhn, H.J. (1985) The Osmotic Response of Salmon Louse, *Lepeophtheirus salmonis* (Copepoda: Caligidae), During the Transition from Sea-Water to Fresh-Water. *Journal of Comparative Physiology B-Biochemical Systemic and Environmental Physiology* 155, 357–365.

Hamoutene, D., Mitchell, J.S., Murray, H.M., Eaves, A., Marshall, K., Belley, R. and George, S. (2016) The Effect of Light Regimen on Settlement Patterns of Sea Lice, *Lepeophtheirus salmonis*, on Atlantic Salmon, *Salmo salar*, Post-Smolts While Taking into Account Fish Size and Fin Erosion in a Static Tank System. *Aquaculture* 465, 1–6.

Hamre, L., Bui, S., Oppedal, F., Skern-Mauritzen, R. and Dalvin, S. (2019) Development of the Salmon Louse *Lepeophtheirus salmonis* Parasitic Stages in Temperatures Ranging from 3 to 24°C. *Aquaculture Environment Interactions* 11, 429–443.

Hamre, L.A., Glover, K.A. and Nilsen, F. (2009) Establishment and Characterisation of Salmon Louse (*Lepeophtheirus salmonis* (Krøyer, 1837)) Laboratory Strains. *Parasitology International* 58, 451–460.

Hamre, L.A. and Nilsen, F. (2011) Individual Fish Tank Arrays in Studies of *Lepeophtheirus salmonis* and Lice Loss Variability. *Diseases of Aquatic Organisms* 97, 47–56.

Heuch, P.A. (1995) Experimental Evidence for Aggregation of Salmon Louse Copepodids (*Lepeophtheirus salmonis*) in Step Salinity Gradients. *Journal of the Marine Biological Association of the UK* 75, 927–939.

Heuch, P.A., Parsons, A. and Boxaspen, K. (1995) Diel Vertical Migration: A Possible Host-Finding Mechanism in Salmon Louse (*Lepeophtheirus salmonis*) Copepodids? *Canadian Journal of Fisheries and Aquatic Sciences* 52, 681–689.

Heuch, P.A. and Karlsen, H.E. (1997) Detection of Infrasonic Water Oscillations by Copepodids of *Lepeophtheirus salmonis* (Copepoda: Caligida). *Journal of Plankton Research* 19, 735–747.

Heuch, P.A., Doall, M.H. and Yen, J. (2007) Water Flow around a Fish Mimic Attracts a Parasitic and Deters a Planktonic Copepod. *Journal of Plankton Research* 29, I3–I16.

Hevrøy, E.M., Boxaspen, K., Oppedal, F., Taranger, G.L. and Holm, J.C. (2003) The Effect of Artificial Light Treatment and Depth on the Infestation of the Sea Louse *Lepeophtheirus salmonis* on Atlantic Salmon (*Salmo salar* L.) Culture. *Aquaculture* 220, 1–14.

Hogans, W.E. and Trudeau, D.J. (1989) Preliminary Studies on the Biology of Sea Lice, *Caligus elongatus*, *Caligus curtus* and *Lepeophtheirus salmonis* (Copepoda: Caligoida) Parasitic on Cage-Cultured Salmonids in the Lower Bay of Fundy. *Canadian Technical Report of Fisheries and Aquatic Sciences* 1715, 14.

Hull, M.Q., Pike, A.W., Mordue, A.J. and Rae, G.H. (1998) Patterns of Pair Formation and Mating in an Ectoparasitic Caligid Copepod *Lepeophtheirus salmonis* (Krøyer 1837): Implications for Its Sensory and Mating Biology. *Philosophical Transactions of the Royal Society of London Series B: Biological Sciences* 353, 753–764.

Jaworski, A. and Holm, J.C. (1992) Distribution and Structure of the Population of Sea Lice, *Lepeophtheirus salmonis* Krøyer, on Atlantic Salmon, *Salmo salar* L., under Typical Rearing Conditions. *Aquaculture and Fisheries Management* 23, 577–589.

Johnson, S.C. and Albright, L.J. (1991a) Development, Growth, and Survival of *Lepeophtheirus salmonis* (Copepoda: Caligidae) under Laboratory Conditions. *Journal of the Marine Biological Association of the UK* 71, 425–436.

Johnson, S.C. and Albright, L.J. (1991b) The Developmental Stages of *Lepeophtheirus salmonis* (Krøyer, 1837) (Copepoda: Caligidae). *Canadian Journal of Zoology* 69(4), 929–950.

Jones, S., Kim, E. and Dawe, S. (2006) Experimental Infections with *Lepeophtheirus salmonis* (Krøyer 1837) on Three-spine Sticklebacks, *Gasterosteus Aculeatus* L., and Juvenile Pacific Salmon, *Oncorhynchus* Spp. *Journal of Fish Diseases* 29, 489–495.

Kabata, Z. and Hewitt, G.C. (1971) Locomotory Mechanisms in Caligidae (Crustacea: Copepoda). *Journal of the Fisheries Research Board of Canada* 28, 1143–1151.

Komisarczuk, A.Z., Grotmol, S. and Nilsen, F. (2017) Ionotropic Receptors Signal Host Recognition in the Salmon Louse (*Lepeophtheirus salmonis*, Copepoda). *PLoS ONE* 12, 18.

Lincoln, R.J. (1970) A Laboratory Investigation into the Effects of Hydrostatic Pressure on the Vertical Migration of Planktonic Crustacea. *Marine Biology* 6, 5–11.

Ljungfeldt, L.E.R., Quintela, M., Besnier, F., Nilsen, F. and Glover, K.A. (2017) A Pedigree-Based Experiment Reveals Variation in Salinity and Thermal Tolerance in the Salmon Louse, *Lepeophtheirus salmonis*. *Evolutionary Applications* 10, 1007–1019.

Longhurst, A.R. (1976) Vertical Migration. In: Cushing, D.H. and Walsh, J.J. (eds) *The Ecology of the Seas*. W.B. Saunders Co, Philadelphia. pp. 116–137.

Lyse, A.A., Stefansson, S.O. and Fernö, A. (1998) Behaviour and Diet of Sea Trout Post-Smolts in a Norwegian Fjord System. *Journal of Fish Biology* 52, 923–936.

Nelson, E.J., Robinson, S.M.C., Feindel, N., Sterling, A., Byrne, A. and Ang, K.P. (2018) Horizontal and Vertical Distribution of Sea Lice Larvae (*Lepeophtheirus salmonis*) in and around Salmon Farms in the Bay of Fundy, Canada. *Journal of Fish Diseases* 41, 885–899.

Nordtug, T., Kvæstad, B. and Hagemann, A. (2021) Responses and Preferences of Salmon Louse (*Lepeophtheirus salmonis* Krøyer 1836) Copepodids to Underwater Artificial Light Sources. *Aquaculture* 532.

Novales Flamarique, I., Browman, H.I., Belanger, M. and Boxaspen, K. (2000) Ontogenetic Changes in Visual Sensitivity of the Parasitic Salmon Louse *Lepeophtheirus salmonis*. *Journal of Experimental Biology* 203, 1649–1657.

Ohtsuka, S., Nishida, Y., Hirano, K., Fuji, T., Kaji, T., Kondo, Y., Komeda, S., Tasumi, S., Koike, K. and Boxshall, G.A. (2021) The Cephalothoracic Sucker of Sea Lice (Crustacea: Copepoda: Caligidae): The Functional Importance of Cuticular Membrane Ultrastructure. *Arthropod Structure & Development* 62, 101046.

Øines, O., Simonsen, J.H., Knutsen, J.A. and Heuch, P.A. (2006) Host Preference of Adult *Caligus elongatus* Nordmann in the Laboratory and Its Implications for Atlantic Cod Aquaculture. *Journal of Fish Diseases* 29, 167–174.

Oppedal, F., Bui, S., Stien, L.H., Overton, K. and Dempster, T. (2019) Snorkel Technology to Reduce Sea Lice Infestations: Efficacy Depends on Salinity at the Farm Site, but Snorkels Have Minimal Effects on Salmon Production and Welfare. *Aquaculture Environment Interactions* 11, 445–457.

Øverli, Ø., Nordgreen, J., Mejdell, C.M., Janczak, A.M., Kittilsen, S., Johansen, I.B. and Horsberg, T.E. (2014) Ectoparasitic Sea Lice (*Lepeophtheirus salmonis*) Affect Behavior and Brain Serotonergic Activity in Atlantic Salmon (*Salmo salar* L.): Perspectives on Animal Welfare. *Physiology & Behavior* 132, 44–50.

Overton, K., Samsing, F., Oppedal, F., Dalvin, S., Stien, L.H. and Dempster, T. (2018) The Use and Effects of Hydrogen Peroxide on Salmon Lice and Post-Smolt Atlantic Salmon. *Aquaculture* 486, 246–252.

Pike, A.W. and Wadsworth, S.L. (1999) Sea Lice on Salmonids: Their Biology and Control. In: Baker, J. R., Muller, R. and Rollinson, D. (eds) *Advances in Parasitology* Vol 44. Elsevier Academic Press Inc, San Diego, California. pp. 233–237.

Poulin, R. (2013) Explaining Variability in Parasite Aggregation Levels among Host Samples. *Parasitology* 140, 541–546.

Ritchie, G. (1997) The Host Transfer Ability of *Lepeophtheirus salmonis* (Copepoda: Caligidae) from Farmed Atlantic Salmon, *Salmo salar* L. *Journal of Fish Diseases* 20, 153–157.

Ritchie, G., Mordue, A.J., Pike, A.W. and Rae, G.H. (1996a) Morphology and Ultrastructure of the Reproductive System of *Lepeophtheirus salmonis* (Krøyer, 1837) (Copepoda: Caligidae). *Journal of Crustacean Biology* 16, 330–346.

Ritchie, G., Mordue, A.J., Pike, A.W. and Rae, G.H. (1996b) Observations on Mating and Reproductive Behaviour of *Lepeophtheirus salmonis*, Krøyer 1837 (Copepoda: Caligidae). *Journal of Experimental Marine Biology and Ecology* 201, 285–298.

Samsing, F., Solstorm, D., Oppedal, F., Solstorm, F. and Dempster, T. (2015) Gone with the Flow: Current Velocities Mediate Parasitic Infestation of an Aquatic Host. *International Journal for Parasitology* 45, 559–565.

Scott, A. (1901) On the Fish Parasites, *Lepeophtheirus* and *Lernaea*. *Rep. Lanes Sea Fish Lab.* 9, 63–94.

Sievers, M., Oppedal, F., Ditria, E. and Wright, D.W. (2019) The Effectiveness of Hyposaline Treatments against Host-Attached Salmon Lice. *Scientific Reports* 9, 6976.

Skern-Mauritzen, R. and Mikkelsen, T.N. (2021) The Information Continuum Model of Evolution. *Biosystems* 209, 104510.

Skern-Mauritzen, R., Torrissen, O. and Glover, K.A. (2014) Pacific and Atlantic *Lepeophtheirus salmonis* (Krøyer, 1838) Are Allopatric Subspecies: *Lepeophtheirus salmonis salmonis* and *L. salmonis oncorhynchi* subspecies novo. *BMC Genetics* 15, 1–9.

Skern-Mauritzen, R., Sissener, N.H., Sandvik, A.D., Meier, S., Sævik, P.N., Skogen, M.D., Vågseth, T., Dalvin, S., Skern-Mauritzen, M. and Bui, S. (2020) Parasite Development Affect Dispersal Dynamics; Infectivity, Activity and Energetic Status in Cohorts of Salmon Louse Copepodids. *Journal of Experimental Marine Biology and Ecology* 530–531, 151429.

Skern-Mauritzen, R., Malde, K., Eichner, C., Dondrup, M., Furmanek, T., Besnier, F., Komisarczuk, A.Z., Nuhn, M., et al. (2021) The Salmon Louse Genome: Copepod Features and Parasitic Adaptations. *Genomics* 113, 3666–3680.

Sundell, K., Wrange, A.L., Jonsson, P.R. and Blomberg, A. (2019) Osmoregulation in Barnacles: An Evolutionary Perspective of Potential Mechanisms and Future Research Directions. *Frontiers in Physiology* 10, 877.

Sutherland, B.J.G., Jantzen, S.G., Yasuike, M., Sanderson, D.S., Koop, B.F. and Jones, S.R.M. (2012) Transcriptomics of Coping Strategies in Free-Swimming *Lepeophtheirus salmonis* (Copepoda) Larvae Responding to Abiotic Stress. *Molecular Ecology* 21, 6000–6014.

Szetey, A., Wright, D.W., Oppedal, F. and Dempster, T. (2021) Salmon Lice Nauplii and Copepodids Display Different Vertical Migration Patterns in Response to Light. *Aquaculture Environment Interactions* 13, 121–131.

Titelman, J., Varpe, Ø., Eliassen, S. and Fiksen, Ø. (2007) Copepod Mating: Chance or Choice? *Journal of Plankton Research* 29, 1023–1030.

Todd, C.D., Walker, A.M., Hoyle, J.E., Northcott, S.J., Walker, A.F. and Ritchie, M.G. (2000) Infestations of Wild Adult Atlantic Salmon (*Salmo salar* L.) by the Ectoparasitic Copepod Sea Louse *Lepeophtheirus salmonis* Krøyer: Prevalence, Intensity and the Spatial Distribution of Males and Females on the Host Fish. *Hydrobiologia* 429, 181–196.

Todd, C.D., Stevenson, R.J., Reinardy, H. and Ritchie, M.G. (2005) Polyandry in the Ectoparasitic Copepod *Lepeophtheirus salmonis* Despite Complex Precopulatory and Postcopulatory Mate-Guarding. *Marine Ecology Progress Series* 303, 225–234.

Treasurer, J.W. (2002) A Review of Potential Pathogens of Sea Lice and the Application of Cleaner Fish in Biological Control. *Pest Management Science* 58, 546–558.

Treasurer, J. and Bravo, S. (2011) The Spatial Distribution Patterns of *Caligus rogercresseyi* and *C. elongatus* on Atlantic Salmon Hosts (*Salmo salar*). *Aquaculture* 320 (3–4), 154–158.

Tucker, C.S., Sommerville, C. and Wootten, R. (2000) The Effect of Temperature and Salinity on the Settlement and Survival of Copepodids of *Lepeophtheirus salmonis* (Krøyer, 1837) on Atlantic Salmon, *Salmo salar* L. *Journal of Fish Diseases* 23, 309–320.

Ugelvik, M.S., Skorping, A. and Mennerat, A. (2016) Parasite Fecundity Decreases with Increasing Parasite Load in the Salmon Louse *Lepeophtheirus salmonis* Infecting Atlantic Salmon *Salmo salar*. *Journal of Fish Diseases* 40(5), 671–678.

Ugelvik, M.S., Mo, T., Mennerat, A. and Skorping, A. (2017) Atlantic Salmon Infected with Salmon Lice Are More Susceptible to New Lice Infections. *Journal of Fish Diseases* 40, 311–317.

Wootten, R., Smith, J.W. and Needham, E.A. (1982) Aspects of the Biology of the Parasitic Copepods *Lepeophtheirus salmonis* and *Caligus elongatus* on Farmed Salmonids, and Their Treatment. *Proceedings of the Royal Society of Edinburgh* 81(3), 185–197.

Wright, D.W., Oppedal, F. and Dempster, T. (2016) Early-Stage Sea Lice Recruits on Atlantic Salmon Are Freshwater Sensitive. *Journal of Fish Diseases* 39, 1179–1186.

CHAPTER 6

Semiochemical tools for the management of sea lice

A.J. Mordue (Luntz)[1], J. Pino-Marambio[2] and J.A. Pickett[3]

[1]School of Biological Sciences, University of Aberdeen, Scotland, UK
[2]Cargill Innovation Center, Colaco, Calbuco, Chile
[3]School of Chemistry, Cardiff University, Wales, UK

6.1 Introduction

In all areas of food production there is a powerful movement, both legislatively and investment driven, towards more sustainable pest management systems with substantially lower potential risk to the environment and human health. In the management of pests of aquatic food production systems, such innovations lag deployment compared to the more mainstream and research-intensive terrestrial production systems. Here we describe the use of semiochemicals, including pheromones, to manipulate the behaviour of sea lice and reduce infection levels with evidence already of field efficacy. We recommend that with more research and development, semiochemicals would form a major part of the arsenal of methods required for integrated pest management (IPM) strategies for sea lice of farmed salmon. Crustaceans, in general, have a highly sensitive olfactory system that allows them to find food, mates, predators, or rivals by the detection of minute amounts of olfactory molecules in the water that function as a signal in a spatiotemporal fashion.

From the 1980s onwards, the use of repellent and attractive stimuli together has been developed into 'push-pull' strategies for many insect pest species. The knowledge gained from these studies paves the way for developing similar strategies for crustaceans and in this instance for sea lice of farmed salmon.

The main principle of a 'push-pull' strategy is to use a combination of behaviour modifying chemicals to manipulate the pest species away from the protected crop or animal (Saini et al., 2017) and towards other areas containing attractive compounds where they can be trapped for removal or destruction. In addition, if a full 'push-pull' strategy is not possible it can still be valuable to provide one half of the strategy as in the case of the application of repellents to the skin of animals or man to protect from the bites of insects which harbour transmittable diseases.

Chemical cues play an essential role in the life cycle of the salmon louse (*Lepeophtheirus salmonis* (Krøyer, 1837)). Chemicals in the water and on the surface of organisms are involved in the modalities of smell and taste and induce specific behaviours in the louse which lead to successful host location, feeding, mate location, and the evasion of non-host species (Ingvarsdóttir et al., 2002a, 2002b;

Mordue (Luntz), 2003; Pino-Marambio et al., 2007; Jodaa Holm et al., 2016). Chemoreceptors on the first antennae and elsewhere on the body of the louse (distributed chemoreception) are stimulated by chemical cues with the transduction process resulting in increased neuronal activity, changes in behaviour leading to mating, host location, landing, and feeding, the upregulation of genes, and the production of proteins and peptides required for the processes of growth, moulting, and reproduction (Zhou et al., 2004; Sánchez-Gracia et al., 2009; Hara, 2011; Núñez-Acuña et al., 2014; Komisarczuk et al., 2017; D'Onofrio et al., 2020).

Olfaction is the first chemical sense to be employed in the key behavioural traits of host and mate location. Chemicals in the water function at a distance by stimulating the antennal receptors and serve to orientate the animal towards or away from the emitting source. Such chemical signals are termed semiochemicals and include pheromones, a term reserved for chemical signals communicating between members of the same species. For sea lice, semiochemical research began in the 1990s with detailed observations of sea louse behaviour (Ritchie et al., 1996; Hull et al., 1998; Devine et al., 2000) coupled with ultrastructural studies of chemoreceptors, behavioural bioassays, and electrophysiological recording techniques (Hull et al., 1998; Ingvarsdóttir et al., 2002a, 2002b; Fields et al., 2007). These resulted in the collection of unequivocal evidence of the involvement of identifiable host-related chemical cues causing alterations in the swimming behaviour and host location of sea lice, as well as of the involvement of sex pheromones in mate location.

As semiochemicals play such a key role in sea louse behaviour and physiology, there is the potential to develop strategies for the disruption of host and mate location in lice with a view to both preventing settlement on fish, and disrupting mating behaviour, which could form part of an integrated pest management strategy.

6.2 What are semiochemicals?

Chemoreception is the dominant sensory modality for most crustaceans (Schmidt and Mellon, 2011). In the aquatic environment crustaceans are surrounded by a vast array of chemical signals that relate to feeding, mating, and avoidance of predators. There have been many attempts to classify the variety of chemical interactions between organisms to clarify nomenclature and to gain some understanding into the evolutionary processes involved. The overarching term semiochemicals is generally accepted as the term to use with subdivisions providing context and understanding (Wyatt, 2010). Semiochemicals (from the Greek *semeion*, mark or signal) are defined as 'chemicals acquired or produced by the sender, involved in the chemical interactions between organisms' (Nordlund and Lewis, 1976).

The term semiochemical can be subdivided into its different components. Pheromones are semiochemicals used within a species, such as sex pheromones released by pre-adult and virgin adult sea lice which attract males for mating, and allelochemicals are semiochemicals used between species. Allelochemicals are further broken down according to the costs and benefits to the sender and receiver. They are termed kairomones if they benefit the receiver, and allomones if they benefit the sender. Examples relating to sea lice would be host chemicals, such as isophorone, which is released into the water and which acts as a kairomone to the sea louse, *L. salmonis*, attracting it to its host. Non-host chemicals, such as 2-aminoacetophenone from turbot, function as an allomone to *L. salmonis* repelling it from the non-host turbot (Mordue and Birkett, 2009).

6.2.1 Pheromones

The word pheromone was coined from the Greek *pherein*, to carry or transfer, and *hormōn*, to excite or stimulate (Wyatt, 2014), and was defined originally by Karlson and Lüscher (1959). Pheromones are evolved chemical signals between members of the same species. The molecules are emitted by an individual and received by a second individual of the same species, in which they cause a specific reaction, for example a stereotyped behaviour or a developmental process (Karlson and Lüscher, 1959; Wyatt, 2010).

In crustaceans, pheromones can be released to the surrounding water as volatile pheromones (distance pheromones) or bound to the body surface (contact pheromones). Urine has been shown to be the carrier of many such chemical signals in large crustaceans. Pheromones are 'signals' within species, rather than molecules with a physiological purpose, and give a selective advantage to both emitter and receiver. Both the signal production of pheromones and the response itself must have evolved for that purpose.

6.2.2 Kairomones and allomones

Kairomones and allomones are different from pheromones as although the receiver may well have evolved to recognise a specific host 'cue', that cue did not evolve in the emitter to affect the recipient. Such cues are chemicals that elicit a response that is not adaptive to the emitting animal (Wyatt, 2011). An example of a 'cue' here would be 2-aminoacetophenone functioning as an allomone which is repellent or 'host masking' to the louse, but which is part of the turbot 'signature mixture' that is not purely involved in resistance to parasites.

In terrestrial systems, plants release a variety of volatile organic compounds (for example, terpenoids and fatty acid derivatives). These compounds are used for distinct functions for plants such as self-defence mechanisms and for their own metabolic processes (Nyasembe and Torto, 2014). These chemicals have a specific function in communication between species, or between members of the same species promoting or deterring interactions between plants and insects. Semiochemicals are organic compounds used by insects to convey specific chemical messages that modify behaviour or physiology (Pitts et al., 2014; El-Sayed, 2019). Typically, semiochemicals are lipophilic, and not usually highly volatile or water soluble and can have a wide range of molecular weights being classified only according to function. It should be noted that within this classification the same molecule could function as a pheromone in one species and as a kairomone or allomone in another, depending on the species.

6.2.3 Chemoreceptors

In aquatic organisms, chemical communication, specifically olfaction, needs to take place in water solution rather than in the gas phase as in terrestrial organisms. Although chemical communication evolved first in aquatic organisms, much terminology around chemical ecology relates to terrestrial systems where the modalities of olfaction and taste are clearly delineated. In crustaceans it is now accepted to list two modes of chemoreception characterised as 'olfaction' and 'distributed chemoreception'. The former relates to chemoreceptors on the antennae which send their axons to the olfactory lobe of the brain and the latter relates to the distribution of chemosensory receptors on the antennae and around the body which send their sensory neurons to different glomeruli in the brain from olfactory receptors (Harzsch and Krieger, 2018).

Olfaction is mediated through specialised sensilla on the distal article of the first antennae in sea lice, the aesthetascs. These sensilla resemble olfactory sensilla in insects and are innervated by many highly branched olfactory receptor neurons (ORNs). The walls of the aesthetascs are permeable which allows entry to a range of organic and inorganic molecules transported to specific receptor sites on the neurons by odour binding proteins (OBPs) and/or chemosensory proteins (CSPs). Excitation of the neurons leads to action potentials travelling via the antennal nerve to the brain (Hull et al., 1998; Zhou et al., 2004; D'Onofrio et al., 2020). Phylogenetic analysis indicates that *Caligus rogercresseyi* (Bravo & Boxshall 2000) express a CSP (CrogCSP) which is clustered in a main micro-crustacean clade. This is consistent with what Vieira and Rozas (2011) reported, as well as Pelosi et al. (2014), and those authors show the different clades of proteins, mainly CSPs. Sequence analysis and homology modelling revealed the presence of four cysteine residues connected among them in two disulphide bridges, which is the main feature of CSPs (Jimenez Benavente, 2017). Aesthetascs do not have mechanosensory neurons. On the other hand, distributed chemoreceptors which are found on both the proximal and distal articles of the first antennae of sea lice and on other parts of the body are bimodal, innervated by both chemoreceptor neurons (CRNs) and mechanoreceptor neurons (MRNs) sending their axons to different lobes of the brain than aesthetascs (Hull et al., 1998; Derby et al., 2016). They contain unbranched dendrites of the CRNs which travel up the length of the hair, which has a distal pore through which molecules may enter. Normally there is only one MRN per hair which terminates at the base of the hair (Hull et al., 1998).

In this chapter we are considering semiochemical strategies for sea lice control and therefore we only review those compounds that fit the definition of a semiochemical. They are olfactory molecules that stimulate the aesthetascs of the first antennae. They are volatile, non-polar organic compounds generally of low molecular weight that have low to high physical volatility dependent on their structure, and which are soluble in water at the concentrations at which they are required for chemical signalling.

6.3 Finding a mate

All free-living copepods use pheromones to locate a mate (Breithaupt and Thiel, 2011). Parasitic copepods are no exception, and it is now well established that chemical cues are involved in both *L. salmonis* and *C. rogercresseyi* mating behaviour (Ritchie et al., 1996; Hull et al., 1998; Devine et al., 2000; Ingvarsdóttir et al., 2002b; Pino-Marambio et al., 2007). Male sea lice are highly mobile and need to successfully locate and land on their host fish to find females. This high mobility results in frequent re-assortment of males on available host fish. Once on the host fish male lice use the body surface of the fish as a substrate for mate location which leads eventually to successful mating.

6.3.1 Mate location

Pre-adult and adult females need to find and settle on the host fish. Once on the fish specific patterns of distribution of females enables their pheromones to be detected by males and mate location and mating to occur. Close observation studies and non-invasive monitoring conducted on lice introduced to salmon held in flumes containing circulating sea water enabled the details of mate location and mating to be fully understood. By monitoring single cohorts of lice through their life cycle (see Chapter 2, this volume, for life cycle figure) on the salmon host, it was demonstrated that protandry occurs in the male (Hull et al., 1998). Males reach the adult stage earlier than females and reach maturity while females moult to the pre-adult II stage, the stage at which most pair formation and pre-copula mate guarding occurs. The distribution of female *L. salmonis* on the host fish is related to their stage of development. Laboratory aquarium experiments showed that most mobile male and female *L. salmonis* were located on the dorsal surface of the fish with pre-adult II females mostly attaching antero dorsally, and prior to pre-copula mainly on the head (Ritchie et al., 1996). These findings were substantiated under sea cage farming conditions, where tagged individual salmon were monitored twice, 15 days apart, from within a population of lice allowed to establish for four months without treatment (Bui et al., 2020a, 2020b). By moving to the head and other dorsal anterior areas, female lice are in a position for long-distance sex pheromones to be effectively deployed. It is common during observations of mate location on salmon to observe adult male lice taking a zigzag track upstream on the fish surface towards the unmated females situated on the head of the fish (A.J. Mordue (Luntz), Aberdeen, 1997, personal communication).

Such behaviours reflect that of insects where male moths fly upwind in a pheromone plume. Once a male moth has located the pheromone plume emitted by the female it flies directly upwind towards the source, adopting a zigzag trajectory flying perpendicular to the direction of the wind (casting) if the signal is lost in order to relocate the pheromone plume and locate the female (Kennedy, 1983). Such behaviour in adult male lice suggests a similar mechanism in these animals too.

6.3.2 Mating behaviour

Flume-based observations revealed details of *L. salmonis* mating behaviour after the female had been located (Ritchie et al., 1996; Hull et al., 1998). The mating behaviour of *L. salmonis* resembles that of other parasitic copepods such as *Lepeophtheirus pectoralis* (O.F. Müller, 1776) (Anstensrud, 1990a, 1990b, 1990c, 1992). Adult male *L. salmonis* locate receptive females preferentially at the pre-adult II stage and at the virgin adult stage. They assume a pre-copula guarding position covering the dorsal genital complex of the female holding her with their second antennae. Mating takes place soon after the moult of the female to the adult stage is completed, the male maintaining his grip on the female by moving to the newly ecdysed cephalothorax of the female that has extended a thin frontal filament to ensure attachment to the fish during moulting. Mating itself is completed by the transfer of spermatophores approximately 24 h post-moult and lasts for several minutes with the male ventral to the female manipulating the genital complex of the female to maintain close contact. A post-copula position like that of the pre-copula position is taken up for several hours while the male secures the cemented spermatophores to the female. After this stage adult mated females are no longer attractive to adult males, presumably due to a switching off of pheromone production as has been reported in newly mated insects (Ritchie et al., 1996).

In the Copepoda, as in all crustaceans, the mating process is indicative of the use of pheromones and contact semiochemicals. Chemical communication between males and pre-adult and adult females was suggested for *L. pectoralis* (Anstensrud, 1990b). Mate searching and testing behaviours in the presence of pre-adult and adult females were demonstrated for *L. salmonis* males (Ritchie et al., 1996) along with clustering of males around in-copular pairs. Hull et al. (1998) showed definitively that adult male behaviour is affected by the presence of chemical cues from pre-adult and virgin adult females. Ablation of the distal tip of the first antennae of adult males significantly reduced successful mating, due to poor mate location and a lack of pair formation. It also reduced the ability of males to find a host suggesting that the effect on host and mate location would be due to a combination of a loss in perception of long-range and short-range olfactory cues together with a loss of contact stimulus perception with the removal of both aesthetascs and mechanoreceptor sensory hairs (Hull et al., 1998).

6.3.3 Sex pheromones

The role of sex pheromones in *L. salmonis* and *C. rogercresseyi* mating behaviour was investigated by Ingvarsdóttir et al. (2002b) and Pino-Marambio et al. (2007) respectively with the use of Y-tube behavioural assays. Ingvarsdóttir et al. (2002b) demonstrated and confirmed the presence of sex pheromones in pre-adult and virgin adult female *L. salmonis* in that adult males showed increased kinesis and positive rheotaxis, seen as increased locomotion and directed swimming towards the females and conditioned water. By analysing total swimming behaviour of adult male *C. rogercresseyi* for 10 min in a Y-tube apparatus (Fig. 6.1) (Pino-Marambio et al., 2007) it was demonstrated in this species also that adult males were attracted to water conditioned with juvenile and adult females.

Solid phase extraction of water conditioned with pre-adult II virgin female *L. salmonis* followed by vacuum distillation revealed that adult males responded positively to the distillate containing small, lipophilic organic molecules but not to the residue (Ingvarsdóttir et al., 2002b). This indicates that long-distance sex pheromones of sea lice are of a comparable molecular weight range and similar physical properties as volatile semiochemicals of terrestrial organisms. This discovery is unsurprising in evolutionary terms as the aquatic system of olfaction predates this system in terrestrial

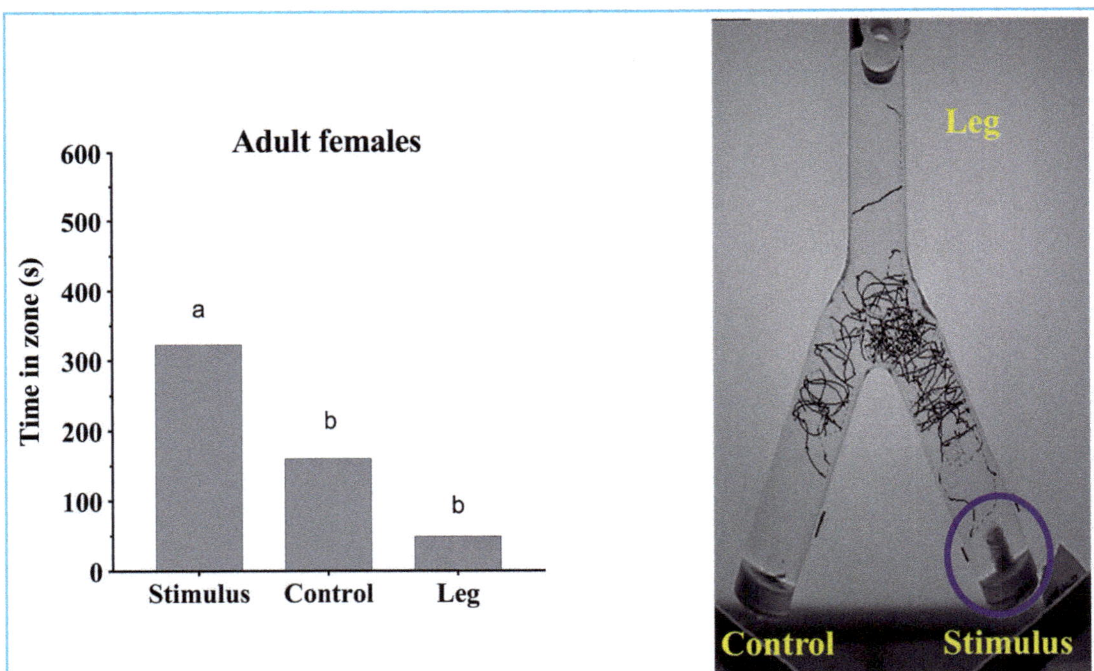

Fig. 6.1 Preference bioassays of adult males of *C. rogercresseyi* for chemical cues released by conspecific adult females. Different letters indicate significant differences based on Conover-Inman's test ($p < 0.05$). Six females were placed in the stimulus arm (in circle). (Adapted from Pino-Marambio, 2007.)

communication. The types of essentially lipophilic compounds used in both ecosystems provide similar concentrations, that is, in water determined by solubility and in air by volatility, which are in turn determined by related physical parameters (Bailey et al., 2006).

The identification of sex pheromones in crustacea has proved to be extremely difficult and lags well behind that of insects where a large number of pheromones have been identified and many used in 'push-pull' strategies for pest control. This lag has probably been aggravated by searches being made for water-soluble signals which are not, as explained above, the main types of semiochemical in aquatic ecosystems. It is hoped that by explaining this issue in this context that the search for aquatic pheromones and other semiochemicals as largely lipophilic molecular entities will be stimulated.

In larger crustaceans sex pheromones often appear to be released in the urine. Due to its release site being near the mouth, it is difficult to separate out key components from a complex mix of different molecules, and to relate these to a suitable bioassay method where mating cues are not confused with food cues. It is often difficult to distinguish between feeding, agonistic, and mating behaviours when designing bioassays and visual and tactile cues may also be of significant importance in finding a mate. Nevertheless, potential sex pheromones have been claimed as being identified (Glesson, 1982; Kamio et al., 2000, 2002; Kamio, 2009; Hardege et al., 2002, 2011; Hardege and Terschak, 2011; Zhang et al., 2020, 2011). Several authors have designed highly specific bioassays for male responses to female sex pheromones, such as males of the hermit crab *Pagurus filholi* (De Man, 1887). Males perform an assessment behaviour towards females as a preliminary step to pre-copulatory guarding (Okamura et al., 2017). Such assessment behaviour involving manipulation of the female prior to guarding is distinctive from responses of males to food cues and odours of predators. This has been used in a reliable bioassay system that has enabled the sex pheromone of *P. filholi* to be characterised as a low molecular weight polar compound of less than 1 kDa in size (Okamura et al., 2017). Further work utilising this bioassay should enable full identification and characterisation of *P. filholi* sex pheromone.

Equally, using a grasping behaviour-guided sponge assay in the crab *Telmessus cheiragonus* (Tilesius, 1815) Kamio et al. (2000) clearly indicated the presence of a sex pheromone in female urine. This specific male behaviour, whereby male crabs grasp sponges treated with pre- and post-moult female urines, was observed at significantly higher percentages than with controls, indicating that only pre- and post-moult female urines possess pheromonal activity. This study indicates that the urine of pre- and post-moult females of *T. cheiragonus* contains sex pheromone to which conspecific males responded by grasping sponges containing female urine. In addition, Kamio et al. (2002) reported a copulation pheromone in the crab *T. cheiragonus*. This pheromone was released from post-moult females for more than 21 days after moulting and evokes copulation behaviour in male crabs. The physicochemical characterisation of this pheromone indicated that it was a water-borne pheromone with a molecular weight of less than 1 kDa.

Successful mating is essential to make the right choice at the correct time. Sex pheromones play a significant role to coordinate mating behaviour. In many crustaceans, including lobsters and shrimps, reproduction starts from initial attraction to formation of a mating pair eventually leading to mating. Zhang et al. (2011) examined the hypothesis that contact pheromones on the female body surface of the shrimp *Lysmata boggessi* (Rhyne & Lin, 2006) were lipophilic in nature, with similar chemical properties to those found in insect cuticular hydrocarbon contact cues. Through chemical analyses and behavioural assays, it was shown that newly moulted *L. boggessi* contained a bouquet of lipophilic odour compounds. (Z)-9-octadecenamide was the key odour with hexadecanamide and methyl linoleate enhancing the bioactivity of the pheromone blend. However, these compounds are probably too close to primary animal metabolism to function as pheromones or other semiochemicals delivering precise behavioural cues at realistic physiological concentrations.

6.3.4 Epideictic pheromones

Other pheromones may well be involved in both mating behaviour of sea lice and in epideictic behaviours on and between host fish. In aquarium tanks male *L. salmonis* were attracted to fish infected with other male lice (Stephenson, 2012). Considerable host switching occurs from fish to fish particularly by adult males (Ritchie et al., 1996; Hull et al., 1998; Stephenson, 2012; Bui et al., 2020b). In flume conditions males switched host three and a half times as much as mated females (Hull et al., 1998). Such behaviours would aid both distribution of male lice between fish and aid the search for virgin females that had already settled on salmon.

6.3.5 Other attraction cues

In Y-tube bioassays *L. salmonis* adult males showed significantly increased activity in the presence of other males and in the presence of mated females but did not exhibit directed movement towards or away from the stimulus. Also, pre-adult II females showed a strong directional response towards adult males suggesting an attraction to them (Ingvarsdóttir et al., 2002b). These latter interactions suggest more than one type of pheromone might be involved in mate location of *L. salmonis* and in dispersal on fish, but much work remains to be done to elucidate their meaning.

6.4 Finding a host

It is essential for parasitic organisms to find their host before development can occur. Sea lice have free-living larvae which along with mobile adults have developed the ability to locate and recognise their fish host in the open environment. *L. salmonis* is an oligophagous species which feeds only on a small number of closely related salmonid fish and has developed a strong evolutionary association with its host. Use of specific olfactory cues from the host along with the physical cues of light, salinity, and water currents enable host location and landing to occur. *C. rogercresseyi* is a more generalist parasitic sea louse of the southern hemisphere but this species also has a range of host fish that support its growth and non-host fish that are avoided. *C. rogercresseyi*, initially misidentified as *Caligus flexispina* (A.G. Lewis, 1964), which is found in the oceanic islands of the Pacific Ocean, was the dominant species found in continental waters on native fish species *Eleginops maclovinus* (G. Cuvier, 1830) and *Odonthestes regia* (Humboldt, 1821). These species of fish inhabit coastal areas and move to estuarine waters for spawning, this being the reason for their close association with salmon aquaculture, where they obtain food (Carvajal et al., 1998). Salmon farming in the Pacific Ocean has enabled a close association to develop between *C. rogercresseyi* and the Atlantic salmon, *Salmo salar* L., which is similar to but more complex than the relationship of *L. salmonis* with its host (Pino-Marambio et al., 2007). These species contrast sharply with *C. elongatus*, a generalist feeder and minor pest of wild and farmed Atlantic salmon, which relies on physical rather than chemical cues to find a host (Smallman, 2009).

Many modalities apart from chemoreception play important roles in host location and other behaviours of lice. However, as this chapter is devoted to semiochemicals they will be mentioned in passing rather than extensively described. Work on the sensory biology of sea lice has shown that visual cues relating to moving shadows and light flashing off a fish as it swims by (Novales Flamarique et al., 2000; Heuch et al., 2006; Fields et al., 2018), together with physical cues such as salinity and water currents (Heuch et al., 2002; Genna et al., 2005; Bricknell et al., 2006; Heuschele and Selander, 2014), are of great importance in enabling sea lice to locate and land on their host. Visual cues operate over a spatial scale of up to tens of metres while hydrodynamic cues in the form of water currents operate over a scale of several millimetres to several centimentres (Fields et al., 2007, 2018). A spatiotemporal hierarchy of sensory input and responses can be envisaged with visual cues detected by copepodids at a distance and hydrodynamic cues at closer distances; however, such cues are non-specific in nature. Intimately involved with these physical cues are chemosensory cues, which provide specificity to the host or non-host related cues. Such olfactory cues also function at long and close distances: up to tens of metres distance at least in salmon farms where large numbers of salmon are present.

Chemoreception plays a defining role in host location, host choice, and landing behaviour in the major parasites of farmed salmon (Mordue (Luntz) and Birkett, 2009; Núñez-Acuña et al., 2014, 2019; Komisarczuk et al., 2017). Olfactory stimuli operate through the range of behaviours from general swimming activity, through activation by host cues and host searching behaviour, to landing on the fish (Fig. 6.2). It is these semiochemicals occurring in the water surrounding the fish that provide the cues leading to host location and which are important as potential candidates for 'push-pull' strategies in sea louse control.

Host location cues were first studied using adult male and female *L. salmonis* (Devine et al., 2000; Ingvarsdóttir et al., 2002a). Kinesis chambers measuring activity by use of a digital tracking system

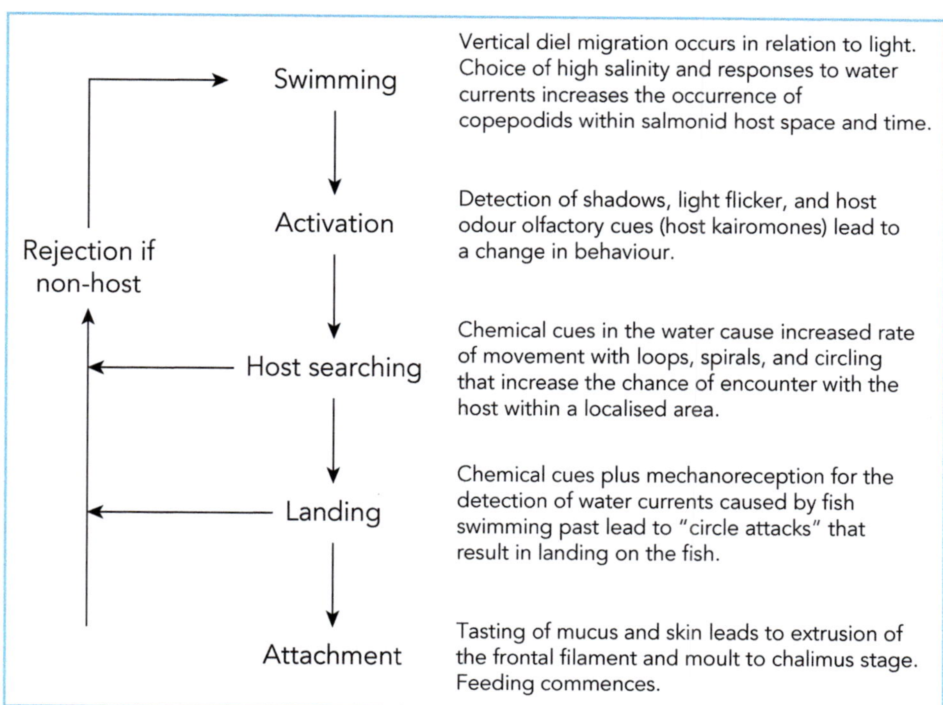

Fig. 6.2 Diagram of proposed behavioural sequence of *Lepeophtheirus salmonis* copepodid during location of a salmon host (adapted from Mordue (Luntz) and Birkett, 2009).

were used to investigate changes in behaviour when lice were presented with host related and non-host related cues. Basic principles were established using water conditioned with a variety of salmon and non-salmon cues including salmon mucus, skin, and flesh. Significant kinetic and rheotactic responses were recorded in adult male *L. salmonis* swimming activity, measured as increased speed of swimming and distance moved in the presence of chemicals from fish material that had been in artificial sea water. Electrophysiological activity in the antennal nerve of *L. salmonis* adult males was demonstrated with whole fish extract (WFE) prepared by soaking salmon flesh, skin, and mucus in filtered sea water (Fields et al., 2007).

Further work developed more realistic conditions for obtaining host location cues by using sea water in which salmon had been allowed to swim (salmon conditioned water, SCW) (Devine et al., 2000; Ingvarsdóttir et al., 2002a). By collecting sea water in which a fish of known mass was allowed to swim for a constant period of time, a standard concentration of fish olfactory cues emanating from the host could be created and used for behavioural studies, electrophysiological studies, and extraction protocols to identify the biologically active compounds within. The development of Y-tube behavioural arenas, horizontally placed for adults and vertically placed for copepodids, in which both kinesis and taxis could be measured enabled both activation and directional behaviours to be measured (Fig. 6.3). Such conditioned waters altered behaviour in the same way as before and revealed strong and significant activation and positive upstream rheotactic responses in both adult male lice and copepodids to SCW (Devine et al., 2000; Ingvarsdóttir et al., 2002a; Bailey et al., 2006). Likewise, even when in no-flow conditions where directional responses could not be measured, in SCW co-presented with a light flashing source, chemosensory signals within the SCW stimulated swimming in *L. salmonis* copepodids (Fields et al., 2018).

Solid phase extraction (SPE) systems for extracting polar, neutral, and lipophilic molecules from SCW and conditioned waters from other non-host fish species yielded components that were biologically active in behavioural assays. Vacuum distillation can split the fraction into 'volatile' and 'non-volatile' components and revealed that biological activity was found in the 'volatile' fraction of the extracts (Ingvarsdóttir et al., 2002a). The copepodids of *L. salmonis* showed that the larval stage is highly activated and strong directional responses were also seen with SCW, with an extract of SCW prepared by SPE and to a vacuum distillate of the PE extract (Bailey et al., 2006).

Coupled gas chromatography-mass spectrometry (GC-MS) of the 'volatile' fraction demonstrated these compounds as having the necessary features of relatively high lipophilicity and low water solubility and revealed several components of aliphatic and aromatic molecules such as aldehydes, ketones, phenol, and alkyl phenols. Isophorone, a catabolite of carotenoid pigments, and 1-octen-3-ol, a known component of salmon flesh, were identified as minor components and confirmed by peak enhancement on a non-polar HP-1 column (Ingvarsdóttir et al., 2002a) as was the isoprenoid oxidation product 6-methyl-5-hepten-2-one among other components in the GC-MS traces of Bailey et al. (2006) (Table 6.1). These discoveries raise another important issue often not considered for the aquatic ecosystem context, namely that pheromones and other semiochemicals are usually extremely diluted components of an animal's environment but intensely active biologically at the molecular level.

Isophorone and 1-octen-3-ol activate and cause directed movement in *L. salmonis* adult males, with isophorone eliciting strong upstream rheotaxis towards the host odour when tested at physiological concentrations (Ingvarsdóttir et al., 2002a). Electrophysiological responses were recorded from adult males to SCW, turbot conditioned water (TCW), isophorone, and 1-octen-3-ol

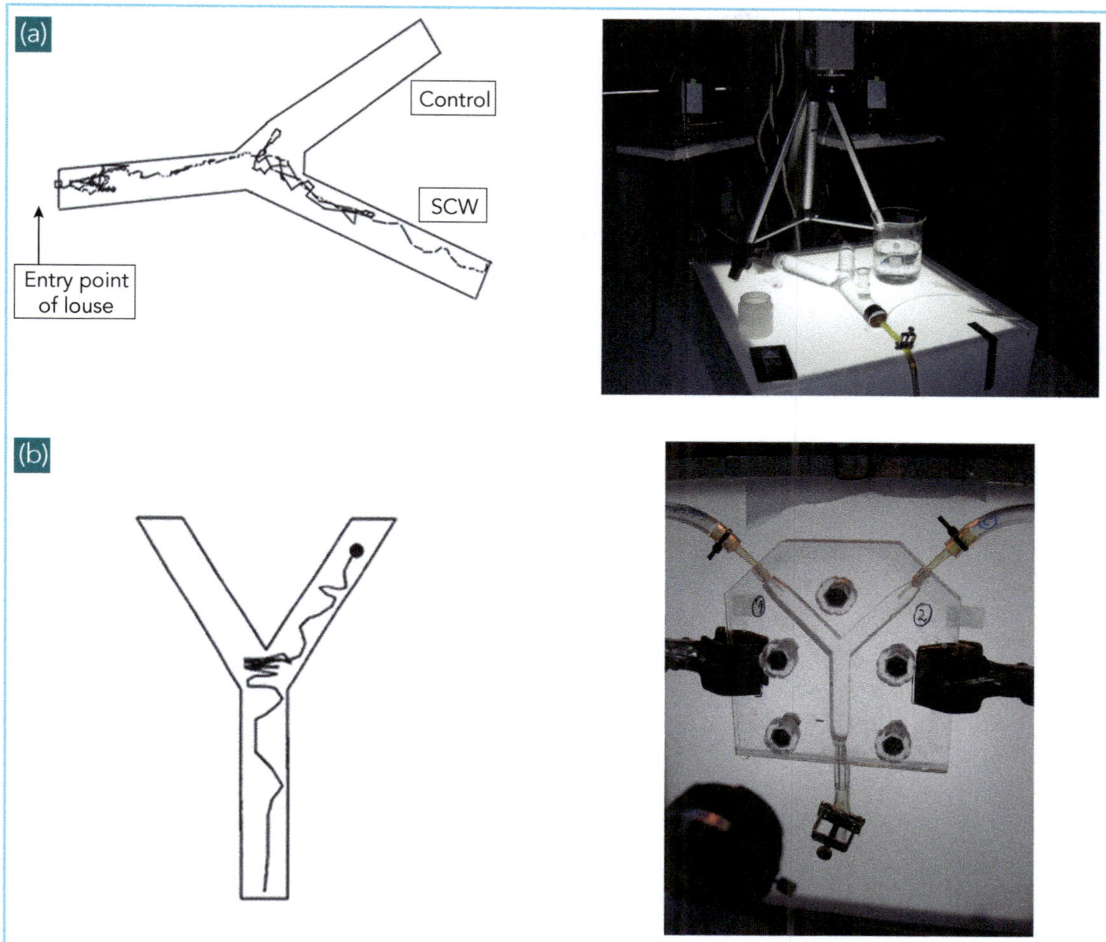

Fig. 6.3 A) Typical track of adult male *L. salmonis* and *C. rogercresseyi* in horizontal Y-tube in response to a choice of salmon conditioned water (SCW) or a sea water control. Length of the apparatus = 30 cm, flow rate of water = 30 ml/min. (Devine et al., 2000; Pino-Marambio et al., 2007). B) Vertical Y-tube arena and louse activity patterns for activation and taxis behaviour in *L. salmonis* and *C. rogercresseyi* copepodids (Devine et al., 2000; Ingvarsdóttir et al., 2002a, 2002b; Bailey et al., 2006; Pino-Marambio et al., 2007). Y-tube, moulded between two sheets of glass, total length = 14 cm, flow rate of water = 2 ml/min.

(Ingvarsdóttir et al., 2002a) confirming the biological activity of these host cues. A comparison of chemicals from both SCW and a non-host fish turbot (*Scophthalmus maximus* L.) showed that isophorone and 6-methyl-5-hepten-2-one were found specifically in the SCW and not in the TCW (Bailey et al., 2006).

Isophorone and 6-methyl-5-hepten-2-one also caused marked activation and upstream rheotaxis in copepodids (Bailey et al., 2006). In *L. salmonis* copepodids there was a noticeable dose-response effect with isophorone showing increasing activity at concentration levels of the compound found in the SCW (around 1 µg/l) (Table 6.1) (Bailey et al., 2006). This was an important finding as both behavioural responses and a dose-response relationship at concentrations that are relevant to both the emitter and recipient of an olfactory cue are key factors in defining a semiochemical.

Table 6.1 Behavioural responses of the sea louse, *L. salmonis*, to host and non-host olfactory stimuli in Y bioassays. Data from Devine et al. (2000); Ingvarsdóttir et al. (2002a); Bailey et al. (2006).

	Stimulus	Adult males			Copepodids	
		Activation	Upstream taxis	Electrophysiology response	Activation	Upstream taxis
Host cues Atlantic salmon (*Salmo salar*)	SCW	✶	✶	✶	✶	✶
	SCW extract	✶	✶		✶	✶
	Vacuum distillation volatiles	✶	✶		✶	✶
	Vacuum distillation non-volatiles	NS	NS			
	3,5,5-Trimethylcyclohex-2-en-1-one (isophorone)	✶	✶	✶	✶	✶
	1-Octen-3-ol	✶	✶			
	6-Methyl-5-hepten-2-one				✶	✶
Non-host cues turbot (*Scophthalmus maximus*)	TCW	✶	NS	✶		
	TCW extract	✶	NS			
Host cues (SCW)	SCW				✶	✶
SWC mixed with non-host cues (from turbot)	SCW + 2-Aminoacetophenone				NS	NS
	SCW + 4-Methylquinazoline				NS	NS

SCW, salmon conditioned water; TCW, turbot conditioned water. Dose response for isophorone is shown in Bailey et al., 2006.

✶: significant response at $p < 0.05 – p < 0.001$; NS: not significant; No entry: not tested.

In adult male *C. rogercresseyi* collected in fish farms, artificial sea water did not elicit any attractant or repellent responses. However, a significant attractant effect was observed, represented as the accumulative time spent (%) in the zone containing salmon odour during the Y-tube bioassays, when SCW from Atlantic salmon was used as the stimulus. This indicates that adult males of *C. rogercresseyi* can specifically identify chemical cues of their host (Pino-Marambio et al., 2007).

In a further behavioural study, a degree of plasticity in behavioural events was seen. First generation adult males of *C. rogercresseyi* reared on rainbow trout, stimulated by conditioned water of three salmonid species (rainbow trout (*Oncorhynchus mykiss* (Walbaum, 1792), Atlantic salmon, and coho salmon (*Oncorhynchus kisutch* (Walbaum, 1792)) were only attracted to rainbow trout conditioned water (Pino-Marambio et al., 2007). The individuals of each generation preferred conditioned water from the fish species they had been reared on as a stimulus source indicating the induction of prior knowledge for host preference (Pino-Marambio, 2007). Although the mechanism for this process is unknown it seems reasonable to propose an epigenetic process to explain this observation.

Similar patterns have been observed in insects where a functional memory for host preference has been demonstrated in several orders, including haematophagous Diptera, where it is known to influence a variety of behaviours, ranging from choice of mate to food location and oviposition site selection (McCall and Kelly, 2002). The F1 progeny of host conditioned mosquitoes *Culex quinquefasciatus* (Say, 1823) did not exhibit the parental preference but were as susceptible to conditioning as their parents. Rearing mosquitoes in infusions of hay or guinea pig faeces produced a similar although less dramatic change, such that the innate propensity for hay infusion could be cancelled by rearing in a guinea pig faeces infusion. The results demonstrated a change in odour preference by *C. quinquefasciatus* following exposure to the odour during development or pupal eclosion, suggesting that some form of larval or early adult imprinting has occurred (McCall and Eaton, 2001).

6.5 Avoidance of non-host

It is just as important for parasites to recognise 'non-host' as it is to recognise 'host' and how this is achieved in sea lice copepodids relates to:

1. the perception of long and short distance volatile allomone chemicals from the non-host fish that repel the lice
2. a lack of, or the masking of, the required positive 'host signal' kairomones that switch on host location behaviour.

6.5.1 Non-host cues

It is well known that copepodids respond to chemical cues from the host fish and the physical cues of water movement as the fish moves through the water resulting in increased swimming speed and directionality and when close to the fish a 'swim burst' and turn ('hooking') behaviour that enables them to land on the fish (Genna et al., 2005; Bailey et al., 2006; Heuch et al., 2006; Pino-Marambio et al., 2007; Mordue (Luntz) and Birkett, 2009; Stephenson, 2012). Landing behaviour is switched on in the presence of host kairomones from Atlantic salmon and is significantly reduced in their absence. In trials with single fish species, Atlantic salmon (*S. salar*), saithe (*Pollachius virens* L.), or Atlantic cod (*Gadus morhua* L.), the number of copepodids landing on the non-host fish saithe or cod was significantly fewer than that on Atlantic salmon (Pert et al., 2009), the lack of host kairomones and the presence of non-host allomones being of importance here.

Turbot has been used as a representative non-host fish in many behavioural studies with both adult males and copepodids of *L. salmonis*. In kinesis experiments turbot conditioned water caused an increase in activity of adult males although this was not seen in Y-tube arenas. However, in neither case did TCW initiate a directional response towards the chemical cue (Devine et al., 2000; Ingvarsdóttir et al., 2002a) (Table 6.1). That adult lice can detect such non-host cues was verified during electrophysiological recordings from male antennae revealing activity to TCW (Ingvarsdóttir et al., 2002a) (Table 6.1). Furthermore *L. salmonis* copepodids were activated by TCW but as shown in Y-tube experiments, upstream positive rheotaxis, with movement into the arm of the Y-tubes containing the fish odour, was seen when only SCW was present and was not seen when TCW along with SCW was presented (Bailey et al., 2006). Copepodids can clearly recognise non-host cues within a mixture of host and non-host odour, and this alters their behaviour to favour the host odour only. This repellence behaviour for non-host cues has also been observed in *C. rogercresseyi* where adult stages showed an avoidance effect with conditioned water of the non-host blenny, *Hypsoblennius sordidus* (E.T. Bennett, 1828) (Pino-Marambio et al., 2007).

By using various extraction methods developed for host cues, different compounds that were found specifically in the TCW extracts were identified as the nitrogen-containing lipophilic compounds, 2-aminoacetophenone (2-AA) and 4-methylquinazoline (Bailey et al., 2006). Both compounds, at realistic physiological concentrations, in the presence of SCW, eliminated activation and directional responses to the SCW (Bailey et al., 2006), strongly suggesting that the presence of non-host cues could influence attraction to host cues (Table 6.1).

These exciting results were taken forward in a small-scale field trial where 2-AA was incorporated into polymer ribbons which were tested in the field for a repellence effect on the numbers of lice settling onto farmed salmon (see section 6.6.1).

6.5.2 Plant cues

The use of intensive monocultures in agriculture facilitates infestation by parasitic insects. To prevent insect orientation towards crops, a mixture of non-host plant odours has been used to neutralise their orientation (Thiery and Visser, 1986). A similar defence mechanism had been observed in some bird species which gather fresh plant material containing secondary products that repel parasites or mask the chemicals used to find the host (Lafuma et al., 2001). The potential for developing botanically derived natural products as novel feed-through repellents for disrupting settlement of the salmon lice *L. salmonis* and *C. rogercresseyi* upon farmed Atlantic salmon, *S. salar*, was investigated using an established laboratory vertical Y-tube behavioural bioassay for assessing copepodid behaviour.

The plant materials used included oils extracted from *Allium sativum* (Amaryllidaceae), *Rosmarinus officinalis* (Lamiaceae), and *Myrica gale* (Myricaceae), and individual components obtained from plant extracts belonging to the *Brassica* genus. Plants in the family Brassicaceae contain secondary metabolites called glucosinolates (GLs) that protect against different disease agents (Hopkins et al., 2009; Jensen et al., 2010). By mechanical action (that is, chewing) or other mechanical processing, the enzyme myrosinase encounters GLs and hydrolyses them into isothiocyanates (ITCs) (Hopkins et al., 2009). These compounds function as insect deterrents and it was thought that their strong pungent flavour may also mask the host smell and obscure host recognition and/or attachment process by sea lice (Skugor et al., 2016). By adding these products to salmon conditioned water, it was shown that *L. salmonis* copepodids lost the ability to respond to salmon cues at extremely low concentrations (picograms per litre, in *L. salmonis* copepodids (O'Shea et al., 2016)). A significant reduction in attractiveness to SCW was observed at a level of around 10 picograms per litre for some compounds from the Brassicaceae. It was surmised that the combination of plant materials at very low concentrations together with SCW would disrupt *L. salmonis* settlement on a host fish (see section 6.6.2).

Caligus rogercresseyi was also exposed to three individual components obtained from plant extracts belonging to the *Brassica* genus. The behavioural response of *C. rogercresseyi* copepodids to SCW was assessed after the addition of three different masking compounds. In these Y-tube assays, the copepodids had a high activity, indicating that the masking compounds did not affect the swimming ability of the lice. Nor was any toxicity seen, confirming a semiochemical role rather than a physiologically based aversion. An Index of Preference (IP = number of visits in stimulus zones/number of visits in the control zone) was calculated to observe if fish conditioned water could elicit any attractant or repellent responses in copepodids of *C. rogercresseyi*. IP = 1 indicated neither avoidance nor preference of the copepodids; IP > 1 indicated that there was a preference for the stimulus; and IP < 1 indicated avoidance of the stimulus or a preference for the control. The IP calculated for the three masking compounds showed that the two highest concentrations (0.01 and 1.0 mg ml^{-1}) reduced the preference of the copepodids for Atlantic salmon. The IP showed that masking compounds were effective in their action of masking the chemical cues from Atlantic salmon (Wadsworth et al., 2009). Subsequent field trials with very low levels of the plant material in the salmon's diet were conducted to see if it was effective in preventing lice settlement under farmed conditions (see section 6.6.2).

6.6 'Push-pull' strategies in sea lice control

The control of sea lice infections of farmed salmon ultimately depends upon a sophisticated integrated pest management (IPM) strategy. The idea of an IPM strategy for sea lice was initiated in Scotland in the 1990s involving universities, industry, and the Scottish Salmon Producers' Organisation (SSPO). This came into the public domain in the conference held in 2001 at the University of Aberdeen, Scotland, in which management practices, the role of veterinary medicines, and biological and behavioural approaches to control were discussed and which laid the foundations for the implementation of IPM strategies for sea lice of farmed salmon (Mordue (Luntz) and Pike, 2002). Since that time many measures have been tested and it is now beginning to be realised that much more complex and simultaneous management strategies are required for the control of sea lice in commercial sea cages. The effectiveness of long-term multiple simultaneous treatments has been demonstrated (Bui et al., 2020b) and the idea of aiming for prevention of infestations rather than treatment after infection is starting to be explored (Barrett et al., 2020). Such approaches promote integrated pest management thinking which requires a proactive approach to predict, prevent, and sustainably treat infestations. The use of 'preventative methods' together with sustainable treatments when necessary is the way forward.

Salmon lice have been a problem for the Atlantic salmon and rainbow trout farming industries in Europe and the Americas for many years. *L. salmonis* infestations are a part of the normal salmon life cycle in sea water on wild and farmed salmon in the Atlantic. The problematic species in South America is *C. rogercresseyi*. Recent discoveries have enhanced knowledge of chemical communication in the salmon louse and there is a potential for semiochemicals to form a component of an IPM strategy. However, the chemical ecology of sea lice is not fully understood, and the lack of knowledge makes semiochemical-based approaches to control less attractive when compared to traditional veterinary medicine treatments. However, such alternative strategies do have potential for the control of salmon lice infestations and would include:

1 the disruption of copepodid host location by the masking of salmon odours
2 using masking compounds in feed
3 the use of synthetic mixtures of pheromones or other attractive volatiles to trap adults and/or copepodids from the close environment.

6.6.1 'Push' approaches – non-host cues

Hastie et al. (2013) reported on three field-based studies that investigated the potential to interfere with natural rates of lice infection in the marine environment, using slow-release formulations of 2-aminoacetophenone (2-AA), identified as one of the non-host recognition semiochemicals from *S. maximus*. Chemical grade 2-AA was melt-formulated into PVC rope to provide a slow and consistent release rate of 0.48 mg/g/day. PVC rope ribbons were fitted onto cylindrical sentinel pens (H =1 m, dia. = 0.9 m) with and without the slow-release formulation of 2-AA. The fish were exposed to challenges of sea lice that occurred naturally in the loch system. Three separate trials were conducted. Trial one involved one long line in loch systems, whereas trials two and three were sea site challenge trials where the sentinel cages were allocated inside production cages (Fig. 6.4).

The first experiment (trial one) showed a 43% reduction in overall prevalence of lice-infected fish in the 2-AA treatment pens. Mean loads of 1.19 lice/fish and 0.52 lice/fish were recorded in the control and treatment pens, respectively (Fig. 6.5).

In trial two, after 11 days the control group was 32% infected (37/116 fish), while the treated group was 12% infected (14/117 fish). In trial three, overall the control and treated groups were 24% and 14% infected, respectively. This infestation represented a 62% (trial two) and 43% (trial three) reduction in overall prevalence of lice-infected fish in the 2-AA treatment pens, respectively. No deleterious effects of 2-AA on fish growth, behaviour, or food intake were observed (Fig. 6.6).

6.6.2 'Push' approaches – masking compounds in feed

Under a controlled infestation trial, a semiochemical masking compound (feed material source of glucosinolates) was added to a conventional feed (Skugor et al., 2016). This feed was used during a pre-infestation period of three weeks before a sea lice challenge with copepodids of *C. rogercresseyi*. During this period, each group was fed with differential diets of 0%, 1%, 2%, or 3% of masking compounds, and 3% of masking compound in combination with an immunostimulant. Fish from each treatment were challenged with *C. rogercresseyi* after 21 days. Fish in all the tested diets with masking compounds showed a decrease of lice infestation, with some groups showing up to a 25% decrease (Núñez-Acuña et al., 2014).

Similar approaches were used to control *L. salmonis* infestations. Masking compounds in the diet (glucosinolates from the Brassicaceae) were used to investigate if Atlantic salmon fed two different doses of masking compounds in the feed would be protected against lice infection. A 25% reduction

Fig. 6.4 Field-based study at Marine Harvest Ardnish, Loch Ailort, Scotland, to investigate the interference of natural rates of sea louse infection in the marine environment, using slow release formulations of 2-aminoacetophenone (2-AA), a non-host cue from turbot, *S. maximus*. Twelve sentinel pens (six treatments (T) and six controls (C)) were placed in a floating long line system (trial one) or in active cages (trials two and three. Design not shown) in which 20 uninfected smolts were placed. Polymer ribbons (dark blue) impregnated with 2-AA with a release rate of 0.48 mg/g/day were attached to the pens. Fish were killed and lice numbers recorded after 39 days (trial one) and 11 days (trials two and three). (From the work of Hastie et al., 2013.)

in lice counts was obtained in the low dose group and a 17% reduction in the high dose group compared to fish fed control feed (Jodaa Holm et al., 2016).

The successes of both non-host cues in salmon pens and plant derived compounds in the diet of salmon leading to a significant reduction of lice on fish of between 17% and 63% is important and should not be underestimated. These figures, as part of several measures to reduce lice numbers to acceptable levels, could be highly significant.

6.6.3 'Pull' components – traps

In insects, monitoring of pest species can involve direct sampling of the commodity itself using visual inspection, general traps to determine if the patch is infested, or direct and specific sampling of the insects dispersing among resource patches using tools such as pheromone traps (Trematerra, 2013). Similarly, semiochemicals derived from salmon may prove useful for the development of odour traps for monitoring lice numbers. Host kairomones such as the already identified host cue isophorone could be used in sea louse traps as could sex pheromones once they have been identified. Further work to define other host kairomonal components and sea lice sex pheromones require to

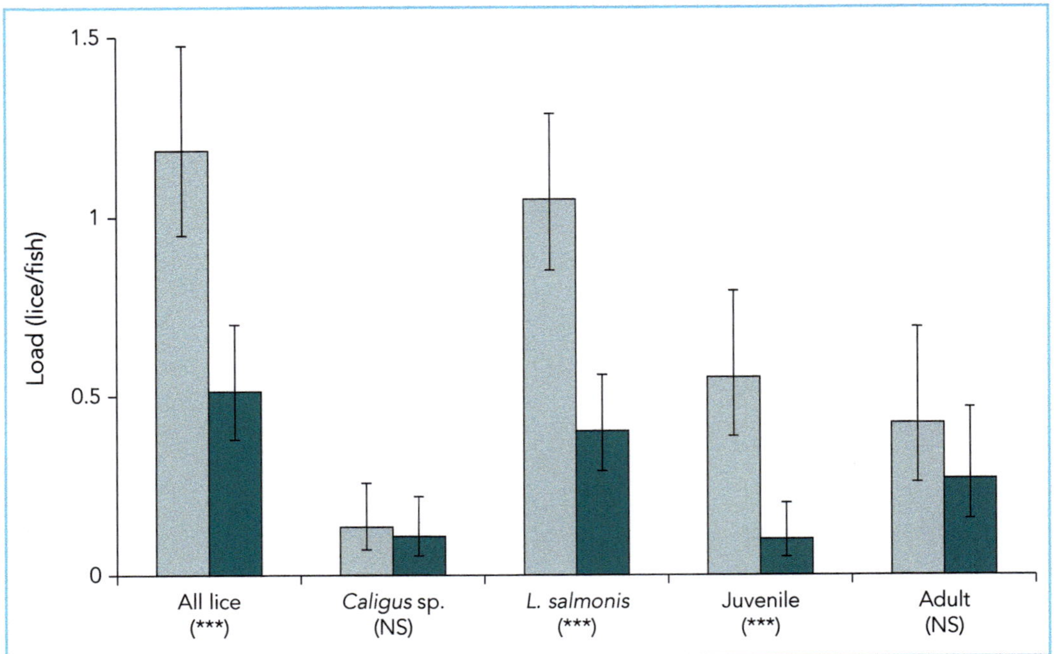

Fig. 6.5 Trial one: Back-transformed predicted mean infection loads (lice/fish) estimated using GLMMs. Control and treatment pen values are represented by light and dark green columns, respectively. 95% CLs are also displayed. Mean loads of juvenile (chalimus, pre-adult) and adult (male, female, gravid) stages of *L. salmonis* are provided in the last two sets of columns on the right (significance: ***p < 0.001, NS = not significant). (From Hastie et al., 2013.)

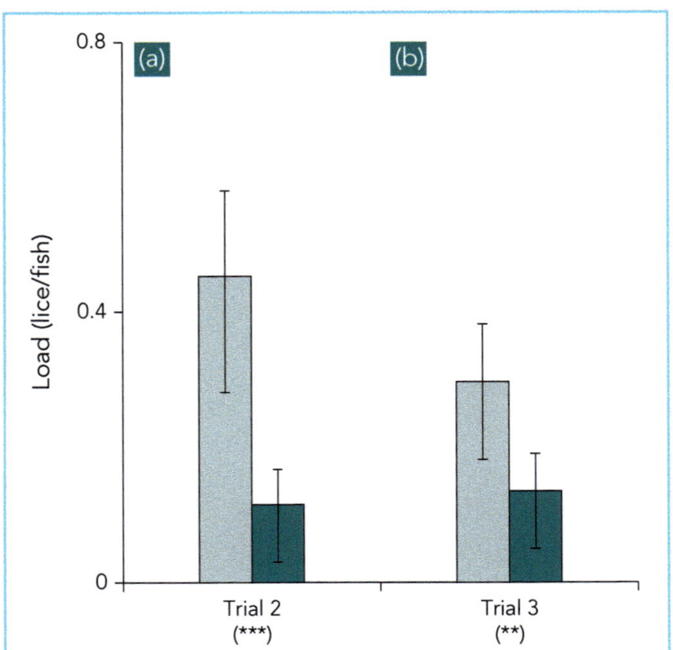

Fig. 6.6 Trials two and three: Back-transformed predicted mean infection loads (lice/fish) estimated using GLMMs. Control and treatment pen values are represented by light and dark green columns, respectively. 95% CLs are also displayed. Mean loads of *L. salmonis* (all stages) are provided (significance: ***p < 0.001, *p < 0.01). (From Hastie et al., 2013.)

be conducted along with strategies to develop traps for sea lice that would be effective in the field. It would be important to combine host kairomones and pheromones together with the physical cues of shadows and light flicker and place them in the context of a moving target to harness all relevant host location cues to effectively corral and remove the parasite population.

6.6.4 'Push-pull' strategies

As already widely demonstrated a 'push-pull' approach in insect pest control is highly successful in resource-poor agricultural food production. The 'push-pull' platform, based on local plants, involves intercropping cereal crops with a forage legume, Fabaceae, and planting a forage grass as a border crop, resulting in substantial grain yield increases, and resulting in over 80% reduction in lepidopterous stemborer infestation of maize (Khan et al., 2014). Protection is given by semiochemicals that are released by the intercrop that repel (push away) the pests while at the same time attracting arthropod parasites, while those released by the border crop attract (pull) the pests to them. A reduction of 82.7% in the average number of larvae per plant and 86.7% in plant damage per plot were observed using this technology of 'push-pull' compared to maize monocrop plots (Khan et al., 2018). This 'push-pull' approach has been readily developed for drought tolerant crops and newly emerging pests (Khan et al., 2018; Midega et al., 2018). It has been demonstrated in other intensive terrestrial livestock farming systems that the application of insect vector behaviour modification to prevent an arthropod entering a space occupied potentially by livestock reduces encounters between the host and vectors. Host orientated behaviour of mosquitoes, ticks, and tsetse flies (*Glossina* species) is not only dependent on kairomones from preferred hosts; there is evidence also of their avoiding non-host individuals by distinguishing them from hosts using specific chemical odours. This evidence was demonstrated when repellents from unpreferred hosts prevented contact between host and vector. Saini et al. (2017) developed a slow-release device (as a neckband) that deployed repellents from water buck, a non-host member of the Bovidae family, that convert cattle into non-hosts for tsetse flies: 'cows in water buck clothing'. This development can be applied as an IPM strategy (push-pull). Thus, the rationale for the 'push-pull' treatments involved slow-release devices (containing spatial repellents) to 'push' the vector away and an insecticide-impregnated visually and olfactorily attractive target to 'pull' and kill the flies (removal trapping). The 'push' comprised the use of recognition semiochemicals from the water buck, which is not considered a host for these tsetse flies. The pull used colour and cow attractants plus deltamethrin for tsetse destruction (Saini et al., 2017). A similar 'push-pull' system to the one against the haematophagous insects that bite livestock was proposed recently for control of malaria mosquitoes that bite humans.

Some of the components of a 'push-pull' strategy for sea lice of farmed salmon have now been established. The use of non-host cues to repel lice is a major advancement in demonstrating the 'push' component of the 'push-pull' in the field (that is, salmon farming pens). Attractants such as host location cues have already been identified to provide the 'pull' and at which destruction of the mobile copepodids of *L. salmonis* needs to take place. Preliminary studies have been made with collaborators at the University of Maine (I. Bricknell group) on a novel idea whereby destruction allows the copepodids to be consumed by farmed mussels (Molloy et al., 2011). Thus, the plan is to use a slow-release formulation of the *L. salmonis* host attractant semiochemicals described above to be released on the off fish pen side of mussel farming ropes taking account of prevailing water currents. The copepodids are then attracted into the mussels and eaten.

6.7 Conclusions

The problems of resistance of sea lice to currently used chemical treatments and the limitation of effective alternative treatments suggest the need for additional methods and the implementation of an integrated strategy for a reduction in sea lice infestations of salmonid species. There are advantages in using effective treatments against all stages of the louse life cycle, and so preventing reproduction and the build-up of populations. This coupled with simultaneous treatments on all farms, such as rest periods or rotation, could reduce louse prevalence and the frequency of chemical treatments.

In this sense, semiochemicals could play an important role in the control of sea lice. Such strategies are based on ethological control methods which cause behavioural changes in the louse. The use of chemical cues would enable all stages of the louse life cycle to be affected leading to the avoidance of encounter with their host, food source, and mating site (Fig. 6.7). The rotation between controlled chemical treatments and a strategy with semiochemicals within an integrated pest management system would increase the efficacy against infestation, as occurs in the control of agricultural pests. These semiochemicals including sex pheromones could be used in integrated pest management, known as 'push-pull' strategies.

Preventing sea lice from entering fish farms in the first place minimises the need for treatments. The identification of sex pheromones and/or aggregation pheromones that mediate the location of potential mates and their use in traps would allow an interruption of the life cycle. By attracting males to traps containing pheromones, a reduction in mating success would ensue with resultant reduced build-up in population levels. In addition, the sensory ability of all stages of lice to locate and identify their hosts could be disrupted. Identified host attractant compounds produced by susceptible species could be used in traps along with the universal attractant of light. Indeed, the use of light systems such as 'Blue Lice' in combination with attractants and other

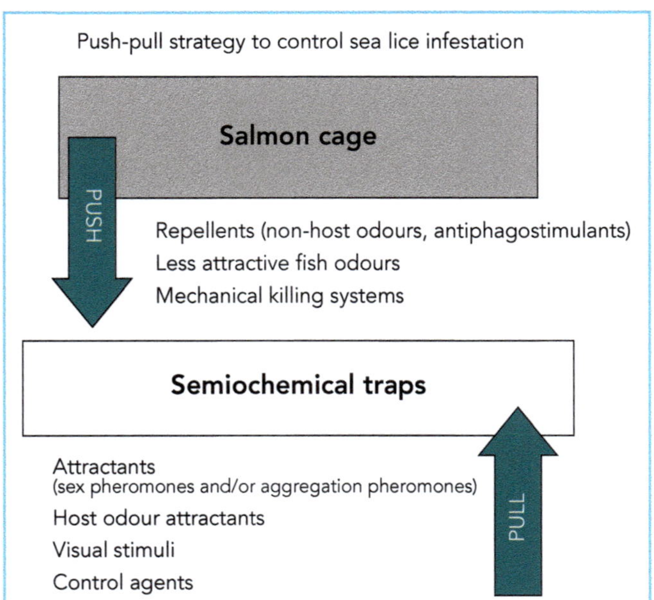

Fig. 6.7 The deployment of semiochemicals as part of a 'push-pull' strategy for control of sea lice (adapted from Mordue (Luntz) and Birkett, 2009).

measures such as sea lice skirts would keep the numbers of sea lice below permitted limits (Blue Lice, 2017).

The use of traps with attractants from more susceptible fish, sex or aggregation pheromones, and visual stimuli would be complemented by the use of less attractive fish odours and/or the use of identified repellents such as non-host fish signals and anti-phagostimulants introduced into the salmon cage. This strategy in combination with other non-chemical technologies would support infestation control. To control lice inside cages, sentinel cages could be installed to attract lice and kill them using chemotherapeutants, mechanical systems, or optical delousing systems with laser such as the stingray system (Stingray Marine Solutions, 2010; Bui et al., 2020a). The latter used in combination with semiochemicals could improve the encounter rate with lice using attractants.

Of course, overuse of semiochemicals in fish lice control will inevitably lead to genetic selection of resistance traits in the lice. However, rather than having to produce a novel veterinary medicine as for conventional control, the lice, needing to preserve the associated behavioural trait for survival, will simply be selected for use of new semiochemicals. Therefore, by the same approach as was applied to the identification of the initially used semiochemicals, the new semiochemicals used by the resistant lice will be identified. The concept of overcoming resistance to pheromone and other semiochemical-based pest control systems has been reviewed as the basis for using evolutionary processes for overcoming resistance to conventional control agents such as insecticides which relates, of course, directly to the veterinary medicines deployed against fish lice (Pickett and Weston, 2018).

References

Anstensrud, M. (1990a) Mating strategies of two parasitic copepods [(*Lernaeocera branchialis* (L.) (Pennellidae) and *Lepeophtheirus pectoralis* (Müller) (Caligidae)] on flounder: polygamy, sex-specific age at maturity and sex ratio. *Journal of Experimental and Marine Biology and Ecology* 136, 141–158.

Anstensrud, M. (1990b) Effects of mating on female behaviour and allometric growth in the two parasitic copepods *Lernaeocera branchialis* (L. 1767) (Pennellidae) and *Lepeophtheirus pectoralis* (Müller, 1776) (Caligidae). *Crustaceana* 59, 245–258.

Anstensrud, M. (1990c) Moulting and mating in *Lepeophtheirus pectoralis* (Copepoda: Caligidae). *Journal of the Marine Biological Association of the United Kingdom* 70, 269–281.

Anstensrud, M. (1992) Mate guarding and mate choice in two copepods, *Lernaeocera branchialis* (L.) (Pennellidae) and *Lepeophtheirus pectoralis* (Müller) (Caligidae), parasitic on flounder. *Journal of Crustacean Biology* 12, 31–40.

Bailey, R., Birkett, M., Ingvarsdóttir, A., Mordue (Luntz), A., O'Shea, B., Pickett, J. and Wadhams, L. (2006) The role of semiochemicals in host location and non-host avoidance by salmon louse (*Lepeophtheirus salmonis*) copepodids. *Canadian Journal of Fisheries and Aquatic Sciences* 63, 448–456.

Barrett, L.T., Oppedal, F., Robinson, N. and Dempster, T. (2020) Prevention not cure: a review of methods to avoid sea lice infestations in salmon aquaculture. *Reviews in Aquaculture* 12, 2527–2543.

Blue Lice (2017) A sustainable approach to sea lice prevention. Available at: https://www.bluelice.no/norsk (accessed 30 January 2022).

Breithaupt, T. and Thiel, M. (eds) (2011) *Chemical Communication in Crustaceans*. Springer, New York.

Bricknell, I.R., et al. (2006) Effect of environmental salinity on sea lice *Lepeophtheirus salmonis* settlement success. *Diseases of Aquatic Organisms* 71(3), 201–212.

Bui, S., Geitung, L., Oppedal, F. and Barrett, L.T. (2020a) Salmon lice survive the straight shooter: a commercial scale sea cage trial of laser delousing. *Preventive Veterinary Medicine* 181, 105063.

Bui, S., Stien, L.H., Nilsson, J., Trengereid, H. and Oppedal, F. (2020b) Efficiency and welfare impact of long-term simultaneous in situ management strategies for salmon louse reduction in commercial sea cages. *Aquaculture* 520, 734934.

Carvajal, J., González, L. and George-Nascimento, M. (1998) Native sea lice (Copepoda: Caligidae) infestation of salmonids reared in netpen systems in southern Chile. *Aquaculture* 166, 241–246.

D'Onofrio, C., Zaremska, V., Zhu, J., Knoll, W. and Pelosi, P. (2020) Ligand-binding assays with OBPs and CSPs. *Methods in Enzymology* 642, 229–258.

Derby, C.D., Kozma, M.T., Senatore, A. and Schmidt, M. (2016) Molecular mechanisms of reception and perireception in crustacean chemoreception: a comparative review. *Chemical Senses* 41, 381–398.

Devine, G., Ingvarsdóttir, A., Mordue, W., Pike A.W, Pickett, J.A., Duce, I. and Mordue (Luntz), A.J. (2000) Salmon lice, *Lepeophtheirus salmonis*, exhibit specific chemotactic responses to semiochemicals originating from the salmonid, *Salmo salar*. *Journal of Chemical Ecology* 26, 1833–1847.

El-Sayed, A. (2019) Pherobase: Semiochemicals – Index. Available at: http://www.pherobase.com/database/compound/compounds-index.php (accessed 2 May 2021).

Fields, D.M., Weissburg, M.J. and Browman, H. (2008) Chemoreception in the salmon louse *Lepeophtheirus salmonis*: an electrophysiology approach. *Diseases of Aquatic Organisms* 78, 161–168.

Fields, D.M., Skiftesvik, A.B. and Browman, H.I. (2017) Behavioural responses of infective-stage copepodids of the salmon louse (*Lepeophtheirus salmonis*, Copepoda: Caligidae) to host-related sensory cues. *Journal of Fish Diseases* 41, 875–884.

Genna, R.L., Mordue, W., Pike, A.W. and Mordue (Luntz) A.J. (2005) Light intensity, salinity, and host velocity influence presettlement intensity and distribution on hosts by copepodids of sea lice, *Lepeophtheirus salmonis*. *Canadian Journal of Fisheries and Aquatic Sciences* 62(12), 2675–2682.

Glesson, R.A. (1982) Morphological and behavioral identification of the sensory structures mediating pheromone reception in the blue crab, *Callinectes sapidus*. *Biological Bulletin* 163, 162–171.

Hara, T.J. (2011) Smell, taste, and chemical sensing | Chemoreception (smell and taste): An introduction. *Encyclopedia of Fish Physiology*. Academic Press, Cambridge, Massachusetts. pp. 183–186.

Hardege, J.D. and Terschak, J.A. (2011) Identification of crustacean sex pheromones. In: Breithaupt, T. and Thiel, M. (eds) *Chemical Communication in Crustaceans*. Springer, New York. pp. 373–392.

Hardege, J.D., Jennings, A., Hayden, D., Müller, C.T., Pascoe, D., Bentley, M.G. and Clare, A.S. (2002) Novel behavioural assay and partial purification of a female-derived sex pheromone in *Carcinus maenas*. *Marine Ecology Progress Series* 244, 179–189.

Hardege, J.D., Bartels-Hardege, H.D., Fletcher, N., Terschak, J.A., Harley, M., Smith, M.A., Davidson, L., Hayden, D., Müller, C.T., Lorch, M., Welham, K., Walther, T. and Bublitz, R. (2011) Identification of a female sex pheromone in *Carcinus maenas*. *Marine Ecology Progress Series* 436, 177–189.

Harzsch, S. and Krieger, J. (2018) Crustacean olfactory systems: a comparative review and a crustacean perspective on olfaction in insects. *Progress in Neurobiology* 161, 23–60.

Hastie, L.C., Wallace, C., Birkett, M.A., Douglas, A., Jones, O., Mordue (Luntz), A.J., Ritchie, G., Pickett, J.A., Webster, J.L. and Bowman, A.S. (2013) Prevalence and infection intensity of sea lice (*Lepeophtheirus salmonis*) on Atlantic salmon (*Salmo salar*) host is reduced by the non-host compound 2-aminoacetophenone. *Aquaculture* 410–411, 179–183.

Heuch, P., Knutsen, J. and Schram, T. (2002) Salinity and temperature effects on sea lice over-wintering on sea trout (*Salmo trutta*) in coastal areas of the Skagerrak. *Journal of the Marine Biological Association of the United Kingdom* 82, 887–892.

Heuch, P.A., Doall, M.H. and Yen, J. (2006) Water flow around a fish mimic attracts a parasitic and deters a planktonic copepod. *Journal of Plankton Research* 29, i3–i16.

Heuschele, J. and Selander, E. (2014) The chemical ecology of copepods. *Journal of Plankton Research* 36(4), 895–913.

Hopkins, R.J., Van Dam, N.M. and Van Loon, J.J.A. (2009) Role of glucosinolates in insect-plant relationships and multitrophic interactions. *Annual Review of Entomology* 54, 57–83.

Hull, M.Q., Pike, A.W., Mordue (Luntz), A.J. and Rae, G.H. (1998) Patterns of pair formation and mating in an ectoparasitic caligid copepod *Lepeophtheirus salmonis* (Krøyer 1837): implications for its sensory and mating biology. *Philosophical Transactions of the Royal Society B: Biological Sciences* 353, 753–764.

Ingvarsdóttir, A., Birkett, M.A, Duce, I., Genna, R.L., Mordue, W., Pickett, J.A., Wadhams, L.J. and Mordue (Luntz) A.J. (2002a) Semiochemical strategies for sea louse control: host location cues. *Pest Management Science* 58, 537–545.

Ingvarsdóttir, A., Birkett, M.A., Duce, I., Mordue, W., Pickett, J.A., Wadhams, L.J., and Mordue (Luntz), A.J. (2002b) Role of semiochemicals in mate location by the parasitic sea louse, *Lepeophtheirus salmonis. Journal of Chemical Ecology* 28, 2107–2017.

Jensen, J., Styrishave, B., Gimseng, A.L. and Hansen, H. (2010) The toxic effects of benzyl glucosinolate and its hydrolysis product, the biofumigant benzyl isothiocyanate, to *Folsomia fimetaria. Environmental Toxicology and Chemistry* 29(2), 359–364.

Jimenez Benavente, H.A. (2017) Identification and in silico structural characterization of a chemosensory protein potentially related to chemical communication in salmon louse, *Caligus rogercresseyi*. Master's thesis, Universidad de La Frontera, Chile.

Jodaa Holm, H., Wadsworth, S., Bjelland, A.K., Krasnov, A., Evensen, O. and Skugor, S. (2016) Dietary phytochemicals modulate skin gene expression profiles and result in reduced lice counts after experimental infection in Atlantic salmon. *Parasites & Vectors* 9, 271.

Kamio, M. (2009) Toward identifying sex pheromones in blue crabs. Using biomarker targeting within the context of evolutionary chemical ecology. *Annals of the New York Academy of Sciences* 1170(1), 456–461.

Kamio, M., Matsunaga, S. and Fusetani, N. (2000) Studies on sex pheromones of the helmet crab, *Telmessus cheiragonus* 1. An assay based on precopulatory mate-guarding. *Zoological Science* 17, 731–733.

Kamio, M., Matsunaga, S. and Fusetani, N. (2002) Copulation pheromone in the crab *Telmessus cheiragonus* (Brachyura: Decapoda). *Marine Ecology Progress Series* 234, 183–190.

Karlson, P. and Lüscher, M. (1959) "Pheromones": a new term for a class of biologically active substances. *Nature* 183, 55–56.

Kennedy, J.S. (1983) Zigzagging and casting as a programmed response to wind-borne odour: a review. *Physiological Entomology* 8, 109–120.

Khan, Z.R., Midega, C.A.O., Pittchar, J.O., Murage, A.W., Birkett, M.A., Bruce, T.J.A. and Pickett, J.A. (2014) Achieving food security for one million sub-Saharan African poor through push-pull innovation by 2020. *Philosophical Transactions of the Royal Society B: Biological Sciences* 369.

Khan, Z.R., Pittchar, J.O., Midega, C.A.O. and Pickett, J.A. (2018) Push-pull farming systems controls fall armyworm: lessons from Africa. *Outlooks on Pest Management* 29, 220–224.

Komisarczuk, A.Z., Grotmol, S. and Nilsen, F. (2017) Ionotropic receptors signal host recognition in the salmon louse (*Lepeophtheirus salmonis*, copepoda). *PLoS ONE* 12(6), 1–18.

Lafuma, L., Lambrechts, M.M. and Raymond, M. (2001) Aromatic plants in bird nests as a protection against blood-sucking flying insects? *Behavioural Processes* 56, 113–120.

McCall, P.J. and Eaton, G. (2001) Olfactory memory in the mosquito *Culex quinquefasciatus. Medical and Veterinary Entomology* 15, 197–203.

McCall, P.J. and Kelly, D.W. (2002) Learning and memory in disease vectors. *Trends in Parasitology* 18, 429–433.

Midega, C.A.O., Pittchar, J.O., Pickett, J.A., Hailu, G.W. and Khan, Z.R. (2018) A climate-adapted push-pull system effectively controls fall armyworm, *Spodoptera frugiperda* (J.E. Smith), in maize in East Africa. *Crop Protection* 105, 10–15.

Molloy, S.D., Pietrak, M.R., Bouchard, D.A. and Bricknell, I. (2011) Ingestion of *Lepeophtheirus salmonis* and the blue mussel *Mytilus edulis. Aquaculture* 311, 61–64.

Mordue (Luntz), A.J. (2003) Arthropod semiochemicals: mosquitoes, midges and sealice. *Biochemical Society Transactions* 31, 128–133.

Mordue (Luntz), A.J. and Birkett, M.A. (2009) A review of host finding behaviour in the parasitic sea louse, *Lepeophtheirus salmonis* (Caligidae: Copepoda). *Journal of Fish Diseases* 32, 3–13.

Mordue (Luntz), A.J. and Pike, A.W. (2002) Proceedings of SCI Symposium: Salmon Farming: towards an integrated pest management strategy for sea lice. In: Mordue (Luntz), A.J. and Pike, A.W. (eds) *Pest Management Science* 58(6), John Wiley & Sons, Ltd, Aberdeen, UK.

Nordlund, D.A. and Lewis, W.J. (1976) Terminology of chemical releasing stimuli in intraspecific and interspecific interactions. *Journal of Chemical Ecology* 2, 211–220.

Novales Flamarique, I., Browman, H.I., Bélanger, M. and Boxaspen, K. (2000) Ontogenetic changes in visual sensitivity of the parasitic salmon louse *Lepeophtheirus salmonis*. *Journal of Experimental Biology* 203, 1649–1657.

Núñez-Acuña, G., Valenzuela-Muñoz, V., Pino Marambio, J., Wadsworth, S. and Gallardo-Escárate, C. (2014) Insights into the olfactory system of the ectoparasite *Caligus rogercresseyi*: molecular characterization and gene transcription analysis of novel ionotropic receptors. *Experimental Parasitology* 145, 99–109.

Núñez-Acuña, G., Gallardo-Escárate, C., Skiftesvik, A.B., Fields, D.M. and Browman, H.I. (2019) Silencing of ionotropic receptor 25a decreases chemosensory activity in the salmon louse *Lepeophtheirus salmonis* during the infective stage. *Gene* 697, 35–39.

Nyasembe, V.O. and Torto, B. (2014) Volatile phytochemicals as mosquito semiochemicals. *Phytochemistry Letters* 8, 196–201.

O'Shea, B., Wadsworth, S., Pino Marambio, J., Birkett, M.A., Pickett, J.A. and Mordue (Luntz), A.J. (2016) Disruption of host-seeking behaviour by the salmon louse, *Lepeophtheirus salmonis*, using botanically derived repellents. *Journal of Fish Diseases* 40, 1–11.

Okamura, S., Kawaminami, T., Matsuura, H., Fusetani, N. and Goshima, S. (2017) Behavioral assay and chemical characters of female sex pheromones in the hermit crab *Pagurus filholi*. *Journal of Ethology* 35, 169–176.

Pelosi, P., Iovinella, I., Felicioli, A. and Dani, F.R. (2014) Soluble proteins of chemical communication: an overview across arthropods. *Frontiers in Physiology* 5, 320.

Pert, C.C., Mordue (Luntz), A.J., Fryer, R.J., O'Shea, B. and Bricknell, I.R. (2009). The settlement and survival of the salmon louse, *Lepeophtheirus salmonis* (Krøyer, 1837), on atypical hosts. *Aquaculture* 288, 321–324.

Pickett, J.A. and Weston, L.A. (2018) Possibilities for rationally exploiting co-evolution in addressing resistance to insecticides, and beyond. *Pesticide Biochemistry and Physiology* 151, 18–24.

Pino-Marambio, J. (2007) Semioquímica que media la interaccion inter- e intra- especifica del copépodo ectoparásito de especies salmonideas *Caligus rogercresseyi* (Copepoda: Caligidae). PhD thesis, Universidad de La Frontera, Chile.

Pino-Marambio, J., Mordue (Luntz), A.J., Birkett, M., Carvajal, J., Asencio, G., Mellado, A. and Quiroz, A. (2007) Behavioural studies of host, non-host and mate location by the sea louse, *Caligus rogercresseyi* (Boxshall & Bravo 2000) (Copepoda: Caligidae). *Aquaculture* 271, 70–76.

Pitts, R.J., Mozūraitis, R., Gauvin-Bialecki, A. and Lempérière, G. (2014) The roles of kairomones, synomones and pheromones in the chemically-mediated behaviour of male mosquitoes. *Acta Tropica* 132, S26–34.

Ritchie, G., Mordue (Luntz), A.J., Pike, A.W. and Rae, G.H. (1996) Observations on mating and reproductive behaviour of *Lepeophtheirus salmonis*, Krøyer (Copepoda: Caligidae). *Journal of Experimental Marine Biology and Ecology* 201, 285–298.

Saini, R.K., Orindi, B.O., Mbahin, N., Andoke, J.A., Muasa, P.N., Mbuvi, D.M., Muya, C.M., Pickett, J.A. and Borgemeister, C.W. (2017) Protecting cows in small holder farms in East Africa from tsetse flies by mimicking the odor profile of a non-host bovid. *PLoS Neglected Tropical Diseases* 11(10), e0005977.

Sánchez-Gracia, A., Vieira, F.G. and Rozas, J. (2009) Molecular evolution of the major chemosensory gene families in insects. *Heredity* 103, 208–216.

Schmidt, M. and Mellon, D. (2011) Neuronal processing of chemical information in crustaceans. In: Breithaupt, T. and Thiel, M. (eds) *Chemical Communication in Crustaceans*. Springer Science & Business, Berlin, Germany. pp. 123–147.

Skugor, S, Holm, H.J., Bjelland, A.K., Pino, J., Evensen, O., Krasnov, A. and Wadsworth, S. (2016) Nutrigenomic effects of glucosinolates on liver, muscle and distal kidney in parasite-free and salmon louse infected Atlantic salmon. *Parasites & Vectors* 9(1), 639.

Smallman, D.R. (2009) The behavioural, chemical and host ecology of two species of copepods (Copepoda: Caligidae) parasitic on Atlantic salmon (*Salmo salar* L.). PhD thesis, University of Aberdeen, Scotland.

Stephenson, J.F. (2012) The chemical cues of male sea lice *Lepeophtheirus salmonis* encourage others to move between host Atlantic salmon *Salmo salar*. *Journal of Fish Biology* 81, 1118–1123.

Stingray Marine Solutions (2010) Optical delousing of salmon and trout using laser. Available at: https://www.stingray.no/delousing-with-laser/?lang=en (accessed 30 January 2022).

Thiery, D. and Visser, J.H. (1986) Masking of host plant odour in the olfactory orientation of the Colorado potato beetle. *Entomologia Experimentalis et Applicata* 41, 165–172.

Trematerra, P. (2013) Aspects related to decision support tools and integrated pest management in food chains. *Food Control* 34, 733–742.

Vieira, F.G. and Rozas, J. (2011) Comparative genomics of the odorant-binding and chemosensory protein gene families across the arthropoda: origin and evolutionary history of the chemosensory system. *Genome Biology and Evolution* 3, 476–490.

Wadsworth, S.L., Gonzalez-Vecino, J.L., Pino-Marambio, J. and Mordue (Luntz), A.J. (2010) Methods and feed compositions for masking of fish semiochemicals WO2011068415A2 World Intellectual Property Organization.

Wyatt, T.D. (2010) Pheromones and signature mixtures: defining species-wide signals and variable cues for identity in both invertebrates and vertebrates. *Journal of Comparative Physiology A: Neuroethology, Sensory, Neural, and Behavioral Physiology* 196(1), 685–700.

Wyatt, T.D. (2011) Pheromones and behavior. In: Breithaupt T. and Thiel M. (eds) *Chemical Communication in Crustaceans*. Springer, New York, NY. pp. 23–38.

Wyatt, T. (2014) *Pheromones and Animal Behavior: Chemical Signals and Signatures* (2nd ed.). Cambridge University Press, Cambridge, UK.

Zhang, D., Terschak, J.A., Harley, M.A., Lin, J. and Hardege, J.D. (2011) Simultaneously hermaphroditic shrimp use lipophilic cuticular hydrocarbons as contact sex pheromones. *PLoS ONE* 6, 17720.

Zhang, D., Liu, X., Harley, M.A. and Hardege, J.D. (2020) Uridine-5'-tri-phosphate is a candidate component of the soluble sex pheromone bouquet in a marine shrimp, *Lysmata wurdemanni*. *Marine Ecology Progress Series* 640, 139–146.

Zhou, J.-J., Zhang, G.-A., Huang, W., Birkett, M.A., Field, L.M., Pickett, J.A. and Pelosi, P. (2004) Revisiting the odorant-binding protein LUSH of *Drosophila melanogaster*: evidence for odour recognition and discrimination. *FEBS Letters* 558, 23–26.

CHAPTER 7

Pathologies associated with sea lice

Hamish Rodger[1], *Ana Herrero*[2] *and Chris Matthews*[3]

[1]VAI Consulting, Kinvara, Co. Galway, Ireland
[2]VAI Consulting, Oban, Scotland
[3]Pharmaq Analytiq, Inverness, Scotland

7.1 Introduction and overview

The parasitic crustacean copepods known as sea lice (*Lepeophtheirus salmonis* (Krøyer, 1837) and *Caligus* spp.) have caused challenges for salmonids in both the wild and farmed fish arenas, but it has been in the marine net pen farming of Atlantic salmon (*Salmo salar* L.) that the issues have been most intense with wide-ranging impacts in fish welfare, farm production, economics, business opportunities and local social licence (Johnson et al., 2004). This chapter reviews the pathologies observed, induced and reported as associated with sea lice infestations on salmonids and will focus on the clinical signs and gross pathology, histopathology, osmoregulatory effects and biochemistry as well as stress responses, host impacts, healing and injuries associated with treatments.

7.2 Clinical signs and gross pathology

There have been reports of heavily sea-lice parasitised wild Atlantic salmon since the late 1930s and early 1940s in which severe skin lesions on the heads of infested fish and mortalities were recorded in returning fish in Eastern Canada (White, 1940). The first cases of serious infestations in farmed salmon in Scotland were observed in 1976 and had been apparent in Norway from the late 1960s (Rae, 1979; Wootten et al., 1982). The majority of the cases were reported associated with *L. salmonis*; however, it is notable that some in both land-based and net pen farms were associated with *Caligus elongatus*, Nordmann 1832. Infestations with caligid copepods were described soon after the founding of the salmonid industry in Chile, with *Caligus rogercresseyi* (Boxshall & Bravo 2000) identified as the dominant species from the 1980s (Boxshall and Bravo, 2000).

Clinical signs observed with lice infestations include an increased frequency of leaping behaviour (Furevik et al., 1988; Grimnes and Jakobsen, 1996), where the fish break the water surface and may collide with net pen structures (although this is not always observed), and grey or pale patches on the skin of the head, operculum and dorsal area (especially behind the dorsal fin) which may become white and eventually red as erosive damage increases in severity, especially over the head and operculum (Fig. 7.1, 7.2 and 7.3). In post smolts black spots may be observed in the skin and fins as a result of attachment sites for the copepodites (Grimnes and Jakobsen, 1996). As the pathology progresses the lesions can extend into the muscle and in severe cases expose cranial cartilage. A decrease in appetite may be observed, which will result in reduced fish growth (Abolofia et al., 2017), with a presentation of anorexia in lesioned fish. Once skin lesions develop, in the absence of treatment, fish become moribund, may rise in the water column close to the surface and become lethargic. Mortalities will then start to occur.

Small haemorrhages may occur in the skin and are most obvious in the ventrum, especially around the vent and behind fins (Fig. 7.4 and 7.5). Punctate skin marks (1–2 mm diameter) may also be

Fig. 7.1 Atlantic salmon with *Lepeophtheirus salmonis* infestation resulting in skin erosion, haemorrhage on operculum and head with traumatic intraocular haemorrhage.

Fig. 7.2 Atlantic salmon with *Lepeophtheirus salmonis* infestation resulting in skin erosion and scale loss behind the adipose fin. A punctate attachment site for a chalimus-stage louse can be seen as a pale spot on the fin.

Fig. 7.3 Marine-farmed *Oncorhynchus mykiss* with *Lepeophtheirus salmonis* associated skin pathology presenting grossly as grey-white areas on dorsum.

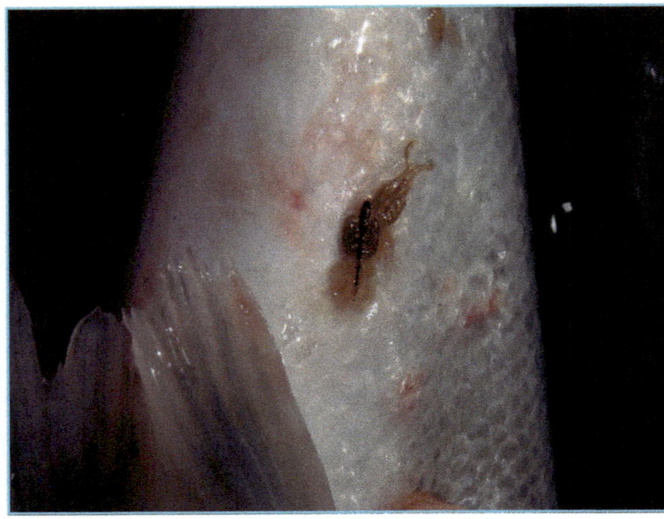

Fig. 7.4 Atlantic salmon with *Lepeophtheirus salmonis* infestation resulting in skin erosion and haemorrhage on ventrum posterior to the pelvic fins.

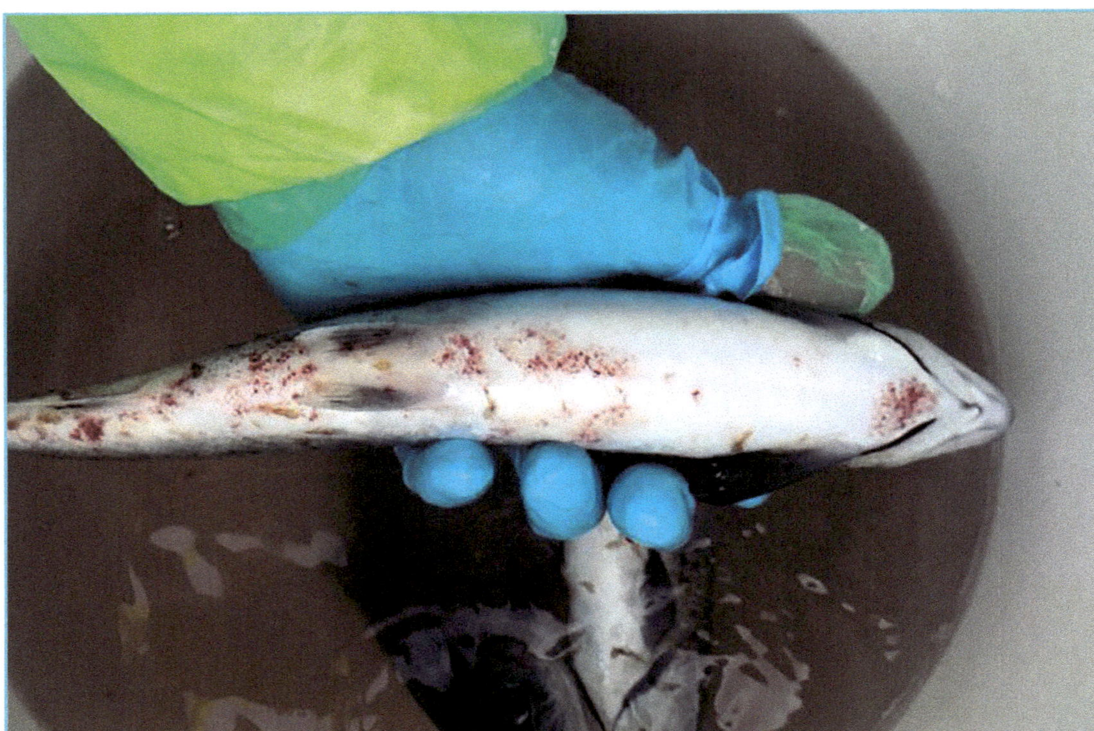

Fig. 7.5 Atlantic salmon with *Caligus elongatus* infestation resulting in skin erosion and haemorrhage on ventrum.

seen on the fins, and can be obvious on the adipose and dorsal fins. All life stages of the parasites that are on the fish are visible to the eye, although the chalimus stages can be easily overlooked, especially against the darker dorsal skin and fins (Fig. 7.6). The attached chalimus stages of the lice are often seen on the fins or ventrum and attachment sites may present as a ring of pigmented tissue with a depressed white centre (Jones et al., 1990). Bron et al. (1991) demonstrated that in Atlantic salmon experimentally challenged with *L. salmonis* 69% of chalimus were recovered from the fins, 21% from the gills and 18% from the body at ten days post-exposure. The centre of the attachment site may then slough leaving the small punctate mark. Mobile stages of the lice (pre-adults and adults) will browse over the body surface and may be seen distant from skin lesions as well as in the mouth and gills. Atrophy and loss of the distal portions of the gill filaments have been reported as a gross finding in the gills of Atlantic and chinook salmon (*Oncorhynchus tshawytscha* (Walbaum, 1792)) (Johnson and Albright, 1992). Fish with sea-lice skin lesions, if taken into freshwater, for instance in the case of broodstock, can develop fungal (*Saprolegnia* sp.) infections, with fungal plaques obvious as white to cream-coloured raised areas. Erosion of the dorsal fin has also been associated with heavy infestations, giving rise to exposure of the fin rays. Fin erosion of Atlantic salmon, with the pectoral, anal and pelvic fins also affected, has been induced with challenges with *C. rogercresseyi* (González et al., 2020) and, as with *L. salmonis*, the chalimus of *C. rogercresseyi* have been shown to have a settling preference for the fins (Araya et al., 2012).

Returning wild sea trout (*Salmo trutta*) post smolts with natural infestations of all stages of *L. salmonis*, and sampled from rivers, presented with no skin damage but severe dorsal fin erosion, and the degree of fin erosion correlated with the level of chalimus burden (Dawson, 1998).

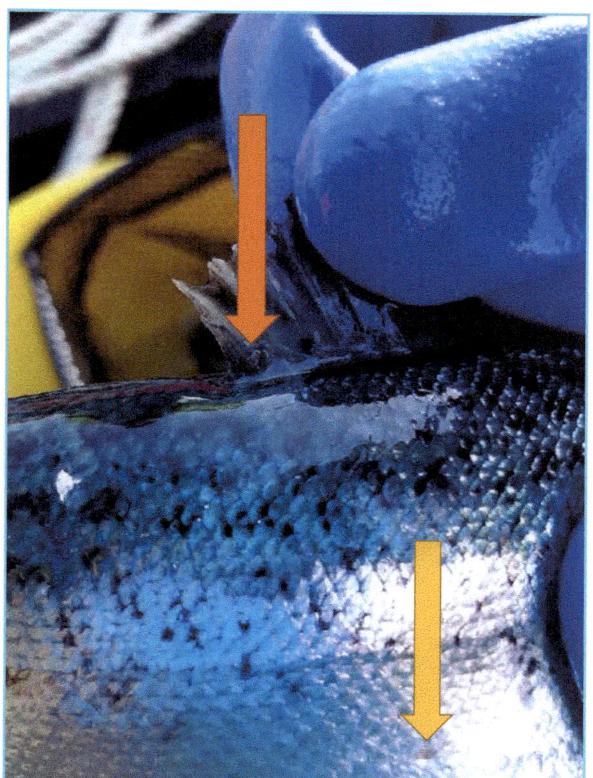

Fig. 7.6 Punctate marks on dorsal fin of Atlantic salmon as caused by chalimus stage of *Lepeophtheirus salmonis* (orange arrow) and some chalimus present on flank (yellow arrow).

Clinical signs of other infectious conditions such as bacterial (vibriosis and piscirickettsiosis) and viral diseases (pancreas disease) may also start to appear in some cases where such pathogens are endemic with the parasitic infestation increasing the susceptibility to pathogen-related clinical disease (Wootten et al., 1982; Rodger and Mitchell, 2007; Arriagada et al., 2019).

7.3 Histopathology

Sea lice predominantly feed on host mucus, tissues and blood. The skin histopathology of *L. salmonis* infested Atlantic salmon has been described where the lice feeding areas present with focal epidermal loss and necrosis, dermal oedema, mild inflammation and haemorrhage. In scaled areas there may be resorption or loss of the scales. The histopathology in Atlantic salmon associated with the pre-adult and adult stages of *L. salmonis* includes oedema and intercellular fissures in the epidermis between basal cells and overlying tissue (Jónsdóttir et al., 1992). The areas of inflammation may be present in the dermis anterior to the parasite and are considered to correspond to the raised areas of host tissue around the area of attachment. The inflammation can extend through the dermal layers and into the hypodermis and muscular tissue. Fibrosis may also be observed consistent with a chronic inflammatory reaction (Jónsdóttir et al., 1992). Johnson and Albright (1992) have described the comparative susceptibility and variation in fin and gill histopathology with different salmonid species to experimental challenge with *L. salmonis*. Histology of gills revealed progressive epithelial erosion, hyperplasia and fusion of gill lamellae in chinook and Atlantic salmon. There was also proliferation of the mucous cells in the hyperplastic epithelium in the gills of the chinook salmon, but this was not observed in the Atlantic salmon. Inflammation in both species was predominantly neutrophils with a few lymphocytes. In the fins of chinook, Atlantic and coho salmon (*Oncorhynchus kisutch* (Walbaum, 1792)), following challenge, the predominant histopathology described was partial to complete erosion of the epidermis with mild inflammation in the dermis of the coho salmon, but no obvious inflammation in chinooks or Atlantics. In coho salmon at the feeding sites for the parasites on fins there was epithelial hyperplasia, which in some cases encapsulated the copepods with inflammatory cells. The lesions extended through the dermis exposing fin rays with necrosis, haemorrhage and inflammation of the dermis. This contrasted with the limited response in both Atlantic and chinook salmon fins despite the breach in epidermis from feeding and exposure of dermis and fin rays. A mild inflammation was observed in the dermis of some fish. From the work by Johnson and Albright (1992) naïve coho salmon appear to be more resistant to *L. salmonis* challenge than chinook or Atlantic salmon and they demonstrated the most well-developed histopathological response in the level of epithelial hyperplasia and inflammation. Nolan et al. (1999), in further mobile stage *L. salmonis* challenge studies in Atlantic salmon, demonstrated that there was increased apoptosis and necrosis in the skin epidermis and a decrease in mucous cells. Chalimus (or mobile stages) are infrequently observed on salmon gills in farm situations; however, a number of groups have reported gill infestations with chalimus in experimental challenges, in addition to attachment to fins.

The location of histopathology induced by the stages of sea lice which reside on salmonids varies with the life stage, for instance that associated with chalimus being predominantly on the fins (Fig. 7.7). The histopathology associated with healing lesions of attachment sites of chalimus stages of *L. salmonis* has been described with attachment filament fragments observed in the epidermis and hyperplasia around the attachment and focal areas of necrosis. In the dermis, fibrosis, inflammatory cell infiltration

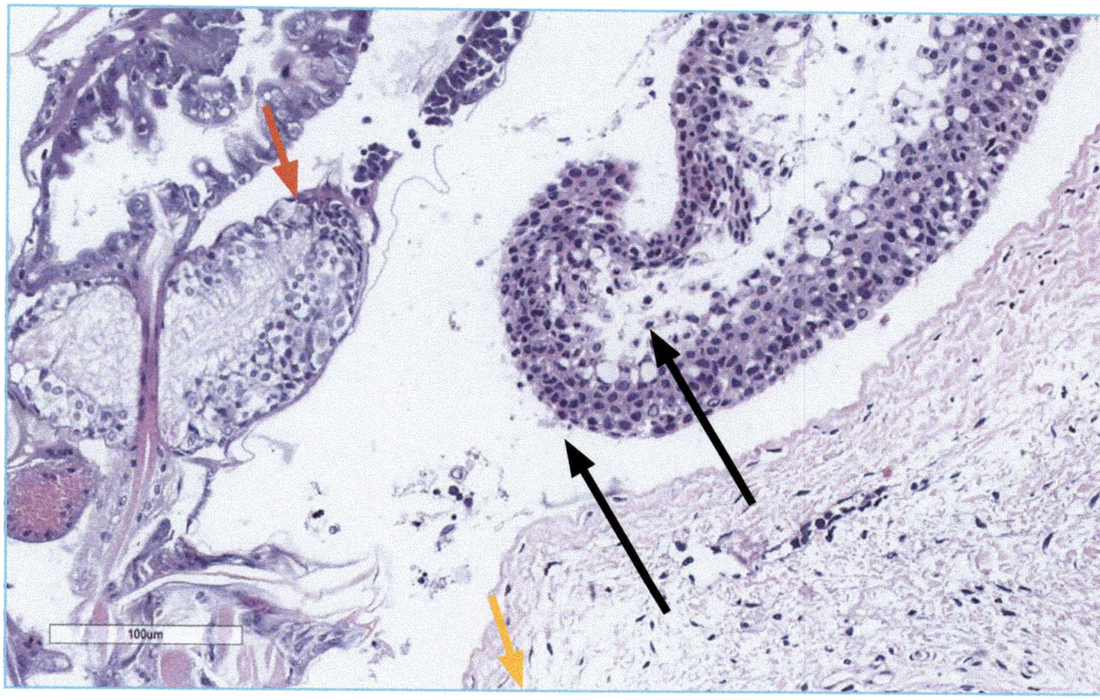

Fig. 7.7 Histological section of *Lepeophtheirus salmonis* chalimus (left of image) on pectoral fin of *S. salar* with epidermal perforation and focal necrosis (arrows) (H&E).

and increased melanisation of the stratum granulosum was reported (Jones et al., 1990). In the gills of Atlantic salmon, chalimus stage *L. salmonis* may induce local haemorrhage, aneurysms, focal lamellar hyperplasia and a low level of inflammation in the gill filament (Fig. 7.8 and 7.9).

Histopathology has been described in the tissues of rainbow trout (*Oncorhynchus mykiss* (Walbaum, 1792)) experimentally challenged with *C. rogercresseyi* (Rojas et al., 2018) with epidermal erosion, dermal hyperplasia, inflammation in the stratum compactum and an increase in number of melanophores and epidermal mucous cells all reported. Rojas et al. (2018) also reported gill changes described as hyperplasia and fusion of the lamellae and increased quantities of inflammatory and mucous cells. There were also minor changes described in the intestine, such as disorganised epithelium and increased numbers of secretory cells, which the authors attributed to a systemic inflammatory response in the trout to parasite infestation.

7.4 Osmoregulatory effects and biochemistry

Atlantic salmon post smolts have been shown to have significant physiological changes as a result of experimental infestation with *L. salmonis*, with significantly increased plasma chlorides and decreased blood total protein, albumin and haematocrit (Grimnes and Jakobsen, 1996). These changes became most obvious when the lice developed to pre-adult stages (25 days post-infestation at 10°C). This can be attributed to the severe mechanical damage that the more mature and mobile life stages of the parasite induce on the fish skin, especially the salmon head and operculum which are devoid of scales.

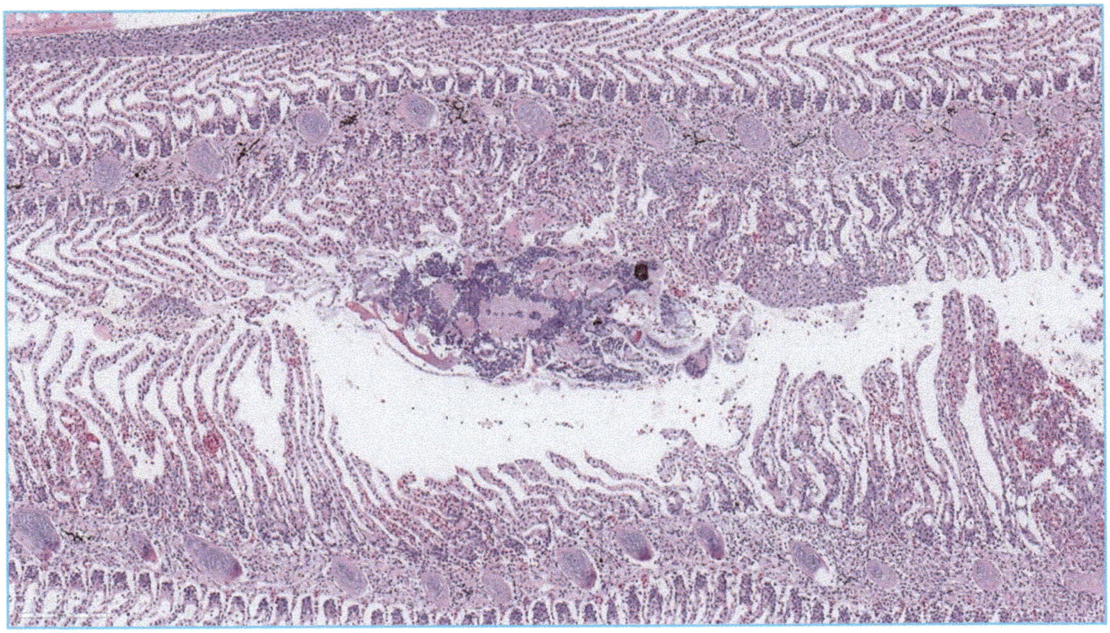

Fig. 7.8 Histological section of Atlantic salmon gills with attached caligid chalimus and associated pathology consisting of aneurysms, haemorrhage, hyperplasia and lamellar fusion as well as low level inflammatory response in the filament (H&E).

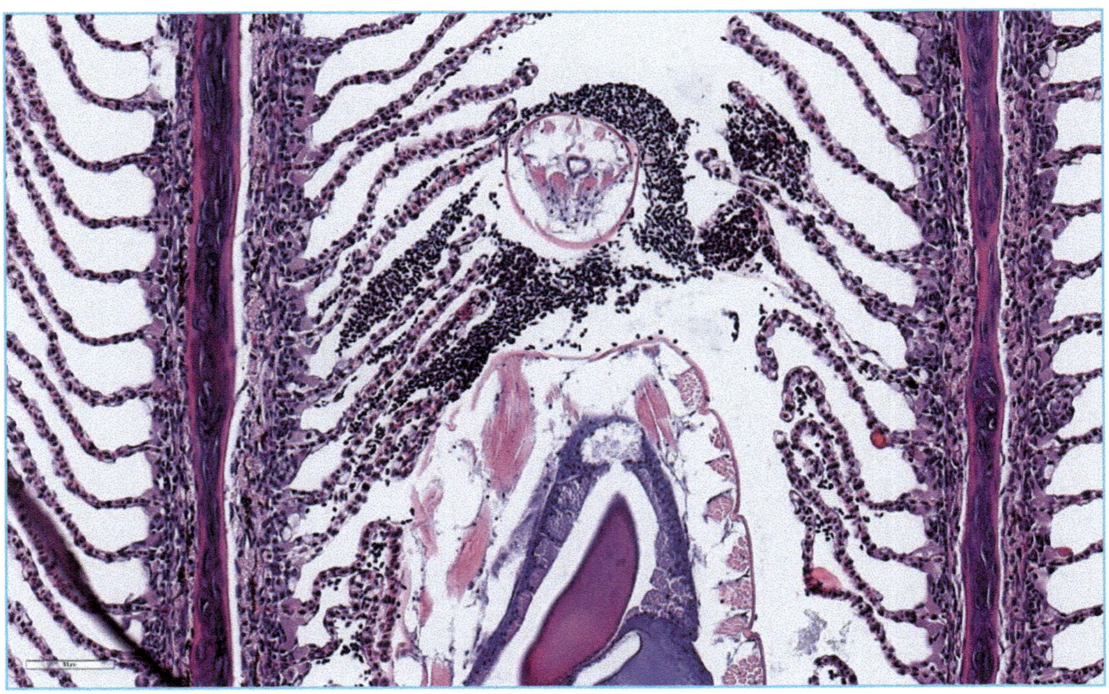

Fig. 7.9 Histopathology of chalimus stage of *Lepeophtheirus salmonis* in gills of *S. salar* with associated lamellar haemorrhage (H&E).

The reduction in haematocrit may be due to leakage of erythrocytes after cell membrane damage. Similar findings were observed in experimentally challenged Atlantic salmon post smolts by Dawson et al. (1999) where at 21 days post-challenge with *L. salmonis* copepodites (at 13.9–16.0°C) there was a significant decrease in haematocrit, total protein and serum albumin, and elevated blood chlorides. This coincided with the establishment of the pre-adult and adult lice and resultant skin damage. This study also demonstrated significant decreases in plasma alkaline phosphatase and lactate dehydrogenase at day 21 post-challenge. However, in the Dawson et al. (1999) study by day 30 post-challenge many of the blood parameters had shown signs of recovery. Wild sea trout (*Salmo trutta*) with natural infestations of *L. salmonis* presented with severe fin erosion, and those fish with mobile stage lice (pre-adults and adults) had significant reductions in blood total protein, serum albumin and cholesterol compared to those with no lice or with chalimus stages only (Dawson, 1998). This sea trout study also found that there was no significant difference between the blood parameters of the fish with attached-stage lice only (chalimus) and uninfected fish. Experimental work by Wagner et al. (2003) has shown that Atlantic salmon critical swimming speeds are significantly lower in higher *L. salmonis* infected fish compared to controls and that after swimming the blood chloride levels increase significantly; they considered that sublethal lice infections (0.1 lice/g fish) compromised the overall fitness of the salmon, as assessed via cardiac output. In addition to elevated blood chloride levels, Nolan et al. (1999) demonstrated significantly elevated gill Na^+/K^+-ATPase activities in pre-adult and adult *L. salmonis*-challenged Atlantic salmon post smolts.

7.5 Stress responses

Research into the stress responses of fish to lice infestations have shown that exposure of rainbow trout to infective-stage *L. salmonis* results in significantly higher blood cortisol after four hours of confinement than in uninfected fish (Ruane et al., 2000). The attachment and feeding mechanisms of mobile lice give rise to skin mucus changes, damage and a stress response with cortisol release (Wagner et al., 2008). Infestations of pre-adult and adult *L. salmonis* in Atlantic salmon are associated with significant increases in cortisol and glucose levels and prolonged infestation leads to the effects from a chronic stress response as well as the physical and osmotic impacts which lead to mortality (Bowers et al., 2000). The elevated gill Na^+/K^+-ATPase activity in pre-adult and adult *L. salmonis* challenged Atlantic salmon post smolts previously mentioned (Nolan et al., 1999) is an indirect consequence of the stress response and has an osmoregulatory impact through induced chloride cell apoptosis. These effects in the gill were proposed as the result of the generalised stress response, rather than direct damage from the parasites. As highlighted by Nolan et al. (1999) lice infestation is stressful for the salmon and at higher lice levels gives rise to compromised skin and gill integrity, plus stress-related immunosuppression. Plasma prostaglandin E_2 (PGE_2), an anti-inflammatory molecule, plus trypsin and cortisol were shown to be significantly elevated in Atlantic salmon experimentally challenged with *L. salmonis*, with the PGE_2 appearing to subside as juvenile attached lice develop to mature stages yet cortisol levels increased further (Fast et al., 2006). In addition, head kidney macrophage populations change in lice-affected Atlantic salmon with a temporary decrease in respiratory burst and phagocytic activities leading to a decreased ability of macrophages to respond to bacterial challenges (Mustafa et al., 2000a). González Gómez et al. (2016) have also shown that low level infestations (five to six lice per fish) of *C. rogercresseyi* induce biochemical effects (elevated glucose, decreased protein and triglycerides) and stress responses (elevated cortisol) in Atlantic salmon.

Following the release of cortisol into the blood there are secondary response effects such as the release of glucose, an increase in heart and respiratory rate, and other physiological changes which all act to increase the metabolic rate and muscle energy levels thus enabling escape or attack (the fight-or-flight response). When cortisol remains elevated the tertiary or longer-term response is enabled which leads to a closing down of non-essential body functions resulting in reduced appetite and eventually leading to adverse effects on growth. Other effects include a decrease in the immune response associated with reduced circulating lymphocyte levels and antibody production, as well as changes in behaviour, all of which can result in lower survival. Iversen and Eliassen (2014) demonstrated if salmon are stressed before a management procedure such as vaccination then poorer outcomes and welfare result, such as higher levels of fin erosion and higher mortalities. It has also been demonstrated experimentally that stressful procedures, such as netting or handling salmon, can increase the risk of infestation with *L. salmonis* copepodites (Delfosse et al., 2021).

7.5.1 Host impacts and susceptibility to other diseases

There is little evidence of a significant tissue response in Atlantic salmon at the lice attachment or feeding sites, in contrast to that shown in coho salmon where there is a marked inflammatory and hyperplastic response. However, *L. salmonis* can affect the fish stress response and cause effects outside the attachment area as shown by Nolan et al. (2000) with increased gill ATPase four hours after challenge in rainbow trout as well as effects on skin and gill epithelium. Increased levels of lysozyme and alkaline phosphatase have been reported in the mucus of Atlantic salmon infected with *L. salmonis*, as well as the presence of plasma proteins. PGE_2 has been identified in *L. salmonis* secretions (Fast et al., 2004) and can downregulate Atlantic salmon inflammatory gene expression and may increase the availability of blood where lice are feeding. The inflammatory cell response seen in coho salmon, in contrast, appears to counter this process rapidly enabling the fish to shed themselves of the parasites. The physical and stress-associated effects on epithelial surfaces of skin and gills coupled with immunosuppression from elevated cortisol means that infestation with sea lice increases the susceptibility of the fish to other diseases. Examples include increased susceptibility to:

1. the microsporidian, *Loma salmonae* (Putz, Hoffman & Dunbar, 1965), in marine reared rainbow trout (Mustafa et al., 2000b) infested with *L. salmonis*
2. infectious salmon anaemia virus (ISAV), in Atlantic salmon infested with *L. salmonis* (Barker et al., 2019)
3. *Piscirickettsia salmonis*, in farmed salmonids infested with *C. rogercresseyi* (Lhorente et al., 2014; Arriagada et al., 2019)
4. *Moritella viscosa* in Atlantic salmon challenged with *L. salmonis* (Carvalho et al., 2020).

There is also experimental work to indicate that Atlantic salmon blood monocytes isolated from *L. salmonis*-infested fish have an inability to upregulate antiviral genes at the early time of salmonid alphavirus (SAV) infection *in vitro*, resulting in higher levels of virus replication (Gamil et al., 2020). These findings highlight another possible impact of lice infestation on SAV infections and are consistent with epidemiological findings (Rodger and Mitchell, 2007).

It has been suggested that host-associated microbiota in the skin and gut are highly significant in parasitic disease outcomes in animals. The skin surface microbiota of *Salmo salar* changes significantly when compared to control fish in parasitic challenge studies, with such changes proposed to promote pathology via proliferation of endogenous bacterial pathogens or via decreased colonisation resistance to exogenous opportunistic pathogens (Llewellyn et al., 2017). The same authors also demonstrated that high *L. salmonis* burdens are associated with metagenomic profiles containing pathogenic bacteria such as *Pseudomonas, Tenacibaculum, Flavobacterium* and *Vibrio* species. The microbiota present upon caligid sea lice themselves has been proposed as a potential reservoir of both fish pathogens and genes for antimicrobial resistance (Gonçalves et al., 2020).

Farmed Atlantic salmon populations are subject to a wide range of gill disease challenges, particularly between late summer and early winter (Herrero et al., 2018). The localised proliferative or hyperplastic gill pathology associated with attached juvenile sea lice is unlikely to be of such severity as to significantly affect respiratory capacity and so cause clinical gill disease in its own right. The clinical effects of host responses reported following sea lice settlement such as significantly reduced haematocrit are not understood, but any resultant deficit in oxygen carrying capacity might be reasonably speculated to ultimately contribute to loss of the respiratory reserve required to cope with seasonal gill challenges. Further, behavioural changes driven by sea lice parasitism, for example populations of lethargic individuals aggregating at the cage surface or slower swimming speeds, are likely to augment the course of certain gill diseases such as amoebic gill disease (AGD), given that the causative agent *Neoparamoeba perurans* shows variable depth distribution within the cage under certain circumstances (Wright et al., 2017). Control methods for sea lice such as lice skirts also occasionally appear to accelerate the course of advanced AGD infections, probably by affecting water hydrodynamics in the upper part of the cage.

A low-level humoral antibody response has been reported in Atlantic salmon with a natural *L. salmonis* infestation (Grayson et al., 1991) and although a rapid local and systemic reaction to the parasite develops it does not result in substantial levels of protection (Tadiso et al., 2011). A biphasic molecular response has been reported in Atlantic salmon to *L. salmonis,* where the initial inflammatory response (interleukin (IL)-1β, IL-1R, IL-12, CD4/8, tumour necrosis factor α, PGE_2) and so on) to attached stage copepodites and chalimus I is then followed by a downregulation of genes with a second spike after the development of mobile lice stages (Tadiso et al., 2011; Fast, 2014). Large fluctuations in splenic proteases with unknown roles have also been reported (Tadiso et al., 2011). Research has indicated that immune responses to *L. salmonis* infection involve limited activation of T helper (Th) cell mediated immunity with a bias towards Th2 responses. Differences in host responses between resistant and susceptible salmonid species also support the view that the character, timing and strength of the immune response play an important role for the outcome of the infection (Holm et al., 2015). A large variation in lice counts has been reported within resistant and susceptible families of Atlantic salmon and the best inverse correlation with infection level has been seen for the prototypical Th1 genes, including members of the interferon pathways (Holm et al., 2015). The gene expression response of Atlantic salmon skin transcriptome to *C. rogercresseyi* has been studied and used to classify family groups of salmon according to the traits of resistance (high versus low lice burden) and growth during infestation (Robledo et al., 2018).

7.5.2 Sea lice as potential carriers of fish pathogens

Different bacteria and viruses have been isolated from sea lice parasitising salmonids. For instance, *T. maritimum, Pseudomonas fluorescens* and *Vibrio* spp. were isolated from the carapace and gut of *L. salmonis* parasitising apparently healthy farmed Atlantic salmon (Barker et al., 2009) and ISAV was detected in the gut of *L. salmonis* obtained from infected fish (Nylund et al., 1993). Sea lice probably acquire fish pathogens by feeding on mucus, skin and blood of hosts; however, the role that caligid sea lice may play in fish disease as a biological or mechanical vector is poorly understood. Novak et al. (2016) found that in certain experimental circumstances *L. salmonis* could transmit *Aeromonas salmonicida* subsp. *salmonicida* between infected fish and smaller naïve individuals. Under laboratory conditions, *C. rogercresseyi* can potentially act as a mechanical vector of ISAV for Atlantic salmon and transmit the virus 48 hours after being separated from an infected host (Oelckers et al., 2014). Similarly, experimental transmission of infectious haematopoietic necrosis virus (IHNV) to naïve fish was possible through the contact of *L. salmonis* previously exposed to IHNV infected fish (Jakob et al., 2011). Conversely, *P. salmonis* was detected by molecular methods in the carapace of sea lice up to one hour after removal from infected farmed salmon affected by piscirickettsiosis but experimental challenges failed to transmit the bacterium from *C. rogercresseyi* to naïve fish (Labra et al., 2020).

The microsporidian *Desmozoon lepeophtherii* has a complex life cycle that involves both the sea lice and Atlantic salmon (Freeman and Sommerville, 2009; Nylund et al., 2010). The route of infection of the microsporidian in salmon or louse is still unknown, but clinical disease has been associated with the microsporidian in Atlantic salmon (Matthews et al., 2013; Weli et al., 2017). It is likely that *L. salmonis* ingests spores while feeding on the epithelial cells of the skin and gills of salmon. Presence of lice in the salmon population does not seem to be a requisite for the microsporidian to infect fish, but this does not exclude the possibility that waterborne spores previously released by the lice could infect salmon (Sveen et al., 2012).

7.5.3 Injuries associated with lice treatments

In addition to the physical, physiological and osmotic damage that sea lice can have on salmonids, treatment-related mortalities can also be significant as has been reported in Norway (Overton et al., 2019; Walde et al., 2021). All sea lice bath, physical or thermal treatments involve crowding of fish (Fig. 7.10), and in some cases pumping, which induces stress and the risk of low dissolved oxygen and fin, eye, gill and skin damage from trauma (Fig. 7.11). Exposure to hydrogen peroxide in bath treatments has been associated with gill pathology, especially when the water temperature is above 14°C (Kiemer and Black, 1997). In mechanical delousing methods (for example, flushing systems), gill bleeding and scale loss have been commonly reported during operations (Overton et al., 2019).

Mortality and serious lesions associated with thermal sea-lice treatments have been reported after industrial treatments in Norway and the histopathology most commonly observed includes gill haemorrhage, scale and skin loss, haemorrhage and vacuolation of thymic tissue, degeneration of nasal epithelium and brain haemorrhage (Poppe et al., 2018) (Fig. 7.11). However, experimental exposure of sedated fish to water at 34°C for 30 s did not cause any statistically significant increase in acute lesions, except for an increase in minor fin lesions, considered to be caused by the strong behavioural reaction of fish when exposed to warm water (Moltumyr et al., 2021). There remain gaps in our knowledge of the various histopathologies and welfare impacts associated with mechanical, thermal and water pressure removal methods for sea lice removal at controlled experimental trial level.

Fig. 7.10 Crowding of salmon for pumping to wellboat for sea lice treatment.

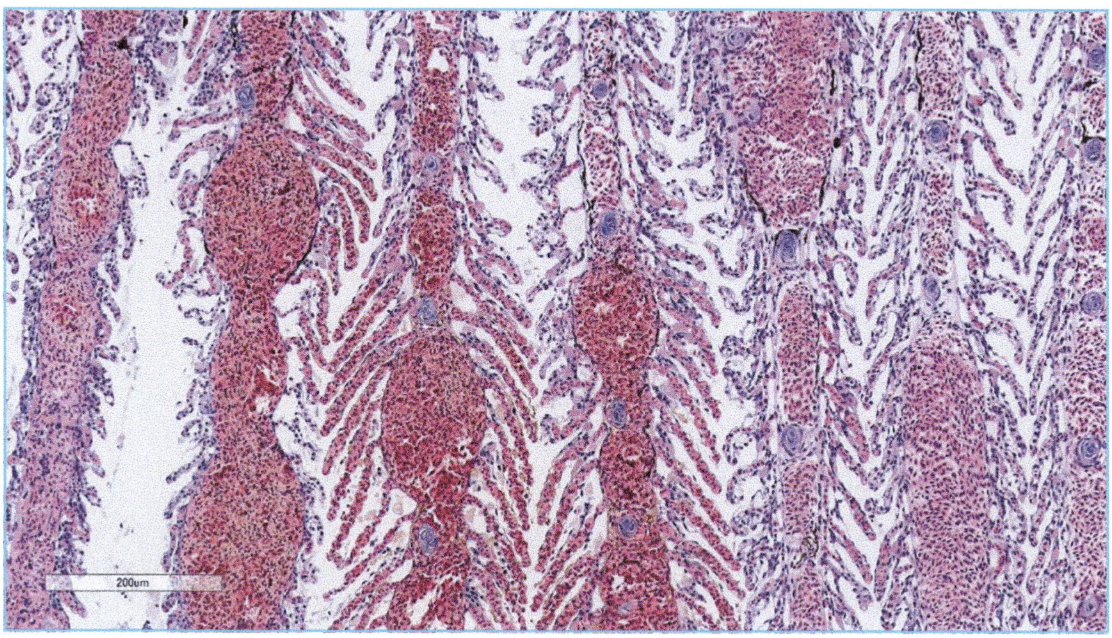

Fig. 7.11 Histopathology of *S. salar* gills following sea lice bath treatment showing extensive haemorrhaging in gill filaments and loss of normal structure (H&E).

Summary

In summary, the stages of sea lice that attach or feed on salmonids induce gross and histopathological changes although some species, such as coho salmon, can reject the challenges within one or two weeks. The lice-induced pathology, as well as the osmoregulatory and stress-related responses, can be fatal if untreated. In addition, infestation with lice increases the susceptibility of salmonids to other infectious diseases including other parasites, bacteria and viruses. The low-level immune response in Atlantic salmon to *L. salmonis* although rapid, as a local and systemic reaction, does not result in substantial levels of protection. Sea lice have been shown to be potential carriers of specific pathogens and in experimental conditions may be involved in pathogen transfer but their relevance as vectors remains to be determined. Control measures and treatments for sea lice, while effective at clearing the parasites, can potentially impact on fish health and the development of other diseases.

References

Abolofia, J., Wilen, J.E. and Asche, F. (2017) The cost of lice: quantifying the impacts of parasitic sea lice on farmed salmon. *Marine Resource Economics* 32, 3.

Araya, A., Mancilla, M., Lhoernte, J.P., Neira, R. and Gallardo, J.A. (2012) Experimental challenges of Atlantic salmon *Salmo salar* with incremental levels of copepodids of sea louse *Caligus rogercresseyi*: effects on infestation and early development. *Aquaculture Research* 43, 1904–1908.

Arriagada, G., Hamilton-West, C., Nekouei, O., Foerster, C., Müller, A., Lara, M. and Gallardo-Escárate, C. (2019) *Caligus rogercresseyi* infestation is associated with *Piscirickettsia salmonis*-attributed mortalities in farmed salmonids in Chile. *Preventive Veterinary Medicine* 171, 104771.

Barker, D.E., Braden, L.M., Coombs, M.P. and Boyce, B. (2009) Preliminary studies on the isolation of bacteria from sea lice, *Lepeophtheirus salmonis*, infecting farmed salmon in British Columbia, Canada. *Parasitology Research* 105, 1173–1177.

Barker, S.E., Bricknell, I.R., Covello, J., Purcell, S., Fast, M.D., Wolters, W. and Bouchard, D.A. (2019) Sea lice, *Lepeophtheirus salmonis* (Krøyer 1837), infected Atlantic salmon (*Salmo salar* L.) are more susceptible to infectious salmon anemia virus. *PLoS ONE* 14 (1), e0209178.

Boxshall, G.A. and Bravo, S. (2000) On the identity of the common *Caligus* (Copepoda: Siphonostomatoida: Caligidae) from salmonid netpen systems in southern Chile. *Contributions to Zoology* 69, 137–146.

Bowers, J.M., Mustafa, A., Speare, D.J., Conboy, G.A., Brimacombe, M., Sims, D.E. and Burka, J.F. (2000) The physiological response of Atlantic salmon, *Salmo salar* L., to a single experimental challenge with sea lice, *Lepeophtheirus salmonis*. *Journal of Fish Diseases* 23, 165–172.

Bron, J.E., Sommerville, C., Jones, M. and Rae, G.H. (1991) The settlement and attachment of early stages of the salmon louse *Lepeophtheirus salmonis* (Copepoda: Caligidae) on the salmon host, *Salmo salar*. *Journal of Zoology (London)* 224, 201–212.

Carvalho, L.A., Whyte, S.K., Braden, L.M., Purcell, S.L., Manning, A.J., Muckle, A. and Fast, M.D. (2020) Impact of co-infection with *Lepeophtheirus salmonis* and *Moritella viscosa* on inflammatory and immune responses of Atlantic salmon (*Salmo salar*). *Journal of Fish Diseases* 43, 459–473.

Dawson, L.H.J. (1998) The physiological effects of salmon lice (*Lepeophtheirus salmonis*) infections on returning post-smolt sea trout (*Salmo trutta* L.) in western Ireland, 1996. *ICES Journal of Marine Science* 55, 193–200.

Dawson, L.H.J., Pike, A.W., Houlihan, D.F. and McVicar, A.H. (1999) Changes in physiological parameters and feeding behaviour of Atlantic salmon *Salmo salar* infected with sea lice *Lepeophtheirus salmonis*. *Diseases of Aquatic Organisms* 35, 89–99.

Delfosse, C., Pageat, P., Lafont-Leculle, C., Asproni, P., Chabaud, C., Cozzi, A. and Bienboire-Frosini, C. (2021) Effect of handling and crowding on the susceptibility of Atlantic salmon (*Salmo salar* L.) to *Lepeophtheirus salmonis* (Krøyer) copepodids. *Journal of Fish Diseases* 44, 327–336.

Fast, M.D. (2014) Fish immune responses to parasitic copepod (namely sea lice) infection. *Developmental and Comparative Immunology* 43, 300–312.

Fast, M.D., Ross, N.W., Craft, C.A., Locke, S.J., MacKinnon, S.L. and Johnson, S.C. (2004) *Lepeophtheirus salmonis*: characterisation of prostaglandin E_2 in secretory products of the salmon louse by RP-HPLC and mass spectrometry. *Experimental Parasitology* 107, 5–13.

Fast, M.D., Muise, D.M., Easy, R.E., Ross, N.W. and Johnson, S.C. (2006) The effects of *Lepeophtheirus salmonis* infections on the stress response and immunological status of Atlantic salmon (*Salmo salar*). *Fish and Shellfish Immunology* 21, 228–241.

Freeman, M.A. and Sommerville, C. (2009) *Desmozoon lepeophtherii* n. gen., n. sp., (Microsporidia: Enterocytozoonidae) infecting the salmon louse *Lepeophtheirus salmonis* (Copepoda: Caligidae). *Parasites & Vectors* 2, 1–15.

Furevik, D.M., Huse, I., Bjordal, Å. and Fernø, A. (1988) Surface activity of Atlantic salmon (*Salmo salar*) in net pens. *International Council for the Exploration of the Sea* ICES C.M. 1988/F, 19.

Gamil, A.A.A., Gadan, K., Gislefoss, E. and Evensen, Ø. (2020) Sea lice (*Lepeophtheirus salmonis*) infestation reduces the ability of peripheral blood monocytic cells (PBMCs) to respond to and control replication of salmonid alphavirus in Atlantic salmon (*Salmo salar* L.). *Viruses* 12, 1450.

Gonçalves, A.T., Collipal-Matamal, R., Valenzuela-Muñoz, V., Nuñez-Acuña, G., Valenzuela-Miranda, D. and Gallardo-Escárate, C. (2020) Nanopore sequencing of microbial communities reveals the potential role of sea lice as a reservoir for fish pathogens. *Scientific Reports* 10: 2895.

González Gómez, M.P., Marín Arribas, S.L. and Vargas-Chacoff, L. (2016) Stress response of *Salmo salar* (Linnaeus 1758) facing low abundance infestation of *Caligus rogercresseyi* (Boxshall & Bravo 2000), an object in the tank, and handling. *Journal of Fish Diseases* 39, 853–865.

González, M.P., Marín, S.L., Mancilla, M., Cañon-Jones, H. and Vargas-Chacoff, L. (2020) Fin erosion of *Salmo salar* (Linnaeus 1758) infested with the parasite *Caligus rogercresseyi* (Boxshall & Bravo 2000). *Animals* 10, 1166.

Grayson, T.H., Jenkins, P.G., Wrathmell, A.B. and Harris, J.E. (1991) Serum responses to the salmon louse, *Lepeophtheirus salmonis* (Krøyer, 1838), in naturally infected salmonids and immunised rainbow trout, *Oncorhynchus mykiss* (Walbaum), and rabbits. *Fish and Shellfish Immunology* 1, 141–155.

Grimnes, A. and Jakobsen, P.J. (1996) The physiological effects of salmon lice infection on post-smolt of Atlantic salmon. *Journal of Fish Biology* 48, 1179–1194.

Herrero, A., Thompson, K.D., Ashby, A., Rodger, H.D. and Dagleish, M.P. (2018) Complex gill disease: an emerging syndrome in farmed Atlantic salmon (*Salmo salar* L.). *Journal of Comparative Pathology* 163, 23–28.

Holm, H., Santi, N., Kjøglum, S., Perisic, N., Skugor, S. and Evensen, Ø. (2015) Difference in skin immune responses to infection with salmon louse (*Lepeophtheirus salmonis*) in Atlantic salmon (*Salmo salar* L.) of families selected for resistance and susceptibility. *Fish and Shellfish Immunology* 42, 384–394.

Iversen, M.H. and Eliassen, R.A. (2014) The effect of allostatic load on hypothalamic-pituitary-interrenal (HPI) axis before and after secondary vaccination in Atlantic salmon postsmolts (*Salmo salar* L.). *Fish Physiology and Biochemistry* 40, 527–538.

Jakob, E., Barker, D.E. and Garver, K.A. (2011) Vector potential of the salmon louse *Lepeophtheirus salmonis* in the transmission of infectious haematopoietic necrosis virus (IHNV). *Diseases of Aquatic Organisms* 97, 155–165.

Johnson, S.C. and Albright, L.J. (1992) Comparative susceptibility and histopathology of the response of naïve Atlantic, chinook and coho salmon to experimental infection with *Lepeophtheirus salmonis* (Copepoda: Caligidae). *Disease of Aquatic Organisms* 14, 179–193.

Johnson, S.C., Treasurer, J.W., Bravo, S., Nagasawa, K. and Kabata, Z. (2004) A review of the impact of parasitic copepods on marine aquaculture. *Zoological Studies* 43, 229–243.

Jones, M.W., Sommerville, C. and Bron, J. (1990) The histopathology associated with the juvenile stages of *Lepeophtheirus salmonis* on the Atlantic salmon, *Salmo salar* L. *Journal of Fish Diseases* 13, 303–310.

Jónsdóttir, H., Bron, J.E., Wootten, R. and Turnbull, J.F. (1992) The histopathology associated with the preadult and adult stages of *Lepeophtheirus salmonis* on the Atlantic salmon, *Salmo salar* L. *Journal of Fish Diseases* 15, 521–527.

Kiemer, M.C. and Black, K.D. (1997). The effects of hydrogen peroxide on the gill tissues of Atlantic salmon, *Salmo salar* L. *Aquaculture* 153, 181–189.

Labra, Á., Bravo, S. and Marshall, S.H. (2020) Defining the role of *Caligus rogercresseyi* in transmission and spreading of *Piscirickettsia salmonis*. *Aquaculture* 528, 735489.

Lhorente, J.P., Gallardo, J.A., Villanueva, B., Carabaño, M.J. and Neira, R. (2014) Disease resistance in Atlantic salmon (*Salmo salar*): coinfection of the intracellular bacterial pathogen *Piscirickettsia salmonis* and the sea louse *Caligus rogercresseyi*. *PLoS ONE* 9(4), e95397.

Llewellyn, M.S., Leadbeater, S., Garcia, C., Sylvain, F.E., Custodio, M., Ang, K.P., Powell, F., Carvalho, G.R., Creer, S., Elliott, J. and Derome, N. (2017) Parasitism perturbs the mucosal microbiome of Atlantic salmon. *Scientific Reports* 7, 43465.

Matthews, C.G.G., Richards, R.H., Shinn, A.P. and Cox, D.I. (2013) Gill pathology in Scottish farmed Atlantic salmon, *Salmo salar* L., associated with the microsporidian *Desmozoon lepeophtherii* Freeman & Sommerville, 2009. *Journal of Fish Diseases* 36, 861–869.

Moltumyr, L., Gismervik, K., Gu, J., Gåsnes, S.K., Kristiansen, T.S., Rønnestad, I., Nilsson, J. and Stien, L.H. (2021) Does the thermal component of warm water treatment inflict acute lesions on Atlantic salmon (*Salmo salar*)? *Aquaculture* 532, 736048.

Mustafa, A., MacWilliams, C., Fernandez, N., Matchett, K., Conboy, G.A. and Burka, J.F. (2000a) Effects of sea lice (*Lepeophtheirus salmonis* Krøyer, 1837) infestation on macrophage functions in Atlantic salmon (*Salmo salar* L.). *Fish and Shellfish Immunology* 10, 47–59.

Mustafa, A., Speare, D.J., Daley, J., Conboy, G.A. and Burka, J.F. (2000b) Enhanced susceptibility of seawater cultured rainbow trout, *Oncorhynchus mykiss* (Walbaum), to the microsporidian *Loma salmonae* during a primary infection with the sea louse, *Lepeophtheirus salmonis*. *Journal of Fish Diseases* 23, 337–341.

Nolan, D.T., Reilly, P. and Wendelaar Bonga, S.E. (1999) Infection with low numbers of the sea louse *Lepeophtheirus salmonis* induces stress-related effects in postsmolt Atlantic salmon (*Salmo salar*). *Canadian Journal of Fisheries and Aquatic Sciences* 56, 947–959.

Nolan, D.T., Ruane, N.M., van der Heijden, Y., Quabius, E.S., Costelloe, J. and Wendelaar Bonga, S.E. (2000) Juvenile *Lepeophtheirus salmonis* (Krøyer) affect the skin and gills of rainbow trout *Oncorhynchus mykiss* (Walbaum) and the host response to a handling procedure. *Aquaculture Research* 31, 823–833.

Novak, C.W., Lewis, D.L., Collicutt, B., Verkaik, K. and Barker, B.E. (2016) Investigations on the role of the salmon louse, *Lepeophtheirus salmonis* (Caligidae), as a vector in the transmission of *Aeromonas salmonicida* subsp. *salmonicida*. *Journal of Fish Diseases* 39, 1165–1178.

Nylund, A., Wallace, C. and Hovland, T. (1993) The possible role of *Lepeophtheirus salmonis* (Krøyer) in the transmission of infectious salmon anaemia. In: Boxshall, G. A. and Defaye, D. (eds) *Pathogens of Wild and Farmed Fish: Sea Lice*. Ellis Horwood, New York. pp. 367–373.

Nylund, S., Nylund, A.R.E., Watanabe, K., Arnesen, C.E. and Karlsbakk, E. (2010) *Paranucleospora theridion* n. gen., n. sp. (Microsporidia, Enterocytozoonidae) with a life cycle in the salmon louse (*Lepeophtheirus salmonis*, Copepoda) and Atlantic salmon (*Salmo salar*). *Journal of Eukaryotic Microbiology* 57, 95–114.

Oelckers, K., Vike, S., Duesund, H., Gonzalez, J., Wadsworth, S. and Nylund, A. (2014) *Caligus rogercresseyi* as a potential vector for transmission of infectious salmon anaemia (ISA) virus in Chile. *Aquaculture* 420, 126–132.

Overton, K., Dempster, T., Oppedal, F., Kristiansen, T.S., Gismervik, K. and Stien, L.H. (2019) Salmon lice treatments and salmon mortality in Norwegian aquaculture: a review. *Reviews in Aquaculture* 11, 1398–1417.

Poppe, T.T., Dalum, A.S., Røislien, E., Nordgreen, J. and Helgesen, K.O. (2018) Termisk nehandling av laks. *Norsk Veterinærtidsskrift* 130 (3), 148–156.

Rae, G.H. (1979) On the trail of the sea lice. *Fish Farmer* 2, 22–25.

Robledo, D., Gutiérrez, A.P., Barria, A., Yáñez, J.M. and Houston, R.D. (2018) Gene expression response to sea lice in Atlantic salmon skin: RNA sequencing comparison between resistant and susceptible animals. *Frontiers in Genetics* 9, 287.

Rodger, H. and Mitchell, S. (2007) Epidemiological observations of pancreas disease in Ireland. *Journal of Fish Diseases* 30, 157–167.

Rojas, V., Sánchez, D., Gallardo, J.A. and Mercado, L. (2018) Histopathological changes induced by *Caligus rogercresseyi* in rainbow trout (*Oncorhynchus mykiss*). *Latin American Journal of Aquatic Research* 46, 843–848.

Ruane, N.M., Nolan, D.T., Rotillan, J., Costelloe, J. and Wendelaar Bonga, S.E. (2000) Experimental exposure of rainbow trout *Oncorhynchus mykiss* (Walbaum) to the infective stages of the sea louse *Lepeophtheirus salmonis* (Krøyer) influences the physiological response to an acute stressor. *Fish and Shellfish Immunology* 10, 451–463.

Sveen, S., Øverland, H., Karlsbakk, E. and Nylund, A. (2012) *Paranucleospora theridion* (Microsporidia) infection dynamics in farmed Atlantic salmon *Salmo salar* put to sea in spring and autumn. *Diseases of Aquatic Organisms* 101, 43–49.

Tadiso, T.M., Krasnov, A., Skugor, S., Afanasyev, S., Hordvik, I. and Nilsen, F. (2011) Gene expression analyses of immune responses in Atlantic salmon during early stages of infection by salmon louse (*Lepeophtheirus salmonis*) revealed bi-phasic responses coinciding with the copepod-chalimus transition. *BMC Genomics* 12, 141.

Wagner, G.N., McKinley, R.S., Bjørn, P.A. and Finstad, B. (2003) Physiological impact of sea lice on swimming performance of Atlantic salmon. *Journal of Fish Biology* 62, 1000–1009.

Wagner, G.N., Fast, M.D. and Johnson, S.C. (2008) Physiology and immunology of *Lepeophtheirus salmonis* infections of salmonids. *Trends in Parasitology* 24, 176–183.

Walde, C.S., Jensen, B.B., Pettersen, J.M. and Stormoen, M. (2021) Estimating cage-level mortality distributions following different delousing treatments of Atlantic salmon ***(Salmo salar)*** in Norway. *Journal of Fish Diseases* 44, 899–912.

Weli, S.C., Dale, O.B., Hansen, H., Gjessing, M.C., Rønneberg, L.B. and Falk, K. (2017) A case study of *Desmozoon lepeophtherii* infection in farmed Atlantic salmon associated with gill disease, peritonitis, intestinal infection, stunted growth and increased mortality. *Parasites and Vectors* 10, 370.

White, H.C. (1940) "Sea lice" (*Lepeophtheirus*) and death of salmon. *Journal of the Fisheries Research Board of Canada* 5, 172–175.

Wootten, R., Smith, J.W. and Needham, E.A. (1982) Aspects of the biology of the parasitic copepods *Lepeophtheirus salmonis* and *Caligus elongatus* on farmed salmonids, and their treatment. *Proceedings of the Royal Society of Edinburgh* 81B, 185–197.

Wright, D.W., Nowak, B., Oppedal, F., Bridle, A. and Dempster, T. (2017) Free-living *Neoparamoeba perurans* depth distribution is mostly uniform in salmon cages, but reshaped by stratification and potentially extreme fish crowding. *Aquaculture Environment Interactions* 9, 269–279.

CHAPTER 8

Sea lice physiology

David Fields[1], Robert Morefield[2] and Ian Bricknell[2]

[1]Bigelow Laboratory for Ocean Sciences, Maine, USA
[2]Aquaculture Research Institute and School of Marine Sciences, University of Maine, USA

8.1 Introduction

The three major species of sea lice, *Lepeophtheirus salmonis* (Krøyer, 1837), *Caligus elongatus*, von Nordmann, 1832, and *Caligus rogercresseyi* (Boxshall & Bravo, 2000), are the most important copepod parasites in aquatic fish production, yet very little research has been conducted on their physiology. At the time of writing, a Web of Science search using the key terms "Physiology" and "*Lepeophtheirus salmonis*" returns 31 results (out of a population of 1319 publications that use *Lepeophtheirus salmonis* as a key term, or 2.3%). This figure is further compounded when those 31 publications that use "Physiology" and "*Lepeophtheirus salmonis*" are interrogated. Only seven of these manuscripts look at the physiology of the sea louse *Lepeophtheirus salmonis*; the rest investigate the effect the parasite has on the physiology of its salmonid host, Atlantic salmon (*Salmo salar* L.). There are also physiological studies conducted on a variety of other salmonid species including *Oncorhynchus* spp. as well as *Salmo* and *Salvelinus* spp. A similar picture is found by repeating the search using the terms *Caligus elongatus* or *Caligus rogercresseyi*, with only about 0.5% of all papers published on the big three sea lice species of concern investigating some aspect of the louse's physiology rather than the impact the parasite has on its host.

This is not as bleak as it seems. Although the physiology of the Siphonostomatoida is relatively poorly studied when compared to other aspects of their biology, some aspects of their physiology can be inferred from those of other ecdysozoans/arthropods. An example is the pathways involved in moulting/ecdysis. Additionally, the considerable number of studies that have been carried out looking at parasiticides targeting sea lice can throw some light on the physiology of sea lice. As most of these compounds started life as insecticides to protect terrestrial crops, their targets and mechanisms of action have generally been well characterized. This is important as, by inference, if an insecticide inhibits a certain enzyme in one organism, there is a high probability that it will inhibit the same enzyme in sea lice. Hence, drug specificity can lead to inference about the presence or absence of particular metabolic pathways in *L. salmonis*. For example, the anti-louse drug Calicide® (teflubenzuron) is a benzoylphenyl urea chitin synthesis inhibitor preventing the correct function of glycosyltransferase chitin synthase, which in turn precludes new cuticle from forming successfully (Branson et al., 2000; Ritchie et al., 2002; Campbell et al., 2006; Toksen et al., 2009; Eichner et al., 2015a, 2015b, 2015c). Teflubenzuron is particularly effective against lepidopteran pests of terrestrial crops. However, it seems to function in a similar way in *L. salmonis*, especially impacting the chalimus stages. This suggests that glycosyltransferase chitin synthase plays a key role in ecdysis of the attached stages in *L. salmonis*.

Most siphonostomatoid copepods have a siphon-like mouth tube and most employ a frontal filament, at least in the larval stages, to attach to their hosts. These attributes have contributed to their great diversity (approximately 2200 species in 39 families) and success. Sea lice (caligids) are well-known fish parasites and are very damaging to fishes in cage culture. Atlantic salmon are particularly susceptible to the salmon louse (*L. salmonis*). In parasitic copepods, the infective larva is, with rare exceptions, the first copepodid stage, and life cycles are direct, involving only a single host. In *L. salmonis*, the nauplius phase is reduced to two lecithotrophic stages and has uniseriate egg strings in which disc-shaped eggs are closely packed into one row extending the length of the string. Most nauplii have a single pair of modified caudal setae known as balancers (although their function is unknown). The life cycles of sea lice have been the topic of much research and debate of late. This attention is due to the damage that they cause cage-cultured fishes, particularly salmonids (Lafferty et al., 2015). Knowledge of the correct life cycle is critical in determining when to treat for the damaging fish-associated stages.

(For a more comprehensive review of the siphonostomatoid copepods see Chapter 2, this volume.)

Few areas have attracted direct research, in terms of sea louse physiology. However, some attention has been paid to their osmoregulation and tolerance of low salinities, iron uptake, digestion, lipid metabolism, immune-modulation, and chemoreception.

In this chapter we plan to review the current status of sea lice physiology research and highlight areas that are rapidly emerging as key areas for further research.

8.2 The cuticle

As mentioned in the introduction there have been some inferences drawn as to the enzymes involved in cuticular synthesis, not least through the effect of benzoylphenyl urea chitin synthesis inhibitors on glycosyltransferase chitin synthase (Chavez-Mardones et al., 2016). Better known is the structure of the cuticle of the sea louse, in particular *L. salmonis* (Bron et al., 2000), as well as considerable research having been conducted on transcriptomics and endocrinology of moulting in *L. salmonis* (Eichner et al., 2008, 2015a, 2015b, 2015c, 2018; Harðardóttir et al., 2019).

Although the study by Bron et al. (2000) specifically looked at the cuticle of the chalimus II stage of *L. salmonis*, it is very comprehensive in nature and identified that the cuticle was fairly typical of other free-living and parasitic copepodids, such as the cuticle of *Lernaeocera branchialis* (L.) or *Cletocamptus retrogressus*, Schmankevitsch, 1875 (Bresciani, 1986; Smith and Whitfield, 1988). The semi-transparent cuticle consists of four layers of epicuticle with a distinct fuzzy outer layer, which is separated from the inner and outer epicuticle by a distinct layer of "transitional" procuticle. Beneath the cuticle is a single layer of epidermal tissues which is rich in integumental glands and chromatophores. In the chalimus II stage, at least, there were no transcuticular canals or pores observed although sensilla have been seen emerging through the cuticle in adult sea lice, especially around the antennae and antennule (Overgard et al., 2016). It is reasonable to conclude that the cuticle of *L. salmonis* is not highly specialized and is functionally closer to free-living copepods than it is to some of the specialized cuticles that have evolved in parasitic crustaceans that have a highly modified body plan.

The structure of the cuticle of sea lice is poorly studied and more research needs to be done especially in regard to the planktonic and the mobile stages, and the relationship of sensilla to the structure of the cuticle, as well as more basic studies on the diffusion rate of oxygen through the cuticle and its basic biomechanics.

8.2.1 Cuticular transcriptomics

Generally, the transcriptome of sea lice is not well researched. Two of the earliest key studies for *L. salmonis* were conducted by Eichner et al. (2015a, 2015b, 2015c) and Harðardóttir et al. (2019). These studies identified six genes associated with chitin synthesis and three genes associated with enzymes that digest old cuticle during ecdysis. The six genes found in the *L. salmonis* genome were LsCHS1, LsCHS2, LsGFAT, LsGNA1, LsAGM, and LsUAP. These genes have a very high level of homology with the functional enzymes found in in the chitin synthesis pathway. Three further enzymes, the chitinases LsChi1, LsChi2, and LsChi4, were characterized by Eichner et al. (2015a, 2015b, 2015c) where they were shown to be important in digesting and weakening the old cuticle prior to the onset of ecdysis.

The transcriptome of *Caligus rogercresseyi* has been less well studied. The most comprehensive study was carried out by Chavez-Mardones et al. (2016). This research looked at the transcriptome of *C. rogercresseyi* that had been treated with azamethiphos (Bayer) and deltamethrin (AlphaMax®). Although these studies were looking at the gene expression that was enhanced or suppressed during treatment with these drugs, numerous genes that have considerable homology with *L. salmonis* were identified. The genes included chitinase, ecdysone receptors, and cuticle proteins 6, 7, 18.6, 19, and 19.8 as well as numerous other genes associated with ecdysis in other arthropods.

A full discussion of the transcriptome of sea lice can be found in Chapter 12, this volume. It is clear that the study of the transcriptomics of ecdysis and cuticle formation have a significant role to play in understanding the physiology of sea lice and allowing the development of safer more targeted antiparasitics.

8.2.2 Respiration

Gas exchange within copepods is considered to occur directly through the exocuticle of the animal. It is likely that areas of thinner cuticle may serve as more specialized zones for gas exchange. In particular, the swimming appendages, which have both a high surface area and relatively thin cuticle, may serve a respiratory function with their movements through the water. In swimming stages, this supports increased gas exchange by thinning boundary layers to allow more rapid gas diffusion. In the more voluminous parts of the body (cephalosome and metasome), gas exchange is augmented by actively pumping fluid in the space below the outer integument. There are relatively few data sets on the oxygen consumption rates of any of the stages (Thompson et al., 2019).

8.2.3 Haemocyanin dogma

There has been a general assumption that sea lice, like many other arthropods, use haemocyanin as the primary oxygen transporting protein in the haemolymph. This assumption was based on the fact haemocyanin is the second most common oxygen transporting molecule in metazoan animals and is very common in crustaceans. However, no systematic investigation of the oxygen transporting molecules of sea lice, or even the protein profiles, of the haemolymph has ever been carried out and there is no evidence that any of the three important species of sea lice *L. salmonis, C. rogercresseyi* and *C. elongatus* utilize haemocyanin.

The recent sequencing of the genome of *L. salmonis* indiates that there are no genes present that code for haemocyanin, suggesting that this protein is absent, at least in *L. salmonis* and *C rogercresseyi.* If there are no genes for haemocyanin then what protein is used to transport oxygen in the haemolymph to the tissues? Fox (Fox, 1945) first identified haemoglobin in parasitic copepods and it is know from many crustacean clades, including parasitic copepods, Malacostracans, Cladocera and Branciopoda (Burmester, 2015). Given that haemoglobin is the most common oxygen transporting protein found in metazoans it is quite feasible that the parasitic copepods have evolved haemoglobin, especially given the ready access they have to the haem molecule from their hosts. However, this question will not be solved until a full genomic and proteonomic analysis is complete and simply remains a hypothesis for now.

8.3 Gut morphology

The anatomy and histology of the alimentary canal have been described for a range of copepod species belonging to different orders. In the Siphonostomatoida the alimentary canal consists of a short esophagus, a midgut, and a short hindgut (Scott, 1901; Nylund et al., 1992). The esophagus and foregut are derived from ectoderm and so are lined with chitin. The ectodermal origin of the foregut allows the development of setae and other cuticular features, resulting in a structure that can be adapted to process a wide variety of food types. Similarly, the hindgut in crustaceans is also derived from ectodermal tissue and lined with chitin. In the harpacticoid copepod *Tigriopus californicus*, the hindgut epithelium is highly invaginated, and the abundance of mitochondria in the cells suggests that they are likely involved in water transport (Sullivan and Bisalputra, 1980). The midgut, where final digestion and absorption occurs, can be divided into three successive zones (Arnaud et al., 1978; Hallberg and Hirche, 1980) containing a cecum, a stomach, and an intestine. Broadly, three different enterocytes have been identified within the epithelium lining the midgut (cf. Arnaud et al., 1978): B-cells, F-cells, and R-cells. The F-cells have digestive function while the B-cells appear to have both digestive and excretory function. The R-cells are assumed to have an absorptive function. Much of our understanding of the copepod gut comes from work on calanoid copepods that are primarily herbivorous and very little is known about the ultrastructure of enterocytes in the Siphonostomatoida.

8.4 Digestive physiology

The physiological basis of digestion in sea lice is poorly understood with only a few dedicated studies having been conducted (for example, that by Andrade-Salas, 1997). It has been suggested that adult sea lice, being carnivorous, might have a similar digestive process to other carnivorous arthropods. This would suggest that they rely upon an acidic pH in the foregut that allows proteolytic enzymes (peptidases) such as trypsins to function, hydrolyzing host proteins that have been ingested during feeding. Unfortunately, there is virtually no evidence backing up the hypothesis that either *L. salmonis* or *C. rogercresseyi* have an acid-based digestion.

Vigneau (2006) found a unique aminopeptidase that was reported as being a "… unique aminopeptidase which has an affinity profile not observed in any other aminopeptidases, requiring zinc in its active site. It has a weakly hydrophobic core, making it require a saline environment …". Such a unique aminopeptidase seems to have evolved to function optimally in a high saline environment such as seawater. Its basic enzymology, however, certainly does not suggest that it has evolved to function in a high pH and low salinity environment. Subsequent transcriptomic research has also been very useful in identifying potential vaccine antigen candidates (see Chapter 20), especially in the chalimus and larval stages (Hamilton et al., 2018). Although the delivery and optimization of a sea louse vaccine remains some way away this approach holds a lot of potential for future vaccines.

Vigneau (2006) proposed that the digestive tract of sea lice has to have a high salinity environment to permit the unique aminopeptidase to function and, in view of the observation of possible osmocytes, at the transition between posterior midgut and hindgut (Bron et al., 2000). The aminopeptidase functioned optimally at 740–1200 milliosmoles l^{-1}, far higher than the osmolarity in salmon plasma, which is typically around 330–350 milliosmoles l^{-1}. It could be suggested that the salts found in the digestive tract were obtained through seawater ingress via the anus or through ingestion during feeding on the host epidermis. The louse mouthparts are not highly specialized, for example to penetrate a blood vessel, as in other members of the Siphonostomatoida clade. It can be hypothesized that as the digestive tract of *L. salmonis* has a high osmotic potential (Bricknell and Pietrak, 2022, in review) due to the ingestion or drinking of seawater, then the pH will be high because of the buffering capacity of the ingested seawater. It is also unlikely that lice secrete hydrochloric acid to reduce the pH of the digestive tract as this would be energetically very expensive and Bron et al. (1993) did not identify any parietal cells or parietal cell analogues in their survey of the digestive tract of *L. salmonis*.

The environment of the gut lumen, while poorly characterized, is nevertheless extremely important in terms of its potential relevance to the consistent failure of vaccines targeted against the adult louse. For antibodies, immune cells, and defenses to function in the digestive tract of a sea louse it would be expected that the environment is similar to that seen in the host. It would be expected for the components of the immune system to function optimally or near optimally the environment would need to be close to pH 7.4, with an osmolarity of 350 milliosmoles l^{-1}. Unfortunately, *L. salmonis* seems to have a pH that is closer to seawater (Bricknell and Pietrak, 2022, in review; and later in this chapter) with an osmolarity that is marginally lower than sea water when the gut is empty, and

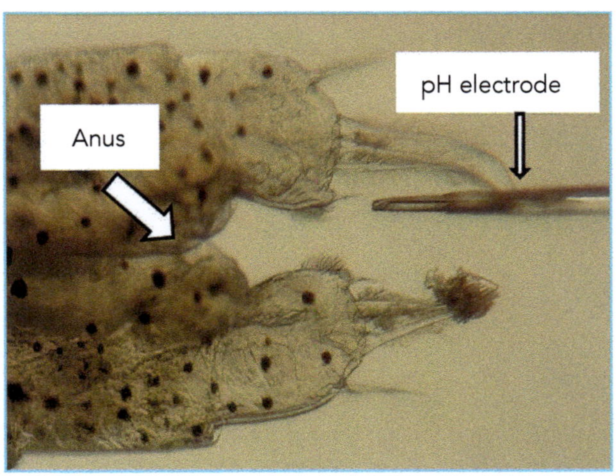

Fig. 8.1 Micro pH electrode being inserted into the distal digestive tract of *L. salmonis* (D. Fields, 2022).

around 850 milliosmoles l^{-1} after feeding. Therefore, digestion is sufficiently slow to allow antibody binding or defensin function. Given the limited data for the physical environment inside the digestive tract of lice it can be assumed that it is a hostile environment for the ingested component of the salmon's immune system to function. Any antibodies produced by a vaccine raised against sea louse gut antigens would have to work in an extremely high osmotic and pH environment (Bricknell et al., 2002) with several proteases trying to digest these proteins before they can damage the gut lining (Vigneau, 2006; Hamilton et al., 2018).

Blood ingestion is, nevertheless, observed in all stages from copepodid onwards, with the likelihood of observing blood in the gut scaling with the size of the stage involved. This may simply reflect the change of depth of feeding penetration and hence blood abundance with size or, in the case of adult females at least, a specific blood feeding behavior. Brandal et al. (1976) first suggested that sea lice, particularly the females, feed on blood and proposed that a blood meal was required for successful egg production, such as is the case for anautogenous mosquitoes such as *Aedes aegypti* (Tsuji et al., 1990). Of the more recent studies looking at the requirement of heme or a blood meal for egg development, the results remain inconclusive. It is clear that sea lice do not always have clearly detectable levels of blood in their digestive tract (Bricknell and Pietrak, 2022, in review), unlike true hematophagous arthropods such as fleas, ticks, and mosquitoes and some other members of the Siphonostomatoida such as *Lernaeocera branchialis* (L.). Certainly, there is a basic physiological requirement for iron, even if it is only involved in electron transfer pathways. Some studies have suggested that host iron restriction occurs as a response to a sea louse infection in salmonids (Valenzuela-Munoz and Gallardo-Escarate, 2017). Typically, this is represented by a rapid increase in serum iron chelators such as ferritin and transferrin in the fish (Trosse et al., 2015; Valenzuela-Munoz et al., 2021). Two publications (Trosse et al., 2015; Heggland et al., 2019) suggest that iron, from hemoglobin and ferritin, is required by females for successful egg development. However, the facultative or obligate nature of this requirement for iron in the diet is still unclear.

The importance of gut pH cannot be overstated. Copepod digestion occurs through well-regulated enzymatic activity along their linear digestive tract (Hassett and Landry, 1983). Assuming that the pH optima of copepod digestive enzymes are like those of other crustaceans (with significantly longer gut lengths), the optimal gut pH would need to span a wide range of pH 2–9 (Mayzaud et al., 1994). Maximum amylase activity is in the range of pH 5–7.5 (Dutta et al., 2009), many extracellular proteases secreted into the digestive system perform best in highly acid environments of pH 1–2, and many intracellular proteases and secreted metalloproteases function at higher pH values of pH 5–7.5.

Lipases, in contrast, show the greatest efficiency in slightly basic conditions at or near pH 8. Such a wide span in the pH optima of the different enzymes suggests the existence of localized variations within the gut or a mix of intracellular and extracellular digestion. It is difficult to imagine how such a large gradient in pH can be maintained in a gut of length 5–10 mm in the adults, let alone the extremely short gut of 1.0 mm in the parasitic copepodid stage. That said, some termite species can maintain a gradient in pH of three log orders over a 300 μm span (Brune et al., 1995). So, it is feasible that some aspect of digestion occurs intracellularly inside vacuoles.

There are very few direct measurements of the gut pH of copepods. Dye studies of the alimentary tract of the calanoid copepod *Calanus helgolandicus* suggest a gradient in pH of pH 6.9–8.0 between the foregut and the hindgut (Pond et al., 2012). However, these authors recognized that the dye approach they utilized may have averaged the pH signal over a wide range of spatial scales, missing very small changes in local pH they sought to measure. More recent measurement using micro-sensors (pH and oxygen) found pH values of 5.5 in the lower gut (20% of the entire intestinal tract) in large copepods (*Calanus helgolandicus*) and a pH of 6.8 in smaller animals of the same species. Although they sampled only the lower 20% of the intestinal tract, they did find two distinct pH zones. Within the region of the urosome they found the pH value to drop by around 1.2 (pH 6.6) over the first 600 μm and a further drop in the pH value by 1.3 (to pH 5.5) in the hindgut. This zonation of pH suggests functional division within the gut. The pH of the digestive tract of sea lice has been investigated using a small (100 μm) pH electrode inserted in the anus of recently fed females and males (Figs. 8.1 and 8.2). The pH ranges from pH 6.5 near the hindgut and anus, to pH 7.0 in the foregut, near the mouth.

8.5 Osmoregulation

Very little has been written about osmoregulation in sea lice, although the earliest papers on salinity tolerance date back to the mid-1980s (Wootten et al., 1982; Hahnenkamp and Fyhn 1985). These papers both noted that adult *L. salmonis* have a low tolerance to a hyposaline environment, with Wootten et al. (1982) noting that there was an impact on *L. salmonis* survival below 29 parts per thousand (ppt). Hahnenkamp and Fyhn (1985) reported that *L. salmonis* died within 24 hours of being removed from a fish and placed in freshwater, while lice attached to salmon could survive for about seven days in freshwater. This may suggest that ions could be obtained directly from the host at least for a brief period of time in a hypoosmotic environment, being ingested in tissue via the mouth tube.

A series of studies by Heuch et al. throughout the late 1990s and early 2000s (Heuch et al., 1995, 2002, 2003, 2009) investigated the use of oceanic salt lenses by *L. salmonis* in inshore Norwegian ecosystems. It was discovered that sea lice larvae aggregate at haloclines. Initially this was proposed as a mechanism that allows larval sea lice to avoid the low salinity, low density surface water, which floats on top of more saline, higher density, deeper water. A study by Bricknell et al. (2006) subsequently showed that salinity had a marked impact on settlement success and copepodid survival. Salinities below 29 ppt caused a significant decline in settlement success and survival, suggesting typical estuarine salinities effectively prevented louse settlement. This was noteworthy as estuaries have

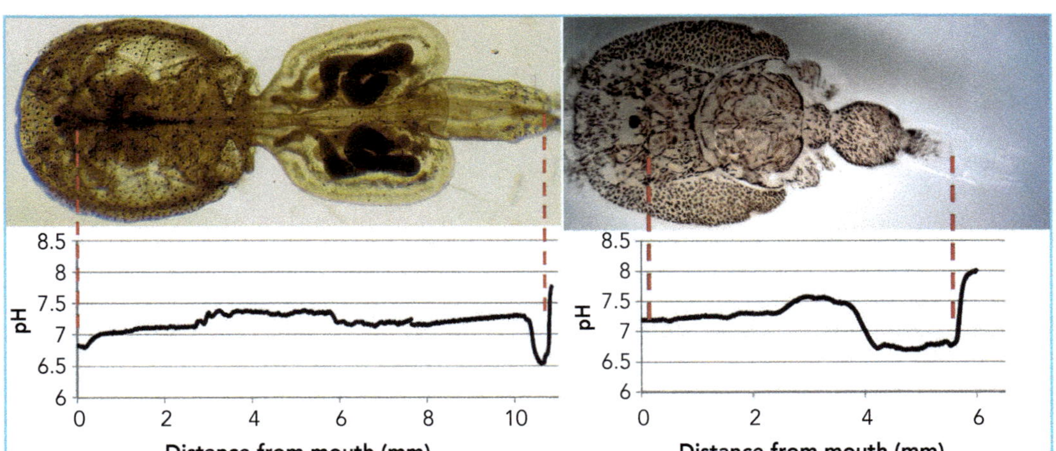

Fig. 8.2 pH levels seen in the digestive tract of *L. salmonis* as determined by a micro pH electrode (Fields and Browman, unpublished findings).

been proposed as a major site of sea lice infection, especially of migratory smolt of Atlantic salmon and sea trout (*Salmo trutta* L.) (Thorstad et al., 2015). However, if the hydrology of the embayment causes the salinity to fall below the key value of 29 ppt for any significant period during a tidal cycle, then it is likely that settlement will not occur (Bricknell et al., 2006).

It is reasonable to consider all aquatic arthropods as being animals mostly encased in an impermeable box sealed with the waxy cuticle. Diffusion of ions can occur in a few limited permeable external areas, such as the arthrodial membranes at limb joints. However, the arthrodial membranes are likely to be highly impermeable too as they consist of highly tanned proteins.

Microscopic reviews of the transcuticular cells of the cuticle and arthrodial membranes of sea lice have not identified any transcuticular mitochondrial rich ion transporting cells (MRITCs) as seen in other crustaceans (Doyle and McMahon 1995; Watts et al., 1996; McMahon and Doyle, 1997; Henry and Watts, 2001; Brooks and Mills, 2006) and aquatic insects (Bricknell and Potts, 1988). The probability is high that ion uptake and excretion occurs in the digestive tract of sea lice. In a major histological survey of the digestive tract of *L. salmonis*, Bron et al. (1993) identified a small cluster of cells at the distal end of the hindgut that were proposed to be osmocytes and involved in ion transport and osmoregulation. However, no further studies have investigated the anatomy of these potential osmocytes, and there have been no studies on ion flux in isotonic, hyposaline, and hypersaline environments that sea lice could potentially encounter in the wild.

It is surprising that it took about 160 years since the first description of *L. salmonis* before the osmolarity of the hemolymph was measured (Andrade-Salas, 1997; Bricknell and Pietrak, 2022, in review) and that study did not look at the ionic composition of sea lice hemolymph, just the osmolarity. Bricknell and Pietrak (2022, in review) demonstrated that the osmotic potential of the hemolymph was close to that of seawater (1000 mOsm kg^{-1}) in starved sea lice and about 20% (approximately 820 mOsm kg^{-1}) lower in sea lice that had been actively feeding on salmon. The latter value found by Bricknell and Pietrak was in close agreement (878 mOsm kg^{-1}) with that reported by Andrade-Salas in his PhD thesis in 1997. These data throw up some interesting points concerning the apparent dilution of the contents of the digestive tract in feeding *L. salmonis*, suggesting that the contents of feeding animals consists of about 28% host material (Bricknell and Pietrak, 2022, in review). These findings support the conclusion that *L. salmonis* is a stenohaline osmoconformer (Wootten et al., 1982; Hahnenkamp and Fyhn, 1985; Heuch et al., 2002; Bricknell et al., 2006), this being typical of other members of copepoda (Villate, 1991; Chew and Chong, 2011; Xu and Gao, 2011; Svetlichny et al., 2017). When combined with reports of a critical survival salinity around 29 ppt, it seems safe to assume that *L. salmonis* is, at best, a poor osmoregulator.

8.6 Nutrition and metabolic pathways

8.6.1 Lipid storage

The endogenous reserves that embryos, nauplii, and free-living copepodids are reliant upon are principally energy-rich neutral lipids. Neutral lipids consist of various forms of fatty acids which are made of at least four carbons with a carboxylic acid group at the end of the macromolecule, an arrangement conserved across taxa (Gonzalez et al., 2016). The high energy content of lipids makes them an ideal storage medium compared to proteins and carbohydrates, typically containing twice the energy content (Lee et al., 2006). Consequently, energy storage in the form of neutral lipids is a ubiquitous strategy in zooplankton, where it provides energy for periods of low food supply, vertical migration, reproduction, and predator avoidance, among others.

The available energetic content of neutral lipids in *L. salmonis* copepodids has been estimated at being comparable to the copepodid stage of free-living, non-parasitic copepods in winter (Tucker et al., 2002). Of the four types of lipid storage known in marine zooplankton, triacylglycerol and wax esters are the most predominant (Lee et al., 2006). Parasitic sea lice have been shown to predominantly utilize triacyglycerols (TAG) as storage lipids (Lee, 1975; Tocher et al., 2000). Triacylglycerols are commonly used as the main energy reserve in marine mammals, fish, and krill species but are relatively uncommon in copepods which primarily use wax esters (Tocher et al., 2010). The relative advantages of each lipid class mostly lie in utilization rates and regulation. TAGs can be quickly hydrolyzed to provide short-term energy and contain more energy per molecule, where wax esters are utilized more slowly due to tight regulation through lipases, a beneficial attribute for diapause (Tucker et al., 2000; Lee et al., 2006). The use of TAGs by parasitic sea lice supports the hypothesis that wax esters are primarily used for long-term energy storage as the short copepodid stages require the constant energy reserves obtained by lipid droplets in the eggs. However, there is still considerable discussion among copepodologists regarding the benefits of different lipid storage types and their evolutionary origins. It should be noted that in contrast to *L. salmonis*, which is a more sedentary species with a limited host range, the generalist *C. elongatus* appears to display specific lipid storage structures in adults that appear similar to the oil sacs of free-living calanoid copepods and may reflect a lifestyle that allows longer periods in the plankton (J. Bron, personal communication).

8.7 Growth

Growth of *L. salmonis*, *C. elongatus*, and *C. rogercresseyi* follows the classic copepod life cycle (see Chapter 2 and Chapter 3, this volume, for the life cycle figures of each species). In the case of these species there is direct hatching from an egg that gives rise to two successive nauplius stages, an infective copepodid stage, then two chalimus stages that are physically attached to the host by means of a frontal filament. According to species there may be an intervening pre-adult stage, which is then followed by the adult stage. During this period, the animals increase in size by over 25 times from a 300–350 µm freshly hatched nauplii to a 12–15 mm-long adult.

8.8 Reproduction

8.8.1 *Sea lice reproduction and life history*

Vitellogenesis, the production of yolk, appears to be initiated long before female *L. salmonis* reach maturity, with evidence showing that this process begins in the chalimus phase of the female life cycle, where associated genes (vitellogenin 1 and 2) are overexpressed in both chalimus and adult female stages (Gonçalves et al., 2014). After a female louse matures, continuous egg production is initiated in the ovaries, from which the eggs migrate through oviducts to the genital complexes where lipids are incorporated (Dalvin et al., 2011). The eggs mature into vitellogenic oocytes and are fertilized as they are extruded into a pair of external egg strings (Khan et al., 2018). Each of these egg strings can contain from 50 to 700 individual eggs which hatch sequentially from the distal end and moult within approximately 18 hours into free-swimming nauplii (Wootten et al., 1982; Costello, 2006). As adult females have been shown to be capable of extruding up to 11 separate clutches of egg strings (Hamre et al., 2009), it is apparent that they are capable of very high fecundity relative to free-living copepod species.

One of the most important metrics of egg quality is the accumulation of lipids, as these reserves must fuel growth through both naupliar stages and into the infective copepodid stage (Taccardi et al., 2021). The longevity of the infective copepodid stage has a significant impact on the likelihood of the parasite locating and infecting a host. As larval lice age through the successive moults of the free-living phase, their reserves decline. In calorimetric research conducted on the planktonic stages of *L. salmonis*, Tucker et al. (2000) showed a steady rate of decline in energy stores with time, this rate remaining steady until the copepodid stage day five when a sharp decline was observed. The energy reserves utilized for development to the copepodid stage are substantial, as the non-host-searching naupliar stages account for approximately 25–50% of their planktonic life (Samsing et al., 2016). The infective period, the copepodid stage, lasts for a comparatively short time, with Johnson and Albright (1991) demonstrating survival at only two to eight days while Gravil (1996) determined they could survive for 9–17 days. The survival time is highly dependent upon water temperature. When energy reserves decline, the ability of sea lice to infect a host becomes compromised, as Tucker et al. (2000) observed when older copepodids aged seven days were shown to have a reduced ability to infect and settle on a host. The reduced infection capability when energy reserves are depleted may not be solely a function of metabolic capacity. Sea lice are known, like many specialist parasites, to immunomodulate their hosts. Inflammatory responses to the parasite are significantly reduced in fish infected by both feeding adult and attached chalimus-stage sea lice (Johnson and Albright, 1992). Experimental evidence has definitively shown that *L. salmonis* secrete prostaglandin E_2 to aid invasion of host tissues through immunomodulation (Fast et al., 2007). Prostaglandin E_2 is a potent vasodilator and is known to downregulate pro-inflammatory cytokines (Tabouret et al., 2003). Prostaglandin E_2 synthesis requires lipids as it is a derivative of omega-6 fatty acid ARA (Tocher et al., 2010). Feeding in sea lice begins as the copepodid moults into the chalimus stage. Therefore, in order to immunomodulate the host and successfully feed, enough lipids must be available to synthesize immune suppressants. These results may explain the reduced ability of sea lice to infect a host at an advanced copepodid age.

An interesting series of experiments using RNAi to inhibit sodium/potassium ATPase (Na^+/K^+-ATPase) in female *L. salmonis* gave a set of unexpected results. Classically Na^+/K^+-ATPase is associated with osmoregulation and is found in elevated levels along the basal membrane of MRITCs. In a study by Komisarczuk et al. (2018), RNAi knockdown of the Na^+/K^+-ATPase caused a series of major malformations to develop in the ovaries with very pathological non-viable eggs being produced. This suggest that Na^+/K^+-ATPase plays a significant role in the transport of materials into the developing egg and disruption of Na^+/K^+-ATPase has a marked effect on reproductive success.

8.9 Muscle and locomotion

There are no studies that look directly at the structure and function of sea lice muscle tissues. However, there are several studies that have looked indirectly at the function and structure of muscle tissue in sea lice. Some of the earliest high quality images of *L. salmonis* muscle tissue can be found in James Bron's PhD thesis (1993), as well as in more recent studies such as those by Komisarczuk et al. (2019)

and Schram (2000b). In these studies the muscle cells seem to be predominantly striated muscle and not dissimilar to the muscle cells described in other copepods (Schram, 2000a).

8.10 Temperature responses

8.10.1 Metabolism and overwintering

Modulation of the copepodid metabolism may allow for some extension of the essential infective copepodid stage and has potential as a strategy for enduring the reduced temperatures of higher latitudes. Temperature has a major regulatory effect on sea lice egg production and viability. Heuch et al. (2000), Ritchie et al. (1993), and Gravil (1996) found that egg strings sampled at a low temperature (7°C) had a greater total length and contained more eggs on average than those sampled at a higher temperature (12°C). However, the individual lower-temperature egg strings were of lower quality as they were smaller and had a higher percentage of non-viable unhatched eggs. The limiting low temperature for eggs to hatch and develop into the copepodid stage appears to be 3°C or below, above which development time increased along with temperature (Boxaspen and Næss, 2000). This has been confirmed *in situ* on salmon aquaculture farms where infection in fish is positively correlated with temperature (Boxaspen, 1997). It seems clear that low temperatures affect egg production and larval viability negatively. However, infection and egg production persist over the winter months. The overwintering strategies employed by sea lice are currently unestablished. What is known is that sea lice endure the cold winter temperatures, which in some regions such as off the coast of New England and Eastern Canada can be as low as -2°C, and populations rise under less adverse environmental conditions each spring. Considerable attention has been paid to the lipid content of wintering sea lice ova (Taccardi et al., 2021). However, less attention has been given to the possibility of an induced seasonal reduction in metabolic rates independent of the reduced temperature related metabolism.

Heuch et al. (2000) also found that generation times were strongly affected by temperature. At 12°C the time from infecting a fish as copepodid to the F1 generation was 30 days; at 7°C the generation time was over three times longer, at 99 days. Photoperiod has also been implicated in altering reproductive output. Egg size in *L. salmonis* has been shown to correlate more closely with photoperiod than temperature, indicating seasonal non-temperature related effects (Ritchie et al., 1993).

Diapause is a state of genetically programmed developmental arrest in response to adverse environmental conditions that has been tied to photoperiod, temperature, and food availability (Christie et al., 2013). As many as 70% of copepods of the mesozooplankton are known to enter diapause, especially in areas of climatic extremes (Record et al., 2018). The "couch potato" protein (CPO) is a key biomolecule involved in regulating diapause and nervous system development within insects. Recently, putative homologs have been discovered in the copepod *Calanus finmarchicus* (Christie et al., 2013). Upregulation of the CPO gene was found in the diapause preparatory copepodid stage of *C. finmarchicus*. While sea lice are not known to enter a full quiescent diapause stage or undergo developmental arrest, elements of diapause-like induced states may have been conserved, as the CPO gene has recently been identified in *C. rogercresseyi* (Gallardo-Escárate et al., 2014). In this study, CPO levels were highest in the nauplius I–II and copepodid stage compared to chalimus, pre-adults, and adults. Notably, after settling onto a host and emerging into the chalimus stage, CPO expression levels of this gene dropped ten-fold, possibly indicative of metabolic rationing in the copepodid stage. Furthermore, downregulation of a key lipid metabolism component gene (triacylglycerol lipase) has been detected in the copepodid stage of *L. salmonis*, suggesting a role in non-temperature related metabolic alteration (Gonçalves et al., 2014). These results support a potential metabolic regulation mechanism for seasonally induced conservation of finite endogenous lipids and would provide a rich area of study in future research.

8.11 Endocrinology

8.11.1 *Ecdysis*

Completion of the sea lice life cycle is highly dependent on the process of moulting, or ecdysis. Ecdysis, the mechanism for growth in arthropods, is tightly regulated through a complex series of ecdysteroid hormones and their derivatives. Ecdysteroids are synthesized from the lipid cholesterol, which is an essential nutrient in crustaceans as they are incapable of its *de novo* synthesis (Ravid et al., 1999). In consequence, sources of cholesterol are restricted to diet and endogenous reserves. Lipid analysis of *L. salmonis* eggs has shown that a substantial portion of the yolk lipids are cholesterol, as egg strings collected from wild salmon show that 7% of total yolk lipids are cholesterols. Notably, eggs obtained from farmed salmon aquaculture showed 9–13% of cholesterol (Tocher et al., 2010). While this increase in cholesterol content may be a result of aquaculture, with farms having healthier fish through optimized nutrition, it may alternatively be an induced microevolutionary response, selecting for increased cholesterol inclusion (thus rapid and ensured ecdysal growth) rather than increased triacylglycerol (extended copepodid longevity). As salmon farms have a high density of non-migrating hosts, selection may favor rapidity to infection over increased time for host locating. Microevolution has been suggested to have occurred in comparisons of sea lice from farmed fish and sea lice from wild fish. Mennerat et al. (2017) found that farm-derived lice produced more eggs in their first clutch and fewer eggs later on than their wild cousins. Additionally, farm-derived lice suffered earlier mortalities than wild populations. These results suggest that farm-derived lice populations have been selected for increased early investment in reproduction at the expense of later fecundity and survival.

References

Andrade-Salas, O. (1997) Feeding and digestion in the ectoparasitic copepod *Lepeophtherius salmonis* (Krøyer 1837). (Publication Number 2844) PhD thesis, University of Stirling, Stirling, UK.

Arnaud, J., Brunet, M. and Mazza, J. (1978) Studies on midgut of *Centropages typicus* (Copepod, Calanoid). 1. Structural and ultrastructural data. *Cell and Tissue Research* 187(2), 333–353.

Boxaspen, K. (1997) Geographical and temporal variation in abundance of salmon lice (*Lepeophtheirus salmonis*) on salmon (*Salmo salar* L.). *ICES Journal of Marine Science* 54(6), 1144–1147.

Boxaspen, K. and Næss, T. (2000) Development of eggs and the planktonic stages of salmon lice (*Lepeophtheirus salmonis*) at low temperatures. *Contributions to Zoology* 69, 51–55.

Brandal, P.O., Egidius, E. and Romslo, I. (1976) Host blood – major food component for parasitic copepod *Lepeophtheirus salmonis* Krøyer, 1838 (Crustacea-Caligidae). *Norwegian Journal of Zoology* 24, 341–343.

Branson, E.J., Ronsberg, S.S. and Ritchie, G. (2000) Efficacy of teflubenzuron (Calicide®) for the treatment of sea lice, *Lepeophtheirus salmonis* (Krøyer 1838), infestations of farmed Atlantic salmon (*Salmo salar* L.). *Aquaculture Research* 31(11), 861–867.

Bresciani, J. (1986) The fine structure of the integument of free-living and parasitic copepods – a review. *Acta Zoologica* 67(3), 125–145.

Bricknell, I.R. and Pietrak, M. (2022) A Glimpse into the Bio-Physical Environment of the Sea Louse (*Lepeophtheirus salmonis* (Krøyer 1837)) Digestive Tract. *Aquaculture* In review.

Bricknell, I.R. and Potts, W.T.W. (1988) The effects of acidification on the coniform mitochondria rich ion transporting cells of *Plecoptera*. *Proceedings of the Royal Microscopic Society* 23, 107–108.

Bricknell, I.R., Bisset, P.F. and Bowden, T.J. (2002) The function of Atlantic salmon (*Salmo salar* L.) antibodies under extremes of pH and osmolarity. *Fish and Shellfish Immunology* 13, 215–219.

Bricknell, I.R., Dalesman, S.J., O'Shea, B., Pert, C.C. and Luntz, A.J.M. (2006) Effect of environmental salinity on sea lice *Lepeophtheirus salmonis* settlement success. *Diseases of Aquatic Organisms* 71, 201–212.

Bron, J.E. (1993) A study of the biology and behaviour of the copepodid larvae of the salmon louse *Lepeophtheirus salmonis* (Krøyer, 1837) (Copepoda: Caligidae). (Publication Number 3258) PhD thesis, University of Stirling, Stirling, UK.

Bron, J.E., Sommerville, C. and Rae, G.H. (1993) The functional morphology of the alimentary tract of larval stages of the parasitic copepod *Lepeophtheirus salmonis*. *Journal of Zoology* 230, 207–220.

Bron, J.E., Shinn, A.P. and Sommerville, C. (2000) Ultrastructure of the cuticle of the chalimus larva of the salmon louse *Lepeophtheirus salmonis* (Krøyer, 1837) (Copepoda: Caligidae). *Contributions to Zoology* 69(1–2), 39–49.

Brooks, S.J. and Mills, C.L. (2006) Gill Na^+/K^+-ATPase in a series of hyper-regulating gammarid amphipods. Enzyme characterisation and the effects of salinity acclimation. *Comparative Biochemistry and Physiology* 144A, 24–32.

Brune, A., Emerson, D. and Breznak, J.A. (1995) The termite gut microflora as an oxygen sink: microelectrode determination of oxygen and pH gradients in guts of lower and higher termites. *Applied and Environmental Microbiology* 61, 2681–2687.

Burmester, T. (2015) Evolution of respiratory proteins across the pancrustacea. *Integrative and Comparative Biology* 55, 792–801.

Campbell, P.J., Hammell, K.L., Dohoo, I.R. and Ritchie, G. (2006) Randomized clinical trial to investigate the effectiveness of teflubenzuron for treating sea lice on Atlantic salmon. *Diseases of Aquatic Organisms* 70(1–2), 101–108.

Chavez-Mardones, J., Valenzuela-Munos, V. and Gallardo-Escarate, C. (2016) In silico transcriptome analysis of cuticle-related genes associated with delousing drug responses in the sea louse *Caligus rogercresseyi*. *Aquaculture* 450, 123–135.

Chew, L.L. and Chong, V.C. (2011) Copepod community structure and abundance in a tropical mangrove estuary, with comparisons to coastal waters. *Hydrobiologia* 666, 127–143.

Christie, A.E., Roncalli, V., Lona, P.B., McCoole, M.D., King, B.L., Bucklin, A., Hartline, D.K. and Lenz, P.H. (2013) In silico characterization of the insect diapause-associated protein couch potato (CPO) in *Calanus finmarchicus* (Crustacea: Copepoda). *Comparative Biochemistry and Physiology D-Genomics & Proteomics* 8, 45–57.

Costello, M.J. (2006). Ecology of sea lice parasitic on farmed and wild fish. *Trends in Parasitology* 22(10), 475–483.

Dalvin, S., Frost, P., Loeffen, P., Skern-Mauritzen, R., Baban, J., Ronnestad, I. and Nilsen, F. (2011) Characterisation of two vitellogenins in the salmon louse *Lepeophtheirus salmonis*: Molecular, functional and evolutional analysis. *Diseases of Aquatic Organisms* 94, 211–224.

Doyle, J.E. and McMahon, B.R. (1995) Effects of acid exposure in the brine shrimp *Artemia franciscana* during development in seawater. *Comparative Biochemistry and Physiology Part A: Physiology* 112, 123–129.

Dutta, P.K., Tripathi, S., Mehrotra, G.K. and Dutta, J. (2009) Perspectives for chitosan based antimicrobial films in food applications. *Food Chemistry* 114, 1173–1182.

Eichner, C., Frost, P., Dysvik, B., Jonassen, I., Kristiansen, B. and Nilsen, F. (2008) Salmon louse (*Lepeophtheirus salmonis*) transcriptomes during post molting maturation and egg production, revealed using EST-sequencing and microarray analysis. *BMC Genomics* 9(15), 126.

Eichner, C., Dalvin, S., Skern-Mauritzen, R., Malde, K., Kongshaug, H. and Nilsen, F. (2015a) Characterization of a novel RXR receptor in the salmon louse (*Lepeophtheirus salmonis*, Copepoda) regulating growth and female reproduction. *BMC Genomics* 16(22), 81.

Eichner, C., Hamre, L.A. and Nilsen, F. (2015b) Instar growth and molt increments in *Lepeophtheirus salmonis* (Copepoda: Caligidae) chalimus larvae. *Parasitology International* 64(1), 86–96.

Eichner, C., Harasimczuk, E., Nilsen, F., Grotmol, S. and Dalvin, S. (2015c) Molecular characterisation and functional analysis of LsChi2, a chitinase found in the salmon louse (*Lepeophtheirus salmonis*, Krøyer 1838). *Experimental Parasitology* 151, 39–48.

Eichner, C., Dondrup, M. and Nilsen, F. (2018) RNA sequencing reveals distinct gene expression patterns during the development of parasitic larval stages of the salmon louse (*Lepeophtheirus salmonis*). *Journal of Fish Diseases* 41(6), 1005–1029.

Fast, M.D., Johnson, S.C., Eddy, T.D., Pinto, D. and Ross, N.W. (2007) *Lepeophtheirus salmonis* secretory/excretory products and their effects on Atlantic salmon immune gene regulation. *Parasite Immunology* 29(4), 179–189.

Fox, H.M. (1945) The oxygen affinities of certain invertebrate haemoglobins. *Journal of Experimental Biology* 21, 161–165.

Gallardo-Escárate, C., Valenzuela-Muñoz, V., Nuñez-Acuña, G., Chávez-Mardones, J. and Maldonado-Aguayo, W. (2014) Transcriptome analysis of the couch potato (CPO) protein reveals an expression pattern associated with early development in the salmon louse *Caligus rogercresseyi*. *Gene* 536(1), 1–8.

Gonçalves, A.T., Farlora, R. and Gallardo-Escarate, C. (2014) Transcriptome survey of the lipid metabolic pathways involved in energy production and ecdysteroid synthesis in the salmon louse *Caligus rogercresseyi* (Crustacea: Copepoda). *Comparative Biochemistry and Physiology B-Biochemistry & Molecular Biology* 176, 9–17.

Gonzalez, M.T., Castro, R., Munoz, G. and Lopez, Z. (2016) Sea lice (Siphonostomatoida: Caligidae) diversity on littoral fishes from the South-Eastern Pacific coast determined from morphology and molecular analysis, with description of a new species (*Lepeophtheirus confusum*). *Parasitology International* 65, 685–695.

Gravil, H.R. (1996) Studies on the biology and ecology of the free swimming larval stages of *Lepeophtheirus salmonis* (Krøyer, 1838) and *Caligus elongatus* Nordmann, 1832 (Copepoda: Caligidae). PhD thesis, University of Stirling, Stirling, UK.

Hahnenkamp, L. and Fyhn, H.J. (1985) The osmotic response of salmon louse, *Lepeophtheirus salmonis* (Copepoda: Caligidae), during the transition from sea-water to fresh-water. *Journal of Comparative Physiology B-Biochemical Systemic and Environmental Physiology* 155, 357–365.

Hallberg, E. and Hirche, H.J. (1980) Differentiation of midgut in adults and over-wintering copepodids of *Calanus finmarchicus* (Gunnerus) and *Calanus helgolandicus*. *Journal of Experimental Marine Biology and Ecology* 48(3), 283–295.

Hamilton, S., McLean, K., Monaghan, S.J., McNair, C., Inglis, N.F., McDonald, H., Adams, S., Richards, R., Roy, W., Smith, P., Bron, J., Nisbet, A.J. and Knox, D. (2018) Characterisation of proteins in excretory/secretory products collected from salmon lice, *Lepeophtheirus salmonis*. *Parasites & Vectors* 11(9), 294.

Hamre, L.A., Glover, K.A. and Nilsen, F. (2009) Establishment and characterisation of salmon louse (*Lepeophtheirus salmonis* (Krøyer 1837)) laboratory strains. *Parasitology International* 58(4), 451–460.

Harðardóttir, H.M., Male, R., Nilsen, F., Eichner, C., Dondrup, M. and Dalvin, S. (2019) Chitin synthesis and degradation in *Lepeophtheirus salmonis*: molecular characterization and gene expression profile during synthesis of a new exoskeleton. *Comparative Biochemistry and Physiology Part A: Molecular & Integrative Physiology* 227, 123–133.

Hassett, R.P. and Landry, M.R. (1983) Effects of food-level acclimation on digestive enzyme activities and feeding behavior of *Calanus pacificus*. *Marine Biology* 75(1), 47–55.

Heggland, E.I., Eichner, C., Stove, S.I., Martinez, A., Nilsen, F. and Dondrup, M. (2019) A scavenger receptor B (Cd36)-like protein is a potential mediator of intestinal heme absorption in the hematophagous ectoparasite *Lepeophtheirus salmonis*. *Scientific Reports* 9, 14.

Henry, R.P. and Watts S.A. (2001) Environmentally mediated carbonic anhydrase reduction in the gills of euryhaline crustaceans. *Journal of Experimental Biology* 204, 991–1002.

Heuch, P.A., Parsons, A. and Boxaspen, K. (1995) Diel vertical migration – a possible host-finding mechanism in salmon louse (*Lepeophtheirus salmonis*) copepodids. *Canadian Journal of Fisheries and Aquatic Sciences* 52, 681–689.

Heuch, P.A., Nordhagen, J.R. and Schram, T.A. (2000) Egg production in the salmon louse [*Lepeophtheirus salmonis* (Krøyer)] in relation to origin and water temperature. *Aquaculture Research* 31(11), 805–814.

Heuch, P.A., Knutsen, J.A., Knutsen, H. and Schram, T. (2002) Salinity and temperature effects on sea lice over-wintering on sea trout (*Salmo trutta*) in coastal areas of the Skagerrak. *Journal of the Marine Biological Association of the United Kingdom* 82, 887–892.

Heuch, P.A., Revie, C.W. and Gettinby, G. (2003) A comparison of epidemiological patterns of salmon lice, *Lepeophtheirus salmonis*, infections on farmed Atlantic salmon, *Salmo salar* L., in Norway and Scotland. *Journal of Fish Diseases* 26, 539–551.

Heuch, P.A., Olsen, R.S., Malkenes, R., Revie, C.W., Gettinby, G., Baillie, M., Lees, F. and Finstad, B. (2009) Temporal and spatial variations in lice numbers on salmon farms in the Hardanger Fjord 2004–06. *Journal of Fish Diseases* 32, 89–100.

Johnson, S.C. and Albright, L.J. (1991) Development, growth, and survival of *Lepeophtheirus salmonis* (Copepoda: Caligidae) under laboratory conditions. *Journal of the Marine Biological Association of the United Kingdom* 71(2), 425–436.

Johnson, S.C. and Albright, L.J. (1992) Comparative susceptibility and histopathology of the response of naive Atlantic, chinook, and coho salmon to experimental infection with *Lepeophtheirus salmonis* (Copepoda: Caligidae). *Diseases of Aquatic Organisms* 14(3), 179–193.

Khan, M.T., Dalvin, S., Waheed, Q., Nilsen, F. and Male, R. (2018) Molecular characterization of the lipophorin receptor in the crustacean ectoparasite *Lepeophtheirus salmonis*. *PLoS ONE* 13(4), 1–23.

Komisarczuk, A.Z., Kongshaug, H. and Nilsen, F. (2018) Gene silencing reveals multiple functions of Na^+/K^+-ATPase in the salmon louse (*Lepeophtheirus salmonis*). *Experimental Parasitology* 185, 79–91.

Komisarczuk, A. Z., Kongshaug, H., Li, M. and Nilsen, F. (2019) RNAi-mediated myosuppressin deficiency affects muscle development and survival in the salmon louse (*Lepeophtheirus salmonis*). *Scientific Reports* 9(1), 6944–6944.

Lafferty, K.D., Harvell, C.D., Conrad, J.M., Friedman, C.S., Kent, M.L., Kuris, A.M., Powell, E.N., Rondeau, D. and Saksida, S.M. (2015) Infectious diseases affect marine fisheries and aquaculture economics. *Annual Review of Marine Science* 7(1), 471–496.

Lee, R.F. (1975) Lipids of parasitic copepods associated with marine fish. *Comparative Biochemistry and Physiology, B: Comparative Biochemistry* 52(3), 363–364.

Lee, R.F., Hagen, W. and Kattner, G. (2006) Lipid storage in marine zooplankton. *Marine Ecology Progress Series* 307(1863), 273–306.

Mayzaud, P., Biggs, D.C. and Roche-Mayzaud, H. (1994) Short term variability of metabolic and digestive enzyme activity in naturally occurring populations of adult copepod *Acartia clausia*. *Ecoscience* 1, 195–207.

McMahon, B.R. and Doyle, J.E. (1997) Acid exposure in euryhaline environments: ion regulation and acid intolerance in larval and adult *Artemia franciscana*. *Hydrobiologia* 352, 1–7.

Mennerat, A., Ugelvik, M.S., Håkonsrud Jensen, C. and Skorping, A. (2017) Invest more and die faster: the life history of a parasite on intensive farms. *Evolutionary Applications* 10(9), 890–896.

Nylund, A., Okland, S. and Bjorknes, B. (1992) Anatomy and ultrastructure of the alimentary canal in *Lepeophtheirus salmonis* (Copepoda, Siphonostomatida). *Journal of Crustacean Biology* 12(3), 423–437.

Overgard, A.C., Hamre, L.A., Harasimczuk, E., Dalvin, S., Nilsen, F. and Grotmol, S. (2016) Exocrine glands of *Lepeophtheirus salmonis* (Copepoda: Caligidae): distribution, developmental appearance, and site of secretion. *Journal of Morphology* 277(12), 1616–1630.

Pond, D.W., Tarling, G.A., Ward, P. and Mayor, D.J. (2012) Wax ester composition influences the diapause patterns in the copepod *Calanoides acutus*. *Deep Sea Research Part II: Topical Studies in Oceanography* 59–60 (Supplement 1): 93–104.

Ravid, T., Tietz, A., Khayat, M., Boehm, E., Michelis, R. and Lubzens, E. (1999) Lipid accumulation in the ovaries of a marine shrimp *Penaeus semisulcatus* (de haan). *Journal of Experimental Biology* 202(13), 1819–1829.

Record, N.R., Ji, R., Maps, F., Varpe, Ø., Runge, J.A., Petrik, C.M. and Johns, D. (2018) Copepod diapause and the biogeography of the marine lipidscape. *Journal of Biogeography* 45(10) 2238–2251.

Ritchie, G., Mordue, A.J., Pike, A.W. and Rae, G.H. (1993) The reproductive output of *Lepeophtheirus salmonis* adult females in relation to seasonal variability of temperature and photoperiod. In: Boxshall, G.A. and Defaye, D. (eds) *Pathogens of Wild and Farmed Fish: Sea Lice*. Ellis Horwood, New York. pp. 153–165.

Ritchie, G., Ronsberg, S.S., Hoff, K.A. and Branson, E.J. (2002) Clinical efficacy of teflubenzuron (Calicide®) for the treatment of *Lepeophtheirus salmonis* infestations of farmed Atlantic salmon *Salmo salar* at low water temperatures. *Diseases of Aquatic Organisms* 51(2), 101–106.

Samsing, F., Oppedal, F., Dalvin, S., Johnsen, I., Vagseth, T. and Dempster, T. (2016) Salmon lice (*Lepeophtheirus salmonis*) development times, body size, and reproductive outputs follow universal models of temperature dependence. *Canadian Journal of Fisheries and Aquatic Sciences* 73, 1841–1851.

Schram, T.A. (2000a) The egg string attachment mechanism in *Hatschekia hippoglossi* (Guerin-Meneville, 1837) (Copepoda: Hatschekiidae). *Sarsia* 85(2), 151–156.

Schram, T.A. (2000b) The egg string attachment mechanism in salmon lice *Lepeophtheirus salmonis* (Copepoda: Caligidae). *Contributions to Zoology* 69(1–2), 21–29.

Scott, A.P. (1901) *Lepeophtheirus and Lernaea. Liverpool Marine Biology Committee Memoirs on Typical British Marine Plants and Animals, Vol VI.* Williams & Norgate, London, UK.

Smith, J.A. and Whitfield, P.J. (1988) Ultrastructural studies on the early cuticular metamorphosis of adult female *Lernaeocera branchialis* (L.) (Copepoda: Pennellidae). *Hydrobiologia* 167, 607–616.

Sullivan, D.S. and Bisalputra, T. (1980) The morphology of a harpacticoid copepod gut – a review and synthesis. *Journal of Morphology* 164(1), 89–105.

Svetlichny, L., Hubareva, E. and Isinibilir, M. (2017) Comparative trends in respiration rates, sinking and swimming speeds of copepods *Pseudocalanus elongatus* and *Acartia clausi* with comments on the cost of brooding strategy. *Journal of Experimental Marine Biology and Ecology* 488, 24–31.

Tabouret, G., Bret-Bennis, L., Dorchies, P. and Jacquiet, P. (2003) Serine protease activity in excretory–secretory products of *Oestrus ovis* (Diptera: Oestridae) larvae. *Veterinary Parasitology* 114(4), 305–314).

Taccardi, E.Y., Bricknell, I.R. and Hamlin, H.J. (2021) Seasonal progression of embryo size and lipid reserves in sea lice *Lepeophtheirus salmonis* collected from salmon farms. *Marine Ecology Progress Series* 664, 79–86.

Thompson, C.R.S., Fields, D.M., Bjelland, R.M., Chan, V.B.S., Durif, C.M.F., Mount, A., Runge, J.A., Shema, S.D., Skiftesvik, A.B. and Browman, H.I. (2019) The planktonic stages of the salmon louse (*Lepeophtheirus salmonis*) are tolerant of end-of-century pCO_2 concentrations. *PeerJ* 7(23), e7810.

Thorstad, E.B., Todd, C.D., Uglem, I., Bjorn, P.A., Gargan, P.G., Vollset, K.W., Halttunen, E., Kalas, S., Berg, M. and Finstad, B. (2015) Effects of salmon lice *Lepeophtheirus salmonis* on wild sea trout *Salmo trutta* – a literature review. *Aquaculture Environment Interactions* 7(2), 91–113.

Tocher, J.A., Dick, J.R., Bron, J.E., Shinn, A.P. and Tocher, D.R. (2010) Lipid and fatty acid composition of parasitic caligid copepods belonging to the genus *Lepeophtheirus*. *Comparative Biochemistry and Physiology – B Biochemistry and Molecular Biology* 156(2), 107–114.

Toksen, E., Nemli, E. and Cankurt, M. (2009) The effect of teflubenzuron on the control of *Lernanthropus kroyeri* (van Beneden, 1851) (Lernanthropidae) infestations in cultured sea bass, *Dicentrarchus labrax* (Linnaeus, 1758). *Bulletin of the European Association of Fish Pathologists* 29(6), 205–209.

Trosse, C., Kongshaug, H., Dondrup, M. and Nilsen, F. (2015) Characterisation of iron regulatory protein 1A and 1B in the blood-feeding copepod *Lepeophtheirus salmonis*. *Experimental Parasitology* 157, 1–11.

Tsuji, N., Okazawa, T. and Yamamura, N. (1990) Autogenous and anautogenous mosquitos – a mathematical analysis of reproductive strategies. *Journal of Medical Entomology* 27(4), 446–453.

Tucker, C.S., Sommerville, C. and Wootten, R. (2000) An investigation into the larval energetics and settlement of the sea louse, *Lepeophtheirus salmonis*, an ectoparasitic copepod of Atlantic salmon, *Salmo salar*. *Fish Pathology* 35, 137–143.

Tucker, C.S., Norman, R., Shinn, A.P., Bron, J.E., Sommerville, C. and Wootten, R. (2002) A single cohort time delay model of the life-cycle of the salmon louse *Lepeophtheirus salmonis* on Atlantic Salmon *Salmo salar*. *Fish Pathology* 37, 107–118.

Valenzuela-Munoz, V. and Gallardo-Escarate, C. (2017) Iron metabolism modulation in Atlantic salmon infested with the sea lice *Lepeophtheirus salmonis* and *Caligus rogercresseyi*: a matter of nutritional immunity? *Fish and Shellfish Immunology* 60, 97–102.

Valenzuela-Munoz, V., Benavente, B.P., Casuso, A., Leal, Y. and Gallardo-Escarate, C. (2021) Chimeric protein IPath® with chelating activity improves Atlantic salmon's immunity against infectious diseases. *Vaccines* 9(4), 361.

Vigneau, A.J. (2006) Digestive enzymes of the salmon louse *(Lepeophtheirus salmonis)*: implications for vaccine development. PhD thesis, University of Aberdeen, Aberdeen, UK.

Villate, F. (1991) Annual cycle of zooplankton community in the Abra Harbor (Bay of Biscay) – abundance, composition and size spectra. *Journal of Plankton Research* 13, 691–706

Watts, S.A., Yeh, E.W. and Henry, R.P. (1996) Hypoosmotic stimulation of ornithine decarboxylase activity in the brine shrimp *Artemia franciscana*. *Journal of Experimental Zoology* 274, 15–22.

Wootten, R., Smith, J.W. and Needham, E.A. (1982) Aspects of the biology of the parasitic copepods *Lepeophtheirus salmonis* and *Caligus elongatus* on farmed salmonids, and their treatment. *Proceedings of the Royal Society of Edinburgh Section B-Biological Sciences* 81, 185–197.

Xu, Z.L. and Gao, Q. (2011) Optimal salinity for dominant copepods in the East China Sea, determined using a yield density model. *Chinese Journal of Oceanology and Limnology* 29, 514–523.

CHAPTER 9

A standardised generic framework of sea lice model components for application in coupled hydrodynamic-particle models

Alexander G. Murray[1], Samuel Shephard[2], Lars Asplin[3], Thomas Adams[4], Bjørn Ådlandsvik[3], Alejandro Gallego[1], Michael Hartnett[5], Ingrid Askeland Johnsen[3], Simon R.M Jones[6], Meadhbh Moriarty[1], Stephen Nash[5], Campbell C. Pert[1], Berit Rabe[1] and Patrick G. Gargan[2]

[1]Marine Scotland Science, Aberdeen, UK
[2]Inland Fisheries Ireland, City West Campus, Dublin, Ireland
[3]Institute of Marine Research, Norway
[4]Scottish Association of Marine Science, Oban, UK
[5]National University of Ireland, Galway, Ireland
[6]Fisheries and Oceans Canada, Nanaimo, British Columbia, Canada

9.1 Introduction: sea lice and their transmission

Salmon production, at $17.1 billion in 2018, is the second largest component of global aquaculture by value (FAO, 2020). Although production is increasing, several issues act to limit the sustainability of this growth (Taranger et al., 2015). Sea lice are considered to have serious impacts on sustainable marine salmon aquaculture in almost all areas where salmon are farmed (Pike and Wadsworth, 1999; Jones and Beamish, 2011), and are the most focused on pathogen of salmon for scientific research (Murray et al., 2016). These ectoparasitic copepods are a dominant factor in limiting the maximum sustainable biomass of farmed salmon (Taranger et al., 2015). Lice are consequently the main limiting factor for growth in the salmonid farming industry under the Norwegian 'traffic light' management system and are key considerations for salmon aquaculture planning, not only in Norway, but also in most countries with marine salmonid aquaculture. In Europe, a particular concern is the salmonid specialist *Lepeophtheirus salmonis* (Krøyer, 1837), but the more generalist *Caligus* lice species such as *Caligus elongatus* (von Nordmann, 1832) in the North Atlantic, *C. clemensi* (Parker & Margolis 1964) in Pacific Canada and especially *C. rogercresseyi* (Boxshall & Bravo 2000) in Chile cause significant infestations on marine salmon farms (Hemmingsen et al., 2020).

Since these *Caligus* lice have similar planktonic dispersion to *L. salmonis*, any dispersion modelling described here also applies to these species. However, parameter values and infestation sources may differ. This may include biological parameters such as tolerance of low salinity and swimming depths. Even within *L. salmonis*, there are significant genetic differences between Atlantic and Pacific populations (Todd et al., 2004), thus parameter values should be adjusted to reflect the species and area described within a study.

Control of sea lice is complicated by planktonic larval stages of several days duration, which allow transmission over considerable distances with ocean currents. The potential for long-distance transport may result in interaction between farms in a neighbourhood and between farmed and wild salmonid stocks (Shephard et al., 2016), requiring area management approaches to control of infestation. The first few days of the planktonic stage are as non-infective nauplii, which indicates a species adapted to spreading.

Transmission can also spread lice that are resistant to treatments, influencing the dynamics of emergence of resistance (McEwan et al., 2015). Since resistance to medicines is an increasing threat to effective control of sea lice (Aaen et al., 2015), and evidence for environmental contamination is increasing (Bloodworth et al., 2019), an understanding of transmission dynamics is important for effective control in both the short and long term.

Although averaged infestation pressure can be described using distance relationships (Aldrin et al., 2013; Salama et al., 2016), interaction among lice on different host populations depends on local currents (Salama et al., 2013). Consequently certain adjacent populations may have relatively little interaction, while other populations could interact over relatively long distances. Interaction may also be unidirectional owing to prevailing currents, so if population A can infest population B, it does not necessarily mean B can infest A (Adams et al., 2015).

These factors mean that, to understand the interactions of lice on different host populations, it is necessary to describe both the dynamics of water movements and the biology of the larval lice. The approach used to do this is coupled hydrodynamic-particle modelling (Fig. 9.1). This coupled modelling approach is used for a variety of processes in the ocean (Paris et al., 2013), such as dispersal of pollutants, distribution dynamics of plankton, and dispersal of eggs and larvae of fish and other animals (Christensen et al., 2018).

Once attached, the lice develop and mature on their hosts (Hamre et al., 2013) and ovigerous females become a source of new eggs. However, such population modelling is not covered in this chapter on dispersal modelling.

Coupled models have been used for many years to describe sea lice dispersal (Fig. 9.1) in countries such as Norway, Scotland, Canada and the Faroe Islands (Asplin et al., 2004; Murray and Gillibrand, 2006; Jones and Beamish, 2011; Salama and Rabe, 2013; Kragesteen et al., 2018). The modelling approach is continuing to develop (Asplin et al., 2014; Johnsen et al., 2016; Samsing et al., 2016a, 2016b; Salama et al., 2018), for example the modelling of large-scale regional dispersal (Adams et al., 2016), mapping the influenced geographical areas (Sandvik et al., 2016) and models are used for network analysis and strategic management (Samsing et al., 2018). These models have been validated using sentinel cages, plankton trawls and counts from wild sea trout (Salama et al., 2013, 2018; Sandvik et al., 2016; Skarđhamar et al., 2018; Myksvoll et al., 2018). Sea lice dispersal models are used as an applied tool in aiding sea lice management, notably in the Norwegian 'traffic light' system, which implements risk-based spatial planning and management. Because of their applied role, there has been considerable investment in the development of reliable and accurate models.

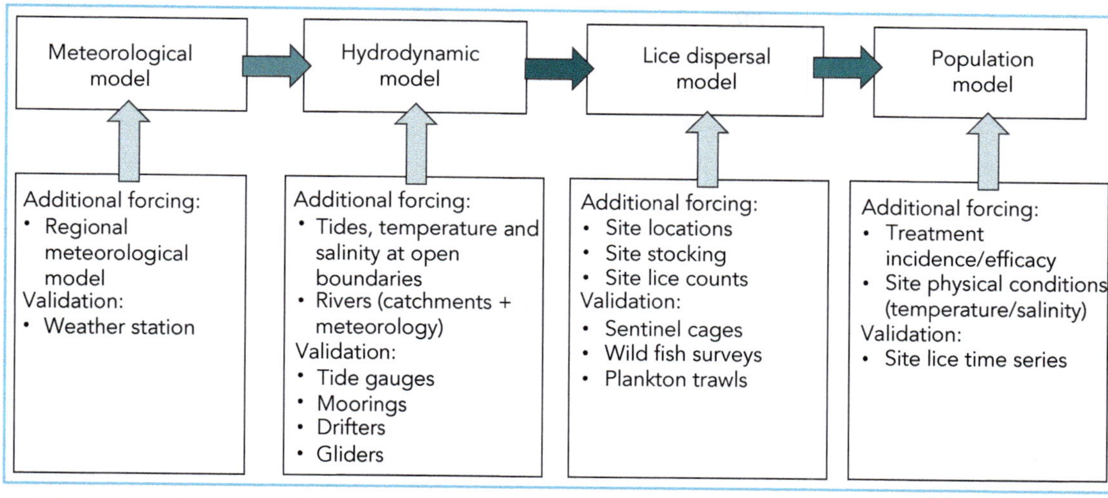

Fig. 9.1 Coupling of models to drive sea lice models for informing sea lice controls. Meteorological models drive hydrodynamic models through winds and precipitation. The hydrodynamic model moves lice particles in a lice dispersal model, which in turn can be used to drive models of lice population dynamics.

9.2 Aim of a standardised generic framework

Different institutions in various countries have developed the models currently used, and differences in approaches may result from this evolution. However, there are many common challenges, particularly in northwest Europe. Here similar ecosystems, with wild sea trout (*Salmo trutta*, L.) and Atlantic salmon (*S. salar*, L.) present, often support marine farmed Atlantic salmon and rainbow trout (*Onchorhynchus mykiss*, Walbaum), although under a range of population densities, temperatures and other environmental factors. Lessons can be (and are) shared in the development of models within this region and more widely with countries such as Canada and Chile, which have major salmonid industries and corresponding sea lice issues.

The issues driving sea lice dispersal, and the modelling methods used, are therefore often similar in different salmon farming areas. Differences between modelling approaches may exist for valid reasons. However, standardisation makes sense unless there are strong reasons to use area-specific or topic-specific models. Standards can be flexible, simulating the local environment explicitly (for example, consideration of local water temperatures is important) by using process rate parameter values that vary with local environment.

Reasons for standardisation include:

A Outputs from model runs carried out by different organisations can be compared directly, and ensure that differences are due to differences in the system, not the modelling approach. This facilitates data sharing.

B An accepted standard model means that parties seeking to model sea lice in a specific area do not have to justify their individual decisions about how they approach lice biology (although this also means the standards must be well validated, and standards need to be questioned to consider if there are reasons a particular model should deviate from them).

C Field validation of models is necessary, but it is expensive and time consuming. An accepted standard means that there is less need for specific validation in each individual case of application. A model that has been shown to work in multiple regions can be applied with reasonable confidence in other regions for which data are lacking; for example, when a model is applied for the whole Scottish west coast it is not possible to validate it in every local area. Validation is even more problematic for the much longer coast of Norway and considerable resources are applied in this cause.

Experiences of scientists from Marine Scotland Science (MSS), the Scottish Association for Marine Science (SAMS), the Norwegian Institute of Marine Research (IMR) and Inland Fisheries Ireland (IFI) were discussed at a workshop in Aberdeen supported by the Marine Association of Science and Technology for Scotland (MASTS). At this meeting, the components required of a coupled model were discussed and common approaches were established, and areas where further work was required for successful standardisation were identified. This meeting preceded the EU-funded NASCO Licetrack project, within which the objective of standardisation of sea lice larval modelling was more fully developed in a collaboration of IFI, MSS and IMR.

There are alternative modelling approaches to the application of coupled hydrodynamic-particle models for inferring sea lice infestations on a farm and subsequent dispersal to other farms and into the wider environment. In predicting individual farm infestation pressure, average abundance of gravid female lice with a two-week lag may be informative (for example, for *C. rogercresseyi* (Kristoffersen et al., 2013)). Interaction between farms on average may be estimated statistically from sea lice counts on farms to assess strength of interaction with seaway distance (Aldrin et al., 2013) or assessed from average dispersal in coupled models (Salama et al., 2016). Distances of interaction of lice on farms and wild fish populations can also be estimated from lice count data (Shephard et al., 2016). Stage-structured models have been developed to better quantify the sea lice infestation pressure by estimating the time lag structures representing growth and recruitment in various water temperatures for different life stages, and using non-linear functions to account for seaway distance to weight the interaction between various farms (Aldrin et al., 2019). While multivariate, autoregressive state-space models provide an intuitive framework for assessing sources of infestation pressures by invoking state variables to account for correlations, so factors like water temperature and various treatments can be accounted for explicitly in the model (Elghafghuf et al., 2020). Simple descriptions of interaction decaying with distance can be used to develop general plans for the management of sea lice (Murray and Moriarty, 2021). All approaches highlighted lack detail of local hydrodynamic factors that affect specific local patterns of interaction. These models do not account for transport by currents of the non-infectious naupliar phases; their movement introduces an unknown, and often substantial, spatial error in the estimated location of the core of the kernel of predicted copepodids, and thus infectious pressure.

Therefore, we aim here to define a basic generic framework for coupled sea lice models which can be driven by an appropriate hydrodynamic model that provides the currents at horizontal, vertical and temporal resolution appropriate to the sea lice model (and the aims of its use, for example strategic large-scale or local interactions between farms). For the hydrodynamic model, we do not attempt to specify how the model is implemented, merely to describe the required resolution of inputs and outputs. An agreed standard description of a biological model is possible, although there is some need for work on vertical swimming behaviour (and this has implications for the appropriate hydrodynamic output).

In developing a standard framework, we aim to describe a model that is simple to implement and computationally efficient. Increased complexity leads to more parameters, which increases model variation that is dependent on parameter uncertainty. The model is a stripped down model based on the principle from Occam's razor that a model should be as simple as possible, but no simpler.

Process functions and parameters (Tables 9.1, 9.2 and 9.3) are taken from models developed and used elsewhere (for example, Asplin et al., 2004; Amundrud and Murray, 2009; Adams et al., 2015), from reviews that have synthesised a very large literature on sea lice biology (for example, Johnson and Albright, 1991; Pike and Wadsworth, 1999; Stien et al., 2005; Jones and Beamish, 2011; Johnsen et al., 2016; Brooker et al., 2018), and from a rapidly growing literature on sea lice on salmon (Murray et al., 2016).

9.3 Standards for the hydrodynamic model

Hydrodynamic models need to resolve in three dimensions the currents that transport sea lice at an appropriate resolution in time and space for the question being addressed (Table 9.1). The exact approach used is not important to the transport of lice, as long as the output meets this appropriate resolution. For example, when modelling at the scale of individual farm interactions in hydrographically complex inshore waters, a fine horizontal resolution is required (100–200 m or finer). In areas offshore or remote from the focus area, coarser resolution are acceptable. These different scales can be resolved by the use of hydrodynamic models of variable resolution ('unstructured grid') (Chen et al., 2011) or nesting of more conventional rectangular grid models. Coarser resolution is also appropriate for some uses, such as strategic modelling of the spread of lice between areas (Wolf et al., 2016), as long as the absence of any fine-scale processes not represented in the model does not significantly influence the results (an aspect that should be investigated by model validation against field data, model sensitivity analyses and/or model inter-model comparison exercises). The duration of the planktonic stages for lice of potentially more than two weeks must be considered.

Adequate vertical resolution is also required. In particular, a surface layer of a few metres depth is particularly important because sea lice tend to locate near the surface (although this may depend on light, salinity and species of louse, see later section on vertical movements of particle), but also as the stratification of the water can cause great differences in the physical environment over short distances here. This layer may be resolved in some detail, for example in the Hardanger Fjord model (Asplin et al., 2014),

Table 9.1 A summary of basic properties of outputs and inputs.

Property	Scale	Comment
Horizontal resolution	≤100 m (at areas of interest)	Variable grid avoids computational overhead.
Vertical resolution	Surface layer or layers	Depends on lice behaviour.
Currents, u, v, w		Velocities in X, Y, Z dimensions at specific time and location.
Turbulence		Stochastic
Temporal	≤1 hour	Resolve tidal currents and rapid shifts in wind condition.
Winds	≤1 km resolution	Resolve topographic steering/channelling of the wind.
Freshwater	Major rivers and diffuse runoff inputs, as relevant.	

where 15 of 35 vertical levels are in the top 10 m. Temperature and salinity are important for lice development and behaviour (see sections 9.5 and 9.6). Larval lice may attempt to avoid a low salinity surface layer by swimming vertically to a deeper layer, and so representation of haloclines is particularly important. Thermoclines may also need to be resolved. Under typical estuarine flows, this deeper layer may be moving in the opposite direction to an outflowing surface layer and it will be less impacted by wind stress. Since the wind-driven current has a logarithmically decaying profile from the surface and downwards, several computational levels in the upper few metres are needed to capture this properly.

In Scotland, a simple approach is considered appropriate for Loch Torridon (Amundrud and Murray, 2009), because relatively small rivers and limited spring snow melt, together with exposure to wind and tides, mean strong haloclines can be ignored, and a surface layer of only a few metres' depth is required to drive sea lice movement. Deeper layers are required for the hydrodynamic simulation, but their outputs are not required for the lice model and the vertical resolution structure of subsurface layers therefore is required to simulate hydrodynamics and need not take lice into account. However, if more complex lice behaviour is modelled, such as vertical migration or freshwater layer avoidance, then the sea lice model will require output of all the layers relevant to this behaviour (that is, covering the depths that contain most sea lice) and vertical resolution must be fine enough to account for sea lice behaviour.

Model temporal scales need to be fine enough to generate output currents at least at hourly resolution, in order to simulate particle transport and to resolve tidal currents and changing winds. Resolution also depends on purpose. For example, strategic level interactions based on 'climatological' model outputs, which simulate general conditions, do not require resolution of short-term changes. Cumulative effects of short-term variability should be parameterised in such cases. It should be recognised that such models omit rare events and tend to underestimate maximum dispersal distances.

The hydrodynamic model will require forcing at scales appropriate to resolving behaviour at the scales of interest. This includes boundary forcing by tides, and often by nesting the model domain within coarser models that cover larger areas. It also requires wind forcing. Given the topographic complexities of most areas in which salmon are farmed (with surrounding hills and mountains), the wind should ideally be resolved at a local scale that is finer than the typical output from national meteorological offices, and may therefore require local wind models. Freshwater inputs are also important, and flows from larger rivers may have to be explicitly described either through simulations or measurements. In addition, in some areas diffuse runoff from land may be more important than large rivers, so this needs to be considered when considering freshwater forcing. It could also be that direct rainfall will be of some importance.

As well as the u, v and w current velocities, particles are also moved by turbulence. This is a stochastic diffusive process ζ. This can be simulated as a function of the current velocities or using a fixed turbulent diffusion coefficient (Amundrud and Murray, 2007). Alternatively, this can be a direct output from the hydrodynamic model.

9.4 Standards for the particle model framework

The hydrodynamic model drives the movements of particles. The biological model of sea lice particles (Fig. 9.2) has two classes of properties. First, the models describing rates (egg production, maturation, mortality), which are relatively simple to implement and have no interaction with the transport model. Second, the model properties related to lice movement, particularly vertical movement but

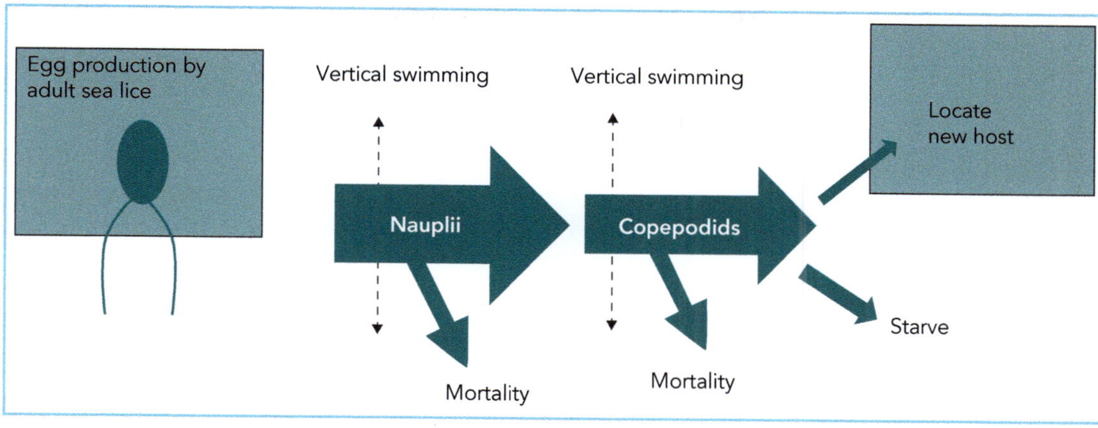

Fig. 9.2 Generic biological model framework. Adult female lice produce eggs that hatch releasing non-infectious nauplii, which mature into infectious copepodids. Larval lice are transported by currents generated by a hydrodynamic model, and have some vertical swimming capacity to control their depth. During these planktonic phases, the larval lice are subject to mortality and eventually will starve if they do not find a host. Larval numbers decay with time for general mortality; starvation is qualitatively different as it limits maximum larval lifespan.

also handling of boundaries (open boundaries or land boundaries). Each particle also needs a location to be defined in the model domain (X, Y and Z), an age and a value for their weighting or number of lice it represents. Particles represent clusters of identical individuals, or 'super individuals', because it is not computationally feasible to represent individual sea lice by individual particles. This element is only required if processes affecting lice numbers, such as settlement or mortality, are being implemented in the model.

9.5 Biological rates

In this section we derive standardised models for biological processes in lice development; parameters are summarised in Table 9.2. It should be noted that the great majority of references to sea lice biology and behaviour concern *L. salmonis*, and this we use to identify standard values for the modelling. Different species of *Caligus* lice may require different parameters, although the basic biological processes, and hence modelling framework, apply.

9.5.1 Production

Production of larval lice (L) from a salmonid population (such as a salmon farm) depends on the number of ovigerous female lice per fish (F), the number of salmonids in the population (S), the rate of egg production per day (E) and their survival to hatching (x). This relationship is represented as:

$$L = xEFS$$

The value of S should be available from records of numbers of fish on farms, or at least estimated from proxies such as regulatory-consented biomass if the exact number is not available; use of proxies

Table 9.2 Biological rate processes.

Biological process	Parameter	Value
Egg production	L = xEFS	
Source population	S	From licensing data
Ovigerous female lice per fish	F	From surveillance data
Egg production rate	E	30 d^{-1} ovigerous $louse^{-1}$
Egg viability	x	1, except under low salinity
Maturation time	M_t	Temperature dependent
Maximum age	D_t	Temperature dependent
Mortality rate	M	1% hr^{-1}, salinity dependent
Infection probability	J	Not evaluated

reduces accuracy of the estimation of L. Estimates of lice on wild salmonid populations and locations may be more difficult to obtain, although they are likely to contribute relatively little to the numbers of larvae, except in the initiation of infection when lice-free smolts are put to sea. Typically, salmon lice represent a real threat only in areas with many fish farms; hence the contribution from wild salmonids will be minor in these areas. However, lice on wild fish will always be present and a complete removal of lice from a region is impossible.

The lice produced are assigned to particles representing many individuals; these particles are referred to as 'super individuals'. A typical release rate may be five super individual particles per site per hour. The number of modelled particles must be sufficient to simulate dispersal patterns so that the results are independent of the number of particles used (this should be explored by a sensitivity analysis), and their weighting depends on the rate of production of lice at the source, divided by the rate of release for the super individual particles.

9.5.2 Biological rate processes

In the relationship L = xEFS, F represents the ovigerous female lice count measured at farms, which should be applied for sea lice modelling to be quantitatively accurate. Normalised counts can be used to simulate relative roles of farms as lice sources. Ovigerous counts for wild fish are less easy to obtain but as there normally are fewer wild fish in an area compared to the number of farmed fish, the wild fish contribute less to total lice production (Heuch and Mo, 2001; Butler, 2002). As it is relatively low, uncertainty in the number of lice on wild fish is less crucial to total lice production, but is important for initiating infection on farmed fish.

The rate of production of eggs depends on a range of factors including the age and health of the louse and on the environment (Heuch et al., 2000; Stien et al., 2005; Samsing et al., 2016a; Brooker et al., 2018), and has evolved under aquaculture with selection for faster production (Mennerat et al., 2017). A standard 150 eggs per egg string has been derived for Norwegian modelling (Skarđhamar et al., 2018), with egg string production rate varying seasonally with temperature (Brooker et al., 2018). A Behelradek function was derived by Stien et al. (2005) and this is still widely used in models with explicit egg production as a function of temperature (for example, Skarđhamar et al., 2018). Production rates may be varied between scenarios representing different seasons. Standardising egg string production time at ten days gives 30 eggs per day given two egg strings per female.

Viability of eggs seems to be high in seawater, but it may be affected by low salinity (Groner et al., 2016). If modelling in seawater (rather than brackish water), x =1 is an approximation that ignores small losses due to viability problems.

A relative infestation pressure can be simulated without explicit evaluation of E and x, by assuming these parameters are the same for all farms in the model domain (which will be the case unless sophisticated modelling of these is implemented as functions of environmental variation). This can be used to assess the relative proportions of lice at a particular sink location that are coming from different individual sources. This approach has been used in Scottish models (for example, Salama et al., 2013), but does not allow absolute infection rates to be calculated.

Maturing larval sea lice go through two non-infective naupliar phases before reaching the copepodid stage responsible for infestation events. Simulation of the nauplii phases is important because larval lice may be transported considerable distances before they become infectious, which may limit or even prevent self-infestation on farm sites in a largely dispersive environment. The time for maturation to occur is heavily dependent on temperature, with longer times required at low temperatures (Samsing et al., 2016a). Although some functions that have been used miss the effects of very cold waters, or are targeted at production under very cold conditions (Boxaspen and Næss, 2000), a standard should apply across a range of conditions; it should also be simple and robust. The formula used by Stien et al. (2005) is a Behelradek function:

$$M_t = [\beta_1/(T - 10 + \beta_1\beta_2)]^2$$

Where parameters fitted to observations are β_1 = 24.79 and β_2 = 0.525, T = temperature. However, a simpler and more robust approach is to use a fixed number of degree days for maturation to occur; 40 degree days is used as a standard from Norwegian modelling (Samsing et al., 2016a, 2016b; Skardhamar et al., 2018; Myksvoll et al., 2018).

It is generally sufficient to implement temperature-dependent development at a seasonal level, that is, just use a single value of M_t appropriate to seasonal average temperatures for simulations covering dispersal of larvae over a few weeks. This neglects the relatively small changes in temperature occurring over such periods and in different locations within a local area (this is not the case when modelling entire coasts such as Norway's for which a considerable temperature range will exist at any one time). However, the degree day approach can be implemented very simply if a degree days counter is included as an attribute of a lice particle, and offers greater flexibility to cope with unusual meteorological and oceanographic effects, as long as these are adequately replicated by the hydrodynamic model or some external dataset.

On reaching maturation time, particles can either simply switch from nauplii to copepodid, or a probabilistic switch may be used (to allow for biological uncertainty), for example 10% h^{-1} probability of maturing once threshold age is reached. This addition appears to make relatively little difference, so the instantaneous switch which is conceptually simpler and simpler to implement is preferable for standardisation.

9.5.3 Mortality and senescence

Larval lice in the environment are subject to mortality; they are highly sensitive to salinity, with mortality rates increasing rapidly when salinity drops below 29 (Johnson and Albright, 1991; Bricknell et al., 2006). Mortality rate is also temperature-dependent in a way that interacts with salinity. Mortality also depends on interaction with grazers such as zooplankton or mussels (Brooker et al., 2018). However, simulation of these effects requires estimation of the populations of such grazers or some external, empirically derived mortality function, greatly increasing model complexity and uncertainty. In the absence of complexity, we assume that the variation is randomised so the average rate value applies.

Copepodids may, while still viable, lose some of their infectivity as they age and become senescent. This process means that, even when present, older copepodids contribute less than freshly matured ones to the infectious pressure. This effect can simply be treated as part of mortality rate for simulation purposes, for example if 50% died and there was a 50% drop in infectivity among survivors, this can be treated as 75% mortality.

In non-brackish seawater, a mortality rate m standard of 1% h^{-1} (or similar) is widely used in models that have proven effective at simulating larval distributions, for example in Loch Linnhe (Salama et al., 2013) and northern Norway (Johnsen et al., 2016). In the absence of detailed knowledge of the actual mortality/predation of planktonic salmon lice, and from the desire to make the modelling as simple as possible (and in the understanding this does not apply in brackish waters), the rate 1% h^{-1} is used as the standard and is considered a normal mortality rate for planktonic organisms of this size (McGurk, 1986).

Larval sea lice use up energy with time, and do not feed, relying instead on stored lipids (Pike and Wadsworth, 1999). This behaviour means that there is a maximum time for which they remain viable. In this characteristic, sea lice differ from, for example, viruses, whose numbers decay exponentially with time at a rate that can continue indefinitely, so they do not have a fixed maximum survival time (other than decay below minimum infectious dose). Maximum survival time is particularly important for strategic modelling of maximum distances of potential connectivity between areas. However, only a very small proportion of larvae would survive to reach this point, so it is likely to have limited significance for local interactions between specific pairs of farms.

9.5.4 Infestation

Infestation is not strictly required for modelling relative infection pressure, but infestation rate among fish is a function of this infestation pressure (that is, the concentration of copepodids in the farm area (Sandvik et al., 2020)). The concentration of copepodids (or relative concentration if only relative production is assessed) can identify relative infestation rates without knowing exact relationship to concentration. It should also be the case that on infestation, larval lice are removed from the system, although this is not actually done in existing models as the effect is extremely small relative to mortality. A parameter rate J is included here; explicit valuation of this rate is required in modelling actual rates of infestation, and also to aid comparison of model results with infestation rates obtained from sentinel cage results.

It is assumed that infestation will be a function of several parameters, including water temperature, current speed, light, salinity and so on (Genna et al., 2005; Samsing et al., 2015) and the age and behaviour of the copepodid (Murray and Moriarty, 2021).

9.6 Vertical and boundary movements of particles

Sea lice particles are moved with the currents generated from the hydrodynamic model. However, particles also have the capacity to actively swim in the vertical (Johnsen et al., 2014, 2016) and sea

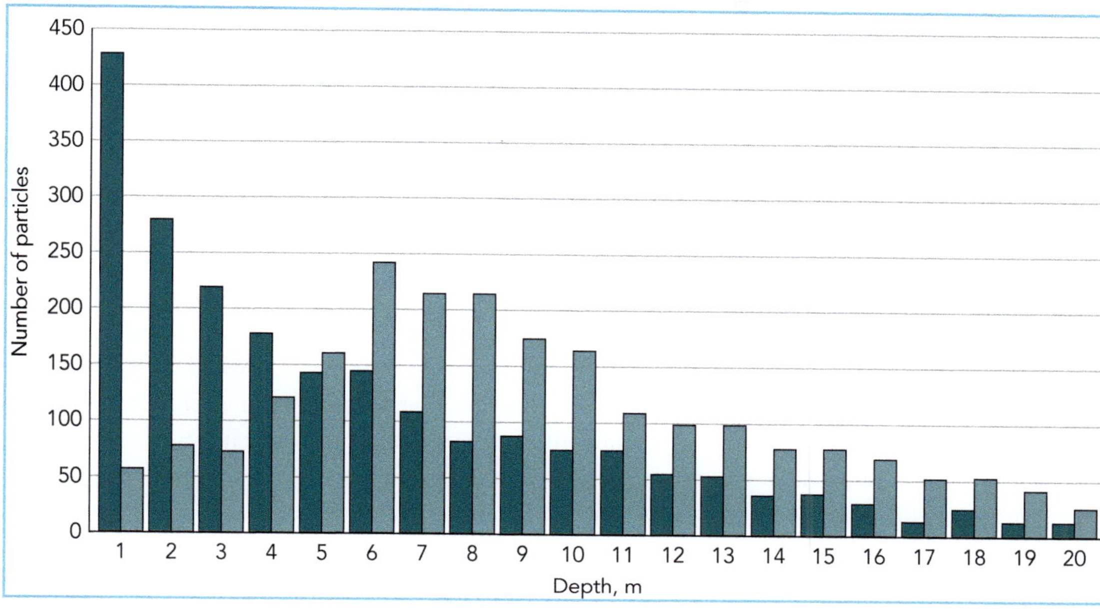

Fig. 9.3 Vertical distribution of copepodids after 20 days under two example scenarios with salinity avoidance (light green bars) and without salinity avoidance (dark green bars). These different depths lead to exposure to very different current regimes.

lice are positively phototactic (Hevrøy et al., 2003). The swimming velocities are relatively small compared to the currents, especially tidal currents, but they are important in that they do enable the lice to move vertically or retain positions in the water column. Even small vertical movements can affect the currents that lice are exposed to, for example counter currents, or enable them to avoid freshwaters (Crosbie et al., 2019) that would be lethal. Active behaviour can thus have substantial effects on lice dispersal (Gillibrand and Willis, 2007; Johnsen et al., 2016). An example is shown from the Hardanger Fjord model with and without vertical swimming behaviour included (Fig. 9.3) to avoid a freshwater inflow event. Vertical avoidance of freshwater layers can keep lice near river mouths, since they avoid being carried down-fjord in a fresh buoyancy-driven surface current. At the same time, mortality is reduced by avoiding low salinity.

In general, larval *L. salmonis* appear to actively attempt to maintain a position near the surface of the water (Salama et al., 2018; Nelson et al., 2018) because they are positively phototactic (Heuch et al., 1995). It is for this reason that technologies such as skirts and snorkels separating salmon from surface waters are effective at controlling lice infestations (Stien et al., 2016; Samsing et al., 2016b; Oppedal et al., 2017). However, there is evidence that *Caligus* lice species may be active below the surface layer, for example *C. elongatus* (á Norđi et al., 2015) or *C. rogercresseyi* (Molinet et al., 2011) owing to different host depth preferences. The simplest behaviour is to assume that lice remain near the surface at all times. This can be simulated simply by moving the particle in the horizontal, but not in the vertical. In such circumstances, lice may become concentrated in downwelling areas/onshore. This situation has been confirmed for Loch Shieldaig (Murray et al., 2011), with high coastal concentrations being associated with onshore winds.

Modelling vertical movement of the lice requires that accurate vertical current data be available, as are data for horizontal currents for the different depths occupied by lice. Vertical behaviour may be modelled as diurnal migration or avoidance of salinity (Crosbie et al., 2019), or as more detailed simulation of vertical swimming such that even if lice attempt to keep near the surface they may not always be able to in times and areas of strong vertical mixing (Johnsen et al., 2016). Incorporating environmental parameters as currents, temperature, salinity or mixing variables from extra layers is conceptually simple, but has added computational storage overheads.

Vertical movement is therefore an area in particular need of further work (Johnsen et al., 2016), and should take into account recent data (Nelson et al., 2018; Salama et al., 2018) before it can be standardised. In areas of Scotland and Ireland without strong haloclines, we can reasonably use the simple, positively buoyant, assumption of lice maintaining themselves in surface waters, particularly in daytime as they are positively phototactic (Heuch et al., 1995; Hevrøy et al., 2003). However, if strong haloclines are present, vertical swimming (Heuch, 1995) is likely to affect both survival and dispersal and so this assumption of the lice being near the surface is not applicable. Neutrally buoyant particle models do not capture the ability of lice to form concentrations associated with oceanographic features (Murray and Salama, 2017) and so are inappropriate for sea lice modelling. We might find that the nauplii stages actually reside deeper than the copepodids (á Norđi et al., 2016), as may *Caligus* species (á Norđi et al., 2016), which potentially will alter the dispersion pattern.

9.6.1 Boundaries: solid or open

Particles that reach the boundary of the model domain must interact with that boundary in some way. When particles reach solid boundaries, representing coastline, they cannot cross the boundary. Instead, they must either stick at the boundary until a current moves them away from the shoreline at a later time, or they could 'bounce'. Sticking is considered preferable to bouncing (Amundrud and Murray, 2007), but in unstructured models, boundary nodes can be set as points of zero velocity and interpolated particle velocity tends to zero as the edge is approached. In traditional rectangular grid models using the 'C' grid (Arakawa and Lamb, 1977), the normal velocity is also zero at land-sea interfaces. This factor reduces the probability that a particle will hit land.

When particles reach open boundaries, for example the boundary between the model area and the open sea outside the model domain, the particle can cross such a boundary. If so, it will be lost from the model domain and the boundary becomes a sink. It is possible to input particles across the boundary, representing input from other areas (or returning particles), but this presents the problem of assessing how many particles to input. Also, the boundary areas are poorly resolved parts of the model, and domain flows across these boundaries, and therefore rates of input, may be poorly represented. The simplest solution is for the boundary to be a sink, but to make it far from the area of interest to the simulation. The use of variable grids that allows an area of coarse resolution to surround the area of interest means boundaries can be made distant without excessive computational effort.

9.7 Particle properties

Each simulated particle carries key information on its location (X, Y and Z coordinates, although in a simple simulation Z may be constant), its age since hatching (including temperature history), status (nauplii or copepodid) and an infectious weighting which depends on the number of larval lice the

Table 9.3 Properties of particles.

Property	Parameter	Comment
Location	X,Y,Z	3-D coordinates (Z fixed in simple form)
Age since hatching	A	
Maturity	M	0 if $A < M_t$, 1 if $A > M_t$
Lice in particle	W	$W_0 = L/P$, decays as A increases

particle represents. These particle properties are summarised in Table 9.3. There may be other useful attributes, for example to calculate rates or for post-processing of results (salinity history and area of origin and other factors).

The weighting of the particle at age $A = 0$ is determined by L/P, where P is the number of particles released per time step. The value of P may itself depend on L, such that when sea lice release rates are high the number of simulated particles released is increased. However, a minimum number of particles must be released in stochastic simulations so that results are not significantly dependent on the number of particles released. With each time step, the particle moves in the X and Y dimensions at rates determined by the hydrodynamic currents that apply at that particular time, interpolated to the particles, specific location and time (if between locations in time and space for which specific hydrodynamic model outputs are available). Random diffusive currents, representing turbulence, also move each particle individually. Particles crossing boundaries behave as described earlier. In the simplest version, the particle does not move vertically and remains near the surface.

The particle ages by the length of the model time step. Particle weighting decays at a rate proportional to 1% h^{-1} over the time step in seawater (see section 9.5). If this gives it an age > M_t, the particle changes from nauplii to copepodid and becomes infectious. If the particle exceeds the maximum age it is no longer considered in the simulation.

9.8 Presenting model outputs

Sea lice particle models produce complex outputs, involving locations of hundreds or thousands of particles at hundreds to thousands of time steps. This structure can make model results difficult to display and so standardised methods of presentation are required to compare results from different dispersal models. As part of the standardisation process, a standardisation of output format is required. (See Barker et al. (2018) for a discussion on format for particle tracking output.) This product will make model-model comparisons more feasible and allow for development of standardised presentation software.

The simplest method of presenting results is to map the location of particles, either as snap-shots of particle distribution at specific times or an aggregated distribution summed over several different time steps, which shows persistent high concentrations and relatively high risk areas. Video presentation can be used to compare movements of concentrations of particles.

Maps of particle positions may be difficult to compare by eye. Furthermore, with particles of different weights and status (nauplii and copepodids), interpretation of associated risk patterns is made more difficult. Therefore, concentrations of larval lice per unit area can be averaged over model domain grid elements by summing the weight (lice each particle represents) for all the particles in a model grid element and then dividing by its area. These concentrations could be plotted with a single value per grid element as a mosaic, or interpolated with respect to adjacent elements to give maps with continuously varying concentrations. An alternative is to use kernel density estimation to produce maps from (super-) particle distributions. Concentrations may be plotted using linear scales or log scales if large ranges in concentration are of interest.

Properties such as vertical distribution present difficulties for presentation of outputs in situations where sea lice are significantly distributed away from the surface, although histograms of lice concentration with depth is one option (Fig. 9.3).

A simple summary output is the proportion of sea lice that travel to a given distance from their source. This distance can be averaged across different model runs and source locations to produce a kernel for estimating relative risk around farms (Salama et al., 2016). Such a kernel is conceptually comparable with those generated by statistical models of risk (Kristoffersen et al., 2014).

The risk of contact between sites or management areas (Adams et al., 2015, 2016; Samsing et al., 2018; Rabe et al., 2020) has been identified, either on the basis of any lice transmission at all or on some threshold level of transmission in simulations, and used to produce networks of contacts between sites or areas. Analysis of such networks identifies areas of high internal connectivity, but also identifies important farms that strongly link the network together and areas that act as barriers or 'firebreaks' between connected areas. Networks of contact (Adams et al., 2015; Salama et al., 2018) could be used to develop models of interacting lice populations, since transport models and population models generally require different time scales (Adams et al., 2015; Salama et al., 2018). Such networks of population models will allow synchronised (or imperfectly synchronised) management practices to be assessed without implementing them full scale. The areas of strongest interaction can define the most appropriate scales for management areas (Adams et al., 2016). These network outputs of dispersal models have great potential to advise aquaculture management, and provide easily understood graphical outputs that define key elements of the aquaculture system.

Another relevant use of sea lice particle model networks is to identify net 'sources' and 'sinks' of lice, that is, locations (for example, fish farms) that are net exporters or importers of particles, respectively. This information can be very useful to identify suitable locations for fish farms or place firebreaks to disrupt a potential route for parasite or disease transmission.

Model outputs need to be compared with observations, and methods of comparison should be standardised. Observed lice counts from fish in sentinel cages or from plankton trawls can be plotted against simulated concentrations from the area of the model domain that corresponds to the sampling locations (Salama et al., 2013). To allow for heterogeneity at levels not captured in the model, the result used may be the maximum within a given range, for example a 500 m distance of the sampling location (Salama et al., 2018). Model results and observations have been compared by matching to defined classes of lice loads (for example, class 1 = zero to one louse per fish, 2 = one to five lice per fish, 3 = five to ten lice per fish and class 4 = more than ten lice per fish). This was done to identify the degree of match or mismatch of predicted and observed lice loads and to generate geographical maps where salmonid fish are likely to be influenced or at risk of harm or death due to lice infestations over a defined time period (Sandvik et al., 2016). Samsing et al. (2016) compared the vertical distribution of lice from the model to observed vertical infestation levels from a scientific trial, and the modelled horizontal concentration of lice to the reported observed levels of lice on fish farm locations. They then calculated how large a proportion of the lice infestations would have been avoided using a snorkel barrier of 5 m depth. Myksvoll et al. (2018) showed that the modelled density of lice from aquaculture locations correlated with the number of lice infestations observed on captured wild sea trout along the Norwegian coast.

9.9 Discussion

The purpose of this chapter is to provide guidance for a generic sea lice model framework where the sea lice, simulated as particles with associated biological and behavioural characteristics, can be driven with the outputs of available hydrodynamic models, provided these have appropriate temporal and spatial resolution in inputs and outputs. The sea lice particle model is intended to be a generic framework that is as robust and simple as possible (but not simpler), so that it can be easily applied in many different situations. A consensus model may not always be suitable for specific applications to specific problems. Alternative functions may therefore be required for non-standard applications.

We have concentrated here on *L. salmonis* as the most important sea louse on salmonid farms in the Northern Hemisphere. However, *C. rogercresseyi* is the most important louse species in Chile (González and Carvajal, 2003), and other *Caligus* species parasitise salmonids elsewhere. These different species, while having similar basic behaviours, may have different environmental tolerances or swimming behaviour that will need to be considered if applying the model to *Caligus* lice.

A standardised generic framework for modelling sea lice dispersal will likely have broad applicability in areas outside the North Atlantic area where salmon are farmed, for example in western Canada where *L. salmonis* is also the dominant lice species, particularly when developing connectivity networks which will assist in the formation of sea lice management zones. However, limited information on the timing, abundance and distribution of migrating juvenile wild Pacific salmon, relative to farm sources of infection, may lessen the applicability of the generic framework as a tool in predicting impacts to wild salmon. In Chile, the sea louse *Caligus rogercresseyi* is a major problem for salmonid aquaculture and, although its life cycle (González and Carvajal, 2003) is similar to *L. salmonis*, some biological parameters are likely to differ between species. In Australia, sea lice do not cause a problem for salmon aquaculture (Murray et al., 2016).

There are existing reviews of the biological functions for the dependence of parameters on the environment (Pike and Wadsworth, 1999; Stien et al., 2005; Samsing et al., 2016; Brooker et al., 2018) and experience in applying models (Johnsen et al., 2014, 2016; Salama et al., 2013, 2018; Adams et al., 2016), so the aim is not to develop new analyses but to derive a model from these existing analyses and applications that can be broadly accepted. Again, a simple model means that there is a minimum of parameters and functions to include, and less disagreement in deriving a consensus. This is not a review of biophysical modelling development or applications either; those are available elsewhere in the literature (North et al., 2009; Gallego, 2011).

In general, existing sea lice dispersal models do use similar approaches. Therefore standardisation on an agreed generic modelling framework is largely a matter of choosing between details of functions rather than between fundamentally different approaches; this makes finding a consensus practicable. A maturation time of 40 degree days is widely used as a default value for *L. salmonis* (Myksvoll et al., 2018) but more information is needed for *Caligus* lice. Mortality is likely to be dominated by predation and probably varies between seasons and areas. Especially in spring bloom systems, the ecosystem is highly variable between seasons. In the lack of detailed information of the primary predators of salmon lice, a constant mortality rate of 1% h^{-1} is the most used practice of today. Egg production also varies with environment (Stien et al., 2005) and the condition of lice, but a rate of 30 viable eggs per day is suggested as default. Thus we identify default values for key parameters consistent with existing standard practice, for use in the absence of local data to tune these to local conditions. One area where there is still considerable uncertainty is the vertical swimming behaviour of sea lice. The details of vertical behaviour are significant in determining horizontal lice distribution (Johnsen et al., 2016), but the data to tie down such behaviour is still limited and apparently contradictory (Salama et al., 2018; Nelson et al., 2018). It is therefore too early to standardise details and, in line with our aim of simplification, the approach of keeping lice relatively close to the surface of the model is recommended in the absence of alternative information; that is, strongly positively buoyant without including further behavioural details. However, it is clear that lice avoid freshwater, and in the presence of a halocline they will seek a depth that avoids low salinity (Crosbie et al., 2019), and therefore this assumption is not valid where haloclines are pronounced. A simple version allowing for low salinity avoidance would be to treat the halocline as though it were the surface and have lice accumulated at or below this depth (Heuch, 1995). It is appropriate for different formulations to be used experimentally in models to assess the impact of vertical swimming on lice particle distributions with respect to that from the simple buoyant model. It is not appropriate to use neutrally buoyant particles in simulations, but standardised models of swimming behaviour are an area for further work once experimental and observational data provide a clearer answer.

Standard models based on a generic framework can be used to compare sea lice infection pressure distributions in different systems with different environments (particularly hydrodynamics and temperature), different levels of production and different distributions of farms. These models can be used to assess the implications of new farms or closures of existing farms, or changing levels of production in farms. As such, a robust and simple standard model that can be agreed a priori (even if it does not include every detail) can be a very useful planning tool, for example to support the implementation of proven management approaches in other jurisdictions. Networks of contact generated by these models can be used to model interacting populations of sea lice. The models already used in Scotland, Norway and Canada use approaches similar to the proposed standard (Jones and Beamish, 2011), as this standardisation draws on decades of modelling work and validation in these and other jurisdictions. The model framework is being applied to sea lice modelling for Killary Harbour, Ireland. This generic model framework will contribute to developing best management practice for sea lice control and has the potential to define a range of production strategies on salmon farms aimed at reducing the presence of sea lice and their negative impacts, both on farmed and wild Atlantic salmon.

Acknowledgements

Work on sea lice model standardisation was funded by the European Union through the North Atlantic Salmon Conservation Organisation (NASCO) under the grants for action programme of the European Maritime and Fisheries Fund, through the LICETRACK Project, coordinated by Inland Fisheries Ireland. Marine Association for Science and Technology Scotland (MASTS) funded an initial meeting.

References

á Norði, G., Simonsen, K., Danielsen, E., Eliasen, K., Mols-Mortensen, A., Christiansen, D.H., Steingrund, P., Galbraith, M. and Pattursson, Ø. (2015) Abundance and distribution of planktonic *Lepeophtheirus salmonis* and *Caligus elongatus* in a fish farming region in the Faroe Islands. *Aquaculture Environment Interactions* 7, 15–27.

á Norði, G., Simonsen, K. and Pattursson, Ø. (2016) A method of estimating *in situ* salmon louse nauplii production at fish farms. *Aquaculture Environment Interactions* 8, 397–405.

Aaen, S.M., Helgesen, K.O., Bakke, M.J., Kaur, K. and Horsberg, T.E. (2015) Drug resistance in sea lice: a threat to salmonid aquaculture. *Trends in Parasitology* 31, 72–81.

Adams, T., Proud, P. and Black, K.D. (2015) Connected networks of sea lice populations: dynamics and implications for control. *Aquaculture Environment Interactions* 6, 273–284.

Adams, T., Aleynik, D. and Black, K.D. (2016) Temporal variability in sea lice population connectivity and implications for regional management protocol. *Aquaculture Environment Interactions* 8, 585–596.

Aldrin, M., Storvik, B., Kristoffersen, A.B. and Jansen P.E. (2013) Space-time modelling of the spread of salmon lice between and within Norwegian marine salmon farms. *PLoS ONE.* Available at: https://doi.org/10.1371/journal.pone.0064039 (accessed 17 January 2022).

Aldrin, M., Jansen, P.A. and Stryhn H. (2019) A partly stage-structured model for the abundance of salmon lice in salmonid farms. *Epidemics* 26, 9–22.

Amundrud, T.L. and Murray, A.G. (2007) Validating particle tracking models of sea lice dispersion in Scottish sea lochs. ICES CM 2007/B:05. Available at: https://www.ices.dk/sites/pub/CM%20Doccuments/CM-2007/B/B0507.pdf (accessed 17 January 2022).

Amundrud, T.L. and Murray, A.G (2009) Modelling sea lice dispersion under varying environmental forcing in a Scottish sea loch. *Journal of Fish Diseases* 32, 27–44.

Arakawa, A. and Lamb, V.R. (1977) Computational design of the basic dynamical processes of the UCLA general circulation model. *Methods in Computational Physics: Advances in Research and Applications* 17, 173–265.

Asplin, L., Boxaspen, K. and Sandvik, A.D. (2004) Modelled distribution of salmon lice in a Norwegian fjord. ICES CM 2004/P:11 Available at: https://www.ices.dk/sites/pub/CM%20Doccuments/2004/P/P1104.pdf (accessed 17 January 2022).

Asplin, L., Johnsen, I.A., Sandvik, A.D., Albretsen, J., Sundfjord, V., Aure, J. and Boxaspen, K.K. (2014) Dispersion of salmon lice in the Hardangerfjord. *Marine Biology Research* 10, 216–225.

Barker, C. (2018) Draft proposal for a netcdf standard for storing/sharing data from particle tracking models. NOAA Emergency Response Division. Available at: https://github.com/NOAA-ORR-ERD/nc_particles/blob/master/nc_particle_standard.md (accessed 17 January 2022).

Bloodworth, J.W., Baptie, M.C., Preedy, K.F. and Best, J. (2019) Negative effects of the sea lice therapeutant emamectin benzoate at low concentrations on benthic communities around Scottish fish farms. *Science of the Total Environment* 669, 91–102.

Boxaspen, K. and Næss, T. (2000) Development of eggs and planktonic stages of sea lice (*Lepeophtheirus salmonis*) at low temperatures. *Contributions to Zoology* 69, 51–55.

Bricknell, I.S., Dalesman, S.J., O'Shea, B., Pert, C.C. and Mordue Luntz, A.J. (2006) Effects of environmental salinity on sea lice *Lepeophtheirus salmonis* settlement success. *Diseases of Aquatic Organisms* 71, 201–212.

Brooker, A.J., Skem-Mauritzen, R. and Bron, J.E. (2018) Production, mortality, and infectivity of planktonic larval sea lice, *Lepeophtheirus salmonis* (Krøyer, 1837): current knowledge and implications for epidemiological modelling. *ICES Journal of Marine Science* 75, 1214–1234.

Butler, J.R.A. (2002) Wild salmonids and sea louse infestations on the west coast of Scotland: sources of infection and implications for the management of marine salmon farms. *Pest Management Science* 58, 595–608.

Chen, C., Beardsley, R.C., Cowles, G., Qi, J., Lai, Z., Gao, G., Stuebe, D., Xu, Q., Xue, P., Ge, J., Ji, R., Hu, S., Tian, R., Huang, H., Wu, L. and Lin, H. (2011) An unstructured grid, finite-volume coastal ocean model. FVCOM user manual. MIT, Cambridge, Mass. Available at: http://fvcom.smast.umassd.edu/wp-content/uploads/2013/11/MITSG_12-25.pdf (accessed 17 January 2022).

Christensen, A., Mariani, P. and Payne, M.R. (2018) A generic framework for individual-based modelling and physical-biological interaction. *PLoS ONE*. Available at: https://doi.org/10.1371/journal.pone.0189956 (accessed 17 January 2022).

Crosbie, T., Wright, D.W., Oppedal, F., Johnsen, I.A., Samsing F. and Dempster T. (2019) Effect of step salinity gradients on salmon lice larval behaviour and dispersal. *Aquaculture Environment Interactions* 11, 181–190.

Elghafghuf, A., Vanderstichel, R., Hammell, L. and Stryhn, H. (2020) Estimating sea lice infestation pressure on salmon farms: Comparing different methods using multivariate state-space models. *Epidemics* 31, 100394.

FAO (2020) FAO yearbook of fishery and aquaculture statistics 2018. Food and Agriculture Organisation of the UN, Rome. Available at: http://www.fao.org/fishery/statistics/yearbook/en (accessed 17 January 2022).

Gallego, A. (2011) Bio-physical models: an evolving tool in marine ecological research. In: Jopp, F., Reuter, H. and Breckling, B. (eds) *Modelling Complex Ecological Dynamics*. Springer, Berlin. pp. 279–290.

Genna, R.L., Mordue, W., Pike, A.W. and Mordue (Luntz), A.J. (2005) Light intensity, salinity, and host velocity influence presettlement intensity and distribution on hosts by copepodids of sea lice, *Lepeophtheirus salmonis*. *Canadian Journal of Fisheries and Aquatic Sciences* 62, 2675–2682.

Gillibrand, P.A. and Willis, K.J. (2007) Dispersal of sea louse larvae from salmon farms modelling the influence of environmental conditions and larval behaviour. *Aquatic Biology* 1, 63–75.

González, L. and Carvajal, J. (2003) Life cycle of *Caligus rogercresseyi* (Copepoda: Caligidae) parasite of Chilean reared salmonids. *Aquaculture* 220, 101–117.

Groner, M.L., McEwan, G.F., Rees, E.E., Gettinby, G. and Revie, C.W. (2016) Quantifying the influence of salinity and temperature on the population dynamics of a marine ectoparasite. *Canadian Journal of Fisheries and Aquatic Sciences* 73, 1281–1291.

Hamre, L.A., Eichner, C., Caipang, C.M.A., Dalvin, S.T., Bron, J.E., Nilsen, F., Boxshall, G. and Skern-Mauritzen, R. (2013) The salmon louse *Lepeophtheirus salmonis* (Copepoda: Caligidae) life cycle has only two chalimus stages. *PLoS ONE*. Available at: doi: 10.1371/journal.pone.0073539 (accessed 17 January 2022).

Hemmingsen, W., MacKenzie, K., Sagerus, K., Remen, M., Bloch-Hansen, K. and Imsland, A.K.D. (2020) *Caligus elongatus* and other sea lice of the genus *Caligus* as parasites of farmed salmonids: a review. *Aquaculture* 522, 735160.

Hevrøy, E.M., Boxaspen, K., Oppedal, F., Taranger, G.L. and Holm, J.C. (2003) The effect of artificial light treatment and depth of infestation of the sea louse *Lepeophtheirus salmonis* on Atlantic salmon (*Salmo salar* L.) culture. *Aquaculture* 220, 1–14.

Heuch, P.A. (1995) Experimental evidence for aggregation of salmon louse copepodids (*Lepeophtheirus salmonis*) in step salinity gradients. *Journal of the Marine Biological Association of the UK* 75, 927–939.

Heuch, P.A. and Mo, T.A. (2001) A model of salmon louse production in Norway: effects of increasing salmon production and public management measures. *Diseases of Aquatic Organisms* 45, 145–152.

Heuch, P.A., Parson, A. and Boxaspen, K. (1995) Diel vertical migration: A possible host-finding mechanism in salmon louse (*Lepeophtheirus salmonis*) copepodids? *Canadian Journal of Fisheries and Aquatic Sciences* 52, 681–689.

Heuch, P.A., Nordhagen, J.R. and Schram, T.A. (2000) Egg production in the salmon louse [*Lepeophtheirus salmonis* (Krøyer)] in relation to origin and water temperature. *Aquaculture Research* 31, 805–814.

Johnsen, I.A., Fiksen, Ø., Sandvik, A.D. and Asplin, L. (2014) Vertical salmon lice behaviour as a response to environmental conditions and its influence on regional dispersion in a fjord system. *Aquaculture Environment Interactions*, 5, 127–141.

Johnsen, I.A., Asplin, L.C., Sandvik, A.D. and Serra-Llinares, R.M. (2016) Salmon lice dispersal in a northern Norwegian fjord system and the impact of vertical movements. *Aquaculture Environment Interactions* 8, 99–116.

Johnson, S.C. and Albright, L.J. (1991) The development stages of *Lepeophtheirus salmonis* (Krøyer, 1837) (Copepoda: Caligidae) *Canadian Journal of Zoology* 69, 929–950.

Jones, S.R.M. and Beamish, R, (2011) *Salmon Lice: an Integrated Approach to Understanding Parasite Abundance and Distribution*. Wiley-Blackwell, Chichester, UK.

Kragesteen, T.J., Simonsen, K., Visser, A.W. and Andersen K.H. (2018) Identifying salmon lice transmission characteristics between Faroese salmon farms. *Aquaculture Environment Interactions* 10, 49–60.

Kristoffersen, A.B., Rees, E.E., Stryhn, H., Ibarra, R., Campisto, J.L., Revie, C.W. and St-Hilaire, S. (2013) Understanding sources of sea lice for salmon farms in Chile. *Preventative Veterinary Medicine* 111, 165–175.

Kristoffersen, A.B., Jimenez, D., Viljugrein, H., Grøntvedt, R., Stein, A. and Jansen, P.A. (2014) Large-scale modelling of salmon lice (*Lepeophtheirus salmonis*) infection pressure based on lice monitoring data from Norwegian salmonid farms. *Epidemics* 9, 31–39.

McEwan, G.F., Groner, M.L., Fast, M.D., Gettinby, G. and Revie, C.W. (2015) Using agent-based modelling to predict the role of wild refugia in the evolution of resistance of sea lice to chemotherapeutants. *PLoS ONE.* Available at: https://doi.org/10.1371/journal.pone.0139128 (accessed 17 January 2022).

McGurk, M. D. (1986) Natural mortality of marine pelagic fish eggs and larvae: role of spatial patchiness. *Marine Ecology Progress Series* 34, 227–242.

Mennerat, A., Ugelvik, M.S., Jensen, C.H. and Skorping, A. (2017) Invest more and die faster: The life history of a parasite on intensive farms. *Evolutionary Applications* 10, 890–896.

Molinet, C., Cáceres, M., González, M.T., Carvajal, J., Asencio, G., Díaz, M., Díaz, P., Castro, M.T. and Codjambassis, J. (2011) Population dynamic of early stages of *Caligus rogercresseyi* in an embayment used for intensive salmon farms in Chilean inland seas. *Aquaculture* 312, 62–71.

Murray, A.G. and Gillibrand, P.A. (2006) Modelling salmon lice dispersal in Loch Torridon, Scotland. *Marine Pollution Bulletin* 53, 128–135.

Murray, A.G. and Moriarty, M. (2021) A simple modelling tool for assessing interaction with host and local infestation of sea lice from salmonid farms on wild salmonids based on processes operating at multiple scales in space and time. *Ecological Modelling* 443, 109459.

Murray A.G. and Salama N.K.G. (2017) Modelling disease in aquatic systems that are spread by processes operating at different temporal and spatial scales: examples from salmon aquaculture. CAB Direct 12 No. 032 1–11 10.1079/PAVSNNR201712032 (accessed 17 January 2022).

Murray, A.G., Amundrud, T.L., Penston, M.J., Pert, C.C. and Middlemas, S.J. (2011) Chapter 2: Abundance and distribution of larval sea lice in Scottish coastal waters. In: Jones, S. and Beamish, R. (eds) *Salmon Lice: an Integrated Approach to Understanding Parasite Abundance and Distribution*. Wiley-Blackwell, Oxford, UK. pp. 51–81.

Murray, A.G., Wardeh, M. and McIntyre, K.M. (2016) Using the H-index to assess priorities for salmon aquaculture. *Preventative Veterinary Medicine* 126, 199–207.

Myksvoll, M.S., Sandvik, A.D., Albretsen, J., Asplin, L., Johnsen, I.A., Karlsen, Ø., Kristensen, N.M., Melsom, A., Skarðhamar, J. and Ådlandsvik, B. (2018) Evaluation of a national operational salmon lice monitoring system – From physics to fish. *PLoS ONE.* Available at: https://doi.org/10.1371/journal.pone.0201338 (accessed 17 January 2022).

Nelson, E.J., Robins, S.M.C., Feindel, N., Sterling, A., Byrne, A. and Pee Ang, K. (2018) Horizontal and vertical distribution of sea lice larvae (*Lepeophtheirus salmonis*) in and around salmon farms in the Bay of Fundy, Canada. *Journal of Fish Diseases* 41, 885–899.

North, E.W., Gallego, A. and Petitgas, P. (2009) Manual of recommended practices for modelling physical - biological interactions during fish early life. *ICES Cooperative Research Report* 295.

Oppedal, F., Samsing, F., Dempster, T., Wright, D.W., Bui, S. and Stien, L.H. (2017) Sea lice infestation levels decrease with deeper 'snorkel' barriers in Atlantic salmon sea-cages. *Pest Management Science* 73, 1935–1943.

Paris, C.B., Helgers, J., van Sebille, E. and Srinivasan, A. (2013) Connectivity modeling system: a probabilistic modeling tool for the multi-scale tracking of abiotic and biotic variability in the ocean. *Environmental Modelling and Software* 42, 47–54.

Pike, A.W. and Wadsworth, S.L. (1999) Sea lice on salmonids: their biology and control. *Advances in Parasitology* 44, 233–337.

Rabe, B., Gallego, A., Wolf, J., O'Hara Murray, R., Stuiver, C., Price, D. and Johnson, H. (2020) Applied connectivity modelling at local to regional scale: The potential for sea lice transmission between Scottish finfish aquaculture management areas. *Estuarine, Coastal and Shelf Science* 238, 106716.

Salama, N.K.G. and Rabe, B. (2013) Developing models for investigating the environmental transmission of disease-causing agents within open-cage salmon aquaculture. *Aquaculture Environment Interactions* 4, 91–115.

Salama, N.K.G., Collins, C.M., Fraser, J.G., Dunn, J., Pert, C.C., Murray, A.G. and Rabe, B. (2013) Development and assessment of a biophysical dispersal model for sea lice. *Journal of Fish Diseases* 36, 323–337.

Salama, N.K.G., Murray, A.G. and Rabe, B. (2016) Simulated environmental transport distances of *Lepeophtheirus salmonis* in Loch Linnhe, Scotland, for informing aquaculture area management structures. *Journal of Fish Diseases* 39, 419–428.

Salama, N.K.G., Dale, A.C., Ivanov, V.V., Cook, P.F., Pert, C.C., Collins, C.M. and Rabe, B. (2018) Using biological-physical modelling for informing sea lice dispersal in Loch Linnhe, Scotland. *Journal of Fish Diseases* 41, 901–919.

Samsing, F., Solstorm, D., Oppedal, F., Solstorm, F. and Dempster, T. (2015) Gone with the flow: current velocities mediate parasitic infestation of an aquatic host. *International Journal for Parasitology* 45, 559–565.

Samsing, F., Oppedal, F., Dalvin, S., Johnsen, I., Vågseth, T. and Dempster, T. (2016a) Salmon lice (*Lepeophtheirus salmonis*) development times, body size, and reproductive outputs follow universal models of temperature dependence. *Canadian Journal of Fisheries and Aquatic Science* 73, 1841–1851.

Samsing, F., Johnsen, I., Stien, L.H., Oppedal, F., Albretsen, J., Asplin, L. and Dempster, T. (2016b) Predicting the effectiveness of depth-based technologies to prevent salmon lice infection using a dispersal model. *Preventative Veterinary Medicine* 129, 48–57.

Samsing, F., Johnsen, I., Dempster, T., Oppedal, F. and Treml, E.A. (2018) Network analysis reveals strong seasonality in the dispersal of a marine parasite and identifies areas for coordinated management. *Landscape Ecology* 32, 1953–1967.

Sandvik, A.D., Bjørn, P.A., Ådlandsvik, B., Asplin, L., Skardhamar, J., Johnsen, I.A., Myksvoll, M. and Skogen, M.D. (2016) Towards a model-based prediction system for salmon lice infestation pressure. *Aquaculture Environment Interactions* 8, 527–542.

Sandvik A.D., Johnsen I.A., Myksvoll M.S., Sævik P.N. and Skogen M.D. (2020) Prediction of the salmon lice infestation pressure in a Norwegian fjord. *ICES Journal of Marine Research* 77, 746–756.

Shephard, S., MacIntyre, C. and Gargan, P. (2016) Aquaculture and environmental drivers of salmon lice infestation and body condition in sea trout. *Aquaculture Environment Interactions* 8, 597–610.

Skardhamar, J., Albretsen, J., Sandvik, A.D., Lien, V.S., Myksvoll, M.S., Johnsen, I.A., Asplin, L., Ådlandsvik, B., Halttunen, E. and Bjørn, P.A. (2018) Modelling salmon lice dispersion and infestation patterns in a sub-arctic fjord. *ICES Journal of Marine Science* 5, 1733–1747.

Stien, A, Bjorn, P.A., Heuch, P.A. and Elston, D.A. (2005) Population dynamics of salmon lice *Lepeophtheirus salmonis* on Atlantic salmon and sea trout. *Marine Ecology Progress Series* 290, 263–275.

Stien, L.H., Dempster, T., Bui, S., Glaropoulos, A., Fosseidengen, J.K., Wright, D.W. and Oppedal, F. (2016) 'Snorkel' sea lice barrier technology reduces sea lice loads on harvest-sized Atlantic salmon with minimal welfare impacts. *Aquaculture* 458, 29–37.

Taranger, G.L., Karlsen, Ø., Bannister, R.J., Glover, K.A., Husa, V., Karlsbakk, E., Kvamme, B.O., Boxaspen, K.K., Bjørn, P.A. Finstad, B., Madhun, A.S., Morton, H.C. and Svåsand, T. (2015) Risk assessment of the environmental impact of Norwegian Atlantic salmon farming. *ICES Journal of Marine Research* 72, 997–1021.

Todd C.D., Walker A.M., Ritchie M.C., Graves A.J. and Walker A.F. (2004) Population genetic differentiation of sea lice (*Lepeophtheirus salmonis*) parasitic on Atlantic and Pacific salmonids: analyses of microsatellite DNA variation among wild and farmed hosts. *Canadian Journal of Fisheries and Aquatic Sciences* 61, 1176–1190.

Wolf, J., Stuiver, C., Price, D., Johnson, H., Gallego, A. and O'Hara Murray, R. (2016) The Scottish shelf model. Part 6: Wider domain and sub-domain integrations. *Scottish Marine and Freshwater Science* 7(8).

CHAPTER 10

Population dynamics of sea lice

Fernando O. Mardones[1,2] *and Manuel Lepe-López*[3,4]

[1]Escuela de Medicina Veterinaria, Facultad de Agronomía e Ingeniería Forestal, Facultad de Ciencias Biológicas y Facultad de Medicina, Pontificia Universidad Católica de Chile, Santiago, Chile
[2]Department of Pediatric Infectious Diseases and Immunology, Facultad de Medicina, Pontificia Universidad Católica de Chile, Santiago, Chile
[3]PhD Program in Conservation Medicine, Universidad Andres Bello, Santiago, Chile
[4]Facultad de Ciencias de la Vida, Universidad Andres Bello, Santiago, Chile

10.1 Introduction

Management of parasites and pests remains a sustainability challenge in many aquaculture sectors, but particularly in intensive salmon production (Naylor et al., 2021). Sea lice of the family Caligidae (Copepoda) are marine ectoparasitic copepods with vast reproductive potential and affect a wide variety of fish species, predominantly salmonids. The estimated global costs of sea lice infections to the salmon industry in 2015 were approximately £700 million, comprising mainly the costs for treatments and to a lesser extent losses in production (Brooker et al., 2018). The most frequently occurring sea louse on salmon in the Northern Hemisphere is *Lepeophtheirus salmonis* (Krøyer 1837), whereas *Caligus elongatus* (Nordmann) is, to some degree, prevalent in Europe and Canada, affecting a variety of salmonid and non-salmonid species. *Caligus elongatus* is only sporadically subject to research because of low prevalence and limited influence on fish morbidity (Aaen et al., 2015). In the Southern Hemisphere *Caligus rogercresseyi* (Boxshall & Bravo 2000) is the principal pathogenic species affecting the Chilean salmon aquaculture industry. In Australia, as is the case in Chile, host-switching from endemic to farmed introduced hosts has been reported, where *Caligus longirostris* was present on farmed Atlantic salmon (*Salmo salar* L.) and rainbow trout (*Oncorhynchus mykiss*) in Tasmania (Nowak et al., 2010). The reported impact of sea lice in Australian salmon aquaculture is almost insignificant.

Among many species of sea lice, this chapter is focused on the population dynamics of *L. salmonis* and *C. rogercresseyi* as they are the dominant species severely affecting the salmon farming industry. Sea lice can cause fish mortality, skin damage, and stress, impacting wild salmonid fish also. The use of antiparasitic treatments to control sea lice may also promote drug resistance. Sea lice infections on farmed salmon are associated with increased infection pressure on adjacent wild salmon stocks, and there is particular concern that juvenile wild salmon newly migrated into the sea are at greatest risk from the adverse effects of salmon louse infection (Marty et al., 2010; Vollset et al., 2018).

10.2 Sea lice population dynamics

Population dynamics studies the size and age composition of populations as dynamic systems, and the biological and environmental processes driving them. A population is defined as a group of individuals of the same species living and interbreeding within a given area. Individuals of the population often rely on the same resources, are subject to similar environmental constraints, and depend on the availability of other members to persist over time. A major component of ecological research focuses on understanding what determines the structure and functioning of sea lice populations and their variation in time and space. There are several processes that occur simultaneously that can affect population size and dynamics. The population size of the sea lice is influenced by the per capita population growth rate, which is the rate at which the population size changes per individual in the population. This growth rate is determined by the stage-specific developmental rates, death, emigration, and migration rates in the population. When the per capita growth rate remains constant, the population can experience exponential growth followed by exponential decline. The maximum per capita growth rate for a population is called the intrinsic rate of increase.

As a population grows in an area, a population may experience the effects of increased densities. In each area, the maximum population size of the species that the environment can sustain is called the carrying capacity, which is determined by the availability of resources including food and habitats. As the density of individuals in a population increases, these individuals must begin competing for limited resources with members of the same species (intra-specific competition) or with members of other species (inter-specific competition). If the population grows indefinitely, the availability of resources will decrease with less being available to sustain the population. The process in which per capita population growth changes as population density changes is referred to as density dependence.

In the following sections, we describe the factors that explain average abundance of the sea lice population and the causes of their fluctuations. We also provide examples using sea lice count data derived from research projects from the Chilean salmon industry.

10.3 Sea lice diversity and developmental stages

Members of the family Caligidae, mainly species of *Caligus* Müller, 1785 and *Lepeophtheirus* Nordmann, 1832, represent a group of siphonostomatoid copepods commonly known as sea lice which have impacted finfish aquaculture worldwide (Lafferty et al., 2015). This family comprises 36 genera and more than 500 valid species, but information on the complete life cycle is only available for 17 species (Maran et al., 2013). These species belong to three genera, that is, *Caligus* (12 species), *Lepeophtheirus* (four species), and *Pseudocaligus* (one species). To date, 125 described species of *Lepeophtheirus* are accepted as valid (Walter and Boxshall, 2022). *Caligus amblygenitalis* and *Caligus chiastos* have been significantly associated with the severity of gross eye damage in southern bluefin tuna farmed in South Australia (Hayward et al., 2009). *Caligus epidemicus* Hewitt, 1971 is an important pathogen in Asia, causing mortality in more than ten marine cultured fish species (Ho, 2000). *Caligus orientalis* Gusev, 1951 often causes serious problems at fish farms in the brackish-water regions of Japan, Taiwan, and China (Nagasawa, 2004). In addition to the impact of *C. rogercresseyi* and *L. salmonis* in salmon farms, there are reported declines of wild fish populations in areas surrounding sea cages associated with sea lice epidemics.

In all caligids, the eggs are carried in linear, uniseriate egg strings and hatch as a free-swimming nauplius (Maran et al., 2013). The free-swimming phase consists of two naupliar stages followed by the copepodid, which is the infective stage that locates the host. Subsequent development on the host appears to be more variable, with different numbers of stages having been reported in varied species of *Caligus* and *Lepeophtheirus*: four or six chalimus stages, zero or two pre-adults, and one adult stage. The life cycle of *L. salmonis* includes eight stages: five free-swimming larval stages (two naupliar stages, one copepodite stage, and two chalimus stages), and three post-larval stages including two pre-adult stages and one adult stage (Schram, 1993; Hamre et al., 2021). Similarly, *C. rogercresseyi* includes eight different development stages: three larval stages (two nauplius and one copepodid), four chalimus, and the adult (González and Carvajal, 2003). The first three stages are planktonic, and the others are parasitic as no pre-adult stage has been observed. The naupliar stages are not infective, while the larval copepodite stage attaches to the salmon and moults into the chalimus stages. The two pre-adult stages are fully motile and may jump between nearby hosts. In the last adult stage, fertilized females develop sets of egg strings which, upon hatching, release a new generation of naupliar stages. Consequently the salmon louse life cycle has only six post-nauplius instars, as in other genera of caligid sea lice and copepods in general (Hamre et al., 2013).

10.4 Host susceptibility

Sea lice primarily infect salmonids belonging to the genera *Salmo*, *Oncorhynchus*, and *Salvelinus* (Johnson and Albright, 1992; Pike and Wadsworth, 1999; Bjørn et al., 2001; Wootten et al., 2011). Susceptibility to *L. salmonis* infection varies among the salmonid hosts, with coho (or Pacific) salmon (*O. kisutch*) and pink salmon (*O. gorbuscha*) being less susceptible than sockeye salmon (*O. nerka*) and Atlantic salmon (Wootten et al., 1982; Johnson and Albright, 1992; Jones and Hargreaves, 2007; Jakob et al., 2013).

Field and experimental studies have shown that the abundance and prevalence of *C. rogercresseyi* adult stages are greater in rainbow trout compared to Atlantic salmon (González et al., 2000; Bravo et al., 2008; Yatabe et al., 2011; Hamilton-West et al., 2012; Valenzuela-Muñoz et al., 2016; Lepe-López et al., 2021b). However, some studies have also observed similarities in *C. rogercresseyi* susceptibility by rainbow trout and Atlantic salmon, arguing differences caused by temperatures greater than 12°C and salinities greater than 25 ppt (Bravo et al., 2014). The susceptibility of rainbow trout to *C. rogercresseyi* could be explained by semiochemicals and mucous cell hypertrophy response (Fast et al., 2002). Besides, there is very low heritability susceptibility to *C. rogercresseyi* in rainbow trout (Bassini et al., 2019). In contrast, coho salmon (as the most *C. rogercresseyi*-resistant species) has a low mucus biochemical response and a different macrophage response mechanism to *C. rogercresseyi* infestation (Valenzuela-Muñoz et al., 2016).

Susceptibility to *L. salmonis* depends on the innate capacity of the fish to mount inflammatory and acute-phase responses in addition to immunomodulation at the site of attachment (Jones et al., 2006; Skugor et al., 2008; Tadiso et al., 2011; Braden et al., 2012). This species also infects a range of other salmonid fish, both farmed and wild, as well as other unrelated fish such as the three-spined stickleback *Gasterosteus aculeatus* (Jones et al., 2006), seabass *Dicentrarchus labrax* (Pert et al., 2006), and saithe *Pollachius virens* (Bruno and Stone, 1990; Lyndon and Toovey, 2001).

Caligus rogercresseyi has been found on native marine fish, including native rock cod *Eleginops maclovinus*, Chilean silverside *Odonthestes regia*, and a flatfish *Paralichthys microps* (Carvajal et al., 1998).

As in most farmed salmon species, the immune responses to *L. salmonis* infestation include inflammatory responses, protease secretion, immunoglobulin deficiency, and T-helper 2 (Th2) responses. However, coho salmon were the most resistant to infection followed by chinook then Atlantic salmon. Infected rainbow trout and Atlantic salmon showed increases in mucus lysozyme activities at one day post injection (dpi) (Fast et al., 2002). Histological sections of coho fins and gills revealed well-developed epithelial hyperplasias and inflammatory responses to the presence of *L. salmonis* (Johnson and Albright, 1992). For *C. rogercresseyi*, liver and muscle carbohydrate intermediate anabolic metabolism in *O. kisutch* increased during infestation, suggesting that this species preferentially uses the glycogenolytic-glucolytic pathway, in contrast with *S. salar* (Vargas-Chacoff et al., 2016). Besides, coho salmon is the more resistant species, presenting a primary Th1 response compared to Atlantic salmon and rainbow trout (Valenzuela-Muñoz et al., 2016). Under both laboratory and field conditions, coho salmon are generally less susceptible to infection with these sea lice species than either rainbow trout or Atlantic salmon (Johnson et al., 2004).

10.5 The abundance and distribution of sea lice

Abundance refers to the average count of sea lice on individuals or a group of sampled salmonids in the wild or farmed conditions. Farm-level sea lice abundances are required to be calculated in many salmon farming countries at regular intervals (that is weekly or monthly) by averaging the pen-level values using specified protocols. Even though sea lice abundance follows a density-dependent process, at salmon farming sites where higher prevalence and intensity of infection has been reported in denser populations of farmed salmon (Jansen et al., 2012), there are factors that modulate abundance of sea lice. Factors influencing this variability are the production of salmon lice eggs (mostly from the farmed fish (Heuch et al., 2005)), the water current that determines the spreading of the planktonic salmon lice, the water temperature determining the growth rates of eggs and the larval stages, and the salinity influencing the salmon lice survival and behavior (Heuch et al., 1995).

A variety of deterministic (Revie et al., 2003; Stien et al., 2005; Robbins et al., 2010; Kristoffersen et al., 2014) and stochastic (Aldrin et al., 2013; Groner et al., 2013) models have been derived to predict sea lice dynamics. For example, statistical regression techniques can be used when identified environmental variables suspected to have an effect on the control of lice populations are investigated as potential risk factors (Revie et al., 2003). Linear models and multiple regression are increasingly being used in quantitative epidemiological studies to formulate and test hypotheses, and to construct statistical models that can explain outcomes related to disease in terms of independent variables. Factors such as level of treatment, type of treatment, cage volume, current speed, loch flushing time, and sea lice levels in the preceding period were found to be important. Some studies combine the monitoring data with a deterministic model to estimate farm production of infectious lice stages, using empirical estimates of the relative risk of salmon lice transmission between farms that depend on inter-farm distances, to estimate the external infection pressure at a farm site (Kristoffersen et al., 2014). More sophisticated stochastic space-time models for *L. salmonis* have shown that about 66% of the expected salmon lice abundance can be attributed to infection within farms, while the remaining 28% may be attributed to infection from neighborhood farms and 6% to non-specified sources of infection (Aldrin et al., 2013).

Factors such as temperature, wind direction, and salinity are significant (Boxaspen, 2006; Bricknell et al., 2006; Amundrud and Murray, 2009), with numerous other variables cited as having the potential to affect sea lice density (Heuch et al., 2000; Revie et al., 2002; Krkošek et al., 2009).

The lifespan of copepods in the water column is highly influenced by water temperature (Groner et al., 2016; Montory et al., 2018). The mean lifespan of free-living stages of *C. rogercresseyi* in the water column is 20 days at 12°C (González and Carvajal, 2003; Montory et al., 2018). The first four days of pelagic life correspond to nauplius larval stages, which moult to a copepodid stage that may live up to 16 days without a host. Early developmental stages (for example, larvae) of marine invertebrates are considered to be the most vulnerable phases in their life cycles (Chaparro et al., 2008; Przeslawski et al., 2015), with the highest mortality rates occurring during planktonic dispersal stages (Rumrill, 1990; Short et al., 2013; Morgan, 2020). This mortality has been associated with physiological stress caused by environmental conditions but also with biological interactions such as predation (Rumrill, 1990; Morgan, 2020), which in turn can threaten population persistence (Allan et al., 1976).

10.6 Reproduction, maturation, and growth rates

The dynamics of the parasites depend on both host abundance and transmission among populations, particularly among wild and domesticated fish (Krkošek, 2017). These transmissions will likely be focused on regional scales around aquaculture activities and wild fish habitats. Recently, Vollset et al. (2018) correlated the estimated infestation pressure of sea lice within farms with louse counts on wild fish, indicating that fish farms play a significant role in the epidemiology of sea lice on wild sea trout. Moreover, these authors demonstrated that a large proportion of the temperature effect on lice counts is mediated through the infection pressure from farms, since these variables interact, increasing the lice counts with warm water (Vollset et al., 2018). In this way, it is possible that the caligid population on Patagonian blenny could play a relevant role in the parasite dynamics of both salmon and native fish, participating as a host and as a source of parasites, respectively.

Factors such as temperature, wind direction, and salinity are significant (Boxaspen and Næss, 2000; Bricknell et al., 2006; Amundrud and Murray, 2009), with numerous other variables cited as having the potential to affect sea lice density (Heuch et al., 2000; Revie et al., 2003; Gillibrand and Willis, 2007). For example, dispersal of planktonic larvae in the water surface is strongly influenced by tidal and wind forces (Murray and Gillibrand, 2006).

Increasing water temperature has an effect on the nauplius and chalimus maturation rates (Stien et al., 2005). On the other hand, increasing water salinity has a proportional effect on egg viability (Johnson and Albright, 1991) while decreasing mortality rates of nauplius, copepodid, chalimus, and adult stages (Bricknell et al., 2006; Connors et al., 2008). Changes in salinity and temperature are reported to affect *L. salmonis* survival and growth rates as the majority of active nauplii died at the copepodid moult stage at salinities of 20 and 25 mg/l, and active copepodids were only obtained at 30 mg/l (Johnson and Albright, 1991).

Temperature, light, and currents are major factors that affect the dispersal of the planktonic stages of both *L. salmonis* and *C. rogercresseyi*, and their survival depends on salinities above 25 psu (Costello, 2006; Brooks, 2009). The probability of larvae occurring at any particular site depends on the number of ovigerous lice present at any one time in a particular area and the distance from, and the prevailing currents to, the point where the larvae are recovered (Costelloe et al., 1998). For *L. salmonis*, the larvae require between 4.2 and 5.8 days to develop to the infective stage at the temperatures typical

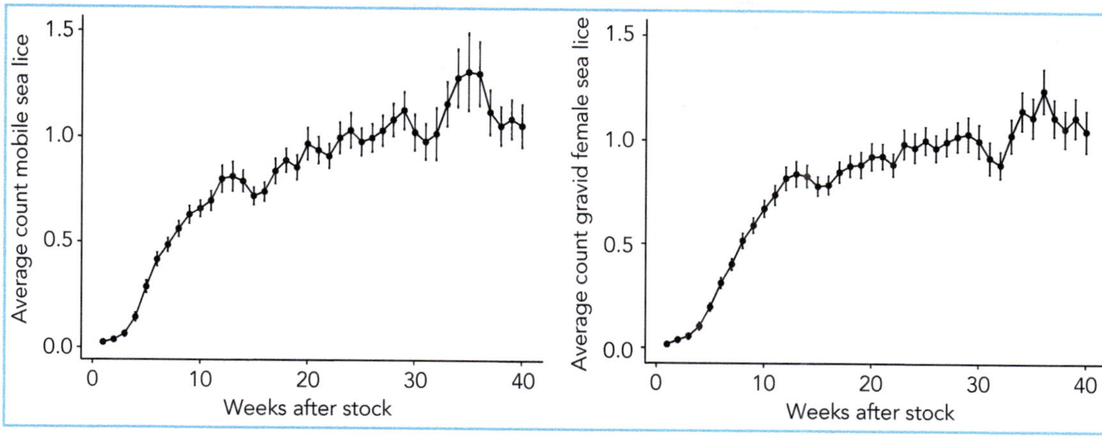

Fig. 10.1 Average counts of mobile (left) and gravid female (right) sea lice from 284 Atlantic salmon farms in southern Chile 2014–2018.

during the time that pink salmon fry are migrating, and do not consistently develop to an infectious stage at salinities less than 30 ppt (Brooks, 2005). Maximal copepodid presettlement occurred at low to medium light, with increasing light levels causing a shift in attachment from the ventral surfaces and fins to the head and dorsal and caudal areas and fins, indicative of a behavioral response to light (Genna et al., 2005). Freshwater inputs force strong seaward-flowing surface currents in the main channels that likely carry imbedded sea lice larvae significant distances during their development to an infective stage (Brooks, 2005). In *C. rogercresseyi* each developmental stage occurs at temperatures between 4°C and 17°C, with the rate of development being a linear function of water temperature (González and Carvajal, 2003). The species is also sensitive to salinities lower than 20% (Bravo et al., 2008). Sea lice counting data from 284 Atlantic salmon farms stocked between 2016 and 2018 in southern Chile and that did not receive any antiparasitic treatment before stocking in the sea showed that infestation rates were strongly modulated by water temperature and salinity as shown in Fig. 10.1.

The developmental time from the extrusion of egg strings to adults is temperature dependent. It takes approximately 40 days for *L. salmonis* for males and 50 days for females at 10°C, but less for *C. rogercresseyi* (Costello, 2006). The total generation time of sea lice depends on water temperature and takes about 98 days at 7°C (Heuch et al., 2000). Under laboratory conditions, the *L. salmonis* adult female can live up to 191 days and produce up to 11 pairs of egg strings each containing a mean number of 285 eggs at 7.2°C (Heuch and Mo, 2001). The nauplius and the infectious copepod stages last about 7 and 12 days at 7°C, respectively (Gislason, 2018). As a result, development of the first pair of egg strings takes about 79 days (98 –191) at 7°C from infection.

10.7 Sea lice mortality rates

Copepodid mortality is important in assessing relative, as well as absolute, distribution of infection pressure. Stien et al. (2005) estimated a mortality rate of 0.22 day^{-1} at a salinity of 30 ppt and with no clear change with temperature. Johnson and Albright (1991) found optimal survival at a salinity of 25 ppt at 10°C. However, in contradiction, Bricknell et al. (2006) found that survival fell steeply between salinities of 26 ppt and 29 ppt, with the 50% mortality time dropping from approximately one day to half a day, giving mortality rates of approximately 0.029 h^{-1} and 0.058 h^{-1}, far higher than the number estimated by Stien et al. (2005). The mortality rate of larval lice is higher in brackish water, with mortality increasing quickly for salinities less than 25 ppt and such that by 12 h mortality was 0.5 h^{-1} (Bricknell et al., 2006). Nauplii may die before maturing to become copepodids, at rates largely dependent on salinity. However, this is not relevant to the estimation of relative infectious pressure in a uniform salinity field. At a salinity of 30 ppt, Stien et al. (2005) estimated mortality rates of 0.17 day^{-1}. Based on this, we assume that nauplii mortality rates can be approximated as equal to copepodid mortality rates and do not vary with salinity. This assumption is valid for the Loch Torridon system, where, aside from directly adjacent to river mouths, the surface salinity is greater than 32 ppt during the summer months (as observed over three years of regular sampling). A weighting factor of 0.01 h^{-1} was applied to each particle to include mortality in the model, regardless of stage, as they aged. Similar treatment of mortality rates can be found in Brown et al. (2005) where particle counts were weighted by a mortality correction. Biological scenarios using mortalities of 0.01–0.06 h^{-1} were investigated earlier (Murray and Amundrud, 2007).

Copepodids are generally distributed at or near the surface of the water (Heuch et al., 1995; Costelloe et al., 1998; Penston et al., 2008) and these observations are supported by data which suggest that infection on salmon held at depths of 0–4 m is approximately an order of magnitude greater than on salmon at 4–8 m or 8–12 m (Hevrøy et al., 2003). In the absence of active motion, copepodid lice tend to sink (Bricknell et al., 2006) such that their near-surface location is actively maintained by a phototactic response (Bron et al., 1993; Heuch et al., 1995) in which speeds of at least 0.1–0.2 cm s^{-1} are possible (Devine et al., 2000). Nauplii have limited swimming and sensory abilities and are positively phototactic (Johannessen, 1978; Heuch et al., 1995). Like copepodids, they have been observed to sink when inactive (Johannessen, 1978), but tend to remain in surface waters while viable, although perhaps to a lesser extent than copepodids. Significantly greater densities of nauplii were recovered at a depth of 5 m in Loch Torridon than at the surface (Penston et al., 2008). There is a paucity of data on *L. salmonis* larvae at depths greater than 5 m, so the potential for lice to be transported below this depth is not known. More accurate estimates for the vertical distribution of planktonic sea lice in day and night conditions are not available. Other models of sea lice dispersion assume an active diurnal migration of the lice. For example, Gillibrand and Willis (2007) assigned vertical motion to their simulated copepodids based on day and night conditions and their model predicts a vertical migration of the center of mass of lice populations that spans 20 m. However, Heuch et al. (1995) found only a 1 m diurnal migration for copepodids sampled in bag experiments in a Norwegian fjord. While the work of Heuch et al. (1995) was limited by the depths of the experimental bags (6 m and 12 m), their results do suggest that assuming a strong vertical migration for sea lice planktonic stages may not be valid. In general, it seems reasonable to suggest that nauplii may be located deeper than copepodids (Heuch et al., 1995; Penston et al., 2008).

The salmon lice pass three pelagic stages in their life cycle. In the first two larval stages, the lice grow and in the third stage become infectious and ready for attaching to a salmonid fish. Typically, the first two stages last between five and ten days, and the third stage about 20 days, depending on the ambient temperature. If the lice fail to find a host, they will die at the end of stage three. Three factors are the most important for the problem of salmon lice on wild salmon in fjords:

1. The production of salmon lice.
2. The distribution of salmon lice within the system.
3. The amount of brackish water in the fjord.

10.8 Infection rate

Sea lice are extraordinarily effective at locating and infecting salmonids and have the potential to establish themselves as a pest disease to salmonid fish (Revie et al., 2009). The ectoparasites use physical (light and salinity) and chemical (kairomones) cues to locate and recognize their host. Another fundamental sensory modality to fish location is mechanoreception through sensory organs, which allows sea lice to detect and land on their host (Mordue (Luntz) and Birkett, 2009; Solé et al., 2021). The infective stage, the copepodid, carries both chemosensory aesthetes and mechanosensory setae on its antennules, indicating that both mechanical and chemical signals may be important in host-finding (Solé et al., 2021). Initial attachment for the copepodid normally occurs on the fish fins where it hooks into the host's tissue. Mainly depending on water temperature, the copepodid undergoes a moult to the first sessile stage in the life cycle, the chalimus, which attaches itself to the fish with a penetrative thread referred to as the frontal filament (Gonzalez-Alanis et al., 2001). Adult *L. salmonis* present several types of sensitive setae on their antennae. Studies have found that salmon lice copepodids are sensitive to low frequency water accelerations such as those produced by a swimming fish (Heuch and Karlsen, 1997; Heuch et al., 2007).

Infection with sea lice occurs almost immediately after juvenile salmonids are introduced into sea net pens. The source of this infection is larvae released from adult female parasites. Data gathered from 284 Atlantic salmon farms in southern Chile that were stocked before lufenuron (a new treatment for sea lice control in salmon aquaculture based on a benzoylurea (or benzoylphenylurea; BPU) was approved, showed that the average number of sea lice rapidly increased during the first ten weeks after stocking (Mardones et al., 2018) (Fig. 10.1). Lufenuron has been reported as a highly effective drug against sea lice by a significant reduction in moulting in *L. salmonis* larvae (Poley et al., 2018).

Sea lice abundances are near zero in areas without salmon farms, but juvenile wild salmon sampled near salmon farms are infected (Harte et al., 2017). The most immature life stages dominate the lice population suggesting that the source of lice is a stationary, local salmonid population. Once infected, lice dynamics are dominated by site reproduction and subsequent dispersal (Adams et al., 2015). Even with identical external infection rates, lice abundances differ dramatically between farms over a production cycle (dependent on oceanographic conditions and resulting between-farm connectivity). Sea lice have been identified as a potential transmission route for the infectious salmon anemia virus (ISAV), acting as a mechanical vector for the ISAV in farmed Atlantic salmon and/or exacerbating disease (Valdes-Donoso et al., 2013; Oelckers et al., 2014; Barker et al., 2019). In Chile, it is recognized that infection with sea lice can predispose fish to the development of other diseases such as infectious pancreatic necrosis, bacterial kidney disease, and salmonid rickettsial septicemia (Johnson et al., 2004).

Median natural infection intensity of adult female *L. salmonis* on sea trout may be as low as zero to three lice per fish at a prevalence of ca. 0–20% in spring, rising to four to eight lice per fish at higher prevalence in the autumn. Sea trout may acquire 10–20 adult female lice per fish in farming areas and in years with moderate infection pressure, towards the end of summer (Heuch et al., 2005). Prevalence of sea lice *C. rogercresseyi* in fish farms has been reported in one study as 53.4% and an average sea lice abundance of 11.8 per fish (Hamilton-West et al., 2012). Previously, the mean adult count was 3.38 lice per fish (Zagmutt-Vergara et al., 2005). Infection rate suggests that *C. rogercresseyi* is not controlled and juvenile phases are persistent among sites (Mancilla-Schulz et al., 2019).

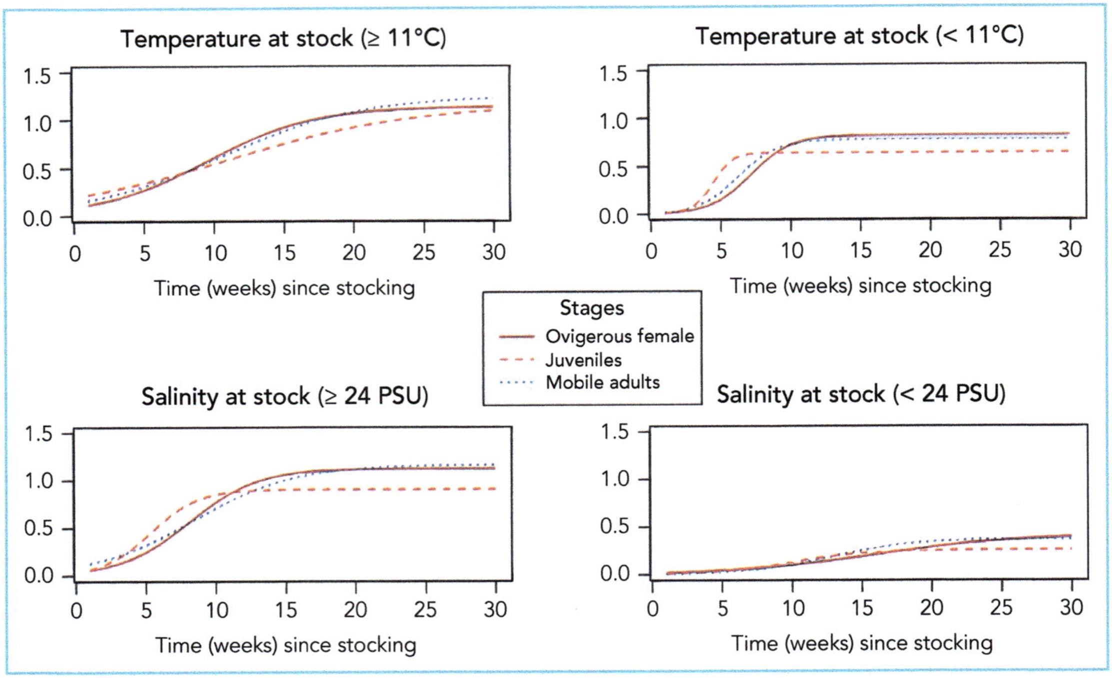

Fig. 10.2 Fitted logistic growth curves for average numbers of *Caligus rogercresseyi* for various stages and environmental conditions from 284 Atlantic salmon farms in southern Chile (2016–2018).

Growth curves were fitted to logistic functions to evaluate the effect of water temperature and salinity from a subset of Atlantic salmon farms in southern Chile untreated before transfer and during the first 30 weeks for three different developmental stages, namely juveniles, mobile adults, and ovigerous females (Mardones et al., 2018). In these scenarios, the most significant factor to minimize sea lice counts was stocking at salinities below 24 psu (Fig. 10.2).

10.9 Sampling sea lice

Different methods are employed for sampling fish and sea lice stages around the world. Little instruction exists on a scheme that can be used routinely by fish farmers in estimating the number of wild or captive fish to sample. In the case of salmon farming, the sampling methods should therefore be to trigger intervention before the female sea lice become gravid (egg-bearing) (Treasurer and Pope, 2000). In wildlife studies, the sampling methods are aimed at estimating the sea lice infestation level and its relationship with the parasitic dynamics in salmon farming or vice versa. Sea lice numbers varied significantly between cages on a farm, indicating that fish from several cages should be sampled to give a representative estimate of the mean. In practice, it is rarely possible to sample more than 20–30 fish in routine counts and the predicted level of accuracy is in the range of 40–50%. A detailed plan for sampling cages of the farms is presented in Treasurer and Pope (2000) based on the level of accuracy of a sample mean and depending on the variability of the data. However, to facilitate monitoring it is recommended that the numbers of mobile adult phases or gravid females are collected, and that these counts are combined with the condition of the epidermis of the fish during counts. The treatment intervention criteria for the threshold for *Caligus* spp. is higher due to less severe petechial hemorrhaging.

Although genera *Lepeophtheirus* and *Caligus* show differences in their life cycle (for example, *C. rogercresseyi* lacks the pre-adult stage), counting focuses on planktonic or parasitic stages (González and Carvajal, 2003; Hamre et al., 2013). The planktonic phases include nauplii and copepodid. The parasitic phases constitute: chalimus, pre-adult, and mobile adult males and females, and gravid females. Planktonic phase counts are usually directed at the infective pressure potential of a fish species, while mobile and adult phase counts are focused on parasitic infection processes on fish hosts. Three estimations can be used to quantify sea lice infestation: prevalence, abundance, and intensity (Jeong and Revie, 2020). Prevalence is defined as the number of hosts infected with one or more individuals of a particular parasite species (or taxonomic group) divided by the number of hosts examined for that parasite species. Abundance is defined as the number of individuals of a particular parasite in/on a single host regardless of whether the host is infected. Intensity is defined as the number of individuals of a particular parasite species in a single infected host (Bush et al., 1997; Rózsa et al., 2000).

In an observational plankton survey (naupliar stages and copepodid) the largest number of localities are selected to cover a geographic site of interest with weekly, biweekly, or monthly (depending on time scale) samplings from near-surface (1 m) water. Plankton samples are collected using a Neuston net by towing slowly and horizontally. The depths of plankton samples depended on topography and can vary typically by 0 m, 1 m, and 5 m. The sample contents are transferred to a sample bottle and fixed in a 4% formaldehyde solution or preserved in alcohol (McKibben and Hay, 2004; Penston et al., 2004, 2008). Skarhamar et al. (2019) verified that conventional methods for plankton sampling are efficient even in fjords with low salinity concentrations combined with patchy distributions near the surface.

In observational parasitic studies (non-mobile: chalimus, and mobile: pre-adult, adult male, adult female, and gravid female) on wild fish, specimens are sampled under conditions similar in space and time to plankton surveys (Costelloe et al., 1998; Aksnes et al., 1997; Adams et al., 2015). Fish samples are collected by electro-fishing or using floating gill nets within 3 km of the river mouths but never in the estuary throughout a tidal cycle. The nets are continuously observed, and fish are removed at least every third hour. To minimize the possible loss of sea lice, the fish are cut from gill nets, placed in individually tagged plastic bags, and frozen. Beach nets and seine nets are also used for fish capture.

Since the life cycle of sea lice is temperature dependent, the frequency of sampling can be varied weekly or every two weeks. Weekly sampling should be carried out depending on the season of the year and at temperatures above 10°C (Skarhamar et al., 2019). The samplings could be done every two weeks in some locations with low lice abundance and also in winter. A basic count should differentiate adults from larval stages. However, a detailed count is more useful, including gravid females, adult females, adult males, pre-adult, and chalimus phases combined in one category (juveniles). The numbers of chalimus or juvenile phases in the field were found to be only half of laboratory counts on average (Treasurer and Pope, 2000).

A standard cage and a minimum of five random ones (usually three or four) are selected for farmed fish. In many farms ten fish are removed by dip net in each cage and transferred to an anesthetic bath for removing and counting sea lice stages and detached parasites are also counted from the bath (Tucker et al., 2002). Sampling of fish by lifting and splitting the net is time-consuming and therefore attracting fish to the surface with feed and removing fish with a net are frequent practices.

10.10 Number of fish to sample

The selection of cages for sampling should be at random, taking fish of representative health. However, if there is a cage with heavily infected fish (personal observation) it should also be selected for sampling in the random group. On subsequent occasions, the heavily infected cage can be sampled again if it is not included in the random sample. Dislodged lice should be counted in the bucket after the final count and added to the overall total count for the cage. At the farm level, approximately 40 individuals are collected (ten fish per cage) (Saksida et al., 2007; Mancilla-Schulz et al., 2019). The observational field efforts to collect fish addressing parasitic phases of the sea lice present ranges from 30 to 100 individuals per site (Rees et al., 2015).

Guidelines as to the suitable sampling approaches needed to achieve accurate sea lice load estimates remain largely underdeveloped, in part due to the lack of wide-ranging data on sea lice infestation distributions on farmed salmon under variable real-world settings. However, it has been demonstrated that sampling a smaller number of fish from a larger number of pens resulted in a more accurate estimate of abundance. Jeong and Revie (2020) proposed that the level of accuracy that can be achieved in practice depends on many factors and an acceptable level of accuracy will, by definition, be subjective and vary according to the purpose for which the estimation is being conducted. Nevertheless, the minimum sample sizes needed to detect a given difference in sea lice prevalence before and after treatment are roughly congruent to values representing a total of 40 sampled fish (four fish from ten pens) and avoiding fewer than four fish per pen.

10.11 Stock and farm density effects on sea lice epizootics

Attached louse counts remained low when neighboring farms were in the first year of the production cycle but rose substantially throughout the season in the second year. In contrast to attached lice, higher counts of mobile lice were generally seen at lower water temperatures. However, this difference may have been attributable to more heavily infested fish being removed from the population. Temperature had an indirect positive effect on mobile louse counts by increasing the infection pressure, which in turn raised the mobile louse counts. Mobile louse counts rose steadily during the year when neighboring farms were in the first year of the production cycle and stayed high throughout the second year. In conclusion, the estimates of the infection pressure effect on louse counts, along with a biennial pattern emerging due to the production cycle of fish farms, suggested in one locality that fish farms play a role in the parasite infestations on wild sea trout (Vollset et al., 2018).

Several field and modelling studies have examined copepodid populations in intertidal zones and have shown that the planktonic stages can be transported tens of kilometers from their source by tides and currents (Costello, 2006). The infectious stage of *L. salmonis* could be successfully identified close to river mouths using plankton sampling. Copepodids were present in high numbers over a sustained period and were abundant in the early spring at times when salmon and sea trout smolts migrated to sea (McKibben and Hay, 2004).

10.12 Population dynamics and sampling considerations

Authorities in salmon farming countries have created legislation and guidelines stipulating the maximum number of sea lice per fish on a farm before treatment or control measurements can be conducted to avoid excessive cross-farm infections. These guidelines are also designed in some countries for the protection of wild salmon species and the environment. We present a brief description of the main countries with regulations of this type.

10.12.1 Norway

Farmers are instructed to count sea lice (*L. salmonis* and *C. elongatus*) on farmed salmonids at regular intervals and report the highest mean count during a month to the Norwegian Food Safety Authority (NFSA) [www.mattilsynet.no]. The mean count of sea lice from a sample of ten fish from multiple net pens is reported. The regulations specify maximum thresholds for sea lice abundance on farms and an obligatory reporting system for sea lice counts. Maximum threshold abundances are set at 0.5 adult females (or three mobile adults) per fish in the period 1st January through to 31st August, and one adult female (or five mobile adults) per fish in the period 1st September through to 31st December (Aldrin et al., 2013). It is mandatory to report key statistics on farmed fish stocks, fish health-related statistics, and water temperature measured at 3 m depth (Heuch et al., 2003). These limits were lowered further in 2017 to 0.5 gravid female but 0.2 in the outward migration of salmonids. The Institute of Marine Research (IMR) [www.hi.no] also coordinates a national salmon louse monitoring program designed to monitor the infestation pressure on wild salmon and trout. Sampling is performed at several regular and irregular stations along the Norwegian coast using traps or gill nets, staying at each site for approximately two weeks (Myksvoll et al., 2018).

10.12.2 Chile

The current regulation for sea lice count (*C. rogercresseyi*) is the Specific Sanitary Program for the Surveillance and Control of Caligidosis by the Secretariat of Fisheries and Aquaculture (Subpesca, http://www.subpesca.cl/) and the National Fisheries and Aquaculture Service (Sernapesca, http://www.sernapesca.cl). This program specifies that each farm site must report weekly mean abundance of parasites per fish based on a sample of 40 fish collected from four randomly selected cages (Jeong and Revie, 2020). The sampling procedure consisted of netting ten fish from each cage and placing them in an anesthetic bath. Once anesthetized, the number of lice on each fish was counted, including gravid females, mobile adult females and males, and juvenile stages (chalimus). Detached parasites were also counted in each category. Maximum threshold abundances are set to at least three gravid females per fish to trigger chemical control measures (Arriagada and Marín, 2018). In addition, each farm reports on the number of fish per site (thousands), average fish weight (kg), biomass per site (tons), water salinity (ppt), and temperature (°C) (Mancilla-Schulz et al., 2019).

10.12.3 Scotland

Marine Scotland's Fish Health Inspectorate (FHI) [www.gov.scot] requires reporting of mean abundance of adult female sea lice (*L. salmonis* and *C. elongatus*) per fish with minimum numbers used as three in the period from July 2017 to June 2019, and two in the period from June 2019 onwards (McKenzie et al., 2004). A minimum of 25 fish per week are sampled irrespective of the number of stocked cages on each farm (FHI, 2019). Thus, when a farm reaches the allowed limit the FHI will increase the level of monitoring of that site and continue to do so until the mean abundance of adult females per fish is reduced to below the reporting level. In addition, exceeding a level of eight adult females might result in enforcement action, including the potential to require a reduction in biomass.

10.12.4 Canada

Fisheries and Oceans Canada's Aquaculture Management Directorate (DFO, www.dfo-mpo.gc.ca/aquaculture) requests monitoring for sea lice on farmed salmon depending on coast location. On the Pacific coast, the objective is to maintain a farm level sea lice (*L. salmonis* and *C. elongatus*) load of fewer than three mobile phases (pre-adults and adults) per fish during the general period of wild pink and chum salmon smolt seaward migration (March to June). Farm operators must take measures to reduce sea lice levels within the next 48 hours. At least three cages per farm are assessed, once or twice per month (twice per month during the spring period), with 20 fish sampled per cage. The objective of on-farm sea lice monitoring on the Atlantic coast is the early detection of sea lice infestation load greater than zero. In the Atlantic provinces, sampling typically consists of evaluating a small number of fish (usually five) from each of four to six cages within a farm (DFO, 2014). Cage selection can be random or based on the position of the cages at the farm, for example with those on the flood side being selected preferentially (Westcott et al., 2004). These selected cages are sampled at each sampling event.

10.12.5 Faroe Islands

The Aquaculture Research Station of the Faroes (Fiskaaling) regularly counts the number of sea lice in the attached development stages (chalimus, mobiles, and adult) for the salmon farms in the Faroe Islands. The lice-counting regulations require counts to be made by Fiskaaling on each site about every 14 days, where a sample of ten fish is examined from each cage. The previously permitted sea lice limit was a mean number of two adult female lice per fish, which has now been reduced to one and a half adult females per fish (Gislason, 2018; Kragesteen et al., 2019). Counts do not include the planktonic stages (chalimus) and there is no distinction between *L. salmonis* and *C. elongatus* in the monitoring program.

10.13 Data handling of sea lice counts

The count data of sea lice plankton surveys gives a number at various sites *i*, denoted by *Ni* from a volume of water *Vi*. One option to describe these data is to use density *Ni/Vi* as a response variable and work with a Gaussian distribution as a poor approach ignoring the differences in volumes (Zuur et al., 2009). On some occasions, for further analysis about the concentration of planktonic phases at certain sites, the average density of the larvae of all the locations is taken at each respective site or the simple ratios as the total density of nauplii found at an individual site divided by the total density of copepods found at that site (Peterson et al., 2004). Nonetheless, not all sites are sampled at the same frequency or time, and their area can present a high variability and overdispersion of the observations. The foregoing implies limitations for linear models and parametric statistical tests. Besides, some non-parametric tests could have sensitivity consequences for the sea lice count data. Different authors have reported that sea lice count data can be approached by statistical modelling techniques that consider the exponential families of Poisson and/or negative binomial.

The Poisson distribution applies to events that occur randomly over individuals in a situation where the probability of an event occurring is a constant from individual to individual and independent of the occurrence of other events. If sea lice randomly choose their host, the statistical distribution should follow a Poisson distribution, where the variance would be approximately equal to the mean. However, it has often been observed that sea lice aggregate on some portion of available hosts, leading to an overdispersed parasite count distribution (that is, where the variance is significantly greater than the mean due to the variation of counts between fish and between cages). The negative binomial distribution often gives a good fit to the frequency distribution of events whose probability of occurring varies from individual to individual in a specific manner and where the probability of occurrence increases when other events have happened. Perhaps in small samples, the Poisson distribution can present a better fit because the variance needs to be equal to the mean.

The first statistical option available to address count-type data distributions is generalized linear models (GLMs). In addition to providing a link function to improve fit to the data, GLMs can deal with heterogeneity (the differences in variation may cause serious trouble in linear regression and analysis of variance models). Sometimes when the underlying question is to determine which covariates are driving a sea lice system, it should be necessary to capture the shape of the relationship through generalized additive models (GAMs). Penston et al. (2008) use GAMs with negative binomial distribution to investigate the influence of station, depth, and time (production week) on nauplius and copepodid densities on the north-west coast of Scotland (Penston et al., 2008).

Usually when count data are analyzed using GLMs and GAMs three potential problems arise: longitudinal observations at each locality, a correlation between near localities data, and large variations of explicatory variables. Zuur et al. (2010) suggest a guide for exploration of outlier values, the collinearity of the variables, and the graphic relationship between the parasitic counts with the explanatory variables. It is also necessary to involve regression techniques using biological knowledge for the selection, validation, and interpretation of models in the framework of multimodel inference (Burnham and Anderson, 2004). Nonetheless, on some occasions, there are a large number of zeros in the data (absence observations) or non-linear relational ones in the modelling process. The zero-inflated models and additive models can be used for these cases. Jansen et al. (2012) use zero-inflated negative binomial models to study the effect of densities of farmed salmonids on farm levels of sea lice count in Norway (Jansen et al., 2012).

Data are often transformed in ecological studies to satisfy parametric test assumptions. Nevertheless, log-transformation or square root transformation implies that the mechanisms are multiplicative on the scale of the raw data. O'Hara and Kotze (2010) simulated data from a negative binomial distribution and compared the outcome of fitting models that were transformed, finding that the transformations performed poorly, except when the dispersion was small, and the mean counts were large. Studies in Chile (Lepe-López et al., 2021a,b) have shown that negative binomial distribution may be appropriate when investigating the effect of several variables in the abundance of *C. rogercresseyi* (Fig. 10.3).

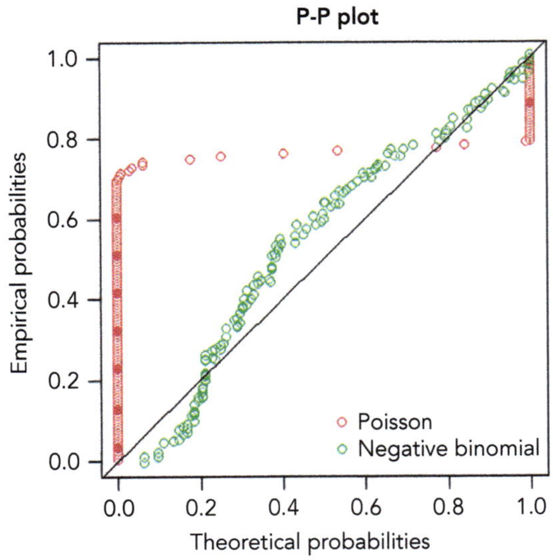

Fig. 10.3 The plot of the probabilities of each fitted distribution (x-axis, Poisson in red and negative binomial in green) against the empirical probabilities (y-axis) gives an example of better fit of the negative binomial distribution to counts of ovigerous female *Caligus rogercresseyi* in marine farms in southern Chile (2016–2018).

10.14 Analyzing zeros and overdispersion values in samples from sea lice in salmon farming

As in most parasite infestations, sea lice counts follow a right skewed distribution, and it is assumed that lice follow a Poisson or negative binomial distribution (Treasurer and Pope, 2000; Krkošek et al., 2005). While the Poisson distribution assumes a simple random distribution of a count, the negative binomial can be seen as an extension of this distribution, where a variance on the mean is allowed, thus giving a broader distribution than the Poisson. The lack of conformation to the Poisson distribution may be real, owing to sampling error (Krkošek et al., 2005; Revie et al., 2005; Gettinby et al., 2011) or to a shift in the population after using control strategies such as wrasse that target certain stages of the lice population.

Sometimes sea lice count data have far more zeros than would be expected for Poisson or negative binomial distribution (Fig. 10.4). Zero inflation data can give better estimated parameters and standard errors in statistics modelling, avoiding the real overdispersion into the analysis (Zuur et al., 2010). It is necessary to know if these zero data signify that sea lice are not present because the habitat is not suitable, or reflects poor sampling practices, seasonal fluctuations, or the effect of treatments (these types of zeros should be removed). It is necessary to say that the zeros signify the absence of the parasite prior to data analysis (true zeros). Two models can be applied to analyze zero-inflated data: two-part and mixture models. In two-part models the first step is considered as zeros versus non-zeros to model the probability that a zero value is observed, and the second step is non-zero observations are modelled with a truncated (because the distributions are zero truncated, they cannot produce zeros). The zeros are modelled in mixture models as coming from two different processes: the binomial process and the count process. Selection of one of these models is based on knowledge of sea lice biology and sampling.

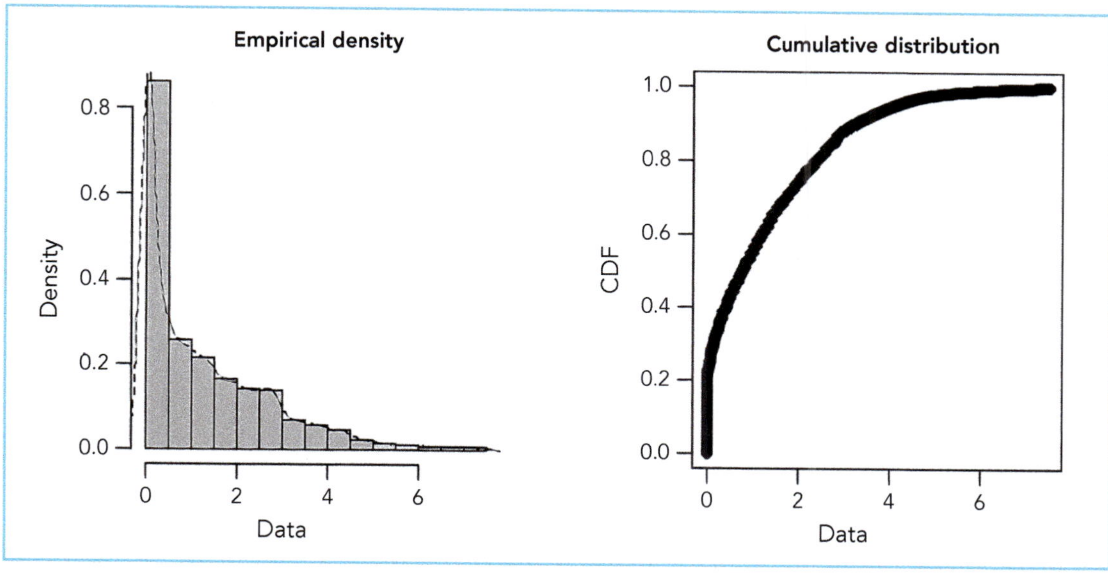

Fig.10.4 Histogram and CDF plots of an empirical distribution (Poisson or negative binomial) of counts of ovigerous female *Caligus rogercresseyi* in marine farms in southern Chile (2016–2018). The number of zeros dominates the distribution of counts.

Acknowledgements

This study was funded by a contract (EPIVET project 201707240090), which was financed by SalmonChile and Sernapesca through the Fondo Inversion Estrategica (FIE-2015-914 V014) from the Ministry of Economy. The National Fund for Scientific and Technological Development of the Government of Chile (FONDECYT) also provided partial funding, through the grant: FONDECYT regular N° 1191675.

References

Aaen, S.M., Helgesen, K.O., Bakke, M.J., Kaur, K. and Horsberg, T.E. (2015) Drug resistance in sea lice: a threat to salmonid aquaculture. *Trends in Parasitology* 31(2), 72–81.

Adams, T.P., Proud, R. and Black, K.D. (2015) Connected networks of sea lice populations: dynamics and implications for control. *Aquaculture Environment Interactions* 6(3), 273–284.

Aksnes, D.L., Miller, C.B., Ohman, M.D. and Wood, S.N. (1997) Estimation techniques used in studies of copepod population dynamics – a review of underlying assumptions *Sarsia* 82(4), 279–296.

Aldrin, M., Storvik, B., Kristoffersen, A.B. and Jansen, P.A. (2013) Space-time modelling of the spread of salmon lice between and within Norwegian marine salmon farms. *PLoS ONE* 8(5), e64039.

Allan, J.D., Kinsey, T.G. and James, M.C. (1976) Abundances and production of copepods in the Rhode river subestuary of Chesapeake Bay. *Chesapeake Science* 17(2), 86–92.

Amundrud, T.L. and Murray, A.G. (2009) Modelling sea lice dispersion under varying environmental forcing in a Scottish sea loch. *Journal of Fish Diseases* 32(1), 27–44.

Arriagada, G.A. and Marín, S.L. (2018) Evaluating the spatial range of the effect of synchronized antiparasitic treatments on the abundance of sea lice *Caligus rogercresseyi* (Boxshall & Bravo 2000) in Chile. *Aquaculture Research* 49(2), 816–831.

Barker, S.E., Bricknell, I.R., Covello, J., Purcell, S., Fast, M.D., Wolters, W. and Bouchard, D.A. (2019) Sea lice, *Lepeophtheirus salmonis* (Krøyer 1837), infected Atlantic salmon (*Salmo salar* L.) are more susceptible to infectious salmon anemia virus. *PLoS ONE* 14, e0209178.

Bassini, L.N., Lhorente, J.P., Oyarzún, M., Bangera, R., Yáñez, J.M. and Neira, R. (2019) Genetic parameters for *Piscirickettsia salmonis* resistance, sea lice (*Caligus rogercresseyi*) susceptibility and harvest weight in rainbow trout (*Oncorhynchus mykiss*). *Aquaculture* 510, 276–282.

Bjørn, P.A., Finstad, B. and Kristoffersen, R. (2001) Salmon lice infection of wild sea trout and Arctic char in marine and freshwaters: The effects of salmon farms. *Aquaculture Research* 32(12), 947–962.

Boxaspen, K. (2006) A review of the biology and genetics of sea lice. *ICES Journal of Marine Science* 63(7), 1304–1316.

Boxaspen, K. and Næss, T. (2000) Development of eggs and the planktonic stages of salmon lice (*Lepeophtheirus salmonis*) at low temperatures. *Contributions to Zoology* 69. Available at: https://brill.com/view/journals/ctoz/69/1-2/article-p51_5.xml. (accessed 7 May 2021).

Braden, L.M., Barker, D.E., Koop, B.F. and Jones, S.R. (2012) Comparative defense-associated responses in salmon skin elicited by the ectoparasite *Lepeophtheirus salmonis*. *Comparative Biochemistry and Physiology Part D Genomics Proteomics* 7(2), 100–109.

Bravo, S., Pozo, V. and Silva, M.T. (2008) The tolerance of *Caligus rogercresseyi* to salinity reduced in southern Chile. *Bulletin of the European Association of Fish Pathologists* 28(5), 198–206.

Bravo, S., Silva, M.T. and Treasurer, J. (2014) Factors affecting the abundance of *Caligus rogercresseyi* (Boxshall & Bravo 2000) on farmed salmonids in Chile in the period 2006–2007. *Aquaculture* 434, 456–461.

Bricknell, I.R., Dalesman, S.J., O'Shea, B., Pert, C.C. and Mordue Luntz, A.J. (2006) Effect of environmental salinity on sea lice *Lepeophtheirus salmonis* settlement success. *Diseases of Aquatic Organisms* 71(3), 201–212.

Bron, J.E., Sommerville, C. and Rae, G.H. (1993) Aspects of the behaviour of copepodid larvae of the salmon louse *Lepeophtheirus salmonis* (Krøyer, 1837). In: Boxshall, G.A. and Defaye, D. (eds) *Pathogens of Wild and Farmed Fish: Sea Lice*. Ellis Horwood Ltd, Chichester, UK. pp. 125–142.

Brooker, A.J., Skern-Mauritzen, R. and Bron, J.E. (2018) Production, mortality, and infectivity of planktonic larval sea lice, *Lepeophtheirus salmonis* (Krøyer, 1837): current knowledge and implications for epidemiological modelling. *ICES Journal of Marine Science* 75, 1214–1234.

Brooks, K.M. (2005) The effects of water temperature, salinity, and currents on the survival and distribution of the infective copepodid stage of sea lice (*Lepeophtheirus salmonis*) originating on atlantic salmon farms in the Broughton Archipelago of British Columbia, Canada. *Reviews in Fisheries Science* 13(3), 177–204.

Brooks, K.M. (2009) Considerations in developing an integrated pest management programme for control of sea lice on farmed salmon in Pacific Canada. *Journal of Fish Diseases* 32(1), 59–73.

Brown, C.A., Jackson, G.A., Holt, S.A. and Holt, G.J. (2005) Spatial and temporal patterns in modeled particle transport to estuarine habitat with comparisons to larval fish settlement patterns. *Estuarine Coastal and Shelf Science* 64, 33–46.

Bruno, D.W. and Stone, J. (1990) The role of saithe, *Pollachius virens* L., as a host for the sea lice, *Lepeoptheirus salmonis* Krøyer and *Caligus elongatus* Nordmann. *Aquaculture* 89(3–4), 201–207.

Burnham, K.P. and Anderson, D.R. (2004) Multimodel inference: Understanding AIC and BIC in model selection. *Sociological Methods and Research* 33(2), 261–304.

Bush, A.O., Lafferty, K.D., Lotz, J.M. and Shostak, A.W. (1997) Parasitology meets ecology on its own terms: Margolis et al. revisited. *The Journal of Parasitology* 83(4), 575–583.

Carvajal, J., González, L. and George-Nascimento, M. (1998) Native sea lice (Copepoda: Caligidae) infestation of salmonids reared in netpen systems in southern Chile. *Aquaculture* 166(3–4), 241–246.

Chaparro, O.R., Cubillos, V.M., Montiel, Y.A., Paschke, K.A. and Pechenik, J.A. (2008) Embryonic encapsulation and maternal incubation: Requirements for survival of the early stages of the estuarine gastropod *Crepipatella dilatata*. *Journal of Experimental Marine Biology and Ecology* 365(1), 38–45.

Connors, B.M., Juarez-Colunga, E. and Dill, L.M. (2008) Effects of varying salinities on *Lepeophtheirus salmonis* survival on juvenile pink and chum salmon. *Journal of Fish Biology* 72, 1825–1830.

Costelloe, M., Costelloe, J., O'Donohoe, G., Coghlan, N.J., Oonk, M. and Heijden, Van der (1998) Planktonic distribution of sea lice larvae, *Lepeophtheirus salmonis*, in Killary Harbour, West Coast of Ireland. *Journal of the Marine Biological Association of the United Kingdom* 78(3), 853–874.

Costello, M.J. (2006). Ecology of sea lice parasitic on farmed and wild fish. *Trends in Parasitology* 22(10), 475–483.

Devine, G.J., Ingvarsdottir, A., Mordue, W., Pike, A.W., Pickett, J., Duce, I.A. and Mordue, A.J. (Luntz) (2000) Salmon lice, *Lepeophtheirus salmonis*, exhibit specific chemotactic responses to semiochemicals originating from the salmonid, *Salmo salar*. *Journal of Chemical Ecology* 26, 1833–1847.

DFO (2014) Sea Lice Monitoring and Non-Chemical Measures, *DFO Canadian Science Advisory Secretariat Science Advisory Report* p. 19. Available at: http://waves-vagues.dfo-mpo.gc.ca/Library/363891.pdf. (accessed 17 July 2021).

Fast, M.D., Ross, N.W., Mustafa, A., Sims, D.E., Johnson, S.C., Conboy, G.A., Speare, D.J., Johnson, G. and Burka, J.F. (2002) Susceptibility of rainbow trout *Oncorhynchus mykiss*, Atlantic salmon *Salmo salar* and coho salmon *Oncorhynchus kisutch* to experimental infection with sea lice *Lepeophtheirus salmonis*. *Diseases of Aquatic Organisms* 52(1), 57–68.

FHI (2019) The regulation of sea lice in Scotland. Available at: https://www.gov.scot/publications/fish-health-inspectorate-sea-lice-information/ (accessed 17 October 2021).

Genna, R.L., Mordue, W., Pike, A.W. and Mordue, A.J. (2005) Light intensity, salinity, and host velocity influence presettlement intensity and distribution on hosts by copepodids of sea lice, *Lepeophtheirus salmonis*. *Canadian Journal of Fisheries and Aquatic Sciences* 62(12), 2675–2682.

Gettinby, G., Robbins, C., Lees, F. and Heuch, P.A. (2011) Use of a mathematical model to describe the epidemiology of *Lepeophtheirus salmonis* on farmed Atlantic salmon *Salmo salar* in the Hardangerfjord, Norway. *Aquaculture* 320(3–4),164–170.

Gillibrand, P.A. and Willis, K.J. (2007) Dispersal of sea louse larvae from salmon farms: modelling the influence of environmental conditions and larval behaviour. *Aquatic Biology* 1, 63–75.

Gislason, H. (2018) Statistical modelling of sea lice count data from salmon farms in the Faroe Islands. *Journal of Fish Diseases* 41(6), 973–993.

González, L. and Carvajal, J. (2003) Life cycle of *Caligus rogercresseyi* (Copepoda: Caligidae) parasite of Chilean reared salmonids. *Aquaculture* 220, 101–117.

González, L., Carvajal, J. and George-Nascimento, M. (2000) Differential infectivity of *Caligus flexispina* (Copepoda: Caligidae) in three farmed salmonids in Chile. *Aquaculture* 183, 13–23.

Gonzalez-Alanis, P., Wright, G.M., Johnson, S.C. and Burka, J.F. (2001) Frontal filament morphogenesis in the salmon louse *Lepeophtheirus salmonis*. *Journal of Parasitology* 87, 561–574.

Groner, M.L., Cox, R., Gettinby, G. and Revie, C. (2013) Use of agent-based modelling to predict benefits of cleaner fish in controlling sea lice, *Lepeophtheirus salmonis*, infestations on farmed Atlantic salmon, *Salmo salar* L. *Journal of Fish Diseases* 36(3), 195–208.

Groner, M.L., et al. (2016) Lessons from sea louse and salmon epidemiology. *Philosophical Transactions of the Royal Society B: Biological Sciences* 371(1689).

Hamilton-West, C., Arriagada, G., Yatabe, T., Valdes, P., Herve-Claude, L.P. and Urclay, S. (2012) Epidemiological description of the sea lice (*Caligus rogercresseyi*) situation in southern Chile in August 2007. *Preventive Veterinary Medicine* 104(3–4), 341–345.

Hamre, L.A., Eichner, C., Caipang, C.M.A., Dalvin, S.T., Bron, J.E., Nilsen, F., Boxshall, G. and Skern-Mauritzen, R. (2013) The salmon louse *Lepeophtheirus salmonis* (Copepoda: Caligidae) life cycle has only two chalimus stages. *PLoS ONE* 8(9), e73539.

Hamre, L.A., Oldham, T., Oppedal, F., Nilsen, F. and Glover, K.A. (2021) The potential for cleaner fish-driven evolution in the salmon louse *Lepeophtheirus salmonis*: Genetic or environmental control of pigmentation? *Ecology and Evolution* 11(12), 7865–7878.

Harte, A.J., Bowman, A.S., Salama, N.K.G. and Pert, C.C. (2017) Factors influencing the long-term dynamics of larval sea lice density at east and west coast locations in Scotland. *Diseases of Aquatic Organisms* 123(3), 181–192.

Hayward, C.J., Bott, N.J. and Nowak, B.F. (2009) Seasonal epizootics of sea lice, *Caligus* spp., on southern bluefin tuna, *Thunnus maccoyii* (Castelnau), in a long-term farming trial. *Journal of Fish Diseases* 32, 101–106.

Hevrøy E.M., Boxaspen K., Oppedal F., Taranger G.L. and Holm J.C. (2003) The effect of artificial light treatment and depth on the infestation of the sea louse *Lepeophtheirus salmonis* on Atlantic salmon (*Salmo salar* L.) culture. *Aquaculture* 220, 1–14.

Heuch, P.A. and Karlsen, E. (1997) Detection of infrasonic water oscillations by copepodids of *Lepeophtheirus salmonis* (Copepoda: Caligidae). *Journal of Plankton Research* 19, 735–747.

Heuch, P.A. and Mo, T.A. (2001) A model of salmon louse production in Norway: effects of increasing salmon production and public management measures. *Diseases of Aquatic Organisms* 45(2), 145–152.

Heuch, P.A., Parsons, A. and Boxaspen, K. (1995) Diel vertical migration: a possible host-finding mechanism in salmon louse (*Lepeophtheirus salmonis*) copepodids? *Canadian Journal of Fisheries and Aquatic Sciences* 52(4), 681–689.

Heuch, P.A., Nordhagen, J.R. and Schram, T.A. (2000) Egg production in the salmon louse [*Lepeophtheirus salmonis* (Krøyer)] in relation to origin and water temperature. *Aquaculture Research* 31, 805–814.

Heuch, P.A, Revie, C.W. and Gettinby, G. (2003) A comparison of epidemiological patterns of salmon lice, *Lepeophtheirus salmonis*, infections on farmed Atlantic salmon, *Salmo salar* L., in Norway and Scotland. *Journal of Fish Diseases* 26(9), 539–551.

Heuch, P.A., Bjorn, P.A., Finstad, B., Holst, J.C., Asplin, L. and Nilsen, F. (2005) A review of the Norwegian "National Action Plan Against Salmon Lice on Salmonids": The effect on wild salmonids. *Aquaculture* 246(1–4), 79–92.

Heuch, P.A., Doall, M.H. and Yen, J. (2007) Water flow around a fish mimic attracts a parasitic and deters a planktonic copepod. *Journal of Plankton Research* 29, i3–i16.

Ho, J.-S. (2000) The major problem of cage aquaculture in Asia relating to sea lice. In: Liao, I.C. and Lin, C.K. (eds) *Cage Aquaculture in Asia. Proceedings of the First International Symposium on Cage Aquaculture in Asia.* Manila and World Aquaculture Society, Bangkok. pp. 13–19.

Jakob, E., Sweeten, T., Bennett, W. and Jones, S.R. (2013) Development of the salmon louse *Lepeophtheirus salmonis* and its effects on juvenile sockeye salmon *Oncorhynchus nerka*. *Diseases of Aquatic Organisms* 106, 217–227.

Jansen, P.A., Kristoffersen, A.B., Viljugrein, H., Jimenez, D., Aldrin, M. and Stien, A. (2012) Sea lice as a density-dependent constraint to salmonid farming. *Proceedings Biological Society* 279(1737), 2330–2338.

Jeong, J. and Revie, C.W. (2020) Appropriate sampling strategies to estimate sea lice prevalence on salmon farms with low infestation levels. *Aquaculture* 518, 734858.

Johannessen, A. (1978) Early stages of *Lepeophtheirus salmonis* (Copepoda: Caligidae). *Sarsia* 63, 169–176.

Johnson, S.C. and Albright, L.J. (1991) The developmental stages of *Lepeophtheirus salmonis* (Krøyer, 1837) (Copepoda: Caligidae). *Canadian Journal of Zoology* 69(4), 929–950.

Johnson, S.C. and Albright, L.J. (1992) Comparative susceptibility and histopathology of the response of naive Atlantic, chinook and coho salmon to experimental infection with *Lepeophtheirus salmonis* (Copepoda: Caligidae). *Diseases of Aquatic Organisms* 14, 179–193.

Johnson, S.C., Treasurer, J., Bravo, S., Nagasawa, K. and Kabata, Z. (2004) A review of the impact of parasitic copepods on marine aquaculture. *Zoological Studies* 43(2), 229–243.

Jones, S.R. and Hargreaves, N.B. (2007) The abundance and distribution of *Lepeophtheirus salmonis* (Copepoda: Caligidae) on pink (*Oncorhynchus gorbuscha*) and chum (*O. keta*) salmon in coastal British Columbia. *Journal of Parasitology* 93, 1324–1331.

Jones, S.R., Prosperi-Porta, G., Kim, E., Callow, P. and Hargreaves, N.B. (2006) The occurrence of *Lepeophtheirus salmonis* and *Caligus clemensi* (Copepoda: Caligidae) on three-spine stickleback *Gasterosteus aculeatus* in coastal British Columbia. *Journal of Parasitology* 92(3), 473–480.

Kragesteen, T.J., Simonsen, K., Visser, A.W. and Andersen, K.H. (2019) Optimal salmon lice treatment threshold and tragedy of the commons in salmon farm networks. *Aquaculture* 512, 734329.

Kristoffersen, A.B., Jimenez, D.,Viljugrein, H., Grondtvedt, R., Stien, O. and Jansen, P.A. (2014) Large scale modelling of salmon lice (*Lepeophtheirus salmonis*) infection pressure based on lice monitoring data from Norwegian salmonid farms. *Epidemics* 9, 31–39.

Krkošek, M. (2017) Population biology of infectious diseases shared by wild and farmed fish. *Canadian Journal of Fisheries and Aquatic Science* 999, 1–9.

Krkošek, M., Lewis, M.A. and Volpe, J.P. (2005) Transmission dynamics of parasitic sea lice from farm to wild salmon. *Proceedings of the Royal Society B: Biological Sciences* 272(1564), 689–696.

Krkošek, M., Morton, A., Volpe, J.P. and Lewis, M.A. (2009) Sea lice and salmon population dynamics: effects of exposure time for migratory fish. *Procedings Biological Society* 276(1668), 2819–2828.

Lafferty, K.D., Harvell, C.D., Conrad, J.M., Friedman, C.S., Kent, M.L., Kuris, A.M., Powell, E.N., Rondeau, D. and Saksida, S.M. (2015) Infectious diseases affect marine fisheries and aquaculture economics. *Annual Review Marine Science* 7, 471–496.

Lepe-López, M., Escobar-Dodero, J., Zimin-Veselkoff, N., Azat, C. and Mardones, F.O. (2021a) Assessing the present and future habitat suitability of *Caligus rogercresseyi* (Boxshall & Bravo 2000) for salmon farming in southern Chile. *Frontiers in Veterinary Sciences* 7, 615039.

Lepe-López, M., Escobar-Dodero, J., Rubio, D., Alvarez, J., Zimin-Veselkoff, N. and Mardones, F.O. (2021b) Epidemiological factors associated with *Caligus rogercresseyi* infection, abundance, and spatial distribution in southern Chile. *Frontiers Veterinary Science* 8, 595024.

Lyndon, A.R. and Toovey, J.P.G. (2001) Occurrence of gravid salmon lice (*Lepeophtheirus salmonis* (Krøyer)) on saithe (*Pollachius virens* (L.)) from salmon farm cages. *Bulletin of the European Association of Fish Pathologists* 21(2), 84–85.

Mancilla-Schulz, J., Marin, S.L. and Molinet, C. (2019) Dynamics of *Caligus rogercresseyi* (Boxshall & Bravo 2000) in farmed Atlantic salmon (*Salmo salar*) in southern Chile: Are we controlling sea lice? *Journal of Fish Diseases* 42(3), 357–369.

Maran, B.A., Moon, S.Y., Ohtsuka, S., Oh, S-Y., Soh, H.Y., Myoung, J-G., Iglikowska, A. and Boxshall, G.A. (2013) The caligid life cycle: New evidence from *Lepeophtheirus elegans* reconciles the cycles of *Caligus* and *Lepeophtheirus* (Copepoda: Caligidae). *Parasite* 20(1).

Mardones, F.O., Rubio, D., Alba, A., Lepe-López, M., Escobar-Dodero, J. and Zimin-Veselkoff, N. (2018) Endemic-epidemic modeling of areal count time series of sea lice and mortalities due to piscirickettsiosis in farmed Atlantic salmon in Chile. *12th International Conference on Sea Lice, Punta Arenas, Chile.* INTESAL, Puerto Montt, 27.

Marty, G.D., Saksida, S.M. and Quinn II, T.J. (2010) Relationship of farm salmon, sea lice, and wild salmon populations. *Proceedings National Academy of Sciences USA* 107(52), 22599–22604.

McKenzie, E., Gettinby, G., McTart, K. and Revie, C.W. (2004) Time-series models of sea lice *Caligus elongatus* (Nordmann) abundance on Atlantic salmon *Salmo salar* L. in Loch Sunart, Scotland. *Aquaculture Research* 35(8), 764–772.

McKibben, M.A. and Hay, D.W. (2004) Distributions of planktonic sea lice larvae *Lepeophtheirus salmonis* in the inter-tidal zone in Loch Torridon, Western Scotland in relation to salmon farm production cycles. *Aquaculture Research* 35(8), 742–750.

Montory, J.A., Cumillaf, J.P., Cubillos, V.M., Paschke, K., Urbina, M.A. and Gebauer, P. (2018) Early development of the ectoparasite *Caligus rogercresseyi* under combined salinity and temperature gradients. *Aquaculture* 486, 68–74.

Mordue (Luntz), A.J. and Birkett, M.A. (2009) A review of host finding behaviour in the parasitic sea louse, *Lepeophtheirus salmonis* (Caligidae: Copepoda). *Journal of Fish Diseases* 32, 3–13.

Morgan, S.G. (2020) Life and death in the plankton: larval mortality and adaptation. *Ecology of Marine Invertebrate Larvae*. Boca Raton, FL (USA).

Murray, A.G. and Amundrud, T.L. (2007) Using coupled biophysical particle tracking models of sea lice in Loch Torridon. In: *Proceedings of the International Congress on Modelling and Simulation*. Christchurch, New Zealand, December 2007, pp. 2848–2853.

Murray, A.G. and Gillibrand, P.A. (2006) Modelling salmon lice dispersal in Loch Torridon, Scotland. *Marine Pollution Bulletin* 53, 128–135.

Myksvoll, M.S., et al. (2018) Evaluation of a national operational salmon lice monitoring system–From physics to fish. *PLoS ONE* 13(7), 1–25.

Nagasawa, K., (2004) Sea lice, *Lepeophtheirus salmonis* and *Caligus orientalis* (Copepoda: Caligidae), of wild and farmed fish in sea and brackish waters of Japan and adjacent regions: a review. *Zoological Studies* 43, 173–178.

Naylor, R.L., et al. (2021) A 20-year retrospective review of global aquaculture. *Nature* 591 (7851), 551–563.

Nowak, B.F., Bryan, J. and Jones, S.R. (2010) Do salmon lice, *Lepeophtheirus salmonis*, have a role in the epidemiology of amoebic gill disease caused by *Neoparamoeba perurans*? *Journal of Fish Diseases* 33(8), 683–687.

Oelckers, K., Vike, S., Duesund, H., Gonzalez, J., Wadsworth, S. and Nylund, A. (2014) *Caligus rogercresseyi* as a potential vector for transmission of infectious salmon anaemia (ISA) virus in Chile. *Aquaculture* 420, 126–132.

O'Hara, R.B. and Kotze, D.J. (2010) Do not log-transform count data. *Methods in Ecology and Evolution* 1(2), 118–122.

Penston, M.J., et al. (2004) Observations on open-water densities of sea lice larvae in Loch Shieldaig, Western Scotland. *Aquaculture Research* 35(8), 793–805.

Penston, M.J., Miller, C.P., Zuur, A., Davies, I.M., et al. (2008) Spatial and temporal distribution of *Lepeophtheirus salmonis* (Krøyer) larvae in a sea loch containing Atlantic salmon, *Salmo salar* L., farms on the north-west coast of Scotland. *Journal of Fish Diseases* 31(5), 361–371.

Pert, C.C., Urquhart, K. and Bricknell, I.R. (2006) The sea bass (*Dicentrarchus labrax* L.): A peripatetic host of *Lepeophtheirus salmonis* (Copepoda: Caligidae)? *Bulletin of the European Association of Fish Pathologists* 26(4), 163–165.

Peterson, A.T., Bauer, J.T. and Mills, J.N. (2004) Ecologic and geographic distribution of filovirus disease. *Emerging Infectious Diseases* 10(1), 40–47.

Pike, A.W. and Wadsworth, S.L. (1999) Sealice on salmonids: their biology and control. *Advances in Parasitology* 44, 233–337.

Poley, J.D., Braden, L.M., Messmer, A.M., Igboeli, O.O., Whyte, S.K., Macdonald, A., Rodriguez, J., Gameiro, M., Rufener, L., Bouvier, J., Wadowska, D.W., Koop, B.F., Hosking, B.C. and Fast, M.D. (2018) High level efficacy of lufenuron against sea lice (*Lepeophtheirus salmonis*) linked to rapid impact on moulting processes. *International Journal for Parasitology: Drugs and Drug Resistance* 8, 174–188.

Przeslawski, R., Byrne, M. and Mellin, C. (2015) A review and meta-analysis of the effects of multiple abiotic stressors on marine embryos and larvae. *Global Change Biology* 21(6), 2122–2140.

Rees, E.E., St-Hilaire, S. and Jones, S. (2015) Spatial patterns of sea lice infection among wild and captive salmon in western Canada. *Landscape Ecology* 30(6), 989–1004.

Revie, C.W., Gettinby, G., Treasurer, J.W., Rae, G. and Clark, N. (2002) Temporal, environmental and management factors influencing the epidemiological patterns of sea lice (*Lepeophtheirus salmonis*) infestations on farmed Atlantic salmon (*Salmo salar*) in Scotland. *Pest Management Science* 58(6), 576–584.

Revie, C.W., Gettinby, G., Treasurer, J.W. and Wallace, C. (2003) Identifying epidemiological factors affecting sea lice *Lepeophtheirus salmonis* abundance on Scottish salmon farms using general linear models. *Diseases of Aquatic Organisms* 57(1–2), 85–95.

Revie, C.W., Robbins, Gettinby, G., Kelly, L. and Treasurer, J.W. (2005) A mathematical model of the growth of sea lice, *Lepeophtheirus salmonis*, populations on farmed Atlantic salmon, *Salmo salar* L., in Scotland and its use in the assessment of treatment strategies. *Journal of Fish Diseases* 28(10), 603–613.

Revie, C.W., Dill, L., Finstad, B. and Todd, C. (2009) Sea Lice Working Group Report. *NINA Special Report* 39. 117.

Robbins, C., Gettinby, G., Lees, F., Baillie, M., Wallace, C. and Revie, C.W. (2010) Assessing topical treatment interventions on Scottish salmon farms using a sea lice (*Lepeophtheirus salmonis*) population model. *Aquaculture* 306(1–4), 191–197.

Rózsa, L., Reiczigel, J. and Majoros, G. (2000) Quantifying parasites in samples of hosts. *The Journal of Parasitology* 86(2), 228–232.

Rumrill, S.S. (1990) Natural mortality of marine invertebrate larvae. *Ophelia* 32(1–2), 163–198.

Saksida, S., Karreman, G.A., Constantine, J. and Donald, A. (2007) Differences in *Lepeophtheirus salmonis* abundance levels on Atlantic salmon farms in the Broughton Archipelago, British Columbia, Canada. *Journal of Fish Diseases* 30(6), 357–366.

Schram, T. (1993) Supplementary descriptions of the developmental stages of *Lepeophtheirus salmonis* (Krøyer, 1837) (Copepoda: Caligidae). In: Boxshall, G.A. and Defaye, D. (eds) *Pathogens of Wild and Farmed Fish: Sea Lice.* Ellis Horwood, Chichester, UK. pp. 30–50.

Short, J., Metaxas, A. and Daigle, R.M. (2013) Predation of larval benthic invertebrates in St George's Bay, Nova Scotia. *Journal of the Marine Biological Association of the United Kingdom* 93(3), 591–599.

Skarhamar, J., Fagerli, F.N., Reigstad, M., Sandvik, A.D. and Bjørn, P. (2019) Sampling planktonic salmon lice in Norwegian fjords. *Aquaculture Environment Interactions* 11, 701–715.

Skugor, S., Glover, K.A., Nilsen, F. and Krasnov, A. (2008) Local and systemic gene expression responses of Atlantic salmon (*Salmo salar* L.) to infection with the salmon louse (*Lepeophtheirus salmonis*). *BMC Genomics* 9, 498.

Solé, M., Lenoir, M., Fortuño, J.-M., De Vreese, S., van der Schaar, M. and André, M. (2021) Sea lice are sensitive to low frequency sounds. *Journal of Marine Science and Engineering* 9, 765.

Stien, A., Bjørn, P., Heuch, P.A. and Elston, D. (2005) Population dynamics of salmon lice *Lepeophtheirus salmonis* on Atlantic salmon and sea trout. *Marine Ecology Progress Series* 290, 263–275.

Tadiso, T.M., Krasnov, A., Skugor, S., Afanasyev, S., Hordvik, I. and Nilsen, F. (2011) Gene expression analyses of immune responses in Atlantic salmon during early stages of infection by salmon louse (*Lepeophtheirus salmonis*) revealed bi-phasic responses coinciding with the copepod-chalimus transition. *BMC Genomics* 12, 141.

Treasurer, J.W. and Pope, J.A. (2000) Selection of host sample number and design of a monitoring programme for ectoparasitic sea lice (Copepoda: Caligidae) on farmed Atlantic salmon, *Salmo salar*. *Aquaculture* 187, 247–260.

Tucker, C.S., Sommerville, C. and Wootten, R. (2002) Does size really matter? Effects of fish surface area on the settlement and initial survival of *Lepeophtheirus salmonis*, an ectoparasite of Atlantic salmon *Salmo salar*. *Diseases of Aquatic Organisms* 49(2), 145–152.

Valdes-Donoso, P., Mardones, F.O., Jarpa, M., Ulloa, M., Carpenter, T.E. and Perez, A.M. (2013) Co-infection patterns of infectious salmon anaemia and sea lice in farmed Atlantic salmon, *Salmo salar* L., in southern Chile (2007–2009). *Journal of Fish Diseases* 36, 353–360.

Valenzuela-Muñoz, V., Boltaña, S. and Gallardo-Escárate, C. (2016) Comparative immunity of *Salmo salar* and *Oncorhynchus kisutch* during infestation with the sea louse *Caligus rogercresseyi:* An enrichment transcriptome analysis. *Fish and Shellfish Immunology* 59, 276–287.

Vargas-Chacoff, L., et al. (2016) Atlantic salmon (*Salmo salar*) and Coho salmon (*Oncorhynchus kisutch*) display differential metabolic changes in response to infestation by the ectoparasite *Caligus rogercresseyi*. *Aquaculture* 464, 469–479.

Vollset, K.W., Qviller, L., Bjørnar, S., Barlaup, B.T. and Dohoo. I. (2018) Parasitic sea louse infestations on wild sea trout: separating the roles of fish farms and temperature. *Parasite Vectors* 11(1), 609.

Walter, T.C. and Boxshall, G.A. (eds) (2013) *Lepeophtheirus* von Nordmann, 1832. *World Copepoda Database*. Accessed through http://www.marinespecies.org/copepoda/aphia.php.

Westcott, J.D., Hammell, K.L. and Burka, J.F. (2004) Sea lice treatments, management practices and sea lice sampling methods on Atlantic salmon farms in the Bay of Fundy, New Brunswick, Canada. *Aquaculture Research* 35(8), 784–792.

Wootten, R., Smith, J.W. and Needham, E. (1982) Aspects of the biology of the parasitic copepods *Lepeophtheirus salmonis* and *Caligus elongatus* on farmed salmonids, and their treatment. *Proceedings of the Royal Society of Edinburgh, Section B: Biological Sciences* 81, 185–197.

Wootten, R., Smith, J.W. and Needham, E.A. (2011) Aspects of the biology of the parasitic copepods *Lepeophtheirus salmonis* and *Caligus elongatus* on farmed salmonids, and their treatment. *Proceedings of the Royal Society of Edinburgh. Section B: Biological Sciences*. Edinburgh, UK. pp. 185–197.

Yatabe, T., Arriagada, G., Hamilton-West, C. and Urcelay, S. (2011) Risk factor analysis for sea lice, *Caligus rogercresseyi*, levels in farmed salmonids in southern Chile. *Journal of Fish Diseases* 34, 345–354.

Zagmutt-Vergara, F.J., Carpenter, T.E., Farver, T.B. and Hedrick, R.P. (2005) Spatial and temporal variations in sea lice (Copepoda: Caligidae) infestations of three salmonid species famed in net pens in southern Chile. *Diseases of Aquatic Organisms* 64(2), 163–173.

Zuur, A.F., Ieno, E.N., Walker, N., Saveliev, A.A. and Smith, G.M. (2009) *Mixed Effects Models and Extensions in Ecology with R*. Springer, New York.

Zuur, A.F., Ieno, E.N. and Elphick, C.S. (2010) A protocol for data exploration to avoid common statistical problems. *Methods in Ecology and Evolution* 1(1), 3–14.

CHAPTER 11

Regulation and authorization of veterinary medicines for sea lice

Stacey Pulver

US Food and Drug Administration

11.1 Introduction

Veterinary medicines are important for disease treatment, sedation, and anesthesia for safe animal handling, and for production needs. Sea lice are the most biologically and economically damaging parasites to the salmonid farming industry worldwide (Costello, 2009; Abolofia et al., 2017), and a variety of tools are needed as part of the solution for treating and controlling this devastating parasite. The management of sea lice relies, in part, on the judicious use of veterinary medicines. Sea lice outbreaks can be financially devastating (Costello, 2009), but so can improper use of veterinary medicines. Improper use can lead to parasites developing resistance to the medicines, harmful tissue residues, and environmental harm to non-target animals and ultimately to the ecosystem. Veterinary medicines are usually part of a larger management plan that incorporates many different strategies (for example, mechanical control, cleaner fish, and improved husbandry) to combat this costly disease and leads to a sustainable system. It is important to understand the laws, regulations, and guidelines of the country in which the drug is being used and, if applicable, where the fish will be imported.

Obtaining a marketing authorization for a veterinary medicine through the competent authority requires submission of data, which can take time and be expensive. However, a rigorous evaluation process ensures that the authorized veterinary medicine is safe for the target animal and the people that eat the animal; is effective; is manufactured to a high quality; has complete and truthful labeling; and has had the impact on the environment of its use assessed. The evaluation process is a systematic way to evaluate the safety and effectiveness of veterinary medicines. The final step in the evaluation process is for the competent authority to grant the marketing authorization and is critical to the availability and accountable use of all classes of drugs used in animal production systems. The use of veterinary medicines in animals should be judicious and according to the label, laws, or regulations of the country in which they are being used.

This chapter focuses on the process for obtaining a marketing authorization for veterinary medicines and considerations for veterinary medicine administration. For the majority of this chapter (except where it is more appropriate to use country specific terms), the author uses the term veterinary medicine and intends to use it synonymously with animal drug or veterinary drug. The European

Union (EU) typically uses the term veterinary medicine, whereas the United States (US) uses the term animal drug, and still other countries may use other terms, such as chemicals, pharmaceuticals, or veterinary drugs. Similarly, the terminology for the processes through which these products are evaluated, or approved, differs between countries. The author intends to use marketing authorization to refer to the approval to market a veterinary medicine within a competent authority. In the EU, this is termed the marketing authorization, but in the US, this is referred to as the drug approval process, and in other countries may be referred to as a license, or registration process.

If a product is not regulated as a veterinary medicine, readers may want to seek out a specific country's laws pertaining to how that product is regulated (for example, pesticides or vaccines). Laws, regulations, and policies change over time, so it is important for readers to consult the competent authority for the most current information.

11.2 Jurisdiction

The laws of each country or regulatory authority will dictate how veterinary medicines are regulated, and which governmental agency oversees the process for marketing authorization. The following section focuses on the countries or competent authorities relevant to the global production of Atlantic salmon.

European Union

In the European Union (EU), medicines can be authorized through different routes, but the process usually involves a partnership between the European Commission, the medicines regulatory authorities in the EU member states and the European Economic Area (EEA), and the European Medicines Agency (EMA). Veterinary medicines and vaccines go through these routes.

All medicines in the EU must be authorized before they can be marketed and made available. There are two main routes for authorizing medicines: a centralized route and a national route. Under the centralized authorization procedure, pharmaceutical companies submit single marketing authorizations to the EMA. This allows the marketing-authorization holder to market the medicine and make it available to patients and healthcare professionals throughout the EU on the basis of a single marketing authorization. EMA's Committee for Medicine Products for Veterinary Use (CVMP) conducts the scientific assessment of the application and provides a recommendation on whether the medicine should be marketed. However, under the EU law EMA has no authority to actually permit marketing in the different EU countries. The European Commission is the authorizing body for all centrally authorized products and makes a legally binding decision based on EMA's recommendation. Once granted by the European Commission, the centralized marketing authorization is valid in all EU member states as well as the EEA countries, Iceland, Liechtenstein, and Norway (EMA, 2021a).

While the majority of new, innovative medicines are evaluated by EMA and authorized by the European Commission in order to be marketed in the EU, most generic medicines, and medicines available without a prescription are assessed and authorized at the national level in the EU. Each member state has its own national authorization procedures. If a sponsor wishes to request marketing authorization in several EU member states for a medicine that is outside the scope of the centralized procedure, it may use a mutual-recognition procedure, or a decentralized procedure. The data requirements and standards governing the authorization of medicines are the same in the EU, irrespective of authorization route (EMA, 2021a).

At the time of publication, the United Kingdom (UK) was preparing for regulation changes due to the UK's withdrawal from the EU, but the Veterinary Medicines Regulations remain in force. In the coming years, it is anticipated that the UK will publish a new set of Veterinary Medicines Regulations (https://www.gov.uk/guidance/vmd-information-hub). As such, this chapter includes EU information without specific reference to the UK, except where appropriate. As the situation changes in the UK, readers should reference the UK government website and other sources for updates.

Chile

In Chile, the Servicio Agricola y Ganadero (SAG or Agricultural and Animal Husbandry Service) is part of the Ministry of Agriculture and companies wishing to sell veterinary medicines must present effectiveness and safety data to SAG.

The production and sale of veterinary drugs, as well as use of experimental drugs, are covered by the Regulation on Veterinary Drugs, implementing the Animal Health and Protection Law (FAO, 2021b).

Canada

In Canada, the Health Canada Veterinary Drugs Directorate (VDD) oversees veterinary drug approval whereas the Pest Management Regulatory Agency (PMRA) regulates pesticides. The VDD regulates in-feed drug treatments, while the PMRA regulates bath (immersion) products. Veterinary biologics are regulated by the Canadian Food Inspection Agency (CFIA, 2020). Every veterinary drug sold in Canada must first be approved by the VDD and issued a Drug Identification Number (DIN). Through VDD, Health Canada evaluates and monitors the safety, quality, and effectiveness, sets standards, and promotes the prudent use of veterinary drugs administered to food-producing and companion animals (Health Canada, 2021).

United States

In the United States, the Food and Drug Administration (FDA) is responsible for regulating animal drugs, animal devices, and food/feed for animals. The US Department of Agriculture (USDA) has authority over veterinary biologics (for example, vaccines), while the Environmental Protection Agency (EPA) has authority over pesticides, insecticides, fungicides, and algaecides. The intended use of a product in the US determines whether it is regulated as a drug. A drug is defined in part, as "articles intended for use in the diagnosis, cure, mitigation, treatment, or prevention of disease in man or other animals" or "articles (other than food) intended to affect the structure or any function of the body of man or other animals" (Federal Food, Drug, and Cosmetic Act). The FDA has regulatory authority over drugs used to treat, control, and prevent sea lice infestations on fish.

For sea lice drugs to be legally marketed in the US, they must be the subject of a new animal drug approval (NADA), conditionally approved NADA, or abbreviated NADA (that is, generic animal drugs).

11.3 Development of new veterinary medicines

It is the responsibility of the applicant (also referred to as "sponsor") of the veterinary medicine to conduct effectiveness, safety, and manufacturing studies, submit the data to the competent authority for review, and ultimately obtain the approval to market the veterinary medicines. This entire process from research and development (R & D) of a particular compound to reaching full marketing authorization of a product can take many years and in the US cost in the order of US$30 million (https://ahi.org/approval-and-regulation-of-animal-medicines/). Whether an applicant takes a medicine to market is often dependent on the potential return on investment for the medicine and a global development strategy. In some countries, veterinary medicine approvals for aquaculture are not lucrative resulting in limited approved options for farmers, veterinarians, and other personnel. The countries with the largest salmonid aquaculture industry will undoubtedly provide the biggest market for sea lice veterinary medicines. Veterinary medicine development and approval in a small market may still occur when the global market is considered.

Hunter et al. (2011) described the phases of veterinary medicine (drug) development as consisting of the discovery phase, development phase, and registration process. The post-authorization phase, which continues indefinitely for the life of the product, is also an important phase. The discovery phase is where the applicant will first conduct a marketing assessment to identify unmet animal health needs that can be addressed and result in a satisfactory return on investment. This phase often includes proof-of-concept studies, preclinical studies (for example, pilot studies or laboratory *in vitro* studies), and assessment of the ease of manufacturing of the product. Should a veterinary medicine make it past this phase, it would enter the development phase where additional assessment of the market would be obtained, and additional studies to characterize the drug substance and manufacturing capability would be conducted. During this phase, the applicant may reach out to the competent authority to officially begin the drug approval or marketing authorization process. The registration phase may begin when studies are conducted to the standards of country requirements. During the registration phase, all the necessary data to support the approval of the veterinary medicine are submitted to the competent authority. The authority then determines if there are enough data to support the approval of the veterinary medicine. If a positive outcome on review of these data occurs, and the product is approved, the applicant may begin legally marketing the product for use according to the approved labeling. The post-authorization phase may involve monitoring of the drug in commercial settings for pharmacovigilance (that is, adverse drug events, changes in or lack of effectiveness) and continued monitoring of manufacturing processes and drug stability. Additionally, applicants may continue to seek approval for new uses for the product.

11.4 Marketing authorization process

The requirements for obtaining a marketing authorization for a veterinary medicine and the standard of conduct for the studies needed will vary by country or competent authority. Countries typically publish guidelines to help applicants through the authorization process and these should be consulted before embarking on studies. Additionally, there are internationally recognized guidelines that provide uniform recommendations and standards for those companies seeking approvals. For example, the International Cooperation on Harmonization of Technical Requirements for Registration of Veterinary Medicinal Products (VICH) publishes guidelines that have been agreed upon by regulatory agencies in the EU, US, and Japan. The governments of Canada, Australia, New Zealand, and South Africa are observers to the VICH process and may also choose to adopt VICH guidelines. Where applicable, VICH guidelines are referenced through this chapter. It is also recommended that facilities conducting studies use their own specific standard operating procedures (SOPs) to provide assurance of quality data. To obtain an approval of a veterinary medicine to treat sea lice, an applicant generally submits data to support the following major sections: effectiveness, target animal safety, environmental impact, human food safety, and manufacturing. Each of these are briefly described below. The order in which these data are developed varies across projects and is not typically dictated by the country or competent authority. Key to a smooth development of a veterinary medicine through the registration process is open communication between the applicant and the regulatory authorities.

Testing of unapproved veterinary medicines

Regulatory authorities may require special approval(s) (for example, permit, certificate, or authorization) to carry out a field trial (that is, testing) of an investigational veterinary medicine in animals. These special approvals would typically need to be in place before the start of the trial. Table 11.1 provides the name of the permit that may be required (or exemption) in several countries. The interested party should contact the appropriate regulatory authority for specific details on how to obtain one.

Effectiveness

An applicant is typically required to demonstrate that the veterinary medicine has the effect it purports or is represented to have under the proposed conditions of use suggested in the labeling. Generally, applicants conduct one or more studies evaluating the medicine under the proposed conditions of use

Table 11.1 Authorities responsible for permits for unapproved drugs.

Country/regulatory authority	Permit or application name	Website
United Kingdom	Animal Test Certificate	https://www.gov.uk/guidance/animal-test-certificates
Ireland	Health Products Regulatory Authority license	http://www.hpra.ie/homepage/veterinary/regulatory-information/veterinary-clinical-field-trials
Canada	Emergency Drug Release Program	https://www.canada.ca/en/health-canada/services/drugs-health-products/special-access/veterinary-drug.html
United States	Investigational New Animal Drug Exemption	https://www.fda.gov/media/75922/download
Norway	Special Exemption	https://www.legemiddelsok.no/
Faroe Islands (Denmark)	Application for Clinical Trials of Veterinary Medicinal Products in Animals	https://laegemiddelstyrelsen.dk/en/licensing/clinical-trials/trials-in-animals/~/media/A5A14B9080B94D3EB08FAF4392C8A4C2.ashx

and may include both laboratory and field studies. The main purpose of documenting the effectiveness is to prove the effect of a veterinary medicine for the proposed indication and at the proposed dosage. Studies should be conducted under appropriate conditions according to the proposed conditions of use of the product. Clinical studies (that is, field studies) intended to test the effectiveness of antiparasitic medicines are interpretable only if the fish and sea lice represent the conditions of use, the dosing is accurate, and the study design is valid. Most studies will be conducted under field conditions, but under certain circumstances it may be acceptable to conduct laboratory studies. It is best to collaborate with the competent authority to determine the most appropriate study or studies for evaluation of effectiveness.

Effectiveness studies are typically conducted according to the principles of Good Clinical Practice (GCP). GCP is an international standard developed under VICH and provides a unified standard for the EU, Japan, and the US to facilitate the mutual acceptance of clinical data by the relevant regulatory authorities (VICH GL9, 2001). It is incumbent for the protocol to be developed based on sound science and quality standards, and if appropriate, to be agreed upon between the investigator, applicant, and regulatory authority.

To the author's knowledge there is no international guidance providing standards specific for evaluating the effectiveness of anti-sea lice veterinary medicines. However, there are two documents that do focus on aquatic animals and provide broader focus to the evaluation of medicines used in fish. The first is the EMA document titled "Guideline on demonstration of target animal safety and efficacy of veterinary medicinal products intended for use in farmed fish" (EMA, 2011). As the title suggests, the scope is for all veterinary medicinal products, not only antiparasitic medicines, and does focus on farmed fish. The second is the publication of the "World Association for the Advancement of Veterinary Parasitology (WAAVP) guideline for testing the efficacy of ectoparasiticides for fish" (Sommerville et al., 2016). The WAAVP is an internationally recognized scientific leader in parasitology and develops guidelines for the evaluation of effectiveness of antiparasitics in various animal groups and provided the first specific antiparasitic guidelines available for aquatic animals. Both documents are valued references and provide a high-level overview of the considerations for planning and execution of studies designed to assess the effectiveness of drugs for fish; however, neither focus specifically on the sea lice disease.

One of the key elements of veterinary medicine development is identifying the appropriate indication (that is, claim) to pursue because this drives the study design for effectiveness and safety studies. The indication will be supported by data that can be measured or observed, has utility to the end user, and is valid. In the case of sea lice, typically the medicine is killing or halting the development of some or all stages of the parasite living on the fish. The effectiveness studies thus may measure the sea lice counts and identify the species and the life stages in which the drug has an effect. The timing of the measurement is important and expected to be determined a priori. If the veterinary medicine is expected to have a residual protection (persistent effect), this will also be measured during effectiveness studies, and may appear in the indication or labeling of the product as the protected period. The test animals and the parasite species should be representative of the conditions in which the medicine would be used. A study should use methods to limit bias which may include incorporating control animals, masking, and randomization, and employ qualified personnel. Additionally, it is important to discuss the study design of the field studies and the details of the protocol development with the competent authority. They may be able to provide input on experimental design, dosing schemes, statistical methods, and sampling procedures.

Target animal safety (TAS)

The TAS evaluation typically includes review of data, scientific literature, and/or other information that assesses the safety of the veterinary medicine in fish under the conditions prescribed, recommended, or suggested in the proposed labeling in the target species or group of animals. The purpose of the studies conducted is to evaluate possible toxic effects of the medicine on treated fish in a controlled setting with the intent of extrapolating those findings to the target population at large. TAS data also are collected during field effectiveness studies.

International harmonization of standards for TAS studies were developed under VICH GL43 and implemented in 2009 (VICH GL43, 2009). Although the scope of the guidance states that it does not provide information for the design of TAS studies in aquatic animals, the recommendations provide a base standard that could be adapted for use in fish and, when followed, confer an expectation of a high-quality standard. Similar to other parts of the regulatory veterinary medicine application, it is necessary to communicate regularly with the regulatory authorities to ensure a smooth process of evaluating TAS.

The specific studies and information needed to assess TAS is highly dependent on the characteristics of the particular animal drug. However, one of the most common studies that is conducted, and also described in VICH GL43, is called a margin of safety study. A margin of safety study is essentially an overdose study, evaluating the highest proposed dose (that is, 1X) and levels above the proposed labeled dose (for example, 3X and 5X) for three times the labeled duration, in comparison to a non-treated control group (VICH GL43, 2009). For fish studies, endpoints generally include physical examinations; daily health observations; and necropsies with histopathology at the study's termination. Hematology and blood chemistry may also be incorporated into the study, but may depend on the country requirements, the drug being evaluated, and the size of the fish. Additional animal safety studies may be appropriate for a particular drug, depending on the conditions of use and the drug's characteristics. For example, if the product is intended for use in reproducing animals, reproductive safety studies may be necessary.

Manufacturing

Before a veterinary medicine can be marketed, it must meet certain manufacturing standards to ensure consistent quality of the product. A review of this information would confirm the identity, strength, quality, and purity of the product. A full description of the methods, facilities, and controls used for the manufacturing, processing, and packaging of the animal drug would be reviewed. Data to support expiry periods and storage conditions would be provided to support product labeling statements. Many countries use the term good manufacturing practices (GMP) to refer to a standard that ensures drugs are of consistent high quality and are appropriate for their intended use. GMP standards may vary between competent authorities. Inspections of the manufacturing facilities may also be required before a drug can be marketed. It is best to collaborate with the appropriate competent authorities, and consult the appropriate laws, regulations, and/or guidance documents to determine what is needed to satisfy the manufacturing requirements of veterinary medicines.

Human food safety

Veterinary medicines approved for use in food-producing species undergo a human food safety evaluation to ensure that food derived from treated animals is safe for human consumption.

Generally, this evaluation reviews data and information submitted to demonstrate that residues of veterinary medicines in the edible tissues of treated animals are safe. There are three components that are evaluated during review of human food safety: toxicology, residue chemistry, and microbial food safety (FDA, 2018).

Toxicology data and information are used to identify and characterize any hazard association with exposure to veterinary medicine residues. These data and information may be used to establish the acceptable daily intake (ADI). This value, usually expressed in micrograms or milligrams of drug that can be safely consumed per kilogram of body weight per day, represents the amount of drug that can be safely consumed by a person daily for a lifetime. This ADI, together with food consumption values for edible tissues, is then used to determine safe concentrations of residues in edible tissue. The safe concentration is the amount of total residue of a veterinary medicine that can be consumed from each edible tissue every day for up to the lifetime of a human without exposing the human to residues in excess of the ADI (FDA, 2018). Safe concentrations, expressed as parts per million (ppm) or parts per billion (ppb), are used for maximum residue limit (MRL) or tolerance determinations.

A target tissue and a marker residue are selected from the information provided about the veterinary medicine metabolism in the target animal. Marker residue depletion data also are collected to demonstrate the depletion of the marker residue post-treatment to below the safe level (for example, MRL or tolerance) and to provide data for calculating a withdrawal period. A withdrawal period is the interval between the time of the last administration of a veterinary medicine and the time when the animal can be slaughtered for food and can be safely consumed. The withdrawal period is typically presented in calendar days or degree days. Degree days is determined by multiplying the water temperature each day with the number of days. For example, ten days at 5°C equals 50 degree days (Chezik et al, 2014). Veterinary medicine applicants in the EU, Japan, and US may reference VICH GL57 to assist with designing residue studies in aquatic species (VICH GL57, 2020).

Along with providing information about the quantity and nature of residues, the applicant also needs to provide a validated analytical method for measuring tissue residues. The analytical method is used to ensure that tissue residues do not exceed the legal safe concentration (for example, MRL or tolerance) after the withdrawal period has been completed. Veterinary medicine applicants in the EU, Japan, and US may reference VICH GL49 to guide them in proper validation of analytical methods (VICH GL49, 2015).

The third part of the human food safety section is to evaluate the risk of emergence or selection of antimicrobial-resistant bacteria of public health concern (for example, *Salmonella* spp., *Campylobacter* spp., *Escherichia coli*) in or on animals treated with any veterinary medicine with antimicrobial properties. Any concern that the use of the veterinary medicine will promote resistance among bacteria of public health concern in or on food-producing animals which may lead to human infection will be reviewed during the course of the approval process (FDA, 2021b).

Maximum residue limits, tolerances, and import tolerances

In most countries, the term maximum residue limit is used, the exception being the US, where the term tolerance is used. *CODEX ALIMENTARIUS* (*CODEX*) defines the MRL for veterinary drug residues as the maximum concentration of residue legally tolerated in a food product obtained from an animal that has received a veterinary medicine. In the EU, MRLs must be established in advance of the marketing authorization for veterinary medicines. Substances that do not have EU MRLs cannot be used to treat food-producing species in the EU.

US FDA defines a tolerance as the maximum concentration of a marker residue, or other residue indicated for monitoring, which can legally remain in a specific edible tissue of a treated animal (FDA, 2018). These tolerances are set for animal drugs approved for use in the US. In the US, it is unlawful to import animal-derived food that bears or contains residues of an unapproved new animal drug unless an import tolerance has been established for residues of that drug and any residues do not exceed that tolerance. Therefore, import tolerances can be established for residues of such unapproved animal drugs present in imported, animal-derived food products. Data on such drugs would need to be submitted to the US FDA to establish such an import tolerance.

MRLs or tolerances are established by the competent authority for purposes of monitoring residues in food. While efforts are made to harmonize, sometimes the conclusions of each country result in establishing different MRLs or tolerances for veterinary drugs. The differences in the animal drug approval systems, husbandry practices, food consumption patterns, methodology used to measure residues, and legal framework vary among countries and contribute to this potential difference in MRLs or tolerances. Within a country, following the labeling directions including the withdrawal period should prevent producers or veterinarians from violating the set tolerance or MRL. However, knowledge of the tolerance or the MRL set for individual countries is important when exporting the fish or fish fillets to that country. Table 11.2 provides resources where these tolerances or MRLs can be found.

Environmental impact

Prior to approval, a veterinary medicine may undergo an environmental risk assessment based on the expected use. This evaluation could include the amount of veterinary medicine expected to be put into the environment, the medicine's potential toxicity to aquatic life, and the potential effects the medicine's use will have on the environment. Depending on the jurisdiction, the risk assessment may also identify whether there is a need for specific environmental risk-mitigation measures, and if needed, could ensure appropriate labeling is in place for these measures. In the countries harmonized to VICH (that is, EU, Japan, and US), a Phase II Environmental Assessment is often needed for veterinary medicines in aquaculture (VICH GL38, 2005).

Table 11.2 Resources where tolerances or MRLs can be found.

Residue limit type	Location to access current limits
Maximum residue limit (MRL) (set by *CODEX*)	http://www.fao.org/fao-who-codexalimentarius/codex-texts/maximum-residue-limits/en/
Tolerance (set by US FDA)	21 Code of Federal Regulations Part 556 Electronic Code of Federal Regulations: www.ecfr.gov/ and to Part 556 https://www.ecfr.gov/current/title-21/chapter-I/subchapter-E/part-556?toc=1
Import tolerance (set by US FDA)	www.fda.gov/animal-veterinary/import-exports/import-tolerances

User safety

Veterinary medicine applications typically require documentation to show the potential risks which may result from the exposure of human beings to the veterinary medicine before its application to the animal (for example, storage or preparation), during its application, or after its application (for example, contact through treated animals) (EMA, 2010). The assessment of the user safety typically addresses situations resulting from normal conditions of use and from foreseeable accidents. It usually does not address deliberate misuse, but this aspect may be addressed when the categorization of the product is reviewed (for example, controlled substances). The EMA has guidance that explains that the user safety assessment would review information on toxicity of the active ingredient or final product, exposure assessment, and risk assessment. This assessment may lead to risk communication and mitigation provided on labeling of the product, in addition to other options that could be considered for risk control (EMA, 2010).

Post-authorization requirements for veterinary medicines

Once a veterinary medicine is approved and marketed, it is important to continue post-approval monitoring to assure the continued safety and effectiveness of the veterinary medicine (FDA, 2021c; EMA, 2021b). This would include monitoring for adverse drug experiences and changes in or lack of effectiveness, as well as continued monitoring for product quality through manufacturing data. Additionally, it may include revising product labeling as necessary and submitting pharmacovigilance reports.

Fish welfare (regulatory studies)

The assessment of animal welfare is important in all aspects of the veterinary medicine authorization process. Sommerville et al. (2016) provide important considerations for fish husbandry and care during the studies to evaluate effectiveness of antiparasitics. Humane euthanasia of fish should also be done correctly at the end of studies. Veterinary medicine applicants should seek out the laws and guidelines of the regulatory authority to determine whether there are local requirements for fish welfare. Member states of the EU should incorporate the 3Rs (that is, replacement, reduction, and refinement) testing approach into regulatory testing (EMA, 2016). Finally, communicating often with the regulatory authorities and having a clear understanding of what is needed to acquire a marketing authorization, will reduce the number of studies, and ultimately the number of fish needed.

11.5 Considerations for drug administration

Once a veterinary medicine reaches approval or full marketing authorization, it is the responsibility of the farmer, veterinarian, and other personnel to assure its proper use. When and how a veterinary medicine gets used may depend on the labeling conditions of use, the veterinary medicine availability, the health of the animals, the sea lice thresholds set for each country, the presence of parasite resistance, and the resources available to carry out the veterinary medicine administration. Additionally, the impact on organic certification of the product, ethics, and other factors may also be considered in choosing an appropriate treatment. The ultimate goal of treatment would be to achieve the greatest impact on the parasite population while being safe to the animal and retaining the lowest rate of resistance development.

Veterinary medicines that have been assessed and approved to treat for sea lice, and those available in various countries, are presented in Chapter 14 of this volume.

11.6 Regulatory lice numbers (sea lice threshold)

The determination to treat for sea lice is usually balanced by a host of circumstances; however, the administration of veterinary medicines to treat sea lice may be dictated by law, regulation, or guidance within each country. Sea lice levels on salmon farms are generally under comprehensive surveillance throughout the seawater production period (Aaen et al., 2015). Infection thresholds within the industry are set conservatively in most, but not all, farming regions with a focus on protecting wild populations of salmon from the impacts of farm-based spill over (Poley et al., 2018). Some governments regulate treatment by setting mandatory maximum thresholds for the amount of mature or motile lice per farmed fish at production sites (Luthman et al., 2019). Table 11.3 provides a summary of sea lice trigger levels for treatment in various countries.

Table 11.3. Sea lice trigger levels for treatment in various countries.

Country	Treatment threshold	Monitoring
Norway[1]	0.2 adult female lice on average per fish during the smolt spring migration period	Weekly when water temperatures are at least 4°C
	0.5 adult female lice on average per fish, all other times	Every 14 days when water temperatures are less than 4°C
Chile[2]	Three motile lice per fish	
Ireland[3]	0.3–0.5 ovigerous lice per fish; number of mobile lice also inform treatment threshold (March to May)	Twice per month, March to May Once per month other times, except December, January
	Two ovigerous lice per fish (all other times during the year)	One inspection during December to January
Scotland[4]	Two adult female lice per fish (reporting threshold) Six adult female lice per fish (intervention threshold)	Weekly, mandatory reporting to Marine Scotland
Canada (British Columbia)[5]	Average of three motile *L. salmonis* per fish (March to June)	
Faroe Islands[6]	1.5 adult female lice per fish	
US	No regulatory threshold set	No regulatory monitoring required

1 https://lovdata.no/dokument/SF/forskrift/2012-12-05-1140.

2 Luthman et al., 2019.

3 The Department of Agriculture, Fisheries & Food 2008. A strategy for improved pest control on Irish salmon farms.

4 https://www.gov.scot/publications/understanding-relative-cost-effectiveness-sea-lice-management-measures-farmed-salmon-production-scotland/pages/5/#:~:text=Scottish%20Government%20regulations%20include%20a,both%20during%20a%20weekly%20count.

5 https://www.dfo-mpo.gc.ca/about-notre-sujet/publications/infographics-infographies/lice-pou-eng.html.

6 Gislason, H. 2018.

11.7 Veterinary medicine delivery

Keys to a successful treatment include the proper diagnosis of disease, assurance of an accurate dose to each animal, and the practicality of the delivery method when used in a commercial setting. The treatment options available may also be dependent on the country and the cost of the treatment. The most common methods of delivery in commercial settings include oral or in-feed drugs and bath or immersion drugs. While injection delivery can be used to administer fish treatments, this method is not often employed for sea lice treatment in net pens because it is stressful, usually requires sedation, is labor intensive and expensive, and would be very time and resource consuming if done in net pens. It is more often used for vaccine administration or spawning drugs when fish are still housed in the hatchery.

The majority of sea lice drug applications occur in the net pens or on wellboats. The choice between the two will be dictated by the drug employed, the directions on the drug's label, labor available, weather and water conditions, access to wellboats, and environmental discharge limitations. Application of a bath treatment requires a known volume of water to treat. Net pen applications would require a tarpaulin skirt (partial enclosure) or tarpaulin bag (full enclosure) for treatment. Delivery of the bath treatment to the tank or net pen may involve various delivery methods including pumps, pipes, and application to multiple locations. A wide margin of safety (toxicity) is recommended owing to the time it takes for these procedures. Application of an oral or in-feed treatment requires a known biomass of fish in the net pens, and monitoring of feeding behavior during treatment.

Hatchery application of certain veterinary medicines may also be an option for some countries. If an option, medicines would be administered prior to seawater transfer of salmon smolts and provide long-term protection against sea lice infestations.

Wellboats have been successfully used to treat sea lice on a commercial scale, having the advantage of containment, collection of dislodged parasites before fish return to the net pen, and improved accuracy of medicine delivery. They still require consent from regulatory authorities for treatment and for the discharge of treated water into a natural body of water. The economic cost and relative shortage of wellboats in some countries may make this a limited option.

Discharge limitations and other environmental considerations

The use of investigational or approved veterinary medicines may result in discharge to fresh or marine receiving waters, and thus may be subject to pollution legislation (Sommerville et al., 2016). In addition to federal regulations, some farms may have local regulations that will need to be obeyed.

11.8 Unapproved and off-label use

The use of authorized veterinary medicines is often legally required; however, when a farmer, veterinarian, or other aquatic animal health professional decides to use an unapproved veterinary medicine (that is, one not approved or authorized for any use), they have the potential to put their safety, the fish's safety, the environment's safety, and their marketed fish product at risk. Unapproved medicines have not been tested for effectiveness or safety (including human food safety), may not be manufactured to quality standards, and their use may have trade implications.

However, it may be necessary to use a veterinary medicine in a manner that is not in accordance with the label or the approved conditions. When this occurs, it is being used in an extra-label use (ELU) manner, sometimes referred to as off-label. As described earlier in the chapter, the process through which the veterinary medicine is evaluated and ultimately granted marketing authorization is designed to show that it is effective and safe for a specific use under specific conditions. Whether this off-label use is allowed, and how it is allowed, is dependent on the laws and regulations of the country in which it is being used. Decisions to utilize ELU may include lack of approved products to address specific needs, lack of effectiveness of approved products, or adjustments to dosing regimens to accommodate the pathogen or animal's health status. In food-producing species, when the decision is made to administer a drug product in an ELU manner, there are potential health and food safety risks involved. These include violative drug residues in food products, violative residues in the environment, potential for adverse reactions, and potential for trade restrictions.

In the EU, off-label use refers to "the veterinary medicinal product that is not in accordance with the summary of the product characteristics (SPC), including the misuse and serious abuse of the product." The EU recognizes clinical situations in which off-label use of a product is necessary and appropriate. For the EU, the details permitting use are published in Articles 10 and 11 of the Directive (known as "the cascade"). A veterinary surgeon may allow for off-label use in accordance with the cascade (EMA, 2018).

In Canada, off-label use refers to the use or intended use of a drug in a manner not in accordance with the label. It also includes the use of unapproved veterinary medicines. In Canada, it is not confined to veterinarians and may be performed by intermediate health professions and laypersons. However, Health Canada recommends that off-label use in food-producing animals be done by a licensed veterinarian within the context of a valid Veterinarian-Client-Patient Relationship (VCPR) (Health Canada, 2014).

In the US, only licensed veterinarians are allowed to prescribe extra-label uses of certain approved animal and human drugs within the context of a valid VCPR and other specific conditions described in the regulations (21 Code of Federal Regulations (CFR) 530). FDA's requirements for ELU in animals limits this use to situations where an animal's health is threatened or where the animal may suffer or die without treatment (FDA, 2021a).

11.9 Veterinary medicine resistance and judicious use

Veterinary medicines are an important tool for treating sea lice but should be used judiciously. However, this historically has not been the case, and this has led, in part, to global resistance of sea lice to many of the available medicines (Aaen et al., 2015; Poley et al., 2018). Widespread resistance of sea lice to veterinary medicines has increased the need to develop new treatment methods (biological, prophylactic, and new medicines) and tools to avoid increased losses due to sea lice disease and to ensure a sustainable salmon farming industry in the future. The expression "resistant sea lice" is mainly applied for populations showing a high degree of reduced sensitivity in bioassays (Aaen et al., 2015). However, while a therapeutic failure in the field could be caused by reduced sensitivity to the medicine, it can also be caused by suboptimal treatment regimens (insufficient drug dispersal, poor feeding management), and poor farm management (husbandry).

To combat resistance, the practice of using combination veterinary medicine treatments of two or more products has also been tried. However, the competent authority may need to approve the combination for use to ensure effectiveness and safety to the fish and humans, ensuring an assessment of the impact to the environment (Jackson et al., 2018).

The importance of judicious use of veterinary medicines cannot be understated and in 2019, the Food and Agricultural Organization of the United Nations (FAO) published a technical guideline for responsible fisheries titled, "Recommendations for prudent and responsible use of veterinary medicines in aquaculture" (FAO, 2019). The guidelines provide recommendations for governments, producers, and animal health professionals to help combat global resistance and provide a "One Health" approach to minimize the emergence and spread of resistance to veterinary medicines.

Development of integrated pest management (IPM) or integrated sea lice management (ISLM) techniques on farms is also important to reduce or slow the spread of resistance. The FAO defines IPM as:

> The careful consideration of all available pest control techniques and subsequent integration of appropriate measures that discourage the development of pest populations and keep pesticides and other interventions to levels that are economically justified and reduce or minimize risks to human health and the environment. IPM emphasizes the growth of a healthy crop with the least possible disruption to agro-ecosystems and encourages natural pest control mechanisms. (FAO, 2021a)

ISLM requires a balanced and robust approach that may include, but is not limited to, drug rotation, fallowing, synchronized treatments within geographic zones, cleaner fish, delousing lasers, plankton shielding skirts, and modified cages. These, and others, are all integral to ISLM and ultimately allow the farm to decrease the amount of veterinary medicine use, but options may be dependent on availability to the farm, or in the country.

11.10 Fish welfare (on farm)

Research on fish welfare is not straightforward and has been constantly evolving over the last several decades. Agreement on the definition of fish welfare (including whether fish feel pain and are capable of suffering) and how to objectively measure fish welfare drive this controversy (Volpato, 2009). However, fish welfare is an important issue for the industry because proper care of fish may improve production efficiency, quality, and quantity (Ashley, 2006), all while ensuring the public confidence in the products they are consuming. Fish welfare may be regulated by the authorities, or there may be guidelines that are recommended. In 2005, the Council of Europe adopted a recommendation on the welfare of farmed fish (COE, 2005). In that document, specific to drug use, it states: "No substance other than those given for therapeutic or prophylactic purposes shall be administered to an animal unless it has been demonstrated by scientific knowledge or established experience that the effect of the substance is not detrimental to the welfare, including health, of the animals." With regard to sea lice, this implies that just because a product is known to be effective in treating sea lice, if the water temperature may create a situation where the drug cannot be safely administered and would put the welfare of the animals at risk, then an alternate method should be used. Additionally, the Council of Europe's recommendations stated: "The routine use of medicines as part of a management system to compensate for poor hygienic conditions, poor management practices, or to mask signs of poor welfare such as pain and distress shall not be allowed." Although the prevention of sea lice entirely is challenging, on-farm husbandry procedures to reduce the impact of sea lice (for example, fallowing), should be employed before the use of veterinary medicines or other compounds. Although fish welfare research and current understanding is not without controversy, improvements in husbandry (for example, stocking density, diet), farm management, and the handling, transport, and slaughter of fish will benefit aquaculture productivity and likely fish welfare (Ashley, 2006).

11.11 Summary

The process for obtaining a marketing authorization for veterinary medicines is designed to be rigorous and provide assurance that the veterinary medicine is safe, effective, and quality manufactured. Users should ensure they are administering veterinary medicines within the context of laws and regulations of their country (or potential importing country), and have mitigated risk to humans, the environment, and the fish. Appropriate use also entails following the labeling of the product and understanding how it should be used and the expected results. While there may be limited options for sea lice treatment and control, it is important when reaching for a veterinary medicine for a sea lice treatment, to reach for those that are approved or have marketing authorization. All use of veterinary medicines should be used according to judicious use principles and have the welfare of the fish at the forefront of the considerations.

References

Aaen, S.M., Helgesen, K.O., Bakke, M.J., Kaur, K. and Horsberg, T.E. (2015) Drug resistance in sea lice: a threat to salmonid aquaculture. *Trends in Parasitology* 31(2), 72–81.

Abolofia, J., Wilen, J.E. and Ashe, F. (2017) The cost of lice: quantifying the impacts of parasitic sea lice on farmed salmon. *Marine Resource Economics* 32(3), 329–349.

Ashley, P.A. (2006) Fish welfare: Current issues in aquaculture. *Applied Animal Behavior Science* 3–4(104), 199–235.

CFIA (2020) Veterinary biologics. Available at: https://inspection.canada.ca/animal-health/veterinary-biologics/eng/1299159403979/1320545281259 (accessed 28 June 2021).

Chezik, K.A., Lester, N.P. and Venturelli, P.A. (2014) Fish growth and degree-days I: selecting a base temperature for a within-population study. *Canadian Journal of Fisheries and Aquatic Sciences* 71(1), 47–55.

Costello, M. (2009) The global economic cost of sea lice to the salmonid farming industry. *Journal of Fish Diseases* (32), 115–118.

Council of Europe (COE) (2005) Standing Committee of the European Convention for the Protection of Animals Kept for Farming Purposes: Recommendation Concerning Farmed Fish. Available at: https://www.coe.int/t/e/legal_affairs/legal_co-operation/biological_safety_and_use_of_animals/Farming/Rec%20fish%20E.asp (accessed 28 June 2021).

EMA (2010) Guideline on user safety for pharmaceutical veterinary medicinal products. Available at: https://www.ema.europa.eu/en/documents/scientific-guideline/guideline-user-safety-pharmaceutical-veterinary-medicinal-products_en.pdf (accessed 16 November 2021).

EMA (2011) Guideline on demonstration of target animal safety and efficacy of veterinary medicinal products intended for use in farmed finfish. Available at: https://www.ema.europa.eu/en/demonstration-target-animal-safety-efficacy-veterinary-medicinal-products-intended-use-farmed (accessed 28 June 2021).

EMA (2016) Guideline on the principles of regulatory acceptance of 3Rs (replacement, reduction, refinement) testing approaches. Available at: https://www.ema.europa.eu/en/documents/scientific-guideline/guideline-principles-regulatory-acceptance-3rs-replacement-reduction-refinement-testing-approaches_en.pdf (accessed 28 June 2021).

EMA (2018) Reflection paper on off-label use of antimicrobials in veterinary medicine in the European Union. Available at: https://www.ema.europa.eu/en/documents/scientific-guideline/reflection-paper-label-use-antimicrobials-veterinary-medicine-european-union-first-version_en.pdf (accessed 28 June 2021).

EMA (2021a) Authorisation of medicines. Available at: https://www.ema.europa.eu/en/about-us/what-we-do/authorisation-medicines (accessed 28 June 2021).

EMA (2021b) Veterinary post-authorisation Q&A. Available at: https://www.ema.europa.eu/en/veterinary-regulatory/post-authorisation/veterinary-post-authorisation-qa-introduction (accessed 30 November 2021).

FAO (2018) *CODEX ALIMENTARIUS* Maximum Residue Limits (MRLs). Available at: https://www.fao.org/fao-who-codexalimentarius/codex-texts/maximum-residue-limits/en/ (accessed 14 January 2022).

FAO (2019) FAO Technical Guidelines for Responsible Fisheries. Aquaculture Development: Recommendations for prudent and responsible use of veterinary medicines in aquaculture. Available at: http://www.fao.org/3/ca7029en/ca7029en.pdf (accessed 28 June 2021).

FAO (2021a) Integrated Pest Management. Available at: http://www.fao.org/agriculture/crops/thematic-sitemap/theme/pests/ipm/en/ (accessed 28 June 2021).

FAO (2021b) National Aquaculture Legislation Overview – Chile. Available at: http://www.fao.org/fishery/legalframework/nalo_chile/en (accessed 28 June 2021).

FDA (2018) General Principles for Evaluating the Human Food Safety of New Animal Drugs Used in Food-Producing Animals. Available at: https://www.fda.gov/regulatory-information/search-fda-guidance-documents/cvm-gfi-3-general-principles-evaluating-human-food-safety-new-animal-drugs-used-food-producing (accessed 29 June 2021).

FDA (2021a) Animal Medicinal Drug Use Clarification Act of 1994 (AMDUCA). Available at: https://www.fda.gov/animal-veterinary/guidance-regulations/animal-medicinal-drug-use-clarification-act-1994-amduca (accessed 28 June 2021).

FDA (2021b) Antimicrobial Resistance. Available at: https://www.fda.gov/animal-veterinary/safety-health/antimicrobial-resistance (accessed September 27, 2021).

FDA (2021c) Data Elements for Submission of Veterinary Adverse Event Reports to the Center for Veterinary Medicine. Available at: https://www.fda.gov/regulatory-information/search-fda-guidance-documents/cvm-gfi-188-data-elements-submission-veterinary-adverse-event-reports-center-veterinary-medicine (accessed 30 November 2021).

FDA (2021d) Import Tolerances. Available at: https://www.fda.gov/animal-veterinary/import-exports/import-tolerances (accessed 14 January 2022).

FDA (2022) Code of Federal Regulations Title 21 Food and Drugs Part 558. Available at: https://www.ecfr.gov/current/title-21/chapter-I/subchapter-E/part-558?toc=1 (accessed 14 January 2022).

Health Canada (2014) Extra-Label Drug Use (ELDU) in Animals. Available at: https://www.canada.ca/en/health-canada/services/drugs-health-products/veterinary-drugs/extra-label-drug-use.html (accessed 28 June 2021).

Health Canada (2021) Regulating veterinary drugs. Available at: https://www.canada.ca/en/health-canada/corporate/mandate/regulatory-role/what-health-canada-regulates-1/veterinary-drugs.html (accessed 28 June 2021).

Hunter, R.P., Shryock, T.R., Cox, B.R., Butler, R.M. and Hammelman, J.E. (2011) Overview of the animal health drug development and registration process: an industry perspective. *Future Medicinal Chemistry* 3(7), 881–886.

Jackson, D., Moberg, O., Stenevik Djupevag, E.M., Kane, F. and Hareide, H. (2018) The drivers of sea lice management policies and how best to integrate them into a risk management strategy: An ecosystem approach to sea lice management. *Journal of Fish Diseases* 41, 927–933.

Luthman, O., Jonell, M. and Troell, M. (2019) Governing the salmon farming industry: Comparison between national regulations and the ASC salmon standard. *Marine Policy* 106, 103534.

Poley, J.D., Braden, L.M., Messmer, A.M., Igboeli, O.O., Whyte, S.K., Macdonald, A., Rodriguez, J., Gameiro, M., Rufener, L., Bouvier, J., Wadowska, D.W., Koop, B.F., Hosking, B.C. and Fast, M.D. (2018) High level efficacy of lufenuron against sea lice (*Lepeophtheirus salmonis*) linked to rapid impact on moulting processes. *IJP: Drugs and Drug Resistance* 8, 174–188.

Sommerville, C., Endris, R., Bell, T.A., Ogawa, K., Buchmann, K. and Sweeney, D. (2016) World Association for the Advancement of Veterinary Parasitology (WAAVP) guideline for testing the efficacy of ectoparasiticides for fish. *Veterinary Parasitology* 219, 84–99.

VICH GL9 (2001) Good Clinical Practice. Available at: https://www.ema.europa.eu/en/documents/scientific-guideline/vich-gl9-good-clinical-practices-step-7_en.pdf, or https://www.fda.gov/regulatory-information/search-fda-guidance-documents/cvm-gfi-85-vich-gl9-good-clinical-practice (accessed 28 June 2021).

VICH GL38 (2005) Environmental Impact Assessment for Veterinary Medicinal Products: Phase II Guidance. Available at: https://www.ema.europa.eu/en/documents/scientific-guideline/vich-gl38-environmental-impact-assessments-veterinary-medicinal-products-vmps-phase-ii_en.pdf (accessed 28 June 2021).

VICH GL43 (2009) Target Animal Safety for Veterinary Pharmaceutical Products. Available at: https://www.fda.gov/media/70438/download, or https://vichsec.org/en/guidelines/pharmaceuticals/pharma-safety/pharma-target-animal-safety.html (accessed 28 June 2021).

VICH GL49 (2015) Studies to Evaluate the Metabolism and Residue Kinetics of Veterinary Drugs in Food Producing Animals: Validation of Analytical Methods Used in Residue Depletion Studies. Available at: https://www.ema.europa.eu/en/documents/scientific-guideline/vich-gl49-studies-evaluate-metabolism-residue-kinetics-veterinary-drugs-food-producing-animals_en.pdf (accessed 29 June 2021).

VICH GL57 (2020) Studies to Evaluate the Metabolism and Residue Kinetics of Veterinary Drugs in Food Producing Animals: Marker Residue Depletion Studies to Establish Product Withdrawal Periods in Aquatic Species. Available at: https://www.ema.europa.eu/en/documents/scientific-guideline/vich-gl57-studies-evaluate-metabolism-residue-kinetics-veterinary-drugs-food-producing-species_en.pdf (accessed 29 June 2021).

Volpato, G. (2009) Challenges in assessing fish welfare. *ILAR Journal* 50(4), 329–337.

CHAPTER 12

Sea lice genomics

Cristian Gallardo-Escárate, Valentina Valenzuela-Muñoz, Gustavo Núñez-Acuña and Diego Valenzuela-Miranda

Interdisciplinary Center for Aquaculture Research (INCAR), University of Concepción, Concepción, Chile

12.1 The genome: an open book to explore the biology of sea lice

The genomic era has increased the research effort to uncover how an organism's genome, and specifically the transcriptome, is modulated after interplaying with pathogenic microorganisms, pollutants, and environmental stressors. Next-generation sequencing technologies are currently utilized for the expression profiling and the discovery of novel gene signaling pathways that are relevant for biological processes such as reproduction, growth, immune response, and metabolism. Notably, the reduction in costs and time required for these technologies have changed in a way allowing their broad application to improve plant and animal production. Indeed, one of the biggest bottlenecks related to the high variation in morphology and performance observed among cultivated organisms is now accessible at an individual level allowing the understanding of the molecular bases that control phenotypic plasticity. The scientific information derived from functional genomics studies highlighted the relevance of incorporating omics approaches in areas such as fish production, where prevalent pathogens threaten the development of sustainable aquaculture. Here, the sea lice *Caligus rogercresseyi* (Boxshall & Bravo 2000) and *Lepeophtheirus salmonis* (Krøyer 1837) are marine ectoparasites and a primary concern for salmon aquaculture worldwide (Gallardo-Escárate et al., 2019a, 2019b). In this context, the sustained growth of salmon farming has led to extensive use of delousing drugs to control lice infestation, generating risk for the emergence of resistance against different chemotherapeutics. Here, genome research has intensively studied the transcriptome signatures of lice species in response to pyrethroids and organophosphates, giving relevant molecular knowledge regarding drug metabolism in sea lice. However, while this information was revealed by canonical gene pathways involved in xenobiotic metabolism, several studies suggest an active role of non-coding RNAs as key regulators in the response to delousing drugs. This approach based on RNAs that do not code for proteins opens novel perspectives to control infestation by ectoparasites, making them interesting targets for the development of novel delousing drug strategies. However, the systematic study of the coding/non-coding RNA interactions and the genome evolution of lice species requires high-quality reference genomes. Hence the construction of chromosome-level genomes will be one of the most challenging milestones for sea lice biology research.

12.1.1 *Caligus rogercresseyi genome*

A reference genome can serve as a foundation for studies of intra- and interpopulation variation, providing insights into the biogeography, local adaptation, and responses at the molecular level of marine species facing global climate change. A reference genome also serves as a catalog for the coding sequences and other functional components of an organism's DNA. The genome is the starting point for a wide range of studies into ecological, conservation, evolutionary, and population genomics. Furthermore, as the practical and financial barriers to *de novo* assembly are reduced and high-quality references are reported for an increasing number of taxa, the field of comparative genomics will flourish, delivering new insights into the patterns and processes underpinning sea lice biology.

The chromosome-level reference genome assembly for *C. rogercresseyi* was constructed through long-read sequencing and scaffolding by Hi-C technique. Herein, eight single-molecular real-time cells in the PacBio Sequel platform were used to generate 38.32 Gb of high-quality DNA genome information. These long subreads were assembled with the Canu V1.5 package (Koren et al., 2017) using default parameters, yielding a draft genome for the sea louse equivalent to 727 Mb with contig N50 of 43,366 bp and 35.55 GC%. The draft genome was assembled with CANU into 25,608 contigs. The size of the genome assembly made by CANU was comparable with previous genome sizes reported for closely related species (Gregory, 2019; Jeong et al., 2020). However, the manual curation of a subset of contigs revealed bacterial DNA contamination. To reduce the bacterial DNA contamination, all contigs assembled in CANU were firstly filtered against NCBI prokaryotic reference sequence database and then against the reference *C. rogercresseyi* transcriptome. For the first filter, BLASTx was applied, while a mapping approach was implemented for the second filter. Taking advantage of the two filters, we removed all the contigs with a significant match to bacterial DNA, reducing the number of contigs produced by the draft genome from 25,608 to 17,711 contigs. Here, the first version of the *C. rogercresseyi* genome was estimated in 505.27 Mb. Notably, the DNA contamination produced by the natural microbiota found in *C. rogercresseyi* was round 30%. This fact remarks the importance of the microbiota in louse biology, revealing putative associations with the pathogenesis of this ectoparasite.

In vivo Hi-C is a technique that maps physical DNA-DNA proximity across the entire genome (Ghurye et al., 2019; Pal et al., 2019). The method was introduced as a genome-wide version of its predecessor, chromosome conformation capture (3C) (Dekker et al., 2002), and has been used as a powerful tool in chromosome-scale genome assembly of many animals in recent years. In this study, Hi-C experiments and data analysis on adult females were used for the chromosome assembly of the sea louse *C. rogercresseyi*. Here, two Hi-C libraries were prepared and sequenced, resulting in around 100 times coverage and around 238 million 150-bp paired-end reads. The Hi-C analysis evidenced that 46.70% of high-quality reads analyzed showed inter-contig signals or Cis-close position (less than 10 kbp on the same contig), and an additional 5.32% of sequence reads revealed a Cis-far conformation (greater than 10 kbp on the same contig). To order and orient the 17,711 contigs Hi-C reads were aligned using Bowtie2 (Langmead and Salzberg, 2012) and scaffolding performed using Proximo (Phase Genomics, Seattle, WA, USA). The Juicebox tool (Durand et al., 2016) for visual inspection and manual corrections were applied, removing 7897 scaffolds that were microbe-sized and disconnected from the rest of the assembly. Herein, the first version of chromosome-level *C. rogercresseyi* assembly was constructed as a useful genomic resource for sea louse genome research and to develop novel control strategies applied to salmon aquaculture. To visualize the scaffold's length construction, the *in vivo* Hi-C data were used to generate 21 pseudo-chromosomes assembled

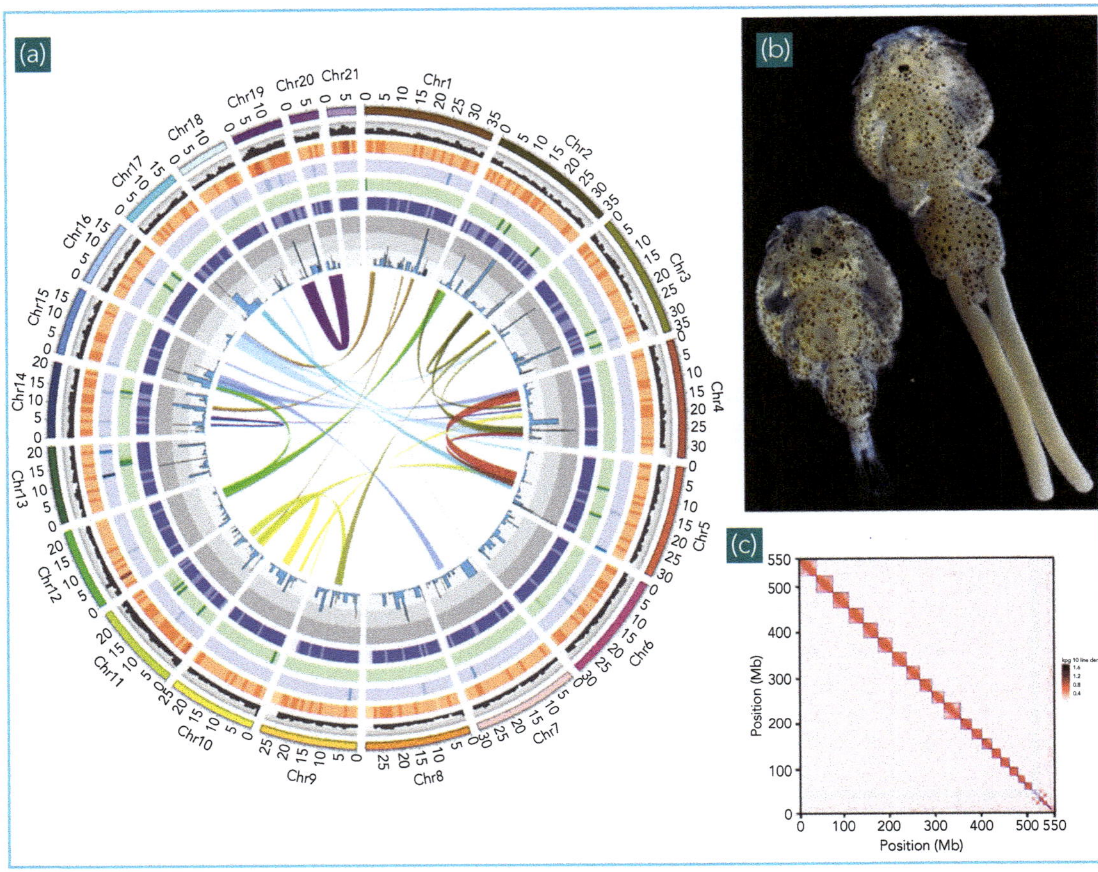

Fig. 12.1 The sea louse *Caligus rogercresseyi* genome. A) Circos plot showing the genomic features for the 21 pseudo-chromosomes. From outside to inside: GC content, gene density, lncRNA density, miRNA density, repetitive elements, transcriptome, and synteny. The chromosome size is shown in Mb scale. B) Adult female (right) and adult male (left) at ten times magnification. C) The *C. rogercresseyi* genome contact matrix using Hi-C data. The red squares represent the draft scaffold. The color bar illuminates the Hi-C contact density in the plot.

with PacBio consensus long DNA reads (Fig. 12.1). The largest scaffold was assembled from 1235 contigs, a size of 36.77 Mb. Meanwhile, the smallest scaffold was 7.98 Mb in length and consisted of 396 original contigs. Notably, the number of contigs in scaffolds was 16,931 (100% of all contigs in chromosome clusters, 95.6% of all contigs) and 505.27 Mb of genome size (100% of all length in chromosome cluster, 97.33% of all sequence length).

12.1.2 Repetitive element and non-coding gene annotation in the C. rogercresseyi genome

Repetitive elements and non-coding genes in the sea louse genome were annotated by homologous comparison and *ab initio* prediction. RepeatMasker (Bedell et al., 2000) was used for homologous comparison by searching against the Repbase database and RepeatModeler (Bao et al., 2015). According to these analyses, about 269.83 Mb of repeat sequences were annotated, which accounted for 51.9% of the whole genome. Herein, DNA transposons, LINE, and LTR transposable elements were identified. Useful genome information for population genetic studies is the identification of simple sequence repeats (SSRs) or microsatellites. The mining of SSRs revealed that the *C. rogercresseyi* genome has 441,494 SSR sequences, where 65.76% represent dinucleotide motifs. The total of SSR sequences accounted for 1.39% of the whole genome, and the genome distribution was correlated with the chromosome size (Fig. 12.2 A). Furthermore, SSRs type dinucleotides, and specifically, the motifs AC/GT were the most abundant, representing 65.76% of the total microsatellite sequences (Fig. 12.2 B). Trinucleotides and tetranucleotides were found in 32.46% of the SSRs sequences.

12.1.3 Protein-coding genes prediction and functional annotation in the *C. rogercresseyi* genome

For the identification of protein-coding genes, two approaches were employed for the sea louse genome, including homologous comparison and *ab initio* prediction. For homologous comparison,

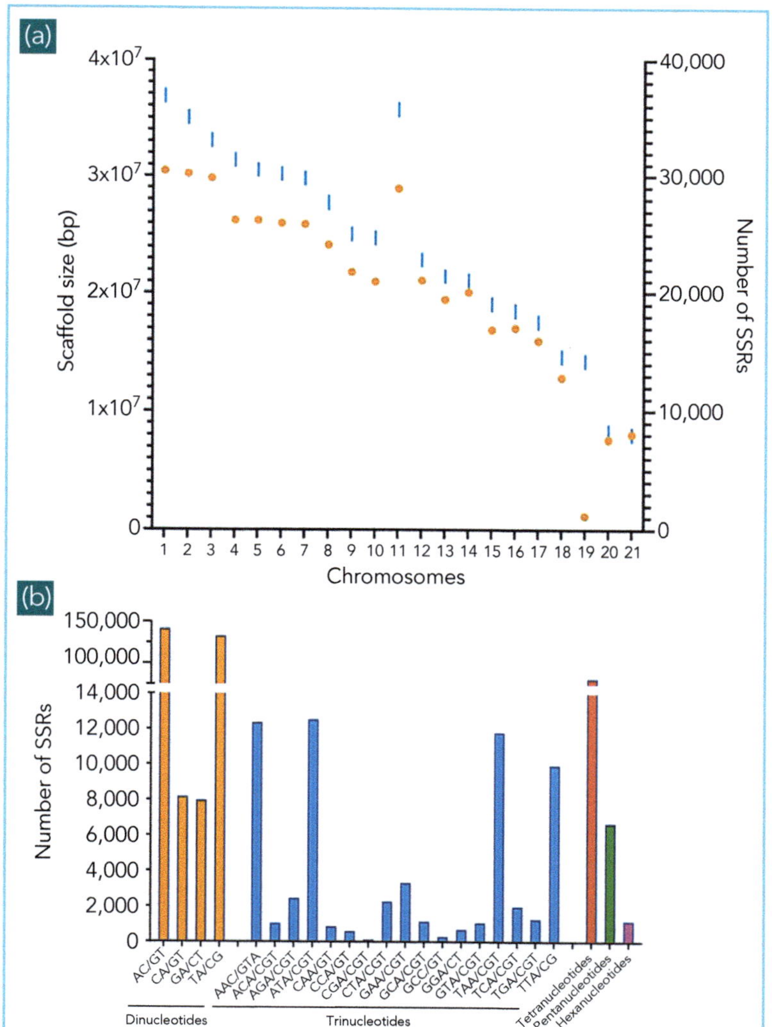

Fig. 12.2 Simple sequence repeats (SSRs) identified in *C. rogercresseyi*. A) Distribution of SSR. Vertical blue lines and orange dots represent the scaffold size (bp) and the number of SSRs, respectively. B) SSR types identified in the sea louse genome.

the protein sequences from *Caenorhabditis elegans* (GCA_000002985.3), *Drosophila melanogaster* (GCA_000001215.4), and *Daphnia pulex* (GL732539.1) genomes were extracted using the respectively published genomes and aligned against the sea louse genome using TBLASTn (e-value < 1e-5). The sequence structure of each candidate gene was predicted using GeneWise (Birney et al., 2004). For *ab initio* prediction, five tools were used to predict protein-coding genes using the Genome Sequence Annotation Server "GenSAS" (Humann et al., 2019). Specifically, Augustus, Braker, GeneMarkES, SNAP, and GlimmerM were used with default parameters. Finally, a non-redundant reference gene set was generated using EvidenceModeler (EVM) and PASA2 tools (Haas et al., 2008). Taken together 25,510 protein-coding genes were identified. Additionally, 437 tRNAs were predicted using tRNAscan-SE, and 39 rRNA genes were annotated using RNAmmer via GenSAS. For non-coding RNAs with putative regulatory roles, 5774 miRNAs and 6308 long-ncRNAs were identified and annotated within the *C. rogercresseyi* genome using transcriptome sequencing data. For functional annotation, the predicted proteins within the sea louse genome were searched by homology against four databases of InterPro (Finn et al., 2017), GO (Harris et al., 2004), KEGG KO (Kanehisa et al., 2014), and Swissprot (Uniprot, 2019). Overall, 88.05%, 68.85%, 64.02%, and 91.02% of genes matched entries in these databases, respectively.

12.2 Unraveling the sea lice metagenomics through genome sequencing

Besides the well-known negative impact of lice in salmon aquaculture, novel evidence suggests the presence of a large and diverse microbiota associated with *C. rogercresseyi* (Gonçalves et al., 2020). However, the biological role of these microorganisms or their potential importance in salmon aquaculture remains largely elusive. Over the last decade the idea that microbes play pivotal roles in the development of almost all living organisms has gained consensus (Hahn and Dheilly, 2016). Microbes can drive physiological and evolutionary changes due to their interaction with their host (Rosenberg and Zilber-Rosenberg, 2011). Particularly in parasites, the microbiota can actively participate during the infective process by producing toxins, redirecting the host immune system, or increasing the parasite's virulence (Chaston et al., 2011; Blaxter and Koutsovoulos, 2014; Dheilly et al., 2015; Hahn and Dheilly, 2016). Herein, the parasite's microbiota has been proposed as a target to develop alternative therapeutic strategies to cope with infestations (Hoerauf et al., 1999; Fichorova et al., 2012). Vector-borne diseases are another major issue regarding parasite-associated microbes in arthropods, responsible for transmitting viruses, bacteria, and protozoa among vertebrate hosts (Kalluri et al., 2007). Copepods are widely colonized by bacteria (Hansen and Bech, 1996; Huq et al., 1983; Dumontet et al., 1996), among which pathogenic species are usually identified (Vezzulli et al., 2010; Heidelberg et al., 2002). Known fish pathogens have been found in other caligid parasitic copepods, such as *Vibrio* spp. in *C. lalandei* (Sepulveda et al., 2017). Likewise, it has been suggested that the sea louse *L. salmonis* could act as a vector of several diseases, including infectious hematopoietic necrosis virus (IHNV), the infectious salmon anemia virus (ISAV) (Barker et al., 2019), and furunculosis (Jakob et al., 2011; Nese and Enger, 1993; Nylund et al., 1993). The study of Gonçalves et al. (2020) identified bacterial pathogens present in *C. rogercresseyi*-associated microbiota, including *Vibrio*, *Tenacibaculum*, and *Aeromonas* species (Gonçalves et al., 2020). However, these species' genomic features are poorly understood, hindering the biological role of the sea lice microbiota and its potential impact on salmon aquaculture.

Previous efforts have been made to understand the microbial communities associated with *L. salmonis* (Barker et al., 2009; Dalvin et al., 2020). However, these approaches were technically limited based on the need to isolate and culture each bacterial species that required identification or the lack of taxonomical resolution at the species or even family level. Long nanopore reads and computational tools have recently allowed the full-length 16s rRNA sequencing of the sea lice-associated microbiota, evidencing the geographical variations and the core microbiota at species-specific levels (Gonçalves et al., 2020) (Fig. 12.3). Here, chromosome proximity ligation (Hi-C) coupled with long-read sequencing is a valuable tool to reconstruct individual bacterial genomes, uncovering novel insights associated with the sea lice microbiome. Thus, the results evidenced the assembly of almost complete genomes from a complex sample with low degrees of cross-contamination. The complete genomes belonged to species from the *Tenacibaculum* and *Dokdonia* genera, which were previously described as abundant components of salmon and sea lice microbiota (Barker et al., 2009; Dalvin et al., 2020; Gonçalves et al., 2020). As expected, our results evidence that each bacterial genome cluster's completeness degree might be directly related to the abundance of each bacterium within the sample.

Beyond identifying bacterial pathogens, the approach allowed the characterization of 4911 coding genes in the *T. ovolyticum* genome. This key information can be useful for the application of reverse

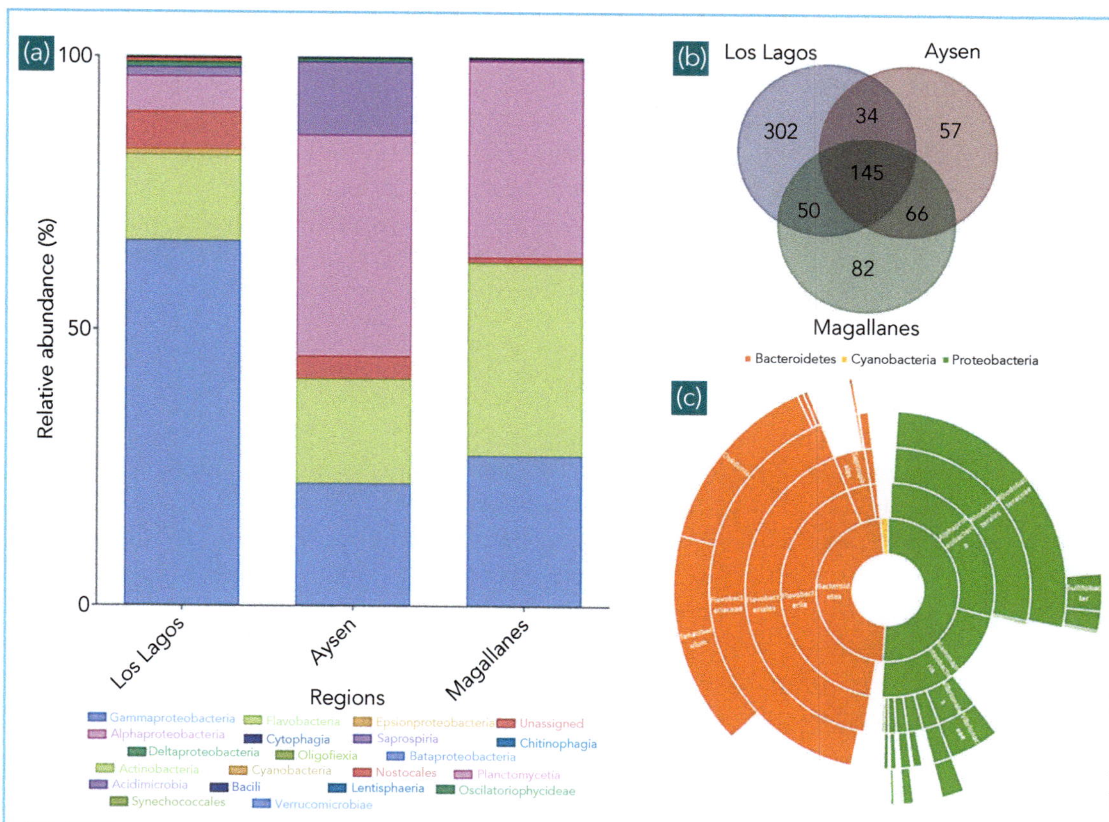

Fig. 12.3 A) Relative abundance of *C. rogercresseyi*-associated microbiota in Chilean aquaculture localities. B) *C. rogercresseyi* core microbiota. Venn diagram representing the taxonomical features identified in Chile. C) Taxonomical distribution of taxa included in the 145 taxa shared by sea lice microbial community of the three regions based on the average relative abundance.

vaccinology, which has been used to cope with vector-borne parasite diseases (Lew-Tabor and Valle, 2016). Using genomic information allows the rapid discovery of potential antigens, being a promising methodology for developing new fish vaccines (Ma et al., 2019). Virulence factors were also identified in the *T. ovolyticum* genome, including the Dot/Icm type IVB secretion system, 60K heat shock proteins (HtpB), and catalases (katA) genes, among others, which have been suggested as key factors during the infection of different fish pathogens (Gomez et al., 2013; Perez-Pascual et al., 2017; Valenzuela-Miranda et al., 2020; Yu et al., 2019). Thus, this genomic information availability can also be used to infer pathogenic mechanisms employed by potential fish pathogens present in *C. rogercresseyi* microbiota.

It is known that arthropod-associated microbiota play pivotal roles in their host biology, where metabolism is crucial (Esposti and Romero, 2017; Zientz et al., 2004). As one of the most represented genome clusters in sea lice microbiota, the authors suggest further investigation of the possible roles of *Dokdonia* sp. on *C. rogercresseyi* biology (Gonçalves et al., 2020). Recent evidence has shown that *Dokdonia* sp. constitute part of the core microbiota associated with sea lice obtained from different geographical areas in Chile, and it was argued that this genus might fulfill roles during the infective process of sea lice because of the secretory capacity of bioactive compounds from this bacteria (Gonçalves et al., 2020). Notably, proximity ligation and deconvolution analysis allowed decoding the *Dokdonia* sp. and *C. rogercresseyi* genomes in parallel (Gallardo-Escárate et al., 2021), finding novel molecular interactions and putative mechanisms of metabolic mutualism in sea lice. The results evidenced that *C. rogercresseyi* lacks essential biosynthetic genes associated with valine, leucine, and isoleucine biosynthesis and degradation, which were in contrast present in the *Dokdonia* sp. genome. The absence of these genes in the *C. rogercresseyi* genome suggests that sea lice cannot biosynthesize these amino acids; therefore, they should be obtained through feeding. However, another possibility is that *C. rogercresseyi* might explode the metabolic machinery present in its core associated microbiota to obtain these amino acids. Metabolic interdependence is a widespread phenomenon between bacteria and arthropods that has been previously reviewed by Zientz et al. (2004).

12.3 Life cycle transcriptomes of the sea lice

The life cycle of sea lice comprises both free-swimming planktonic and parasitic phases. The planktonic phase consists of two nauplii stages and one copepodid stage, the latter being the first infective stage. The planktonic stages of *L. salmonis* have around seven days to find a host, although this time is temperature, and species, dependent (Stucchi et al., 2011). The copepodid finds its host by following behavioral traits, including gradients of host-derived semiochemicals (for example, kairomones), and vibrations of host origin (Heuch et al., 1995; Heuch and Karlsen, 1997; Bailey et al., 2006). The copepodid attaches to the fish skin using its hooked pair of antennae; this process is mediated by chemoreceptors in the copepodid's antennules (Gresty et al., 1993). During its developmental process, the parasite gradually increases in size and is always attached to the fish. Once on the fish, the copepodid molts into chalimus I and develops a frontal filament to attach itself to the fish (Johnson and Albright, 1991; González and Carvajal, 2003). The *C. rogercresseyi* parasite molts into three more chalimus stages (II, III, and IV) while attached to the fish, while in the case of *L. salmonis* (González and Carvajal, 2003), recent research has reported only two chalimus stages (I and II) before the louse develops to the pre-adult stage (Hamre et al., 2013). The *C. rogercresseyi* species lacks these pre-adult stages, maturing from chalimus IV into adult stages (González and

Carvajal, 2003). Sexual dimorphism can be seen at the chalimus III stage; however, it is clearer at chalimus IV. In the case of *L. salmonis*, two pre-adult stages have been described after the chalimus stages and before the adult stage (Johnson and Albright, 1991; Boxshall and Bravo, 2000; González and Carvajal, 2003).

Major challenges facing transcriptomic research in non-model organisms are increasing the speed and accuracy in discovering new genes and metabolic pathways, as well as determining how gene transcription variations are regulated by specific DNA polymorphisms. Understanding the transcriptome is essential for interpreting the functional elements of the genome, for revealing the molecular constituents of cells and tissues, and for understanding complex biological processes such as growth, reproduction, and immune response (Wang et al., 2009). Next-generation sequencing (NGS) technologies offer the opportunity to generate genome-wide sequence data sets for a reasonable cost and time (Vera et al., 2008; Marguerat and Bahler, 2010; Wilstermann and Vidal, 2013). Although these powerful and rapidly evolving technologies have only been available for a few years, they are already making substantial contributions to the understanding of genome expression and regulation under different conditions. A popular application of NGS is species transcriptome generation, which affords direct access to the coding sequences of many genes and information on their relative expression levels (Bellin et al., 2009; Wilhelm and Landry, 2009; Neira-Oviedo et al., 2011). NGS transcriptome data analysis is therefore a useful source for mining molecular markers such as SNPs and EST-SSRs (Sahasrabudhe et al., 1998; Everett et al., 2011; Seeb et al., 2011; Ritdachyeng et al., 2013).

The transcriptome analysis can be applied to identify relevant gene expression profiles across early larval development. It could be hypothesized that this developmental phase is representative of the major physiological changes during the life cycle of copepods. For instance, a study related to peptidergic signaling in *C. finmarchicus* reported that the highest expression levels from six stages (embryo, early nauplius, late nauplius, early copepodid, late copepodid, and adult) are seen in the naupliar and copepodid stages, while the lowest levels are present in embryos and adult females (Christie et al., 2013b). Specifically, in the copepodid stage, host-seeking behavior has been displayed by *L. salmonis* during its infectious stage, including moving towards river mouths and maintaining location in haloclines during salmon migrations (Brooks, 2005). Consequently, the copepod must be able to cope with abiotic stress conditions (Sutherland et al., 2012) and respond to the inflammatory defense mechanisms at the site of salmon parasite attachment (Tadiso et al., 2011). Furthermore, relatively little is known concerning sex differentiation and its endocrine control in crustaceans, and most available data have been obtained in decapods (Rodriguez et al., 2007). However, transcriptome sequencing studies have facilitated the discovery of novel sex-related genes, which thus far have suggested pivotal transcriptional differences between female and male adults (Eichner et al., 2008; He et al., 2012).

The reference *C. rogercresseyi* transcriptome revealed stage-specific gene clusters (Gallardo-Escárate et al., 2014), which were differentially expressed between the nauplius I–II, copepodid, and chalimus stages (Fig. 12.4). Interestingly, some upregulated genes were mainly associated with metalloproteinase, arginine kinase, and cuticle protein, which evidence participation in the digestion of intake proteins, tissue development, cuticle remodeling, and specific cleavage events to activate or inactivate proenzymes and bioactive peptides (Singh, 2010). Concerning nervous system development in copepodids, some relevant genes such as nicotinic acetylcholine receptor, flotillin, synaptotagmin, allotostatin, frequenin, and opsin were highly overexpressed (Gallardo-Escárate et al., 2014). These results are congruent with previous studies of transcriptome profiles reported

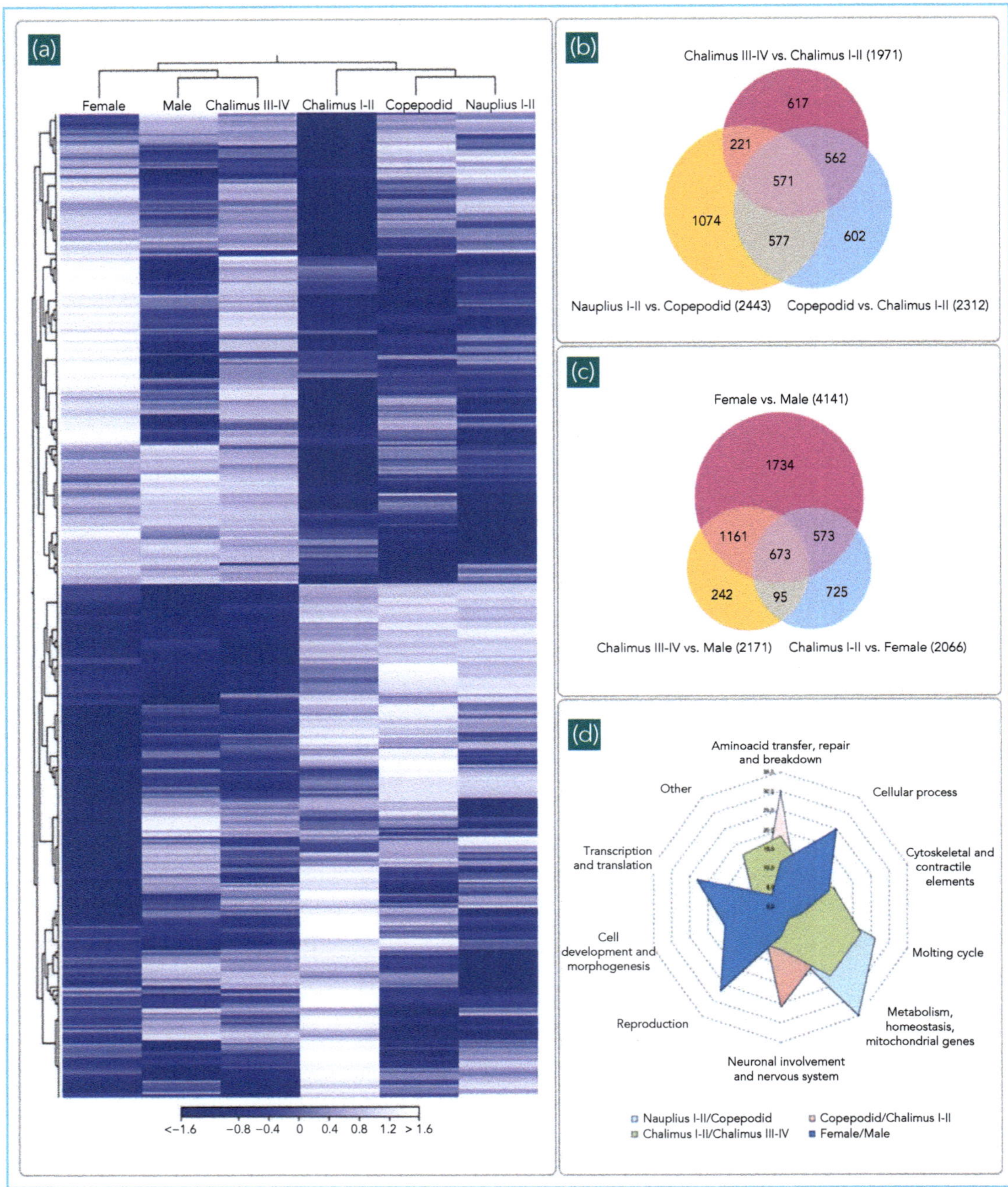

Fig. 12.4 The reference *Caligus rogercresseyi* transcriptome. A) Expression profiling of coding genes during the life cycle. The heatmap was based on transcripts per million (TPM) calculation and hierarchical clustering on Manhattan distances with average linkage. White colors mean upregulated coding genes, blue colors downregulated genes. B) to C) Venn diagrams showing shared and unique genes expressed among the six developmental stages. D) GO enrichment of stage-specific genes annotated for key biological processes differentially expressed. The radar plot represents the comparison between two developmental stages according to the *C. rogercresseyi* life cycle.

for copepodid stages (Christie et al., 2008, 2013a, 2013b). Furthermore, a study by Wilson and Hartline (2011) demonstrated high peripheral and central nervous system development in individuals transitioning from the nauplius to copepodid stage. The upregulation of allotostatin could be associated with the regulation of juvenile hormone production, or, more interestingly, with recent findings where the activation of neurons, or neuroendocrine cells, that expressed the neuropeptide allotostatin modulated feeding behavior in *Drosophila*, including increased food intake and enhanced behavioral responsiveness to nutrients or molecular clues (Hergarden et al., 2012). It is important to note that these effects on feeding behavior could be related to changes induced by the start of the parasitic phase in the salmon louse. Furthermore, investigations of opsin function outside of vertebrate systems have long been focused on arthropod visual pigments (Porter et al., 2007), indicating that copepods possess a sensory apparatus sensitive to different wavelengths that could have implications during the host-finding process, especially in the copepodid stage (Aarseth and Schram, 1999).

For *C. rogercresseyi* chalimus stage, trypsin, alpha-amylase, and carboxypeptidase genes were overregulated (Gallardo-Escárate et al., 2014). Peptidases from the different families may be involved in a wide range of cellular and biological processes, thus making it difficult to infer specific functions across salmon louse development. Host blood has been reported as a food component for the salmon louse *L. salmonis* (Brandal et al., 1976). Blood degradation in several hematophagous organisms has been shown to require the catalysis of several peptidases (Kvamme et al., 2004, 2005). Of the regulated peptidases in the present study, the most overregulated was trypsin, a secretory endopeptidase within the serine protease superfamily. This superfamily includes important digestive enzymes that constitute a major part of digestive fluids and act as activators of other digestive enzymes. These results are congruent with previous reports on the interaction between parasitic copepods and salmon hosts (Fast et al., 2003; Firth et al., 2000).

In addition, genes showing differential transcription from female and male adults were highly modulated (Gallardo-Escárate et al., 2014). For female adults, sex-related genes such as vitellogenins and estradiol 17-beta-dehydrogenase were identified (Gallardo-Escárate et al., 2014). However, effects of the vertebrate-like steroid hormones on reproductive processes, such as oocyte maturation in crustaceans, remain unresolved. Vitellogenins are the major yolk proteins in most invertebrates, and several different vitellogenins typically give rise to vitelline granules in mature eggs (Belles and Maestro, 2005). The role of multiple vitellogenin genes in some organisms, such as insects, is unknown (Tufail and Takeda, 2008) despite those proteins with domain structures similar to vitellogenins also being involved in other developmental processes, such as in the regulation of osmolarity, immunity, and clotting (Sappington and Raikhel, 1998). The present data showed a wide diversity of vitellogenins, including LsVit1 and LsVit2 as reported in *L. salmonis* (Dalvin et al., 2011), and several vitellogenin-like proteins. Moreover, genes associated with cell development, including homologues of vasa, homeobox, Argonaute, cell division protein kinase, and centromere-associated protein, were also specifically expressed in female adults. For salmon louse males, transcription activity related to cytoskeletal and molting cycles, as well as with sperm proteins, was mainly overregulated. The *C. rogercresseyi* transcriptome study observed upregulation of actin, troponin, myosin, and cuticle proteins, an important part of cuticle formation during the final molt (Gallardo-Escárate et al., 2014).

In recent years, the advent of novel biotechnological techniques has allowed for exploring new strategies of pest control in model and non-model species (Christiaens et al., 2010; Horowitz and Ishaaya, 2013; Van Leeuwen et al., 2013; Hui et al., 2013). However, improving the current understanding of the biological and molecular bases of reproduction in sea lice is required, especially given that manipulating the mechanisms controlling sex determination could lead to novel methods for sea lice control (Carmichael et al., 2013). The *C. rogercresseyi* transcriptome identified sex-related transcripts and reported the occurrence of SNPs in several differentially expressed contigs that annotated for sex-related genes. These results showed a slight imbalance in favor of the male sequences. This tendency has been previously reported in the sequences of other organisms such as the lake sturgeon, *Acipenser fulvescens*, (Hale et al., 2010) where 3332 transcripts were expressed in females and 4008 were expressed in males; in the nematode *Oesophagostomum dentatum* (Cantacessi et al., 2010), where 3451 transcripts were found in females and 10,344 in males; and in the abalone, *Haliotis rufescens*, (Valenzuela-Muñoz et al., 2012) where 1296 and 2254 transcripts were found to be significantly expressed in female and male tissues, respectively.

Gene ontology (GO) annotations for the *C. rogercresseyi* contigs revealed a significant difference in the number of expressed transcripts between male and female individuals, where there was an overall predominance of female over male transcripts. This gender-biased transcriptional activity is probably associated with the biosynthesis required for accumulating lipids and manufacturing eggs, as well as with the enlargement of the genital segment and abdomen that occurs during the maturation of females in *C. rogercresseyi* (González and Carvajal, 2003). Coinciding with the present results, Eichner et al. (2008) described a female-specific upregulation of genes during egg production in the salmon louse *L. salmonis*. The transcript expression pattern obtained in the present work closely resembles the GO results obtained by Ning et al. (2013) using a 454 GS FLX platform for the copepod *Calanus sinicus*, where the most represented categories also corresponded to binding and catalytic activity (molecular functions), cell parts (cellular components), the metabolic process, biological regulation, and cellular processes (biological processes).

The *C. rogercresseyi* transcriptome was examined to identify those genes that might be functionally associated with sex determination and sex differentiation. The detected genes included sex-determining region Y chromosome (*SRY*)-box containing protein (Sinclair et al., 1990), *Dmrt3* (Raymond et al., 1998; Kopp, 2012; Smith et al., 1999), *FOXL2* (Adell and Muller, 2004; Shimeld et al., 2010; Ma et al., 2012; Veitia, 2010), *FEM1* (Doniach and Hodgkin, 1984; Hodgkin, 1987; Spence et al., 1990), *VASA* (Lasko and Ashburner, 1988), and *17 beta-HSD* (Mindnich et al., 2004), among other potential candidates. In addition, other putative sex-related genes were also found for the first time in *C. rogercresseyi*, such as *ZFY1* (Nagamine and Carlisle, 1997), *Argonaute* (Modzelewski et al., 2012), *Piwi* (Deng and Lin, 2002; Cox et al., 2000), *SPAG6* (Sapiro et al., 2000), and FmrFamide (Walker et al., 2009). Since conclusive data regarding the mechanism of sex determination and the molecular control of reproduction in *C. rogercresseyi* is currently not available, further studies are required to understand the molecular function of the putative reported sex-related genes.

12.3.1 The aquaporin family genes in C. rogercresseyi

The *in silico* scrutiny of the *C. rogercresseyi* transcriptome allowed the identification of ten putative sequences with close homology with the aquaporin family of proteins (Farlora et al., 2016), as well as reported aquaporin sequences in *L. salmonis* (Stavang et al., 2015), a related parasitic copepod from the same Caligidae family (Fig. 12.5 A). To maintain a coherent nomenclature with up-to-date existent phylogenetic analyses in this taxon, they were termed Aqp12L1, Aqp12L1_v1, Aqp12L2, Glp1_v1, Glp2, Glp3_v1, Glp3_v2, PripL_v1, PripL_v2, and Bib. Out of these ten sequences, eight exhibited complete ORFs and were selected for posterior structural predictions. The *C. rogercresseyi*

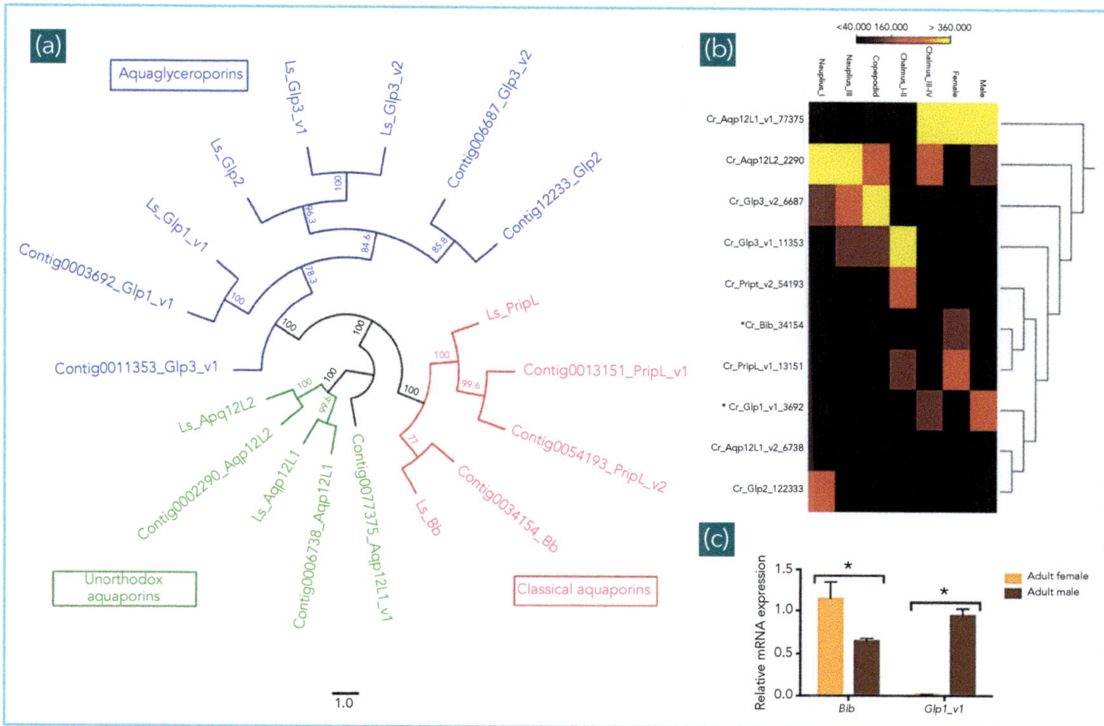

Fig. 12.5 A) Phylogenetic tree for aquaporin family proteins in the Caligidae family (Copepoda: Siphonostomatoida). The numbers at each node indicate the % of bootstrapping using the Jukes-Cantor genetic distance model and the neighbor-joining method. B) Transcriptomic expression of contigs annotated as aquaporin family genes at different developmental stages in *C. rogercresseyi*. C) Gene expression of the sequences marked with an asterisk *Cr-Bib and *Cr-Glp1_v1 validated using qPCR in adult female and male individuals.

aquaporin family ORFs were between 765 and 1989 bp in length, coding for proteins of 255–663 amino acids. Alignments of the sequences as well as secondary and tertiary structural predictions revealed that *C. rogercresseyi* aquaporins share a common general structure, comprised of six transmembrane domains connected by five loops (A–E), with the N- and C- termini located in the cytoplasm, as has been described in mammalian aquaporins (Krane and Goldstein, 2007; Verkman and Mitra, 2000). Highly conserved NPA (Asn-Pro-Ala) motifs were also detected in the segments H3 and H7, forming short hydrophobic helices that penetrate halfway into the membrane from opposite sides. The largest protein was Bib, having a deduced length of 633 amino acids, an isoelectric point (pI) of 8.63, and a predicted molecular weight of 71.1 kD. It harbors a long N-terminus of 131 amino acids and an even longer C-terminus of 336 amino acids. Its subdomain structure is very similar to both *Drosophila* and *L. salmonis* Bib (Rao et al., 1990; Stavang et al., 2015). The smallest aquaporin in *C. rogercresseyi* was PripL_v1 with 255 amino acids, an isoelectric point (pI) of 8.87, and a predicted molecular weight of 27 kD.

To get insight into the expression patterns of the previously identified aquaporin family genes during the ontogeny of the sea lice *C. rogercresseyi*, *in silico* transcription expression using RNA-seq data from different developmental stages and sexes were evaluated (Farlora et al., 2016). Aquaporin transcripts were identified from nauplius I, nauplius II, copepodid, chalimus I–II, chalimus

III–IV, adult female, and adult male samples. Overall, the aquaporin genes exhibited a developmentally regulated expression, with the different genes showing distinctive expression patterns in the gene expression heatmap (Fig. 12.5 B). At early pre-infective stages (from nauplius I to copepodid) Aqp12L2 showed a high expression that was reduced in the chalimus I–II stage, and then increased again in chalimus III–IV. In adults, the expression of this transcript was slightly higher in males as compared with female samples. Glp3_v2 gene was also preferentially expressed at pre-infective stages, exhibiting an increasing pattern of expression from nauplius I until copepodid, and then decreasing in the more advanced developmental stages. Similarly, Glp3_v1 gene was expressed at nauplius II and copepodid stages, with a peak of expression at chalimus I–II stage, decreasing its expression afterward. Interestingly, Aqp12L1_v1 transcript showed higher expression values at chalimus III–IV and adult stages in both sexes, but reduced expression at pre-infective and chalimus I–II stages. In addition, a cluster of genes comprised of PripL_v2, Bib, PripL_v1, Glp1_v1, and Aqp12L1 showed comparatively lower expression mainly restricted to post-infective stages, except for Glp2 which was mostly expressed at nauplius I stage. Importantly, within this cluster of genes, Glp1_v1 exhibited a male-biased pattern of expression in adults as has been reported earlier using RNA-seq data for the same transcript which was originally described as *Cr-Aqp3*. In the present work, we further confirmed this sex-specific expression by qPCR analysis (Fig. 12.5 C). Stavang et al. (2015) analyzed the expression of the *L. salmonis* Glp1_v1 homolog, which was reported in pre-adult II and adult males only, as detected by RT-PCR using specific primers. In this species, the authors found a second shorter splice variant of the gene, Glp1_v2, expressed in pre-adult and adult stages of both sexes. Notably, a homolog for the Glp1_v2 gene in *C. rogercresseyi* transcriptome was absent (Farlora et al., 2016). Another gene exhibiting sex-specific expression was Bib. The expression analysis revealed a female-biased expression pattern in adults using both *in silico* and qPCR analyses (Fig. 12.5 C). Interestingly, this gene has been associated with the morphogenesis of the ovarian follicle cells in *Drosophila* (Ruohola et al., 1991), therefore this sex-biased gene might be also involved in the oogenesis of *C. rogercresseyi*.

The functions of aquaporins in the reproductive system and gametogenesis of vertebrates have been previously reviewed (Huang et al., 2006; Skoblina, 2008), and they have been previously associated with fluid homeostasis in teleosts (Chauvigne et al., 2011).

12.4 Genomics of host recognition in the sea lice

For ectoparasites with free-living life stages, such as the salmon louse *L. salmonis*, host-finding is a process that is critical to complete the life cycle (Vinson, 1975). Thus, there has been strong adaptive-evolutionary pressure for this species to develop reliable and sophisticated links to their hosts (Lindsey and Lasker, 1974). Salmon lice use a combination of mechanical, visual, and chemical sensory cues to identify and locate potential hosts (Fields et al., 2018). However, chemical signals may confer the highest specificity for host recognition (Mordue (Luntz) and Birkett, 2009). The detection of chemical cues released by the host organism (kairomones) provides an adaptive advantage to the parasite. To detect kairomones, salmon lice have evolved chemical reception systems (Genna et al., 2005; Mordue (Luntz) and Birkett, 2009), including chemical receptors found on the antennules located in the first antennae (Ingvarsdóttir et al., 2002). Antennules are responsive to chemical cues associated with salmon hosts (Hull et al., 1998; Fields et al., 2007). This suggests the presence of kairomone receptor proteins in these olfactory structures. While general suites of chemical signals have been previously investigated (Fields et al., 2007), to our knowledge, specific compounds that invoke physiological and behavioral changes in the salmon lice are poorly studied.

The first description of kairomone receptor proteins in sea lice species was the finding of ionotropic receptor (IR) genes in *C. rogercresseyi*, which were responsive to the presence of kairomone-blocking compounds (Núñez-Acuña et al., 2014a, 2016a). These receptors were also linked to neurotransmission gene pathways and had expression patterns related to the presence of antiparasitic xenobiotic drugs (Núñez-Acuña et al., 2016c). Furthermore, new IR genes in the salmon louse, *L. salmonis*, that are related to olfactory transduction have been described, as demonstrated by *in situ* hybridization and RNA-interference techniques (Komisarczuk et al., 2017). These IR genes are the only receptor proteins with known functions related to olfaction that have been found in any marine invertebrate species (Stepanyan et al., 2006; Corey et al., 2013). The role of these receptors in olfaction was predicted from their structural similarity to IRs found in insects that have known olfactory function (Touhara and Vosshall, 2009). Despite the diversity of genes associated with the chemosensory system in the sea lice, the kairomones that they can identify remain poorly known.

Salmon odor containing unidentified chemical cues triggers altered swimming behavior (related to host-seeking) in *L. salmonis* (Devine et al., 2000; Ingvarsdóttir et al., 2002), and in *C. rogercresseyi*, which was attracted by the presence of salmon host odor (Pino-Marambio et al., 2007). *L. salmonis* also exhibit chemoreceptive activity in the presence of salmon extracts from the mucus in the skin tissue, which contains its odor (Fields et al., 2007). However, the specific chemical cues that sea lice detect were not identified in these studies. The first attempt to identify host-derived attractant molecules was the annotation of cDNA libraries from the specific sites where the salmon is being infected by sea lice, which showed upregulation of various peptides and proteins (Tasumi et al., 2015). Peptides related to the antimicrobial peptide (AMP) class of cathelicidins appear to be involved in the sea lice host-recognition system (Núñez-Acuña et al., 2016c). These AMPs have been tested by *in vitro* assays, producing significant effects on frontal filament formation (which favor sea lice infection) and activation of chemosensory-related genes.

Núñez-Acuña et al. (2018) exposed *L. salmonis* copepodids to Atlantic salmon cathelicidin-2 peptide (Cath-2), which triggered a broad differential transcriptomic response that was concentration-dependent (Fig. 12.6 A–C). This also implied that more biological processes were differentially expressed at the highest Cath-2 concentration. This is consistent with the assumption that Cath-2 is a host signal for *L. salmonis* – a high concentration would indicate host proximity and the parasite would respond by activating numerous gene pathways. These results are analogous to those reported in other species, especially insects, which exhibit a broad transcriptomic response when exposed to kairomone (Gardiner et al., 2008; Hoang et al., 2015; Eyres et al., 2017). In aquatic environments, most of the studies on gene expression in response to chemical cues have been conducted on *Daphnia* spp., with results analogous to ours: an upregulation of various genes (Schwarzenberger et al., 2009; Miyakawa et al., 2010). Most of the differentially expressed transcripts with GO annotation for molecular function in our study belonged to genes related to the formation of the cuticle structure. This is consistent with a previous study in which copepodids of the sea lice species *Caligus rogercresseyi* were exposed to Cath-2, and cuticle-related genes were upregulated (Núñez-Acuña et al., 2016c). These genes are related to the molting events (Chávez-Mardones et al., 2016), suggesting that the animals are activating these genes to accelerate their development in response to the detection of a chemical cue indicating that a host is nearby.

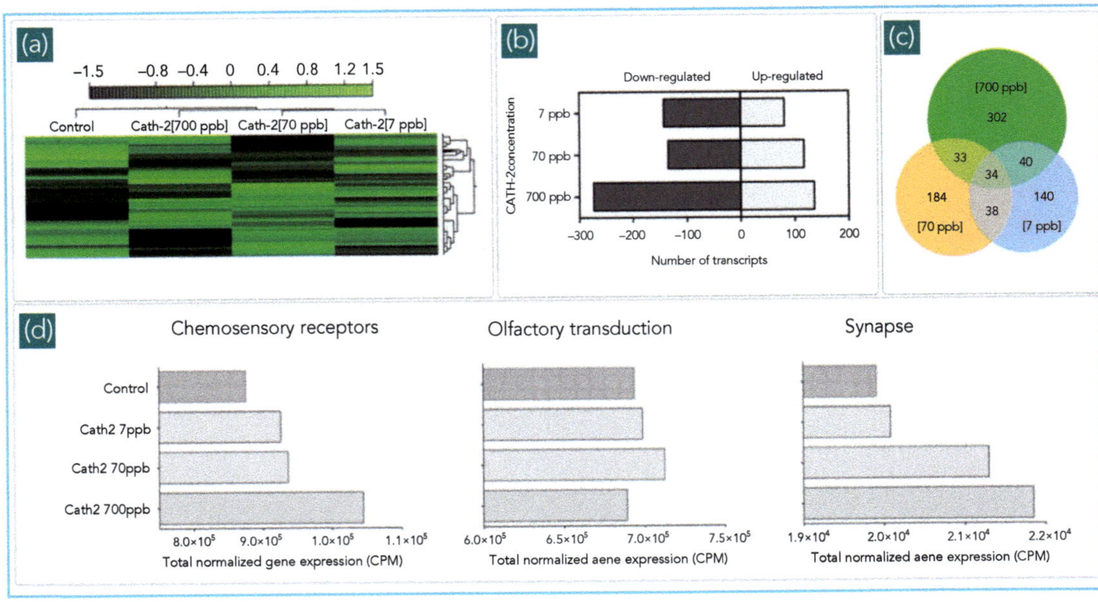

Fig. 12.6 Transcriptomic analyses of *Lepeophtheirus salmonis* copepodids exposed to different concentrations of cathelicidin-2. A) Heatmap showing gene expression according to the concentration of the peptide. B) to C) The number of significantly expressed transcripts (|fold change| > 4; $p < 0.05$) in copepodids at each of the concentrations of the peptide against the control group. D) Top-enriched gene ontology (GO) terms according to normalized expression values (CPM) of selected transcripts.

According to GO criteria of cellular components, Núñez-Acuña et al. (2018) identified that most of the differentially expressed genes code for proteins located in the integral component of the cellular membrane (Fig. 12.6 D). This is consistent with gene expression results because chemosensory receptors, such as ionotropic receptors, are in this cellular compartment. Congruently, a concentration-dependent gene expression level was observed in these chemosensory-related genes, being consistent with a previous study in the sea lice species *C. rogercresseyi* exposed to the same peptide (Núñez-Acuña et al., 2016c). An upregulation of the ionotropic receptor 25a gene in a Cath-2 concentration-dependent trend was also observed. This gene is one of the most studied olfactory receptors in sea lice species and other invertebrates (Rytz et al., 2013). On the other hand, synapse-related genes were upregulated also in a Cath-2 concentration-dependent trend. These molecular findings were also consistent with electrophysiological and swimming behavior analyses, strengthening the hypothesis that Cath-2 peptide is acting as an attractant chemical involved in the host-recognition process of sea lice species. Notably, a study conducted by Núñez-Acuña et al. (2019) using double-strand RNA interference (dsRNA) oligonucleotides for gene silencing of IR25a revealed behavioral responses associated with host recognition after lice exposure to Cath-2 and light, respectively. The dsIR25a group decreased expression levels of IR25a by more than sevenfold regarding the control group. This group was also 26% slower than the control group (control swimming speed was 69 mm/s, while the treated group was 51 mm/s). Since the swimming activity of salmon lice copepodids is associated with the activation of the chemosensory system, these results indicate that the *L. salmonis* chemosensory perception system was not fully activated due to gene silencing.

12.5 The hidden nature of the non-coding RNAs in the sea lice biology

Advances in molecular biology, primarily based on next-generation sequencing technologies, have tremendously changed our understanding of how the transcriptome is modulated and how the non-coding information is pivotal for key biological processes. One of these non-coding RNAs, denoted microRNAs (miRNAs), were first reported to regulate development timing in the nematode *Caenorhabditis elegans* (Lee et al., 1993). Since then, miRNA sequences have been reported in a wide variety of organisms, from animals to viruses, and these sequences are already publicly available through the miRBase database (Griffiths-Jones et al., 2008; Kozomara and Griffiths-Jones, 2014). The main function of miRNAs is regulating gene expression at the post-transcriptional level by adding switch controls to complex cell signaling pathways associated with developmental timing. For instance, miRNAs have been shown to be essential in obtaining optimal expression levels of genes and in regulating the transcripts of target genes involved in the different stages and sites of ontogeny (Kim and Nam, 2006). However, although research has been conducted on miRNAs for 20 years, much is still unknown (Ambros, 2004), such as in regards to miRNA biogenesis and the dynamic interactions of miRNAs with target genes. Moreover, one of the major challenges facing future miRNA research will be relating the discovery of new miRNAs to specific biological processes.

The biogenesis of miRNAs in animals is a complex and multistep process starting in the nucleus, passing through many post-transcriptional modifications, and ending in the cytoplasm. The canonical maturation pathway (Makarova and Kramerov, 2007; Zhang et al., 2007; Chawla and Sokol, 2011), which is similar to protein-coding genes, initiates at transcription (mostly by RNA polymerase II or III), thus generating a primary miRNA (pri-miRNA). The pri-miRNA is characterized by a hairpin RNA structure recognized by the nuclear RNAse III enzyme Drosha and its cofactor DGCR8 in vertebrates or Pasha in invertebrates. These proteins work within a complex of several proteins known as the microprocessor. The microprocessor cleaves the pri-miRNA to generate a shorter hairpin about 70 nucleotides in length called pre-miRNA. Pre-miRNA, an intermediate stage, is exported from the nucleus to the cytoplasm via exportin-5. In the cytoplasm, a second RNAse III enzyme, Dicer, makes the pair of cuts that defines the other end of the miRNA, generating a siRNA-like duplex known as the miR/miR* duplex. This duplex is incorporated into the RNA inducing silencing complex (RISC) in which the Argonaute-1 protein is the main component. The miRNA* passenger strand is then degraded, and the miRNA guide strand leads the RISC complex to the target mRNA. Assembly of the mature, single-stranded miRNA (~22 nucleotides in length) from the duplex into the RISC completes miRNA biogenesis. Recently, Valenzuela-Miranda et al. (2015) identified the biogenesis pathway of miRNAs in the marine copepod ectoparasite *C. rogercresseyi* evidencing that RNA silencing mechanisms can be found during the life cycle of this sea louse species (Fig. 12.7).

So far, transcriptome databases have been reported for *C. rogercresseyi* (Gallardo-Escárate et al., 2014) and *L. salmonis* (Eichner et al., 2018). Importantly, transcripts annotated as *metalloproteinase*, *arginine kinase*, and the *cuticle protein* were primarily upregulated in early developmental stages, evidencing participation in the digestion of intake proteins, tissue development, cuticle remodeling, and specific cleavage events that activated or inactivated proenzymes and bioactive peptides (Singh, 2010). Regarding the development of the nervous system, some relevant genes such as the *nicotinic acetylcholine receptor*, *flotillin*, *synaptotagmin*, *allatostatin*, and *opsin* were also highly expressed in copepodids and congruent with previous studies (Christie et al., 2008, 2013a, 2013b).

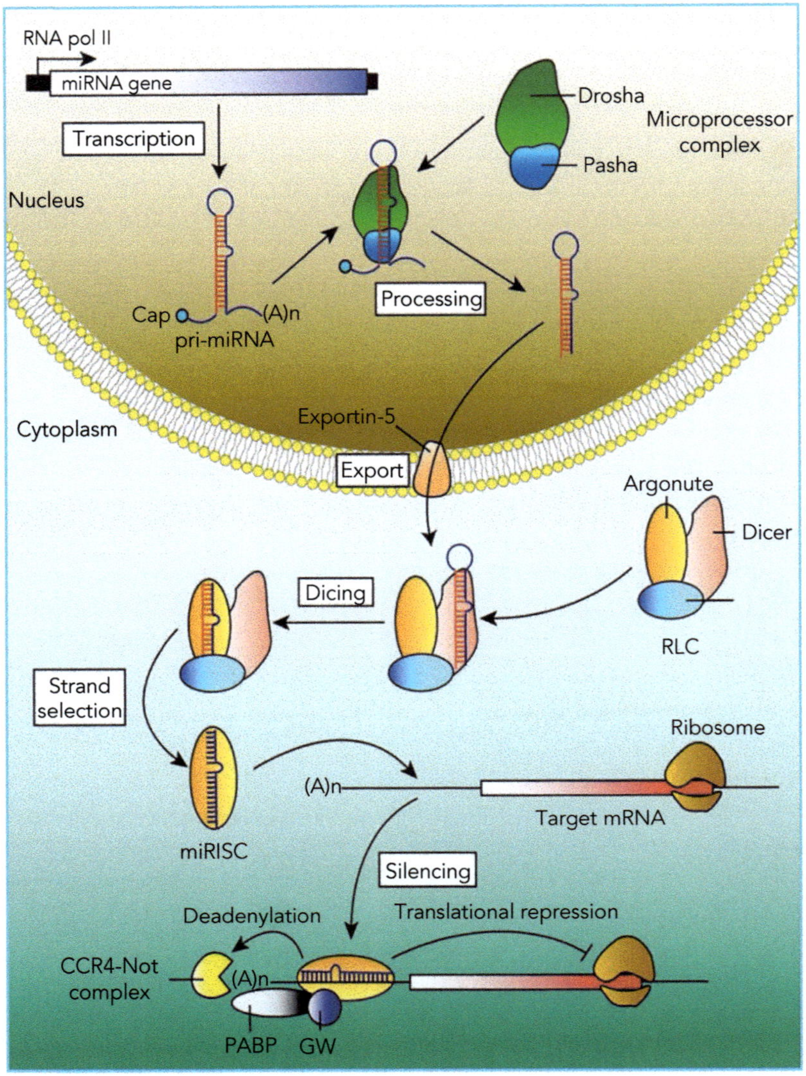

Fig. 12.7 Schematic representation of the post-transcriptional gene silencing mediated by miRNA in *C. rogercresseyi*.

With respect to the *C. rogercresseyi* miRNome (Gallardo-Escárate et al., 2017), the results showed a high number of differentially expressed miRNAs during early developmental stages belonging to the bantam, mir-8, mir-9, mir-996, and mir-184 families (Fig. 12.8). Here, miR-8 has been identified as a regulator of insulin signaling in *Drosophila* fat body and has been shown to be putatively involved in regulating body size and promoting cellular growth; meanwhile large-scale analysis of miRNAs expression revealed that miR-9 is highly enriched in both the developing and mature nervous system of vertebrates and insects (Heimberg et al., 2010). Functional analyses in model species have highlighted a prominent role of miR-9 in regulating the behavior of neural progenitors, as well as the differentiation of some neuronal populations (Coolen et al., 2013). With respect to miRNA-184, it was originally identified by expression cloning from the small RNA fraction of *Drosophila* embryos but is conserved from insects to humans (Aravin et al., 2003). It has been reported in embryos, larvae, and adults, and its expression pattern displays dynamic changes at different embryo stages, suggesting that miR-184 may have an important function during the ontogeny (Li et al., 2011; Faggins, 2012).

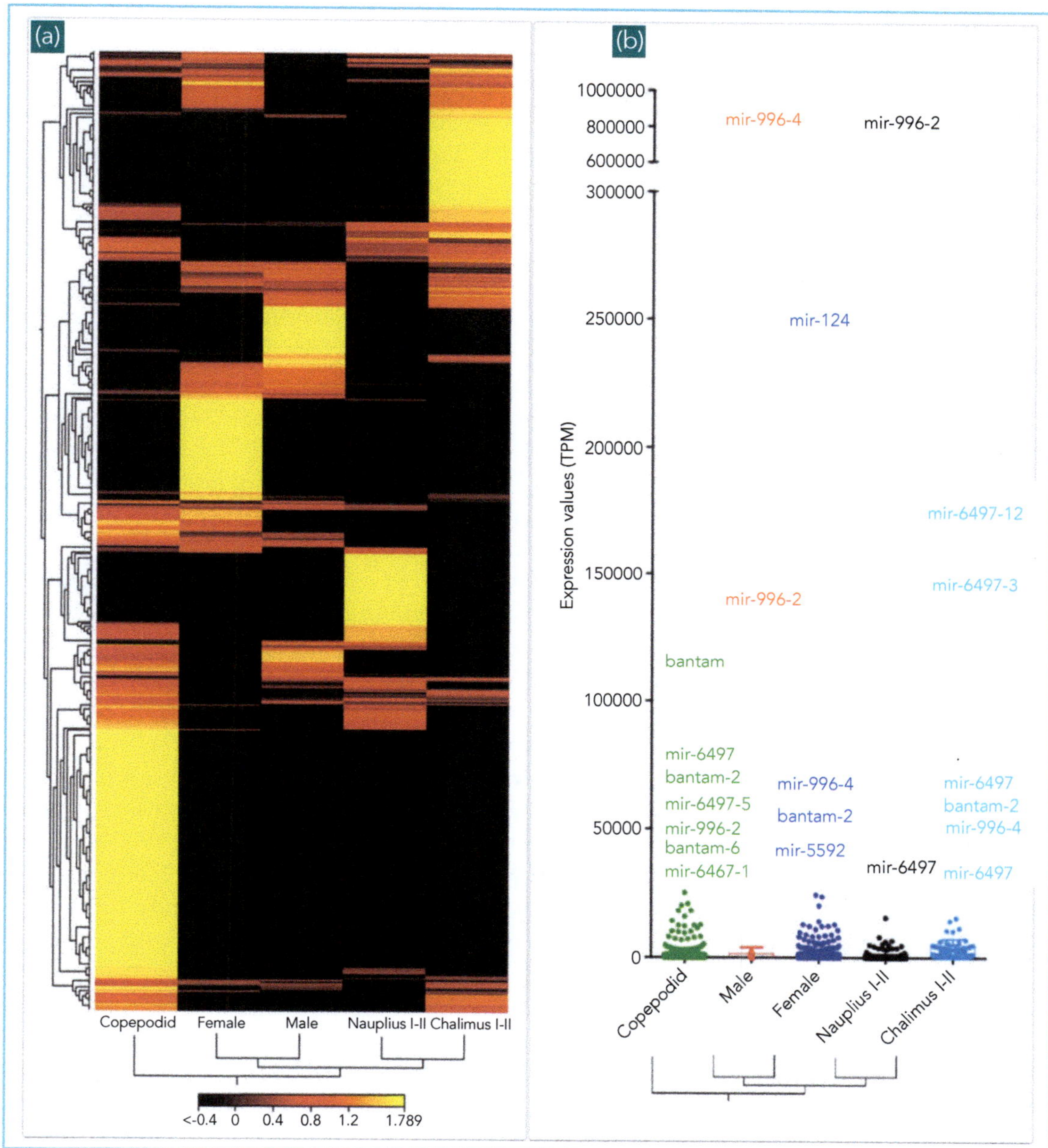

Fig. 12.8 Transcriptome profiles of miRNAs during the developmental stages of the sea louse *C. rogercresseyi*. A) Heatmap of miRNAs through the ontogenesis. B) Transcription expression values of highly regulated miRNAs.

Furthermore, a central question in arthropods is how chemoreception is mediated through transmembrane-domain receptors and which molecular mechanisms provide "fine-tuning" regulation. Interestingly, two small non-coding RNAs closely located at the genome level in *Drosophila*, mir-279 and mir-996, have emerged as key regulators of olfactory neurons (Sun et al., 2015). In turn, these results can be associated with the increased expression of olfactory-related genes during the infective stage of copepodid in *C. rogercresseyi* (Núñez-Acuña et al., 2014a). Moreover, several transcripts related to *cuticle proteins, metalloproteinases,* and *carbonic anhydrases* were putatively found with miRNA binding regions for bantam, mir-8, mir-100, and mir-2765, suggesting a key role of these miRNAs during the molting process in sea lice (Gallardo-Escárate et al., 2017).

For the chalimus stage, *trypsin, alpha amylase,* and *carboxypeptidase* genes were upregulated in *C. rogercresseyi*. Peptidases from different families may be involved in a wide range of cellular and biological processes, making it difficult to infer specific functions throughout sea louse development (Fast et al., 2003; Firth et al., 2000). For instance, blood degradation has been reported in the salmon louse *L. salmonis* (Brandal et al., 1976), associating its peptidase activity with a hematophagous behavior (Kvamme et al., 2004, 2005). With respect to miRNAs reported with putative post-transcriptional regulation on peptidases, Jayachandran et al. (2013) identified a miRNA, har-miR-2002b, that is specifically expressed during larval stages of the cotton bollworm, *Helicoverpa armigera*. Importantly, a trypsin-like serine protease (*Ha-TLP*) was identified as a target of har-miR-2002b. *In vivo* and *in vitro* studies revealed that in the presence of the miRNA mimics, the *Ha-TLP* transcript/protein levels were significantly reduced. In contrast, inhibition of har-miR-2002b led to significant upregulation of *Ha-TLP* transcript levels. Gallardo-Escárate et al. (2017) evidenced several miRNAs such as bantam, mir-8, mir-100, mir-81, and let-7 with peptidases as predicted target genes.

Genes showing differential transcription expression between female and male adult sea lice have been reported (Farlora et al., 2015; Núñez-Acuña and Gallardo-Escárate, 2015; Núñez-Acuña et al., 2019). For female adults, sex-related genes such as *vitellogenins* and *estradiol 17-beta-dehydrogenase* were identified. However, the effects of the vertebrate-like steroid hormones on reproductive processes, such as oocyte maturation in crustaceans, remain unidentified. Vitellogenins are the major yolk proteins in most invertebrates, and several different vitellogenins typically give rise to vitelline granules in mature eggs (Belles and Maestro, 2005). The role of multiple vitellogenin genes in some organisms, such as insects, is unknown (Tufail and Takeda, 2008). The data obtained for *C. rogercresseyi* showed several vitellogenin-like proteins and a wide diversity of vitellogenins (Farlora et al., 2014), including *LsVit1* and *LsVit2*, as reported in *L. salmonis* (Dalvin et al., 2011) (Fig. 12.9 A–B). Moreover, genes associated with cell development, including homologues of vasa, homeobox, Argonaute, cell division protein kinase, and the centromere-associated protein, were also primarily expressed in female adults. The study of sex-biased expression of gene products in model species has produced a number of insightful observations (Marco et al., 2013). However, this effect may be different to that observed for protein-coding genes, as proteins and miRNAs differ in their evolutionary dynamics. For instance, gene duplication is the main mechanism by which novel protein-coding genes emerge, whereas a majority of miRNAs emerge by *de novo* formation within existing transcripts (Campo-Paysaa et al., 2011). Consequently, novel miRNAs are more likely to be lost than protein-coding genes in a short evolutionary period. Although miRNAs have been extensively studied in *Drosophila* (Marco, 2014), the effect of sex-biased expression in miRNA in non-model species remains largely unexplored. Our results show sex-biased expression in two small RNAs, mir-996-4 and mir-124 for male and female sea lice, respectively. Interestingly, the same miRNAs have been reported in *Drosophila*, but inversely expressed, where mir-996 has been associated to females and the mir-124 to male flies (Marco, 2014; Weng et al., 2013). Furthermore, bioinformatic target prediction suggests that mir-996-4 in *C. rogercresseyi* can potentially knock down *vitellogenin I–II* and *vasa* genes in male adults (Fig. 12.9 C–D).

The study conducted by Núñez-Acuna and Gallardo-Escárate (2019) revealed 211 upregulated miRNAs in males and 60 in females in the sea louse *L. salmonis*. Most miRNAs associated with males were the bantam, mir-263a, and mir-276 families, while in females they were associated with the

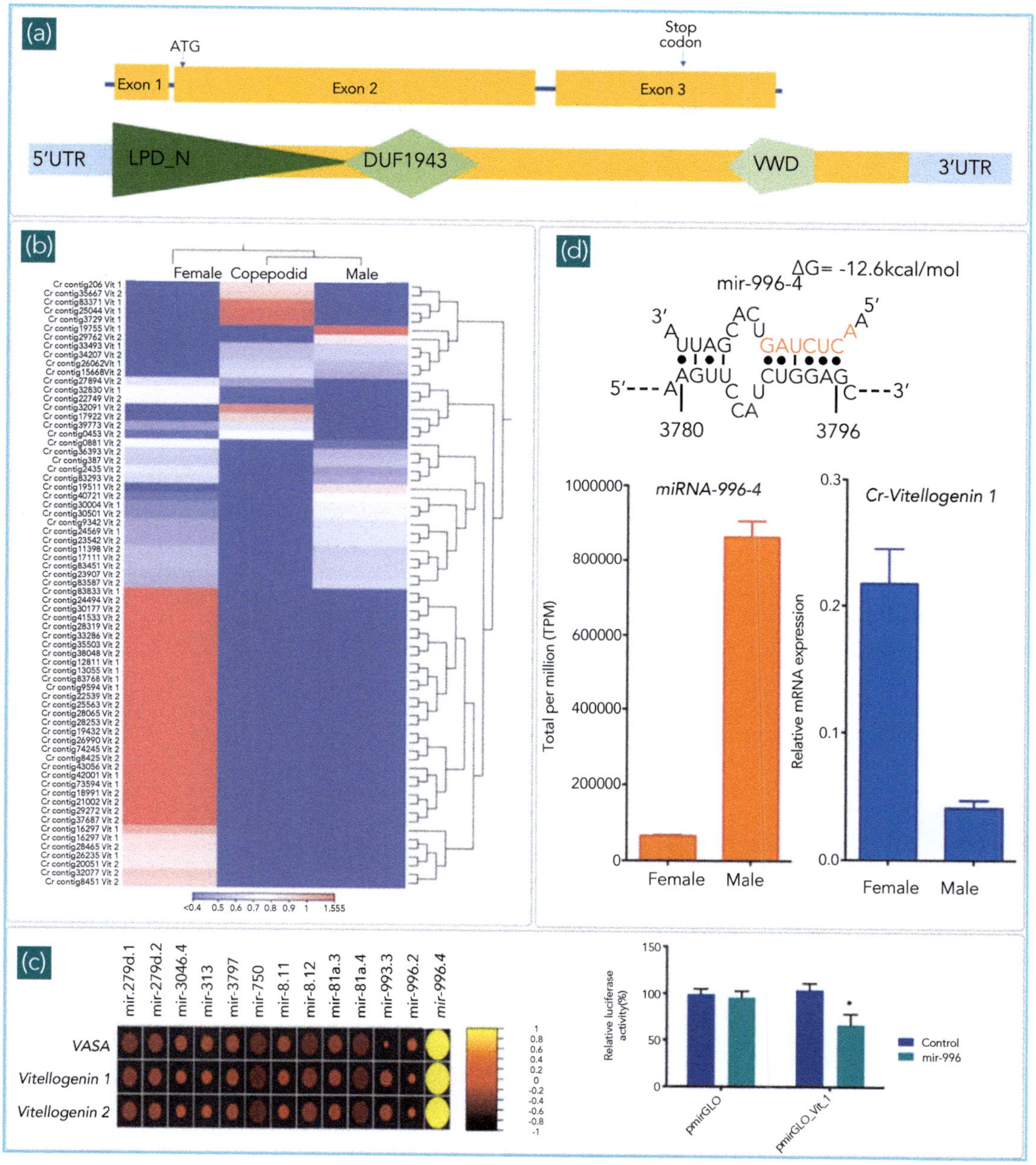

Fig. 12.9 A) Genomic characterization of vitellogenin 1, showing the exon 1, exon 2, and exon 3. The mature mRNA showed the motifs LPD_N, DUF1943, and VWD. B) Transcription profiling for vitellogenin 1 and 2 isoforms evaluated in copepodid, adult males, and females. C) Correlation expression analyses for Vasa and vitellogenin 1 and 2 with putative miRNAs. Luciferase assay was also evaluated using pmirGLO-3'UTR for vitellogenin 1 to corroborate the gene-silencing interaction. D) Transcription expression mir-996 and vitellogenin 1 for adult female and male. DNA interaction between mir-996 and 3'UTR.

bantam, mir-1, and mir-100 families, among others. A heatmap of upregulated miRNAs in males showed three clusters: the first related to the mir-276 family, which comprise largely overregulated miRNAs; the second to the bantam family; and the third to the mir-263a and bantam families, where both presented slighter expression differences than the first one. Upregulation of miRNAs associated with the female group comprised two clusters, revealing the first one with slight expression differences that are more related to the bantam, mir-100, and mir-6497 families, and another with stronger expression differences related to the bantam family. Furthermore, a total of 1455 differentially expressed genes with significant changes (fold change > |4|; $p < 0.05$) were found in both sexes: 691 were female-biased and 764 male-biased (Fig. 12.10 A). A heatmap comprising these significant genes revealed four differentiated clusters: one with slight upregulation in males, another with strong upregulation in males, another with slight upregulation in females, and the last one with strong differences in the female group. Enrichment analyses of annotated mRNAs by GO evidenced that the most abundant biological processes in differentially expressed genes were related to sex and reproduction, such as male gonad development, female meiosis I, female meiotic division, and gastrulation (Fig. 12.10 B). Other enriched GO terms were related to cell differentiation such as cell proliferation, growth, mitotic cell cycle, positive regulation of cell migration, and negative regulation of cell differentiation.

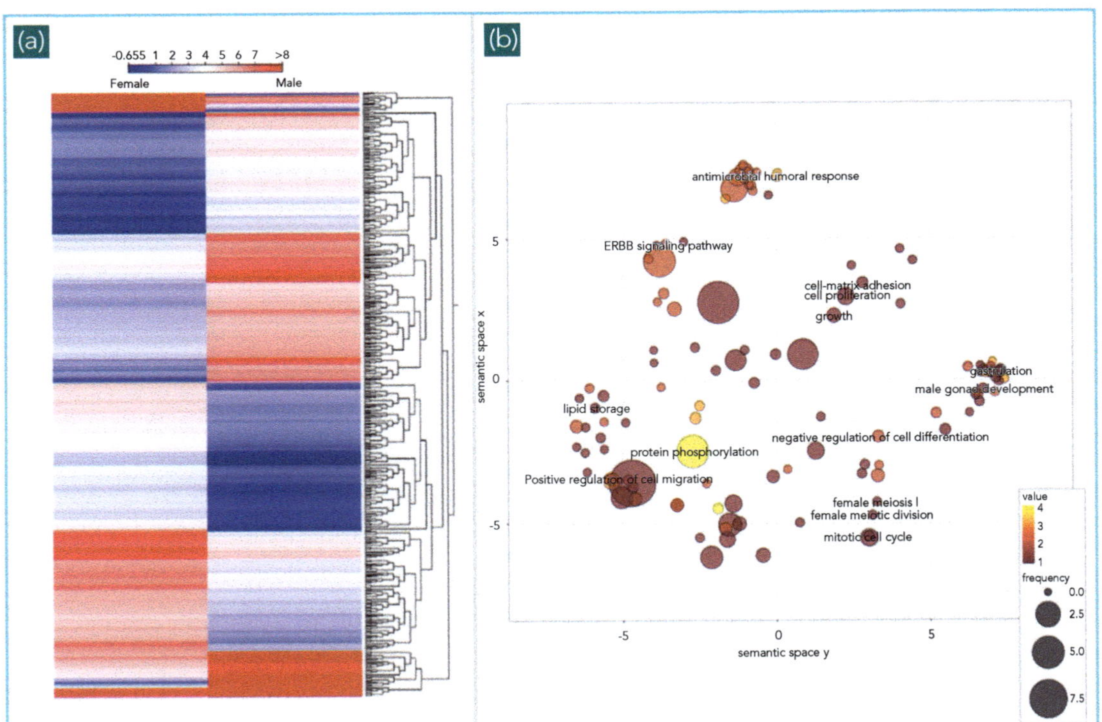

Fig. 12.10 A) Transcriptome profiles of mRNAs associated to adult stages in *L. salmonis*. Heatmap for hierarchical clustering by Manhattan's distances and average linkage. B) Enrichment analyses of significantly expressed annotated mRNAs in *L. salmonis*. REVIGO analyses were conducted based on number of GO categories found for biological processes. Value scales correspond to the number of times of each GO term, and frequency indicates the number of GO terms in each category.

Finally, it is important to note the high number of bantam variants in the *C. rogercresseyi* miRNome comprising 215 isomiRs (Gallardo-Escárate et al., 2017). Bantam was the first miRNA characterized in *Drosophila* and was identified in gain-of-function screens for genes promoting organ growth (Hipfner et al., 2002; Brennecke et al., 2003). Further research in imaginal tissues indicated that bantam is an effector of several signaling pathways during development, namely the Hippo, Notch, Dpp, and epidermal growth factor receptor (EGFR) pathways (Oh and Irvine, 2011). Deletion of the bantam locus results in pupae with a 50% reduction in size and associated with unusual levels of ecdysone at the time of metamorphosis (Boulan et al., 2013). Thus, bantam plays a key role in regulating growth at the systemic level. However, the authors' results suggest a novel function putatively related to immune evasion in *C. rogercresseyi* during the infestation in Atlantic salmon. Using bioinformatic approaches, we evidenced that bantam binds with lower duplex free energy on the 3′-UTRs of the most highly modulated transcripts in Atlantic salmon in response to sea lice infestation such as macrophage activation, Th1/Th2 pathways, and TLRs genes. Notably, several parasite species secrete extracellular vesicles or exosomes that play an important role in pathogen-host interactions (Mathivanan et al., 2010). These vesicles are synthesized through reverse budding of the late endosomal membrane, resulting in the formation of multivesicular bodies to be released into the extracellular environment (Colombo et al., 2014). Here, exosomes comprise functional proteins, lipids, and miRNAs, which can transduce signals in target cells. Interestingly, a recent study evidenced that in one of the causative agents of bilharzia, *Schistosoma japonicum* (Katsurada, 1904), exosome-like vesicles are involved in the transfer of their miRNA cargo to the host. Here, bantam emerges among the top five most expressed miRNAs (Zhu et al., 2016).

12.6 Beyond proteins, genes, and mutations of the sea lice drug response

To control sea lice infections, treatments principally comprise delousing drugs such as emamectin benzoate, deltamethrin, cypermethrin, and azamethiphos. However, research in recent years has shown a decrease in the effectiveness of these treatments (Aaen et al., 2015). This resistance to insecticides in invertebrates is generally linked to changes in gene regulation, which is in turn connected to ion channels dependent on voltage-gating, enzymes involved in esterase activity, and detoxifying molecules, among others (Ffrench-Constant et al., 2004). Among the mechanisms which influence a reduction in the sensitivity to xenobiotics are a reduced capacity of the chemical to penetrate the cuticle of the organism, an increase in metabolic processes that participate in the degradation of xenobiotics, and a decreased sensitivity of target molecules (Sevatdal et al., 2005; Lanning et al., 1996). As such, the main barrier against the absorption of insecticides is the cuticle (Vontas et al., 2007). A thickening of the cuticle has been observed in arthropods exposed to pyrethroids, which is directly related to xenobiotic resistance (Wood et al., 2010; Koganemaru et al., 2013; Lin et al., 2012). Microarray analysis of the mosquito *Anopheles stephensi* revealed an increased expression of the precursor gene for the cuticle-1 protein in strains resistant to insecticides (Vontas et al., 2007). Similar results have been reported by our research group where high modulation of genes associated with cuticle formation in *C. rogercresseyi* females exposed to deltamethrin and azamethiphos were evidenced (Chávez-Mardones et al., 2016).

Enzymes involved in esterase activity have been reported as players in the metabolism of xenobiotics (Villani et al., 1983) and even catalyzing the direct degradation of pyrethroids (Xiong et al., 2014).

Gallardo-Escárate's research group have identified trypsin and chymotrypsin in *C. rogercresseyi* and evaluated the transcriptional modulation of these after exposure to deltamethrin and azamethiphos. High-throughput transcriptome sequencing identified 44 putative trypsin-like and seven putative chymotrypsin-like transcripts in *C. rogercresseyi* that showed differentiated transcriptional modulation after drug exposure. Together, these results suggest that trypsin-like transcripts in *C. rogercresseyi* might play a role in metabolizing delousing drugs (Valenzuela-Miranda and Gallardo-Escárate, 2016a). Furthermore, cytochrome P450 is one of the principal actors in the metabolism of pyrethroids, and high modulation has been observed in resistant strains of arthropods such as *Drosophila* (Wheelock and Scott, 1992); the mosquitoes *A. aegypti* (Bariami et al., 2012) and *A. stephensi* (Vontas et al., 2007); and the salmon louse *L. salmonis* (Sevatdal et al., 2005).

In *C. rogercresseyi*, the expression of *cytochrome P450* was significantly increased in adults treated with azamethiphos (Valenzuela-Muñoz et al., 2015a), and this expression was greater in males than females, a behavior that was also observed in *Drosophila* (Le Goff et al., 2006). Moreover, among the detoxification mechanisms employed by resistant organisms are the ABC transporters, which transport the drugs along the membrane, and which participate in processes of detoxifying pyrethroids, organophosphates, and avermectins, among other chemicals generally used in the control of invertebrates (Pouliot et al., 1997; James and Davey, 2009; Pohl et al., 2012). An increased modulation of the ABC transporters was observed in populations of pyrethroids-resistant *A. aegypti*, thus marking these genes as potential indicators of a reduction in sensitivity to this type of chemical (Bariami et al., 2012). From the transcriptome database of *C. rogercresseyi*, 57 putative members of ABC protein sequences have been identified and phylogenetically classified into the eight subfamilies described for ABC transporters. RNA-seq analysis conducted in adult sea lice exposed to delousing drugs revealed high transcription levels of ABCB and ABCC subfamilies (Valenzuela-Muñoz et al., 2015b).

Pyrethroids cause changes in the activation of sodium channels, maintaining these open so as to generate the depolarization of the neuronal membrane affecting motor ability in insects, thus leading to paralysis or death (Hemingway et al., 2004). In strains of pyrethroids-resistant *A. stephensi*, a reduced expression of genes belonging to the sodium/calcium transporter family was observed (Vontas et al., 2007), suggesting an association between these channels and pyrethroids during the neurotoxic effect of the xenobiotic. This effect has also been observed in sea lice, where pyrethroids affect the sodium channels and, consequently, the nerve endings in adults (Grant, 2002; Burridge et al., 2010). For instance, in response to deltamethrin, female *C. rogercresseyi* showed increased expression levels of the $Na^{+}K^{+}$-ATPase transporter in comparison to the control group, a result that was not present in males that instead showed no significant variation. On the other hand, organophosphates act on acetylcholinesterase (AChE), which is required for the generation of nervous impulses. The inhibition of this enzyme was shown to generate an accumulation of acetylcholine, resulting in an over-stimulation of the nervous system and muscular fibers that ultimately led to paralysis and death (Burridge et al., 2010; Walsh et al., 2007). Furthermore, to determine whether delousing drugs can shift the transduction of neurotransmission signals, glutamatergic synapse-related transcripts were also evaluated in *C. rogercresseyi*.

Significant differences were observed between azamethiphos and deltamethrin regarding gene regulation, including solute carrier family 8, members 3 (*SN1*) and 2 (*GLS*); protein phosphatase (*PP2B*); *mitogen-activated protein kinase 1*; *phospholipase D*; and the mGluR regulators ankyrin repeat domain (*SHANK*) and *Homer*. Regarding other receptor proteins, solute carrier family 1, member 2 (*EEAT*); ionotropic kainate receptors (*KA*); and *mGluR 3–5* were modulated by deltamethrin,

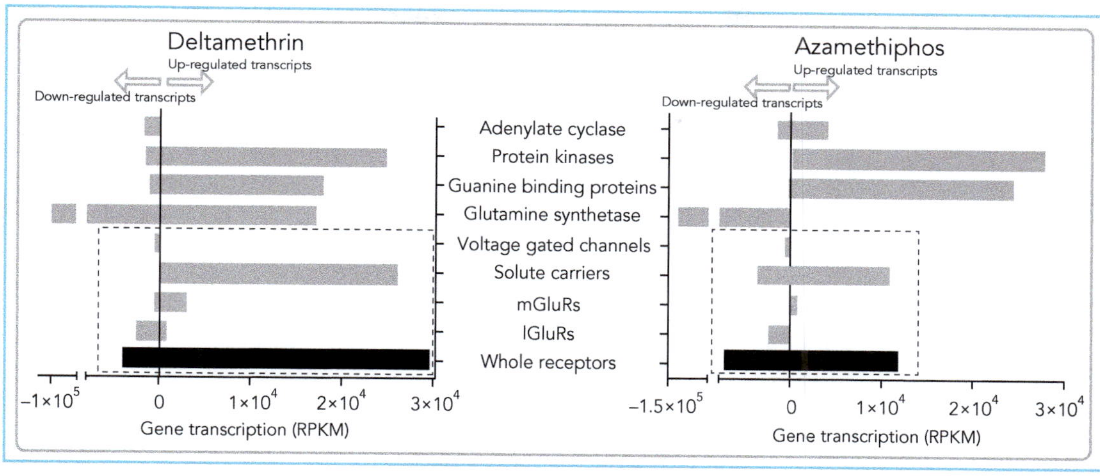

Fig. 12.11 Glutamatergic synapse-related gene expression grouped by molecular functions. Gray bars: gene family; black bar: all receptors' gene families. Dotted rectangle: gene families related to membrane.

while the 7-transmembrane glutamate receptor domain containing protein (*7mGluR*) and ionotropic *NMDAR* receptors were modulated by azamethiphos. Although both drugs notably downregulated glutamine synthetase, also known as glutamate-ammonia ligases (GLNs), some contigs related to this gene were upregulated by deltamethrin (Fig. 12.11). This coordinated response indicates that delousing drugs can drive directionally juxtaposed variations in the mRNA abundances of the glutamatergic pathway (Núñez-Acuña et al., 2016b).

With respect to gene mutations, single nucleotide polymorphisms (SNPs) in genes involved in xenobiotic metabolism have been related to insecticide resistance in arthropods. Such a mutation has been observed in the mosquitoes *A. gambiae* and *A. coluzzii*, species in which a variation of G119S in the acetylcholinesterase gene generated a change of the amino acid Ser for Gly, resulting in a resistance response to the insecticides used in the control of this parasite, among which were organophosphates (Essandoh et al., 2013). In the New World screw-worm fly, *Cochliomyia hominivorax*, a G137D mutation was observed in the carboxylesterase gene in a resistant population to organophosphate (Da Silva et al., 2011). In sea lice, notable evidence has been reported for *L. salmonis*, where a standard TaqMan assay for rapid and high-throughput screening of resistance markers has been developed. The pyrethroid-marker test targets a resistance-associated single nucleotide polymorphism in the cytochrome b gene in the mitochondrial genome (Jensen et al., 2017). On this basis, the parasites can be classified as either carrying the resistance marker (R) or not (S). Furthermore, the same research group proposed an organophosphate-marker test that targets the mutation Phe362Tyr in the AChE1a gene of *L. salmonis* described by Kaur et al. (2016). Based on this assay, each parasite could be classified as homozygote wild type (SS; Phe362/Phe362), heterozygote (RS; 362Tyr/Phe362), or homozygote mutated (RR; 362Tyr/362Tyr).

Núñez-Acuña et al. (2014b) reported a total of 3439 SNP variants identified from *de novo* assembly on the *C. rogercresseyi* transcriptome. Around 63.5% of total SNPs were mainly identified as transitions (A/G or C/T), while less than 36.5% were found to be transversions (A/C, G/T, and C/G). KAAS annotation of putative KEGG pathways involved in the resistance to pesticides was performed, and from here 17 candidate genes were selected with more than 20 polymorphisms. The cell

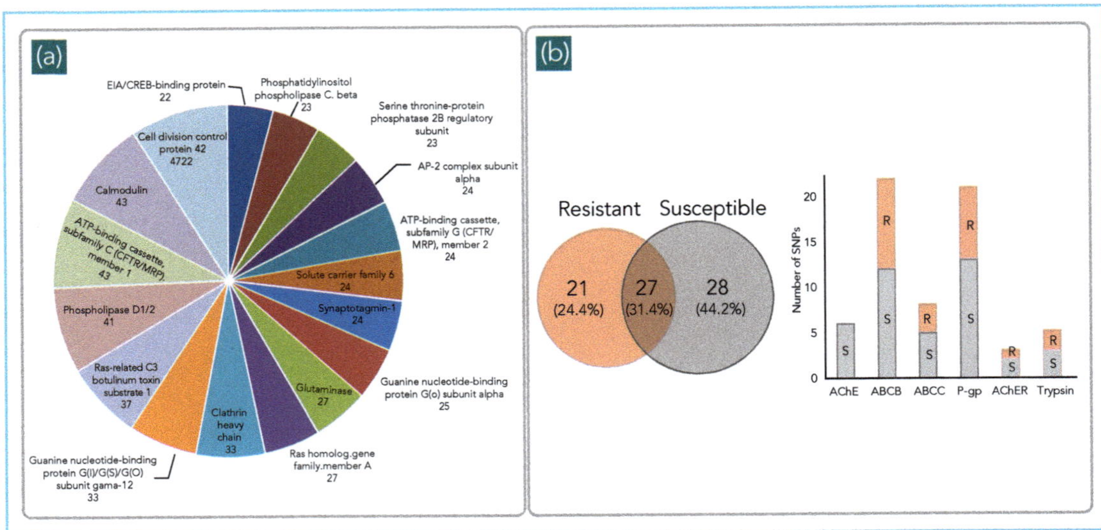

Fig. 12.12 A) Most polymorphic contigs related to nervous system and ABC transporters. The numbers represent the count of SNPs found in the respective transcript. B) Number of SNPs found in some drug-ADME genes associated with resistant or susceptible populations.

division control protein 42 gene, a component of the neurotrophic signaling pathway, was the most polymorphic and presented 47 SNPs. Following this, 43 SNPs were present in the ATP-binding cassette subfamily C member 2 gene, which corresponded to an ABC transporter, and the calmodulin gene, which was present in the dopaminergic synapse, long-term potentiation, and the neurotrophic signaling pathway. Finally, the Phospholipase D1/2 gene, a component of the glutamatergic synapse, presented 41 SNPs (Fig. 12.12 A).

Most of the variations in the individual response to drug action are not only explained by genetic polymorphisms as single mutations in target genes. Indeed, the most important differences in these variations are explained by the action of complex physiological and genetic components, including drug metabolism, or the biotransformation process that normally takes place by the action of specialized enzymatic systems (Peng and Zhong, 2015). In these systems, there are some well-characterized genes and proteins related to the absorption, degradation, metabolism, and excretion of toxicant compounds, called drug-ADME genes. Here, three phases of enzymatic reactions, including oxidation, reduction, and hydrolysis (phase I); conjugation reactions (phase II); and uptake and excretion (phase III) (Meyer et al., 2013; Schröder et al., 2013) have been proposed. In phase I enzymes, cytochrome p450 families of proteins are one of the most important.

These genes are crucial for interindividual response to drugs, allowing the definition of four different phenotypes of humans responding to drugs, ranging from ultra-rapid metabolizer individuals to poor metabolizers (Ingelman-Sundberg and Sim, 2010). In the phase II enzymes, glutathione-S-transferases (GST) are the most described as one of the critical enzymes due to this role in the metabolism of drugs to inactive molecules, showing upregulation in resistant individuals (Colmegna et al., 2017; Tew, 2016). ATP-binding cassette (ATP) transporters and solute carrier (SLC) genes are the two most relevant families of proteins in phase III, which comprise a vast number of proteins acting as transporters and efflux pumps for toxicant compounds (Colmegna et al., 2017). Many of these proteins are key players in individual response to drugs leading to pharmacological resistance, particularly the multidrug-resistance protein (MDR), an ABC transporter also described as P-glycoprotein, which confers resistance to a wide array of drugs (Szakács et al., 2006). In general, drug-ADME genes are highly variable at the genetic level, where many SNPs in some of these genes are responsible for the interindividual response to different drugs (Daly, 2010; Giacomini et al., 2007; Ma et al., 2002; Sirugo et al., 2019). Notably, SNPs associated with drug-ADME genes identified in *C. rogercresseyi* have been identified in resistant lice strains to pesticides (Fig. 12.13 B).

Beyond mRNAs and their gene mutations involved in sea lice drug resistance, emerging classes of non-coding RNAs including miRNAs, intron–exon junctions, repetitive RNAs, and lncRNAs have revealed the role of non-coding RNAs as key players during drug metabolism in parasites. Each class of RNA performs different endogenous functions, providing a variety of opportunities and challenges for drug discovery. Here, we believe that the understanding of the ncRNAs dynamics involved in the response of *C. rogercresseyi* to delousing drugs highlights the opportunity to design novel treatments against this ectoparasite.

12.6.1 The hidden non-coding nature of drug resistance

Studies have shown that the miRNome has a role in drug exposure-responses and pharmacological resistance in several species (Haenisch and Cascorbi, 2012). The molecular mechanism involving miRNAs includes regulating the gene expression of key genes for drug response. For instance, miRNA-451 in humans targets the drug transporter gene *ABCB1* and is inversely correlated to resistance to doxorubicin drugs (Kovalchuk et al., 2008). However, this resistance is not exclusive to humans or vertebrate species. The diamondback moth *Plutella xylostella* (L.) exhibits two miRNAs (miR-7a and miR-8519) that regulate the *ryanodine receptor* gene, conferring resistance to diamide insecticides and increasing the activity of detoxication enzymes related to pharmacological resistance (Li et al., 2015). In this same study, small-RNA sequencing was used to identify the miRNAs involved in responses to chlorantraniliprole insecticide exposure, which are also involved in the upregulation in detoxification genes such as *cytochrome p450* isoforms (Etebari et al., 2018). After exposure to the widely used pyrethroid pesticide deltamethrin, the mosquito *Culex pipiens* L. showed changes in 142 miRNAs after drug exposure and the regulation of key detoxification genes in deltamethrin-sensitive and deltamethrin-resistant strains (Hong et al., 2014). Specifically, the conserved miR-278-3p is involved in this species' pyrethroid resistance through the regulation of *CYP6AG11* gene expression (Lei et al., 2015). Although cumulative studies on arthropods have revealed that pharmacological resistance involves diverse biological processes, the role of the miRNome in regulating key molecular signaling pathways remains unresolved.

The previous *C. rogercresseyi* miRNome identified 663 unique annotated miRNAs (Gallardo-Escárate et al., 2017), while a recent study reported 2776 annotated miRNAs (unpublished data). This difference is mostly attributable to the amount of small-RNA reads used for the miRNome characterization (around 45 million reads vs. almost 260 million reads) due to the use of a sequencer with a higher yield (Illumina NovaSeq 6000 vs. MiSeq). Current studies being conducted to characterize the full genome of *C. rogercresseyi* at a chromosome-scale level will help clarify these results and produce a final, agreed upon, miRNome for this species. The miRNome differentially expressed included almost 30% of all identified miRNAs, suggesting that the response to this drug in *C. rogercresseyi* is not restricted to a few miRNAs. Moreover, in the case of mRNAs, modulation involved a systemic or global response.

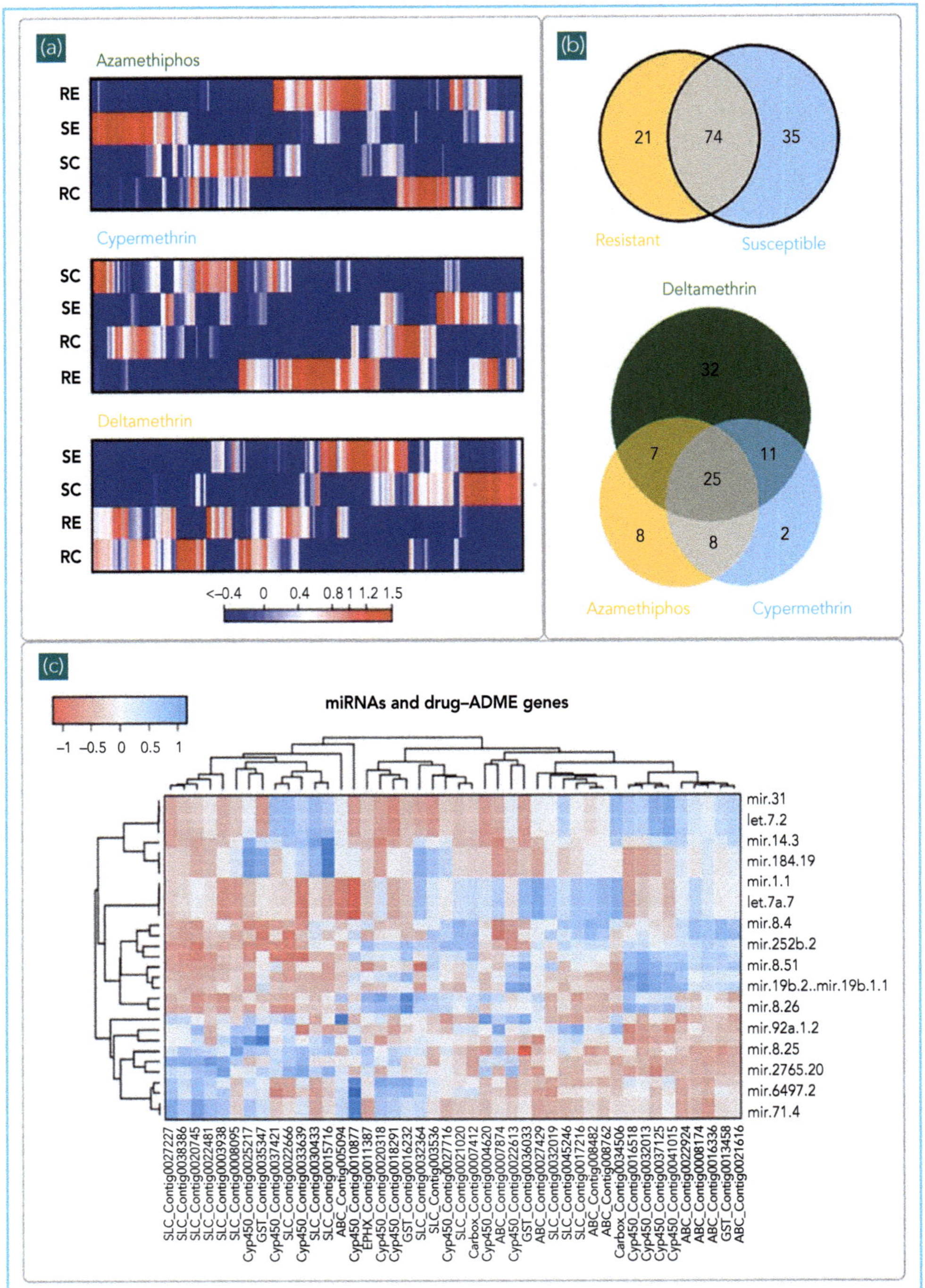

Fig. 12.13 Correlation matrix for miRNAs and drug absorption, distribution, metabolism, and excretion (ADME) gene expression levels. A) Heatmaps for drug ADME genes. B) Venn diagrams with the number of differentially expressed drug ADME genes (|fold change| > 4, $p < 0.01$) of resistant vs. susceptible lice exposed to different drugs. C) Heatmap based on the Pearson's correlation values between the TPM values in all the samples of differentially expressed miRNAs and mRNAs (|fold change| > 4, $p < 0.01$).

Interestingly, the 779 differentially expressed miRNAs revealed in this study featured clustered expression profiles in the lice exposed to each drug (Fig. 12.13). Some clusters were more strongly associated with exposure to the drug (control vs. exposed samples), while others related it to the population of origin (resistant vs. susceptible lice). This is quite interesting because it suggests that putative targets from a specific biological process might act simultaneously to respond to drug exposure. Further validation of specific miRNAs and targets could elucidate this claim, and *in vitro* techniques could confirm the miRNA–mRNA interactions suggested in this study. Most of the differentially expressed miRNAs varied in comparisons between the control vs. drug-exposed samples and not in the resistant vs. susceptible population comparisons. This difference was expected because previous studies have shown that drug exposure profoundly impacts the condition of parasites and because a previous study presented mRNA sequences with the same trend (Núñez-Acuña et al., 2020). However, the miRNAs that were differentially expressed according to resistant/susceptible populations are unique and essential because they could be involved in molecular functions conferring resistance to these drugs.

Some of the differentially expressed miRNAs obtained in this study had potential binding sites in drug–ADME genes involved in pharmacological resistance through drug metabolism, distribution, or excretion from the cell (Schröder et al., 2013). These groups of genes act in enzymatic reactions grouped into three phases. Phase I is related to the oxidation, reduction, and hydrolysis of drugs, including *cytochrome p450* gene families. One of the family of genes present in phase II is *glutathione-S-transferase*, and it is related to conjugation reactions of foreign drugs in the cell (Tew, 2016; Chávez-Mardones and Gallardo-Escárate, 2015). Phase III corresponds to the uptake and excretion of drugs in the cell, where transporter proteins such as *ATP-binding cassette transporters* (ABC transporters) and *solute carrier* (SLC) are the most common in different taxa (Tew, 2016; Colmegna et al., 2017). Some of these genes were studied in the model species used in this research on *C. rogercresseyi*, evidencing a potential role in drug response and emerging pharmacological resistance to organophosphate and pyrethroid compounds (Chávez-Mardones and Gallardo-Escárate, 2015; Núñez-Acuña et al., 2016b; Valenzuela-Muñoz et al., 2015a, 2015b). The expression, correlation, and sequence homology analyses conducted on the miRNAs strongly suggest the post-transcriptional regulation of drug–ADME genes during the three mentioned phases, where these non-coding RNAs might modulate the drug responses of sea lice through the repression of this cluster of genes. This finding is particularly interesting for mir-184, let-7a, and other families with potential targets for exposure to the three drugs.

Conversely, long non-coding RNAs (lncRNAs) are endogenous, capped, and polyadenylated mRNA-like RNAs longer than 200 nucleotides and transcribed by RNA polymerase II; nevertheless, lncRNAs do not encode for proteins (Moran et al., 2012, Mercer et al., 2009). The molecular characterization of lncRNAs in different model species has revealed that these transcripts share some common features. As compared to mRNAs, lncRNAs are generally shorter, less enriched in guanine-cytosine (GC), and expressed at lower levels, and tend to conform fewer secondary structures (Yang and Zhang, 2015; Mu et al., 2016; Boltaña et al., 2016; Etebari et al., 2015; Huang et al., 2012; Pauli et al., 2012). Despite these common features, few degrees of sequence conservation between closely related species have been reported (Pauli et al., 2012; Etebari et al., 2015; Mu et al., 2016; Paneru et al., 2016). Although the first functional lncRNAs were characterized in the 1990s (Brown et al., 1992; Brockdorff et al., 1992), the massive identification of lncRNAs has only recently been achieved through advances in large-scale transcriptomics. As a result, few lncRNAs have been functionally characterized, and, thus, the mechanisms whereby these transcripts can modulate gene networks are still unclear. Cumulative evidence suggests that lncRNAs can regulate protein production through chromatin remodeling, transcriptional control, or post-transcriptional processing by gene silencing (Mercer et al., 2009). This wide range of actions means lncRNAs could be involved in regulating multiple biological processes (Taylor et al., 2015; Zhao and Lin, 2015; Imamura and Akimitsu, 2014). Recent evidence suggests that lncRNAs are also triggered by different stress factors, including genotoxic agents and insecticides (Mizutani et al., 2012; Etebari et al., 2015). This body of evidence indicates that lncRNAs might fulfill a pivotal role in the response to xenobiotic compounds. Consequently, lncRNAs could be putative targets for understanding the drug response and might be candidates for developing novel mitigation strategies against insecticide-treated parasites.

The growing evidence for the regulatory role of ncRNAs in different biological processes has increased the scientific interest in these transcripts. However, ncRNAs research in marine species is still limited and has been mainly associated with functions in immune responses (Boltaña et al., 2016; Mu et al., 2016; Paneru et al., 2016; Tarifeno-Saldivia et al., 2017). Notably, our research group has recently published a study focusing on the identification of lncRNAs in *C. rogercresseyi* and on the evaluation of the transcriptional modulation during exposure to the delousing drugs deltamethrin and azamethiphos. Through a custom pipeline, 1835 putative lncRNAs that shared features of other previously reported lncRNAs were identified (Valenzuela-Miranda et al., 2017). Furthermore, transcriptional responses of *C. rogercresseyi* coding and non-coding RNAs were also compared during exposure to antiparasitic drugs. Venn diagrams were constructed for each condition considering the differentially expressed transcripts (fold-change > 2; $p < 0.05$). Results evidenced a similar distribution of differentially expressed mRNAs and lncRNAs in both treatments, with higher numbers of transcripts exclusively regulated in the group exposed to azamethiphos (Fig. 12.14 A). However, regarding the intensity and direction of the change (upregulated or downregulated), lncRNAs were strongly downregulated after delousing drug exposure (Fig. 12.14 B–C), but since this study was not designed to reveal how lncRNAs are differentially expressed between resistant and susceptible strains of sea lice no further conclusions could be made.

Regarding sequence homology, few *C. rogercresseyi* lncRNAs were effectively mapped against closely related genomes, even with other sea lice species. In contrast to mRNAs, lncRNAs did not

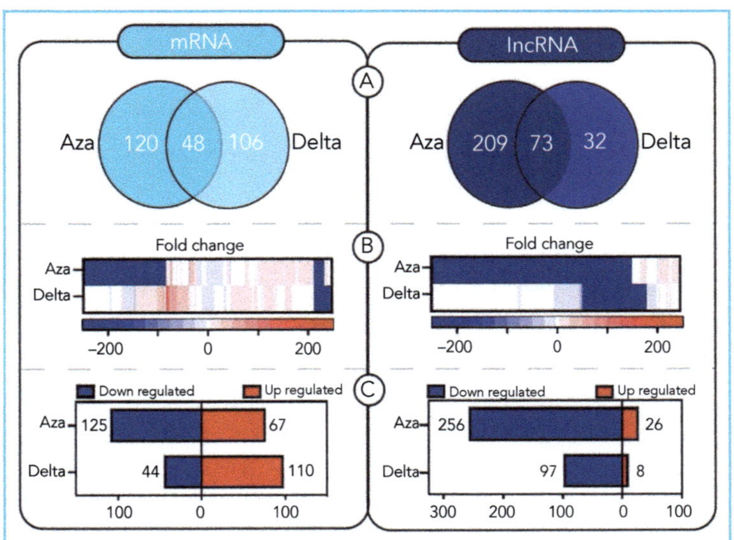

Fig. 12.14 Transcriptional regulation of mRNAs and lncRNAs. A) Venn diagram representing the distribution of differentially expressed transcripts exclusive and unique to each drug. B) Heatmap representing the fold change value for each differentially expressed transcript, ranging from −200 (blue) to 200 (red). C) Distribution of upregulated (red) and downregulated (blue).

share the high sequence conservation between closely related species. However, it has been shown that this lack of conservation does not imply a lack of function (Johnsson et al., 2014). For instance, 47 different orthologues in *Drosophila* species have been reported for the lncRNA roX, and this loss of sequence conservation could be offset by a dynamic change in the respective binding sites (Quinn et al., 2016). Moreover, the intensity and direction (up- or downregulated) of expressed lncRNAs exhibit a tendency to increase in expression after stimulation or compared with the control. These results are congruent with previously reported evidence suggesting that insecticide-resistant strains of *P. xylostella* deploy a large number of lncRNAs against different toxins (Etebari et al., 2015). This result contrasts with the author's previous result of lncRNAs in *C. rogercresseyi* exposed to deltamethrin and azamethiphos, where lower transcription expression was observed (Valenzuela-Miranda et al., 2017). However, this fact could evidence that the population used at that study belongs to an uncharacterized susceptible strain. However, a novel study in this species with characterized resistant and susceptible strains confirmed the specific modulation of lncRNAs expression and interactions with coding RNAs, and suggested a role in pharmacological resistance against azamethiphos, deltamethrin, and cypermethrin drugs (Núñez-Acuña et al., 2020). Finally, co-expression and co-localization approaches have been previously used to identify possible interactions between coding and non-coding RNAs (Etebari et al., 2015; Valenzuela-Miranda and Gallardo-Escárate, 2016b; Al-Tobasei et al., 2016). This analysis is based on the assumption that lncRNAs regulate genes proximal to the respective binding site (Villegas and Zaphiropoulos, 2015).

Acknowledgements

The authors would like to thank an anonymous reviewer during the preparation of this chapter and FONDAP #15110027 awarded by ANID-Chile.

References

Aaen, S.M., Helgesen, K.O., Bakke, M.J., Kaur, K. and Horsberg, T.E. (2015) Drug Resistance in Sea Lice: A Threat to Salmonid Aquaculture. *Trends in Parasitology* 31, 72–81.

Aarseth, K.A. and Schram, T.A. (1999) Wavelength-Specific Behaviour in *Lepeophtheirus salmonis* and *Calanus finmarchicus* to Ultraviolet and Visible Light in Laboratory Experiments (Crustacea: Copepoda). *Marine Ecology Progress Series* 186, 211–217.

Adell, T. and Muller, W.E. (2004) Isolation and Characterization of Five Fox (Forkhead) Genes from the Sponge Suberites *Domuncula*. *Gene* 334, 35–46.

Al-Tobasei, R., Paneru, B. and Salem, M. (2016) Genome-Wide Discovery of Long Non-Coding RNAs in Rainbow Trout. *PLoS ONE* 11(2), e0148940.

Ambros, V. (2004) The Functions of Animal MicroRNAs. *Nature* 431, 350–355.

Aravin, A.A., Lagos-Quintana, M., Yalcin, A., Zavolan, M., Marks, D., Snyder, B., Gaasterland, T., Meyer, J. and Tuschl, T. (2003) The Small RNA Profile During *Drosophila* Melanogaster Development. *Developmental Cell* 5, 337–350.

Bailey, R.J.E., Birkett, M.A., Ingvarsdóttir, A., Mordue, A.J., Mordue, W., O'Shea, B., Pickett, J.A. and Wadhams, L.J. (2006) The Role of Semiochemicals in Host Location and Non-Host Avoidance by Salmon Louse (*Lepeophtheirus salmonis*) Copepodids. *Canadian Journal of Fisheries and Aquatic Sciences* 63, 448–456.

Bao, W.D., Kojima, K.K. and Kohany, O. (2015) Repbase Update, a Database of Repetitive Elements in Eukaryotic Genomes. *Mobile DNA* 6, 11.

Bariami, V., Jones, C.M., Poupardin, R., Vontas, J. and Ranson, H. (2012) Gene Amplification, ABC Transporters and Cytochrome P450s: Unraveling the Molecular Basis of Pyrethroid Resistance in the Dengue Vector, *Aedes aegypti*. *PLoS Neglected Tropical Diseases* 6(6), e1692.

Barker, D.E., Braden, L.M., Coombs, M.P. and Boyce, B. (2009) Preliminary Studies on the Isolation of Bacteria from Sea Lice, *Lepeophtheirus salmonis*, Infecting Farmed Salmon in British Columbia, Canada. *Parasitology Research* 105, 1173–1177.

Barker, S.E., Bricknell, I.R., Covello, J., Purcell, S., Fast, M.D, Wolters, W. and Bouchard, D.A. (2019) Sea Lice, *Lepeophtheirus salmonis* (Krøyer 1837), Infected Atlantic Salmon (*Salmo salar* L.) Are More Susceptible to Infectious Salmon Anemia Virus. *PLoS ONE* 14(1), e0209178.

Bedell, J.A., Korf, I. and Gish, W. (2000) Maskeraid: A Performance Enhancement to Repeatmasker. *Bioinformatics* 16, 1040–1041.

Belles, X. and Maestro, J.L. (2005) Endocrine Peptides and Insect Reproduction. *Invertebrate Reproduction & Development* 47, 23–37.

Bellin, D., Ferrarini, A., Chimento, A., Kaiser, O., Levenkova, N., Bouffard, P. and Delledonne, M. (2009) Combining Next-Generation Pyrosequencing with Microarray for Large-Scale Expression Analysis in Non-Model Species. *BMC Genomics* 10, 555.

Birney, E., Clamp, M. and Durbin, R. (2004) Genewise and Genomewise. *Genome Research* 14, 988–995.

Blaxter, M. and Koutsovoulos, G. (2014) The Evolution of Parasitism in Nematoda. *Parasitology* 25, 1–14.

Boltaña, S., Valenzuela-Miranda, D., Aguilar, A., Mackenzie, S. and Gallardo-Escárate, C. (2016) Long Noncoding RNAs (lncRNAs) Dynamics Evidence Immunomodulation During ISAV-Infected Atlantic Salmon (*Salmo salar*). *Scientific Reports* 6, 22698.

Boulan, L., Martin, D. and Milan, M. (2013) Bantam miRNA Promotes Systemic Growth by Connecting Insulin Signaling and Ecdysone Production. *Current Biology* 23, 473–478.

Boxshall, G.A. and Bravo, S. (2000) On the Identity of the Common Caligus (Copepoda: Siphonostomatoida: Caligidae) from Salmonid Netpen Systems in Southern Chile. *Contributions to Zoology* 69, 137–146.

Brandal, P., Egidius, E. and Romslo, I. (1976) Host Blood: A Major Food Component for the Parasitic Copepod *Lepeophtheirus salmonis* Krøyer, 1838 (Crustacea: Caligidae). *Norwegian Journal of Zoology* 24, 341–343.

Brennecke, J., Hipfner, D.R., Stark, A., Russell, R.B. and Cohen, S.M. (2003) Bantam Encodes a Developmentally Regulated MicroRNA That Controls Cell Proliferation and Regulates the Proapoptotic Gene Hid in *Drosophila*. *Cell* 113, 25–36.

Brockdorff, N., Ashworth, A., Kay, G.F., Mccabe, V.M., Norris, D.P., Cooper, P.J., Swift, S. and Rastan, S. (1992) The Product of the Mouse Xist Gene is a 15 Kb Inactive X-Specific Transcript Containing no Conserved Orf and Located in the Nucleus. *Cell* 71, 515–526.

Brooks, K.M. (2005) The Effects of Water Temperature, Salinity, and Currents on the Survival and Distribution of the Infective Copepodid Stage of Sea Lice (*Lepeophtheirus salmonis*) Originating on Atlantic Salmon Farms in the Broughton Archipelago of British Columbia, Canada. *Reviews in Fisheries Science* 13, 177–204.

Brown, C.J., Hendrich, B.D., Rupert, J.L., Lafreniere, R.G., Xing, Y, Lawrence, J. and Willard, H.F. (1992) The Human Xist Gene – Analysis of a 17 Kb Inactive X-Specific RNA That Contains Conserved Repeats and Is Highly Localized within the Nucleus. *Cell* 71, 527–542.

Burridge, L., Weis, J.S., Cabello, F., Pizarro, J. and Bostick, K. (2010) Chemical Use in Salmon Aquaculture: A Review of Current Practices and Possible Environmental Effects. *Aquaculture* 306, 7–23.

Campo-Paysaa, F., Semon, M., Cameron, R.A., Peterson, K.J. and Schubert, M. (2011) MicroRNA Complements in Deuterostomes: Origin and Evolution of MicroRNAs. *Evolution & Development* 13, 15–27.

Cantacessi, C., Jex, A.R., Hall, R.S., Young, N.D., Campbell, B.E., et al. (2010) A Practical, Bioinformatic Workflow System for Large Data Sets Generated by Next-Generation Sequencing. *Nucleic Acids Research* 38(17), e171.

Carmichael, S.N., Bekaert, M., Taggart, J.B., Christie, H.R., Bassett, D.I., et al. (2013) Identification of a Sex-Linked Snp Marker in the Salmon Louse (*Lepeophtheirus salmonis*) Using Rad Sequencing. *PLoS ONE* 8, e77832.

Chaston, J.M., Suen, G., Tucker, S.L., Andersen, A.W., Bhasin, A., et al. (2011) The Entomopathogenic Bacterial Endosymbionts Xenorhabdus and Photorhabdus: Convergent Lifestyles from Divergent Genomes. *PLoS ONE* 6(11), e27909.

Chauvigne, F., Zapater, C. and Cerda, J. (2011) Role of Aquaporins During Teleost Gametogenesis and Early Embryogenesis. *Frontiers Physiology* 2, 66.

Chávez-Mardones, J. and Gallardo-Escárate, C. (2015) Next-Generation Transcriptome Profiling of the Salmon Louse *Caligus rogercresseyi* Exposed to Deltamethrin (Alphamax™): Discovery of Relevant Genes and Sex-Related Differences. *Marine Biotechnology* 17, 793–810.

Chávez-Mardones, J., Valenzuela-Muños, V. and Gallardo-Escárate, C. (2016) In Silico Transcriptome Analysis of Cuticle-Related Genes Associated with Delousing Drug Responses in the Sea Louse *Caligus rogercresseyi*. *Aquaculture* 450, 123–135.

Chawla, G. and Sokol, N.S. (2011) MicroRNAs in *Drosophila* Development. *International Review of Cell and Molecular Biology* 286, 1–65.

Christiaens, O., Iga, M., Velarde, R.A., Rouge, P. and Smagghe, G. (2010) Halloween Genes and Nuclear Receptors in Ecdysteroid Biosynthesis and Signalling in the Pea Aphid. *Insect Molecular Biology* 19, Suppl 2, 187–200.

Christie, A.E., Sousa, G.L., Rus, S., Smith, C.M., Towle, D.W., Hartline, D.K. and Dickinson, P.S. (2008) Identification of a-Type Allatostatins Possessing – Yxfgi/Vamide Carboxy-Termini from the Nervous System of the Copepod Crustacean *Calanus finmarchicus*. *General and Comparative Endocrinology* 155, 526–533.

Christie, A.E., Fontanilla, T.M., Nesbit, K.T. and Lenz, P.H. (2013a) Prediction of the Protein Components of a Putative *Calanus finmarchicus* (Crustacea, Copepoda) Circadian Signaling System Using a *de novo* Assembled Transcriptome. *Comparative Biochemistry and Physiology D-Genomics & Proteomics* 8, 165–193.

Christie, A.E., Roncalli, V., Wu, L.S., Ganote, C.L., Doak, T. and Lenz, P.H. (2013b) Peptidergic Signaling in *Calanus finmarchicus* (Crustacea, Copepoda): In Silico Identification of Putative Peptide Hormones and Their Receptors Using a *de novo* Assembled Transcriptome. *General and Comparative Endocrinology* 187, 117–135.

Colmegna, B., Morosi, L. and D'Incalci, M. (2017) Molecular and Pharmacological Mechanisms of Drug Resistance: An Evolving Paradigm. *Mechanisms of Drug Resistance in Cancer Therapy*. Springer, Switzerland. pp. 1–12.

Colombo, M., Raposo, G. and Thery, C. (2014) Biogenesis, Secretion, and Intercellular Interactions of Exosomes and Other Extracellular Vesicles. *Annual Review of Cell and Developmental Biology* 30, 255–289.

Coolen, M., Katz, S. and Bally-Cuif, L. (2013) Mir-9: A Versatile Regulator of Neurogenesis. *Frontiers in Cellular Neuroscience* 7, 220.

Corey, E., Bobkov, Y., Ukhanov, K. and Ache, B.W. (2013) Ionotropic Crustacean Olfactory Receptors. *PLoS ONE* 8, e60551.

Cox, D.N., Chao, A. and Lin, H. (2000) Piwi Encodes a Nucleoplasmic Factor Whose Activity Modulates the Number and Division Rate of Germline Stem Cells. *Development* 127, 503–514.

Da Silva, N.M., De Carvalho, R.A. and De Azeredo-Espin, A.M.L. (2011) Acetylcholinesterase CDNA Sequencing and Identification of Mutations Associated with Organophosphate Resistance in *Cochliomyia hominivorax* (Diptera: Calliphoridae). *Veterinary Parasitology* 177, 190–195.

Dalvin, S., Frost, P., Loeffen, P., Skern-Mauritzen, R., Baban, J., Ronnestad, I. and Nilsen, F. (2011) Characterisation of Two Vitellogenins in the Salmon Louse *Lepeophtheirus salmonis*: Molecular, Functional and Evolutional Analysis. *Diseases of Aquatic Organisms* 94, 211–224.

Dalvin, S., Skaftnesmo, K.O., Malde, K., Glover, K.A., Leong, J.S., Koop, B.F., Edvardsen, R.B. and Sandlund, N. (2020) Microbial Communities Associated with the Parasitic Copepod *Lepeophtheirus salmonis*. *Marine Genomics* 49, 100688.

Daly, A.K. (2010) Genome-wide Association Studies in Pharmacogenomics. *Nature Reviews Genetics* 11, 241.

Dekker, J., Rippe, K., Dekker, M. and Kleckner, N. (2002) Capturing Chromosome Conformation. *Science* 295, 1306–1311.

Deng, W. and Lin, H. (2002) Miwi, a Murine Homolog of Piwi, Encodes a Cytoplasmic Protein Essential for Spermatogenesis. *Developmental Cell* 2, 819–830.

Devine, G.J., Ingvarsdóttir, A., Mordue, W., Pike, A.W., Pickett, J., Duce, I. and Mordue, A.J. (2000) Salmon Lice, *Lepeophtheirus salmonis*, Exhibit Specific Chemotactic Responses to Semiochemicals Originating from the Salmonid, *Salmo salar*. *Journal of Chemical Ecology* 26, 1833–1847.

Dheilly, N.M., Poulin, R. and Thomas, F. (2015) Biological Warfare: Microorganisms as Drivers of Host-Parasite Interactions. *Infection, Genetics and Evolution* 34, 251–259.

Doniach, T. and Hodgkin, J. (1984) A Sex-Determining Gene, *Fem-1*, Required for Both Male and Hermaphrodite Development in *Caenorhabditis elegans*. *Developmental Biology* 106(1), 223–235.

Dumontet, S., Krovacek, K., Baloda, S.B., Grottoli, R., Pasquale, V. and Vanucci, S. (1996) Ecological Relationship between *Aeromonas* and *Vibrio* spp. and Planktonic Copepods in the Coastal Marine Environment in Southern Italy. *Comparative Immunology Microbiology and Infectious Diseases* 19, 245–254.

Durand, N.C., Robinson, J.T., Shamim, M.S., Machol, I., Mesirov, J.P., Lander, E.S. and Aiden, E.L. (2016) Juicebox Provides a Visualization System for Hi-C Contact Maps with Unlimited Zoom. *Cell Systems* 3, 99–101.

Eichner, C., Frost, P., Dysvik, B., Jonassen, I., Kristiansen, B. and Nilsen, F. (2008) Salmon Louse (*Lepeophtheirus salmonis*) Transcriptomes During Post Molting Maturation and Egg Production, Revealed Using Est-Sequencing and Microarray Analysis. *BMC Genomics* 9(126).

Eichner, C., Dondrup, M. and Nilsen, F. (2018) RNA Sequencing Reveals Distinct Gene Expression Patterns During the Development of Parasitic Larval Stages of the Salmon Louse (*Lepeophtheirus salmonis*). *Journal of Fish Diseases* 41, 1005–1029.

Esposti, M.D. and Romero, E.M. (2017) The Functional Microbiome of Arthropods. *PLoS ONE* 12(5), e0176573.

Essandoh, J., Yawson, A.E. and Weetman, D. (2013) Acetylcholinesterase (Ace-1) Target Site Mutation 119s Is Strongly Diagnostic of Carbamate and Organophosphate Resistance in *Anopheles gambiae* Ss and *Anopheles coluzzii* across Southern Ghana. *Malaria Journal* 12, 404.

Etebari, K., Furlong, M.J. and Asgari, S. (2015) Genome Wide Discovery of Long Intergenic Non-Coding RNAs in Diamondback Moth (*Plutella xylostella*) and Their Expression in Insecticide Resistant Strains. *Scientific Reports* 5, 14642.

Etebari, K., Afrad, M., Tang, B., Silva, R., Furlong, M. and Asgari, S. (2018) Involvement of MicroRNA Mir-2b-3p in Regulation of Metabolic Resistance to Insecticides in *Plutella xylostella*. *Insect Molecular Biology* 27, 478–491.

Everett, M.V., Grau, E.D. and Seeb, J.E. (2011) Short Reads and Nonmodel Species: Exploring the Complexities of Next-Generation Sequence Assembly and Snp Discovery in the Absence of a Reference Genome. *Molecular Ecology Resources* 11, 93–108.

Eyres, I., Duvaux, L., Gharbi, K., Tucker, R., Hopkins, D., Simon, J.J.C., Ferrari, J., Smadja, C.M. and Butlin, R.K. (2017) Targeted Re-Sequencing Confirms the Importance of Chemosensory Genes in Aphid Host Race Differentiation. *Molecular Ecology* 26, 43–58.

Faggins, A. (2012) Mir-184 Targets Tramtrack69 to Control Development and Proliferation of Glial Cells in the *Drosophila* Third Instar Larval Brain. *Journal of Neurogenetics* 26, 25–26.

Farlora, R., Araya-Garay, J. and Gallardo-Escárate, C. (2014) Discovery of Sex-Related Genes through High-Throughput Transcriptome Sequencing from the Salmon Louse *Caligus rogercresseyi*. *Marine Genomics* 15, 85–93.

Farlora, R., Núñez-Acuña, G. and Gallardo-Escárate, C. (2015) Prohibitin-2 Gene Reveals Sex-Related Differences in the Salmon Louse *Caligus rogercresseyi*. *Gene* 564, 73–80.

Farlora, R., Valenzuela-Muñoz, V., Chávez-Mardones, J. and Gallardo-Escárate, C. (2016) Aquaporin Family Genes Exhibit Developmentally-Regulated and Host-Dependent Transcription Patterns in the Sea Louse *Caligus rogercresseyi*. *Gene* 585, 119–127.

Fast, M.D., Burka, J.F., Johnson, S.C. and Ross, N.W. (2003) Enzymes Released from *Lepeophtheirus salmonis* in Response to Mucus from Different Salmonids. *Journal of Parasitology* 89, 7–13.

Ffrench-Constant, R.H., Daborn, P.J. and Goff, G.L. (2004) The Genetics and Genomics of Insecticide Resistance. *Trends in Genetics* 20, 163–170.

Fichorova, R.N., Lee, Y., Yamamoto, H.S., Takagi, Y., Hayes, G.R., et al. (2012) Endobiont Viruses Sensed by the Human Host – Beyond Conventional Antiparasitic Therapy. *PLoS ONE* 7, e48418.

Fields, D.M., Weissburg, M.J. and Browman, H.I. (2007) Chemoreception in the Salmon Louse *Lepeophtheirus salmonis*: An Electrophysiological Approach. *Diseases of Aquatic Organisms* 78, 161–168.

Fields, D., Skiftesvik, A. and Browman, H. (2018) Behavioural Responses of Infective-Stage Copepodids of the Salmon Louse (*Lepeophtheirus salmonis*, Copepoda: Caligidae) to Host-Related Sensory Cues. *Journal of Fish Diseases* 41, 875–884.

Finn, R.D., Attwood, T.K., Babbitt, P.C., Bateman, A., Bork, P., et al. (2017) Interpro in 2017 – Beyond Protein Family and Domain Annotations. *Nucleic Acids Research* 45, D190–D199.

Firth, K.J., Johnson, S.C. and Ross, N.W. (2000) Characterization of Proteases in the Skin Mucus of Atlantic Salmon (*Salmo Salar*) Infected with the Salmon Louse (*Lepeophtheirus salmonis*) and in Whole-Body Louse Homogenate. *Journal of Parasitology* 86, 1199–1205.

Gallardo-Escárate, C., Valenzuela-Muñoz, V. and Núñez-Acuña, G. (2014) RNA-Seq Analysis Using *de novo* Transcriptome Assembly as a Reference for the Salmon Louse *Caligus rogercresseyi*. *PLoS ONE* 9(4), e92239.

Gallardo-Escárate, C., Valenzuela-Muñoz, V., Boltana, S., Nuñes-Acuña, G., Valenzuela Miranda, S., et al. (2017) The *Caligus rogercresseyi* Mirnome: Discovery and Transcriptome Profiling During the Sea Lice Ontogeny. *Agri Gene* 4, 8–22.

Gallardo-Escárate, C., Arriagada, G., Carrera, C., Gonçalves, A.T., Núñez-Acuña, G. et al. (2019a) The Race between Host and Sea Lice in the Chilean Salmon Farming: A Genomic Approach. *Reviews in Aquaculture* 11, 325–339.

Gallardo-Escárate, C., Valenzuela-Muñoz, V., Núñez-Acuña, G., Carrera, C., Gonçalves, A.T., Valenzuela-Miranda, D., Benavente, B.B.P. and Roberts, S. (2019b) Catching the Complexity of Salmon-Louse Interactions. *Fish and Shellfish Immunology* 90, 199–209.

Gallardo-Escárate, C., Valenzuela-Muñoz, V., Nuñez-Acuña, G., Valenzuela-Miranda, D., Gonçalves, A.T., et al. (2021) Chromosome-Scale Genome Assembly of the Sea Louse *Caligus rogercresseyi* by Smrt Sequencing and Hi-C Analysis. *Scientific Data* 8(60).

Gardiner, A., Barker, D., Butlin, R.K., Jordan, W.C. and Ritchie, M.G. (2008) *Drosophila* Chemoreceptor Gene Evolution: Selection, Specialization and Genome Size. *Molecular Ecology* 17, 1648–1657.

Genna, R.L., Mordue, W., Pike, A.W. and Mordue, A.J. (2005) Light Intensity, Salinity, and Host Velocity Influence Presettlement Intensity and Distribution on Hosts by Copepodids of Sea Lice, *Lepeophtheirus salmonis*. *Canadian Journal of Fisheries and Aquatic Sciences* 62, 2675–2682.

Ghurye, J., Rhie, A., Walenz, B.P., Schmitt, A., Selvaraj, S., Pop, M., Phillippy, A.A.M. and Koren, S. (2019) Integrating Hi-C Links with Assembly Graphs for Chromosome-Scale Assembly. *PLoS Computational Biology* 15(8), e1007273.

Giacomini, K.M., Brett, C.M., Altman, R. B., Benowitz, N.L., Dolan, M.E., et al. (2007) The Pharmacogenetics Research Network: From SNP Discovery to Clinical Drug Response. *Clinical Pharmacology & Therapeutics* 81, 328–345.

Gomez, F.A., Tobar, J.A., Henriquez, V., Sola, M., Altamirano, C. and Marshall, S.H. (2013) Evidence of the Presence of a Functional Dot/Icm Type Iv-B Secretion System in the Fish Bacterial Pathogen *Piscirickettsia salmonis*. *PLoS ONE* 8(1), e54934.

Gonçalves, A.T., Collipal-Matamal, R., Valenzuela-Muñoz, V., Núñez-Acuña, G., Valenzuela-Miranda, D. and Gallardo-Escárate, C. (2020) Nanopore Sequencing of Microbial Communities Reveals the Potential Role of Sea Lice as a Reservoir for Fish Pathogens. *Scientific Reports* 10, 2895.

González, L. and Carvajal, J. (2003) Life Cycle of *Caligus rogercresseyi*, (Copepoda: Caligidae) Parasite of Chilean Reared Salmonids. *Aquaculture* 220, 101–117.

Grant, A.N. (2002) Medicines for Sea Lice. *Pest Management Science* 58, 521–527.

Gregory, T.R. (2019) Animal Genome Size Database. Available at: http://www.genomesize.com (accessed March 2021).

Gresty, K.A., Boxshall, G.A. and Nagasawa, K. (1993) Antennulary Sensors of the Infective Copepod Larva of the Salmon Louse *Lepeophtheirus salmonis* (Copepoda: Caligidae). In: Boxshall, G.A. and Defaye, D. (eds) *Pathogens of Wild and Farmed Fish: Sea Lice*. Ellis Horwood, Chichester, UK. pp. 83–98.

Griffiths-Jones, S., Saini, H.K., Van Dongen, S. and Enright, A.J. (2008) Mirbase: Tools for MicroRNA Genomics. *Nucleic Acids Research* 36, D154–D158.

Haas, B.J., Salzberg, S.L., Zhu, W., Pertea, M., Allen, J.E., Orvis, J., White, O., Buell, C.C.R. and Wortman, J.J.R. (2008) Automated Eukaryotic Gene Structure Annotation Using Evidencemodeler and the Program to Assemble Spliced Alignments. *Genome Biology* 9, R7.

Haenisch, S. and Cascorbi, I. (2012) miRNAs as Mediators of Drug Resistance. *Epigenomics* 4, 369–381.

Hahn, M.A. and Dheilly, N.M. (2016) Experimental Models to Study the Role of Microbes in Host-Parasite Interactions. *Frontiers in Microbiology* 7.

Hale, M.C., Jackson, J.R. and Dewoody, J.A. (2010) Discovery and Evolution of Candidate Sex-Determining Genes and Xenobiotics in the Gonads of Lake Sturgeon (*Acipenser fulvescens*). *Genetica* 138, 745–756.

Hamre, L.A., Eichner, C., Caipang, C.M.A., Dalvin, S.T., Bron, J.E., Nilsen, F., Boxshall, G. and Skern-Mauritzen, R. (2013) The Salmon Louse *Lepeophtheirus salmonis* (Copepoda: Caligidae) Life Cycle Has Only Two Chalimus Stages. *PLoS ONE* 8(9), e73539.

Hansen, B. and Bech, G. (1996) Bacteria Associated with a Marine Planktonic Copepod in Culture. 1. Bacterial Genera in Seawater, Body Surface, Intestines and Fecal Pellets and Succession During Fecal Pellet Degradation. *Journal of Plankton Research* 18, 257–273.

Harris, M.A., Clark, J., Ireland, A., Lomax, J., Ashburner, M., et al. (2004) The Gene Ontology (GO) Database and Informatics Resource. *Nucleic Acids Research* 32, D258–261.

He, L., Wang, Q., Jin, X.K., Wang, Y., Chen, L.L., Liu, L.L.H. and Wang, Y. (2012) Transcriptome Profiling of Testis During Sexual Maturation Stages in Eriocheir Sinensis Using Illumina Sequencing. *PLoS ONE* 7(3), e33735.

Heidelberg, J.F., Heidelberg, K.B. and Colwell, R.R. (2002) Bacteria of the Gamma-Subclass Proteobacteria Associated with Zooplankton in Chesapeake Bay. *Applied and Environmental Microbiology* 68, 5498–5507.

Heimberg, A.M., Cowper-Sallari, R., Semon, M., Donoghue, P.C.J. and Peterson, K.J. (2010) MicroRNAs Reveal the Interrelationships of Hagfish, Lampreys, and Gnathostomes and the Nature of the Ancestral Vertebrate. *Proceedings of the National Academy of Sciences of the United States of America* 107, 19379–19383.

Hemingway, J., Hawkes, N.J., Mccarroll, L. and Ranson, H. (2004) The Molecular Basis of Insecticide Resistance in Mosquitoes. *Insect Biochemistry and Molecular Biology* 34, 653–665.

Hergarden, A.C., Tayler, T.D. and Anderson, D.J. (2012) Allatostatin-a Neurons Inhibit Feeding Behavior in Adult *Drosophila*. *Proceedings of the National Academy of Sciences of the United States of America* 109, 3967–3972.

Heuch, P.A. and Karlsen, E. (1997) Detection of Infrasonic Water Oscillations by Copepodids of *Lepeophtheirus salmonis* (Copepoda: Caligida). *Journal of Plankton Research* 19, 735–747.

Heuch, P.A., Parsons, A. and Boxaspen, K. (1995) Diel Vertical Migration: A Possible Host Finding Mechanism in Salmon Louse (*Lepeophtheirus salmonis*) Copepodids? *Canadian Journal of Fisheries and Aquatic Sciences* 52, 681–689.

Hipfner, D.R., Weigmann, K. and Cohen, S.M. (2002) The Bantam Gene Regulates *Drosophila* Growth. *Genetics* 161, 1527–1537.

Hoang, K., Matzkin, L.M. and Bono, J.M. (2015) Transcriptional Variation Associated with Cactus Host Plant Adaptation in *Drosophila mettleri* Populations. *Molecular Ecology* 24, 5186–5199.

Hodgkin, J. (1987) Sex Determination and Dosage Compensation in *Caenorhabditiselegans*. *Annual Review of Genetics* 21, 133–154.

Hoerauf, A., Nissen-Pahle, K., Schmetz, C., Henkle-Duhrsen, K., Blaxter, M.L., et al. (1999) Tetracycline Therapy Targets Intracellular Bacteria in the Filarial Nematode *Litomosoides sigmodontis* and Results in Filarial Infertility. *The Journal of Clinical Investigation* 103, 11–18.

Hong, S., Guo, Q., Wang, W., Hu, S., Fang, F., et al. (2014) Identification of Differentially Expressed MicroRNAs in *Culex Pipiens* and Their Potential Roles in Pyrethroid Resistance. *Insect Biochemistry and Molecular Biology* 55, 39–50.

Horowitz, A.R. and Ishaaya, I. (2013) *Advanced Technologies for Managing Insect Pests: An Overview*. Springer, Dordrecht.

Huang, H.F., He, R.H., Sun, C.C., Zhang, Y., Meng, Q.X. and Ma, Y.Y. (2006) Function of Aquaporins in Female and Male Reproductive Systems. *Human Reproduction Update* 12, 785–795.

Huang, W., Long, N.Y. and Khatib, H. (2012) Genome-Wide Identification and Initial Characterization of Bovine Long Non-Coding RNAs from Est Data. *Animal Genetics* 43, 674–682.

Hui, J.H.L., Bendena, W.G. and Tobe, S.S. (2013) Future Perspectives for Research on the Biosynthesis of Juvenile Hormones and Related Sesquiterpenoids in Arthropod Endocrinology and Ecotoxicology. In: Devillers, J. (ed.) *Juvenile Hormones and Juvenoids: Modeling Biological Effects and Environmental Fate*. CRC Press, Boca Raton, Florida. pp. 15–30.

Hull, M.Q., Pike, A.W., Mordue, A.J. and Rae, G.H. (1998) Patterns of Pair Formation and Mating in an Ectoparasitic Caligid Copepod *Lepeophtheirus salmonis* (Krøyer 1837): Implications for Its Sensory and Mating Biology. *Philosophical Transactions of the Royal Society of London. Series B: Biological Sciences* 353, 753–764.

Humann, J.L., Lee, T., Ficklin, S. and Main, D. (2019) Structural and Functional Annotation of Eukaryotic Genomes with Gensas. *Gene Prediction: Methods and Protocols* 1962, 29–51.

Huq, A., Small, E.B., West, P.A., Huq, M.I., Rahman, R. and Colwell, R.R. (1983) Ecological Relationships between Vibrio-Cholerae and Planktonic Crustacean Copepods. *Applied and Environmental Microbiology* 45, 275–283.

Imamura, K. and Akimitsu, N. (2014) Long Non-Coding RNAs Involved in Immune Responses. *Front Immunology* 5, 573.

Ingelman-Sundberg, M. and Sim, S.C. (2010) Pharmacogenetic Biomarkers as Tools for Improved Drug Therapy; Emphasis on the Cytochrome P450 System. *Biochemical and Biophysical Research Communications* 396, 90–94.

Ingvarsdóttir, A., Birkett, M.A., Duce, I., Genna, R.L., Mordue, W., Pickett, J.A., Wadhams, L.J. and Mordue, A.J. (2002) Semiochemical Strategies for Sea Louse Control: Host Location Cues. *Pest Management Science* 58, 537–545.

Jakob, E., Barker, D.E. and Garver, K.A. (2011) Vector Potential of the Salmon Louse *Lepeophtheirus salmonis* in the Transmission of Infectious Haematopoietic Necrosis Virus (IHNV). *Diseases of Aquatic Organisms* 97, 155–165.

James, C.E. and Davey, M.W. (2009) Increased Expression of ABC Transport Proteins Is Associated with Ivermectin Resistance in the Model Nematode *Caenorhabditis elegans*. *International Journal for Parasitology* 39, 213–220.

Jayachandran, B., Hussain, M. and Asgari, S. (2013) An Insect Trypsin-Like Serine Protease as a Target of MicroRNA: Utilization of MicroRNA Mimics and Inhibitors by Oral Feeding. *Insect Biochemistry and Molecular Biology* 43, 398–406.

Jensen, E.M., Sevatdal, S., Bakke, M.J., Kaur, K. and Horsberg, T.E. (2017) A Selection Study on a Laboratory-Designed Population of Salmon Lice (*Lepeophtheirus salmonis*) Using Organophosphate and Pyrethroid Pesticides. *PLoS ONE* 12.

Jeong, C.-B., Lee, B.-Y., Choi, B.-S., Kim, M.-S., Park, J.C., et al. (2020) The Genome of the Harpacticoid Copepod Tigriopus Japonicus: Potential for Its Use in Marine Molecular Ecotoxicology. *Aquatic Toxicology* 222, 105462.

Johnson, S.C. and Albright, L.J. (1991) The Developmental Stages of *Lepeophtheirus salmonis* (Krøyer, 1837) (Copepoda: Caligidae). *Canadian Journal of Zoology* 69, 929–950.

Johnsson, P., Lipovich, L., Grandér, D. and Morris, K.V. (2014) Evolutionary Conservation of Long Non-Coding RNAs; Sequence, Structure, Function. *Biochimica et Biophysica Acta (BBA) – General Subjects* 1840, 1063–1071.

Kalluri, S., Gilruth, P., Rogers, D. and Szczur, M. (2007) Surveillance of Arthropod Vector-Borne Infectious Diseases Using Remote Sensing Techniques: A Review. *PLoS Pathogens* 3, 1361–1371.

Kanehisa, M., Goto, S., Sato, Y., Kawashima, M., Furumichi, M. and Tanabe, M. (2014) Data, Information, Knowledge and Principle: Back to Metabolism in Kegg. *Nucleic Acids Research* 42, D199–205.

Kaur, K., Jansen, P.A., Aspehaug, V.T. and Horsberg, T.E. (2016) Phe362tyr in AChE: A Major Factor Responsible for Azamethiphos Resistance in *Lepeophtheirus salmonis* in Norway. *PLoS ONE* 11(2), e0149264.

Kim, V.N. and Nam, J.W. (2006) Genomics of MicroRNA. *Trends in Genetics* 22, 165–173.

Koganemaru, R., Miller, D.M. and Adelman, Z.N. (2013) Robust Cuticular Penetration Resistance in the Common Bed Bug (*Cimex lectularius L.*) Correlates with Increased Steady-State Transcript Levels of Cpr-Type Cuticle Protein Genes. *Pesticide Biochemistry and Physiology* 106, 190–197.

Komisarczuk, A.Z., Grotmol, S. and Nilsen, F. (2017) Ionotropic Receptors Signal Host Recognition in the Salmon Louse (*Lepeophtheirus salmonis*, Copepoda). *PLoS ONE* 12, e0178812.

Kopp, A. (2012) Dmrt Genes in the Development and Evolution of Sexual Dimorphism. *Trends Genetics* 28, 175–184.

Koren, S., Walenz, B.P., Berlin, K., Miller, J.R., Bergman, N.H. and Phillippy, A.M. (2017) Canu: Scalable and Accurate Long-Read Assembly Via Adaptive K-Mer Weighting and Repeat Separation. *Genome Research* 27, 722–736.

Kovalchuk, O., Filkowski, J., Meservy, J., Ilnytskyy, Y., Tryndyak, V.P., Vasyl'f, C. and Pogribny, I.I.P. (2008) Involvement of MicroRNA-451 in Resistance of the Mcf-7 Breast Cancer Cells to Chemotherapeutic Drug Doxorubicin. *Molecular cancer therapeutics* 7, 2152–2159.

Kozomara, A. and Griffiths-Jones, S. (2014) Mirbase: Annotating High Confidence MicroRNAs Using Deep Sequencing Data. *Nucleic Acids Research* 42, D68–D73.

Krane, C.M. and Goldstein, D.L. (2007) Comparative Functional Analysis of Aquaporins/Glyceroporins in Mammals and Anurans. *Mammalian Genome* 18, 452–462.

Kvamme, B.O., Skern, R., Frost, P. and Nilsen, F. (2004) Molecular Characterisation of Five Trypsin-Like Peptidase Transcripts from the Salmon Louse (*Lepeophtheirus salmonis*) Intestine. *International Journal for Parasitology* 34, 823–832.

Kvamme, B.O., Kongshaug, H. and Nilsen, F. (2005) Organisation of Trypsin Genes in the Salmon Louse (*Lepeophtheirus salmonis*, Crustacea, Copepoda) Genome. *Gene* 352, 63–74.

Langmead, B. and Salzberg, S.L. (2012) Fast Gapped-Read Alignment with Bowtie 2. *Nature Methods* 9, 357–354.

Lanning, C.L., Fine, R.L., Corcoran, J.J., Ayad, H.M., Rose, R.L. and Abou-Donia, M.B. (1996) Tobacco Budworm P-Glycoprotein: Biochemical Characterization and Its Involvement in Pesticide Resistance. *Biochimica et Biophysica Acta (BBA) – General Subjects* 1291, 155–162.

Lasko, P.F. and Ashburner, M. (1988) The Product of the *Drosophila* Gene Vasa Is Very Similar to Eukaryotic Initiation Factor-4a. *Nature* 335, 611–617.

Le Goff, G., Hilliou, F., Siegfried, B.D., Boundy, S., Wajnberg, E., Sofer, L., Audant, P. and Feyereisen, R. (2006) Xenobiotic Response in *Drosophila melanogaster*: Sex Dependence of P450 and Gst Gene Induction. *Insect Biochemistry and Molecular Biology* 36, 674–682.

Lee, R.C., Feinbaum, R. L. and Ambros, V. (1993) The C-Elegans Heterochronic Gene Lin-4 Encodes Small RNAs with Antisense Complementarity to Lin-14. *Cell* 75, 843–854.

Lei, Z., Lv, Y., Wang, W., Guo, Q., Zou, F., et al. (2015) Mir-278-3p Regulates Pyrethroid Resistance in *Culex pipiens pallens*. *Parasitology Research* 114, 699–706.

Lew-Tabor, A.E. and Valle, M.R. (2016) A Review of Reverse Vaccinology Approaches for the Development of Vaccines against Ticks and Tick Borne Diseases. *Ticks and Tick-Borne Diseases* 7, 1236–1237.

Li, P., Peng, J.J., Hu, J.B., Xu, Z.X., Xie, W. and Yuan, L.D. (2011) Localized Expression Pattern of Mir-184 in *Drosophila*. *Molecular Biology Reports* 38, 355–358.

Li, X., Guo, L., Zhou, X., Gao, X. and Liang, P. (2015) miRNAs Regulated Overexpression of Ryanodine Receptor Is Involved in Chlorantraniliprole Resistance in *Plutella xylostella* (L.) *Scientific Reports* 5, 1–9.

Lin, Y., Jin, T., Zeng, L. and Lu, Y. (2012) Cuticular Penetration of B-Cypermethrin in Insecticide-Susceptible and Resistant Strains of *Bactrocera dorsalis*. *Pesticide Biochemistry and Physiology* 103, 189–193.

Lindsey, J. and Lasker, R. (1974) Chemical Signals in the Sea: Marine Allelochemicals and Evolution. *Fishery Bulletin* 72, 1.

Ma, J., Bruce, T. J., Jones, E.M. and Cain, K.D. (2019) A Review of Fish Vaccine Development Strategies: Conventional Methods and Modern Biotechnological Approaches. *Microorganisms* 7(11), 569.

Ma, K., Qiu, G., Feng, J. and Li, J. (2012) Transcriptome Analysis of the Oriental River Prawn, *Macrobrachium nipponense* Using 454 Pyrosequencing for Discovery of Genes and Markers. *PLoS ONE* 7, e39727.

Ma, M.K., Woo, M.H. and Mcleod, H.L. (2002) Genetic Basis of Drug Metabolism. *American Journal of Health-System Pharmacy* 59, 2061–2069.

Makarova, J.A. and Kramerov, D.A. (2007) Noncoding RNAs. *Biochemistry – Moscow* 72, 1161–1178.

Marco, A. (2014) Sex-Biased Expression of MicroRNAs in *Drosophila melanogaster*. *Open Biology* 4. DOI:10.1098/rsob.140024 (accessed March 2021)

Marco, A., Kozomara, A., Hui, J., Emery, A.M., Rollinson, D., Griffths-Jones, S. and Ronshaugen, M. (2013) Sex-Biased Expression of MicroRNAs in *Schistosoma mansoni*. *PLoS Neglected Tropical Diseases* 7(9), e2402.

Marguerat, S. and Bahler, J. (2010) RNA-Seq: From Technology to Biology. *Cellular and Molecular Life Sciences* 67, 569–579.

Mathivanan, S., Ji, H. and Simpson, R.J. (2010) Exosomes: Extracellular Organelles Important in Intercellular Communication. *Journal of Proteomics* 73, 1907–1920.

Mercer, T.R., Dinger, M.E. and Mattick, J.S. (2009) Long Non-Coding RNAs: Insights into Functions. *Nature Reviews Genetics* 10, 155–159.

Meyer, U.A., Zanger, U.M. and Schwab, M. (2013) Omics and Drug Response. *Annual Review of Pharmacology and Toxicology* 53, 475–502.

Mindnich, R., Möller, G. and Adamski, J. (2004) The Role of 17 Beta-Hydroxysteroid Dehydrogenases. *Molecular and Cellular Endocrinology* 218, 7–20.

Miyakawa, H., Imai, M., Sugimoto, N., Ishikawa, Y., Ishikawa, A., et al. (2010) Gene Up-Regulation in Response to Predator Kairomones in the Water Flea, *Daphnia pulex*. *BMC Developmental Biology* 10, 45.

Mizutani, R., Wakamatsu, A., Tanaka, N., Yoshida, H., Tochigi, N., et al. (2012) Identification and Characterization of Novel Genotoxic Stress-Inducible Nuclear Long Noncoding RNAs in Mammalian Cells. *PLoS ONE* 7(4), e34949.

Modzelewski, A.J., Holmes, R. J., Hilz, S., Grimson, A. and Cohen, P.E. (2012) Ago4 Regulates Entry into Meiosis and Influences Silencing of Sex Chromosomes in the Male Mouse Germline. *Developmental Cell* 23, 251–264.

Moran, V.A., Perera, R. J. and Khalil, A.M. (2012) Emerging Functional and Mechanistic Paradigms of Mammalian Long Non-Coding RNAs. *Nucleic Acids Research* 40, 6391–6400.

Mordue (Luntz), A.J. and Birkett, M.A. (2009) A Review of Host Finding Behaviour in the Parasitic Sea Louse, *Lepeophtheirus salmonis* (Caligidae: Copepoda). *Journal of Fish Diseases* 32, 3–13.

Mu, C., Wang, R., Li, T., Li, Y., Tian, M., et al. (2016) Long Non-Coding RNAs (lncRNAs) of Sea Cucumber: Large-Scale Prediction, Expression Profiling, Non-Coding Network Construction, and lncRNA-MicroRNA-Gene Interaction Analysis of lncRNAs in *Apostichopus japonicus* and *Holothuria glaberrima* During Lps Challenge and Radial Organ Complex Regeneration. *Marine Biotechnology (NY)* 18, 485–499.

Nagamine, C.M. and Carlisle, C. (1997) Zfy2/1 Fusion Gene Fails to Replicate Zfy1 Expression Pattern in Fetal Gonads. *Genomics* 43, 397–398.

Neira-Oviedo, M., Tsyganov-Bodounov, A., Lycett, G.J., Kokoza, V., Raikhel, A.S. and Krzywinski, J. (2011) The RNA-Seq Approach to Studying the Expression of Mosquito Mitochondrial Genes. *Insect Molecular Biology* 20, 141–152.

Nese, L. and Enger, O. (1993) Isolation of *Aeromonas salmonicida* from Salmon Lice *Lepeophtheirus salmonis* and Marine Plankton. *Diseases of Aquatic Organisms* 16, 79–81.

Ning, J., Wang, M.X., Li, C.L. and Sun, S. (2013) Transcriptome Sequencing and *de novo* Analysis of the Copepod *Calanus sinicus* Using 454 Gs Flx. *PLoS ONE* 8(5), e63741.

Núñez-Acuña, G. and Gallardo-Escárate, C. (2015) Two Novel Male-Associated Peroxinectin Genes Are Downregulated by Exposure to Delousing Drugs in *Caligus rogercresseyi*. *Gene* 557, 98–102.

Núñez-Acuña, G. and Gallardo-Escárate, C. (2019) Characterization of the Salmon Louse *Lepeophtheirus salmonis* miRNome: Sex-Biased Differences Related to the Coding and Non-Coding RNA Interplay. *Marine Genomics* 45, 38–47.

Núñez-Acuña, G., Valenzuela-Muñoz, V., Marambio, J.P., Wadsworth, S. and Gallardo-Escárate, C. (2014a) Insights into the Olfactory System of the Ectoparasite *Caligus rogercresseyi*: Molecular Characterization and Gene Transcription Analysis of Novel Ionotropic Receptors. *Experimental Parasitology* 145, 99–109.

Núñez-Acuña, G., Valenzuela-Muñoz, V. and Gallardo-Escárate, C. (2014b) High-Throughput SNP Discovery and Transcriptome Expression Profiles from the Salmon Louse *Caligus rogercresseyi* (Copepoda: Caligidae). *Comparative Biochemistry and Physiology Part D: Genomics and Proteomics* 10, 9–21.

Núñez-Acuña, G., Vera-Bizama, F., Boltaña, S., Hawes, C., Marambio, J.P., Wadsworth, S. and Gallardo-Escárate, C. (2016a) In-Feed Additives Modulate Ionotropic Receptor Genes from the Sea Louse *Caligus rogercresseyi*: A Comparative Analysis in Two Host Salmonid Species. *Aquaculture* 451, 99–105.

Núñez-Acuña, G., Boltaña, S. and Gallardo-Escárate, C. (2016b) Pesticides Drive Stochastic Changes in the Chemoreception and Neurotransmission System of Marine Ectoparasites. *International Journal of Molecular Sciences* 17(6), 700.

Núñez-Acuña, G., Marambio, J.P., Valenzuela, T., Wadsworth, S. and Gallardo-Escárate, C. (2016c) Antimicrobial Peptides from *Salmon salar* Skin Induce Frontal Filament Development and Olfactory/Cuticle-Related Genes in the Sea Louse *Caligus rogercresseyi Aquaculture*. 464, 171–177.

Núñez-Acuña, G., Gallardo-Escárate, C., Fields, D.M., Shema, S., Skiftesvik, A.B., Ormazábal, I. and Browman, H.I. (2018) The Atlantic Salmon (*Salmo salar*) Antimicrobial Peptide Cathelicidin-2 Is a Molecular Host-Associated Cue for the Salmon Louse (*Lepeophtheirus salmonis*). *Scientific Reports* 8, 13738.

Núñez-Acuña, G., Gallardo-Escárate, C., Skiftesvik, A.B., Fields, D.M. and Browman, H.I. (2019) Silencing of Ionotropic Receptor 25a Decreases Chemosensory Activity in the Salmon Louse *Lepeophtheirus salmonis* During the Infective Stage. *Gene* 697, 35–39.

Núñez-Acuña, G., Sáez-Vera, C., Valenzuela-Muñoz, V., Valenzuela-Miranda, D., Arriagada, G. and Gallardo-Escárate, C. (2020) Tackling the Molecular Drug Sensitivity in the Sea Louse *Caligus rogercresseyi* Based on mRNA and lncRNA Interactions. *Genes* 11, 857.

Nylund, A., Wallace, C. J. and Hovland, T. (1993) The Possible Role of *Lepeophtheirus salmonis* (Krøyer) in the Transmission of Infectious Salmon Anaemia. In: Boxshall, G.A. and Defaye, D. (eds) *Pathogens of Wild and Farmed Fish: Sea Lice*. Ellis Horwood, Chichester, UK. pp. 363–378.

Oh, H. and Irvine, K.D. (2011) Cooperative Regulation of Growth by Yorkie and Mad through Bantam. *Developmental Cell* 20, 109–122.

Pal, K., Forcato, M. and Ferrari, F. (2019) Hi-C Analysis: From Data Generation to Integration. *Biophysical Reviews* 11, 67–78.

Paneru, B., Al-Tobasei, R., Palti, Y., Wiens, G.D. and Salem, M. (2016) Differential Expression of Long Non-Coding RNAs in Three Genetic Lines of Rainbow Trout in Response to Infection with *Flavobacterium psychrophilum*. *Scientific Reports* 6, 36032.

Pauli, A., Valen, E., Lin, M.F., Garber, M., Vastenhouw, N.L., et al. (2012) Systematic Identification of Long Noncoding RNAs Expressed During Zebrafish Embryogenesis. *Genome Research* 22, 577–591.

Peng, L. and Zhong, X. (2015) Epigenetic Regulation of Drug Metabolism and Transport. *Acta Pharmaceutica Sinica B* 5, 106–112.

Perez-Pascual, D., Lunazzi, A., Magdelenat, G., Rouy, Z., Roulet, A., et al. (2017) The Complete Genome Sequence of the Fish Pathogen *Tenacibaculum maritimum* Provides Insights into Virulence Mechanisms. *Frontiers in Microbiology* 8.

Pino-Marambio, J., Mordue (Luntz), A.J., Birkett, M., Carvajal, J., Asencio, G., et al. (2007) Behavioural Studies of Host, Non-Host and Mate Location by the Sea Louse, *Caligus rogercresseyi* Boxshall & Bravo, 2000 (Copepoda: Caligidae). *Aquaculture* 271, 70–76.

Pohl, P., Klafke, G., Júnior, J.R., Martins, J.R., Silva Vaz, I., Jr. and Masuda, A. (2012) ABC Transporters as a Multidrug Detoxification Mechanism in *Rhipicephalus* (*Boophilus*) *microplus*. *Parasitology Research* 111, 2345–2351.

Porter, M.L., Cronin, T.W., Mcclellan, D.A. and Crandall, K.A. (2007) Molecular Characterization of Crustacean Visual Pigments and the Evolution of Pancrustacean Opsins. *Molecular Biology and Evolution* 24, 253–268.

Pouliot, J.-F.S., L'heureux, F.S., Liu, Z., Prichard, R.K. and Georges, E. (1997) Reversal of P-Glycoprotein-Associated Multidrug Resistance by Ivermectin. *Biochemical Pharmacology* 53, 17–25.

Quinn, J.J., Zhang, Q.F.C., Georgiev, P., Ilik, I.A., Akhtar, A. and Change, H.Y. (2016) Rapid Evolutionary Turnover Underlies Conserved lncRNA-Genome Interactions. *Genes & Development* 30, 191–207.

Rao, Y., Jan, L.Y. and Jan, Y.N. (1990) Similarity of the Product of the *Drosophila* Neurogenic Gene Big Brain to Transmembrane Channel Proteins. *Nature* 345, 163–167.

Raymond, C.S., Shamu, C.E., Shen, M.M., Seifert, K.J., Hirsch, B., Hodgkin, J. and Zarkower, D. (1998) Evidence for Evolutionary Conservation of Sex-Determining Genes. *Nature* 391, 691–695.

Ritdachyeng, E., Manaboon, M., Tobe, S.S. and Singtripop, T. (2013) Possible Roles of Juvenile Hormone and Juvenile Hormone Binding Protein on Changes in the Integument During Termination of Larval Diapause in the Bamboo Borer Omphisa Fuscidentalis. *Physiological Entomology* 38, 183–191.

Rodriguez, E. M., Medesani, D.A. and Fingerman, M. (2007) Endocrine Disruption in Crustaceans Due to Pollutants: A Review. *Comparative Biochemistry and Physiology a-Molecular & Integrative Physiology* 146, 661–671.

Rosenberg, E. and Zilber-Rosenberg, I. (2011) Symbiosis and Development: The Hologenome Concept. *Birth Defects Research Part C – Embryo Today: Reviews* 93, 56–66.

Ruohola, H., Bremer, K.A., Baker, D., Swedlow, J.R., Jan, L.Y. and Jan, Y.N. (1991) Role of Neurogenic Genes in Establishment of Follicle Cell Fate and Oocyte Polarity During Oogenesis in *Drosophila. Cell* 66, 433–449.

Rytz, R., Croset, V. and Benton, R. (2013) Ionotropic Receptors (IRs): Chemosensory Ionotropic Glutamate Receptors in *Drosophila* and Beyond. *Insect Biochemistry and Molecular Biology* 43, 888–897.

Sahasrabudhe, P.V., Tejero, R., Kitao, S., Furuichi, Y. and Montelione, G.T. (1998) Homology Modeling of an Rnp Domain from a Human RNA-Binding Protein: Homology-Constrained Energy Optimization Provides a Criterion for Distinguishing Potential Sequence Alignments. *Proteins-Structure Function and Genetics* 33, 558–566.

Sapiro, R., Tarantino, L.M., Velazquez, F., Kiriakidou, M., Hecht, N.B., Bucan, M. and Strauss, J.F. (2000) Sperm Antigen 6 Is the Murine Homologue of the *Chlamydomonas reinhardtii* Central Apparatus Protein Encoded by the PF16 Locus. *Biology of reproduction* 62, 511–518.

Sappington, T.W. and Raikhel, A.S. (1998) Molecular Characteristics of Insect Vitellogenins and Vitellogenin Receptors. *Insect Biochemistry and Molecular Biology* 28, 277–300.

Schröder, A., Klein, K., Winter, S., Schwab, M., Bonin, M., Zell, A. and Zanger, U. (2013) Genomics of ADME Gene Expression: Mapping Expression Quantitative Trait Loci Relevant for Absorption, Distribution, Metabolism and Excretion of Drugs in Human Liver. *The Pharmacogenomics Journal* 13, 12–20.

Schwarzenberger, A., Courts, C. and Von Elert, E. (2009) Target Gene Approaches: Gene Expression in *Daphnia magna* Exposed to Predator-Borne Kairomones or to Microcystin-Producing and Microcystin-Free *Microcystis aeruginosa*. *BMC Genomics* 10, 527.

Seeb, J.E., Carvalho, G., Hauser, L., Naish, K., Roberts, S. and Seeb, L.W. (2011) Single-Nucleotide Polymorphism (SNP) Discovery and Applications of SNP Genotyping in Nonmodel Organisms. *Molecular Ecology Resources* 11, 1–8.

Sepulveda, F.A., Torres, J.F., Infante, C.D. and González, M.T. (2017) Potential Role of Ectoparasites (*Zeuxapta seriolae* and *Caligus lalandei*) in the Transmission of Pathogenic Bacteria in Yellowtail Kingfish *Seriola lalandi*, Inferred from Cultivable Microbiota and Molecular Analyses. *Journal of Fish Diseases* 40, 979–985.

Sevatdal, S., Fallang, A., Ingebrigtsen, K. and Horsberg, T.E. (2005) Monooxygenase Mediated Pyrethroid Detoxification in Sea Lice (*Lepeophtheirus salmonis*). *Pest Management Science* 61, 772–778.

Shimeld, S.M., Boyle, M.J., Brunet, T., Luke, G.N. and Seaver, E.C. (2010) Clustered Fox Genes in Lophotrochozoans and the Evolution of the Bilaterian Fox Gene Cluster. *Development Biology* 340, 234–248.

Sinclair, A., Berta, P., Palmer, M., Hawkins, J., Griffiths, B. and Palmer, M. (1990) A Gene from the Human Sex-Determining Region Encodes a Protein With. *Nature* 346, 240–244.

Singh, B. (2010) Matrix Metalloproteinases – an Overview. *Research and Reports in Biology* 1, 1–20.

Sirugo, G., Williams, S.M. and Tishkoff, S.A. (2019) The Missing Diversity in Human Genetic Studies. *Cell* 177, 26–31.

Skoblina, M. (2008) Aquaporins in Gametogenesis of Vertebrate Animals. *Russian Journal of Developmental Biology* 39, 35–41.

Smith, C.A., Mcclive, P.J., Western, P.S., Reed, K.J. and Sinclair, A.H. (1999) Evolution: Conservation of a Sex-Determining Gene. *Nature* 402, 601–602.

Spence, A.M., Coulson, A. and Hodgkin, J. (1990) The Product of *Fem-1*, a Nematode Sex-Determining Gene, Contains a Motif Found in Cell Cycle Control Proteins and Receptors for Cell-Cell Interactions. *Cell* 60, 981–990.

Stavang, J.A., Chauvigné, F., Kongshaug, H., Cerdà, J., Nilsen, F. and Finn, R.N. (2015) Phylogenomic and Functional Analyses of Salmon Lice Aquaporins Uncover the Molecular Diversity of the Superfamily in Arthropoda. *BMC Genomics* 16, 618.

Stepanyan, R., Day, K., Urban, J., Hardin, D.L., Shetty, R.S., Derby, C.C.D., Ache, B.B.W., McClintock, T.S. and Har-, D.D.H. (2006) Gene Expression and Specificity in the Mature Zone of the Lobster Olfactory Organ. *Physiological Genomics* 25, 224–233.

Stucchi, D.J., Guo, M., Foreman, M.G., Czajko, P., Galbraith, M., Mackas, D.D.L. and Gillibrand, P. (2011) Modeling Sea Lice Production and Concentrations in the Broughton Archipelago, British Columbia. In: Jones, S. and Beamish, R. (eds) *Salmon Lice: An Integrated Approach to Understanding Parasite Abundance and Distribution.* John Wiley & Sons Ltd, Chichester, UK. pp. 117–150.

Sun, K.L., Jee, D., De Navas, L.F., Duan, H. and Lai, E.C. (2015) Multiple *in vivo* Biological Processes Are Mediated by Functionally Redundant Activities of *Drosophila* Mir-279 and Mir-996. *PLoS Genetics* 11(6), e1005245.

Sutherland, B.J.G., Jantzen, S.G., Yasuike, M., Sanderson, D.S., Koop, B.F. and Jones, S.R.M. (2012) Transcriptomics of Coping Strategies in Free-Swimming *Lepeophtheirus salmonis* (Copepoda) Larvae Responding to Abiotic Stress. *Molecular Ecology* 21, 6000–6014.

Szakács, G., Paterson, J.K., Ludwig, J.A., Booth-Genthe, C. and Gottesman, M.M. (2006) Targeting Multidrug Resistance in Cancer. *Nature Reviews Drug Discovery* 5, 219.

Tadiso, T.M., Krasnov, A., Skugor, S., Afanasyev, S., Hordvik, I. and Nilsen, F. (2011) Gene Expression Analyses of Immune Responses in Atlantic Salmon During Early Stages of Infection by Salmon Louse (*Lepeophtheirus salmonis*) Revealed Bi-Phasic Responses Coinciding with the Copepod-Chalimus Transition. *BMC Genomics* 12, 141.

Tarifeno-Saldivia, E., Valenzuela-Miranda, D. and Gallardo-Escárate, C. (2017) In the Shadow: The Emerging Role of Long Non-Coding RNAs in the Immune Response of Atlantic Salmon. *Developmental and Comparative Immunology* 73, 193–205.

Tasumi, S., Norshida, I., Boxshall, G.A., Kikuchi, K., Suzuki, Y. and Ohtsuka, S. (2015) Screening of Candidate Genes Encoding Proteins Expressed in Pectoral Fins of Fugu *Takifugu rubripes*, in Relation to Habitat Site of Parasitic Copepod *Caligus fugu*, Using Suppression Subtractive Hybridization. *Fish and Shellfish Immunology* 44, 356–364.

Taylor, D.H., Chu, E.T.J., Spektor, R. and Soloway, P.D. (2015) Long Non-Coding RNA Regulation of Reproduction and Development. *Molecular Reproduction and Development* 82, 932–956.

Tew, K.D. (2016) Glutathione-Associated Enzymes in Anticancer Drug Resistance. *Cancer Research* 76, 7–9.

Touhara, K. and Vosshall, L.B. (2009) Sensing Odorants and Pheromones with Chemosensory Receptors. *Annual Review of Physiology* 71, 307–332.

Tufail, M. and Takeda, M. (2008) Molecular Characteristics of Insect Vitellogenins. *Journal of Insect Physiology* 54, 1447–1458.

Uniprot, C. (2019) Uniprot: A Worldwide Hub of Protein Knowledge. *Nucleic Acids Research* 47, D506–D515.

Valenzuela-Miranda, D. and Gallardo-Escárate, C. (2016a) *Caligus rogercresseyi* Serine Proteases: Transcriptomic Analysis in Response to Delousing Drugs Treatments. *Aquaculture* 465, 65–77.

Valenzuela-Miranda, D. and Gallardo-Escárate, C. (2016b) Novel Insights into the Response of Atlantic Salmon (*Salmo salar*) to *Piscirickettsia salmonis*: Interplay of Coding Genes and lncRNAs During Bacterial Infection. *Fish and Shellfish Immunology* 59, 427–438.

Valenzuela-Miranda, D., Núñez-Acuña, G., Valenzuela-Muñoz, V., Asgari, S. and Gallardo-Escárate, C. (2015) MicroRNA Biogenesis Pathway from the Salmon Louse (*Caligus rogercresseyi*): Emerging Role in Delousing Drug Response. *Gene* 555, 231–241.

Valenzuela-Miranda, D., Etebari, K., Asgari, S., and Gallardo-Escárate, C. (2017) Long Noncoding RNAs: Unexplored Players in the Drug Response of the Sea Louse *Caligus rogercresseyi*. *Agri Gene* 4, 1–7.

Valenzuela-Miranda, D., Valenzuela-Munoz, V., Núñez-Acuña, G. and Gallardo-Escarate, C. (2020) Long-term Serial Culture of *Piscirickettsia salmonis* Leads to a Genomic and Transcriptomic Reorganization Affecting Bacterial Virulence. *Aquaculture* 2020, 529.

Valenzuela-Muñoz, V., Bueno-Ibarra, M.A. and Escárate, C.G. (2012) Characterization of the Transcriptomes of *Haliotis rufescens* Reproductive Tissues. *Aquaculture Research* 1–15.

Valenzuela-Muñoz, V., Chavez-Mardones, J. and Gallardo-Escárate, C. (2015a) RNA-Seq Analysis Evidences Multiple Gene Responses in *Caligus rogercresseyi* Exposed to the Anti-Salmon Lice Drug Azamethiphos. *Aquaculture* 446, 156–166.

Valenzuela-Muñoz, V., Sturm, A. and Gallardo-Escárate, C. (2015b) Transcriptomic Insights on the ABC Transporter Gene Family in the Salmon Louse *Caligus rogercresseyi*. *Parasites & Vectors* 8, 1–14.

Van Leeuwen, T., Dermauw, W., Grbic, M., Tirry, L. and Feyereisen, R. (2013) Spider Mite Control and Resistance Management: Does a Genome Help? *Pest Management Science* 69, 156–159.

Veitia, R.A. (2010) Foxl2 Versus Sox9: A Lifelong "Battle of the Sexes". *Bioessays* 32, 375–380.

Vera, J.C., Wheat, C.W., Fescemyer, H.W., Frilander, M.J. and Crawford, D.L. (2008) Rapid Transcriptome Characterization for a Nonmodel Organism Using 454 Pyrosequencing. *Molecular Ecology* 17, 1636–1674.

Verkman, A. and Mitra, A.K. (2000) Structure and Function of Aquaporin Water Channels. *American Journal of Physiology – Renal Physiology* 278, F13–F28.

Vezzulli, L., Pruzzo, C., Huq, A. and Colwell, R.R. (2010) Environmental Reservoirs of *Vibrio cholerae* and Their Role in Cholera. *Environmental Microbiology Reports* 2, 27–33.

Villani, F., White, G.B., Curtis, C.F. and Miles, S.J. (1983) Inheritance and Activity of Some Esterases Associated with Organophosphate Resistance in Mosquitoes of the Complex of *Culex pipiens* L. (Diptera: Culicidae). *Bulletin of Entomological Research* 73, 153–170.

Villegas, V.E. and Zaphiropoulos, P.G. (2015) Neighboring Gene Regulation by Antisense Long Non-Coding RNAs. *International Journal of Molecular Sciences* 16, 3251–3266.

Vinson, S.B. (1975) Biochemical Coevolution between Parasitoids and Their Hosts. In: Price, P.W. (ed.) *Evolutionary Strategies of Parasitic Insects and Mites*. Springer US, Boston, MA. pp. 14–48.

Vontas, J., David, J., Nikou, D., Hemingway, J., Christophides, G., Louis, C. and Ranson, H. (2007) Transcriptional Analysis of Insecticide Resistance in *Anopheles stephensi* Using Cross-Species Microarray Hybridization. *Insect Molecular Biology* 16, 315–324.

Walker, R.J., Papaioannou, S. and Holden-Dye, L. (2009) A Review of Fmrfamide-and Rfamide-Like Peptides in Metazoa. *Invertebrate Neuroscience* 9, 111–153.

Walsh, T.K., Lyndon, A.R. and Jamieson, D.J. (2007) Identification of cDNAs Induced by the Organophosphate Trichlorphon in the Parasitic Copepod *Lepeophtheirus salmonis* (Copepoda: Caligidae). *Pesticide Biochemistry and Physiology* 88, 26–30.

Wang, Z., Gerstein, M. and Snyder, M. (2009) RNA-Seq: A Revolutionary Tool for Transcriptomics. *Nature Reviews Genetics* 10, 57–63.

Weng, R.F., Chin, J.S.R., Yew, J.Y., Bushati, N. and Cohen, S.M. (2013) Mir-124 Controls Male Reproductive Success in *Drosophila*. *Elife* 2, e00640.

Wheelock, G.D. and Scott, J.G. (1992) The Role of Cytochrome P450ipr in Deltamethrin Metabolism by Pyrethroid-Resistant and Susceptible Strains of House Flies. *Pesticide Biochemistry and Physiology* 43, 67–77.

Wilhelm, B.T. and Landry, J.R. (2009) RNA-Seq-Quantitative Measurement of Expression through Massively Parallel RNA-Sequencing. *Methods* 48, 249–257.

Wilson, C.H. and Hartline, D.K. (2011) Novel Organization and Development of Copepod Myelin. I. Ontogeny. *Journal of Comparative Neurology* 519, 3259–3280.

Wilstermann, A. and Vidal, S. (2013) Western Corn Rootworm Egg Hatch and Larval Development under Constant and Varying Temperatures. *Journal of Pest Science* 86, 419–428.

Wood, O., Hanrahan, S., Coetzee, M., Koekemoer, L. and Brooke, B. (2010) Cuticle Thickening Associated with Pyrethroid Resistance in the Major Malaria Vector *Anopheles funestus*. *Parasites & Vectors* 3, 67.

Yang, J.R. and Zhang, J. (2015) Human Long Noncoding RNAs Are Substantially Less Folded Than Messenger RNAs. *Molecular Biology and Evolution* 32, 970–977.

Yu, Y.X., Zhang, Z., Wang, Y.G., Liao, M.J., Rong, X.J., Li, B., Wang, K., Chen, J. and Zhang, H. (2019) Complete Genome Sequence of *Photobacterium damselae* Subsp. *Damselae* Strain Sspd1601 Isolated from Deep-Sea Cage-Cultured *Sebastes schlegelii* with Septic Skin Ulcer. *International Journal of Genomics* 2019, 4242653.

Xiong, C.R., Fang, F.J., Chen, L., Yang, Q.G., He, J., et al. (2014) Trypsin-Catalyzed Deltamethrin Degradation. *PLoS ONE* 9(3), e89517.

Zhang, B.H., Wang, Q.L. and Pan, X.P. (2007) MicroRNAs and Their Regulatory Roles in Animals and Plants. *Journal of Cellular Physiology* 210, 279–289.

Zhao, X.Y. and Lin, J.D. (2015) Long Noncoding RNAs: A New Regulatory Code in Metabolic Control. *Trends in Biochemical Sciences* 40, 586–596.

Zhu, L.H., Liu, J.T., Dao, J.W., Lu, K., Li, H., et al. (2016) Molecular Characterization of *S. japonicum* Exosome-Like Vesicles Reveals Their Regulatory Roles in Parasite-Host Interactions. *Scientific Reports* 6, 25885.

Zientz, E., Dandekar, T. and Gross, R. (2004) Metabolic Interdependence of Obligate Intracellular Bacteria and Their Insect Hosts. *Microbiology and Molecular Biology Reviews* 68, 745–770.

CHAPTER 13

Salmon-louse interaction and immunological consequences for the host

Mark Fast and Laura Braden

Department of Pathology and Microbiology, Atlantic Veterinary College, Charlottetown, PEI, Canada

13.1 Introduction

Ectoparasitic copepods are an unusually successful group of parasites with numerous unique adaptions to their specific host species, allowing for rather complex ecological, physiological, and immunological interactions. Most representatives of this group are marine in nature, providing rather distinct boundaries for their life history interactions with anadromous hosts. The circumpolar distribution of *Lepeophtheirus salmonis* (Krøyer 1837), commonly referred to in the Northern Hemisphere as the 'salmon louse', and its impacts on wild and farmed fish in both the Pacific and Atlantic Oceans offer a singular insight into host-parasite interactions, where different hosts can be compared across differing environments while keeping the parasite end of the equation 'relatively constant.' Furthermore, there is a good deal of field data from wild and farmed sources that can be compared against laboratory experiments in order to validate and/or determine whether controlled exposure studies are indeed representative of the natural interaction.

Lepeophtheirus salmonis has diverged into two subspecies: *L. salmonis salmonis* for those found in the Atlantic, and *L. salmonis oncorhynchi* subspecies *novo* for those from the Pacific (Skern-Mauritzen et al., 2014). This divergence must be considered when examining host-pathogen interactions, as these subspecies are consistent with independent co-evolution of the parasite and have occurred in salmonid species within their respective oceans over the last 2.5–11 million years (Yazawa et al., 2008). Historically, the North Pacific and North Atlantic have been drastically different environments, and this has necessitated the need for different aquaculture industry treatment control measures. For example, the refugia population of wild salmonids is larger than the cultured population of Atlantic salmon in Pacific North America, whereas the wild population is endangered and rare in Atlantic North America, or less dense in Europe than cultured populations.

Although the prevalence of *L. salmonis* on these salmonid populations, be they wild, cultured, sentinel, or other, approaches 100% in both the Pacific and Atlantic Oceans (Butler, 2002; Beamish et al., 2005; Jackson et al., 2013; Frederick, 2018; Sandvik et al., 2020), the frequency of *L. salmonis* on these hosts in the field belay the differential susceptibility of host species examined in the laboratory. Johnson and Albright (1992a), Fast et al. (2002), and Jones et al. (2007) were the first to describe differential rejection in the laboratory of *L. salmonis* by salmonids, with coho (*Oncorhynchus kisutch* (Walbulm 1792)) and pink salmon (*O. gorbuscha* (Walbulm 1792)) showing divergent resistant mechanisms compared to Atlantic salmon and other susceptible hosts. This was later shown to be impacted by size and life stage in pink salmon by Sutherland et al. (2011) and Braden et al. (2015a) respectively. These and supplementary studies on sockeye salmon (*O. nerka*) (Jakob et al., 2013; Long et al., 2018) show general agreement with field assessments of returning adult salmon in Central British Columbia by Beamish et al. (2005), where average intensities on coho and chinook salmon (*O. tschawyscha* (Walbulm 1792)) were significantly lower than lice intensities on chum (*O. keta* (Walbulm 1792)), sockeye, and pink salmon.

Host response to infection is a combination of many factors including the age-stage of the host, environmental conditions (in particular temperature/salinity), and the life stage of the parasite the host encounters. While there have been numerous studies on both systemic and localized effects on *L. salmonis* infection, perhaps our most comprehensive understanding has occurred through large-scale transcriptomics studies and their examination of the host-parasite interaction over time and space (Skugor et al., 2008; Tadiso et al., 2011; Sutherland et al., 2014; Braden et al., 2015b, 2016, 2022 – in review; Holm et al., 2016; Umasuthan et al., 2020). However, it is difficult to draw general conclusions on pathways associated with true resistance, as most data are derived from the more ecologically threatened and economically viable/important species Atlantic salmon. Furthermore, the host-parasite relationship in these studies is assessed at different time points, within different tissues, and in some cases using differing exposure methods. Thus, the goal of this review will be to focus more on initiation of the host response in the skin/fin, and then to consider other sources of information for a more complete understanding of this host-parasite interaction.

13.2 Wound healing in salmon and its alteration through sea lice infection

Sentinel studies on *L. salmonis* (Pert et al., 2014; Frederick, 2018) have shown that infection pressure on naïve salmon is almost completely due to the infectious copepodid stage. Others have shown that, while swim/water flow speed can impact settlement location, the most common areas of initial infection on salmon are the fins (Bron et al., 1991; Samsing et al., 2016). Within minutes of copepodid exposure and settlement, salmonids across susceptibilities quickly exhibit erratic behaviors such as increased jumping, rolling, and changes in burst swim patterns (Samsing et al., 2016; Bui et al., 2016). Interestingly, cohabitation studies have demonstrated that initial settlement is similar between susceptible (Atlantic salmon, rainbow trout) and resistant hosts (coho salmon) up to 24 h post-exposure (Fast et al., 2002). Because of this commonality, we must first understand the normal response to skin/fin trauma (that is, wound healing) and how this is regulated to determine what is 'normal'," what happens due to mechanical damage and/or parasite virulence, and how this changes between different hosts.

Wound healing in fish, similar to other vertebrates, involves four main steps: re-epithelialization, inflammation, cell proliferation with granulation tissue formation, and tissue remodeling. Unique to both amphibians and fish is that keratocytes (epithelial cells) migrate to cover the wound from the intact surrounding tissue. In Atlantic salmon, this may occur within a few hours for small wounds but is dependent on temperature, stress, and nutrition, among other things (Andrews et al., 2015; Jensen et al., 2015). A recent study by Sveen et al. (2019) characterized the short- and long-term wound repair mechanisms in Atlantic salmon smolts by excising 5 mm biopsies of skin from behind the dorsal fin and following the molecular and histological progression over time. The location of these wounds and environmental conditions such as temperature and salinity make this work an excellent resource of normal healing mechanisms in Atlantic salmon to compare salmon lice infection against. As would be expected the healing process was dominated early on by re-epithelialization and hemostasis within the first one to three days post-wounding (dpw) (Sveen et al., 2019). At 1 dpw, coagulation (*antithrombin, urokinase, serpine1*), fibrin clot formation, and extracellular matrix adhesion (*fibronectin* and *integrins*) were induced along with epithelial migration (*epigen precursor, unconventional myosins*), cell to cell adherence (*epithelial cadherins*), and components of the basement membrane (*laminin* and *hyaluronidase-2*). Wounding was also accompanied by hyperpigmentation at the wound edges and spreading of pigmented bodies at the wound surface. This was followed by the expression of the most upregulated genes in the study, *matrix metalloproteinase 9* and *13*, which increased 128-fold by 3 dpw. The authors observed migrating keratocytes and mucous cells at the wound margin and suggested the formation of fibrin-like structures initially protect and provide a surface aiding in epithelial sliding/migration over the wound. Transcription of acute phase proteins (for example, *serum amyloid A*), eicosanoid metabolism genes (for example, *leukotriene hydrolase*), lymphocyte specific genes (for example, *cd4*), neutrophil and other leucocyte chemotactic factors such as *macrophage inflammatory protein* and *leukocyte cell-derived chemotaxin*, and inflammatory cytokine receptors (*il-1r2* and *il-6rα*) peaked in the skin at 1–3 dpw, which was followed by infiltration of inflammatory cells into the wound and accompanied by tissue degradation between three and seven days post wounding (Sveen et al., 2019).

The early stages of *L. salmonis* infection on coho salmon fins shows many similarities to the normal skin healing process in Atlantic salmon, both histologically and transcriptomically. The initial work by Johnson and Albright (1992a) was carried out on coho salmon with *L. salmonis oncorhynchii nova*, whereas Fast et al. (2002) and the recent work by Braden et al. (2016, 2022 – in review) involved infecting coho salmon with the Atlantic subsp. *L. salmonis salmonis*. Yet despite the parasite subsp. and host size differences across studies, similar gross and histological descriptions of the host response to the parasite were reported followed by >95% reductions in lice within 10–15 days post-infection (dpi) (Johnson and Albright, 1992a; Fast et al., 2002; Braden et al., 2016, 2022 – in review). Coho salmon also reject *Caligus rogercresseyi* (Boxshall & Bravo 2000), the parasitic copepod species impacting Chilean salmon aquaculture, with the same intensity and over the same one- to two-week time frame (Vera-Bizama et al., 2015; Valenzuela-Muñoz et al., 2016, 2017, 2018). And unlike pink salmon, which enter seawater at a much smaller/earlier life stage (200–300 mg; reviewed in Groot and Margolis, 1991) and develop resistance to *L. salmonis* with increasing growth/size (≥ 700 mg), resistance in coho salmon appears to be as intensive and successful in rejecting *L. salmonis* within 24 h of entry/acclimation to seawater (at 10–12 g, more than two lice/g) as it is after one month in seawater (at 20 g, more than two lice/g) (Jones et al., 2008; Sutherland et al., 2011; Braden et al., 2016, 2022 – in review). All of these lines of evidence suggest a conserved and extremely effective response by coho salmon to ectoparasitic challenge from caligid copepods.

Within 24 h after exposure, moderate inflammation with multifocal granulocytic infiltration is observed in coho salmon fin at the louse attachment site, accompanied by submucosal and intravascular granulocyte infiltration in affected gills. This continues through 4 dpi, where significant melanin deposition is observed in the fin and epidermal hyperplasia is initiated. Granulocyte infiltration is reduced, and epidermal hyperplasia intensified and more commonly observed by 7 dpi, and this is accompanied by eosinophilic granular cell infiltration in the gill arch subcutis. Full encapsulation of the parasite in some cases occurs by 10 dpi (Johnson and Albright, 1992a; Braden et al., 2016, 2022 – in review). The coho salmon transcriptome response to *L. salmonis* copepodids is localized to the attachment site, is initiated within 24 h, and is primarily composed of significant induction of innate (chitin/lectin) sensing and inflammatory pathways which peaks by 7 dpi. Whereas protein synthesis and transport are the common pathways elicited at 1 dpi, acute inflammation, T-cell differentiation, apoptosis, and lymphocyte differentiation are all strongly impacted by 4 dpi in the fin. During this period (1–4 dpi) and right through to 10 dpi, cellular activation, degranulation, chemotaxis, and acute inflammation are hallmarks of the coho salmon fin response to *L. salmonis* (Braden et al., 2016, 2022 – in review) (Fig. 13.1).

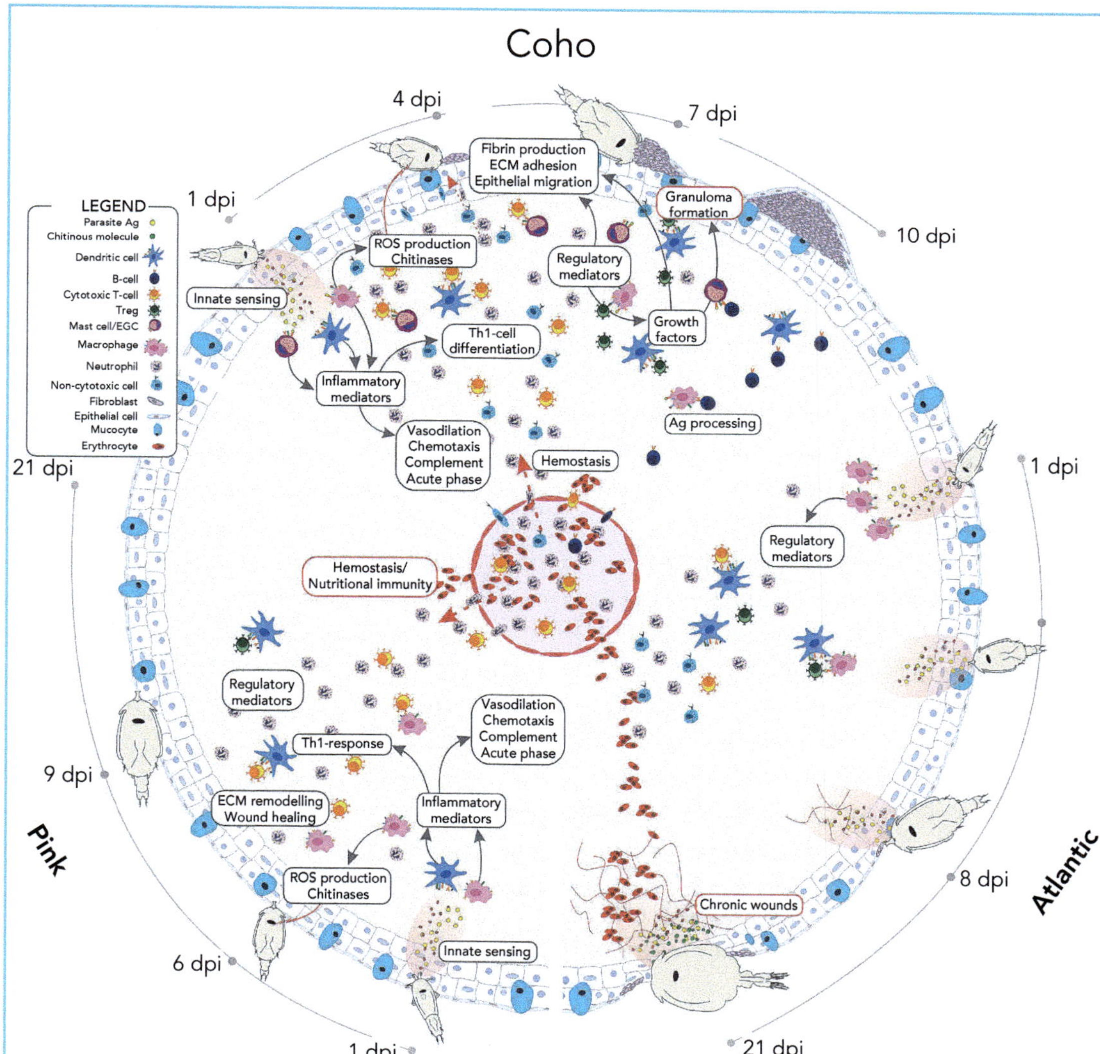

Fig. 13.1 Interaction diagram of salmonid cell signaling and regulation over time in response to *Lepeophtheirus salmonis* infection. Individual host response representations are based on integration of transcriptomic and histological studies on coho, pink, and Atlantic salmon fin/skin studies (Sutherland et al., 2014; Braden et al., 2015b, 2016, 2017, 2022 – in review; Umasuthan et al., 2020).

Copepodid settlement involves insertion of the second antennae and first maxilla into the host epithelium to initiate feeding so the parasite can moult into a chalimus I larvae and secure its attachment through the frontal filament (Bron et al., 1991). As in other crustaceans, these structures contain an exocuticle which is made up of chitin, sclerotized protein, hard minerals (calcite or carbonate and so on), phospholipids, and sugars like glucose, mannose, glucosamine, and others. Therefore, the host epithelium is likely exposed to these exoskeletal components immediately upon infection. Indeed, within the first 24 h of infection, mannose binding lectin (*mbl*), macrophage mannose receptor (*mrc1*), macrophage inducible Ca^{2+}-dependent lectin (*mincle/clec4e*), and several other pattern recognition receptors (PRRs) that recognize mannose, chitin, or other sugar moieties are strongly induced in coho salmon fins (Braden et al., 2016, 2022 – in review) (Fig. 13.1). As stated by Braden et al. (2022 – in review) this strongly supports chitin and/or sugar moiety sensing/binding as part of the initial response in coho salmon.

Chitin has been shown to modulate innate immune responses of keratinocytes, mainly through TLR-2, resulting in upregulation of cytokines/chemokines (CXCL8) (Koller et al., 2011), and chitin stimulates pro-inflammatory cytokines like IL-1β, IL-12, and TNFα from macrophages contributing to Th1-biased responses, at least partially mediated through the macrophage mannose receptor (Shibata et al., 1997, 2000; Lee et al., 2011). Under breeding selection for resistance in Atlantic salmon families, Robledo et al. (2018) observed *mrc1* and *clec4e* in the skin positively associated with *C. rogercresseyi* abundance on the fins and Holm et al. (2016) found these genes associated with modest *L. salmonis* reductions in skin of Atlantic salmon fed functional diets.

The fast-acting nature of this response to a parasite (likely due to pre-formed mediators), accompanied by cellular activation, degranulation, chemotaxis, and acute inflammation in the fins of coho salmon throughout the 1–10 dpi period, supports a mast cell (MC)-like driven hypersensitivity. Interestingly, Braden et al. (2015b) observed pre-formed mediators in mucous cells as well, positively staining with antibodies for $IL1\beta^+$ protein and significantly higher populations of $IL1\beta^+$- and $TNF\beta^+$-lymphocytes at the site of adult *L. salmonis* attachment in coho compared to Atlantic salmon as early as 24 hpi. MC degranulation has been shown to be induced by lectins and provides an activation loop for neutrophil recruitment (Moreno et al., 2003). In fact, Braden et al. (2016, 2022 – in review) describe GSEA analysis which suggests significant MC-like involvement in the coho salmon response, including significant upregulation of FcγR. While MCs have not been fully characterized in teleosts, eosinophilic granule cells (EGC) have been described to carry out MC-like function in fish (Reite and Evensen, 2006). The response in coho salmon, however, does not show consistent EGC proliferation and/or trafficking to the infected skin tissue in all cases, but rather, they are a common feature in the skin and gill epithelium of coho salmon, possibly more so than in susceptible species like Atlantic salmon (Fig. 13.2).

Granulomatous inflammation and EGC infiltration have been previously described as responding to dead/remnant copepods and the chitinous structures therein, in fish hosts of lernaepodid and penellid copepods, *Salmincola californiensis* (Dana, 1852) and *Penella balaenopterae* (Koren & Danielssen, 1877) respectively (Roberts et al., 2004; Mazzariol et al., 2011). In active infections of these parasitic copepods, however, these responses may be maladaptive as they are also related to host fibrous tissue formation, and eventual formation of epithelioid cells that encapsulate the bulla or anchors of the parasite, effectively enhancing their holdfast.

While the transcriptome of Atlantic salmon fin samples has not been assessed at coordinated early time points to make true comparisons with coho salmon possible, Tadiso et al. (2011) did show transient increases in the skin of a few mannose binding receptors and C-type lectins. Yet these were

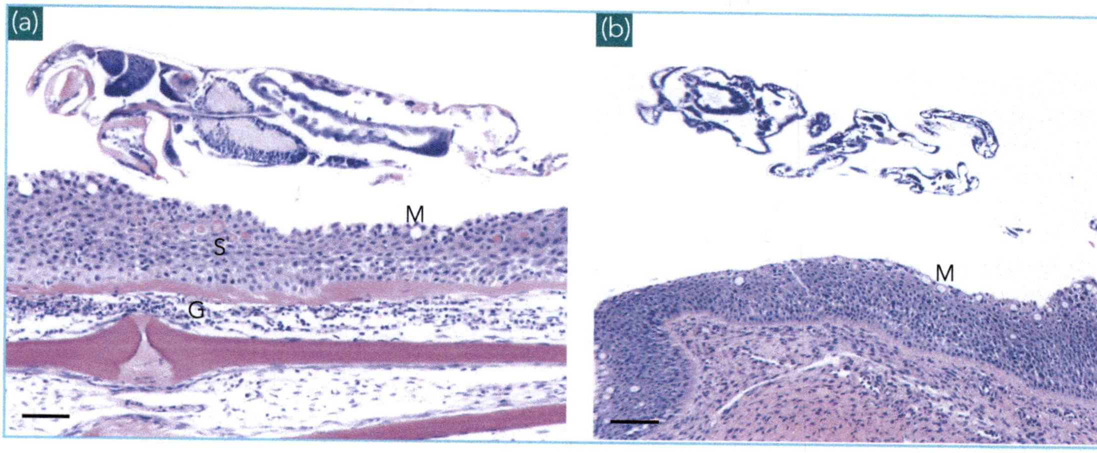

Fig. 13.2 Skin epithelium section of coho (A) and Atlantic (B) salmon showing general architecture under larval (chalimus) *Lepeophtheirus salmonis* infection. (M) mucous cell; (G) granulocyte infiltration; (S) granulocyte infiltration.

small and largely absent after 3 dpi. Acute phase genes (serum amyloid A, C1q complement factors), interferons, extra cellular matrix (ECM), and tissue remodeling matrix metalloproteases (MMPs) were all downregulated, and no lymphocyte response was measured, during these initial 1–5 dpi (Tadiso et al., 2011). Whereas coho salmon display acute inflammation, epithelial hyperplasia, and evidence of wound healing processes as described by Sveen et al. (2019), there is little to no comparable reaction observed in Atlantic salmon other than parasite-derived epithelial erosion (Johnson and Albright, 1992a; Covello et al., 2012). By 7 dpi (60–70 dd), we are able to make direct comparisons of fin responses across salmonids. This is a very important time point as this represents the transition of copepod to chalimus I life stages in the parasite, which generally occurs at ca. 50 dd, and also represents an inflection point in the bi-phasic response of Atlantic salmon to sea lice (Skugor et al., 2008; Tadiso et al., 2011).

13.3 Differential host response following moult to chalimus I

Umasuthan et al. (2020) examined *L. salmonis* attached and adjacent fin sites in Atlantic salmon at 7 dpi, using similar methods to Braden et al. (2016, 2022 – in review), and while the first utilized microarrays and the latter RNA-sequencing, with some differences in differential analysis, these studies used the same subspecies of lice, temperature, and sampling regime, presenting a unique opportunity to contrast the transcriptome response to lice with normal wound repair. Comparing coho responses (Braden et al., 2016, 2022 – in review) to Atlantic responses (Umasuthan et al., 2020) we find approximately a two to one ratio of differential gene expression (Fig. 13.3), with a shared dysregulation of ca. 180 genes. Aside from the processes already mentioned in coho salmon fin, responses at 7 dpi were also characterized by exogenous antigen processing, leucocyte chemotaxis, coagulation, TLR2 signaling, response to TNF, and TCR/FcR signaling (that is, T and B-cell lymphocytes) (Braden et al., 2016, 2022 – in review). In contrast, pattern recognition and chemoattraction to the site of injury are comparatively absent in Atlantic salmon at 7 dpi. Pro-inflammatory cytokines/receptors (il-1β, il12b, litaf, cxcr1, cxcr4), chemokines (il8, ccl2, ccl4, ccl20), and complement factors

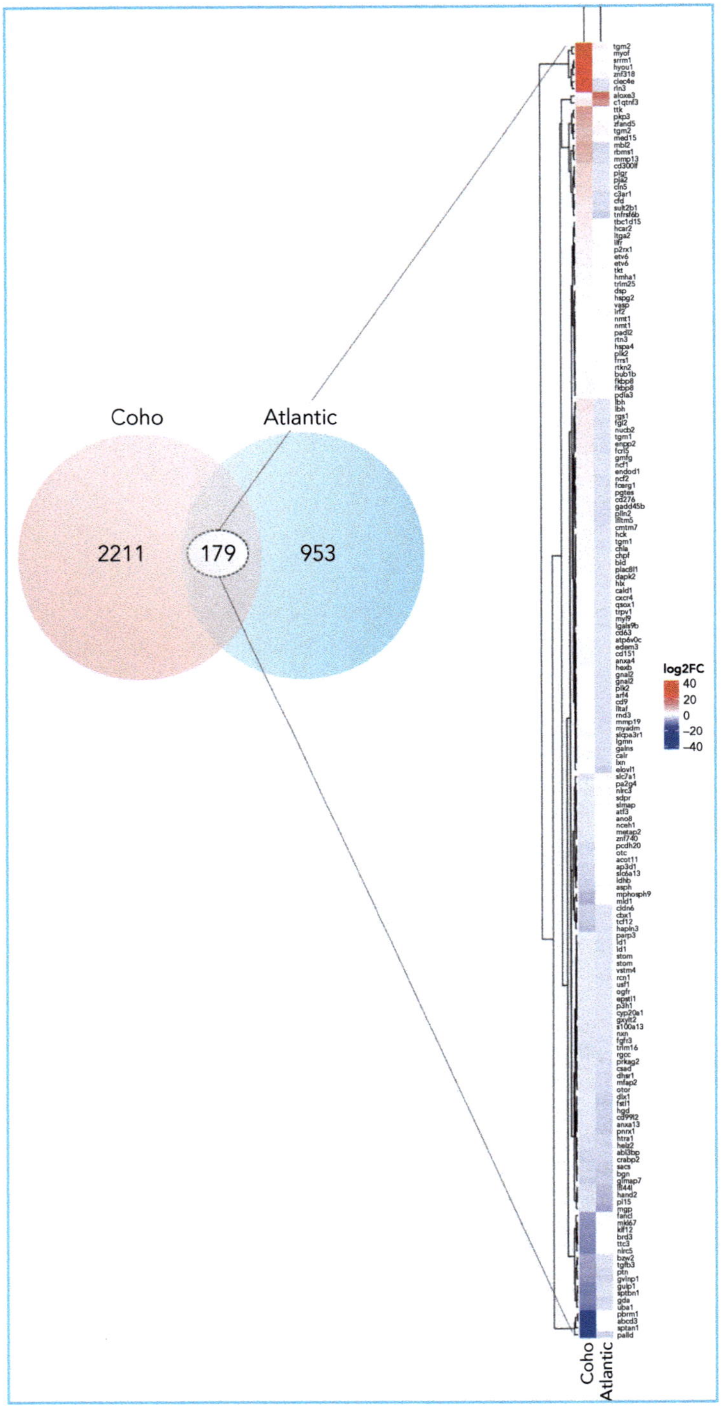

Fig. 13.3 Comparison of Atlantic (blue; differentially regulated genes) and coho (pink; differentially regulated genes) salmon fin transcriptome responses at seven days post-infection (Braden et al., 2016, 2022 – in review; Umasuthan et al., 2020). Shared differential genes (179) are identified in Venn diagram and individual heatmap highlighting fold change (FC) differences across species for individual genes.

(c3aR, c5aR, c4, c6, c1qb), which are essential signals for setting up gradients at sites of injury to attract phagocytic and antigen presenting cells (neutrophils, macrophages, dendritic cells, and so on), are all highly induced in coho salmon fins but downregulated in Atlantic salmon fins at this point (Table 13.1). Many of these same pro-inflammatory and tissue repair genes (*il1β, il8, tnfα, cox2, c/ebpβ, mmp9, mmp13*) have also been reported in coho salmon skin infected with adult *L. salmonis oncorhynchii nova* (Braden et al., 2015b) and Atlantic salmon (*il1β, mmp9*) fed functional diets resulting in reduced *L. salmonis salmonis* infection (Covello et al., 2012; Cai et al., 2022). Rolling and endothelial cell adhesion of immune effector cells into the affected tissue also appears to be divergent between coho and Atlantic salmon fins at 7 dpi. Adhesion molecules (kcad), vascular cell adhesion molecule (vcam1), intercellular adhesion molecule (icam), and integrins (itgb1) are all upregulated in coho salmon and downregulated in Atlantic salmon as is transglutaminase-2, which strongly binds fibronectin and is important in ECM stabilization and cell adhesion. Hence, cellular adhesion is also important in re-epithelialization as well as tissue remodeling. Two strongly upregulated genes in the fin of coho salmon at 7 dpi were *myof* which regulates VEGF and *pkp3* which regulates epithelial cell-cell adhesions. Furthermore, two of the most important factors involved in salmonid wound healing, mmp 9 and mmp13 (Sveen et al., 2019), were strongly induced in coho salmon fin, as were mmp14, mmp17, mmp19, mmp23, and mmp25. In contrast, mmp2, mmp13, mmp19, and mmp23 were downregulated in Atlantic salmon fins (Table 13.1). This divergent response in tissue remodeling and wound repair may represent a key piece of what is driving chronic pathology in the latter. Of the 24 different MMPs, each has their own substrate specificity and functionality that under homeostasis act in concert to regulate cell-cell and cell-matrix interactions by releasing cytokines and growth factors, and controlling cellular effector crosstalk (reviewed in Krishnaswamy et al. (2017)). Hence, development of chronic wounds, as observed in Atlantic salmon, is a consequence of downregulation of both MMPs (for example, mmp2) and inhibitors that control them (for example, timp3). In the absence of these critical regulators of tissue repair, cellular migration, and cutaneous immunity at the attachment site of the parasite, chronic wounds persist. Importantly, the fin transcriptome in coho salmon is representative of an encapsulative response and as such, the prominent and dramatic overexpression of ECM and tissue remodeling components should be considered in that context.

The exact mechanism by which coho salmon immune effector cells negatively impact the louse remains unknown. Similar to what has been previously suggested for pink salmon (Sutherland et al., 2014), coagulation (*fgl2, f3, f5, plat, plau*) cascades initiated by coho salmon may reduce the nutritional content of the louse meal, and superoxide production by neutrophils/granulocytes (*ncf1, ncf2*) would likely damage exoskeletal structures of the louse, as would chitinases (*chia*) and lysozymes, which have been shown to be highly induced in coho salmon mucus late in infection with *L. salmonis* (Fast et al., 2002). The louse transcriptome at these same time points can give us an understanding of the louse side of the relationship with a resistant host, and dual-RNA sequencing in Braden et al. (2016, 2022 – in review) allowed louse transcripts to be matched with host fin transcripts simultaneously. At 1 and 4 dpi, chitin synthesis, development, homeostasis, antioxidant, ECM degradation, and virulence genes were all overexpressed in the louse. At 4 dpi, the addition of immune and coagulative pathways in the louse also appear to be induced. This contrasts with the 7 and 10 dpi transcriptome of *L. salmonis*, which is limited to dysregulation of ECM degradation and coagulation genes.

Other studies on *L. salmonis* and *C. rogercresseyi* have shown key chitin synthesis pathway genes are cyclical, some increasing and others decreasing within the transition between moults (Eichner et al., 2015; Poley et al., 2018; Michaud et al., 2019; Braden et al., 2020a). In Braden et al. (2016, 2022 – in review), chitin synthesis genes upregulated at 1 and 4 dpi are no longer impacted by 7–10 dpi

Table 13.1 Categories of select differentially expressed genes associated with the fin response in resistant (coho salmon) and susceptible (Atlantic salmon) hosts during *Lepeophtheirus salmonis* infection. Transcriptomic data for coho salmon were obtained from Braden et al. (2016, 2022 – in review) while data for Atlantic salmon were obtained from Umasuthan et al. (2020).

Category	Gene	Expression		Description
		Coho	Atlantic	
Cellular effector markers	*cd209e/dcsign*	4.87	−1.85	PRR for mannose on DCs; initiation of primary immune response; presents Ag to T-cells
	cd63	0.88	−1.26	Involved in MC degranulation
	cd276	2.38	−1.79	T-cell mediated immunity
	hck	2.13	−1.54	Regulator of innate immune responses
	pigr	5.94	−1.62	Transcytosis of Ig across epithelial mucosa
	fcerg1	2.46	−1.82	Mediates allergic inflammatory signaling in mast cells
	mbl	9.58	−2.57	PRR for mannose residues; activates complement
	chia	2.03	−1.71	PRR that recognizes chitin; degrades into smaller immunogenic fragments
	cd300lf	5.93	−1.49	Inhibitory receptor for mast cells
	clec4e/mincle	24.9	−1.57	PRR that recognizes fungi and bacteria; interacts with FcRg/FCεR1γ
	fcrl5	3.54	−1.90	B-cell development
	mrc1	8.86	nc	Macrophage phagocytic receptor for bacteria, fungi; response to chitin
	cd22	9.44	nc	B-cell interactions; positive and negative regulation of BCR signaling
Inflammation	*ptges*	2.36	−1.72	PGE2 production; key role in early inflammation
	alox5ap	2.77	nc	Required for leukotriene biosynthesis
	alox5	2.01	nc	Arachidonate metabolism
	ccr1	3.04	nc	Receptor for C-C chemokines
	ccl2	2.34	nc	Chemotactic for monocytes, basophils
	ccl4	4.70	nc	Inflammatory; antiviral
	ccl13	2.36	nc	Chemotactic for monocytes, lymphocytes, basophils, and eosinophils
	ccl19	−2.61	nc	Chemotactic for T and B cells
	ccl20	3.10	nc	Chemotactic for DCs, T and B cells in epithelial mucosa
	ccl21	−3.40	nc	Inhibits hemopoiesis
	ccl25	−1.45	nc	T-cell development
	cxcr1	7.41	nc	Receptor for IL8 on neutrophils
	cxcr4	1.53	−1.62	Tissue regeneration
	cxcl11	2.86	nc	Chemotactic for activated T-cells; skin immunity
	cxcl13	4.25	nc	Chemotactic for B-cells
	cxcl14	1.84	nc	Chemotactic for neutrophils, DCs
	il1b	3.57	nc	Pro-inflammatory cascades
	ilr2a	5.54	nc	Decoy receptor for IL1β; neutralizes IL1β
	il6ra	1.87	nc	Regenerative and anti-inflammatory; drives T-cells to Th_{17} via DC signaling
	il7ra	1.48	nc	Receptor for IL7 and TSLP; growth factor for lymphoid cells
	il8	4.40	nc	Chemotactic for neutrophils, basophils, T-cells; activates neutrophils

Table 13.1 (Continued)

Category	Gene	Expression		Description
		Coho	Atlantic	
	il10rb	1.29	nc	Receptor for IFN-stimulated genes; antiviral
	il12b	4.26	nc	Growth factor for activated T and NK cells; enhances cytotoxicity; stimulate IFNγ production
	il12rb2	0.58	nc	Proliferation of NK and T cells; promotes T_h1 differentiation
	il13ra1	1.23	nc	Binds IL4/14 with low affinity
	il13ra2	0.91	3.48	Binds IL13 with high affinity; does not bind IL4; decoy receptor to regulate excessive IL13 signaling
	il17d	–1.74	nc	Induces IL6, IL8, and GM-CSF from endothelial cells
	il17rd	–1.42	1.95	Inhibitor of fibroblast growth factor
	il17rc	1.34	nc	Receptor for IL17A/F; TRAF6 mediated production of cytokine, antimicrobial peptides in mucosal epithelium
	il21r	7.29	nc	Receptor for IL21; immunoregulatory; inhibition of DC activation
	il31r	1.60	nc	Mediates IL31-associated itch; skin immunity
	il34	–1.66	nc	Promotes survival, proliferation of monocytes; pro-inflammatory cytokine release
	lgals9b	1.20	–1.68	Suppresses T_h17 signaling, promotes Treg differentiation
	aloxe3	2.54	16.72	Epidermal barrier formation
	pja2	5.51	–1.34	Positive regulation of TLR2 signaling
	cd151	0.62	–1.57	Regulator of cell activation, proliferation, migration, and adhesion
	enpp2	3.54	–1.79	Production of lysophosphatidic acid; lipid mediator of proliferation, migration, and survival; wound healing
	fgl2	3.86	–1.81	Secreted by T-cells; coagulant activity
	lxn	0.76	–2.04	Inhibitor of tissue carboxypeptidases
	litaf	1.04	–1.99	Regulator of TNFα, CCL2, CXCL1, IL10 by lysosomal degradation
	nlrc5	–9.17	1.45	Negative regulator of NFκB pathways
	c1qt1	–1.90	nc	Anti-inflammatory adipokine
	c1qt3	4.07	12.03	Anti-inflammatory adipokine
Tissue remodeling	*mmp2*	nc	–4.44	Cleaves collagens and gelatins; involved in wound healing
	mmp9	4.94	nc	ECM degradation; leukocyte migration; cleaves collagens and gelatins
	mmp13	7.96	–1.94	ECM degradation; wound healing, tissue remodeling
	mmp14	2.09	nc	Activates MMP2; degrades collagens; positive regulator of cell growth and activation
	mmp17	3.48	nc	Degrades fibrin; activates membrane bound GFs and TNFα
	mmp19	1.19	–2.12	ECM degradation
	mmp23	1.40	–1.63	Protease
	mmp25	3.35	nc	Activates MMP2
	fbn1	9.43	–1.91	Tissue and ECM homeostasis
	sult2b1	4.13	–2.09	Epidermal cholesterol metabolism, regulates epidermal proliferation
	tgm1	3.69	–1.70	Cross-links epidermal proteins
	myof	26.70	1.34	Plasmalemma repair; regulates VEGF
	kcad	10.05	nc	Epithelial cell-cell adhesions

Table 13.1 (Continued)

Category	Gene	Expression		Description
		Coho	Atlantic	
Growth & apoptosis	*mid1*	−1.04	1.34	Ubiquitination and subsequent regulation of IGBP1
	gadd45b	2.45	−2.10	Regulates growth and apoptosis
	rbms1	8.87	−1.79	DNA binding protein; DNA replication
	ttc3	−8.51	1.64	Regulator of AKT signaling
	pbrm1	−23.09	1.40	Negative regulator of cell proliferation
	tnfrsf6b	4.49	−3.61	Decoy receptor for toxic ligands; protects from apoptosis
	bid	1.80	−1.44	Induces caspases and apoptosis
	fgf10	7.08	1.51	Role in wound healing
	hgfac	2.78	−2.19	Activates hepatocyte growth factor (HGF); potent mitogen
	met	3.45	nc	Receptor for HGF
	hbegf	2.74	nc	Fibroblast and smooth muscle mitogen; binds to EGF; macrophage-mediated cellular proliferation
	tgfb1	1.29	nc	Tissue remodeling; activation of MMPs; promotion of Th17 or Treg; cell migration
	vegfaa	3.56	nc	Growth factor; angiogenesis, vasculogenesis, endothelial cell proliferation; cell migration; inhibitory of apoptosis
	vegfc	3.91	nc	Growth factor; angiogenesis, vasculogenesis, endothelial cell proliferation; cell migration; inhibitory of apoptosis
Cellular adhesion	*vcam1*	3.19	−1.82	Leukocyte endothelial cell adhesion
	mag	4.10	−3.28	Mediates interactions between myelinating cells
	kcam	2.49	nc	Epithelial cell adhesion
	cdh1	2.91	nc	Cell-cell, mobility and proliferation of epithelial cells
	cdh26	4.79	nc	Integrin binding; modulates CD4 T cell activation
	dsg2	4.03		Mediates cell-cell adhesion
	epcam	0.74	nc	Physical interaction between epithelial cells and lymphocytes at mucosal epithelium
	itga2	3.94	1.33	Receptor for laminin, collagens, fibronectins; cell-cell adhesion in ECM
	itgb2	1.10	−1.43	ICAM receptor; NK cell cytotoxicity; adhesion/transmigration of leukocytes
Cytotoxic molecules	*aep1*	6.12	nc	Pore-forming molecular, cytotoxic cellular effector
	ncf1	3.12	−1.53	Superoxide production in neutrophils
	ncf2	3.08	−1.89	Superoxide production in neutrophils
Hemostasis	*fa5*	23.68	nc	Regulator of hemostasis
	plat	4.15	−1.56	Converts plasminogen to plasmin; tissue degradation and remodeling
	serpine1	3.05	nc	Inhibitor of plat; regulates fibrinolysis; stimulates keratinocyte migration in cutaneous wound repair
	hmox	1.17	nc	Iron sequestration
	f3	4.46	−2.29	Initiates coagulation; key regulator of hemostasis
Complement and acute phase	*c3ar1*	5.12	−2.16	Receptor for C3a; stimulates chemotaxis, granule enzyme release, superoxide production
	c5ar1	3.39	nc	Receptor for C5a; stimulates chemotaxis, granule enzyme release, superoxide production
	c4bpa	4.41	−1.81	Classical complement activation

Table 13.1 (Continued)

Category	Gene	Expression		Description
		Coho	Atlantic	
	cfh	5.36	−2.81	Modulates complement activation; mediates adhesion of neutrophils to pathogens
	c1qb	7.99	nc	Activation of Ab-mediated complement
	c4b	−1.78	nc	Classical complement
	c4	−2.41	nc	Classical complement cascades; binds to Ig
	c6	1.91	nc	Membrane attack complex
	cfd	4.55	−2.15	Lectin complement pathway activation
	saa1	0.86	nc	Major acute phase protein; promotes mast cell adhesion in ECM
	saa5	2.97	nc	Binds fibrinogen and promotes fibrin formation
	sap	1.86	nc	Inhibits granulocyte adhesion; binds IL8; mediate phagocytosis
	crp	nc	−2.61	Limits neutrophil recruitment to site of injury; mediates phagocytosis
	trfe	2.43	nc	Iron binding

despite the preparation to moult to chalimus II. ECM degradation/tissue remodeling likely plays an important part in preparing for the moult as well, but the fact that MMPs are dysregulated at all time points and unaccompanied by chitin synthesis pathway impacts (Braden et al., 2016, 2022 – in review) suggest these genes may also be responding to other stimuli such as cuticular damage from host responses (that is, chitinases, lysozyme, superoxides, aerolysins) or perhaps a virulence mechanism along with coagulation pathways to impact host epithelial integrity and parasite feeding. Sutherland et al. (2017) found *setd7*, a transcription factor for collagenases similar to *mmp9*, also upregulated in lice infecting fish fed an immunostimulant diet and associated with greater host skin inflammation and reduced lice loads. Upregulation of antioxidant genes by the louse may also be in response to ROS production by host granulocytes, which were common at the *L. salmonis* feeding site on coho salmon fins.

Coho salmon fin also expressed significantly enhanced redox homeostasis and response to oxidative stress genes at 10 dpi. This was accompanied by significant melanin deposition, which interacts with free radicals like superoxide anions, suggesting the need for strong antioxidant activity by the host and also perhaps the louse at these sites. Reactive oxygen species generation appears to be conserved in the host-parasitic-lice relationship, as the blue mussel, *Mytilus edulis* L., differentially expresses ROS pathways in the 'resistant' phenotype of hosts, alongside chitin digestion genes and cuticle proteins (that is, *ECM protein peroxidasin*) being differentially regulated in *Myticola intestinalis* Steuer, 1902, (suborder: Poecilostomatoida) (Feis et al., 2018).

Coho and Atlantic salmon fins show strong differential transcriptomic responses paralleled by functionally different cellular responses to *L. salmonis*, all within a seven day time frame. Work on pink salmon juveniles, which appear to be the next most resistant host species to *L. salmonis*, suggests some similarities and differences in pathways involved in reducing lice infection. An important caveat here is that infection studies in pink salmon involve *L. salmonis oncorhynchii nova*, and direct comparison is not possible to the Atlantic/coho work with *L. salmonis salmonis*, as there may be subspecies differences in the dynamics of the host-parasite relationship, despite no gross differences being apparent in coho salmon response to these different subspecies. Jones et al. (2007) first described transient increases in *il-8* and *tnf-α* at 7 and 14 dpi, respectively, in the skin during targeted profiling of inflammatory transcripts. This was accompanied by mixed leucocyte infiltration at the attachment site in gills and fins; however, this did not differ between the resistant pink and susceptible chum salmon in response.

Furthermore, no encapsulation of juvenile parasites was observed in pink salmon in any laboratory exposures as has been described in coho salmon (Jones et al., 2008; Sutherland et al., 2011, 2014). Sutherland et al. (2014) showed that low numbers of lice occurred in pink salmon within 3 dpi, though as these fish were cohabitated with chum and Atlantic salmon, were unable to determine whether the low level of infection was due to active host rejection, a lower affinity of copepodid attraction toward pink salmon juveniles compared to chum or Atlantic salmon, or a combination of these factors. Whereas in all cases, be they cohabitation or single habitation lab studies, there is significant settlement of lice on coho salmon by 24 hpi followed by a steady decline of infection through 72 h to 10–14 dpi, resulting in nearly 100% loss (Johnson and Albright, 1992a; Fast et al., 2002; Braden et al., 2016, 2022 – in review), the pink salmon studies show significant reductions from 7 to 14 days (>50% in most cases) without the early time point resolution, and generally do not achieve the nearly complete loss of lice by 14 dpi (Jones et al., 2008; Sutherland et al., 2011, 2014). Sutherland et al. (2014) also examined comparative fin responses of pink, chum, and Atlantic salmon at 6 dpi; however, little transcriptome-wide change was observed in pectoral fins, while targeted expression profiles again showed elevated inflammatory markers (*il-1β* and *tnf-α*) in pink salmon only. Increased expression of head kidney (hk) thioredoxin in pink salmon also matches the switch to an antioxidant state in resistant species to protect against the oxidative stress induced by strong inflammatory responses to the parasite (Sutherland et al., 2014).

Systemic assessment through hk analysis also suggested delays in acute phase responses (APR) in Atlantic salmon relative to pink salmon, and APR in decline by 9 dpi in pink. Aside from hepcidin-1, serotransferrin and other iron homeostatic components were dysregulated in head kidneys specific to pink salmon in the 3–9 dpi time frame, suggesting nutritional immunity and sequestration of host iron/nutrients play a unique role in pink salmon resistance to *L. salmonis*. Similar to coho salmon fins, pink salmon hk expression of c-type lectin domain family 4 member M (*clec4m*) and acidic mammalian chitinase (*amcase; chia*) was induced in contrast to susceptible species in Sutherland et al. (2014), as well as *amcase/chia* being one of the most highly regulated genes in response to louse infection in 0.7 g pink salmon (Sutherland et al., 2011). The involvement of chitinases may also positively enhance immunological priming and inflammation at the attachment site, as chitosan and smaller chitin metabolites themselves can be further immunogenic. Taken together, these studies demonstrate rapid (within 24–72 h) and sustained (through the louse moult to chalimus) sensing of the louse both locally and systemically, through chitin, mannose, and other non-self PRRs, initiates acute inflammation at the louse attachment site resulting in potentiating acute phase protein and complement release in resistant species. Whether this is driven by epithelial cell sensing, mast cell-like EGC degranulation, or a combination of these and other signals, recruited neutrophil/mixed leucocytes, oxidative stress, lectin binding, and cytotoxic molecules likely contribute to negative consequences on louse health.

Resistance mechanisms appear to significantly diverge between coho and pink salmon specifically within the formation of granulomatous encapsulation that is unique to coho. Nutritional immunity through coagulation and/or heme regulation is also a feature of resistant hosts with lymphocyte activation (in particular Th1) and trafficking, tissue remodeling, antioxidant, and epithelial growth/recovery responses. This transition from acute inflammation to recovery and homeostasis is similar in most respects to 14 days post-wounding (Sveen et al., 2019) marking the transition between early and late healing in Atlantic salmon. In Atlantic salmon/susceptible species to salmon lice, inflammation does not occur over these time frames, likely resulting in a lack of granulation tissue and/or elements of ECM stabilization (fibronectin, collagen, laminin, proteoglycans) which is required for dermal repair (Sveen et al., 2019). Innate immune function/gene regulation in resistant hosts is also complementary to normal wound healing processes which observe oxidative burst complexes (*ncf1*), APR and complement signals, and eicosanoid production which mostly subside within 14 days and lead to activation of B and T cells (*SRC-like-adapter, SH3 protein, lymphocyte cytosolic protein 1, cd80-like protein, cd274,* and so on). For this reason, it is proposed that it is imperative to examine these differences between resistance/susceptibility in salmonids within the first 10–14 days as measurements at >14 d are in essence measuring the tolerance/adaptation to chronic impacts of the lice in susceptible hosts, with the confounding variables of the louse becoming mobile and producing higher quantities of immunomodulatory molecules, aggregating in the head/tail regions, and driving significant glucocorticoid stress, which takes off at this stage (reviewed in Johnson and Fast, 2004; Wagner et al., 2008). Transition from Th1 to Th2 can be exemplified in the significant Th1-biased responses observed during infection with the attached larval lice in Braden et al. (2016, 2022 – in review), while in coho salmon infected with mobile adult *L. salmonis*, activated MHIIβ+, IL1β+, and TNFα+ cells were observed in coho salmon skin at the louse attachment site, accompanied by subsequent increases in *il4* and *tgfβ* (Braden et al. 2015b).

13.4 Subversion of immunity – co-infections

Analysis of pink and coho salmon shows the divergence of responses to *L. salmonis* from the susceptible phenotype occurs within the first few days, and subsequent processes 14 d and beyond are a result of this inability to mount an appropriate initial response in Atlantic salmon. Fast et al. (2006a, 2006b) and Skugor et al. (2008) show earlier upregulation of antigen presentation, alterative macrophage, and immunoglobulin-like genes systemically, which likely drive the Th2 response observed later on with increases in *il10* and *tgfβ* expression. Without the initial Th1 inflammatory impact that appears to produce an inhospitable environment for the louse and initiates proper wound healing to take place, a dysregulated Th2-type response in Atlantic salmon contributes to chronic wounds (Skugor et al., 2008). This is likely further exacerbated by cortisol production and release in the susceptible phenotype which generally occurs at the louse moult to pre-adult, and further complicates the inflammatory/healing process of the skin (Johnson and Albright, 1992b; reviewed in Johnson and Fast, 2004).

In heavy infections of *L. salmonis*, cortisol-mediated immunosuppression may occur. However, as recently reviewed in Braden et al. (2020b), ectoparasites including *L. salmonis* commonly secrete a cocktail of pharmacologically active factors into the host to manipulate the feeding site, prevent clotting, and interfere with wound healing. Based on host species, and possibly parasite subspecies, the milieu of these secretions appears to change (Fast et al., 2003; Braden et al., 2017). For a complete review Braden et al. (2020b) identified over 35 factors from five major classes of protein that facilitate evasion of host immune system and assist in feeding and attachment. Collagenases and serine proteases predicted to interact with fibronectin, ECM regeneration, and collagen deposition have been identified and found to be enhanced in *L. salmonis* secretome when feeding on more susceptible hosts

(Braden et al., 2017, 2020b). As noted above, the complement system is highly induced in coho salmon from both recognition proteins C1q (classical pathway) and mannose binding lectin (lectin pathway). The opposite is true for susceptible hosts like Atlantic salmon. Inhibition of the complement cascade is a common method of circumventing host responses in numerous host-parasite relationships.

Schistosomes are hypothesized to evade complement through rapid outer surface turnover, antibody ingestion, and production of complement receptor/decoys, whereas in ixodid ticks, anti-complement proteins are secreted from the saliva and enhance feeding (Schroeder et al., 2009). *L. salmonis* has also been shown to enhance production of C3-degrading molecules such as hypodermin B during feeding on susceptible hosts (Braden et al., 2017, 2020b). Complement plays a role in defense against all types of pathogens, not just parasites but viruses and bacteria as well. C3 plays a central role in activating the classical and alternative complement pathways, and its deficiency can cause significant bacteremia in mammals (Lachmann, 1975; Lawrenz et al., 2003; Rottgerding and Kraiczy, 2020).

In Atlantic salmon sea cage culture, *Moritella viscosa* is a common bacterial infection and the causative agent of winter ulcer disease. *Moritella viscosa* infection has been shown to induce inflammatory and complement component C3 gene expression in kidneys and skin of Atlantic salmon during infection (Løvoll et al., 2009; Parrish et al., 2020). Parrish et al. (2020) further showed that under co-infection with *L. salmonis*, *M. viscosa* lesion sites displayed significant increases in *c3*, whereas *L. salmonis* attachment/feeding sites exhibited a significant inhibition of *c3* gene expression. Both Parrish et al. (2020) and Carvalho et al. (2020), additionally showed that prior infection with *L. salmonis* significantly increased mortality associated with *M. viscosa* infection, and this was exacerbated in high intensity infections of *L. salmonis* (n>30 adult lice; <200 g salmon smolts).

In this work, Atlantic salmon with bacterial-induced lesions were mainly observed in the high lice co-infection group and showed inhibited inflammatory (*il-8*, *hepc1*, and so on) and wound healing gene (*mmp9*) responses in the skin (Carvalho et al., 2020). These authors hypothesized that strong inflammatory and APR signals early during co-infection were markers of future successful resolution of co-infection. While *M. viscosa* does not appear to use the louse lesions as a portal of entry (Carvalho et al., 2020), downregulation of complement and other inflammatory responses have been described both at and away from sites of *L. salmonis* infection in the skin (Carvalho et al., 2020; Parrish et al., 2020; Umasuthan et al., 2020), suggesting immunomodulation extending beyond the feeding site, and in the absence of cortisol-mediated immunosuppression. Modulation of macrophage and T-cell responses in Atlantic salmon discussed above are in line with descriptions of altered macrophage/leucocyte populations in kidneys of *L. salmonis* infected salmon, with reduced phagocytic capacity, respiratory burst, and MH class I expression (Fast et al., 2002, 2006a, 2006b), again in the absence of cortisol-mediated immunosuppression. Taken together, these studies suggest *L. salmonis* infection results in an altered immunological response of the host, which results in an immunological environment conducive to particular co-infections.

Irrespective of the efficacy of host responses in resolving *L. salmonis*/copepod infections, and/or the contribution from louse derived secretions, this altered immunological environment exists across all salmonid infections with sea lice. Resistant hosts, for instance, show upregulation of host response to lice at both the local and systemic levels, and this coincides with suppressed antiviral responses (Sutherland et al., 2014; Braden et al., 2022 – in review). Holm et al. (2016) found that while dietary phytochemicals were able to stimulate antiviral responses in Atlantic salmon, following infection with *L. salmonis* these were significantly inhibited. Similarly, Barker et al. (2019) showed downregulation of antiviral genes (for example, *mh class I, galectin 9, TRIM, isg15, mx1*) in the head kidney following single infection with *L. salmonis*, and subsequently significantly higher mortality in co-infected salmon with ISAV, than ISAV infection without lice. In Cai et al. (2022) weighted gene co-expression network analysis (WGCNA) revealed genes in the skin of lice/ISAV co-infected Atlantic salmon, positively correlated with lice load/susceptibility, were interferon and wound healing related. Type I interferons (IFN-α/β) bind to IFN-α/ß receptors, leading to induction of antiviral proteins such as Mx, interferon-stimulated gene 15 (*isg15*), and protein kinase R (Haller et al., 2006). These same authors found pre-exposure to *L. salmonis* increased the susceptibility of Atlantic salmon to a secondary exposure with ISAV due to compromised antigen presentation and T cell differentiation.

Although Cai et al. (2022) found fatty acid (FA) enriched diets (1.0% FA) had a significant effect on lice reduction, they did not significantly protect the fish during co-infection. Going further, Parrish et al. (2020) found that Atlantic salmon fed diets enriched with CpG, which showed significantly increased relative percent survival (RPS) against ISAV compared to control diets (>30% RPS), lost all ISAV-related protection when fish were co-infected with *L. salmonis*, even at infection levels that may be at or below industry treatment trigger levels (that is, one to three adult lice/fish). This altered immunological environment conducive to co-infection created by *L. salmonis* is also in agreement with other parasitic copepod infections in salmon. Lhorente et al. (2014) reported that lice infection (*C. rogercresseyi*) reduced the resistance of Atlantic salmon to the bacterial pathogen *Piscirickettsia salmonis* (*Psal*) and Figueroa et al. (2017) found reduced survival in vaccinated Atlantic salmon that were co-infected with *C. rogercresseyi* and *Psal*, the co-infected group in this study having significantly increased bacterial load and clinical signs of disease.

The impact of sea lice in a co-infection scenario, however, may also work in the opposite way. While we have discussed the diversion and/or subversion of immune responses due to lice, despite the commonality with which lice infect salmon fins, a common finding in *L. salmonis* infections are the lack of copepodid settlement/development at inflamed fin sites (Fast, personal observations). As rapid inflammation and wound repair are important corollaries of louse rejection, prior damage/erosion or induction of inflammation to fin sites may impact *L. salmonis* infection. In fact, Parrish et al. (2020) observed that the order of co-infection impacted outcomes, such that Atlantic salmon infected first with *M. viscosa*, while having increased subsequent louse abundances, exhibited no additive or synergistic impacts on clinical disease or survival compared to single infection with *M. viscosa*. In contrast those infected with *L. salmonis* first had significantly lower survival than both *M. viscosa* only, or *M. viscosa* then *L. salmonis* infected groups. The mechanisms behind this are currently being studied, but again may involve the normal inflammatory induction in the skin caused by *M. viscosa*.

13.5 Future directions

Despite it being almost 30 years since the seminal first edition of *Pathogens of Wild and Farmed Fish: Sea Lice*, and intensive research on biology and control, the salmon louse (*L. salmonis*) continues to plague wild and farmed salmon today. Heavy antiparasitic use in the 1990s to control this parasite in Atlantic salmon farms has begun to give way to non-medicinal interventions (that is, heated water, biological control, lasers, and other methods) as well as unique cage designs and other husbandry practices. But by comparing natural variation in host species' susceptibility/resistance we see there are vast improvements that could be made if the inherent resistance in coho salmon could be harnessed and applied to Atlantic salmon culture. The salmonid immune system has a potential answer to the sea lice problem, yet vaccines have been unable to show any clinical significance and selective breeding has made little genetic progress despite its relative advantages (Fast and Dalvin, 2020).

Emergence of new technologies has not only allowed us to better understand the mechanisms behind the host-parasite interaction, but with the ability to apply targeted mutation through CRISPR/Cas9 gene editing (GE) offers an exciting if not ethically challenging route to examine this immune potential in susceptible hosts. Current global initiatives are underway to develop a detailed, genetic-function roadmap of resistance and susceptibility across numerous salmonid species' genomes. This would allow key molecular processes and gene networks to be modified through GE to transfer coho/pink salmon resistance traits to susceptible species such as Atlantic salmon. A GE process to develop a more louse-resistant Atlantic salmon must follow responsible research and innovation ethical guidelines, and would need to have considerable societal and regulatory review before application within and outside of the laboratory, but provides the opportunity for new insights and solutions to the sea lice problem to be realized. However, if the medicinal intervention and selection of resistant-louse strains over the last two to three decades have taught us anything about the salmon louse, this parasite has an immense propensity for responding quickly to strong anthropogenic selection.

The next decade(s) of salmon louse research will also have to contend with global climate change and its impacts on coastal ocean temperature and dissolved CO_2. Hamre et al. (2019), Godwin et al. (2020), and Medcalf et al. (2020) have all given us an initial peek at the expected shortening of louse generation times and resulting physiological impacts in already susceptible host species, painting an even darker picture for Atlantic salmon; however, what oscillating ocean conditions may mean for resistant phenotypes/species is unknown, and so is the potential for co-evolutionary mechanisms which may enhance louse virulence (Feis et al., 2018; Braden et al. 2020b; Fast and Dalvin, 2020). These will be important areas of expanded effort in the years to come if the economic and ecological impacts of the salmon louse are to finally be curtailed.

References

Andrews, M., Stormoen, M., Schmidt-Posthaus, H., Wahli, T. and Midtlyng, P.J. (2015) Rapid temperature-dependent wound closure following adipose fin clipping of Atlantic salmon *Salmo salar* L. *Journal of Fish Diseases* 38(6), 523–531.

Barker, S.E., Bricknell, I.R., Covello, J., Purcell, S.L., Fast, M.D., Wolters, W. and Bouchard, D.A. (2019) Sea lice (*Lepeophtheirus salmonis*, Krøyer 1837) infected Atlantic salmon (*Salmo salar* L.) are more susceptible to infectious salmon anemia virus. *PLoS ONE* 14, e0209178.

Beamish, R.J., Neville, C.M., Sweeting, R.M. and Ambers, N. (2005) Sea lice on adult Pacific salmon in the coastal waters of Central British Columbia, Canada. *Fisheries Research* 76, 198–208.

Braden, L.M., Barker, D.E., Koop, B.F. and Jones, S.R.M. (2015a) Differential modulation of resistance biomarkers in skin of juvenile and mature pink salmon, *Oncorhynchus gorbuscha* by the salmon louse, *Lepeophtheirus salmonis*. *Fish and Shellfish Immunology* 47, 7–14.

Braden, L.M., Koop, B.F. and Jones, S.R.M. (2015b) Signatures of resistance to *Lepeophtheirus salmonis* include a TH2-type response at the louse-salmon interface. *Developmental and Comparative Immunology* 48, 178–191.

Braden, L.M., Poley, J.D., Byrne, P.J. and Fast, M.D. (2016). Functional transcriptomic characterization of *Lepeophtheirus salmonis*, rejection by coho salmon (*Oncorhynchus kisutch*). 11th International Sea Lice Conference, Westport, Ireland, September 26–28, 2016.

Braden, L.M., Sutherland, B.J.G., Koop, B.F. and Jones, S.R.M. (2017) Enhanced transcriptomic responses in the Pacific salmon louse *Lepeophtheirus salmonis oncorhynchi* to the non-native Atlantic salmon *Salmo salar* suggests increased parasite fitness. *BMC Genetics* 18, 110.

Braden, L.M., Michaud, D., Igboeli, O.O., Dundrop, M., Hamre, L., Dalvin, S., Purcell, S.L., Kongshaug, H., Eichner, C., Nilsen, F. and Fast, M.D. (2020a) Identification of critical enzymes in the salmon louse synthesis pathway as revealed by RNAi-mediated abrogation of infectivity. *International Journal of Parasitology* 50, 10–11.

Braden, L.B., Monaghan, S. and Fast, M.D. (2020b) Salmon immunological defense and interplay with the modulatory capabilities of its ectoparasite *Lepeophtheirus salmonis*. *Parasite Immunology*. Available at: https://doi.org/10.1111/pim.12731 (accessed 31 March 2021).

Bron, J.E., Sommerville, C., Jones, M. and Rae, G.H. (1991) The settlement and attachment of early stages of the salmon louse, *Lepeophtheirus salmonis* (Copepoda: Caligidae) on the salmon host, *Salmo salar*. *Journal of Zoology* 224, 201–212.

Bui, S., Oppedal, F., Stien, L. and Dempster, T. (2016) Sea lice infestation level alters salmon swimming depth in sea-cages. *Aquaculture Environment Interactions* 8, 429–435.

Butler, J.R. (2002) Wild salmonids and sea louse infestations on the west coast of Scotland: sources of infection and implications for the management of marine salmon farms. *Pest Management Science* 58(6), 595–608.

Carvalho, L.A., Whyte, S.K., Braden, L.M., Purcell, S.L., Manning, A.J., Muckle, C.A. and Fast, M.D. (2020) Impact of co-infection with *Lepeophtheirus salmonis* and *Moritella viscosa* on inflammatory and immune responses of Atlantic salmon (*Salmo salar*). *Journal of Fish Diseases* 43, 459–473.

Cai, W., Kumar, S., Umasuthan, N., Caballero-Solares, A., Carvalho, L.A., Whyte, S.K., Purcell, S.L., Gagne, N., Hori, T., Allen, M., Taylor, R.G., Rise, M.L. and Fast, M.D. (2022) Transcriptomic analysis of Atlantic salmon (*Salmo salar*) skin in response to sea lice and infectious salmon anemia virus coinfection under different functional diets. *Frontiers in Immunology*. Volume 22, DOI=10.3389/fimmu.2021.787033 (accessed 23 January 2022).

Covello, J., Friend, S.E., Purcell, S.L., Burka, J.F., Markham, R.J.F., Donkin, A., Groman, D.B. and Fast, M.D. (2012) Orally administered immunostimulants affect inflammatory gene expression and sea lice (*Lepeophtheirus salmonis*) burdens on Atlantic salmon (*Salmo salar*). *Aquaculture* 366–367, 9–16.

Eichner, C., Dalvin, S., Skern-Mauritzen, R., Malde, K., Kongshaud, H. and Nilsen, F. (2015) Characterization of a novel RXR receptor in the salmon louse (*Lepeophtheirus salmonis*, Copepoda) regulating growth and female reproduction. *BMC Genetics* 16, 81.

Fast, M.D. and Dalvin, S. (2020) Global climate change impacts on *Lepeophtheirus salmonis* infections in salmon. In: Woo, P.T.K. (ed.) *Climate Change and Infectious Fish Disease*. CABI, Boston, USA. pp. 471–498.

Fast, M.D., Ross, N.W., Mustafa, A., Sims, D.E., Johnson, S.C., Conboy, G.A., Speare, D.J., Johnson, G. and Burka, J.F. (2002) Susceptibility of rainbow trout *Oncorhynchus mykiss*, Atlantic salmon *Salmo salar* and coho salmon *Oncorhynchus kisutch* to experimental infection with sea lice *Lepeophtheirus salmonis*. *Diseases of Aquatic Organisms* 52, 57–68.

Fast, M.D., Burka, J.F., Johnson, S.C. and Ross, N.W. (2003) Enzymes released from *Lepeophtheirus salmonis* in response to mucus from different salmonids. *Journal of Parasitology* 89, 7–13.

Fast, M.D., Muise, D.M., Easy, R.E., Ross, N.W. and Johnson, S.C. (2006a) The effects of successive *Lepeophtheirus salmonis* infections on the immunological status of Atlantic salmon (*Salmo salar*). *Fish and Shellfish Immunology* 21, 228–241.

Fast, M.D., Ross, N.W., Muise, D.M. and Johnson, S.C. (2006b) Differential gene expression in Atlantic salmon, *Salmo salar*, infected with *Lepeophtheirus salmonis* (Copepoda: Caligidae). *Journal of Aquatic Animal Health* 18, 116–127.

Feis, M.E., John, U., Lokmer, A., Luttikhuizen, P.C. and Wegner, K.M. (2018) Dual transcriptomics reveals co-evolutionary mechanisms of intestinal parasite infections in blue mussels *Mytilus edulis*. *Molecular Ecology* 27, 1505–1519.

Figueroa, C., Bustos, P., Torrealba, D., Dixon, B., Soto, C., Conejeros, P. and Gallardo, J.A. (2017) Coinfection takes its toll: Sea lice override the protective effects of vaccination against a bacterial pathogen in Atlantic salmon. *Scientific Reports* 7(1), 17817.

Frederick, C.A. (2018) The Infection Dynamics of Sea Lice on Atlantic Salmon (*Salmo salar*) in Cobscook Bay, Maine, USA. Electronic Theses and Dissertations. 3018. Available at: https://digitalcommons.library.umaine.edu/etd/3018 (accessed December 2020).

Godwin, S.C., Fast, M.D., Kuparinen, A., Medcalf, K.E. and Hutchings, J.A. (2020) Increasing temperatures accentuate negative fitness consequences of a marine parasite. *Scientific Reports* 10, 18467.

Groot, C. and Margolis, L. (1991) *Pacific Salmon Life Histories*. UBC Press, Vancouver.

Haller, O., Kochs, G. and Weber, F. (2006) The interferon response circuit: Induction and suppression by pathogenic viruses. *Virology* 344, 119–130.

Hamre, L.A., Bui, S., Oppedal, F., Skern-Mauritzen, R. and Dalvin, S. (2019) Development of the salmon louse *Lepeophtheirus* salmonis parasitic stages in temperatures ranging from 3 to 24°C. *Aquaculture Environmental Interactions* 11, 429–443.

Holm, H.J., Wadsworth, S., Bjelland, A-K., Krasnov, A., Evensen, O. and Skugor, S. (2016) Dietary phytochemicals modulate skin gene expression profiles and result in reduced lice counts after experimental infection in Atlantic salmon. *Parasites & Vectors* 9, 271.

Jackson, D., Kane, F., O'Donohoe, P., McDermott, T., Kelly, S., Drumm, A. and Newell, J. (2013) Sea lice levels on wild Atlantic salmon, *Salmo salar* L., returning to the coast of Ireland. *Journal of Fish Diseases* 36(3), 293–298.

Jakob, E., Sweeten, T., Bennett, W. and Jones, S.R.M. (2013) Development of the salmon louse *Lepeophtehirus salmonis* and its effects on juvenile sockeye salmon *Oncorhynchus nerka*. *Diseases of Aquatic Organisms* 106, 217–227.

Jensen, L.B., Wahli, T., McGurk, C., Eriksen, T.B., Obach, A., Waagbo, R., Handler, A. and Tafalla, C. (2015) Effect of temperature and diet on wound healing in Atlantic salmon (*Salmo salar* L.). *Fish Physiology and Biochemistry* 41, 1527–1543.

Johnson, S.C. and Albright, L.J. (1992a) Comparative susceptibility and histopathology of the response of naive Atlantic, chinook and coho salmon to experimental infection with *Lepeophtheirus salmonis* (Copepoda: Caligidae). *Diseases of Aquatic Organisms* 14, 179–93.

Johnson, S.C. and Albright, L.J. (1992b) Effects of cortisol implants on the susceptibility and the histopathology of the responses of naive coho salmon *Oncorhynchus kisutch* to experimental infection with *Lepeophtheirus salmonis* (Copepoda: Caligidae). *Diseases of Aquatic Organisms* 14, 195–205.

Johnson, S.C. and Fast, M.D. (2004) Interactions between sea lice and their hosts. In: Wiegertjes, G.F. and Flik, G. (eds) *Host-parasite Interactions*. BIOS Scientific Publishers Ltd, Oxford, UK. pp. 131–161.

Jones, S.R.M., Fast, M.D., Johnson, S.C. and Groman, D.B. (2007) Differential rejection of salmon lice by pink and chum salmon: disease consequences and expression of pro-inflammatory genes. *Diseases of Aquatic Organisms* 75, 229–238.

Jones, S., Kim, E. and Bennett, W. (2008) Early development of resistance to the salmon louse, *Lepeophtheirus salmonis* (Krøyer), in juvenile pink salmon, *Oncorhynchus gorbuscha* (Walbaum). *Journal of Fish Diseases* 31(8), 591–600.

Koller, B., Muller-Wiefel, A.S., Rupec, R., Korting, H.C. and Ruzicka, T. (2011) Chitin modulates innate immune responses of keratinocytes. *PLoS ONE* 6(2), e16594.

Krishnaswamy, V.R., Mintz, D. and Sagi, I. (2017) Matrix metalloproteinases: The sculptors of chronic cutaneous wounds. *Biochimica Biophysica Acta Molecular Cell Research* 1864, 2220–2227.

Lachmann, P. (1975) Genetics of the complement system. *Journal of Medical Genetics* 12, 372–377.

Lawrenz, M.B., Wooten, R.M., Zachary, J.F., Drouin, S.M., Weis, J.J., Wetsel, R.A. and Norris, S.J. (2003) Effect of complement component C3 deficiency on experimental Lyme borreliosis in mice. *Infection and Immunity* 71, 4432–4440.

Lee, C.G., Da Silva, C.A., Dela Cruz, C.S., Ahangari, F., Ma, B., Kang, M.J., He, C.H., Takyar, S. and Elias, J.A. (2011) Role of chitin and chitinase/chitinase-like proteins in inflammation, tissue remodeling, and injury. *Annual Reviews in Physiology* 73, 479–501.

Lhorente, J.P., Gallardo, J.A., Villanueva, B., Carabaño, M.J. and Neira, R. (2014) Disease resistance in Atlantic salmon (*Salmo salar*): coinfection of the intracellular bacterial pathogen *Piscirickettsia salmonis* and the sea louse *Caligus rogercresseyi*. *PLoS ONE* 9(4), e95397.

Long, A., Garver, K.A. and Jones, S.R.M. (2018) Differential effects of adult salmon lice *Lepeophtheirus salmonis* on physiological responses of sockeye salmon and Atlantic salmon. *Journal of Aquatic Animal Health* 31, 75–87.

Løvoll, M., Wiik-Nielsen. C.R., Tunsjø, H.S., Colquhoun, D., Lunder, T., Sørum, H. and Grove S. (2009) Atlantic salmon bath challenged with *Moritella viscosa* – pathogen invasion and host response. *Fish and Shellfish Immunology* 26(6), 877–884.

Mazzariol, S., Di Guardo, G., Petrella, A., Marsili, L., Fossi, C.M., Leonzio, C. Zizzo, N., Vizzini, S., Gaspari, S., Pavan, G., Podesta, M., Garibaldi, F. and Ferrante, M. (2011) Sometimes sperm whales (*Physeter macrocephalus*) cannot find their way back to the high seas: a multidisciplinary study on a mass stranding. *PLoS ONE* 6(5), e19417.

Medcalf, K.E., Hutchings, J.A., Fast, M.D., Kuparinen, A. and Godwin, S.C. (2020) Warming temperatures and ectoparasitic sea lice impair internal organs in juvenile Atlantic salmon. *Marine Ecology Progress Series* 660, 161–169.

Michaud, D., Cai, W.C., Poley, J.D., Belliveau, P., MacDonald, A.M., Hardottir, H., Dalvin, S. and Fast, M.D. (2019) Impacts of different benzoylureas across various life stages of *L. salmonis*. *19th International Conference on Diseases of Fish and Shellfish*. European Association of Fish Pathologists (EAFP) 219-O, p. 23.

Moreno, A.N., Jamur, M.C., Oliver, C. and Roque-Barreira, M.C. (2003) Mast cell degranulation induced by lectins: effect on neutrophil recruitment. *International Archives of Allergy Immunology* 132(3), 221–230.

Parrish, K.S., Purcell, S.L., Whyte, S.K., Manning, A.J., Carvalho, L., Dalvin, S., Taylor, R.G., Rise, M.L. and Fast, M.D. (2020) Pathological synergies in co-infecting pathogens are impacted by exposure order, and host response to initial infection. *Bulletin of the European Association of Fish Pathologists* 40, 1–2.

Pert, C.C., Fryer, R.J., Cook, P., Kilburn, R., McBeath, S., McBeath, A., Matejusova, I., Urquhart K., Weir, S.J., McCarthy, U., Collins, C., Amundrud, T. and Bricknell, I.R. (2014) Using sentinel cages to estimate infestation pressure on salmonids from sea lice in Loch Shieldaig, Scotland. *Aquaculture Environment Interactions* 5, 49–59.

Poley, J.D., Braden, L.M., Messmer, A.M., Igboeli, O.O., Whyte, S.K., Gameiro, M.D., MacDonald, A.M., Rufener, L., Bouvier, J., Wadowska, D.W., Koop, B.F., Hosking, B. and Fast, M.D. (2018) High level efficacy of lufenuron against sea lice (*Lepeophtheirus salmonis*) linked to rapid impact on moulting processes. *International Journal of Parasitology: Drugs and Drug Resistance* 8, 174–188.

Reite, O.B. and Evensen, O. (2006) Inflammatory cells of teleostean fish: A review focusing on mast cells/eosinophilic granule cells and rodlet cells. *Fish and Shellfish Immunology* 20, 192–208.

Roberts, R.J., Johnson, K.A. and Casten, M.T. (2004) Control of *Salmincola californiensis* (Copepoda: Lernaeapodidae) in rainbow trout, *Oncorhynchus mykiss* (Walbaum): a clinical and histopathological study. *Journal of Fish Diseases* 27, 73–79.

Robledo, D., Gutiérrez, A.P., Barría, A., Yáñez, J.M. and Houston, R.D. (2018) Gene expression response to sea lice in Atlantic salmon skin: RNA sequencing comparison between resistant and susceptible animals. *Frontiers in Genetics* 9, 287.

Rottgerding, F. and Kraiczy, P. (2020) Immune evasion strategies of relapsing fever spirochetes. *Frontiers in Immunology* 23(11), 1560.

Samsing, F., Oppedal, F., Dalvin, S., Johnsen, I., Vagseth, T. and Dempster, T. (2016) Salmon lice (*Lepeophtheirus salmonis*) development times, body size, and reproductive outputs follow universal models of temperature dependence. *Canadian Journal of Fisheries and Aquatic Sciences* 73, 1841.

Sandvik, A.D., Johnsen, I.A., Myksvoll, M.S., Sævik, P.N. and Skogen, M.D. (2020) Prediction of the salmon lice infestation pressure in a Norwegian fjord. *ICES Journal of Marine Sciences* 77, 746–756.

Schroeder, H., Skelly, P., Zipfel, P.F., Losson, B. and Vanderplasschen, A. (2009) Subversion of complement by hematophagous parasites. *Developmental and Comparative Immunology* 33, 5–13.

Shibata, Y., Metzger, W.J. and Myrvik, Q.N. (1997) Chitin particle-induced cell-mediated immunity is inhibited by soluble mannan: mannose receptor-mediated phagocytosis initiates IL-12 production. *Journal of Immunology* 159, 2462–2467.

Shibata, Y., Foster, L.A., Bradfield, J.F. and Myrvik, Q.N. (2000) Oral administration of chitin down-regulates serum IgE levels and lung eosinophilia in the allergic mouse. *Journal of Immunology* 164, 1314–1321, 1323.

Skern-Mauritzen, R., Torissen, O. and Glover, K.A. (2014) Pacific and Atlantic *Lepeophtheirus salmonis* (Krøyer 1838) are allopatric subspecies: *Lepeophtheirus salmonis salmonis* and *L. salmonis oncorhynchi* subspecies *novo*. *BMC Genetics* 15, 32.

Skugor, S., Glover, K.A. and Nilsen, F. (2008) Local and systemic gene expression responses of Atlantic salmon (*Salmo salar* L.) to infection with the salmon louse (*Lepeophtheirus salmonis*). *BMC Genomics* 9, 498.

Sutherland, B.J.G., Jantzen, S.G., Sanderson, D.S., Koop, B.F. and Jones, S.R.M. (2011) Differentiating size-dependent responses of juvenile pink salmon (*Oncorhynchus gorbuscha*) to sea lice (*Lepeophtheirus salmonis*) infections. *Comparative Biochemistry and Physiology – Part D* 6, 213–223.

Sutherland, B.J.G., Koczka, K.W., Yasuike, M., Jantzen, S.G., Yazawa, R. and Koop, B.F. (2014) Comparative transcriptomics of Atlantic *Salmo salar*, chum *Oncorhynchus keta* and pink salmon *O. gorbuscha* during infections with salmon lice *Lepeophtheirus salmonis*. *BMC Genetics* 15, 200.

Sutherland, B.J.G., Covello, J.M., Friend, S.E., Koczka, K.W., Purcell, S.L., MacLeod, T.L., Donovan, B.R., Pino Marambio, J., González Vecino, J.L., Troncoso, J., Koop, B.F., Wadsworth, S.L. and Fast, M.D. (2017) Host-parasite interaction of Atlantic salmon *Salmo salar* with salmon louse *Lepeophtheirus salmonis* during enhanced parasite rejection through immunostimulation. *FACETS* 2, 477–495.

Sveen, L.R., Timmerhaus, G., Krasnov, A., Takle, H., Handeland, S. and Ytteborg, E.(2019) Wound healing in post-smolt Atlantic salmon (*Salmo salar* L.). *Scientific Reports* 9, 3565.

Tadiso, T.M., Krasnov, A., Skugor, S., Afanasyev, S., Hordvik, I. and Nilsen, F. (2011) Gene expression analyses of immune responses in Atlantic salmon during early stages of infection by salmon louse (*Lepeophtheirus salmonis*) revealed bi-phasic responses coinciding with the copepod-chalimus transition. *BMC Genomics* 12, 141.

Umasuthan, N., Xue, X., Caballero-Solares, A., Kumar, S., Westcott, J.D., Chen, Z., Fast, M.D., Skugor, S., Nowak, B.F., Taylor, R.G. and Rise, M.L. (2020) Transcriptomic profiling in fins of Atlantic salmon parasitized with sea lice: evidence for an early imbalance between chalimus-induced immunomodulation and the host's defense response. *International Journal of Molecular Science* 21(7), 2417.

Valenzuela-Muñoz, V. and Gallardo-Escárate, C. (2017) Iron metabolism modulation in Atlantic salmon infested with the sea lice *Lepeophtheirus salmonis* and *Caligus rogercresseyi*: A matter of nutritional immunity? *Fish and Shellfish Immunology* 60, 97–102.

Valenzuela-Muñoz, V., Boltaña, S. and Gallardo-Escárate, C. (2016) Comparative immunity of *Salmo salar* and *Oncorhynchus kisutch* during infestation with the sea louse *Caligus rogercresseyi*: An enrichment transcriptome analysis. *Fish and Shellfish Immunology* 59, 276–287.

Valenzuela-Muñoz, V., Valenzuela-Miranda, D. and Gallardo-Escárate, C. (2018) Comparative analysis of long non-coding RNAs in Atlantic and coho salmon reveals divergent transcriptome responses associated with immunity and tissue repair during sea lice infestation. *Developmental and Comparative Immunology* 87, 36–50.

Vera-Bizama, F., Valenzuela-Muñoz, V., Gonçalves, A.T., Marambio, J.P., Hawes, C., Wadsworth, S. and Gallardo-Escárate, C. (2015) Transcription expression of immune-related genes from *Caligus rogercresseyi* evidences host-dependent patterns on Atlantic and coho salmon. *Fish and Shellfish Immunology* 47, 725–731.

Wagner, G.N., Fast, M.D. and Johnson, S.C. (2008) Physiology and immunology of *Lepeophtheirus salmonis* infections on salmonids. *Trends in Parasitology* 24(4), 175–182.

Yazawa, R., Yasuike, M., Leong, J., von Schalburg, K.R., Cooper, G.A., Beetz-Sargent, M., Robb, A., Davidson, W.S., Jones, S.R.M. and Koop, B.F. (2008) EST and mitochondrial DNA sequences support a distinct Pacific form of salmon louse, *Lepeophtheirus salmonis*. *Marine Biotechnology* 10, 741–749.

CHAPTER 14

Evolution of sea louse (Copepoda: Caligidae) resistance against control agents

Armin Sturm, Claudia Tschesche and James E. Bron

Institute of Aquaculture, University of Stirling

14.1 Introduction

Infestations of farmed Atlantic salmon (*Salmo salar*) with sea lice (Copepoda: Caligidae) constitute one of the major health management challenges for industrial salmon farming. In the North Atlantic, the salmon louse *Lepeophtheirus salmonis* (Krøyer 1837) is the main caligid species affecting farmed Atlantic salmon (Torrissen et al., 2013), with the smaller species *Caligus elongatus* Nordmann 1832 also causing infestations (Jackson et al., 2000). *L. salmonis* further occurs in the North Pacific, where it infects varied species of Pacific salmon as well as farmed Atlantic salmon (Skern-Mauritzen et al., 2014). In Chile, the sea louse species *Caligus rogercresseyi* (Boxshall & Bravo 2000) constitutes a major pathogen for farmed Atlantic salmon (Gallardo-Escárate et al., 2019). Control of sea louse infections at mariculture sites has, until recently, relied heavily on the use of veterinary drugs (Denholm et al., 2002; Jensen et al., 2020). However, due to the high potential of *L. salmonis* to evolve resistance to treatments, management strategies based predominantly on chemical control have proven unsustainable (Aaen et al., 2015).

Integrated pest management (IPM) is a systems-based approach that employs multiple control techniques and has the aim of minimising pest damage/economic losses while at the same time avoiding selection pressures driving resistance. As IPM minimises chemical use, it further reduces adverse environmental impacts of treatments (Alyokhin et al., 2015). IPM strategies for sea louse control, first proposed by Sommerville (1995), combine site management measures with non-medicinal control methods and veterinary drug treatments, while coordinating treatments within management areas (Brooks, 2009; Jackson et al., 2002). IPM approaches further involve the monitoring of sea lice levels and resistance status (Jansen et al., 2016; Jensen et al., 2020) and use knowledge of sea lice population dynamics, larval dispersal and environmental factors to deploy treatments strategically (Adams et al., 2016; Asplin et al., 2020; Groner et al., 2016; Rabe et al., 2020). IPM also relies, for maximum efficacy, upon effective communication within organisations, within and between management areas and with wider stakeholders.

The use of non-medicinal sea louse control approaches has increased dramatically in the last decade, and since 2017 non-medicinal interventions have been more frequently used than chemical treatments (Jensen et al., 2020). Non-medicinal approaches to caligid control include cage designs reducing infection rates (Stien et al., 2016), and co-culture of salmon with different species of cleaner fish (Brooker et al., 2018), as well as thermal, mechanical and freshwater delousing (Overton et al., 2019). Moreover, genetic improvement of farmed Atlantic salmon strains and the use of feed additives with immunostimulatory or lice repellent activities have the potential to contribute to increasing host defences against caligid infections (Holborn et al., 2019; Martin and Król, 2017; O'Shea et al., 2017; Tsairidou et al., 2020).

Despite the decreased of use chemical salmon delousing agents, veterinary drugs remain a significant part of sea lice control strategies. Managing existing drug resistance problems and preventing the emergence of resistance to currently effective treatments is therefore key to successful sea louse management. In the last ten years, considerable progress has been achieved regarding improvement in understanding of the genetic determinants and molecular mechanisms of drug resistance in *L. salmonis* and to a lesser extent *C. rogercresseyi*. The present review focuses on more recent insights, such as progress in the understanding of the molecular mechanisms of resistance, in an attempt to keep overlap with earlier excellent summaries of the topic area to a minimum (Aaen et al., 2015; Coates et al., 2021; Gallardo-Escárate et al. 2019; Torrissen et al., 2013). The aim of this brief review of the current state of understanding of resistance in sea lice is to contribute to the design of strategies minimising the risk of future resistance development.

14.2 Anti-sea louse therapeutants in current use

Pharmaceuticals used for salmon delousing are administered either orally as medicated feeds, or topically through immersion bath treatments. Bath treatments can be carried out either in sea cages, which for this purpose are reduced in volume and enclosed by a tarpaulin before the medical compound is added, or by pumping the fish into a well boat equipped with a system to dose the medicine (Overton et al., 2019). Currently available licensed salmon delousing agents belong to five drug classes, which comprise organophosphates, pyrethroids, avermectins, hydrogen peroxide and benzoylureas (Aaen et al., 2015). In addition, a novel treatment containing the neonicotinoid imidacloprid is currently in the process of obtaining marketing authorisation for Norway.

14.2.1 *Organophosphates*

Organophosphorus compounds constituted the first anti-sea louse drugs to be widely employed. The oral treatment metrifonate, introduced in 1977, showed a low therapeutic index and was replaced by the related compound dichlorvos, which was administered through bath treatments. Dichlorvos was widely used during the 1980s and early 1990s (Torrissen et al., 2013) before being replaced by azamethiphos, another organophosphorus bath treatment showing superior efficacy and occupational safety (Roth et al., 1996). While azamethiphos use in Norway stopped after 1999 due to resistance problems and the compound was not employed for most of the 2000s, it was reintroduced to the market in 2008 due to alternative treatments having become inefficient (Torrissen et al., 2013).

Organophosphate insecticides inhibit the enzyme acetylcholinesterase (AChE), which catalyses the degradation of the neurotransmitter acetylcholine at cholinergic synapses and neuromuscular junctions (Casida and Durkin, 2013). Inhibition of AChE leads to the accumulation of acetylcholine in the synaptic cleft, causing prolonged effects on the postsynaptic membrane and resulting in paralysis. While azamethiphos is well tolerated by Atlantic salmon when applied for 60 min at a concentration of 0.1 mg l^{-1} (Roth et al., 1993), toxicity is observed at concentrations above 0.5 mg l^{-1}, or when the length of treatment with rainbow trout (*Oncorhynchus mykiss*) is extended to 240 min (Intorre et al., 2004).

14.2.2 Pyrethroids

Pyrethrum, a plant extract containing insecticides known as pyrethrins, was introduced as a bath treatment in 1989. In the mid-1990s, pyrethrum was replaced by the chemically related synthetic pyrethroid cypermethrin, which showed a superior efficacy and therapeutic index when used in bath treatments. Another pyrethroid salmon delousing agent, deltamethrin, was introduced in 1998 in Norway and in 2008 in the UK (Denholm et al., 2002; Torrissen et al., 2013).

Both pyrethrins and pyrethroids act on the voltage-gated sodium channel (Na_v) of arthropods, which plays a central role in the initiation and propagation of action potentials (Casida and Durkin, 2013). Binding of pyrethroids to Na_v prolongs channel opening, resulting in membrane depolarisation and disruption of nerve function (Dong et al., 2014). The therapeutic margin of cypermethrin (100x) is higher than that of deltamethrin (3.5x) (Roth, 2000).

14.2.3 Avermectins

Emamectin benzoate is a macrocyclic lactone of the avermectin class, which was introduced as an in-feed salmon delousing treatment in 1999 (Aaen et al., 2015). Prior to resistance development, a seven-day course of treatment with the drug provided protection against sea lice infestation for up to 60 days (Stone et al., 2002).

Avermectins bind and irreversibly activate glutamate-gated chloride channels in arthropods and nematodes, resulting in prolonged effects on membrane potential and blocking of nervous function (Cornejo et al., 2014; Cully et al., 1996; Wolstenholme and Rogers, 2005). The therapeutic margin of emamectin benzoate in salmon is moderate (7x), possibly due to its ability to cross the blood-brain barrier and interact with GABA-gated chloride channels of the central nervous system (Roth, 2000; Sevatdal et al., 2005a).

14.2.4 Hydrogen peroxide

Hydrogen peroxide, an oxidative agent widely used as a disinfectant, was introduced in Norway as a salmon delousing bath treatment in 1990 (Torrissen et al., 2013). The compound is also used as a treatment against infections of Atlantic salmon with the protozoan gill parasite *Neoparamoeba perurans*, the causative agent of amoebic gill disease (AGD) (Wynne et al., 2020). Hydrogen peroxide is a salmon delousing agent of comparatively narrow therapeutic index and its effects are affected by environmental factors such as temperature (Johnson et al., 1993). In the late 1990s, hydrogen peroxide was discontinued due to the availability of new delousing drugs (Torrissen et al., 2013). Ten years later, hydrogen peroxide was reintroduced, as resistance against other available treatments had emerged (Aaen et al., 2015).

Hydrogen peroxide bath treatments were suggested to cause the formation of gas bubbles in the haemolymph of sea lice (Bruno and Raynard, 1994), resulting in pre-adult and adult stages being dislodged, while having no effect on the attachment of chalimus stages (McAndrew et al., 1998). Lice removed by hydrogen peroxide bath treatments can recover and reattach to the fish. After exposure, egg hatching success is decreased in gravid females, as is the development rate of chalimi to further stages (McAndrew et al., 1998; Treasurer and Grant, 1997).

14.2.5 Benzoylureas

The benzoylureas diflubenzuron and teflubenzuron, often termed insect growth regulators (IGRs), are administered as in-feed treatments and were introduced in 1996 in Norway (Torrissen et al., 2013). Both compounds ceased being used for most of the 2000s but have been employed again since 2009 due to efficacy losses of other treatments (Jackson et al., 2017). Further benzoylureas, lufenuron and hexaflumuron, are licensed as salmon delousing agents in Chile.

Benzoylureas interfere with chitin synthesis in arthropods, causing abortive moulting and hatching defects (Grosscurt, 1978; Mulder and Gijswijt, 1973). In *L. salmonis* teflubenzuron prevented development of settled larvae to chalimus and pre-adult stages but had no apparent effects on adult parasites (Branson et al., 2000). Lufenuron, which is administered as an in-feed treatment prior to seawater transfer in order to provide long-term protection against sea lice infections in the marine environment, induces cuticle deformities and an abnormal moulting process (Poley et al., 2018). In insects, the mode of action of benzoylureas has been suggested to involve interaction with chitin synthase 1 (CHS1) (Douris et al., 2016). *L. salmonis* possesses two chitin synthases both of which have unique roles essential for survival and reproduction (Braden et al., 2020; Harðardóttir et al., 2021).

14.2.6 Neonicotinoids

Neonicotinoids, commonly used for control of agricultural pests, are analogues of nicotine and act as agonists of the arthropod neuronal (nicotinic) acetylcholine receptor (nAChR), a ligand-gated ion channel activated by neurotransmitter acetylcholine (Casida and Durkin, 2013). In contrast to the natural ligand acetylcholine, neonicotinoids are not subject to enzymatic breakdown, causing persistent activation of nAChR, overstimulation of cholinergic synapses, hyperexcitation, paralysis and ultimately death. The neonicotinoid imidacloprid, which possesses high activity against *L. salmonis* (Aaen and Horsberg, 2016), is currently in the process of obtaining market authorisation for bath treatments in well boats equipped with a system that removes the compound prior to the release of treatment water to the environment.

14.3 Molecular mechanisms of pesticide resistance in terrestrial arthropods

In terrestrial arthropods, insecticide resistance has been shown to involve two main molecular mechanisms, target site insensitivity and enhanced detoxification (Feyereisen, 1995; Hemingway and Ranson, 2000). In addition, reduced cuticular penetration and changes in behaviour may contribute to resistance (Koganemaru et al., 2013; Ranson et al., 2011).

Target site insensitivity is conferred by point mutations of genes encoding proteins that constitute insecticide target sites, causing changes in specific amino acids that disrupt pesticide binding. Several insecticide classes act on neuronal target sites, such as ion channels and enzymes involved in neurotransmission in arthropods (Casida and Durkin, 2013). These target sites are usually evolutionarily conserved, and structural constraints imposed by their function limit the number of mutations that disrupt insecticide binding while at the same time maintaining normal physiological function. Parallel evolution of target site insensitivity has been reported in different insect species. For example, identical or similar mutations have independently evolved in GABA-gated chloride channels, causing resistance to cyclodiene insecticides (ffrench-Constant et al., 1993), voltage-gated sodium channels, conferring resistance to DDT and pyrethroids (Dong et al., 2014), and acetylcholinesterase, resulting in resistance to organophosphates (Weill et al., 2003). In addition to target site resistance mechanisms affecting pesticides with neuronal targets in arthropods, target site mutations have also been described for insecticides interfering with physiological processes related to growth and development, such as chitin synthesis (Douris et al., 2016).

Pesticide resistance based on enhanced detoxification, also called metabolic resistance, involves broad specificity enzymes and transporters that metabolise foreign organic chemicals to usually less toxic products, sequester foreign compounds or mediate their cellular efflux (Dermauw and van Leeuwen, 2014; Li et al., 2007). Metabolic resistance can therefore affect compounds of different pesticide classes. Biochemical pathways contributing to metabolic resistance involve a number of member-rich gene families, such as the cytochrome P450 monooxygenases (CYPs), carboxylesterases (CaE), glutathione S-transferases (GSTs) and ATP-binding cassette (ABC) transporters. Enhanced activity of enzymes and transporters of the above families in pesticide resistant arthropods can result from mutations that lead to increased gene expression (Daborn et al., 2002), or can be the consequence of gene amplification (that is, result from an increase in gene copy number) (Field et al., 1999). In rare cases, mutation or gene conversion can lead to a gene product of altered specificity, conferring upon resistant arthropods the ability to inactivate a pesticide not metabolised by carriers of the wild type gene (Joussen et al., 2012; Newcomb et al., 1997).

14.3.1 Gene families associated with metabolic resistance

The CYP gene superfamily includes haem-containing enzymes typically acting as monooxygenases, catalysing the transfer of one oxygen atom derived from molecular oxygen into endogenous and foreign chemical substrates (Feyereisen, 2012). Due to their genetic diversity, broad substrate specificity and catalytic versatility, CYPs are involved in pesticide resistance to a wide range of compounds across numerous arthropod species, including organochlorides, organophosphates, carbamates, pyrethroids, neonicotinoids, avermectins and benzoylureas (Feyereisen, 2012).

The diverse CaE gene family comprises a number of highly specialised enzymes, with some members being highly substrate-specific, others showing low substrate specificity and further members being catalytically inactive but nevertheless playing key roles in development and surface recognition (Oakeshott et al., 2005). Catalytically active CaEs are characterised by a conserved amino acid motif called the catalytic triad and typically mediate the hydrolysis of pesticides containing ester groups, such as organophosphates or pyrethroids. CaEs may also mediate resistance by sequestration of pesticides, impairing interaction with their toxicological target sites (Hemingway, 2000).

The family of GSTs conjugate organic compounds containing electrophilic centres with reduced glutathione (GSH), facilitating cellular export (Enayati et al., 2005). GSTs can also catalyse dehydrochlorination reactions and sequester pesticide molecules. In addition, GSTs with glutathione peroxidase activity have been suggested to protect tissues against pesticide induced oxidative damage (Enayati et al., 2005).

The ABC gene family contains transmembrane proteins, some of which act as transporters that mediate the cellular efflux of a wide range of organic chemicals (Dean et al., 2001). Initially discovered as biochemical factors conferring multidrug resistance to cancers (Gottesman et al., 2002), ABC drug transporters play a central role in the biochemical defence against toxicants in metazoans and show high expression levels in tissues involved in excretion and at blood-tissue barriers (Lagas et al., 2009). Roles of ABC transporters in pesticide resistance are increasingly recognised (Dermauw and van Leeuwen, 2014). Enhanced expression of ABC transporters can be associated with arthropod resistance against different pesticide classes (Dermauw and van Leeuwen, 2014) and contributes to the resistance of nematodes against avermectins (Dicker et al., 2011).

14.4 Drug resistance in caligid sea lice and molecular mechanisms underlying resistance

In the North Atlantic, *L. salmonis* shows widespread resistance against the organophosphate azamethiphos, the pyrethroids deltamethrin and cypermethrin, the macrocyclic lactone emamectin benzoate and the non-specific oxidant hydrogen peroxide (Jensen et al., 2020). Similarly, in the Chilean salmon farming industry resistance of *C. rogercresseyi* has been reported for deltamethrin, cypermethrin and emamectin benzoate, with recent data also indicating a loss of efficacy of azamethiphos (Agusti et al., 2016; Arriagada et al., 2020). The current state of knowledge regarding the genetic determinants and molecular mechanisms of resistance against the above compounds is reviewed below. No reports of resistance currently exist in *L. salmonis* with respect to the chitin synthesis inhibitors teflubenzuron and diflubenzuron, perhaps due to their more restricted use to date.

14.4.1 *Organophosphates*

Dichlorvos was introduced as a salmon delousing agent in 1979 and remained the principal treatment against sea lice infections throughout the 1980s (Torrissen et al., 2013). Dichlorvos resistance was first formally reported in the early 1990s (Jones et al., 1992). Replacement of dichlorvos by another organophosphate compound, azamethiphos, in 1994 did not resolve resistance problems, suggesting that resistance against both compounds could be based on a shared mechanism (Roth et al., 1996). Due to losses in efficacy, use of azamethiphos was initially discontinued in 1999; however, due to resistance problems with other treatments, the compound was reintroduced in 2008 (Torrissen et al., 2013). However, organophosphate resistance re-emerged quickly and was widespread by 2014 (Jansen et al., 2016). Scanning the *L. salmonis* genome led to the identification of two genes coding for acetylcholinesterase (AChE), named *ace1a* and *ace1b* (Kaur et al., 2015a), with sequence and tissue expression suggesting that *ace1a* represents the main synaptic form of AChE. Further studies led to the identification of a missense mutation in *ace1a* linked to azamethiphos resistance, which causes an amino acid substitution in the catalytic site of AChE (Phe362Tyr) (Kaur et al., 2015b). Compared to drug susceptible wild type animals, homozygous carriers of the mutation are highly azamethiphos resistant, while heterozygotes show partial resistance (Jensen et al., 2017; Kaur et al., 2015b). Field studies show correlation between the spread of the Phe362Tyr mutation and azamethiphos resistance measured in bioassays, supporting the notion that it is the main determinant of azamethiphos resistance in *L. salmonis* (Kaur et al., 2016).

Following the introduction of azamethiphos as a salmon delousing agent in Chile in 2013, the analysis of farm data showed greater than 90% treatment efficacy for 2013–2014 (Agusti et al., 2016). However, analysis of treatment outcomes from 2015 to 2017 suggest small but significant reductions in the efficacy of azamethiphos in Chile (Arriagada et al., 2020). Analysis of the *C. rogercresseyi* genome led to the identification of two AChE genes called *ace1a* and *ace1b*, of which the former was predicted to localise to synapses (Agusti-Ridaura et al., 2018). When *ace1a* was screened for sequence polymorphisms, a non-synonymous point mutation was identified, which causes an amino acid change in the vicinity of the active site of the enzyme (Phe318Val) and is homologous to an AChE mutation previously reported in organophosphate-resistant mosquitoes. While the Phe318Val mutation was present in all parasites surviving a diagnostic azamethiphos concentration in bioassays, its utility as a resistance marker requires confirmation through further studies (Agusti-Ridaura et al., 2018).

14.4.2 Pyrethroids

The pyrethroids cypermethrin and deltamethrin were introduced as salmon delousing agents in Norway in 1996 and 1998, respectively (Torrissen et al., 2013). Following reports of treatment failures, the pyrethroid susceptibility of *L. salmonis* was assessed in Ireland, Norway and Scotland between 2001 and 2003 (Sevatdal et al., 2005b). Maximal observed resistance ratios, defined as the quotient of median effective concentrations (EC_{50}) in tested and reference strains, were 7.0 for cypermethrin and 11.4 for deltamethrin (Sevatdal et al., 2005b). Deltamethrin bioassays conducted between 2007 and 2012 in Norway (Jensen et al., 2020) document a sudden increase in resistance levels from 2008 onwards, characterised by resistance ratios of 80 and above (recalculated using reference values from Sevatdal et al. (2005b)). In Chile, deltamethrin was introduced as a salmon delousing drug in 2007, followed by market authorisation for cypermethrin in 2009 (Bravo et al., 2014). Assessment of bioassay results, as well as analysis of data on the levels of use of pyrethroids in salmon farming, for the period 2007–2013 provides evidence for the evolution of pyrethroid resistance in *C. rogercresseyi* (Helgesen et al., 2014), which was confirmed in subsequent studies (Agusti et al., 2016).

Pyrethroid resistance often involves mutations of the voltage-gated sodium channel (Na_v), considered the main molecular target site for pyrethroid toxicity in arthropods. So-called knock down resistance (*kdr*) results from non-synonymous mutations of Na_v that affect the ability of pyrethroids to bind to the channel and alter its gating kinetics (Soderlund and Knipple, 2003). In *L. salmonis*, three Na_v homologues named Na_v 1.1, Na_v 1.2 and Na_v 1.3 are recognised (Carmona-Antoñanzas et al., 2019). One missense mutation identified in *L. salmonis* Na_v 1.3 is homologous to a previously characterised *kdr* mutation of insect Na_v (I936V, numbering according to *Musca domestica* Vssc1) (Carmona-Antoñanzas et al., 2019; Usherwood et al., 2007). While expressed in deltamethrin resistant *L. salmonis* strains studied (Carmona-Antoñanzas et al., 2019), the *Na_v 1.3* I936V mutation failed to predict deltamethrin resistance of lice populations causing infestations at Scottish production sites (Tschesche et al., 2021a), suggesting that pyrethroid susceptibility is mostly determined by other genetic determinants.

Genetic crosses between highly deltamethrin resistant and drug susceptible *L. salmonis* revealed that deltamethrin resistance shows a predominantly maternal mode of inheritance and is associated with mitochondrial single nucleotide polymorphisms (SNPs) (Bakke et al., 2018; Carmona-Antoñanzas et al., 2017; Nilsen and Espedal, 2015). The association of mitochondrial mutations with deltamethrin resistance suggests hypothetical mitochondrial target sites of deltamethrin toxicity. Establishing which of several mitochondrial mutations identified are causally linked to the resistance mechanism is complicated by the uniparental inheritance and lack of recombination of mitochondrial genes, which leads to a large number of 'hitchhiking' mutations showing association with a trait while not being functionally linked to it.

The metazoan mitochondrial genome contains 13 genes coding for proteins involved in oxidative phosphorylation (Tjensvoll et al., 2005). In mammalian cells, deltamethrin and other pyrethroids have been reported to disrupt the mitochondrial membrane potential and inhibit oxidative phosphorylation (Braguini et al., 2004; Gassner et al., 1997; Zhang et al., 2007). Consistent with inhibitory effects on mitochondrial ATP synthesis, deltamethrin decreased ATP levels in drug susceptible but not deltamethrin resistant *L. salmonis* (Carmona-Antoñanzas et al., 2017). In addition to their role in energy metabolism, mitochondria are involved in the regulation of apoptosis. Apoptosis is controlled by two main signalling pathways, the intrinsic (mitochondrial) and the extrinsic (death receptor) pathway (Bock and Tait, 2020; Hengartner, 2000), of which the former involves mitochondrial outer membrane permeabilisation, cytochrome c release from mitochondria and caspase activation. Deltamethrin has been shown to induce apoptosis in different mammalian cell types, by mechanisms involving mitochondrial as well as other pathways (Hossain and Richardson, 2011; Kumar et al., 2016). Deltamethrin exposure increased apoptosis in subcuticular tissue, skeletal muscle and central ganglion cells in *L. salmonis* of a drug susceptible strain, but did not do so, or did so to a lesser degree, in a deltamethrin resistant strain (Bakke et al., 2018). Deltamethrin resistant *L. salmonis* further showed lower background levels of apoptosis in subcuticular tissue and skeletal muscle. More research is required to distinguish primary and secondary mitochondrial targets of deltamethrin toxicity in *L. salmonis.*

Exposure to deltamethrin and other pyrethroids has been linked to oxidative stress in many animal tissues (reviewed by: Lu et al., 2019; Yang et al., 2020). In line with these reports, deltamethrin exposure of *C. rogercresseyi* increased the level of transcript expression of a number of genes involved in the defence against oxidative stress including superoxide dismutase, phospholipid-hydroperoxide glutathione peroxidase, ferritin and glutathione S-transferase kappa (Chávez-Mardones and Gallardo-Escárate, 2014). The disruption of mitochondrial function and oxidative stress could be linked in different ways. On the one hand, reactive oxygen species (ROS) generated by pyrethroid metabolism may decrease mitochondrial membrane potential and cause dysfunction of the organelle. On the other hand, mitochondria are the main site of cellular ROS production and mitochondrial damage could lead to massive leak of ROS into the cytoplasm (Lu et al., 2019; Yang et al., 2020).

14.4.3 Emamectin benzoate

Following the UK market authorisation of emamectin benzoate in 2000, the analysis of treatment efficacy data from Scottish salmon production sites between 2002 and 2006 revealed a reduction in the efficacy of the compound over time (Lees et al., 2008). *L. salmonis* strains isolated in 2008–2009 stably express emamectin benzoate hyposensitivity in the absence of drug exposure (Espedal et al., 2013). Between 2000 and 2007, emamectin benzoate was the sole anti-sea louse medicine authorised for the treatment of farmed salmon in Chile by the state agency Sernapesca (Bravo et al., 2014). However, reports of losses of efficacy provide evidence for the evolution of resistance against the drug in *C. rogercresseyi* (Bravo et al., 2008), with recent reports suggesting resistance is common in farming farming-intensive regions of Chile (Agusti et al., 2016).

Transcriptomic responses in emamectin benzoate resistant and susceptible *L. salmonis* populations were studied in two microarray studies performed on Scottish and Canadian isolates (Carmichael et al., 2013; Sutherland et al., 2015). In both studies, resistant lice showed an increased expression of a number of hydrolytic enzymes, whereas in the Scottish study emamectin benzoate resistant

L. salmonis further showed a transcriptional downregulation of the neuronal acetylcholine receptor and a subunit of the GABA gated chloride channel (Carmichael et al., 2013).

Using an array of 6000 SNP markers, Besnier et al. (2014) scanned the genome of *L. salmonis* farm populations from Norway, the Faroe Islands, Shetland, Ireland and the Canadian east coast. This approach resulted in the detection of selective sweeps indicating recent positive selection, with one genomic region affected being linked to emamectin benzoate resistance. A later study investigating Pacific *L. salmonis* populations (Messmer et al., 2018) used a high density (200K) SNP marker array to map putative determinants of emamectin benzoate hyposensitivity to the linkage group previously shown to be involved in resistance in North Atlantic populations (Besnier et al., 2014). Despite the identification of the genome regions involved, the nature of the genes conferring emamectin benzoate resistance remains to be elucidated. Similarly, while the molecular target of emamectin benzoate, the glutamate-gated chloride channel, has been identified and functionally characterised in *C. rogercresseyi* (Cornejo et al., 2014), the mechanisms of resistance of emamectin benzoate in *C. rogercresseyi* are also currently unknown.

14.4.4 Hydrogen peroxide

When resistance problems with organophosphates became common during the 1990s, sea louse control shifted towards hydrogen peroxide (Treasurer and Grant, 1997). However, in 2000 a loss of efficacy of the compound was reported in Scotland (Treasurer et al., 2000). Due to new treatments becoming available around this time, use of hydrogen peroxide stopped for almost a decade, but the drug was reintroduced in 2009 when other salmon delousing agents decreased in efficacy (Torrissen et al., 2013). Resistance of an *L. salmonis* farm isolate to hydrogen peroxide was reported in 2013, with an F1 generation derived from the collected animals generated in the laboratory being almost as resistant as the parental generation (Helgesen et al., 2015). The established hydrogen peroxide resistant *L. salmonis* laboratory strain showed significantly increased expression and activity of the antioxidant enzyme catalase (Helgesen et al., 2017). However, while bioassay results indicated that hydrogen peroxide resistance of the strain was virtually unchanged after four years of laboratory culture in the absence of hydrogen peroxide exposure, catalase expression of the strain was no longer enhanced compared to a reference strain (Agusti-Ridaura et al., 2020), suggesting other genetic factors contributing to hydrogen peroxide resistance. The authors then combined back-crossing and selection experiments with transcriptome sequencing (RNA-seq) techniques to identify further candidate genes differentially expressed between hydrogen peroxide resistant and susceptible *L. salmonis*, which included a number of aquaporins as well as genes potentially involved in the repair of cellular damage, such as a DNA polymerase (Agusti-Ridaura et al., 2020). The study further documented that catalase expression was restored in strains obtained after selection, suggesting potential roles of enzyme induction and/or epigenetic effects in maintaining enhanced expression of the antioxidant enzyme (Agusti-Ridaura et al., 2020).

While hydrogen peroxide was introduced as a salmon delousing bath treatment in Chile in 2007 (Bravo et al., 2010), the compound is less frequently used there than other bath treatments, and at present no reports of decreasing efficacy exist (Marín et al., 2018; Valenzuela-Muñoz et al., 2020). A study investigating the effects of hydrogen peroxide exposure on the transcript expression of antioxidant genes in *C. rogercresseyi* found that glutathione peroxidase, glutathione S-transferase and a sequence annotated as GST-like were induced following drug exposure, while hydrogen peroxide treatment had no effect on catalase and superoxide dismutase transcript abundance

(Chávez-Mardones et al., 2017). Moreover, hydrogen peroxide exposure had sex-specific effects on the transcript expression of genes with putative roles in reproduction, inducing expression of vitellogenin isoform Vg1 and decreasing transcript abundances of the ecdysone and retinoid X receptors in females, while not affecting these transcripts in males (Farlora et al., 2018). A subsequent study employing transcriptome sequencing (RNA-seq) confirmed effects of hydrogen peroxide on genes with roles in antioxidant responses and reproduction (Valenzuela-Muñoz et al., 2020).

14.4.5 Caligid gene families with roles in the biochemical defence against toxicants

Metabolic pesticide resistance in terrestrial arthropods typically involves members of gene families coding for enzymes or transporters with roles in the biochemical defence against toxicants, such as the cytochrome P450 monooxygenases (CYPs), carboxylesterases (CaE), glutathione S-transferase, and ATP-binding cassette (ABC) transporters (Dermauw and van Leeuwen, 2014; Li et al., 2007).

Compared to the number of CYP, CaE and ABC genes typically present in the genomes of free-living arthropods, homology searches of the *L. salmonis* genome revealed that compared to other arthropods the salmon louse possesses low numbers of genes coding for proteins involved in detoxification pathways (Carmona-Antoñanzas et al., 2015; Humble et al., 2019; Tschesche et al., 2021b). Genome scans identified 25 CYP members in *L. salmonis* (Humble et al., 2019), compared to 52 to 75 CYP genes present in the genomes of *D. pulex* and the free-living copepod *Tigriopus japonicus* (Baldwin et al., 2009; Han et al., 2014). Similarly, 21 CaE genes have been detected in *L. salmonis* (Tschesche et al., 2021b), compared to 35 and 51 CaE genes in the genomes of *D. melanogaster* and the mosquito *Anopheles gambiae*, respectively (Claudianos et al., 2006). While the planktonic crustacean *Daphnia pulex* and the fruit fly *Drosophila melanogaster* possess 64 and 65 ABC genes, respectively (Dean et al., 2001; Sturm et al., 2009), only 33 genes encoding ABC proteins were found in *L. salmonis* (Carmona-Antoñanzas et al., 2015). In contrast, 57 sequences coding for ABC members were found in a transcriptome assembly of the sea louse *C. rogercresseyi* (Valenzuela-Muñoz et al., 2015a). However, due to the potential presence of partial sequences, as well as splicing and allelic variants, some ABC genes could be represented more than once in the transcriptome assembly, so that the number of ABC genes in the genome of *C. rogercresseyi* requires confirmation in future studies.

Despite the overall low number of CYP, CaE and ABC genes in *L. salmonis*, the species possesses genes coding for conserved family members having essential physiological roles. For instance, *L. salmonis* exhibits CYPs with roles in ecdysteroid biosynthesis (Humble et al., 2019; Sandlund et al., 2018), CaEs with neuro-developmental functions such as neurotactins, neurolignins, acetylcholinesterases and gliotactins (Kaur et al., 2015a; Tschesche et al., 2021b), different mitochondrial ABCB transporters involved in iron homeostasis, a homologue to the sulfonylurea receptors ABCC8/9, as well as ABCE and ABCF proteins involved in translation (Carmona-Antoñanzas et al., 2015). At the same time, *L. salmonis* shows a low number of genes of ABC subfamilies known to contain drug transporters, and lacks CYP and CaE clades present in insects that contain enzymes associated with xenobiotic metabolism (Carmona-Antoñanzas et al., 2015; Humble et al., 2019; Tschesche et al., 2021b). The deficiency of *L. salmonis* with regard to biochemical defence genes may be related to the evolution of parasitism in caligid sea lice. A reduced number of detoxification genes has also been found in the human body louse *Pediculus humanus humanus* (Lee et al., 2010). Due to their parasitic lifestyle, which involves extended periods of host contact, host-associated stages of both parasites can be expected to have a low risk of exposure to environmental toxins, against which they are protected by the biochemical defence system of the host, which predicts low selection pressure to retain genes involved in detoxification pathways.

Using RT-qPCR, the transcript expression of CYPs was compared between a drug susceptible and a multi-resistant *L. salmonis* strain IoA-02, which shows hyposensitivity to emamectin benzoate, deltamethrin and azamethiphos (Humble et al., 2019). While no differences in constitutive CYP expression could be detected between strains, exposure to deltamethrin or emamectin benzoate induced transcriptional upregulation of certain CYPs (Humble et al., 2019). Upregulation of three CYP transcripts was further found in a microarray study assessing the effects of cypermethrin exposure on *L. salmonis* copepodids (Poley et al., 2016). In a study comparing transcript expression between drug susceptible and emamectin resistant *L. salmonis* using microarray, expression of two CYPs was increased by about two- to three-fold in the resistant strain (Carmichael et al., 2013). In contrast, another microarray study assessing transcript expression in a number of *L. salmonis* strains differing in emamectin benzoate susceptibility did not find upregulation of CYPs in emamectin resistant lice (Sutherland et al., 2015). In *C. rogercresseyi*, exposure to azamethiphos caused moderate upregulation of a transcript showing homology to *Tigriopus japonicus* CYP3038B1 (Valenzuela-Muñoz et al., 2015b).

When expression of catalytically active CaEs was studied in the multi-resistant *L. salmonis* strain IoA-02, transcript levels of the soluble acetylcholinesterase 1b (*ace1b*) increased by 2.84-fold compared to a reference strain. Moreover, exposure to sublethal levels of emamectin benzoate led to transcriptional upregulation of *ace1b* in female parasites (Tschesche et al., 2021b). While transcript expression of the ABC transporter P-glycoprotein (P-gp) did not differ between two *L. salmonis* strains differing in emamectin benzoate susceptibility, treatment of resistant parasites with the drug induced P-gp transcript abundance in male lice (Heumann et al., 2012). In contrast, transcript levels of four ABCC transporters did not differ between strains of different emamectin benzoate susceptibility or between drug-treated and control lice (Heumann et al, 2014).

While usually not considered to be a gene family involved in detoxification, proteases are hydrolases – some of which can catalyse the degradation of pesticides. *L. salmonis* express high levels of proteases, which are involved in many physiological processes including yolk degradation and utilisation in larvae, attachment to the host, digestion of the host's tissue and immunomodulation (Fast et al., 2007; Skern-Mauritzen et al., 2009). Cypermethrin exposure of *L. salmonis* copepodids induced transcript expression of five trypsin-like proteases (Poley et al., 2016), while deltamethrin exposure of *C. rogercresseyi* modulated the expression of trypsin-like and chymotrypsin-like transcripts (Valenzuela-Miranda and Gallardo-Escárate, 2016).

14.4.6 Global changes in gene expression related to drug resistance

Using custom 15K oligonucleotide microarrays, Carmichael et al. (2013) investigated transcriptional differences between adult male *L. salmonis* of two strains of Scottish origin that differed in emamectin benzoate susceptibility. Following 1–3 h of aquatic exposure to 200 µg/l emamectin benzoate, very few transcriptional responses were observed in animals of the resistant strain, compared to mock-treated controls. In the absence of drug exposure, 359 transcripts were found to be differentially regulated between strains in two independent experiments. Gene ontology annotation revealed that genes differing in expression were enriched in functions including calcium binding, chitin metabolism and muscle structure. In addition, two members of the Cys-loop ligand-gated ion channel family, the γ-aminobutyric acid (GABA)-gated chloride channel and the neuronal acetylcholine receptor, showed significant downregulation in the resistant strain. Moreover, in each of the two experiments ~800 additional transcripts showed differential expression not confirmed in the other trial (Carmichael et al., 2013).

L. salmonis transcriptional responses related to emamectin benzoate resistance were assessed in another study using a 38K oligonucleotide microarray (Sutherland et al., 2015), considering pre-adult II parasites of both sexes in one resistant and one susceptible strain isolated from sites located at the Canadian Atlantic coast, as well as one susceptible strain derived from Pacific populations. Parasite sex was the factor with the greatest influence on gene expression, followed by population (Sutherland et al., 2015). Again, emamectin benzoate exposure caused comparatively few changes in gene expression, particularly in resistant lice. The analysis of transcript expression in the absence of chemical exposure in Atlantic lice revealed that genes upregulated in resistant lice were enriched for those annotated as being involved in lipid metabolic processes, responses to chemical stimulus and serine peptidase activity, while functions enriched among genes showing greater expression in susceptible lice included calcium binding, system development and cell differentiation (Sutherland et al., 2015).

Taken together, available studies on emamectin resistance in *L. salmonis* (Carmichael et al., 2013; Sutherland et al., 2015) illustrate that differences in transcriptome responses between strains are complex, and include differential expression of genes that, based on their function, may not be expected to be directly involved in resistance mechanisms. Expression differences in such genes may be due to downstream effects of the primary resistance mechanism or reflect adaptations of the populations to environmental factors. Crossing experiments between strains differing in drug susceptibility followed by selection of resistant and susceptible offspring allows researchers to obtain resistant and susceptible strains sharing the same genetic background, reducing interfering factors related to differences in the strain origin (Agusti-Ridaura et al., 2020).

A study reporting the assessment of transcriptome profiles in *C. rogercresseyi* by RNA-seq (Núñez-Acuña et al., 2020) involved the investigation of susceptible and resistant strains for three drugs, deltamethrin, cypermethrin and azamethiphos. The study revealed complex transcriptomic patterns related to exposure for each of the treatments, as well as transcripts differentially expressed between susceptible and resistant strains. Interestingly, distinct differences were observed in the abundance of both coding and non-coding transcripts, with a number of long non-coding RNAs (lncRNAs) showing a high degree of correlation with specific coding sequences, suggesting roles of lncRNAs in gene regulation (Núñez-Acuña et al., 2020).

14.5 Factors affecting the evolution of resistance

14.5.1 High dose versus low dose selection

In agriculture, insecticides are typically applied at levels well above those tolerated by pests, so that a typical treatment kills close to 100% of a susceptible pest population. Only individuals carrying rare mutations that decrease susceptibility by a large extent will be able to survive. As a result, insecticide resistance emerging from selection in the field is initially typically monogenic and based on mutations having a large effect on susceptibility (ffrench-Constant, 2013; McKenzie and Batterham, 1994).

In contrast, when resistant insect strains were generated experimentally by selection of laboratory populations, the smaller size of the starting population decreases the likelihood of rare mutations being present. Accordingly, laboratory selection usually results in resistant strains in which the resistance mechanism is polygenic (that is, based on different genes, most of which have small effects on susceptibility) (ffrench-Constant, 2013; McKenzie and Batterham, 1994).

High dose strategies are typical for insecticide use for crop protection. As most insecticides interact with arthropod-specific targets that are lacking in plants, adverse side effects on crops are absent. In contrast, many salmon delousing agents, including the bath treatments azamethiphos, deltamethrin and hydrogen peroxide and the in-feed treatment emamectin benzoate, show only moderate differential toxicity between the pest and the protected species (Roth, 2000), necessitating dose limitation to avoid detrimental effects in the host (Aaen et al., 2015). As a result, relatively low levels of resistance in the parasite can become clinically relevant, potentially facilitating the evolution of resistance.

In *L. salmonis*, treatment efficacies against susceptible adult stage lice have been reported as 85–100% for azamethiphos (Jensen et al., 2017; Roth et al., 1996), 70–99% for pyrethroids (Hart et al., 1997; Jensen et al., 2017), 68–98% for emamectin benzoate (Stone et al., 2002) and 85–100% for hydrogen peroxide (Thomassen, 1993), although for the latter compound lice can recover following treatment (McAndrew et al., 1998). Based on the observation of typical treatment efficacies below 100%, one may speculate that selection conditions may have favoured the evolution of polygenic rather than monogenic resistance in sea lice. While the current understanding of genetic mechanisms of drug resistance in sea lice is still incomplete, available data provide evidence for both monogenic and polygenic mechanisms, depending on the compound. Organophosphate resistance has been linked to one major genetic determinant, a target site mutation in the *ace-1a* gene (Kaur et al., 2015b, 2016). Similarly, while the exact mechanism of pyrethroid resistance in *L. salmonis* still awaits clarification, available data suggest that resistance has one major genetic determinant residing in the mitochondrial genome (Carmona-Antoñanzas et al., 2017; Nilsen and Espedal, 2015). A monogenic mechanism further seems the most parsimonious explanation for emamectin resistance in *L. salmonis*, which is associated with a distinct selective sweep and the enrichment of one specific haplotype in different Atlantic populations (Besnier et al., 2014), even though transcriptomic data suggest differential regulation of multiple genes (Carmichael et al., 2013; Sutherland et al., 2015). While increased catalase expression has been suggested as the main mechanism of hydrogen peroxide resistance in *L. salmonis* (Helgesen et al., 2017), a recent study provides support for roles of multiple genes in conferring the resistance phenotype (Agusti-Ridaura et al., 2020).

14.5.2 Physiological costs of resistance

Mutations conferring pesticide resistance can potentially be associated with fitness costs in the absence of the pesticide. Evidence for such costs is provided by studies carried out across a range of arthropod orders. However, other studies of pesticide resistance failed to detect any associated costs (Freeman et al., 2021; Kliot and Ghanim, 2012). From a theoretical perspective, the evolution of pesticide resistance constitutes an adaptation to a new environment. Gene mutations with major phenotypic effects proving beneficial in a new environment can be expected to disrupt the complex interaction of genes established during the adaptation to the ancestral habitat, leading to fitness costs that make the new phenotype less viable in the old environment (Coustau et al., 2000; ffrench-Constant and Bass, 2017). Theoretical considerations further suggest that following the emergence of gene mutations that carry fitness costs, further mutations at other loci can act as 'modifiers' mitigating fitness costs associated with the first mutation.

If alleles conferring pesticide resistance are associated with fitness costs in the absence of chemical exposure, one would expect that after cessation of the use of the relevant compound(s), the resistance alleles would be selected against and would decrease in frequency. For the same reason, resistance alleles carrying fitness costs are predicted to be very rare in pest populations prior to the beginning of pesticide use (Freeman et al., 2021). In the period 2000–2019, the number of chemical delousing treatments administered in Norway peaked in 2015, followed by a drastic decline in chemical usage after 2016 (Jensen et al., 2020). Compared to peak use, the average annual consumption (2017–2019) of azamethiphos and cypermethrin decreased by more than 95%, while that of deltamethrin and hydrogen peroxide declined by 93% and 84%, respectively. The reduction in use of these compounds was mirrored by slight decreases in resistance after 2017, as apparent from bioassay data (Jensen et al., 2020), which could point to a decrease in resistance allele frequencies due to counter-selection. Extensive genotyping results for SNPs associated with resistance to azamethiphos and deltamethrin have been published and provide baselines against which future trends in allele frequency can be measured (Fjørtoft et al., 2017, 2020, 2021; Kaur et al., 2016). The mitochondrial genotype associated with pyrethroid resistance could not be detected in *L. salmonis* collected throughout the North Atlantic in 2000 or 2002 (Fjørtoft et al., 2020), suggesting it was rare prior to selection. A selective sweep associated with resistance to emamectin benzoate has been identified in the genome scans of *L. salmonis* (Besnier et al., 2014). Within the region of the selective sweep, parasites collected throughout the North Atlantic possessed the same resistance associated haplotype, suggesting that emamectin benzoate resistance has evolved from a *de novo* mutation and spread rapidly following widespread use of the drug (Besnier et al., 2014). While no allele frequency data are available for the *ace-1a* gene mutation Phe362Tyr linked to azamethiphos hyposensitivity for time points before resistance emerged (Kaur et al., 2015b), the analysis of SNPs surrounding the mutation showed that multiple haplotypes linked with the resistance allele exist, suggesting that the Phe362Tyr mutation existed before organophosphate agents were introduced and likely represents a neutral allele not subject to positive or negative selection in the absence of treatments (Kaur et al., 2017).

Fitness costs associated with pesticide resistance likely depend on the molecular mechanisms involved (Coustau et al., 2000). Resistance conferred by the energetically expensive constitutive overproduction of detoxification enzymes can be expected to result in fitness costs, while no such costs seem to be associated with regulation of gene expression by induction. Moreover, fitness costs may result from mutations that disrupt pesticide interaction with a target site while also compromising its physiological function. Fitness costs measured in insects include effects of resistance mutations on growth, reproductive output, pre-adult development, production and response to pheromones, and defence against parasitoids (Freeman et al., 2021; Kliot and Ghanim, 2012).

A number of genetic mechanisms mitigating fitness costs of insecticide resistance are documented. In the aphid *Myzus persicae*, carboxylesterase (CaE) overexpression contributes to organophosphate, carbamate and pyrethroid resistance (Bass et al., 2014). Enhanced CaE expression in resistant aphids is the result of serial amplification of esterase genes E4 and FE4, leading to gene copy numbers of up to 80 (Field et al., 1999). Notably, aphids possessing amplified E4 esterase genes can silence their costly expression by DNA de-methylation (Field, 2000). In the Australian blowfly *Lucilia cuprina*, resistance against the organophosphate diazinon is conferred by a missense mutation in the esterase *Rop-1* that decreases its CaE activity while at the same time enabling it to hydrolyse organophosphates (Newcomb et al., 1997). The resistance-associated *Rop-1* allele has fitness costs that are compensated for by a mutation in an epistatic modifier gene that seems to have co-evolved with the mutant *Rop-1* allele (Davies et al., 1996; McKenzie, 1990). In insects, resistance against organophosphates and carbamates based on mutations of the *ace-1* gene that code for pesticide-insensitive acetylcholinesterases are often associated with fitness costs, particularly in homozygotes (Bourguet et al., 2004; Gazave et al., 2001). In two mosquito species, *Culex pipiens* and *Anopheles gambiae*, parallel evolution led to heterogeneous gene duplication of the *ace-1* gene (Assogba et al., 2016; Lenormand et al., 1998), producing a compound heterozygote combining one susceptible and one resistant copy of the gene. Compared to homogenous duplication of the resistant copy of the *ace-1* gene, the heterogeneous allele conferred an intermediate phenotype, characterised by lower resistance but also lower fitness costs (Assogba et al., 2016).

Comparatively little is known about potential fitness costs of drug resistance in *L. salmonis*. No differences in egg production, hatching success and larval survival were observed in a study investigating emamectin benzoate resistant and drug susceptible *L. salmonis* strains (Espedal et al., 2013). While the results demonstrate the absence of fitness costs related to reproductive output and early development under standard husbandry conditions, potential effects on other variables and/or different environmental conditions cannot be excluded. In terrestrial arthropods, the relative fitness costs of resistance are often more pronounced under adverse environmental conditions such as those encountered in winter (Gazave et al., 2001; McKenzie, 1990). In *L. salmonis*, no published studies are available that have investigated the potential involvement of gene copy number in drug resistance. Interestingly, hydrogen peroxide resistance in an *L. salmonis* strain maintained under laboratory culture was initially associated with increased catalase expression and activity (Helgesen et al., 2017). Over several generations of culture in the absence of pesticide exposure, the degree of resistance as well as catalase overexpression of the strain diminished, but was restored upon renewed hydrogen peroxide exposure (Agusti-Ridaura et al., 2020).

14.5.3 Population size and connectivity

Evolution of pesticide resistance in a pest species is favoured by large population sizes, as this increases the likelihood of resistance-conferring mutations being already present or emerging *de novo*. Generation time is a further relevant factor, as species with a short generation time tend to have a greater rate of mutation per year than species with a long generation time (Thomas et al., 2010). In *L. salmonis*, population genetic studies support the hypothesis of a single panmictic population throughout the entire North Atlantic Ocean (Besnier et al., 2014; Glover et al., 2011). The high genetic connectivity of local populations is likely the result of both the planktonic dispersal of salmon lice larvae (Salama et al., 2013), and migratory wild Atlantic salmon originating from rivers located in different regions of the North Atlantic sharing oceanic feeding grounds (Jacobsen et al., 2012). Contributing to the large population size of *L. salmonis* in the North Atlantic is the high number of farmed hosts, which provide a novel habitat for *L. salmonis* that is available throughout the year. Reproductive output and generation time in the species is temperature dependent (Johnson and Albright, 1991). Overall, *L. salmonis* shows a high fecundity, with females producing up to 11 pairs of egg strings over their lifespan, which can reach 191 days at a water temperature of 7.2°C (Heuch et al., 2000). It has been estimated that *L. salmonis* complete between five and seven generations per year in Ireland (Tully, 1992). Taken together, population genetics and life cycle traits of *L. salmonis* are likely major factors explaining the well documented high potential of the species to develop resistance against treatments.

14.5.4 Wild host populations as refuges

In addition to farmed Atlantic salmon, available hosts for *L. salmonis* in the North Atlantic include three species of wild salmonids, Atlantic salmon, sea trout (*Salmo trutta*) and arctic char (*Salvelinus alpinus*). Parasite populations on wild and farmed hosts are in genetic exchange through cross-infection via larvae released into the marine environment. While farmed salmon are subject to sea louse management measures, these are absent for wild salmonid hosts, which therefore provide a refuge where there is no selection for resistance and non-resistant allelic variants can be preserved. In addition, both treatments and regular harvests exert specific selection pressures affecting *L. salmonis* life cycle traits in the farmed environment (Mennerat et al., 2012, 2017).

From a theoretical point of view, refuges in which pesticide selection on pest or parasite populations is absent can slow down resistance development and provide a source of input of non-resistant allelic variants into non-refuge populations. The effect of interactions between caligid populations on farmed and wild salmonids and the resultant effects on resistance development have been explored by different authors using modelling. In a population-based model, the extent of the impact of refuges upon the speed of resistance development was affected by the relative size of treated versus non-treated host populations, the degree of genetic exchange between parasites infecting both types of hosts, and the level of treatments in the farmed habitat (Murray, 2011). In the model, selection for resistance is highest at low to moderate levels of treatment, whereas high levels of treatment are predicted to prevent resistance selection. While this suggestion supports a high dose refuge strategy, an approach that has been initially suggested to prevent resistance formation against *Bacillus thuringiensis* insecticidal toxins expressed by transgenic plants (Gould, 1998), a high degree of genetic exchange between parasite populations can attenuate the efficacy of this approach (Murray, 2011). Only when farm population size is small relative to wild population size is exchange between populations predicted to prevent resistance formation regardless of treatment intensity. A study using agent-based modelling (McEwan et al., 2015) found that when wild salmon populations are of equal or greater number compared to farmed fish, resistance formation can theoretically be prevented. In addition, even small fitness costs of resistance significantly slow down resistance formation. In the North Atlantic, the size of farmed salmonid populations by far outweighs that of wild host populations, whereas in the North Pacific, wild salmonid hosts of *L. salmonis* outnumber farmed Atlantic salmon. Modelling conditions in the Pacific, a study assuming that genetic connectivity between *L. salmonis* populations on farms and lice present on large populations of oceanic salmonids is maintained by migrating local wild salmonids predicts a key role of the local wild hosts in delaying resistance development (Kreitzman et al., 2018). This study further shows that the degree of resistance exhibited by heterozygote carriers of a resistance allele strongly impacts the spread of the resistance.

14.6 Non-medicinal approaches to sea lice control

Over the last decade, a range of non-medicinal sea louse control approaches have been widely implemented at industrial scale (Jensen et al., 2020). Available methodologies can be divided into those that remove host-attached sea lice from farmed fish through physical or biological agents (Overton et al., 2019), those that decrease infection rates by reducing the likelihood of fish encountering infectious sea lice larvae, and those that diminish the ability of settled early infectious stages of sea lice to successfully progress through the life cycle (Barrett et al., 2020).

14.6.1 Most common non-medicinal control technologies

A number of non-medicinal approaches to sea louse control have been developed, which are covered in detail in other chapters of this volume. In brief, systems of thermal, mechanical and freshwater delousing (Overton et al., 2019) employ brief exposure to warm seawater, soft brushes and/or high-pressure water jets to remove sea lice from their hosts. Biological control using so-called cleaner fish involves the co-culture of salmon with lumpsucker (*Cyclopterus lumpus*) or wrasse (Labridae), which remove ectoparasites from other fish (Brooker et al., 2018; Powell et al., 2018). Different preventative approaches are based on the vertical distribution of the infectious *L. salmonis* copepodid larvae, which is characterised by high larval densities in the top layers of the water column (Heuch et al., 1995). Modified cage designs mounting mesh barriers to prevent entry of planktonic lice larvae in the upper range of the cage include snorkel cages and net skirts (Grøntvedt et al., 2018; Stien et al., 2016), while other systems use deep feeding and/or submerged lighting to promote residency of fish in deep water layers where larval densities are low (Frenzl et al., 2014). Sea lice use both chemical and physical cues for host location, with chemoreception playing a key role in identifying the correct host species (Fields et al., 2018; Mordue (Luntz) and Birkett, 2009). A number of plant-derived chemicals have been shown to disrupt host location behaviours in *L. salmonis* (O'Shea et al., 2017) and functional feeds with claimed sea lice repellent activity are commercially available. Further approaches aim at enhancing host immune responses to combat sea lice infestations. In Atlantic salmon, heritable variation in susceptibility to sea lice infestation has been reported (Glover et al., 2005; Kolstad et al., 2005; Gharbi et al., 2015). Currently, a number of research groups are employing selective breeding to improve farmed stock resistance to sea louse infection (Holborn et al., 2019; Tsairidou et al., 2020). In addition to repellent effects reducing larval infection success discussed above, functional feeds can enhance the resistance of host fish to louse infestation by stimulating immune responses (Martin and Król, 2017). While vaccines against sea lice would provide a cost-effective strategy for management (Raynard et al., 2002), there are currently no commercial vaccines available for sea lice and only moderate efficacy has been reported from vaccine trials to date (Contreras et al., 2020; Swain et al., 2020; Valenzuela-Muñoz et al., 2021).

14.6.2 Potential of L. salmonis to genetically adapt to non-medicinal control approaches

The well documented high potential of *L. salmonis* to develop resistance to antiparasitic drugs raises concerns about potential resistance development against non-medicinal control methods. Evolution of resistance to non-chemical methods of pest control has been documented in terrestrial arthropods. From a theoretical point of view, evolution of resistance is theoretically conceivable if treatments are not 100% efficient and if genetic determinants providing differential parasite fitness in the presence of treatments also exist.

Many non-medicinal treatments rely on stressors that are encountered by *L. salmonis* in its natural environment, such as variations in temperature and salinity, and exposure to predation. Accordingly, *L. salmonis* can be expected to have the capacity to physiologically adapt to certain non-chemical treatments, and may exhibit genetic variation in tolerance to treatments both at the individual and the population level.

Bioassays assessing effects of short (2 min) exposure to high temperature on copepodid and pre-adult stage *L. salmonis* revealed significant differences in thermal tolerance among populations originating from different regions of Norway (Andrews and Horsberg, 2021). Another report found thermal tolerance of *L. salmonis* larvae to increase after nauplii had been 'hardened' by exposure to 26°C prior to heat shock treatment (Borchel et al., 2018). Among 37 heat shock protein (HSP) genes identified, nine showed highly induced transcript expression after 1 h of exposure of nauplii to 28°C. Induction of a smaller number of HSP genes was observed after exposure to hydrogen peroxide or low salinity (Borchel et al., 2018). In a study producing full-sibling families derived from two *L. salmonis* strains originating from the north and south of Norway, respectively, survival after heat shock exposure significantly differed among families (Ljungfeldt et al., 2017). While hybrids were at a small advantage at surviving the heat shock, no difference in thermal tolerance was found between the two strains. Taken together, available evidence suggests that both environmental and genetic factors influence susceptibility to warm water exposure in *L. salmonis*.

The tolerance to low salinity exposure was assessed for four *L. salmonis* populations derived from different regions of Norway using bioassays with copepodid and pre-adult II stage parasites. Significant differences among populations were observed for copepodids, but not pre-adult II lice (Andrews and Horsberg, 2020). The above study by Ljungfeldt et al. (2017) also assessed low salinity tolerance among full-sibling families of *L. salmonis*, which were derived from two strains originating from sites characterised by predominantly brackish and full-strength seawater conditions, respectively. While the study found differences in survival after a low salinity challenge among families, treatment and background effects could not be unequivocally separated (Ljungfeldt et al., 2017).

As a result of vertical swimming behaviour in response to environmental cues including light and salinity, the vertical distribution of *L. salmonis* copepodids peaks at shallow depths during the daytime, while extending into deeper layers at night (Heuch, 1995; Heuch et al., 1995). Barrier technologies such as plankton skirts and snorkels reduce infection pressure by preventing *L. salmonis* larvae present in the upper part of the water column from entering mariculture cages, but do not restrict movement of lice at lower depths. Provided depth preference of *L. salmonis* copepodids has a significant hereditary component, widespread use of barrier technologies could lead to a selection for lice against which these control approaches are less effective. Coates and co-workers (Coates et al., 2020) demonstrated that upward swimming copepodids increased after an increase in hydrostatic pressure, and that the intensity of this response differed strongly among families, supporting a contribution of genetic factors to depth preference.

Ballan wrasse removal of *L. salmonis* in laboratory trials was demonstrated to be selective for larger parasites (Leclercq et al., 2014), with observations during trials further suggesting preferential predation on intensely coloured individuals. Anecdotal reports have claimed an increased occurrence of less pigmented phenotypes of *L. salmonis* in farming intensive areas where there is a high usage of cleaner fish (Soltveit, 2018), but currently evidence for this from systematic studies is lacking. Hamre et al. (2021) developed a scoring system to assess *L. salmonis* colour phenotypes and provide evidence for strong environmental effects on coloration. In addition, significant differences between strains in laboratory culture suggest additional roles of genetic factors in determining *L. salmonis* body coloration.

14.7 Utility of control strategies implementing dissimilar treatments

While there is consensus that use of control agents possessing a single mode of action over extended time periods should be avoided, as this enhances the selection pressure driving the evolution of resistance (Denholm et al., 2002), it is less obvious how pesticides of different modes of action should optimally be combined to provide a treatment strategy delaying the evolution of resistance. Bourguet and associates (Bourguet et al., 2013) reviewed studies assessing the performance of four common approaches, of which 'responsive alternation' involves use of a first compound until resistance emerges, and then switching to another product, while 'periodic application' is characterised by cycles of treatments rotating between at least two control agents. Moreover, in a 'mosaic' strategy, use of different compounds is varied on a spatial rather than temporal scale, whereas 'combination' involves the concomitant use of at least two pesticides. Based on the analysis of 29 studies, 'combination' was identified as the best strategy while 'response alternation' performed worst. The reviewed data did not allow a clear conclusion to be reached regarding the utility of the remaining approaches 'periodic application' and 'mosaic', as results differed among studies (Bourguet et al., 2013).

Variants of the above strategies were compared using an agent-based model exploring the use of two delousing agents in a salmon mariculture scenario (McEwan et al., 2016). In the study, treatment programmes corresponding to each of the above four strategies were simulated, assuming that treatments were triggered at caligid densities of two adult lice per fish. For 'periodic application', two further variants were included in the model, both of which assumed treatments being carried out at predetermined time intervals rather than in response to infection severity, with inter-treatment periods being either shorter or longer than the parasite's generation time. When the simulation was run for six production cycles, 'combination' was the best strategy. While inferior to 'combination', the remaining strategies ('responsive alternation', 'mosaic', 'periodic application') performed similarly to one another when treatments were triggered by lice densities, but the variants of 'periodic application' using pre-set delousing schedules were inferior compared to all other strategies (McEwan et al., 2016). In practice, sea louse control in the North Atlantic exclusively relied on only one compound class, organophosphates, from 1981 to 1992, while between 2000 and 2007 two pesticide classes were dominant, the pyrethroids and avermectins (Torrissen et al., 2013).

To explain differences among various strategies of combining dissimilar acting compounds with regard to each strategy's inherent risk to drive the evolution of resistance, the theoretical concept 'degree of treatment heterogeneity' (DTH) has been introduced (Bourguet et al., 2013). DTH denotes the probability of individuals of the controlled pest population carrying resistance genes being brought into contact with more than one pesticide during a time interval considered, which can be within the pest's generation time, or span more than one generation. Assuming that resistance genes for different pesticide classes are all initially rare, the likelihood of any given individual carrying more than one such gene is very low. The 'combination' strategy is characterised by a high DTH, as individuals carrying one type of resistance gene will likely encounter more than one pesticide, often within one generation of the pest. Accordingly, almost all resistance gene carriers in the population will be killed, and selection for resistance is avoided. At the other end of the spectrum, the DTH is low for 'response alternation', where over extended periods of time individuals carrying resistance genes will be confronted by only one pesticide, leading to genes conferring resistance against this pesticide to increase in frequency until resistance emerges at the population level. Overall, successful resistance prevention and management strategies are associated with high DTH (Bourguet et al., 2013).

With regard to sea lice control, it seems worthwhile to try to expand the concept of DTH to be applied to combinations of non-medicinal treatments, as well as strategies encompassing both chemical and alternative control techniques. However, application of DTH to non-chemical control measures will require a good understanding of the selection pressures exerted by such treatments. Moreover, some of the relevant traits discussed in section 14.7 are complex and likely under the control of multiple genes. An in-depth discussion of selection by non-chemical treatments is beyond the scope of this paper and interested readers are referred to a recent excellent review considering the current state of knowledge in the field (Coates et al., 2021).

One factor contributing to the high potential of *L. salmonis* to evolve resistance against treatments is its high potential for dispersal through planktonic larval stages. Increasingly sophisticated hydrodynamic models implementing particle tracking allow prediction of the connectivity between sea lice populations (Salama et al., 2013; Adams et al., 2016; Asplin et al., 2020; Urke et al., 2021), and predictions are improved through new insights into relevant aspects of dispersal and infection biology, such as how infectivity and vertical swimming behaviour develop over time and are affected by environmental variables such as temperature and salinity (Crosbie et al., 2020; Skern-Mauritzen et al., 2020). Mosaic strategies could potentially gain in efficiency when strategically employed based on a detailed understanding of specific patterns of local larval dispersal.

14.8 Summary and conclusions

Until recently, control of *L. salmonis* in salmon farming areas of the North Atlantic and Pacific has relied heavily on the use of a restricted number of chemical control agents, which has led to the evolution of resistance against most available compounds. Similarly, control of the caligid *C. rogercresseyi* at salmon production sites in Chile occurs mainly through treatments with delousing drugs, with resistance being commonly encountered for most compounds in current use. Studies of the molecular basis of resistance in *L. salmonis* suggest that mechanisms are compound class specific and likely based on a limited number of resistance genes. While the molecular mechanisms of drug resistance are at present not fully understood, currently available results provide evidence that *L. salmonis* shares certain mechanisms of pesticide resistance with other arthropods, while other mechanisms appear to be unique to *L. salmonis*. Progress in the availability of both genomic resources and well-characterised laboratory strains for the species *L. salmonis* and *C. rogercresseyi* can be expected to enable more powerful functional genomic and whole genome association, leading to further insights into resistance mechanisms. Questions that are still unresolved include the wider eco-physiological implications of resistance, particularly whether potential fitness costs of resistance exist and under which environmental conditions such costs might have the greatest impacts. The high potential of *L. salmonis* to develop treatment resistance and the high capacity of resistance genes to spread geographically are factors to consider in the design of wider-scale treatment strategies aiming to preserve the efficacy of current treatment approaches. Compared to the situation in the past, where sea lice control relied largely on farm management approaches and upon a restricted number of pesticides, a wide range of control technologies are available today, reducing selection pressure for resistance development. Moreover, increasingly accurate models predicting *L. salmonis* population dynamics and larval dispersal have the potential to contribute to the design of improved control strategies. The main challenge in caligid control is to weaponise these insights to generate updated integrated pest management strategies that optimise combined use of all available control approaches in a manner maximising treatment efficacy while minimising selection pressures driving resistance formation.

References

Aaen, S.M. and Horsberg, T.E. (2016) A screening of multiple classes of pharmaceutical compounds for effect on pre-adult salmon lice *Lepeophtheirus salmonis*. *Journal of Fish Diseases* 39, 1213–1223.

Aaen, S.M., Helgesen, K.O., Bakke, M.J., Kaur, K. and Horsberg, T.E. (2015) Drug resistance in sea lice: a threat to salmonid aquaculture. *Trends in Parasitology* 31, 72–81.

Adams, T.P., Aleynik, D. and Black, K.D. (2016) Temporal variability in sea lice population connectivity and implications for regional management protocols. *Aquaculture Environment Interaction* 8, 585–596.

Agusti, C., Bravo, S., Contreras, G., Bakke, M.J., Helgesen, K.O., Winkler, C., Silva, M.T., Mendoza, J. and Horsberg, T.E. (2016) Sensitivity assessment of *Caligus rogercresseyi* to anti-louse chemicals in relation to treatment efficacy in Chilean salmonid farms. *Aquaculture* 458, 195–205.

Agusti-Ridaura, C., Dondrup, M., Horsberg, T.E., Leong, J.S., Koop, B.F., Bravo, S., Mendoza, J. and Kaur, K. (2018) *Caligus rogercresseyi* acetylcholinesterase types and variants: a potential marker for organophosphate resistance. *Parasites & Vectors* 11, 570.

Agusti-Ridaura, C., Bakke, M.J., Helgesen, K.O., Sundaram, A.Y.M., Bakke, S.J., Kaur, K. and Horsberg, T.E. (2020) Candidate genes for monitoring hydrogen peroxide resistance in the salmon louse, *Lepeophtheirus salmonis*. *Parasites & Vectors* 13, 1–16.

Alyokhin, A., Mota-Sanchez, D., Baker, M., Snyder, W.E., Menasha, S., Whalon, M., Dively, G. and Moarsi, W.F. (2015) The Red Queen in a potato field: Integrated pest management versus chemical dependency in Colorado potato beetle control. *Pest Management Science* 71, 343–356.

Andrews, M. and Horsberg, T.E. (2020) Sensitivity towards low salinity determined by bioassay in the salmon louse, *Lepeophtheirus salmonis* (Copepoda: Caligidae). *Aquaculture* 514, 734511.

Andrews, M. and Horsberg, T.E. (2021) In vitro bioassay methods to test the efficacy of thermal treatment on the salmon louse, *Lepeophtheirus salmonis*. *Aquaculture* 532, 736013.

Arriagada, G., Figueroa, J., Marín, S.L., Arriagada, A.M., Lara, M. and Gallardo-Escárate, C. (2020) First report of the reduction in treatment efficacy of the organophosphate azamethiphos against the sea lice *Caligus rogercresseyi* (Boxshall & Bravo, 2000). *Aquaculture Research* 51, 436–439.

Asplin, L., Albretsen, J., Johnsen, I.A. and Sandvik, A.D. (2020) The hydrodynamic foundation for salmon lice dispersion modeling along the Norwegian coast. *Ocean Dynamics* 70, 1151–1167.

Assogba, B.S., Milesi, P., Djogbénou, L.S., Berthomieu, A., Makoundou, P., Baba-Moussa, L.S., Fiston-Lavier, A.S., Belkhir, K., Labbé, P. and Weill, M. (2016) The ace-1 locus is amplified in all resistant *Anopheles gambiae* mosquitoes: Fitness consequences of homogeneous and heterogeneous duplications. *PLoS Biology* 14, 1–26.

Bakke, M.J., Agusti, C., Bruusgaard, J.C. and Sundaram, A.Y.M. (2018) Deltamethrin resistance in the salmon louse, *Lepeophtheirus salmonis* (Krøyer): Maternal inheritance and reduced apoptosis. *Scientific Reports* 8, 8450.

Baldwin, W.S., Marko, P.B. and Nelson, D.R. (2009) The cytochrome P450 (CYP) gene superfamily in *Daphnia pulex*. *BMC Genomics* 10, 169.

Barrett, L.T., Oppedal, F., Robinson, N. and Dempster, T. (2020) Prevention not cure: a review of methods to avoid sea lice infestations in salmon aquaculture. *Reviews in Aquaculture* 12, 2527–2543.

Bass, C., Puinean, A.M., Zimmer, C.T., Denholm, I., Field, L.M., Foster, S.P., Gutbrod, O., Nauen, R., Slater, R. and Williamson, M.S. (2014) The evolution of insecticide resistance in the peach potato aphid, *Myzus persicae*. *Insect Biochemistry and Molecular Biology* 51, 41–51.

Besnier, F., Kent, M., Skern-Mauritzen, R., Lien, S., Malde, K., Edvardsen, R.B., Taylor, S., Ljungfeldt, L.E.R.R., Nilsen, F. and Glover, K.A. (2014) Human-induced evolution caught in action: SNP-array reveals rapid amphi-atlantic spread of pesticide resistance in the salmon ectoparasite *Lepeophtheirus salmonis*. *BMC Genomics* 15, 937.

Bock, F.J. and Tait, S.W.G. (2020) Mitochondria as multifaceted regulators of cell death. *Nature Reviews Molecular Cell Biology* 21, 85–100.

Borchel, A., Komisarczuk, A.Z., Rebl, A., Goldammer, T. and Nilsen, F. (2018) Systematic identification and characterization of stress-inducible heat shock proteins (HSPs) in the salmon louse (*Lepeophtheirus salmonis*). *Cell Stress and Chaperones* 23, 127–139.

Bourguet, D., Guillemaud, T., Chevillon, C. and Raymond, M. (2004) Fitness costs of insecticide resistance in natural breeding sites of the mosquito *Culex pipiens*. *Evolution* 58, 128–135.

Bourguet, D., Delmotte, F., Franck, P., Guillemaud, T., Reboud, X., Vacher, C., Bordeaux, U. and Walker, A.S. (2013) Heterogeneity of selection and the evolution of resistance. *Trends in Ecology and Evolution* 28, 110–118.

Braden, L., Michaud, D., Igboeli, O.O., Dondrup, M., Hamre, L., Dalvin, S., Purcell, S.L., Kongshaug, H., Eichner, C., Nilsen, F. and Fast, M.D. (2020) Identification of critical enzymes in the salmon louse chitin synthesis pathway as revealed by RNA interference-mediated abrogation of infectivity. *International Journal for Parasitology* 50, 873–889.

Braguini, W.L., Cadena, S.M.S.C., Carnieri, E.G.S., Rocha, M.E.M. and De Oliveira, M.B.M. (2004) Effects of deltamethrin on functions of rat liver mitochondria and on native and synthetic model membranes. *Toxicology Letters* 152, 191–202.

Branson, E.J., Rùnsberg, S.S. and Ritchie, G. (2000) Efficacy of teflubenzuron (Calicide®) for the treatment of sea lice, *Lepeophtheirus salmonis* (Krøyer 1838), infestations of farmed Atlantic salmon (*Salmo salar* L.). *Aquaculture Research* 31, 861–867.

Bravo, S., Sevatdal, S. and Horsberg, T.E (2008) Sensitivity assessment of *Caligus rogercresseyi* to emamectin benzoate in Chile. *Aquaculture* 282, 7–12.

Bravo, S., Treasurer, J., Sepulveda, M. and Lagos, C. (2010) Effectiveness of hydrogen peroxide in the control of *Caligus rogercresseyi* in Chile and implications for sea louse management. *Aquaculture* 303, 22–27.

Bravo, S., Sepulveda, M., Silva, M.T. and Costello, M.J. (2014) Efficacy of deltamethrin in the control of *Caligus rogercresseyi* (Boxshall & Bravo 2000) using bath treatment. *Aquaculture* 432, 175–180.

Brooker, A.J., Papadopoulou, A., Gutierrez, C., Rey, S., Davie, A. and Migaud, H. (2018) Sustainable production and use of cleaner fish for the biological control of sea lice: recent advances and current challenges. *Veterinary Record* 183, 383.

Brooks, K.M. (2009) Considerations in developing an integrated pest management programme for control of sea lice on farmed salmon in Pacific Canada. *Journal of Fish Diseases* 32, 59–73.

Bruno, D.W. and Raynard, R.S. (1994) Studies on the use of hydrogen peroxide as a method for the control of sea lice on Atlantic salmon. *Aquaculture International* 2, 10–18.

Carmichael, S.N., Bron, J.E., Taggart, J.B., Ireland, J.H., Bekaert, M., Burgess, S.T.G., Skuce, P.J., Nisbet, A.J., Gharbi, K. and Sturm, A. (2013) Salmon lice (*Lepeophtheirus salmonis*) showing varying emamectin benzoate susceptibilities differ in neuronal acetylcholine receptor and GABA-gated chloride channel mRNA expression. *BMC Genomics* 14, 408.

Carmona-Antoñanzas, G., Carmichael, S.N., Heumann, J., Taggart, J.B., Gharbi, K., Bron, J.E., Bekaert, M. and Sturm, A. (2015) A survey of the ATP-binding cassette (ABC) gene superfamily in the salmon louse (*Lepeophtheirus salmonis*). *PLoS ONE* 10, e0137394.

Carmona-Antoñanzas, G., Bekaert, M., Humble, J.L., Boyd, S., Roy, W., Bassett, D.I., Houston, R.D., Gharbi, K., Bron, J.E. and Sturm, A. (2017) Maternal inheritance of deltamethrin resistance in the salmon louse *Lepeophtheirus salmonis* (Krøyer) is associated with unique mtDNA haplotypes. *PLoS ONE* 12, e0180625.

Carmona-Antoñanzas, G., Helgesen, K.O., Humble, J.L., Tschesche, C., Bakke, M.J., Gamble, L., Bekaert, M., Bassett, D.I., Horsberg, T.E., Bron, J.E. and Sturm, A. (2019) Mutations in voltage-gated sodium channels from pyrethroid resistant salmon lice (*Lepeophtheirus salmonis*). *Pest Management Science* 75, 527–536.

Casida, J.E. and Durkin, K.A. (2013) Neuroactive insecticides: targets, selectivity, resistance, and secondary effects. *Annual Review of Entomology* 58, 99–117.

Chávez-Mardones, J. and Gallardo-Escárate, C. (2014) Deltamethrin (AlphaMaxTM) reveals modulation of genes related to oxidative stress in the ectoparasite *Caligus rogercresseyi*: Implications on delousing drug effectiveness. *Aquaculture* 433, 421–429.

Chávez-Mardones, J., Asencio, G., Latuz, S. and Gallardo-Escárate, C. (2017) Hydrogen peroxide modulates antioxidant system transcription, evidencing sex-dependent responses in *Caligus rogercresseyi*. *Aquaculture Research* 48, 969–978.

Claudianos, C., Ranson, H., Johnson, R.M., Biswas, S., Schuler, M.A., Berenbaum, M.R., Feyereisen, R. and Oakeshott, J.G. (2006) A deficit of detoxification enzymes: pesticide sensitivity and environmental response in the honeybee. *Insect Molecular Biology* 15, 615–636.

Coates, A., Phillips, B.L., Oppedal, F., Bui, S., Overton, K. and Dempster, T. (2020) Parasites under pressure: salmon lice have the capacity to adapt to depth-based preventions in aquaculture. *International Journal for Parasitology* 50, 865–872.

Coates, A., Phillips, B.L., Bui, S., Oppedal, F., Robinson, N.A. and Dempster, T. (2021) Evolution of salmon lice in response to management strategies: a review. *Reviews in Aquaculture* 13, 1397–1422.

Contreras, M., Karlsen, M., Villar, M., Olsen, R.H., Leknes, L.M., Furevik, A., Yttredal, K.L., Tartor, H., Grove, S., Alberdi, P., Brideseth, B. and de la Fuente, J. (2020) Vaccination with ectoparasite proteins involved in midgut function and blood digestion reduces salmon louse infestations. *Vaccines* 8(1), 1–17.

Cornejo, I., Andrini, O., Niemeyer, M.I., Marabolí, V., González-Nilo, F.D., Teulon, J., Sepúlveda, F.V and Cid, L.P. (2014) Identification and functional expression of a glutamate- and avermectin-gated chloride channel from *Caligus rogercresseyi*, a Southern Hemisphere sea louse affecting farmed fish. *PLoS Pathogens* 10, e1004402.

Coustau, C., Chevillon, C. and ffrench-Constant, R. (2000) Resistance to xenobiotics and parasites: can we count the cost? *Trends in Ecology & Evolution* 15, 378–383.

Crosbie, T., Wright, D.W., Oppedal, F., Dalvin, S., Myksvoll, M.S. and Dempster, T. (2020) Impact of thermoclines on the vertical distribution of salmon lice larvae. *Aquaculture Environment Interactions* 12, 1–10.

Cully, D.F., Paress, P.S., Liu, K.K., Schaeffer, J.M. and Arena, J.P. (1996) Identification of a *Drosophila melanogaster* glutamate-gated chloride channel sensitive to the antiparasitic agent avermectin. *Journal of Biological Chemistry* 271, 20187–20191.

Daborn, P.J., Yen, J.L., Bogwitz, M.R., Goff, G. le, Feil, E., Jeffers, S., Tijet, N., Perry, T., Heckel, D., Batterham, P., Feyereisen, R. and Wilson, T.G. (2002) A single P450 allele associated with insecticide resistance in *Drosophila*. *Science* 297, 2253–2257.

Davies, A.G., Game, A.Y., Chen, Z., Williams, T.J., Goodall, S., Yen, J.L., Mckenzie, J.A. and Batterham, P. (1996) Scalloped wings is the *Lucilia cuprina* homologue and a candidate for the modifier of fitness and asymmetry of diazinon resistance. *Genetics* 143, 1321–1337.

Dean, M., Rzhetsky, A. and Allikmet, R. (2001) The human ATP-binding cassette (ABC) transporter superfamily. *Genome Research* 11, 1156–1166.

Denholm, I., Devine, G.J., Horsberg, T.E., Sevatdal, S., Fallang, A., Nolan, D.V. and Powell, R. (2002) Analysis and management of resistance to chemotherapeutants in salmon lice, *Lepeophtheirus salmonis* (Copepoda: Caligidae). *Pest Management Science* 58, 528–536.

Dermauw, W. and van Leeuwen, T. (2014) The ABC gene family in arthropods: comparative genomics and role in insecticide transport and resistance. *Insect Biochemistry and Molecular Biology* 45, 89–110.

Dicker, A.J., Nisbet, A.J. and Skuce, P.J. (2011) Gene expression changes in a P-glycoprotein (Tci-pgp-9) putatively associated with ivermectin resistance in *Teladorsagia circumcincta*. *International Journal for Parasitology* 41, 935–942.

Dong, K., Du, Y., Rinkevich, F., Nomura, Y., Xu, P., Wang, L., Silver, K. and Zhorov, B.S. (2014) Molecular biology of insect sodium channels and pyrethroid resistance. *Insect Biochemistry and Molecular Biology* 50, 1–17.

Douris, V., Steinbach, D., Panteleri, R., Livadaras, I., Pickett, J.A., Van Leeuwen, T., Nauen, R. and Vontas, J. (2016) Resistance mutation conserved between insects and mites unravels the benzoylurea insecticide mode of action on chitin biosynthesis. *Proceedings of the National Academy of Sciences of the United States of America* 113, 14692–14697.

Enayati, A.A., Ranson, H. and Hemingway, J. (2005) Insect glutathione transferases and insecticide resistance. *Insect Molecular Biology* 14, 3–8.

Espedal, P.G., Glover, K.A., Horsberg, T.E. and Nilsen, F. (2013) Emamectin benzoate resistance and fitness in laboratory reared salmon lice (*Lepeophtheirus salmonis*). *Aquaculture* 416–417, 111–118.

Farlora, R., Valdebenito-Aguayo, F., Valenzuela-Muñoz, V. and Gallardo-Escárate, C. (2018) Hydrogen peroxide treatment modulates the transcription of sex-related genes in the sea lice *Caligus rogercresseyi*. *Journal of Fish Diseases* 41, 921–926.

Fast, M.D., Johnson, S.C., Eddy, T.D., Pinto, D. and Ross, N.W. (2007) *Lepeophtheirus salmonis* secretory/excretory products and their effects on Atlantic salmon immune gene regulation. *Parasite Immunology* 29, 179–189.

Feyereisen, R. (1995) Molecular biology of insecticide resistance. *Toxicology Letters* 82/83, 83–90.

Feyereisen, R. (2012) Insect CYP genes and P450 enzymes. In: Gilbert, L.I. (ed.) *Insect Molecular Biology and Biochemistry*. Elsevier, San Diego, CA, USA. pp. 236–316.

ffrench-Constant, R.H. (2013) The molecular genetics of insecticide resistance. *Genetics* 194, 807–815.

ffrench-Constant, R.H. and Bass, C. (2017) Does resistance really carry a fitness cost? *Current Opinion in Insect Science* 21, 39–46.

ffrench-Constant, R.H., Rocheleau, T.A., Steichen, J.C. and Chalmers, A.E. (1993) A point mutation in a *Drosophila* GABA receptor confers insecticide resistance. *Nature* 363, 449–451.

Field, L.M. (2000) Methylation and expression of amplified esterase genes in the aphid *Myzus persicae* (Sulzer). *Biochemical Journal* 349, 863–868.

Field, L.M., Blackman, R.L., Tyler-Smith, C. and Devonshire, A.L. (1999) Relationship between amount of esterase and gene copy number in insecticide-resistant *Myzus persicae* (Sulzer). *Biochemical Journal* 339, 737–742.

Fields, D.M., Skiftesvik, A.B. and Browman, H.I. (2018) Behavioural responses of infective-stage copepodids of the salmon louse (*Lepeophtheirus salmonis*, Copepoda: Caligidae) to host-related sensory cues. *Journal of Fish Diseases* 41, 875–884.

Fjørtoft, H.B., Besnier, F., Stene, A., Nilsen, F., Bjørn, P.A., Tveten, A.K., Finstad, B., Aspehaug, V. and Glover, K.A. (2017) The Phe362Tyr mutation conveying resistance to organophosphates occurs in high frequencies in salmon lice collected from wild salmon and trout. *Scientific Reports* 7, 1–10.

Fjørtoft, H.B., Nilsen, F., Besnier, F., Espedal, P.G., Stene, A., Bjørn, A., Aspehaug, V.T., Glover, K.A. and Tveten, A. (2020) Aquaculture-driven evolution: distribution of pyrethroid resistance in the salmon louse throughout the North Atlantic in the years 2000–2017. *ICES Journal of Marine Science* 77, 1805–1815.

Fjørtoft, H.B., Nilsen, F., Besnier, F., Stene, A., Tveten, A.K., Bjørn, P.A., Aspehaug, V.T. and Glover, K.A. (2021) Losing the "arms race": Multiresistant salmon lice are dispersed throughout the North Atlantic Ocean. *Royal Society Open Science* 8, 210265.

Freeman, J.C., Smith, L.B., Silva, J.J., Fan, Y., Sun, H. and Scott, J.G. (2021) Fitness studies of insecticide resistant strains: lessons learned and future directions. *Pest Management Science* 77, 3847–3856.

Frenzl, B., Stien, L.H., Cockerill, D., Oppedal, F., Richards, R.H., Shinn, A.P., Bron, J.E. and Migaud, H. (2014) Manipulation of farmed Atlantic salmon swimming behaviour through the adjustment of lighting and feeding regimes as a tool for salmon lice control. *Aquaculture* 424, 183–188.

Gallardo-Escárate, C., Arriagada, G., Carrera, C., Gonçalves, A.T., Nuñez-Acuña, G., Valenzuela-Miranda, D. and Valenzuela-Muñoz, V. (2019) The race between host and sea lice in the Chilean salmon farming: a genomic approach. *Reviews in Aquaculture* 11, 325–339.

Gassner, B., Wüthrich, A., Scholtysik, G. and Solioz, M. (1997) The pyrethroids permethrin and cyhalothrin are potent inhibitors of the mitochondrial complex I. *Journal of Pharmacology and Experimental Therapeutics* 281, 855–860.

Gazave, E., Chevillon, C., Lenormand, T., Marquine, M. and Raymond, M. (2001) Dissecting the cost of insecticide resistance genes during the overwintering period of the mosquito *Culex pipiens*. *Heredity* 87, 441–448.

Gharbi, K., Matthews, L., Bron, J., Roberts, R., Tinch, A. and Stear, M. (2015) The control of sea lice in Atlantic salmon by selective breeding. *Journal of The Royal Society Interface* 12, 20150574.

Glover, K.A., Aasmundstad, T., Nilsen, F., Storset, A. and Skaala, Ø. (2005) Variation of Atlantic salmon families (*Salmo salar* L.) in susceptibility to the sea lice *Lepeophtheirus salmonis* and *Caligus elongatus*. *Aquaculture* 245, 19–30.

Glover, K.A., Stølen, Å.B., Messmer, A., Koop, B.F., Torrissen, O. and Nilsen, F. (2011) Population genetic structure of the parasitic copepod *Lepeophtheirus salmonis* throughout the Atlantic. *Marine Ecology Progress Series* 427, 161–172.

Gottesman, M.M., Fojo, T. and Bates, S.E. (2002) Multidrug resistance in cancer: Role of ATP-dependent transporters. *Nature Reviews Cancer* 2, 48–58.

Gould, F. (1998) Sustainability of transgenic insecticidal cultivars: Integrating pest genetics and ecology. *Annual Review of Entomology* 43, 701–726.

Groner, M.L., Rogers, L.A., Bateman, A.W., Connors, B.M., Frazer, L.N., Godwin, S.C., Krkošek, M., Lewis, M.A., Peacock, S.J., Rees, E.E., Revie, C.W. and Schlaegel, U.E. (2016) Lessons from sea louse and salmon epidemiology. *Philosophical Transactions of the Royal Society B: Biological Sciences* 371, 20150203.

Grøntvedt, R.N., Kristoffersen, A.B. and Jansen, P.A. (2018) Reduced exposure of farmed salmon to salmon louse (*Lepeophtheirus salmonis* L.) infestation by use of plankton nets: Estimating the shielding effect. *Aquaculture* 495, 865–872.

Grosscurt, A.C. (1978) Diflubenzuron: some aspects of its ovicidal and larvicidal mode of action and an evaluation of its practical possibilities. *Pesticide Science* 9, 373–386.

Hamre, L.A., Oldham, T., Oppedal, F., Nilsen, F. and Glover, K.A. (2021) The potential for cleaner fish-driven evolution in the salmon louse *Lepeophtheirus salmonis*: Genetic or environmental control of pigmentation? *Ecology and Evolution* 11, 7865–7878.

Han, J., Won, E.-J., Hwang, D.-S., Shin, K.-H., Lee, Y.S., Leung, K.M.-Y., Lee, S.-J. and Lee, J.-S. (2014) Crude oil exposure results in oxidative stress-mediated dysfunctional development and reproduction in the copepod *Tigriopus japonicus* and modulates expression of cytochrome P450 (CYP) genes. *Aquatic Toxicology* 152, 308–317.

Harðardóttir, H.M., Male, R., Nilsen, F. and Dalvin, S. (2021) Chitin synthases are critical for reproduction, molting, and digestion in the salmon louse (*Lepeophtheirus salmonis*). *Life* 11, 1–24.

Hart, J.L., Thacker, J.R.M., Braidwood, J.C., Fraser, N.R. and Matthews, J.E. (1997) Novel cypermethrin formulation for the control of sea lice on salmon (*Salmo salar*). *Veterinary Record* 140, 179–181.

Helgesen, K.O., Bravo, S., Sevatdal, S., Mendoza, J. and Horsberg, T.E. (2014) Deltamethrin resistance in the sea louse *Caligus rogercresseyi* (Boxhall and Bravo) in Chile: bioassay results and usage data for antiparasitic agents with references to Norwegian conditions. *Journal of Fish Diseases* 37, 877–890.

Helgesen, K.O., Romstad, H., Aaen, S.M. and Horsberg, T.E. (2015) First report of reduced sensitivity towards hydrogen peroxide found in the salmon louse *Lepeophtheirus salmonis* in Norway. *Aquaculture Reports* 1, 37–42.

Helgesen, K.O., Bakke, M.J., Kaur, K. and Horsberg, T.E. (2017) Increased catalase activity – A possible resistance mechanism in hydrogen peroxide resistant salmon lice (*Lepeophtheirus salmonis*). *Aquaculture* 468, 135–140.

Hemingway, J. (2000) The molecular basis of two contrasting metabolic mechanisms of insecticide resistance. *Insect Biochemistry and Molecular Biology* 30, 1009–1015.

Hemingway, J. and Ranson, H. (2000) Insecticide resistance in insect vectors of human disease. *Annual Review of Entomology* 45, 371–391.

Hengartner, M.O. (2000) The biochemistry of apoptosis. *Nature* 407, 770–776.

Heuch, P.A. (1995) Experimental evidence for aggregation of salmon louse copepodids (*Lepeophtheirus salmonis*) in step salinity gradients. *Journal of the Marine Biological Association of the United Kingdom*, 75(4), 927–939.

Heuch, P.A., Parsons, A. and Boxaspen, K. (1995) Diel vertical migration – A possible host finding mechanism in salmon louse (*Lepeophtheirus salmonis*) copepodids. *Canadian Journal of Fisheries and Aquatic Sciences* 52, 681–689.

Heuch, P.A., Nordhagen, J.R. and Schram, T.A. (2000) Egg production in the salmon louse [*Lepeophtheirus salmonis* (Krøyer)] in relation to origin and water temperature. *Aquaculture Research* 31, 805–814.

Heumann, J., Carmichael, S., Bron, J.E., Tildesley, A. and Sturm, A (2012) Molecular cloning and characterisation of a novel P-glycoprotein in the salmon louse *Lepeophtheirus salmonis*. *Comparative Biochemistry and Physiology, Part C* 155, 198–205.

Heumann, J., Carmichael, S.N., Bron, J.E. and Sturm, A. (2014) Isolation and characterisation of four partial cDNA sequences encoding multidrug resistance-associated proteins (MRPs) in the salmon louse *Lepeophtheirus salmonis* (Krøyer, 1837). *Aquaculture* 424–425, 207–214.

Holborn, M.K., Rochus, C.M., Ang, K.P., Elliott, J.A.K., Leadbeater, S., Powell, F. and Boulding, E.G. (2019) Family-based genome wide association analysis for salmon lice (*Lepeophtheirus salmonis*) resistance in North American Atlantic salmon using a 50 K SNP array. *Aquaculture* 511, 734215.

Hossain, M.M. and Richardson, J.R. (2011) Mechanism of pyrethroid pesticide-induced apoptosis: Role of calpain and the ER stress pathway. *Toxicological Sciences* 122, 512–525.

Humble, J.L., Carmona-Antoñanzas, G., McNair, C.M., Nelson, D.R., Bassett, D.I., Egholm, I., Bron, J.E., Bekaert, M. and Sturm, A. (2019) Genome-wide survey of cytochrome P450 genes in the parasite *Lepeophtheirus salmonis* (Krøyer 1837) (salmon louse). *Parasites & Vectors* 12, 563.

Intorre, L., Soldani, G., Cognetti-Varriale, A.M., Monni, G., Meucci, V. and Pretti, C. (2004) Safety of azamethiphos in eel, seabass and trout. *Pharmacological Research* 49, 171–176

Jackson, D., Deady, S., Hassett, D. and Leahy, Y. (2000) *Caligus elongatus* as parasites of farmed salmonids in Ireland. *Contributions to Zoology* 69, 65–70.

Jackson, D., Hassett, D. and Copley, L. (2002) Integrated lice management on Irish salmon farms. *Fish Veterinary Journal* 6, 28–38.

Jackson, D., Kane, F., Hareide, H., Stenevik Djupevåg, E.M. and Moberg, O. (2017) The drivers of sea lice management policies and how best to integrate them into a risk management strategy: An ecosystem approach to sea lice management. *Journal of Fish Diseases* 41, 927–933.

Jacobsen, J.A., Hansen, L.P., Bakkestuen, V., Halvorsen, R., Reddin, D.G., White, J., Maoileidigh, N.O., Russell, I.C., Potter, E.C.E.T., Fowler, M., Smith, G.W., Mork, K.A., Isaksson, A., Oskarsson, S., Karlsson, L. and Pedersen, S. (2012) Distribution by origin and sea age of Atlantic salmon (*Salmo salar*) in the sea around the Faroe Islands based on analysis of historical tag recoveries. *ICES Journal of Marine Science* 69, 1598–1608.

Jansen, P.A., Grøntvedt, R.N., Tarpai, A., Helgesen, K.O. and Horsberg, T.E. (2016) Surveillance of the sensitivity towards antiparasitic bath-treatments in the salmon louse (*Lepeophtheirus salmonis*). *PLoS ONE* 11, e0149006.

Jensen, E.M., Sevatdal, S., Bakke, M.J., Kaur, K. and Horsberg, T.E. (2017) A selection study on a laboratory-designed population of salmon lice (*Lepeophtheirus salmonis*) using organophosphate and pyrethroid pesticides. *PLoS ONE* 12, 1–15.

Jensen, E.M., Horsberg, T.E., Sevatdal, S. and Helgesen, K.O. (2020) Trends in de-lousing of Norwegian farmed salmon from 2000–2019 – Consumption of medicines, salmon louse resistance and non-medicinal control methods. *PLoS ONE* 15, e0240894.

Johnson, S.C. and Albright, L.J. (1991) Development, growth, and survival of *Lepeophtheirus salmonis* (Copepoda: Caligidae) under laboratory conditions. *Journal of the Marine Biological Association of the United Kingdom* 71, 425.

Johnson, S.C., Constible, J.M. and Richard, J. (1993) Laboratory investigations on the efficacy of hydrogen peroxide against the salmon louse *Lepeophtheirus salmonis* and its toxicological and histopathological effects on Atlantic salmon *Salmo salar* and chinook salmon *Oncorhynchus tshawytscha*. *Diseases of Aquatic Organisms* 17, 197–204.

Jones, M.W., Sommerville, C. and Wootten, R. (1992) Reduced sensitivity of the salmon louse, *Lepeophtheirus salmonis*, to the organophosphate dichlorvos. *Journal of Fish Diseases* 15, 197–202.

Joussen, N., Agnolet, S., Lorenz, S., Schone, S.E., Ellinger, R., Schneider, B. and Heckel, D.G. (2012) Resistance of Australian *Helicoverpa armigera* to fenvalerate is due to the chimeric P450 enzyme CYP337B3. *Proceedings of the National Academy of Sciences of the United States of America* 109, 15206–15211.

Kaur, K., Bakke, M.J., Nilsen, F. and Horsberg, T.E., (2015a) Identification and molecular characterization of two acetylcholinesterases from the salmon louse, *Lepeophtheirus salmonis*. *PLoS ONE* 10, e0125362.

Kaur, K., Helgesen, K.O., Bakke, M.J. and Horsberg, T.E. (2015b). Mechanism behind resistance against the organophosphate azamethiphos in salmon lice (*Lepeophtheirus salmonis*). *PLoS ONE* 10, e0124220.

Kaur, K., Jansen, P.A., Aspehaug, V.T. and Horsberg, T.E. (2016) Phe362Tyr in AChE: A major factor responsible for azamethiphos resistance in *Lepeophtheirus salmonis* in Norway. *PLoS ONE* 11, 1–14.

Kaur, K., Besnier, F., Glover, K.A., Nilsen, F., Aspehaug, V.T., Fjørtoft, H.B. and Horsberg, T.E. (2017) The mechanism (Phe362Tyr mutation) behind resistance in *Lepeophtheirus salmonis* pre-dates organophosphate use in salmon farming. *Scientific Reports* 7, 12349.

Kliot, A. and Ghanim, M. (2012) Fitness costs associated with insecticide resistance. *Pest Management Science* 68, 1431–1437.

Koganemaru, R., Miller, D.M. and Adelman, Z.N. (2013) Robust cuticular penetration resistance in the common bed bug (*Cimex lectularius* L.) correlates with increased steady-state transcript levels of CPR-type cuticle protein genes. *Pesticide Biochemistry and Physiology* 106, 190–197.

Kolstad, K., Heuch, P.A., Gjerde, B., Gjedrem, T. and Salte, R. (2005) Genetic variation in resistance of Atlantic salmon (*Salmo salar*) to the salmon louse *Lepeophtheirus salmonis*. *Aquaculture* 247, 145–151.

Kreitzman, M., Ashander, J., Driscoll, J., Bateman, A.W., Chan, K.M.A., Lewis, M.A. and Krkošek, M. (2018) Wild salmon sustain the effectiveness of parasite control on salmon farms: Conservation implications from an evolutionary ecosystem service. *Conservation Letters* 11, 1–13.

Kumar, A., Sasmal, D., Bhaskar, A., Mukhopadhyay, K. and Thakur, A. (2016) Deltamethrin-induced oxidative stress and mitochondrial caspase-dependent signaling pathways in murine splenocytes. *Environmental Toxicology* 31, 808–819.

Lagas, J.S., Vlaming, M.L.H. and Schinkel, A.H. (2009) Pharmacokinetic assessment of multiple ATP-binding cassette transporters: The power of combination knockout mice. *Molecular Interventions* 9, 136–145.

Leclercq, E., Davie, A. and Migaud, H. (2014) Delousing efficiency of farmed ballan wrasse (*Labrus bergylta*) against *Lepeophtheirus salmonis* infecting Atlantic salmon (*Salmo salar*) post-smolts. *Pest Management Science* 70, 1274–1282.

Lee, S.H., Kang, J.S., Min, J.S., Yoon, K.S., Strycharz, J.P., Johnson, R., Mittapalli, O., Margam, V.M., Sun, W., Li, H.-M., Xie, J., Wu, J., Kirkness, E.F., Berenbaum, M.R., Pittendrigh, B.R. and Clark, J.M.

(2010) Decreased detoxification genes and genome size make the human body louse an efficient model to study xenobiotic metabolism. *Insect Molecular Biology* 19, 599–615.

Lees, F., Baillie, M., Gettinby, G. and Revie, C.W. (2008) The efficacy of emamectin benzoate against infestations of *Lepeophtheirus salmonis* on farmed Atlantic salmon (*Salmo salar* L) in Scotland, 2002–2006. *PLoS ONE* 3, e1549.

Lenormand, T., Guillemaud, T., Bourguet, D. and Raymond, M. (1998) Appearance and sweep of a gene duplication: Adaptive response and potential for new functions in the mosquito *Culex pipiens*. *Evolution* 52, 1705–1712.

Li, X., Schuler, M.A. and Berenbaum, M.R. (2007) Molecular mechanisms of metabolic resistance to synthetic and natural xenobiotics. *Annual Review of Entomology* 52, 231–53.

Ljungfeldt, L.E.R., Quintela, M., Besnier, F., Nilsen, F. and Glover, K.A. (2017) A pedigree-based experiment with the salmon louse, *Lepeophtheirus salmonis* reveals variation in salinity and thermal tolerance. *Evolutionary Applications* 1–13.

Lu, Q., Sun, Y., Ares, I., Anadón, A., Martínez, M., Martínez-Larrañaga, M.R., Yuan, Z., Wang, X. and Martínez, M.A. (2019) Deltamethrin toxicity: A review of oxidative stress and metabolism. *Environmental Research 1*70, 260–281.

Marín, S.L., González, M.P., Madariaga, S.T., Mancilla, M. and Mancilla, J. (2018) Response of *Caligus rogercresseyi* (Boxshall & Bravo, 2000) to treatment with hydrogen peroxide: Recovery of parasites, fish infestation and egg viability under experimental conditions. *Journal of Fish Diseases* 41, 861–873.

Martin, S.A.M. and Król, E. (2017) Nutrigenomics and immune function in fish: new insights from omics technologies. *Developmental and Comparative Immunology* 75, 86–98.

McAndrew, K.J., Sommerville, C., Wootten, R. and Bron, J.E. (1998) The effects of hydrogen peroxide treatment on different life-cycle stages of the salmon louse, *Lepeophtheirus salmonis* (Krøyer 1837). *Journal of Fish Diseases* 21, 221–228.

McEwan, G.F., Groner, M.L., Fast, M.D., Gettinby, G. and Revie, C.W. (2015) Using agent-based modelling to predict the role of wild refugia in the evolution of resistance of sea lice to chemotherapeutants. *PLoS ONE* 10, e0139128.

McEwan, G.F., Groner, M.L., Burnett, D.L., Fast, M.D. and Revie, C.W. (2016) Managing aquatic parasites for reduced drug resistance: lessons from the land. *The Royal Society Interface* 13, 20160830.

McKenzie, J.A. (1990) Selection at the dieldrin resistance locus in overwintering populations of *Lucilia cuprina* (Wiedemann). *Australian Journal of Zoology* 38, 469–477.

McKenzie, J.A. and Batterham, P. (1994) The genetic, molecular and phenotypic consequences of selection for insecticide resistance. *Trends in Ecology and Evolution* 9, 166–169.

Mennerat, A., Hamre, L., Ebert, D., Nilsen, F., Dávidová, M. and Skorping, A. (2012) Life history and virulence are linked in the ectoparasitic salmon louse *Lepeophtheirus salmonis*. *Journal of Evolutionary Biology* 25, 856–861.

Mennerat, A., Ugelvik, M.S., Håkonsrud Jensen, C. and Skorping, A. (2017) Invest more and die faster: The life history of a parasite on intensive farms. *Evolutionary Applications* 10, 890–896.

Messmer, A.M., Leong, J.S., Rondeau, E.B., Mueller, A., Despins, C.A., Minkley, D.R., Kent, M.P., Lien, S., Boyce, B., Morrison, D., Fast, M.D., Norman, J.D., Danzmann, R.G. and Koop, B.F. (2018) A 200K SNP chip reveals a novel Pacific salmon louse genotype linked to differential efficacy of emamectin benzoate. *Marine Genomics* 40, 45–57.

Mordue (Luntz), A.J. and Birkett, M.A. (2009) A review of host finding behaviour in the parasitic sea louse, *Lepeophtheirus salmonis* (Caligidae: Copepoda). *Journal of Fish Diseases* 32, 3–13.

Mulder, R. and Gijswijt, M.J. (1973) The laboratory evaluation of two promising new insecticides which interfere with cuticle deposition. *Pesticide Science* 4, 737–745.

Murray, A.G. (2011) A simple model to assess selection for treatment-resistant sea lice. *Ecological Modelling* 222, 1854–1862.

Newcomb, R.D., Campbell, P.M., Ollis, D.L., Cheah, E., Russell, R.J. and Oakeshott, J.G. (1997) A single amino acid substitution converts a carboxylesterase to an organophosphorus hydrolase and confers insecticide resistance on a blowfly. *Proceedings of the National Academy of Sciences of the United States of America* 94, 7464–7468.

Nilsen, F. and Espedal, P.G. (2015) Method for detection of pyrethroid resistance in crustaceans and oligonucleotide sequences useful in detection of pyrethroid resistance. Canadian Patent Application CA 2920588 A1.

Núñez-Acuña, G., Sáez-Vera, C., Valenzuela-Muñoz, V., Valenzuela-Miranda, D., Arriagada, G. and Gallardo-Escárate, C. (2020) Tackling the molecular drug sensitivity in the sea louse *Caligus rogercresseyi* based on mRNA and lncRNA interactions. *Genes* (Basel) 11, 1–19.

Oakeshott, J.G, Claudianos, C., Campbell, P.M., Newcomb, R.D. and Russell, R. (2005) Biochemical genetics and genomics of insect esterases. In: Gilbert, L.I., Iatrou, K. and Gill S.S. (eds) *Comprehensive Molecular Insect Science*. Elsevier, Amsterdam, The Netherlands. pp. 1–73.

O'Shea, B., Wadsworth, S., Pino Marambio, J., Birkett, M.A., Pickett, J.A. and Mordue (Luntz), A.J. (2017) Disruption of host-seeking behaviour by the salmon louse, *Lepeophtheirus salmonis*, using botanically derived repellents. *Journal of Fish Diseases* 40, 495–505.

Overton, K., Dempster, T., Oppedal, F., Kristiansen, T.S., Gismervik, K. and Stien, L.H. (2019) Salmon lice treatments and salmon mortality in Norwegian aquaculture: a review. *Reviews in Aquaculture* 11, 1398–1417.

Poley, J.D., Braden, L.M., Messmer, A.M., Whyte, S.K., Koop, B.F. and Fast, M.D. (2016) Cypermethrin exposure induces metabolic and stress-related gene expression in copepodid salmon lice (*Lepeophtheirus salmonis*). *Comparative Biochemistry and Physiology Part D: Genomics and Proteomics* 20, 74–84.

Poley, J.D., Braden, L.M., Messmer, A.M., Igboeli, O.O., Whyte, S.K., Macdonald, A., Rodriguez, J., Gameiro, M., Rufener, L. and Bouvier, J. (2018) High level efficacy of lufenuron against sea lice (*Lepeophtheirus salmonis*) linked to rapid impact on moulting processes. *International Journal for Parasitology: Drugs and Drug Resistance* 8, 174–188.

Powell, A., Treasurer, J.W., Pooley, C.L., Keay, A.J., Lloyd, R., Imsland, A.K. and Garcia de Leaniz, C. (2018) Use of lumpfish for sea-lice control in salmon farming: challenges and opportunities. *Reviews in Aquaculture* 10, 683–702.

Rabe, B., Gallego, A., Wolf, J., O'Hara Murray, R., Stuiver, C., Price, D. and Johnson, H. (2020) Applied connectivity modelling at local to regional scale: The potential for sea lice transmission between Scottish finfish aquaculture management areas. *Estuarine, Coastal and Shelf Science* 238, 106716.

Ranson, H., N'guessan, R., Lines, J., Moiroux, N., Nkuni, Z. and Corbel, V. (2011) Pyrethroid resistance in African anopheline mosquitoes: what are the implications for malaria control? *Trends in Parasitology* 27, 91–98.

Raynard, R.S., Bricknell, I.R., Billingsley, P.F., Nisbet, A.J., Vigneau, A. and Sommerville, C. (2002) Development of vaccines against sea lice. *Pest Management Science* 58, 569–575.

Roth, M. (2000) The availability and use of chemotherapeutic sea lice control products. *Contributions to Zoology* 69, 109–118.

Roth, M., Richards, R. and Sommerville, C. (1993) Current practices in the chemotherapeutic control of sea lice infestations in aquaculture: a review. *Journal of Fish Diseases* 16, 1–26.

Roth, M., Richards, R.H., Dobson, D.P. and Rae, G.H. (1996) Field trials on the efficacy of the organophosphorus compound azamethiphos for the control of sea lice (Copepoda: Caligidae) infestations of farmed Atlantic salmon (*Salmo salar*). *Aquaculture* 140, 217–239.

Salama, N.K.G., Collins, C.M., Fraser, J.G., Dunn, J., Pert, C.C., Murray, A.G. and Rabe, B. (2013) Development and assessment of a biophysical dispersal model for sea lice. *Journal of Fish Diseases* 36, 323–337.

Sandlund, L., Kongshaug, H., Horsberg, T.E., Male, R., Nilsen, F. and Dalvin, S. (2018) Identification and characterisation of the ecdysone biosynthetic genes neverland, disembodied and shade in the salmon louse *Lepeophtheirus salmonis* (Copepoda: Caligidae). *PLoS ONE* 13, e01941995.

Sevatdal, S., Magnusson, A., Ingebrigtsen, K., Haldorsen, R. and Horsberg, T.E. (2005a) Distribution of emamectin benzoate in Atlantic salmon (*Salmo salar* L.). *Journal of Veterinary Pharmacology and Therapeutics* 28, 101–107.

Sevatdal, S., Copley, L., Wallace, C., Jackson, D. and Horsberg, T.E. (2005b) Monitoring of the sensitivity of sea lice (*Lepeophtheirus salmonis*) to pyrethroids in Norway, Ireland and Scotland using bioassays and probit modelling. *Aquaculture* 244, 19–27.

Skern-Mauritzen, R., Frost, P., Dalvin, S., Kvamme, B.O., Sommerset, I. and Nilsen, F. (2009) A trypsin-like protease with apparent dual function in early *Lepeophtheirus salmonis* (Krøyer) development. *BMC Molecular Biology* 10, 44.

Skern-Mauritzen, R., Torrissen, O. and Glover, K.A. (2014) Pacific and Atlantic *Lepeophtheirus salmonis* (Krøyer, 1838) are allopatric subspecies: *Lepeophtheirus salmonis salmonis* and *L. salmonis oncorhynchi* subspecies novo. *BMC Genetics* 15, 32.

Skern-Mauritzen, R., Sissener, N.H., Sandvik, A.D., Meier, S., Sævik, P.N., Skogen, M.D., Vågseth, T., Dalvin, S., Skern-Mauritzen, M. and Bui, S. (2020) Parasite development affect dispersal dynamics; infectivity, activity and energetic status in cohorts of salmon louse copepodids. *Journal of Experimental Marine Biology and Ecology*, 530–531, 151429.

Soderlund, D.M. and Knipple, D.C. (2003) The molecular biology of knockdown resistance to pyrethroid insecticides. *Insect Biochemistry and Molecular Biology* 33, 563–577.

Soltveit, T. (2018) A clear threat? Transparent lice cleaner fish can't catch. *Fishfarming Expert*, October 2018.

Sommerville, C. (1995) Latest weapons in the war on lice. *Fish Farmer*, March/April, 53–55.

Stien, L.H., Dempster, T., Bui, S., Glaropoulos, A., Fosseidengen, J.E., Wright, D.W. and Oppedal, F. (2016) "Snorkel" sea lice barrier technology reduces sea lice loads on harvest-sized Atlantic salmon with minimal welfare impacts. *Aquaculture* 458, 29–37.

Stone, J., Roy, W.J., Sutherland, I.H., Ferguson, H.W., Sommerville, C. and Endris, R. (2002) Safety and efficacy of emamectin benzoate administered in-feed to Atlantic salmon, *Salmo salar* L., smolts in freshwater, as a preventative treatment against infestations of sea lice, *Lepeophtheirus salmonis* (Krøyer). *Aquaculture* 210, 21–34.

Sturm, A., Cunningham, P. and Dean, M. (2009) The ABC transporter gene family of *Daphnia pulex*. *BMC Genomics* 10, 170.

Sutherland, B.J.G., Poley, J.D., Igboeli, O.O., Jantzen, J.R., Fast, M.D., Koop, B.F. and Jones, S.R.M. (2015) Transcriptomic responses to emamectin benzoate in Pacific and Atlantic Canada salmon lice *Lepeophtheirus salmonis* with differing levels of drug resistance. *Evolutionary Applications* 8, 133–148.

Swain, J.K., Carpio, Y., Johansen, L.H., Velazquez, J., Hernandez, L., Leal, Y., Kumar, A. and Estrada, M.P. (2020) Impact of a candidate vaccine on the dynamics of salmon lice (*Lepeophtheirus salmonis*) infestation and immune response in Atlantic salmon (*Salmo salar* L.). *PLoS ONE* 15, 1–21.

Thomas, J.A., Welch, J.J., Lanfear, R. and Bromham, L. (2010) A generation time effect on the rate of molecular evolution in invertebrates. *Molecular Biology and Evolution* 27, 1173–1180.

Thomassen, J.M. (1993) Hydrogen peroxide as a delousing agent for Atlantic salmon. In: Boxshall, G.A. and Defaye, D. (eds) *Pathogens of Wild and Farmed Fish: Sea Lice*. Ellis Horwood, Chichester, UK. pp. 290–295.

Tjensvoll, K., Hodneland, K., Nilsen, F. and Nylund, A. (2005) Genetic characterization of the mitochondrial DNA from *Lepeophtheirus salmonis* (Crustacea; Copepoda). A new gene organization revealed. *Gene* 353, 218–230.

Torrissen, O, Jones, S., Asche, F., Guttormsen, A., Skilbrei, O.T., Nilsen, F., Horsberg, T.E. and Jackson, D. (2013) Salmon lice – impact on wild salmonids and salmon aquaculture. *Journal of Fish Diseases* 36, 171–194.

Treasurer, J.W. and Grant, A. (1997) The efficacy of hydrogen peroxide for the treatment of farmed atlantic salmon, *Salmo salar* L. infested with sea lice (Copepoda: Caligidae). *Aquaculture* 148, 265–275.

Treasurer, J.W., Wadsworth, S. and Grant, A. (2000) Resistance of sea lice, *Lepeophtheirus salmonis* (Krøyer), to hydrogen peroxide on farmed Atlantic salmon, *Salmo salar* L. *Aquaculture Research* 31, 855–860.

Tsairidou, S., Hamilton, A., Robledo, D., Bron, J.E. and Houston, R.D. (2020) Optimizing low-cost genotyping and imputation strategies for genomic selection in Atlantic salmon. *G3 – Genes, Genomes, Genetics* 10, 581–590.

Tschesche, C., Bekaert, M., Bassett, D.I., Mitchell, C., North, B., Boyd, S., Carmona-Antoñanzas, G., Bron, J.E. and Sturm, A. (2021a) Investigation of deltamethrin resistance in salmon lice (*Lepeophtheirus salmonis*) provides no evidence for roles of mutations in voltage-gated sodium channels. *Pest Management Science* 77, 1052–1060.

Tschesche, C., Bekaert, M., Humble, J.L., Bron, J.E. and Sturm, A. (2021b) Genomic analysis of the carboxylesterase family in the salmon louse (*Lepeophtheirus salmonis*). *Comparative Biochemistry and Physiology Part – C: Toxicology and Pharmacology* 248, 109095.

Tully, O. (1992) Predicting infestation parameters and impacts of caligid copepods in wild and cultured fish populations. *Invertebrate Reproduction and Development* 22, 91–102.

Urke, H.A., Daae, K., Viljugrein, H., Kandal, I., Staalstrøm, A. and Jansen, P.A. (2021) Improvement of aquaculture management practice by integration of hydrodynamic modelling. *Ocean and Coastal Management* 213, 105849.

Usherwood, P.N.R., Davies, T.G.E., Mellor, I.R., O'Reilly, A.O., Peng, F., Vais, H., Khambay, B.P.S., Field, L.M. and Williamson, M.S. (2007) Mutations in DIIS5 and the DIIS4-S5 linker of *Drosophila melanogaster* sodium channel define binding domains for pyrethroids and DDT. *FEBS Letters* 581, 5485–5492.

Valenzuela-Miranda, D. and Gallardo-Escárate, C. (2016) *Caligus rogercresseyi* serine proteases: Transcriptomic analysis in response to delousing drugs treatments. *Aquaculture* 465, 65–77.

Valenzuela-Muñoz, V., Sturm, A. and Gallardo-Escárate, C. (2015a) Transcriptomic insights on the ABC transporter gene family in the salmon louse *Caligus rogercresseyi*. *Parasites & Vectors* 8, 1–14.

Valenzuela-Muñoz, V., Chavez-Mardones, J. and Gallardo-Escárate, C. (2015b). RNA-seq analysis evidences multiple gene responses in *Caligus rogercresseyi* exposed to the anti-salmon lice drug azamethiphos. *Aquaculture* 446, 156–166.

Valenzuela-Muñoz, V., Gallardo-Escárate, A., Sáez-Vera, C., Garcés, F., Bonfatti, J. and Gallardo-Escárate, C. (2020) More than bubbles: In vivo assessment and transcriptome modulation of *Caligus rogercresseyi* and Atlantic salmon exposed to hydrogen peroxide (PARAMOVE®). *Aquaculture* 522, 735170.

Valenzuela-Muñoz, V., Benavente, B.P., Casuso, A., Leal, Y. and Gallardo-Escárate, C. (2021) Chimeric protein ipath® with chelating activity improves Atlantic salmon's immunity against infectious diseases. *Vaccines* 9(4), 1–15.

Weill, M., Luffalla, G., Mogensen, K., Chandre, F., Berthomieu, A., Berticat, C., Pasteur, N., Philips, A., Fort, P. and Raymond, M. (2003) Insecticide resistance in mosquito vectors. *Nature* 423, 136–137.

Wolstenholme, A.J. and Rogers, A.T. (2005) Glutamate-gated chloride channels and the mode of action of the avermectin/milbemycin anthelmintics. *Parasitology* 131, S85–95.

Wynne, J.W., Stratford, C., Slinger, J., Samsing, F., Rigby, M., McCulloch, R., Quezada-Rodriguez, P. and Taylor, R.S. (2020) The interaction between temperature and dose on the efficacy and biochemical response of Atlantic salmon to hydrogen peroxide treatment for amoebic gill disease. *Journal of Fish Diseases* 43, 39–48.

Yang, C., Lim, W. and Song, G. (2020) Mediation of oxidative stress toxicity induced by pyrethroid pesticides in fish. *Comparative Biochemistry and Physiology Part – C: Toxicology and Pharmacology* 234, 108758.

Zhang, S.-Y., Ito, Y., Yamanoshita, O., Yanagiba, Y., Kobayashi, M., Taya, K., Li, C., Okamura, A., Miyata, M., Ueyama, J., Lee, C.-H., Kamijima, M. and Nakajima, T. (2007) Permethrin may disrupt testosterone biosynthesis via mitochondrial membrane damage of Leydig cells in adult male mouse. *Endocrinology* 148, 3941–3949.

PART II

Sea lice control methods

CHAPTER 15

An introduction to sea lice control

Jim Treasurer[1] *and Sandra Bravo*[2]

[1]FAI Farms, Ardtoe, Argyll, UK
[2]Aquaculture Institute, Universidad Austral de Chile, Puerto Montt, Chile

15.1 Introduction

Sea lice have compromised the health, appearance, and condition of farmed salmon from the early days of salmon farming, in particular the species *Lepeophtheirus salmonis* (Krøyer 1837), *Caligus elongatus* Nordmann, and *C. rogercresseyi* (Boxshall & Bravo 2000). Sea lice also impact harvest quality, and susceptibility to other diseases (Torrissen et al., 2013; Øverli et al., 2014; Iversen et al., 2015; Gonzalez et al., 2016). Costello (2006) noted that even in areas with relatively low salmon production, sea lice numbers could increase and require treatment intervention. The *Caligus* species *C. teres* Wilson and *C. rogercresseyi* have been the main threat in Chile from the commencement of salmon farming (Bravo, 2003). Management of sea lice is critical to maintain good fish condition and to prevent morbidity as well as to avoid harvest downgrade and potential environmental impact on wild salmonids. The industry in many countries would like to increase production and the number of salmon farms to meet increasing demand for a healthy food product. However, governments and regulators have indicated that effective sea lice control is critical to upgrading and extending licences and to increase harvest tonnage and aquaculture expansion.

Methods to control sea lice were therefore a priority for both farmed and wild fish, and it was a focus for the salmon farming 'pioneers' (Wootten et al., 1982). Salmon farming companies appointed health managers and veterinarians to manage fish health and to prescribe medicines for sea lice control (Grant, 2002; Rae, 2002). Treatments were carried out on an individual site basis and without wider strategic control policies. However, farms operated within the constraints of animal welfare and veterinarian input from an early date and were therefore governed by veterinarian monitoring and authorised treatment (Hannisdal et al., 2020). Medicines had to be approved by the relevant drugs authority and also discharge to the sea approved by the environmental regulator in each country with regard to discharge consents on a site-by-site basis, although use of a 'cascade system' for licensed products in other domains was possible (Grant, 2002). In contrast, once medicines have been approved by the authority in Chile it is unnecessary to obtain consent from the environmental authority. Protocols and plans for the treatment of sea lice soon developed into a high priority for salmon farm companies. Initially, treatments were limited to medicinal treatments with metriphonate and Neguvon in Norway (Hannisdal et al., 2020) and dichlorvos (DDVP) in Scotland (Rae, 2002). These were applied by bathing the fish by enclosing the net pen with a fully enclosed tarpaulin or a tarpaulin skirt (Rae, 2002; Treasurer, 2018). Skirts did not retain all the medicine for the required treatment period of one hour giving rise to a dilution effect and an abbreviated treatment. These were possibly instrumental in encouraging the development of resistance to medicines and persisted into the 1990s (Treasurer et al., 2000). From the 1990s much experimentation with different treatment approaches and medicines followed, and licences were also pursued for alternative sea lice medicines (Roth et al., 1996). However, although azamethiphos was part of an animal test certificate (ATC) in 1990, it was ten years before a full product licence was obtained (Roth et al., 1996). Likewise, the period from the first trials to full licensing of cypermethrin was over five years.

Many new methods and medicines to control sea lice have been developed over the last 30 years (Hannisdal et al., 2020). This section of the book introduces several chapters that deal with different approaches to controlling sea lice numbers, including medicines both oral and bath, biological control, vaccination, immune stimulants to protect gut and the epidermis, barrier protective systems, and other preventative methods. These will be discussed and compared in this chapter together with a discussion on efficacy of each method, potential strategies for sea lice control, new control methods that are being developed, and novel approaches to sea lice control.

15.2 Sea lice association with other diseases

Fish health has not just been impaired by sea lice, as lice infestation in the 1980s and 1990s was also associated with infectious pancreatic necrosis (IPN) and furunculosis infection. Before the advent of effective furunculosis vaccines, salmon commonly succumbed to furunculosis outbreaks following crowding and holding within a tarpaulin for sea lice treatment. Today, the threat of furunculosis has receded due to effective vaccines but other diseases, such as piscirickettsiosis (especially in Chile), heart diseases, amoebic gill disease (AGD), and to some extent increasing sea temperature, plankton blooms, and reduced water quality, have increased the difficulty of sea lice bath treatments. Adequate oxygenation and minimising handling help to reduce the stress of treatments.

Lepeophtheirus salmonis has been associated with the transmission of bacterial and viral pathogens (such as *Aeromonas salmonicida* (Nese and Enger, 1993); *Tenacibaculum maritimum*; *Pseudomonas fluorescens*; and *Vibrio* spp.) isolated from the outer surface and midgut of sea lice (Barker et al., 2009). In addition, salmonid alphavirus (Petterson et al., 2009) and infectious salmon anaemia virus (Nylund et al., 1991, 1993) have been isolated from *L. salmonis* attached to clinically diseased fish. Similarly, Jakob et al. (2011) documented that adult *L. salmonis* can act as a mechanical vector of infectious haematopoietic necrosis virus (IHNV) and transmit it to unexposed Atlantic salmon through parasitism under experimental conditions. However, it was concluded that *L. salmonis* acts as a mechanical vector rather than a biological vector or reservoir for IHNV. *C. rogercresseyi* was also hypothesised to be a potential vector involved in the horizontal transmission of *Piscirickettsia salmonis*, the most important pathogen infecting the three salmonid species reared in seawater in Chile, but Labra et al. (2020) reported that *C. rogercresseyi* does not act as a biological vector for *P. salmonis*, nor as a mechanical vector. In contrast *C. rogercresseyi* was reported as a potential vector for the transmission of infectious salmon anaemia virus (Oelckers et al., 2014).

15.3 History and development of sea lice control methods

Although much is now known about the biology of sea lice and their effects on farmed and wild salmonids and other species (Costello, 2006; Krkošek, 2013), there has been parallel development from the first days of salmon farming on the control and treatment and management of sea lice numbers in the confines of fish farms (Rae, 2002; Mordue (Luntz) and Pike, 2002). This has taken the form of detailed work on chemotherapy and a range of novel treatments (for example Roth et al., 1996), the development of preventative methods (Barrett et al., 2020a), and the implementation of regional and national control strategies (Grant, 2002; Rae, 2002). The science also included study of resistance to medicines and measuring of resistance (Aaen et al., 2015), monitoring of sea lice numbers, the epidemiology of sea lice infection, understanding patterns of infection, identifying complicating factors, and predicting optimum treatment intervention. In terms of control methods most companies input substantial financial, logistical, and human resources to control sea lice numbers, and it is regarded as the main health problem to farmed salmon with an estimated global cost of US$500 million in 2008, which has significantly increased in recent years. For example, the costs associated with sea lice infection in Norway were reported to be US$436 million in 2011 (Abolofia et al., 2017) and US$320 million in Chile in 2019 (Dresdner et al., 2019). The feeding and growth of the fish are disrupted by sea lice and are included in these costs (Sinnott, 1999). It has been calculated that current costs are in the region of US$1 billion (see Chapter 23). Improvements in salmon growth performance in recent years can be partly attributed to the reduced sea lice loading, better sea lice control, and improved fish health.

The first sea lice epizootics on salmon farms were reported in the 1960s (Wootten et al., 1982). They mentioned that both *L. salmonis* and *C. elongatus* were found on Atlantic salmon and described early control methods. The first method to treat salmon with sea lice was with Neguvon (Brandal and Egidius, 1979) in Norway and Nuvan (dichlorvos) in Scotland (Rae, 2002), and this was later marketed as Aquagard. This is an organophosphate that works by inhibiting acetylcholinesterase and is effective against only pre-adult and adult lice stages (Roth et al., 1996). There were drawbacks relating to human safety, and occupational health monitoring was introduced. Of course, as with other medicines, there was a withdrawal period before the fish could be harvested (Table 15.1). The initial treatments in Europe used cage skirts which formed only a surface layer and the nets were drawn up to 2 m depth below this, and later full tarpaulins that enclosed the whole treatment volume were very successful (Fig. 15.1). However, after a few production cycles there were signs of the development of resistance (Aaen et al., 2015). The safety margins in using Aquagard also seemed lower when fish were compromised with other health issues such as furunculosis or pancreas disease (PD). Mortalities were reported when using bath treatments with Aquagard (Overton et al., 2018), but it was thought this could have been allied to concurrent infection with *Aeromonas salmonicida* as there were no effective vaccines to this disease or IPN in 1990.

Salmon on many farms were partially resistant to dichlorvos products by 1993 and trials commenced with another organophosphate, azamethiphos, with large-scale farm assessment under animal test certificates (Roth et al., 1996). Alternatives had to be found and hydrogen peroxide was assessed (prompted by two companies, Solvay and Brenntag) and licensed from 1993. Peroxide concentration and treatments had to be well judged as treatments are affected by the length of treatment, peroxide concentration, and temperature (Thomassen, 1993; Bruno and Raynard, 1994; Treasurer and Grant, 1997; Kiemer and Black, 1997; Bravo et al., 2010). Initially treatments were

Table 15.1 Minimum residue limit (MRL) and withdrawal period for antiparasitic products used to treat salmon infected with sea lice.

Sea lice medicine	Withdrawal period (degree days)	MRL USA (μg/kg salmon)	MRL EU (μg/kg salmon)	MRL Chile (μg/kg salmon)
Emamectin benzoate	0	absence	100	100
Deltamethrin	10	absence	30	30
Cypermethrin	30	absence	50	50
Azamethiphos	10	absence	0	0
Hydrogen peroxide	0	absence	0	0
Diflubenzuron	300	absence	1000	1000
Teflubenzuron	96	absence	500	400
Lufenuron	1350	absence	–	2050
Hexaflumuron	1923	absence	–	500

Fig. 15.1 Bath treatments of salmon with sea lice.

effective, but resistance did develop over a number of years on some farms (Treasurer et al., 2000). This treatment requires large volumes of hydrogen peroxide, and it was widely noted that sea lice were 'knocked down' by treatment but recovered at the recommended dose of 1500 ppm and could potentially re-infect salmon (Bravo et al., 2010; Marín et al., 2018). However, extensive study in enclosed sea loch systems showed that resettlement of *L. salmonis* did not occur (Treasurer and Grant, 1997), with the exception of moribund or dead fish at the water surface, and sea lice were washed from the cage area. Resettlement was identified as emanating from the fixed chalimus stages that, in the main, were not killed by the peroxide treatment. It has been claimed that resettlement of *C. rogercresseyi* does occur following treatment with hydrogen peroxide in Chile (Bravo et al., 2010) and this may be due to the more active swimming and planktonic behaviour of this species. Cypermethrin and emamectin benzoate (EB) were both licensed from 1998 (Table 15.2). Both were adopted universally and were initially effective, often requiring only two treatments with EB in a production cycle to control sea lice numbers (Treasurer et al., 2002; Bravo, 2003; Lees, 2008). Efficacy was often high, achieving up to 98% reduction in sea lice numbers. However, the use of these medicines declined dramatically in Chile and Europe from 2005 due to reduced efficacy (Bravo et al., 2008a; Agusti et al., 2016). A major shift to new medicines and the re-introduction of stocking of cleaner fish developed in Europe from 2009, especially in Norway. Volumes of these medicines increased in Norway to 1413 kg active ingredient (AI) annually for diflubenzuron, 2028 kg for teflubenzuron, and 1884 kg of azamethiphos (Hannisdal et al., 2020) (Fig. 15.2). The use of diflubenzuron increased in Chile with 3878 kg AI up to 2009. However, azamethiphos was not used in Chile until 2013 when 3207 kg AI was prescribed, and use increased rapidly reaching doses of 15,915 kg AI in 2019 (Fig. 15.2). Diflubenzuron continued to be prescribed in Norway until 2019 but the use of teflubenzuron declined rapidly from 2017. The use of medicines such as cypermethrin and azamethiphos also declined in Norway from 2017, possibly due to the implementation of physical sea lice removal methods. Hydrogen peroxide treatments were also introduced for a second time in Europe from 2009 but its use also declined from 2017 with the advent of these physical control methods.

Hydrogen peroxide was used to treat sea lice in Chile for the first time in 2007 but a requirement to report its use to the health authority was only introduced in 2018. Azamethiphos has been used in increasing volumes in Chile from 2013, and recently lufenuron has been introduced (Fig. 15.2).

The range of medicines utilised in Scotland has been limited since 2012 and this has coincided with the large-scale stocking of cleaner fish and the recent introduction of physical methods such as thermolicers and hydrolicers. Cypermethrin was discontinued from use in Scotland in 2012, and teflubenzuron discontinued in 2013. However, there is still reliance in Norway, Chile, and Scotland on two medicines that have been used for many years – azamethiphos and hydrogen peroxide. For example, the volume of hydrogen peroxide and azamethiphos used to treat salmon with sea lice in Scotland increased from 2012 (Fig. 15.2).

Treatment with hydrogen peroxide has often been viewed by the public as harmful, but it has environmental benefits as no residues are discharged. It may also be used to treat gill related health issues (Table 15.3). Such are the environmental credentials of hydrogen peroxide that the regulatory authority in Scotland, SEPA, does not require notification of the intention to use this medicine, nor national recording of the volumes used. Improvements to peroxide treatments in recent years have involved the pumping of fish into wellboat holds and treating salmon in known volumes of water with hydrogen peroxide (Aquapharma). Wellboat treatments with freshwater have also been carried out (Groner et al., 2019). Maximum stocking density in these treatments depends on the size of fish, and ranges from 45 kg m^{-3} for a 100 g fish to over 100 kg for a 5 kg fish. The tonnage that can be treated on a boat with, for example, two wells depends on the boat and the volume of the wells, but can range from a maximum number of 5000 to 18,000 fish with a mass of, for example, 3 kg. This gives a maximum treatable tonnage range from 15 to 54 tonnes per boat.

Table 15.2 Sea lice medicine use by country and dates of application (* denotes emergency licence).

Product	Method of administration	Dose (of active ingredient)	Norway	Chile	Scotland	Ireland	Faroe Islands	Canada	USA
Trichlorfon (Neguvon®)	bath	0.25 ppm	1974–1986	1981–1985	–	–	1980s	–	–
Dichlorvos (Nuvan®)	bath	1 ppm/1 h	1986–1997	1985–2001	1979–1997	1980–1985	1980s	1990–2000	–
Azamethiphos	bath	100 ppb/30 min	1994–1999 2008 to date	2013 to date	1994 to date	2008 to date*	1992	Atlantic	–
Deltamethrin	bath	0.2 ppb/30 min	1998 to date	2007 to date	2008 to date	2005 to date	1994	2009–2010	–
Cypermethrin	bath	15 ppb/30 min	1996–2017	2010 to date	1997–2011	1999–2009	1994	–	–
Hydrogen peroxide	bath	1,500 ppm/20 min	1993–1997 2009 to date	2007 to date	1993–1999 2009 to date	2009 to date	1991–1992	BC; Atlantic	–to date
Ivermectin	in feed	0.07 mg/kg fish/ week/3 weeks	–	1989–2003	–	1987–2000	1995–1996	Atlantic	–
Emamectin benzoate	in feed	50 µg/kg fish/day/ 7 days	1998 to date	1999 to date	2000 to date	2000 to date	1999 to date	1999– BC; Atlantic	–to date
Teflubenzuron	in feed	10 mg/kg fish/7 days	1996–2001 2009 to date	–	2007–2013	2006–2012	1997 2011 to date	–	–
Diflubenzuron	in feed	6 mg/kg fish/ day/14 days	1996–2000 2009 to date	2009 to date	–	–	1997 2011 to date	–	–
Lufenuron	in feed	5 mg/kg fish/ day/7 days	–	2016 to date	–	–	–	–	–
Hexaflumuron	bath	2 ppm/2 hours	–	2019 to date	–	–	–	–	–

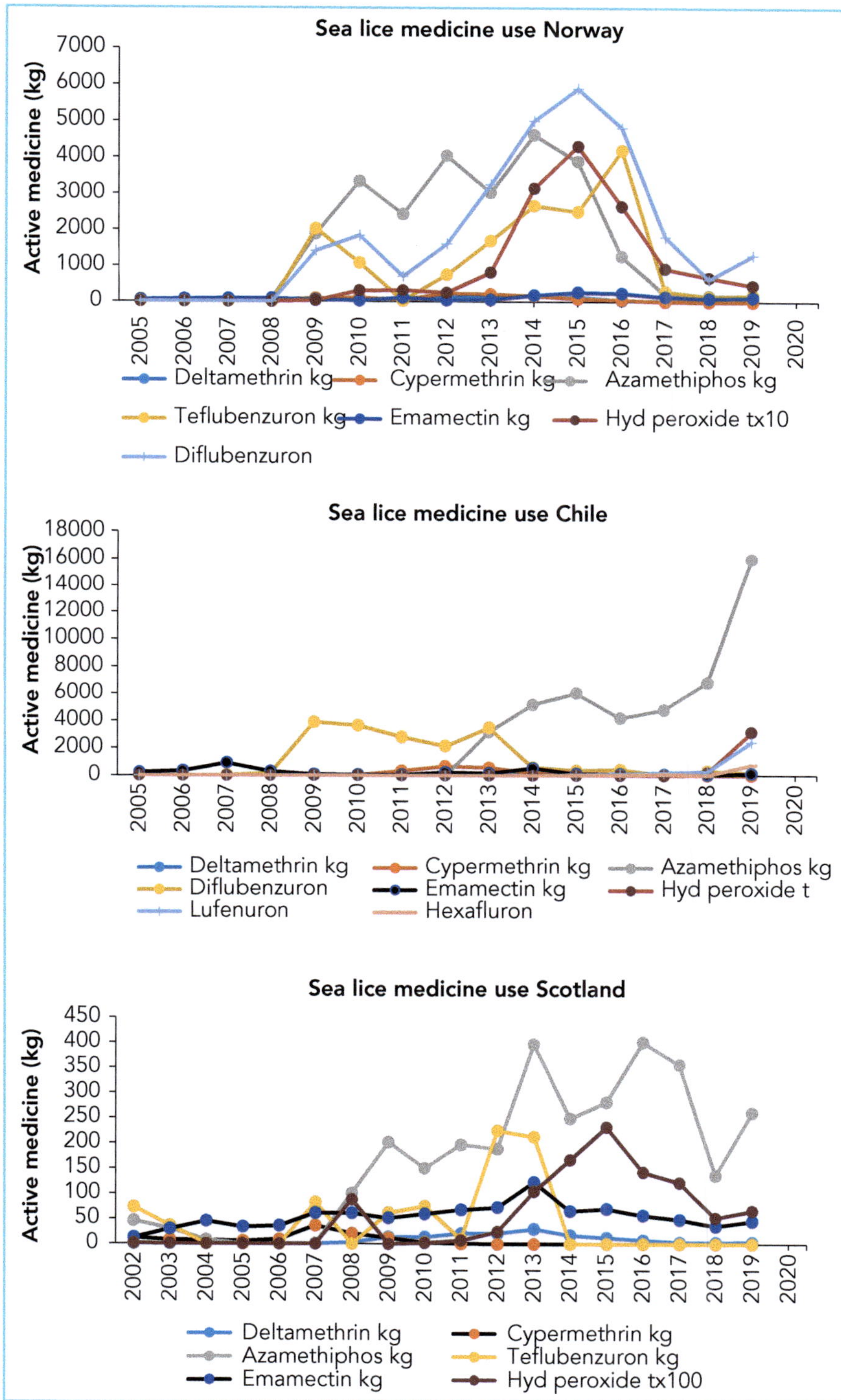

Fig. 15.2 Volumes of sea lice medicine dispensed per year in Norway, Chile and Scotland. Data from Sernapesca, Norwegian Institute of Public Health, SEPA monthly reports, and SEPA/Veterinary Medicines Directorate for hydrogen peroxide use, the VMD on information request from J. Treasurer.

Table 15.3 The mode of action of sea lice medicines (adapted from Aaen et al., 2015).

Medicine	Mode(s) of action	Resistance mechanisms in sea lice	References	Reported resistance
Emamectin benzoate	Activation of glutamate- and GABA-gated chloride channels, thereby reducing the cell's excitability. Mechanisms probably not fully elucidated	Mechanisms unclear. Overexpression of metabolic enzymes, P-gps, and downregulation of GABA-gated chloride channels and neuronal AChE receptors	Treasurer et al., 2012 Sutherland et al., 2014 Carmichael et al., 2013	Chile: Bravo et al., 2008 Norway: Grøntvedt et al., 2014 Canada: Jones et al., 2012, 2013 Scotland: Lees et al., 2008
Benzoylureas (diflubenzuron; teflubenzuron)	Inhibition of chitin synthesis, rendering the parasites unable to detach from their exuviae during moulting	Not reported	Merzendorfer, 2013 Van Leeuwen et al., 2012	Not reported
Azamethiphos	Inhibition of AChEs in cholinergic synapses, leading to excitation and subsequently paralysis	Mutation in one of the genes coding for AChE	Roth et al., 1996 Denholm et al., 2002 Kaur et al., 2015	Norway: Grøntvedt et al., 2014 Chile: Marín et al., 2015
Pyrethroids (deltamethrin; cypermethrin)	Disruption of neuronal signals through interference with VDSCs, leading to excitation and subsequently paralysis	Metabolic resistance most likely	Weston et al., 2013 Sevatdal, 2005	Norway: Sevatdal et al., 2005 Chile (2012): Helgesen et al., 2014
Hydrogen peroxide	Not fully elucidated. Etching action. Gas bubbles in the body rendering the parasites unable to hold on to a surface. Also used against amoebic gill disease	Mechanism not elucidated	Thomassen, 1993 Bruno and Raynard, 1994 Helgesen et al., 2014	Scotland: Treasurer et al., 2000 Norway: Helgesen et al., 2015

15.4 Non-medicinal treatment methods

As there has been widespread loss of sensitivity by sea lice to many medicines other means of controlling sea lice have been developed (Aaen et al., 2015). Non-medicinal methods have therefore emerged as alternatives (Table 15.4) and include treatment with freshwater. Freshwater treatment is normally carried out in wellboats with a treatment time of 12 to 16 h and high treatment efficacy has been recorded in all developmental stages (Paterson, 2018). The freshwater is obtained by abstraction from a known water source or by desalination on the wellboat.

Salmon have been treated in thermal systems on boats or barges with temperatures of up to 28–34°C for 30 seconds. This may present a welfare risk and it was suggested that salmon have aversion to such high temperatures (Gismervik et al., 2019). However, a study by Moltumyr et al. (2021) showed that there was no significant increase in the presence of lesions due to the thermal component of the treatment. Nevertheless, stress associated with the treatments may impact the

Table 15.4 Non-medicinal methods used to control sea lice in each country. E = experimental.

Method	E=experimental					
	Norway	Chile	Scotland	Faroes	Canada	USA
Freshwater	+	+	E	+	–	–
Water flush	+	+	+	+	+	+
Thermal	+	+	+	+	+	+
Brushing (SkaMik)	+	–	+	–	–	–
Laser	E	–	E	E	–	–
Ultrasound	–	–	–	–	–	–
Vacuum	–	–	–	–	–	–

fish. In this thermal treatment system a boat or barge is moored alongside the pen and the salmon are pumped into the thermal system (Fig. 15.3).The fish are then pumped through a dewatering unit and into a treatment chamber filled with warm seawater (Overton et al., 2018). The high temperature detaches the lice from fish and then the salmon are returned to the pen. The water is removed from the thermal unit after treatment and the detached lice are filtered from the water. The system may cause physical damage to the gills and the eyes, and to the thymus and nasal cavity at higher duration of exposure, from 72 to 140 seconds (Poppe et al., 2018). Some studies involving temperatures in excess of 28°C suggested that there could be a pain response and avoidance by salmon.

Flusher systems use high pressure jets of water which are directed from tail to head of the fish to ease the release of lice which are dislodged from the salmon and filtered from the effluent water (Gismervik et al., 2019). Flushers are mounted on a boat or barge adjacent to the salmon pens. The fish are pumped into the treatment unit and the lice are removed mechanically. A development from this is the SkaMik system which incorporates a brusher in the flushing system. Fish are brushed from the head to tail and the sea lice are flushed and brushed from the fish. The system, and especially the brushing, can cause loss of scales and damage to the gills (Hjeltnes et al., 2018). A laser system has also been developed, incorporating a camera to identify individual lice and directing a laser pulse to kill the specific louse once detected. However, there was no significant difference in sea lice numbers between fish in pens with or without laser nodes after a set time (Bui et al., 2020). A large number of laser pulses were delivered to lice numbers, suggesting that a lack of lethality rather than a lack of target detection may have been responsible for low efficacy. There were no adverse effects on salmon welfare indicators using the laser system, for example condition of the skin or eyes.

Mechanical treatments have been effective and there has been a reduction in the volumes of medicines used over the last five years (see Norway and Scotland in Figs 15.2 and 15.4). Only 5% of the total treatments applied in Norway in 2015 to control sea lice were not medicine-based, but non-medicinal alternatives increased to 80% of treatments by 2018 (Fig. 15.4). Non-medicinal treatment accounted for 50% of all treatments in one major salmon farming company in Scotland in 2019. The efficacy of hydrolicer treatments increased after initial use in 2016 with reductions in numbers of adult female lice regularly in excess of 80% in both hydrolicer and thermolicer treatments (Fig. 15.5). The use of mechanical treatments continues to increase in Norway and Scotland and has become the dominant treatment approach. In contrast non-medicinal methods in Chile in 2019 were less than 1% of the total sea lice treatments. This comprised 18 bath treatments with freshwater and

Fig. 15.3 Examples of mechanical treatments, a) A thermolicer system; b) A treatment with hydrogen peroxide in a wellboat.

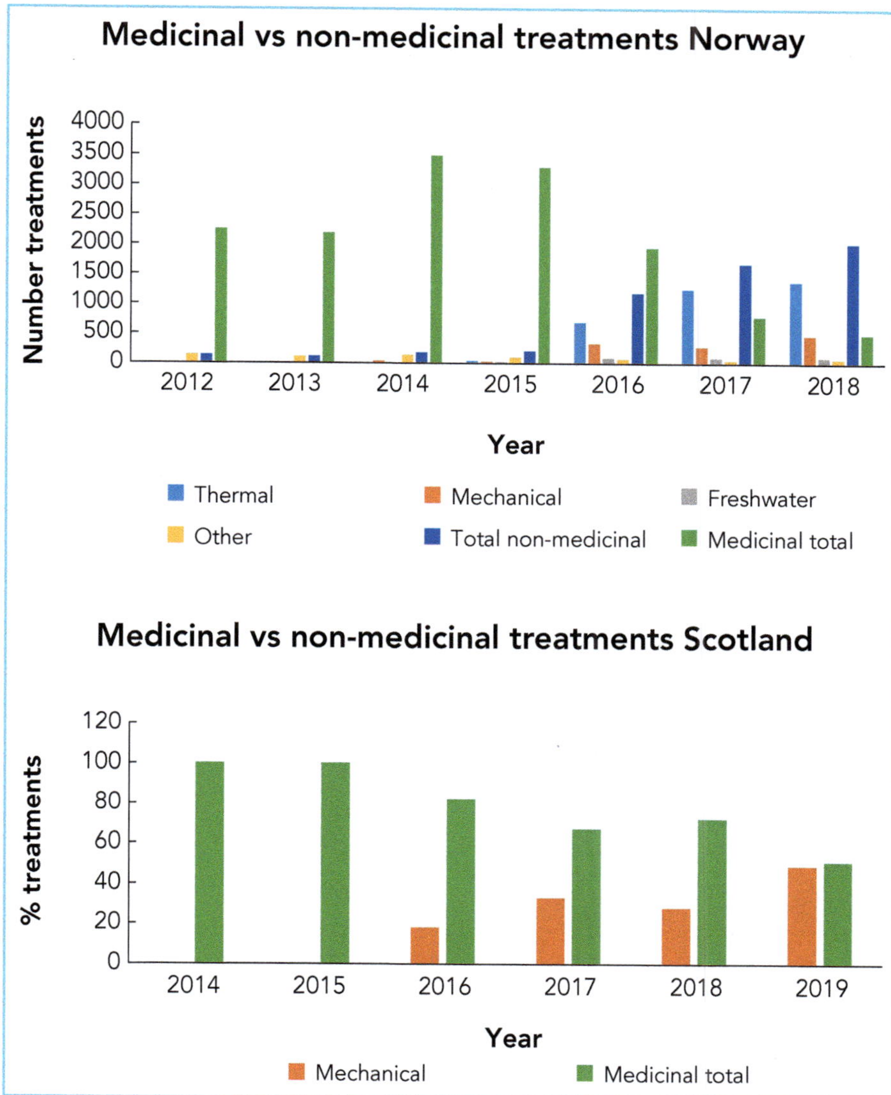

Fig. 15.4 Non-medicinal and chemotherapeutant use in Norway and Scotland.

eight treatments with water flush. In the same year 1270 bath treatments with azamethiphos were applied and 30 with deltamethrin (Sernapesca, 2020). Both flusher and thermal systems have been introduced in Maine, British Columbia, and Atlantic Canada and the number of treatments by these systems is increasing rapidly.

The tonnage of fish that can be treated by mechanical methods in an 8 h day (t h^{-1}) is dependent on the boat capacity. The first equipment pumped around 80–100 t h^{-1}, but the latest equipment with four treatment lines installed in parallel can go up to 200 t h^{-1}. A current comparison of systems is Flatsetsund Engineering AS (FLS) systems which can treat 40–50 t h^{-1} per line, hydrolicers 35–45 t h^{-1} per line, and thermolicers 90 t h^{-1} per line.

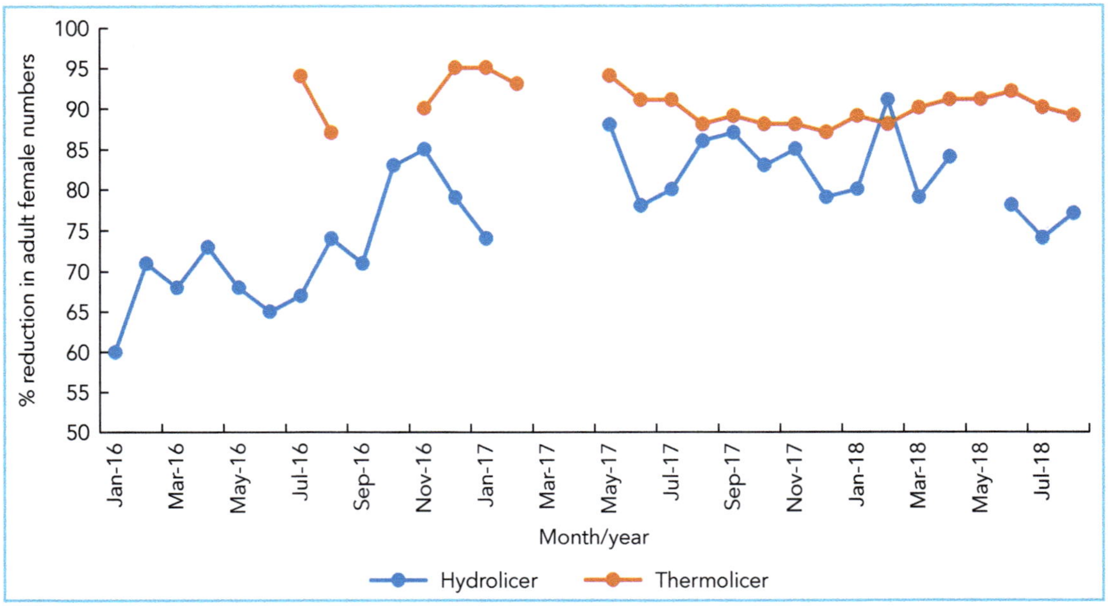

Fig. 15.5 The % reduction in adult female sea lice numbers with hydrolicer and thermolicer in one salmon farming company in Scotland. Mean figures for all treatments in each month are quoted. Figures reworked from published data source (Paterson, 2018).

15.5 Preventative methods

Much has been made of the use of preventative methods in sea lice control. These were introduced from an early stage, initially in Norway and also in Scotland by Marine Harvest and McConnell Salmon in Loch Sunart, west Scotland (Bron et al., 1993; Grant and Treasurer, 1993). This involved close cooperation between several companies in the same hydrographic area and was based largely on a fallowing period of four to six weeks decided from the duration of the life cycle of *L. salmonis* (Bron et al., 1993). This was shown to break the cycle of sea lice infection and maintain lice free salmon up to the first winter in the production cycle in Scotland and Ireland. The minimum time required for a fallow break was assessed as 30 days to ensure no further recruitment by copepodites to the fish (Bron et al., 1993). All farms in the same management area employed an all-in-all-out stocking strategy with coordinated fallowing between farms and companies in the same health management zone. The agreement also required regular sea lice counting, importantly coordinated sea lice treatments between sites and companies in the same area, exchanged sea lice counts, and joint health management coordination. A study from Norway (Guarracino et al., 2018) suggested that fallowing of farms in geographic areas had little effect on sea lice numbers, as reinfestation is rapid from outside of the fallowed zone. However, hydrographic conditions have been considered in Scotland and the effects of fallowing in many sea lochs operated by the same company or with management agreements have shown dramatic success (Bron et al., 1993). In these cases five or six bath treatments were formerly required in the first production year and the frequency was almost eliminated in the 1990s with the first treatment required only by December.

Barrett et al. (2020b) proposed a classification of preventative methods and two categories were defined. The first is reducing encounter. The intention of this approach is the reduction of any opportunity for an encounter between the salmon and the infective copepodid stage. The methods used include the designation of hydrographic and area management zones during a production cycle (Bron et al., 1993; Samsing et al., 2017, 2019). The effects of fallowing have been considerably understated and effective fallowing of farms in the same hydrographic zone can delay the need for sea lice treatment in the first production year (Bron et al., 1993). Barrier methods utilise the behaviour of copepodids to be present in surface layers and a skirt made from permeable plankton mesh can be applied to pens to prevent sea lice accessing the pen (Stien et al., 2016; Bui et al., 2019); oxygen levels may decline with skirts and need to be monitored and managed (Grøntvedt et al., 2018). Behavioural measures are also utilised to keep salmon in lower depths in the water column, such as the use of lighting regimes (Barrett et al., 2020a). There are new obstacles and Coates et al. (2020) reported that *L. salmonis* copepodids may have the capacity to adapt to these preventative measures and can bypass these barriers and infest farms by selection of individuals with a greater preference for a deeper depth distribution. Another method to keep salmon deeper and away from surface copepodids is the snorkel cage which has a layer of netting below the surface and an enclosed tube extending to the water surface. This enables salmon to come to the surface to take in air and fill the swim bladder (Tables 15.5 and 15.6) (Oppedal et al., 2019). The ultimate solution uses closed cages where water is filtered or pumped from deeper water (Nilsen et al., 2017). These closed containment systems have the potential to eradicate lice settlement, but give expensive production costs and deployment issues, may have a higher carbon cost, and present potential welfare issues.

A novel alternative is to create a barrier of air bubbles at the bottom of the net which prevents infective lice entering the net. Barriers can also be created using electrical fences such as the Seafarm Pulse Guard® (SPG) system, which consists of vertical cables that surround the net and power pulses are transmitted horizontally in the water (Bredahl, 2014). Sea lice traps with pheromone attractant have also been assessed but have not yet been deployed extensively (Ingvarsdóttir et al., 2002). The behaviour of salmon can be altered to encourage them to swim in deeper water through the placement of lights or feeder systems deeper in the pen (Hevrøy et al., 2003).

The second preventative category proposed by Barrett et al. (2020b) was that of reducing the attachment by interventions that avoid the post-settlement survival of copepodids. These include

Table 15.5 Preventative methods used to control sea lice in different countries. E = experimental.

Method	Norway	Chile	Scotland	Ireland	Faroes	Canada BC	Canada Atlantic	USA
Breeding	+	+	–	+	–	–	–	–
Bubble curtain	+	E	–	+	–	–	–	–
Cleaner fish	+	E	+	+	+	–	+	+
Deep light/deep feeding	+	–	–	–	–	–	–	–
Electrical fence	E	–	–	–	–	–	–	–
Enclosed cage	E	–	–	–	–	–	–	–
Functional feeds	+	+	E	+	E	–	E	E
Sea lice traps	+	–	–	E	–	–	E	–
Snorkel cage	+	E	E	–	–	–	E	–
Vaccines	E	E	–	–	–	–	–	–
Skirt	+	+	+	+	–	+	E	–
Freshwater lens and skirt	–	–	–	+	–	–	–	–

Table 15.6 Advantages and disadvantages of preventative methods.

Method	Advantages	Disadvantages
Skirt	Prevents the lice free stages entering the net pen Gives some protection Easy to fit to cages	Reduced oxygen in the upper part due to the skirt, requires oxygen supplementation Risk to fish in sudden oxygen depletion Chronic lower oxygen than ambient levels will reduce food consumption and growth rate Great infrastructure round cage Higher costs with fitment and maintenance Greater degree of net fouling in upper strata Still some lice settlement Sea lice selection for deeper distribution possible
Snorkel cage	Can reduce abundance of lice	Does not secure that there will be no lice, and the increment after the first lice will likely be high Higher costs Difficulty in deployment and increased maintenance
Closed cage	Lice-free water No infection, no damage	High operating and investment cost Risk of system failure Welfare issues
Bubble curtain	Prevents lice entering the pen Gives protection against infection	There is not enough evidence about the effectiveness Not entirely effective Experimental in many areas
Electrical fences (SPG)	Prevents lice entering the pen Reduces infection risk	Needs a relatively high expertise to operate, high energy consumption and can be a safety issue No evidence of level of efficacy Still experimental
Sea lice traps	Non harmful method for salmon welfare Easy to set up Environmentally friendly	There is not enough evidence about the effectiveness High cost and maintenance Infrastructure needs to be embedded in cages Experimental
Deep light/deep feeding	Non harmful method for salmon welfare Reduces sea lice infection Easy to implement	There is not enough evidence about the effectiveness of the system
Cleaner fish	A biological method to reduce lice abundance Works continuously No medicinal discharge Environmentally friendly	Concern about cleaner fish welfare Costs Care required to implement vaccination programmes and good health cover, and continuous scrutiny of condition needed
Functional feed	Improves the immune response of salmon Reduces sea lice infection and damage Easy to implement	High cost There is not enough evidence about the effectiveness of the functional feed on lice abundance Still experimental and further development required
Vaccines	A good preventative method Reduces sea lice abundance Environmentally friendly	Difficult to develop an effective vaccine against sea lice High development costs Unproven Partially or marginally protective

Table 15.6 (Continued)

Method	Advantages	Disadvantages
Breeding	Selective breeding Ideal solution if workable Prevents fish damage and reduces treatment costs and disruption Beneficial to the environment	There is no evidence about the effectiveness Further development of selection programmes required Still experimental High investment required and long-term process
Land salmon farm	Prevents lice or predators on the farm Highly controlled environment No sea lice infection	High operating and investment cost Could make operation uneconomic Needs robust and reliable engineering Need to ensure good water quality There is high energy use and potentially high carbon footprint and environmental impact

the use of cleaner fish (Brooker et al., 2018), and functional feeds that reduce attachment success by modifying the mucus layer or modulating the immune response of the skin (Martin and Krol, 2017), and improve the healing response (Dawood et al. 2017; Leclerq et al., 2018). Vaccination against sea lice has been trialled (Raynard et al., 2002), but with limited initial success (see Chapter 20). Selective breeding for resistance to sea lice is also an option that is being investigated (Holborn et al., 2019).

15.6 Know your enemy: measurement of sea lice numbers

Regular monitoring of sea lice numbers on farms is essential for effective sea lice management and is recommended on a weekly basis (Treasurer and Pope, 2000; Revie et al., 2005; Salmon Scotland, 2015). The mean lice numbers are considered in relation to lice number trigger levels and control strategies and national sea lice treatment plans (Tables 15.7 and 15.8). This is often based on communication of sea lice numbers with other companies and taking account of synchronised treatments with other farms and companies in the same hydrographic area (Salmon Scotland, 2015). Discharge consents from regulatory authorities such as SEPA in Scotland need to be complied with and the effects of medicines on the environment monitored. However, in other countries, for example Chile, only authorisation of use of the medicine is required. The sea lice count is carried out by a health manager or a site-based health observer who has received training in sea lice stage identification and counting (Treasurer and Pope, 2000). Fish sample size for sea lice counts varies between country, but Revie et al. (2005) and Treasurer and Pope (2000) recommend five fish should be sampled per pen and suggest that it is more important to sample more pens on the site with at least six suggested (Table 15.8). In Ireland, 20 fish are examined in each of three cages. The sample size varies elsewhere, with higher fish numbers per pen also sampled in Norway (Jackson et al., 2018).

The fish are anaesthetised and mobile stages counted by species. Different countries have different levels of detail on the count, but normally pre-adult lice are grouped together, while adult males and females are counted separately. Ovigerous females are also recorded in some countries. The fixed chalimus stages I and II are then counted by looking methodically at the dorsal surface from head to tail, the ventral area, and then around the fins. Originally it was thought that *L. salmonis* (for life cycle

Table 15.7 Threshold numbers of sea lice to initiate treatment.

Country	Migration period of wild salmonids	Other times of the year	Comments
Canada (Atlantic)	> 3 total mobiles	> 3 total mobiles	
Canada (Pacific)	> 3 total mobiles	> 3 total mobiles	Reduced in 2006 from six mobiles at 'other' times
Chile	no wild salmonids	> 3 gravid females	Three consecutive counts above threshold: premature harvest of fish can be ordered
Ireland	> 0.5 gravid females	> 2 gravid females	Three consecutive counts above threshold: premature harvest of fish can be ordered
Norway	> 0.2 adult females	> 0.5 adult females	If more than 0.5 adult lice for several weeks – farms will be fined daily and ordered to harvest
Scotland	> 0.5 adult females	> 1 adult female	
Faroe Islands	–	> 1.5 adult females	Premature harvest enforced if three consecutive lice counts are above threshold

Table 15.8 Sea lice monitoring schemes practised in each country.

Country	Number pens	Number fish/pen	Frequency of count	Reference
Norway	2–3	20	weekly	http://mattilsynet.no
Chile	4	10	weekly	http://sernapesca.cl
Scotland	4–8	5–10	weekly	CoGP
Ireland	2	30	bi-weekly	Marine Institute
Faroe Islands	all pens	10	bi-weekly	www.hfs.fo
Canada (Atlantic)	4–6	5	bi-weekly	Department of Fisheries and Oceans (2014)
Canada (British Columbia)	3	20	bi-weekly	

see Chapter 2) had four chalimus stages (Schram, 1993), but this has since been confirmed as two stages only (Hamre et al., 2013). The water must be filtered to detect any lice that may have fallen off in the tanks and these lice must be included when calculating the average number of lice on the fish. Copepodids will also settle on the fish gills but these are not counted and higher numbers are often associated with bath challenge settings (Treasurer and Wadsworth, 2004a, 2004b). The monitoring procedure has been standardised in several countries by the use of sea lice count manuals (http://lusedata.no; www.sernapesca.cl).

The monitoring of sea lice on salmon farms (Table 15.8) is an important activity which is regularly carried out to assess infestation levels and to determine the need and timing for treatment. Other objectives are to determine the structure of the developmental stages and to assess the number of adult females, damage to the fish, and treatment efficacy. The sea lice counting system should set out the frequency of sampling, the numbers of salmon to sample in each pen, the number of pens to sample, and how to assess the level of damage to the hosts. While some countries focus on the number of female sea lice all stages should be counted to indicate population structure, as all stages will cause damage. Sampling frequency varies between countries but should be carried out at least weekly as monthly is inadequate. It is important to consider the life cycle length and take account of the seasonal water temperature when deciding the frequency of sampling (Table 15.8). On most farms trained farm staff carry out the sea lice count, as in Norway where lice numbers are reported to the Norwegian Food Safety Authority. In Chile, the count is reported to Sernapesca. Lice counting in Canada is periodically verified by the Ministry of Agriculture, Fisheries and Food in British Columbia (Saksida et al., 2007). Lice counts by farm staff in Scotland are sent to the Scottish Salmon Producers' Organisation and weekly counts are submitted directly to Marine Scotland (FishFarmingExpert, 2021). Government health staff conduct counts in Ireland (O'Donohoe et al., 2020).

15.7 Fish welfare during treatments

Treatments used to reduce or eliminate lice numbers should improve the physical appearance of salmonids and the epidermis and skin lesions should also improve over a few days, depending on the magnitude of damage. Appetite and the extent of jumping activity by the fish are often affected by sea lice (Stone et al., 2002) and these will improve following treatment. Fish are fasted for a period of 24 to 36 hours before treatment to reduce stress and oxygen consumption during enclosure. Oxygen should be supplied during full tarpaulin treatment to maintain oxygen levels at 90 to 100% of saturation. Lifting of nets during treatment should be carried out gradually, evenly and carefully during bath treatment to prevent entrapment in the net, and strong tides should be avoided.

Regardless of the environmental benefits there are welfare concerns about the use of medicines and preventative methods such as snorkel cages that prevent fish swimming to the surface or close containment systems. Mechanical and thermal treatment methods may also produce physical abrasion and scale loss in the fish.

Salmon farmers are also governed in many cases by accreditation and quality schemes such as the RSPCA assured Freedom Food scheme (RSPCA, 2021) which also stipulates monitoring of sea lice numbers and instructs implementation of treatment when numbers of sea lice exceed trigger levels. The welfare of fish is assessed against a welfare scale and the extent of sea lice damage, if any, is recorded. Failure to adequately manage sea lice numbers results in penalties and withdrawal from the certification scheme. A national action plan to tackle salmon lice was introduced in Norway in 1997 (Heuch et al., 2005). In Ireland, a monitoring scheme for salmon farms was adopted in 1991 (Jackson and Minchin, 1993). A protocol for monitoring sea lice was published in 2000 and the Department of Agriculture, Fisheries and Food (DAFF) presented 'A strategy for the improved pest control of Irish salmon farms'.

15.8 Company and veterinary approaches to sea lice control

Fish farming companies instigate methods of monitoring, counting, and treating sea lice in the context of national policies and regulation. Companies, health staff, and veterinarians decide on the type of treatment method within permitted medicinal and mechanical use, with due regard to resistance to specific treatments, and therefore to achieve maximum efficacy. A rapid response to evidence of reduced sensitivity and ongoing assessment of sensitivity should be applied. Treatment must be carried out within environmental consents, and especially taking into account fish welfare. An example of strategic approach operated by Mowi on one of their farms was presented in a public planning application. The lice control strategy is based on the use of lice skirts and stocking of cleaner fish in the first salmon production quarter (Q1), that is in calendar weeks 1 to 13. In-feed treatments are used in week 21, continuing use of cleaner fish through weeks 18–22 of Q2, bath treatments in weeks 27, and mechanical treatment in week 36 of Q3. No treatment other than cleaner fish is required in Q4 and Q1 of the second production year with a bath treatment applied as part of the national treatment strategy in week 13 (Paterson, 2018). This treatment seems late as Wadsworth (1998) recommended treatment in the second year by week 10 as copepodite release and sea lice infection rose rapidly after that date. Thereafter, Paterson (2018) reported that Mowi would use mechanical treatments only with treatment in c. week 8 (production cycle week 68) of Q2, weeks 77 and 88 of Q3, and no treatments would be required in Q4.

Horsberg (2012) proposed four treatment scenarios depending on the level of resistance, from no resistance, to emamectin resistant, to emamectin and pyrethroid resistant, to the most difficult sites which were emamectin, pyrethroid, and azamethiphos resistant. Given that sea lice today are resistant to most medicines this type of strategy may be redundant. These strategies have been overtaken with an increased use of preventative and physical control methods. In areas where sea lice are resistant, and in general use, early control in a production cycle is more commonly applied with preventative methods such as skirts accompanied by cleaner fish use. Later in the first year, and throughout the second year, the priority is on using mechanical methods with little reliance on medicines.

15.9 National sea lice treatment strategies

National control strategies and plans are now a feature in all salmon producing countries and have a great impact in managing sea lice. Companies that sign up to these plans agree to share sea lice counts, coordinate treatments with neighbours, and carry out treatments when target threshold mean sea lice counts hit or exceed average targets of a set number of adult female lice. Generally countries divide sea areas into hydrographic and geographic zones that carry out treatments in synchrony. Compliance with this is voluntary but national regulatory authorities also have statutory duties to enforce sea lice treatment when lice numbers exceed threshold standards.

The national sea lice treatment strategy for Scotland (Table 15.8) was initiated in 1998 following scientific work on sea lice annual dynamics which showed that egg production and infestation were lower over winter at reduced temperatures and salinities (Wadsworth, 1998; Salmon Scotland, 2015). This prompted a strategic approach to sea lice control by synchronising treatments nationally in spring (before week ten) before an increase in egg production and copepodite settlement. This entailed the division of the coastline into hydrographic management areas with spatial separation and break zones, and agreement on treatment criteria (Salmon Scotland, 2015). This was based largely around available bath treatments, which had relatively short efficacy of treatment and were only effective against some of the life cycle stages. Treatment regimens were improved through the use of in-feed medicines that affected all life stages and other non-medicinal approaches such as biological control, mechanical methods, and a range of preventative methods. In recent years additional inputs have been given from the integrated pest management (IPM) approach in reducing the development of resistance and recommending the most efficient medicinal application. Treatments are also focused on minimising lice numbers when wild salmonids are entering seawater. The principles of the national treatment plan in Scotland are incorporated in the CoGP (Salmon Scotland, 2015) and this enables auditing and compliance inspection. The salmon farmers in the management areas are identified and agree to observe the details of the national strategy (Table 15.7). Farm management groups are set up in each area with veterinary involvement to agree the treatment plans, and to coordinate and monitor treatments. The monitoring protocol is set on the basis of scientific and statistical advice (Treasurer and Pope, 2000; Salmon Scotland, 2015).

The limit on infestation levels on all Norwegian salmonid farms ensures less than 0.5 adult female lice on fish (0.2 during the season of out-migration of wild salmonids) (Norwegian Ministry of Fisheries and Coastal Affairs, 2012). In Scotland, an annual meeting of company technical staff (NTS) reviews and discusses sea lice treatment strategy and assesses treatment lice levels and synchrony of spring treatments before sea lice numbers rise in the spring (Salmon Scotland, 2015). It indicates that the pre-adult lice should be monitored to ensure that there is treatment before sea lice become egg bearing. The treatment trigger level is 0.5 adult females between 1st Feb and 30th June and a site average of one adult female between 1st July and 31st January (Table 15.8). Sernapesca in Chile established the concept of 'group of salmonids concessions' (ACS) in 2009. These are salmon farms which are within an area suitable for aquaculture development, in a sector that has epidemiological, oceanographic, operational, or geographical characteristics that justify health management. Agreements were reached on the coordination of fallowing for salmon farms that are part of the concessions group (Sernapesca, 2015).

15.10 Integrated pest management

It is widely recognised that reliance on medicines alone cannot be the future for sea lice control because of an increase in resistance and the prolonged time to develop new licensed medicines (Barrett et al., 2020a). Treating sea lice effectively has proven such a difficult task that several different preventative and treatment methods are required. In 2001 a conference was held to review all sea lice treatment methods and to propose adoption of an integrated pest management (IPM) system for sea lice control (Mordue (Luntz) and Pike, 2002). This IPM approach has been used for many years in terrestrial agriculture (DeBach and Rosen, 1991; Pimentel, 1991) and is seen as critical to sea lice control (Caligus, 1997; Treasurer, 2018).

While prevention will be favoured by farmers, there will be a requirement for medicines if other measures do not control sea lice to suppress the pest effectively, economically, and in an environmentally sound manner. IPM is suited to sea lice control and requires a knowledge of the hosts and the parasites. IPM should include separation of year classes, fallowing between production cycles, surveillance of sea lice numbers, area management rotation of medicine, and determination of efficacy of treatments. From 1998 an IPM approach was adopted by the major salmon farming companies with a mix of approaches to sea lice control.

IPM has been implemented in several countries:

- Norway in 1997 as the national programme to control sea lice (Norwegian Animal Health Authorities).
- Scotland in 1998 through the code of good practices to control sea lice (SQS National Treatment Strategy).
- Ireland in 2002 as single bay management (SBM).

- Canada in 2003; termed integrated pest management (IPM).
- Chile in 2007 through the specific health programme for surveillance and control of caligidosis.

This strategy has been adopted by many salmon farming companies. It is based on a range of preventative methods (Fig. 15.6) including fallowing, single year classes, recently modified cages such as snorkel to reduce infection rate, barrier netting, and closed cage systems. Functional feeds may also be used to boost the immune system, increase gill and skin regeneration, and reduce sea lice loading and damage. Biological control using cleaner fish has also become an integral part of sea lice control. There are a number of aspects of site location that have been shown to affect lice numbers on farms. Characteristics such as water depth, tidal range, currents, bottom type, temperature, and salinity are generally considered as part of site selection and will have an influence on the quality of the site and general health of the fish. The presence of adequate currents can help ensure the infectious copepodid larvae are not retained on site. Sites located in areas of lower salinity are also associated with reduced sea lice infestation pressure.

15.11 When to apply a treatment

Elevated numbers of sea lice impact salmon farms in three ways. One issue is impaired salmon growth and another, in the worst scenario, is that salmon can die of infection either directly or indirectly from secondary infections (Pike and Wadsworth, 1999). The treatment itself can be another issue, as it is both expensive and can have a negative effect on the local environment. Perhaps the most controversial and highly debated and studied issue is the artificially increased infection pressure on wild salmonids (Krkošek et al., 2007; Kristoffersen et al., 2018). However, if treatments are applied when sea

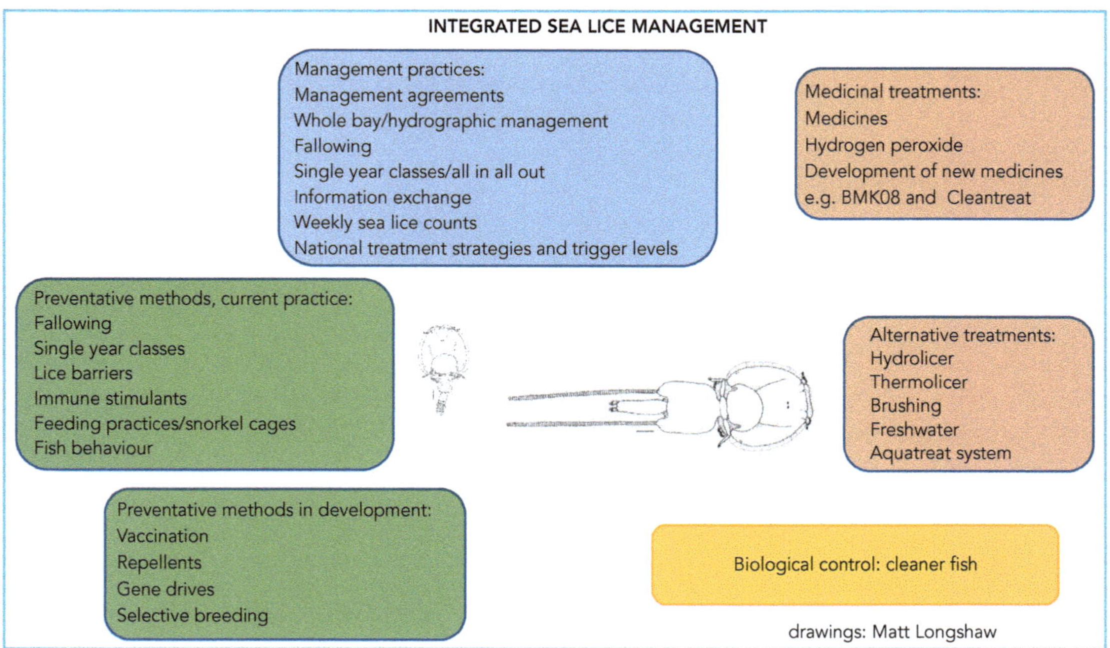

Fig. 15.6 Components of an integrated sea lice management plan.

lice numbers are exceptionally low there may be an increase in the number of treatments, which will increase the development of resistance against the medicines (Kragesteen et al., 2019). Regulations for permitted sea lice limits were introduced in Norway relatively early. These were lowered in 2009 to 0.5 gravid lice per salmon for the first eight months in a production year and one gravid louse for the rest of the year (Table 15.7). The 'code of good practice' (CoGP) in Scotland recommends treatments when lice numbers exceed 0.5 adult lice/salmon during migration periods, and one adult lice/salmon at other times. Weekly counts exceeding an average of two adult female sea lice/salmon were reported (Marine Scotland, 2017) and it is now a legal requirement to notify Marine Scotland weekly of the numbers of sea lice on each farm (FishFarmingExpert, 2021). A monitoring system was introduced in the Faroe Islands in 2009 to reduce the spread of sea lice and its impact on farmed and wild salmon with a limit of two gravid lice per salmon, which was lowered to 1.5 gravid lice per fish in a revision of the regulations in 2017.The regulations implemented by the health authority in Chile in 2007 and updated in 2015 require maintaining the number of gravid females per fish at lower than three, after the application of any antiparasitic treatment. A farm faces mandatory harvesting if there are three reported consecutive notifications of a high lice loading.

15.12 The future of sea lice control, and research and development needs

Aquaculture places a high priority on research and development of sea lice control methods and management systems (Rae, 2002; Barrett et al., 2020). Development of a new medicine can be a long and costly process and, despite several new medicines being approved from 1998 onwards, resistance to these can develop over a short time period (Aaen et al., 2015). The range of medicines available in different countries varies and, in British Columbia and Maine in particular, the range of medicines that can be used has been highly limited. The ability for veterinarians to prescribe unlicensed medicines under the 'cascade system' remains in most salmon producing countries, when no alternatives are available, but this approach is restricted by environmental discharge and medicinal residue regulations and is rarely used. The consequent emphasis on alternative physical and mechanical treatments since 2015 has seen a large increase in treatments being carried out using hydrolicers and thermolicers (Barrett et al., 2020). This led to an 18% reduction in medicine use by Mowi Norway between 2015 and 2018 when 67% of their farms were using non-medicinal treatments (Fig. 15.4).

Cleaner fish have also been widely introduced and have become established in the IPM mix (Powell et al., 2018; Treasurer, 2018; Brooker et al., 2018; Haugland et al., 2020; Imsland et al., 2020). Further development of vaccines for cleaner fish is required to improve survival (Haugland et al., 2018) together with continued introduction of functional feeds and protective cover and hides to improve cleaner fish welfare (Rabadan, 2018). The limitations in cleaner fish techniques have been highlighted (Bui et al., 2020; Gentry et al., 2020; Barrett et al., 2020). However, the concerns and welfare issues are being tackled vigorously (Imsland et al., 2020), capture fisheries for wrasse are being regulated (Salama et al., 2017), and more information is being obtained about the efficacy of cleaner fish use (Imsland et al., 2020).

Much of the future research activity in sea lice control will be in the continued introduction of preventative methods. It should be recognised that good health management and timely initiation of treatments and cooperation between companies will be crucial. As farms become larger and biomasses increase regard should be paid to maximum biomasses and upper limits to sustainable salmon production when these can be accurately and safely assessed. If sea lice control becomes difficult then restrictions and penalties are likely to be applied to the industry. Regulators, both government and environmental, will continue to increase regulatory control. The advent of self-containment and onshore rearing of salmon will be promoted but the higher costs, environmental costs, and potential high risk to stocks through system failures could make these uneconomic and compromising of fish welfare. Sea cage rearing systems are therefore likely to dominate for economic, logistical scaling, and carbon cost reasons for many years.

No single control method for sea lice will be universally successful and there will always be a need for an IPM approach to sea lice control, good management and adaptable response, and good health and veterinary input to treatment decisions. Future IPM strategy will rely on the development of the preventative methods (Fig. 15.6).

References

Aaen, S.M., Helgesen, K.O., Bakke, M.J., Kaur, K. and Horsberg, T.E. (2015) Drug resistance in sea lice: a threat to salmonid aquaculture. *Trends in Parasitology* 31, 72–81.

Abolofia, J., Asche, F. and Wilen, J.E. (2017) The cost of lice: quantifying the impacts of parasitic sea lice on farmed salmon. *Marine Resources Economics* 32, 329-349.

Agusti, C., Bravo, S., Contreras, G., Bakke, M.J., Helgesen, K.O., Winkler, C., Silva, M.T., Mendoza, J. and Horsberg, T.E. (2016) Sensitivity assessment of *Caligus rogercresseyi* to anti-louse chemicals in relation to treatment efficacy in Chilean salmonid farms. *Aquaculture* 458, 195–205.

Barker, D.E., Braden, L.M., Coombs, M.P. and Boyce, B. (2009) Preliminary studies on the isolation of bacteria from sea lice, *Lepeophtheirus salmonis*, infecting farmed salmon in British Columbia, Canada. *Parasitology Research* 105, 1173–1177.

Barrett, L.T., Overton, K., Stien, L.H., Oppedal, F. and Dempster, T. (2020a) Effect of cleaner fish on sea lice in Norwegian salmon aquaculture: a national scale data analysis. *International Journal for Parasitology* 50, 787–796.

Barrett, L.T., Oppedal, F., Robinson, N. and Dempster, T. (2020b) Prevention not cure: a review of methods to avoid sea lice infestations in salmon aquaculture. *Reviews in Aquaculture* 12, 2527–2543.

Brandal, P.O. and Egidius, E. (1979) Treatment of salmon lice (*Lepeophtheirus salmonis* Krøyer 1838) with Neguvon. *Aquaculture* 10, 177–178.

Bravo, S. (2003) Sea lice in Chilean salmon farms. *Bulletin of the European Association of Fish Pathologists* 23(4), 197–200.

Bravo, S., Sevatdal, S. and Horsberg, T.E. (2008) Sensitivity assessment of *Caligus rogercresseyi* to emamectin benzoate in Chile. *Aquaculture* 282, 7–12.

Bravo, S., Treasurer, J., Sepulveda, M. and Lagos, C. (2010) Effectiveness of hydrogen peroxide in the control of *Caligus rogercresseyi* in Chile and implications for sea louse management. *Aquaculture* 303, 22–27.

Bredahl, H. (2014) Electrical fence and use of the same in a fish farm. Patent WO 2014/054951 Al.

Bron, J.E., Sommerville, C., Wootten, R. and Rae, G.H. (1993) Fallowing of marine Atlantic salmon, *Salmo salar* L., farms as a method for the control of sea lice, *Lepeophtheirus salmonis* (Krøyer, 1837). *Journal of Fish Diseases* 16, 487–493.

Brooker, A., Papadopoulou, A., Gutierrez, C., Rey, S., Davie, A. and Migaud, H. (2018) Sustainable production and use of cleaner fish for the biological control of sea lice: recent advances and current challenges. *Veterinary Record* 183, Art No 383.

Bruno, D. and Raynard, R.S. (1994) Studies on the use of hydrogen peroxide as a method for the control of sea lice on Atlantic salmon. *Aquaculture International* 2, 10–18.

Bui, S., Oppedal, F., Nilsson, J., Oldham, T. and Stien, L.H. (2019) Summary and status of deep lights and deep feed use in commercial settings: welfare, behaviour and infestation at three case study sites, Rapport Fra Havforskningen 2019–4. Institute of Marine Research Norway, Bergen.

Bui, S., Geitung, L., Oppedal, F. and Barrett, L. (2020) Salmon lice survive the straight shooter: a commercial scale sea cage trial of laser delousing. *Preventive Veterinary Medicine* 181 (2020) 105063.

Caligus (1997) EU FAIR programme. EcoServe. Issue 2. 20 pp.

Carmichael, S.N., et al. (2013) Salmon lice (*Lepeophtheirus salmonis*) showing varying emamectin benzoate susceptibilities differ in neuronal acetylcholine receptor and GABA-gated chloride channel mRNA expression. *BMC Genetics* 14, 408.

Coates, A., Phillips, B.L., Oppedal, F., Bui, S., Overton, K. and Dempster, T. (2020) Parasites under pressure: salmon lice have the capacity to adapt to depth-based preventions in aquaculture. *International Journal for Parasitology* 50(10–11), 865–872.

Costello, M.J. (2006) Ecology of sea lice parasitic on farmed and wild fish. *Trends in Parasitology* 2006, 475–483.

Dawood, M.A.O., Koshio, S. and Esteban, M.A. (2017) Beneficial roles of feed additives as immune stimulants in aquaculture: a review. *Reviews in Aquaculture* 10, 950–974.

DeBach, P. and Rosen, B. (1991) *Biological Control by Natural Enemies.* Cambridge University Press, New York.

Denholm, I., Devine, G.J., Horsberg, T.E., Sevatdal, S., Fallang, A., Nolan, D.V. and Powell, R. (2002) Analysis and management of resistance to chemotherapeutants in salmon lice, *Lepeophtheirus salmonis* (Copepoda: Caligidae). *Pest Management Science* 58, 528–536.

Department of Fisheries and Oceans (2014) *Sea Lice Monitoring and Non-chemical Measures.* Science Advisory Report 2014/006, Canadian Science Advisory Secretariat. 19 pp.

Dresdner, J., Chávez, C., Quiroga, M., Jiménez, D., Artacho, P. and Tello, A. (2019) *The Impact of Caligus Treatments on Unit Production Costs of Heterogeneous Salmon Farms: Evidence from Chilean Aquaculture.* Intesal Report. 42 pp.

FishFarmingExpert (2021) Scotland: Weekly reporting to be mandatory. Issue 1. Available at: FishFarmingExpert.com (accessed 14 July 2021).

Gentry, K., Bui, S. and Oppedal, F. (2020) Sea lice prevention strategies affect cleaner fish delousing efficacy in commercial Atlantic salmon sea cages. *Aquaculture Environment Interactions* 12, 67–80.

Gismervik, K., Gasnes, S.K., Gu, J., Stien, L.H., Madaro, A. and Nilsson, J. (2019) Thermal injuries in Atlantic salmon in a pilot laboratory trial. *Veterinary and Animal Science* 100081.

Gonzalez, M.P., Munoz, J.L.P., Valerio, V. and Vargas-Chacoff, I. (2016) Effects of the ectoparasite *Caligus rogercresseyi* on *Salmo salar* blood parameters under farm conditions. *Aquaculture* 457, 29–34.

Grant, A. and Treasurer, A. (1993) The effects of fallowing on caligid infestations on farmed Atlantic salmon (*Salmo salar* L.) in Scotland. In: Boxshall, G. and Defaye, D. (eds) *Pathogens of Wild and Farmed Fish: Sea Lice.* Ellis Horwood, London, UK. pp. 255–262.

Grant, A.N. (2002) Medicines for sea lice. *Pest Management Science* 58, 521–527.

Groner, M., Laurin, E., Stormoen, M., Sanchez, J., Fast, M. and Revie, C. (2019) Evaluating the potential for sea lice to evolve freshwater tolerance as a consequence of freshwater treatments in salmon aquaculture. *Aquaculture Environment Interactions* 11, 507–519.

Grøntvedt, R.N., Jansen, P.A., Horsberg, T.E., Helgesen, K. and Tarpai, S. (2014) The surveillance programme for resistance to chemotherapeutants in salmon lice (*Lepeophtheirus salmonis*) in Norway 2014. *Surveillance Programmes for Terrestrial and Aquatic Animals in Norway. Annual Report 2014.* Norwegian Veterinary Institute, Oslo.

Grøntvedt, R.N., Kristoffersen, A.B. and Jansen, P.A. (2018) Reduced exposure of farmed salmon to salmon louse (*Lepeophtheirus salmonis* L.) infestation by use of plankton nets: estimating the shielding effect. *Aquaculture* 495, 865–872.

Guarracino, M., Qviller, L. and Lillehaug, A. (2018) Evaluation of aquaculture management zones as a control measure for salmon lice in Norway. *Diseases of Aquatic Organisms* 130, 1–9.

Hamre, L.A., Eichner, C., Caipang, C.M.A., Dalvin, S.T., Bron, J.E. and Nilsen, F. (2013) The salmon louse *Lepeophtheirus salmonis* (Copepoda: Caligidae) life cycle has only two chalimus stages. *PLoS ONE* 8(9): e73539.

Hannisdal, R., Nøstbakken, O.J., Hove, H., Madsen, L., Horsberg, T.E. and Lunestad, B.J. (2020) Anti-sea lice agents in Norwegian aquaculture; surveillance, treatment trends and possible implications for food safety. *Aquaculture* 521, 735044.

Haugland, G.T., Rønneseth, A. and Wegeland, H.I. (2018) Immunology and vaccinology of lumpfish and wrasse. In: Treasurer, J. (ed.) *Cleaner Fish Biology and Aquaculture Application*. 5M Publications, Sheffield, UK. pp. 258–280.

Haugland, G.T., Imsland, A., Reynolds, P. and Treasurer, J. (2020) Application of biological control: use of cleaner fish. In: Kibenge, S.B. and Powell, M.D. (eds) *Aquaculture Health Management*. Academic Press, London, UK. pp. 319–369.

Helgesen, K.O., Bravo, S., Sevatdal, S., Mendoza, J. and Horsberg, T.E. (2014) Deltamethrin resistance in the sea louse *Caligus rogercresseyi* (Boxhall and Bravo) in Chile: bioassay results and usage data for antiparasitic agents with references to Norwegian conditions. *Journal of Fish Diseases* 37, 877–90.

Helgesen, K.O., Romstad, H., Aaen, S.M. and Horsberg, T.E. (2015) First report of reduced sensitivity towards hydrogen peroxide found in the salmon louse *Lepeophtheirus salmonis* in Norway. *Aquaculture Reports* 1, 37–42.

Heuch, P.A., Bjørn, P.A., Finstad, B., Holst J.C., Asplin, L. and Nilsen, F. (2005) A review of the Norwegian National Action Plan Against Salmon Lice on Salmonids: the effect on wild salmonids. *Aquaculture* 246 (1–4), 79–92.

Hevrøy E.M., Boxaspen K., Oppedal F., Taranger G.L. and Holm J.C. (2003) The effect of artificial light treatment and depth on the infestation of the sea louse *Lepeophtheirus salmonis* on Atlantic salmon (*Salmo salar* L.) culture. *Aquaculture* 220, 1–14.

Hjeltnes, B., Bang-Jensen, B., Bornø, G., Haukaas, A. and Walde, C.S.E. (2018) The health situation in Norwegian aquaculture 2017. *Norwegian Veterinary Institute Report 1b*. Norwegian Veterinary Institute, Oslo, Norway.

Holborn, M.K., Rochus, C.M., Ang, K.P., Elliott, J.A.K., Leadbeater, S. and Powell, F. (2019) Family-based genome wide association analysis for salmon lice (*Lepeophtheirus salmonis*) resistance in North American Atlantic salmon using a 50 K SNP array. *Aquaculture* 511, 734215.

Horsberg, T.E. (2012) Avermectin use in aquaculture. *Current Pharmaceutical Biotechnology* 13, 1095–1102.

Imsland, A.K., Remen, M., Blcoh-Hansen, K., Sagerup, K., Mathisen, R., Myklebust, E.A. and Reynolds, P. (2020) Possible use of lumpfish to control *Caligus elongatus* infestation on farmed Atlantic salmon: a mini review. *Journal of Ocean University of China* 19, 1133–1139.

Ingvarsdóttir, A., Birkett, M.A., Duce, I., Genna, R.L., Mordue, W. and Prickett J.A. (2002) Semiochemical strategies for sea louse control: host location cues. *Pest Management Science* 58, 537–545.

Iversen, A., Hermansen, Ø., Andreassen, O., Brandvik, R.K., Marthinussen, A. and Nystøyl, R. (2015) Cost drivers in salmon farming (In Norwegian: Kostnadsdrivere i lakseoppdrett). Nofima report 41/2015, Trondheim. Available at: https://nofima.no/en/pub/1281065/ (accessed 16 July 2021).

Jackson D. and Minchin, D. (1993) Lice infestations of farmed salmon in Ireland. In: Boxshall, G.A. and Defaye, D. (eds) *Pathogens of Wild and Farmed Fish: Sea Lice*. Ellis Horwood Ltd., London, UK. pp. 188–201.

Jackson, D., Moberg, O., Djupevag, E.M., Kane, F. and Hareide, H. (2018) The drivers of sea lice management policies and how best to integrate them to a risk management strategy: an ecosystem approach to sea lice management. *Journal of Fish Diseases* 41, 927–933.

Jakob E., Barker D.E. and Garver K.A. (2011) Vector potential of the salmon louse *Lepeophtheirus salmonis* in the transmission of infectious haematopoietic necrosis virus (IHNV). *Diseases of Aquatic Organisms* 97, 155–165.

Jones, P.G., Hammell, K.L., Dohoo, I.R. and Revie, C.W. (2012) Effectiveness of emamectin benzoate for treatment of *Lepeophtheirus salmonis* on farmed Atlantic salmon *Salmo salar* in the Bay of Fundy, Canada. *Diseases of Aquatic Organisms* 102, 53–64.

Jones, P.G., Hammell, K.L., Gettinby, G. and Revie, C.W. (2013) Detection of emamectin benzoate tolerance emergence in different life stages of sea lice, *Lepeophtheirus salmonis*, on farmed Atlantic salmon, *Salmo salar*. *Journal of Fish Diseases* 36, 209–220.

Kaur, K., Helgesen, K.O., Bakke, M.J. and Horsberg, T.E. (2015). Mechanism behind resistance against the organophosphate azamethiphos in salmon lice (*Lepeophtheirus salmonis*). *PLoS ONE* 10, e0124220.

Kiemer, M.C.B. and Black, K.D. (1997) The effects of hydrogen peroxide on the gill tissues of Atlantic salmon, *Salmo salar* L. *Aquaculture* 153, 181–189.

Kragesteen T.J., Simonsen K., Visser A.W. and Andersen K.H. (2019) Optimal salmon lice treatment threshold and tragedy of the commons in salmon farm networks. *Aquaculture* 512.

Kristoffersen, A.B., Qviller. L., Helgesen, K.O., Vollset, K.W., Viljugrein, H. and Jansen, P.A. (2018) Quantitative risk assessment of salmon louse-induced mortality of seaward-migrating post-smolt Atlantic salmon. *Epidemics* 23, 19–33.

Krkošek, M., Revie, C.W., Gargan, P.G., Skilbrei, O.T., Finstad, B. and Todd, C.D. (2013) Impact of parasites on salmon recruitment in the Northeast Atlantic Ocean. *Proceedings of the Royal Society B; Biological Sciences* 280, 20122359.

Krkošek, M., Ford, J.S., Morton, A., Lele, S., Myers, R.A. and Lewis M.A. (2007) Declining wild salmon populations in relation to parasites from farm salmon. *Science* 318, 1772–1775.

Labra A., Bravo S. and Marshall S. (2020) Defining the role of *Caligus rogercresseyi* in transmission and spreading of *Piscirickettsia salmonis*. *Aquaculture* 528, 735489. 6 pp.

Leclerq, E., Zerafa, B., Brooker, A.J., Davie, A. and Migaud, H. (2018) Application of passive-acoustic telemetry to explore the behavior of ballan wrasse (*Labrus bergylta*) and lumpfish (*Cyclopterus lumpus*) in commercial Scottish salmon sea-pens. *Aquaculture* 495, 1–12.

Lees, F., Baillie, M., Gettinby, G. and Revie, C.W. (2008) Efficacy of emamectin benzoate against infestations of *Lepeophtheirus salmonis* on farmed Atlantic salmon (*Salmo salar* L) in Scotland, 2002–2006. *PLOS ONE.* 3, e1549.

Marín, S.L., Ibarra, R., Medina, M.H. and Jansen, P.A. (2015) Sensitivity of *Caligus rogercresseyi* (Boxshall & Bravo 2000) to azamethiphos and pyrethroids measured using bioassay tests – a large scale spatial study. *Preventive Veterinary Medicine* 122, 33–41.

Marín, S.L., González, M.P., Madariaga, S.T., Mancilla, M. and Mancilla, J. (2018) Response of *Caligus rogercresseyi* (Boxshall & Bravo 2000) to treatment with hydrogen peroxide: recovery of parasites, fish infestation and egg viability under experimental conditions. *Journal of Fish Diseases* 41, 861–873.

Marine Scotland (2017) The Regulation of Sea Lice in Scotland. Available at: https://www. gov.scot/resource/0051/00516518.pdf (accessed 14 July 2021).

Martin, S.A.M. and Krol, E. (2017) Nutrigenomics and immune function in fish: new insights from omics technologies. *Developmental and Comparative Immunology* 75, 86–98.

Merzendorfer, H. (2013) Chitin synthesis inhibitors: old molecules and new developments. *Insect Science* 20, 121–138.

Moltumyr, L., Gismervik, K., Gu, J., Gåsnes, S.K., Kristiansen, T.S., Rønnestad, I. Nilsson, J. and Stien, L.H. (2021) Does the thermal component of warm water treatment inflict acute lesions on Atlantic salmon *Salmo salar*. *Aquaculture* 736048.

Mordue (Luntz), A.J. and Pike, A.W. (2002) Integrated pest management for sea lice. An introduction. *Pest Management Science* 58, 513–514.

Nese, L. and Enger, Ø. (1993) Isolation of *Aeromonas salmonicida* from salmon lice *Lepeophtheirus salmonis* and marine plankton. *Diseases of Aquatic Organisms* 16, 79–81.

Nilsen, A., Nielsen, K.V., Biering, E. and Bergheim, A. (2017) Effective protection against sea lice during the production of Atlantic salmon in floating enclosures. *Aquaculture* 466, 41–50

Norwegian Ministry of Fisheries and Coastal Affairs (2012) Available at: https:// lovdata.no/dokument/SF/forskrift/2012-12-05-1140#KAPITTEL_1 (accessed 18 July 2021).

Nylund, A., Bjørknes, B. and Wallace, C. (1991) *Lepeophtheirus salmonis*: a possible vector in the spread of diseases on salmonids. *Bulletin of the European Association of Fish Pathologists* 11, 213–216.

Nylund, A., Wallace, C. and Hovland, T. (1993) The possible role of *Lepeophtheirus salmonis* (Krøyer) in the transmission of infectious salmon anaemia. In: Boxshall, G.A. and Defaye, D. (eds) *Pathogens of Wild and Farmed Fish: Sea Lice.* Ellis Horwood, Chichester, UK. pp. 367–378.

O'Donohoe, P., Kane, F., Kelly, S., McDermott, T., D'Arcy, J., Casserly, J., Downes, J.K., McLoughlin, S., Ruane, N. and Jackson, D. (2020) National survey of sea lice (*Lepeophtheirus salmonis* Krøyer and *Caligus elongatus* Nordmann) on fish farms in Ireland – 2019. *Marine Institute, Irish Fisheries Bulletin* 50.

Oelckers, K., Vike, S., Duesund, H., Gonzalez, J., Wadsworth, S. and Nylund, A. (2014) *Caligus rogercresseyi* as a potential vector for transmission of infectious salmon anaemia (ISA) virus in Chile. *Aquaculture* 420–421, 126–132.

Oppedal, F., Bui, S., Stien, L.H., Overton, K. and Dempster, T. (2019) Snorkel technology to reduce sea lice infestations: efficacy depends on salinity at the farm site, but snorkels have minimal effects on salmon production and welfare. *Aquaculture Environment Interactions* 11, 445–457.

Øverli, Ø., Nordgreen, J., Mejdell, C., Janczak, A., Kittilsen, S., Johansen, I. and Horsberg, T. (2014) Ectoparasitic sea lice (*Lepeophtheirus salmonis*) affect behavior and brain serotonergic activity in Atlantic salmon (*Salmo salar* L.): perspectives on animal welfare. *Physiology and Behavior* 132, 44–50.

Overton K., Dempster T., Oppedal F., Kristiansen T.S., Gismervik K. and Stien L.H. (2018) Salmon lice treatments and salmon mortality in Norwegian aquaculture: a review. *Reviews in Aquaculture* 11, 1398–1417.

Paterson, R. (2018) Marine Aquaculture Site. Annex 1: Sea lice management and efficacy statement. sepa.org.uk/media/404305/1002327_annex_1.pdf (accessed 14 July 2021).

Petterson, E., Sandberg, M. and Santi, N. (2009) Salmonid alphavirus associated with *Lepeophtheirus salmonis* (Copepoda: Caligidae) from Atlantic salmon, *Salmo salar* L. *Journal of Fish Diseases* 32, 477–479.

Pike A. and Wadsworth, S. (1999) Sea lice on salmonids: their biology and control. *Advances in Parasitology* 44, 233–337.

Pimentel, D. (1991) Diversification of biological control strategies in agriculture. *Crop Protection* 10, 243–253.

Poppe, T., Dalum, A.S., Røislien, E., Nordgreen, J. and Helgesen, K.O. (2018) Termisk behandling av laks (In Norwegian, English summary). *Norsk Veterinærtidsskrift* (3), 130–155.

Powell, A., Treasurer, J.W., Pooley, C.L., Keay, A.J., Lloyd, R., Imsland, A.K. and Garcia de Leaniz, C. (2018) Use of lumpfish for sea-lice control in salmon farming: challenges and opportunities. *Reviews in Aquaculture* 10, 683–702.

Rabadan, C.G. (2018) Improving the use of wrasse (Labridae) stocked in commercial sea cages with Atlantic salmon (*Salmo salar*) for sea lice control: key features for a successful deployment. In: Treasurer, J. (ed.) *Cleaner Fish Biology and Aquaculture Applications.* 5M Publications, Sheffield, UK. pp. 54–68.

Rae, G.H. (2002) Sea louse control in Scotland, past and present. *Pest Management Science* 58, 515–520.

Raynard R.S., Bricknell I.R., Billingsley P.F., Nisbet A.J., Vigneau A. and Sommerville, C. (2002) Development of vaccines against sea lice. *Pest Management Science* 58, 569–575.

Revie, C.W., Robbins, C., Gettinby, G., Kelly, L. and Treasurer, J.W. (2005) A mathematical model of the growth of sea lice *L. salmonis* populations on farmed salmon in Scotland and its use in the assessment of treatment strategies. *Journal of Fish Diseases* 28, 603–613.

Roth, M., Richards, R.H., Dobson, D.P. and Rae, G.H. (1996) Field trials on the efficacy of the organophosphorus compound azamethiphos for the control of sea lice (Copepoda: Caligidae) infestations on farmed Atlantic salmon (*Salmo salar*). *Aquaculture* 140, 217–239.

RSPCA (2021) RSPCA Welfare Standards for Farmed Atlantic Salmon. RSPCA, Horsham, UK. Available at: http://www.rspca.org.uk/sciencegroup/farmanimals/standards. (accessed 22 July 2021).

Saksida, S., Karreman, G.A., Constantine, J. and Donald, A. (2007) Differences in *Lepeophtheirus salmonis* abundance levels on Atlantic salmon farms in the Broughton Archipelago, British Columbia, Canada. *Journal of Fish Diseases* 30, 357–366.

Salama, N.K.G., Fraser, U. and Mclay, A. (2017) The developing wrasse fishery of Scotland. MASTS ASM, Glasgow.

Salmon Scotland (2015) A Code of Good Practice (CoGP) for Scottish Finfish Aquaculture. Available at: http://thecodeofgoodpractice.co.uk (accessed 19 January 2022).

Samsing F., Johnsen, I., Dempster, T., Oppedal, F. and Treml, E.A. (2017) Network analysis reveals strong seasonality in the dispersal of a marine parasite and identifies areas for coordinated management. *Landscape Ecology* 32, 1953–1967.

Samsing, F., Johnsen, I., Treml, E.A. and Dempster, T. (2019) Identifying 'firebreaks' to fragment dispersal networks of a marine parasite. *International Journal for Parasitology* 49, 277–286.

Schram, T. (1993) Supplementary descriptions of the developmental stages of *Lepeophtheirus salmonis* (Krøyer, 1837) (Copepoda: Caligidae). In: Boxshall, G. and Defaye, D. (eds) *Pathogens of Wild and Farmed Fish: Sea Lice.* Ellis Horwood, London, UK. pp. 30–50.

Sernapesca (2012) Programa de Sanitario Específico de Vigilancia y Control de Caligidosis (PSEVC-Caligidosis). Resolución exenta N°1141, Servicio Nacional de Pesca y Acuicultura, Gobierno de Chile pp. 9.

Sernapesca (2015) Programa de Sanitario Específico de Vigilancia y Control de Caligidosis (PSEVC-Caligidosis). Resolución exenta N°013. Available at: http://www.sernapesca.cl/sites/default/files/importacion/Acuicultura/res_013_2015.pdf (accessed 27 January 2022).

Sevatdal, S. (2005) Monooxygenase mediated pyrethroid detoxification in sea lice (*Lepeophtheirus salmonis*). *Pest Management Science* 61, 772–778.

Sinnott, R. (1999) Cost of sea lice to Scottish salmon farmers. Trouw. *Outlook* 11, 8–10.

Stien, L.H., Dempster, T., Bui, S., Glaropoulos, A., Fosseidengen, J.E. and Wright D.W. (2016) Snorkel sea lice barrier technology reduces sea lice loads on harvest-sized Atlantic salmon with minimal welfare impacts. *Aquaculture* 458, 29–37.

Stone, J., Boyd S., Sommerville, C. and Rae, G.H. (2002) An evaluation of freshwater bath treatments for the control of sea lice, *Lepeophtheirus salmonis* (Krøyer), infections in Atlantic salmon, *Salmo salar* L. *Journal of Fish Diseases* 25, 371–373.

Sutherland, B., Poley, J.D., Igboeli, O.O., Jantzen, J.R., Fast, M.D., Koop, B.F. and Jones, S.R.M. (2014) Transcriptomic responses to emamectin benzoate in Pacific and Atlantic Canada salmon lice *Lepeophtheirus salmonis* with differing levels of drug resistance. *Evolutionary Applications* 8, 133–148.

Thomassen, J.M. (1993) Hydrogen peroxide as a delousing agent for Atlantic salmon. In: Boxshall, G. and Defaye, D. (eds) *Pathogens of Wild and Farmed Fish: Sea Lice.* Ellis Horwood, London, UK. pp. 290–295.

Torrissen, O., Jones, S., Asche, F., Guttormsen, A., Skilbrei, O.T. and Nilsen, F. (2013) Salmon lice impact on wild salmonids and salmon aquaculture. *Journal of Fish Diseases* 36, 171–194.

Treasurer, J. (2018) An introduction to sea lice and the rise of cleaner fish. In: Treasurer, J. (ed.) *Cleaner Fish Biology and Aquaculture Application.* 5M Publications, Sheffield, UK. pp. 3–25.

Treasurer, J. and Grant, A. (1997) The efficacy of hydrogen peroxide for the treatment of farmed Atlantic salmon, *Salmo salar* L. infested with sea lice (Copepoda: Caligidae). *Aquaculture* 148, 265–275.

Treasurer, J. and Pope, J. (2000) Selection of host sample number and design of a monitoring programme for ectoparasitic sea lice (Copepoda: Caligidae) on farmed Atlantic salmon, *Salmo salar. Aquaculture* 187, 247–260.

Treasurer, J.W. and Wadsworth, S.L. (2004a) Interspecific comparison of experimental and natural routes of *Lepeophtheirus salmonis* and *Caligus elongatus* challenge and consequences for distribution of chalimus on salmonids and therapeutant screening. *Aquaculture Research* 35, 773–783.

Treasurer, J.W. and Wadsworth, S. (2004b) Fish gills: important sites of sea lice attachment? *Fish Farmer* 27 (5), 31–33.

Treasurer, J. Wadsworth, S. and Grant, A. (2000) Resistance of sea lice, *Lepeophtheirus salmonis* Krøyer, to hydrogen peroxide on farmed Atlantic salmon, *Salmo salar* L. *Aquaculture Research* 31, 1–6.

Treasurer, J.W., Wallace, C. and Dear, G. (2002) Control of sea lice on farmed Atlantic salmon *S. salar* with the oral treatment emamectin benzoate. *Bulletin of the European Association of Fish Pathologists* 22, 375–380.

Van Leeuwen, T., et al. (2012) Population bulk segregant mapping uncovers resistance mutations and the mode of action of a chitin synthesis inhibitor in arthropods. *Proceedings of the National Academy of Sciences USA* 109, 4407–4412.

Wadsworth, S.L. (1998) The control of sea lice *Lepeophtheirus salmonis* (Krøyer, 1837) (Copepoda: Caligidae) on Atlantic salmon *Salmo salar* L. production sites. PhD thesis, University of Aberdeen.

Weston, D.P., Poynton, H.C., Wellborn, G.A., Lydy, M.J., Blalock, B.J., Sepulveda, M.S. and Colbourne, J.K. (2013) Multiple origins of pyrethroid insecticide resistance across the species complex of a nontarget aquatic crustacean, *Hyallela azteca. Proceedings National Academy Sciences USA* 110, 16532–16537.

Wootten, R., Smith, J.W. and Needham, E.A. (1982) Aspects of the biology of the parasitic copepods *Lepeophtheirus salmonis* and *Caligus elongatus* on farmed salmonids, and their treatment. *Proceedings of the Royal Society of Edinburgh* 81, 185–197.

World Health Organization (1992) Vector resistance to pesticides. Fifteenth report of the WHO expert committee on vector biology and control. Available at: http://apps.who.int/iris/bitstream/10665/37432/1/WHO_TRS_818.pdf (accessed 27 August 2021).

CHAPTER 16

Sensitivity bioassays: understanding the biology of sea lice and its management in salmon farming

Sandra L. Marín[1,2], Tor E. Horsberg[3], Margarita González[4], Kiranpreet Kaur[5] and Celia Agusti-Ridaura[3]

[1]Aquaculture Institute, Universidad Austral de Chile, Puerto Montt, Chile
[2]Research Center Dynamics of High Latitude Marine Ecosystems IDEAL, Universidad Austral de Chile, Valdivia, Chile
[3]Faculty of Veterinary Medicine, Norwegian University of Life Sciences, Norway
[4]Hydrobiological Health Department, Aquaculture Research Division, Fisheries Development Institute (IFOP), Puerto Montt, Chile
[5]Aker BioMarine Antarctic AS, Lysaker, Norway

16.1 Introduction

Lepeophtheirus salmonis (Krøyer 1837) and *Caligus rogercresseyi* (Boxshall & Bravo 2000) are the main caligid species affecting farmed salmonids in the Northern and Southern Hemispheres (Aaen et al., 2015; Arriagada and Marín, 2018). *Caligus elongatus* von Nordmann is also an important species parasitizing salmonid fish in the northern Atlantic Ocean (Igboeli et al., 2014a; Jansen et al., 2012; Agusti-Ridaura et al., 2019), while *Caligus clemensi* (Parker & Margolis) is a generalist also infesting salmonid fish in the northern Pacific Ocean (Hemmingsen et al., 2020). Salmon producing countries have set maximum thresholds for sea lice abundance per fish that should not be exceeded. To implement this the salmon farming industry all around the world has relied mainly on medicinal treatments belonging to a few drug groups: pyrethroids (deltamethrin, cypermethrin), pyrethrum, organophosphates (trichlorfon, dichlorvos, carbamyl, azamethiphos), avermectins (ivermectin, emamectin benzoate), benzoylureas (diflubenzuron, teflubenzuron, and more recently lufenuron), and the oxidizer hydrogen peroxide (Igboeli et al., 2014b). The continuous use of these drugs has resulted in widespread resistance of *L. salmonis* and *C. rogercresseyi* (Aaen et al., 2015; Marín et al., 2015). Resistance can be understood as the selection of a strain or population that is capable of surviving the exposure to a control agent dose that should be lethal to the majority of individuals from a susceptible population (Ffrench-Constant and Roush, 1990).

Bioassays for sensitivity testing of sea lice provide an indication of the degree of susceptibility of sea lice to control agents (with reference to chemotherapeutants, hydrogen peroxide, freshwater, salinity), while avoiding the possibility of the efficacy results being influenced by field factors, such as adverse treatment conditions and inappropriate dosage (Denholm et al., 2002). Chemical agents are the main driving forces for selection of resistant parasites (Aaen et al., 2015) and, given the limited range of active ingredients used to control caligid copepods, appropriate management of them is needed to maintain their efficacy (Lees et al., 2008). This will delay the emergence of resistance (Denholm et al., 2002) and allow early detection of parasite resistance to the chemotherapeutants (Brooks, 2009).

The first bioassay that tested sensitivity of *L. salmonis* to deltamethrin was reported by Sevatdal and Horsberg (2003). The protocol was based on the general bioassay designs for other arthropods by Robertson and Preisler (1992) and modified for sea lice. Laboratory-reared, pre-adult II stage *L. salmonis* were exposed to an increasing gradient of six concentrations for 30 min, the same exposure time as the treatment time indicated by the supplier of deltamethrin, and the effects were recorded 24 h after the finish of the exposure time. Later the methodology was used to develop a handbook to manage resistance of *L. salmonis* where protocols to perform sensitivity bioassays to different control agents were presented, and this was last updated in 2006 (SEARCH, 2006). From the early 2000s, bioassays have been used to systematically monitor parasite response to control agents and to relate parasite response to treatment efficacy and local treatment intensity (Helgesen et al., 2015; Jansen et al., 2016; Jensen et al., 2020). However, as the sensitivity monitoring was becoming more frequent several aspects of the methodology needed standardization, either because they were not included in the handbook or because some aspects had been modified for scientists to solve difficulties (Marín et al., 2018). A bioassay workshop titled “Standardization of traditional bioassay process by sharing best practices” took place during the 11th International Sea Lice Conference held in Westport, Ireland, in 2016. As a result, it was agreed to update the protocols for testing sensitivity. The updated protocol and the technical references supporting the methodology were presented in a new workshop as part of the 12th International Sea Lice Conference in Punta Arenas, Chile, in 2018.

Scientists were also reporting new bioassay methods to test sensitivity of the sea lice species using approaches quite different to those included in the 2006 handbook. These changes were a response to the following difficulties: i) a large number of sea lice is required, (ii) the procedures are labor-intensive and require well-trained observers to determine the condition of the parasite, (iii) they are expensive, and (iv) results are highly sensitive to parasite condition on arrival at the laboratory, and management procedures during the assay (Westcott et al., 2008; Helgesen and Horsberg, 2013; Whyte et al., 2013; Marín et al., 2015). As a result, several one-concentration bioassays have been presented in the literature (Helgesen and Horsberg, 2013; Whyte et al., 2013; Marín et al., 2015), as well as two-concentration bioassays (Grøntvedt et al., 2014–2016; Helgesen et al., 2017–2018, 2019–2020), and one bioassay based on time taken for detecting sea lice response (Carmona-Antoñanzas et al., 2016). Even though simplified bioassays solve some of the limitations of classical bioassays, the main sources of errors remain (for example, fitness of the parasites to be tested; laboratory differences; and inter-observer variation in interpretation of results). Since bioassays with sea lice are difficult to automate, high throughput screening is not feasible. Molecular methods have become a new option for automation.

In spite of the large body of knowledge that has been generated for the different sea lice species, this parasitism is still a major concern for the salmon industry, and it has been suggested that sea lice may limit the growth of the industry (Jansen et al., 2012). Sensitivity testing is one of the available tools for managing this parasitism as it provides information on the spatial and temporal variation of parasite response to the different control agents. The objective of this chapter is to provide an overview of the different approaches to test sea lice sensitivity and the knowledge that has been generated from these approaches, and recommendations associated to the decision-making process for sea lice management, with emphasis on *Caligus rogercresseyi* and *Lepeophtheirus salmonis*.

16.2 Sensitivity testing for sea lice species

16.2.1 Classical bioassays

This type of bioassay has been used to evaluate sea lice species sensitivity to control agents applied by bath and for one applied by fish feed (emamectin benzoate). The objective of classical bioassays is to evaluate the response of the individuals when they are exposed to an increasing gradient of a chemical resulting in deleterious effects. The response of sea lice species has been evaluated mainly through the estimation of the 50% effective concentration (EC_{50}), which is based on the number of affected and/or immobilized individuals (Robertson and Preisler, 1992). This parameter describes the population sensitivity, but not the sensitivity of individual parasites. The methodology to perform classical bioassays consists of exposing individuals to a gradient of at least five different concentrations plus one control. The best fit of data to the model is obtained by correctly choosing the concentration gradient to be used. It should contain one maximum concentration affecting 100% of the exposed individuals but avoid the inclusion of more concentrations causing 100% response (Chapman et al., 1996; Robertson et al., 2009). The most accurate estimation of the EC_{50} is obtained when the two lowest concentrations cause more than 0% but less than 50% efficacy, the two higher concentrations cause more than 50% but less than 100% efficacy, and the highest concentration causes 100% of the expected response (Chapman et al., 1996; Robertson et al., 2009). As a result of this methodology a concentration-response curve is obtained and plotted on a logarithmic *x*-axis (concentration) and a linear *y*-axis (effect) (Fig. 16.1). It should follow a symmetric sigmoid form through which the EC_{50} parameter can be determined. Concentration response is used instead of dose response because dose is not being tested in individuals, but instead they are subjected to solutions that have a compound at a certain concentration. Figure 16.1 shows the results of a hypothetical bioassay in which the 50% effective concentration and the 50% lethal concentration (LC_{50}) were estimated. The concentration gradients for estimating EC_{50} (0–100 ppb) and LC_{50} (0–128 ppb) are different since a lower concentration is required to affect 100% of the sea lice than to kill 100% of them.

Before performing a bioassay with a given control agent it may be necessary to test several concentration gradients because the optimal one may vary across parasite species and within the same species. This is due to differences in the sensitivity between species and among geographical populations of the same species.

Variations in bioassay design were found for exposure time and time for response evaluation (time elapsed between the end of exposure and observation of response), as well as for developmental stages and sea lice generations (Tables 16.1 and 16.2). Other specific procedures may also vary across studies, for example materials for preparing chemotherapeutant dilutions and exposure equipment, and criteria to define parasite condition. Management of parasites during their collection and transportation are also procedures that introduce variation to the overall procedure that needs to be taken into consideration when comparing results (Marín et al., 2018).

A special case of classical bioassay is that where a concentration gradient is replaced by a time interval (time-to-response bioassay) (Carmona-Antoñanzas et al., 2016). This type of bioassay measures the time required until 50% of the lice become affected (showing impaired swimming and/or attachment behavior) (ET_{50}). It is reliable and requires a relatively small number of lice, although may be laborious and time-consuming. In addition, it has the advantage of properly discriminating

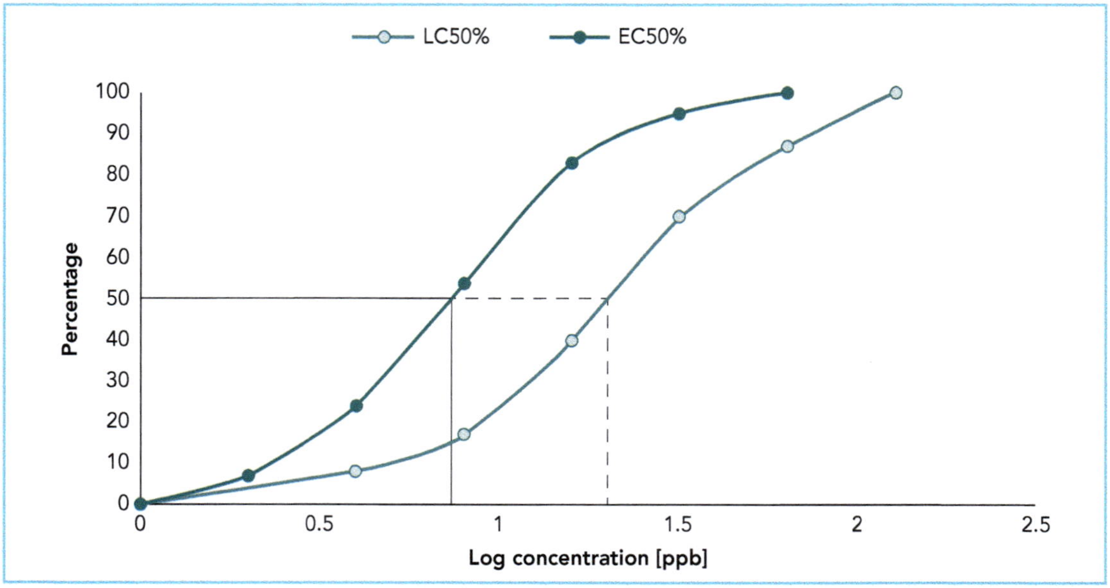

Fig. 16.1 Concentration response curve showing the EC_{50} and LC_{50} values (intersection of the straight lines with the fitted line) for two hypothetical bioassays.

sensitive from resistant lice at the individual louse level, which can be useful when analyzing molecular responses of salmon lice to anti-lice treatments.

Classical bioassays use a small number of sea lice compared to the number of sea lice that are present in a salmon farm. This raises the question about the representation of the sensitivity response estimated for the sample to the sensitivity of the farm population. Helgesen and Horsberg (2013) indicated that representation will depend on the degree of resistance of the population, and that natural variation may also affect the result of a given bioassay. Relating bioassay results with treatment efficacy obtained in farm could contribute to the validation of bioassay results.

Bioassays for sensitivity evaluation have also been performed for *C. elongatus* in Norway demonstrating that it is sensitive to deltamethrin, azamethiphos, hydrogen peroxide, and emamectin benzoate (Agusti-Ridaura et al., 2019). In the Bay of Fundy, Canada, this species was sensitive to emamectin benzoate (Igboeli et al., 2012). In Chile, *L. mugiloides* is a common ectoparasite of the native fish *Eleginops maclovinus* and estimate of EC_{50} for emamectin benzoate was low compared to that reported for *C. rogercresseyi* during the same time interval (Table 16.3) (Bravo et al., 2008a).

16.2.2 Simplified bioassays

This type of bioassay aims to facilitate monitoring of sea lice sensitivity at a large spatial and temporal scale, by reducing the number of sea lice needed to perform the assays. They are usually easier to conduct at a farm site than the classical bioassays.

One-dose bioassay: This type of bioassay was evaluated under laboratory conditions for deltamethrin, azamethiphos, and emamectin benzoate by Helgesen and Horsberg (2013). The sensitivity of four *L. salmonis* strains was firstly identified by traditional six-concentration bioassays (Sevatdal

Table 16.1 Classical bioassay designs and EC_{50} values for *Lepeophtheirus salmonis*. Evaluation: time elapsed between exposure and response observation.

Control agent	Concentration	Exposure time	Evaluation (h)	EC50	Reference
Emamectin benzoate (ppb)	0, 1, 3, 10, 30, 100, 300	24 h	0	25–118	Westcott et al. (2008)
	0, 10, 100, 200, 400, 800	24 h	0	129.4–195.4	Saksida et al. (2013)
	0, 31.3, 62.5, 125, 250, 500	24 h	0	112.3–204.6	Saksida et al. (2013)
	0, 10, 100, 300, 1000, 3000	24 h	0	315.3–457.2	Igboeli et al. (2012)
	0, 5, 15, 25, 50, 100	24 h	24	NA	SEARCH (2006)
Azamethiphos (ppb)	0, 3, 10, 30, 100, 300	60 min	24	4.6–402	Whyte et al. (2016)
	0, 3, 10, 30, 100, 300	60 min	24	NA	SEARCH (2006)
	NA	60 min	24	2.1	Jensen et al. (2017)
	NA	60 min	24	*62	Jensen et al. (2017)
	NA	60 min	24	**>100	Jensen et al. (2017)
Cypermethrin (ppb)	0, 0.15, 0.5, 1.5, 5.0, 15.0	30 min	24	0.22–0.4	Sevatdal et al. (2005)
	0, 0.05, 0.15, 0.5, 1.5, 5.0	60 min	24	0.22–0.4	Sevatdal et al. (2005)
	0, 0.15, 0.5, 1.5, 5.0, 15.0	30 min	24	NA	SEARCH (2006)
Deltamethrin (ppb)	0, 0.03, 0.1, 0.3, 1.0, 3.0	30 min	24	0.09–1.03	Sevatdal et al. (2005)
	0, 0.1, 0.3, 0.6, 1.0, 3.0	30 min	24	0.20–3.03	Whyte et al. (2016)
	0, 0.15, 0.5, 1.5, 5.0, 15.0	30 min	24	NA	SEARCH (2006)
	NA	30 min	24	**0.2	Jensen et al. (2017)
	NA	30 min	24	***3.3	Jensen et al. (2017)
	NA	30 min	24	****58	Jensen et al. (2017)
Hydrogen peroxide (ppm)	0–5000	30 min	0	*216, ***1767	Helgesen et al. (2015)
	0–120	24 h	0	45.9*, 64,7*, ***138	Helgesen et al. (2015)
Salinity (%)	35, 31, 29, 27, 25, 23, 21, 19, 17, 15, 13, 11, 9, 7, 5	24 h	0	11.3–17.5	Andrews and Horsberg (2020)
	35, 26, 20, 15, 10, 5, 0	24 h	0	2.8	Andrews and Horsberg (2020)

* sensitive population to deltamethrin, azamethiphos, and hydrogen peroxide, ** resistant population to deltamethrin and azamethiphos, *** multi resistant population to deltamethrin, azamethiphos, and hydrogen peroxide. NA: not available.

Table 16.2 Classical bioassay designs and EC_{50} values for *Caligus rogercresseyi*. Evaluation refers to the time elapsed between exposure and response observation.

Control agent	Concentrations (ppb)	Exposure time	Evaluation (h)	EC_{50}	Reference
Emamectin benzoate (ppb)	0, 20, 40, 80,120,160, 320	24 h	0	202–870	Bravo et al. (2008)
	0, 20, 40, 80,120,160, 320	24 h	0	#152.2–254.7	Bravo et al. (2010)
	100–700	24 h	0	125–327	Agusti et al. (2016)
Azamethiphos (ppb)	3, 10, 30, 100, 300	1 h	24	4.0–5.6	González et al. (2019)
	0.1–3	24 h	0	0.24–2.36	Agusti et al. (2016)
Cypermethrin (ppb)	0, 0.5, 1.5, 5.0, 50, 100	30 min	24	1.6–3.5	González et al. (2019)
Deltamethrin (ppb)	0, 0.5, 1, 3, 10, 20, 40, 80	40 min	24	0.4–1.0	González et al. (2019)
	0.2–10	24 h	0	0.23–3.34	Agusti et al. (2016)
Hydrogen peroxide (ppm)	0, 152, 321, 483, 662, 825, 1003, 1443, 1749, 2414, 3468	30 min	1 and 24	709.8	Marín et al. (2018)

#different generations of sea lice.

Table 16.3 Classical bioassay designs reported in the literature and EC_{50} values estimated for *C. elongatus* and *L. mugiloides*. Response evaluation was made in all the studies immediately post-exposure.

Control agent	Sea lice species	Concentration	Exposure time (h)	EC_{50}	Reference
Emamectin benzoate (ppb)	*C. elongatus*	0, 5, 15, 40, 80, 160	24	15–31	Agusti-Ridaura et al. (2019)
	C. elongatus	0, 15, 50, 100, 200, 300	24	15–31	Agusti-Ridaura et al. (2019)
	C. elongatus	0, 10, 100, 300, 1000	24	105 (males)	Igboeli et al. (2014b)
	C. elongatus	0, 10, 100, 300, 1000	24	55 (females)	Igboeli et al. (2014b)
	L. mugiloides	0, 20, 40, 80, 120, 160, 320	24	34.2	Bravo et al. (2008)
Azamethiphos (ppb)	*C. elongatus*	0, 0.2, 0.4, 0.8, 1.2, 2	24	0.15–0.20	Agusti-Ridaura et al. (2019)
	C. elongatus	0, 0.05, 0.1, 0.2, 0.3, 0.5	24	0.15–0.20	Agusti-Ridaura et al. (2019)
Deltamethrin (ppb)	*C. elongatus*	0, 0.1, 0.2, 0.4, 0.8, 1.2	24	< 0.1	Agusti-Ridaura et al. (2019)
	C. elongatus	0, 0.01, 0.03, 0.06, 0.1, 0.2	24	< 0.1	Agusti-Ridaura et al. (2019)
Hydrogen peroxide (ppm)	*C. elongatus*	0, 75, 150, 240, 360, 600	0.5	245–267, > 600	Agusti-Ridaura et al. (2019)

and Horsberg, 2003) and small-scale treatments. Then, 24 h bioassays were developed using six or more concentrations for each control agent, adjusting the concentration to the lice strain sensitivity and the length of the exposure time. The main differences between the new 24 h bioassay and the traditional assays were that lice were exposed to each dose for 24 h in one liter of aerated chemical solution and that lice conditions were evaluated immediately after the exposure. The modelling of the concentration-response curves generated from the 24 h bioassays was used to find a single concentration that separated sensitive from resistant lice. This concentration was established at the concentration that gave the largest difference between the lower 80% prediction interval for the sensitive strain and the upper 80% prediction interval for the strain showing reduced sensitivity: 0.04 ppb for deltamethrin, 0.3 ppb for azamethiphos, and 47 ppb for emamectin benzoate. Whyte et al. (2013) used a fixed-concentration bioassay to successfully screen the sensitivity of *L. salmonis* towards emamectin benzoate in Canada. Two experimental solutions were tested: 400 ppb and 800 ppb in pre-adult and adult lice for 24 h. The study showed significant differences in lice sensitivity between geographical regions, chemical concentrations, season, and gender-developmental stages. In Chile, Marín et al. (2015) used one-concentration bioassay for a large-scale spatial study to test the sensitivity of *C. rogercresseyi* to deltamethrin, cypermethrin, and azamethiphos. The chosen concentrations and exposure times for each control agent were those recommended by the suppliers. Results showed that sensitive populations were still present in several geographical areas. In this type of bioassay, results are expressed as percentage of unaffected or affected/immobilized individuals.

Two-dose bioassay: The surveillance program for salmon lice resistance in Norway adapted the Helgesen and Horsberg (2013) protocol by using two-concentration bioassays for monitoring louse sensitivity against four control agents: deltamethrin, azamethiphos, emamectin benzoate, and hydrogen peroxide (Grøntvedt et al., 2014–2016; Helgesen et al., 2017–2018, 2019–2020; Jansen et al., 2016). Lice are exposed to either a low or a high chemical concentration for 24 h (Table 16.4). These chemical concentrations were chosen based on the results of Helgesen and Horsberg (2013), as well as on additional six-concentration bioassays and small-scale clinical trials performed by the authors. The low concentration was targeted as the lowest concentration affecting 80–100% of sensitive parasites. The percentage of salmon lice visibly affected by the exposure to the lowest concentration is then used to estimate the sensitivity status of the salmon lice population. If more than 80% of the lice are affected, the population is considered fully sensitive. The percentage of affected lice at the highest concentration is used as an approximation to the expected outcome of a subsequent farm treatment, when the fish are treated with the dosage recommended by the suppliers and using the approved protocol.

One- or two-concentration bioassays have also been used to efficiently separate sensitive from resistant sea lice populations, and for conducting molecular studies, such as mutation/SNP detection

Table 16.4 Anti-lice chemical concentrations used in the two-dose bioassays for *L. salmonis* in Norway. ppb: µg l^{-1}; ppm: mg l^{-1}.

Control agent	Low concentration	High concentration
Deltamethrin	0.2 ppb	1 ppb
Azamethiphos	0.4 ppb	2 ppb
Emamectin benzoate	100 ppb	300 ppb
Hydrogen peroxide	120 ppm	240 ppm

or differential gene expression, in order to elucidate resistance mechanisms or developing resistance molecular markers (Kaur et al., 2015; Carmona-Antoñanzas et al., 2017; Bakke et al., 2018; Agusti-Ridaura et al., 2020). In the two-concentration bioassays, lice affected at the low concentration would be considered sensitive, and lice unaffected at the high concentration, resistant. However, direct gene expression comparisons between sensitive and resistant lice should be avoided since lice are exposed to different chemical concentrations. To overcome this problem, one-concentration bioassays would be a better approach, choosing in each case the concentration that better separates sensitive from resistant lice (EC_{50} values could be a good starting point to find such a concentration). The downside is that proper separation of sensitive and resistant phenotypes would be challenging. As a possible solution, lice firmly attached to the bottle wall could be considered resistant, and lice completely immobilized could be considered sensitive.

16.3 Bioassay for chitin synthesis inhibitor chemotherapeutants

Harðardóttir et al. (2019) studied the action of four chitin synthesis inhibitors (CSIs), diflubenzuron, hexaflumuron, lufenuron, and teflubenzuron, on larval moult. They performed independent three-concentration bioassays with four replicates, each containing 20–30 larvae. The response was analyzed at seven days post-treatment by counting normal larvae and larvae with incomplete moulting. The EC_{50} values were defined as the concentration where incomplete moulting was observed in half of the treated individuals. The average percentage of incomplete moulting obtained from each of the four replicates was plotted against the concentration of the chemical. Linear regression was performed for each of the three experiments to produce concentration-response curves, and EC_{50} values were calculated from the average values obtained from the trendlines. All four CSIs interfered with the nauplius II moult to copepodids in a concentration-dependent manner. The EC_{50} values for diflubenzuron, hexaflumuron, lufenuron, and teflubenzuron were 93.2, 1.2, 22.4, and 11.7 nM, respectively. The larvae were most sensitive to hexaflumuron and least sensitive to diflubenzuron.

16.4 Non-chemical bioassays

Tolerance can also emerge towards non-medicinal control agents. Freshwater bathing and exposing fish to high temperatures are non-medicinal options being used to reduce sea lice abundance on fish. However, variable treatment efficacies of these options have raised concerns regarding their general applicability and bioassay techniques were developed.

Salinity tolerance: Andrews and Horsberg (2020) developed a methodology to determine the salinity in parts per thousand (‰) at which median survival (EC_{50}) was obtained at 24 h for the copepodid and pre-adult II stages of *L. salmonis* from four geographically separated populations. Furthermore, this method was used to develop a bioassay method for on-site pre-treatment tests. Parasites were exposed to a range of salinities (35‰ to 0‰) for 24 h, survival was observed after 24 h, and results were classified as unaffected/affected *L. salmonis* copepodids. This showed significant differences in tolerance among populations with EC_{50} between 11.3‰ and 17.5‰. No significant differences in tolerance were observed between populations for pre-adult II. Grouping pre-adult II from the populations yielded an EC_{50} of 2.8‰ for both sexes combined. The authors concluded that all stages of *L. salmonis* examined in the study exhibited high tolerances to low salinity and suggested regular monitoring of freshwater sensitivity in areas where this control option is used regularly.

Thermal tolerance: Andrews and Horsberg (2021) tested the efficacy of thermal treatment on *L. salmonis* by exposing copepodids and pre-adult II belonging to seven different populations to an increasing gradient of temperature (expressed as Δ°T) for 2 min, and then returning them to the maintenance temperature of 12°C. Treatment temperatures were calculated as Δ°T, the difference between the maintenance temperature (12°C) and the treatment temperature. Parasites were observed and classified as unaffected/affected immediately following exposure and 24 h post-treatment. The percentage of unaffected copepodids showed differences among populations immediately post-exposure. However, at 24 h post-exposure, copepodids had recovered showing tolerances varying between Δ23°C and Δ26°C, depending on the population. Recovery of pre-adults at 24 h post-exposure was observed even in the larger ΔT (23°C). EC_{50} for pre-adults at 2 min post-exposure was similar among populations (Δ17.1–Δ20 °T), and for copepodids the Δ°T varied between 18.3°C and 21.4°C. The authors recommended that to monitor thermal tolerance and ensure high treatment efficacy, regular *in vitro* bioassays should be performed.

16.5 Molecular methods for sea lice management

Molecular methods study biological processes at fundamental molecular levels and are mostly used to understand the interactions between primary components of cells, DNA and RNA. These methods are easy and could be carried out on fixed/preserved sea lice samples and could be carried out on a relatively small number of samples, due to their high sensitivity and specificity, and can be automated for a high throughput screening. Moreover, unlike bioassays, the protocols for molecular assays could be easily standardized and adapted in different laboratories. Hence, it is very likely reproducible results will be produced from different laboratories using the same molecular methods. However, the molecular diagnostic assays could only be designed if we have knowledge about the underlying mode of action and/or resistance mechanisms, or have a validated biomarker uniquely coupled with resistance. Recently, the unraveling of resistance mechanisms towards different parasiticides against sea lice, such as organophosphates, has led to the development of a successful diagnostic assay, which has been used in different countries like Norway, Scotland, Faroe Islands, Ireland, and Atlantic Canada (Kaur et al., 2016). This diagnostic assay is a simple TaqMan probe method, which is designed to detect the single missense mutation in the gene that codes for an enzyme called acetylcholinesterase. The mutation hinders the interaction between enzyme and organophosphates, and the sea lice survives the treatment (Kaur et al., 2015). The diagnostic assay detects this mutation with 100% sensitivity and specificity and hence could differentiate completely between the sensitive and resistant sea lice.

A similar mutation has been detected in acetylcholinesterase for *Caligus rogercresseyi*. The mutation, however, still needs to be validated for the development of a diagnostic assay (Agusti-Ridaura et al., 2018). Similarly, diagnostic assays have been developed for pyrethroids and hydrogen peroxide. The diagnostic assay for pyrethroids is based on the PCR screening for polymorphisms in the mitochondrial genome (Bakke et al., 2018), whereas the diagnostic assay for hydrogen peroxide is based on the induced expression of the catalase gene (Helgesen et al., 2017) and the downregulation of an aquaporin gene (Agusti-Ridaura et al., 2020). These diagnostic assays are a part of the Norwegian surveillance program to monitor sensitivity of salmon lice (Helgesen et al., 2019).

Molecular diagnostic assays are not yet developed for all control agents used against sea lice; therefore, bioassays are a very valid tool at present. It is recommended when developing molecular markers to provide the EC_{50} values of the lice populations used and the rationale for selecting the tested concentrations. Better understanding of the mode of action of parasiticides and the biochemical pathways underlying resistance in sea lice are warranted, for the development of more molecular tools to determine and control resistance.

16.6 What we have learned from bioassays

The evolution of methods to test sea lice sensitivity to control agents was driven mainly by the loss of efficacy of the control agents and it became urgent to identify the spatial distribution of the sensitive and resistant populations to the different control agents. EC_{50} values cannot be estimated by using one- or two-concentration bioassays. Therefore, the information given by one of those assays is less accurate than the classical six-concentration ones. However, the robustness of the one/two-dose bioassays is dependent on large-scale screening. Large numbers of simplified bioassays in a wide area and/or across time can provide an accurate general overview of the sensitivity status of lice populations. In fact, the results obtained during seven years of the surveillance program in Norway (2013–2019) have shown the usefulness of the two-concentration bioassays for monitoring resistance in *L. salmonis*. The most recent scientific development to test lice sensitivity is based on molecular methods that provide objective markers to discriminate sensitive and resistant sea lice populations. These methods allow large spatial testing but also solve difficulties associated with the considerable number of variation sources associated with classical and one- and two-dose bioassay methods.

Regarding gender differences we have learned that there are similarities between sea lice species in their response to control agents but that the response cannot be extrapolated to all control agents. The mechanisms behind the gender related differences are not yet well understood and may be different among sea lice species. *C. rogercresseyi* males are more sensitive to deltamethrin, cypermethrin, and azamethiphos than females (Marín et al., 2015), although this was not observed in efficacy results from farm treatments (Arriagada et al., 2014). Valenzuela-Muñoz et al. (2014) indicated that the greater expression of the ATP-binding cassette transporter P-glycoprotein observed in females of *C. rogercresseyi* could be linked to a greater resistance to pyrethroids. *L. salmonis* females are also less sensitive than males when exposed to deltamethrin (Whyte et al., 2014; Helgesen et al., 2014). For *L. salmonis*, the larger size of females or their thicker cuticle has been offered as an explanation as to why they have lower sensitivity to deltamethrin compared to males (Whyte et al., 2014). However, *C. rogercresseyi* males and females are similar in size (4.8 mm) (González and Carvajal, 2003). This difference may be explained by the faster depuration rate in females through egg production during the first 24 h (Medina et al., 2002). *C. rogercresseyi* males showed a larger sensitivity to hydrogen peroxide than females at 1 h and 24 h post-exposure, as they had a larger mortality than females (Marín et al., 2017). For emamectin benzoate, *L. salmonis* females are more sensitive than males (Westcott et al., 2008; Whyte et al., 2013; Igboeli et al., 2013). Mechanisms underlying gender differences in sensitivity response need to be better understood, especially considering that females are responsible for producing the sea lice dispersive stage.

We have learned that temperature may have an effect on sea lice sensitivity to emamectin benzoate, but there is no information about temperature effects on the sea lice sensitivity to the other control agents. Westcott et al. (2008) showed that sensitivity to emamectin benzoate was higher during periods of warmer temperatures in *L. salmonis*, which is coincident with the higher efficacy of emamectin benzoate against *L. salmonis* during spring and summer, than that observed for autumn and winter treatments (Lees et al., 2008). The effect of temperature on toxicity has been associated with alterations in chemical penetration, metabolism, altered effectiveness at the target sites, and distribution within the individual (Scott, 1995). Although these studies suggest that toxicity of emamectin benzoate could be affected by temperature the mechanisms underlying the response are unknown for the sea lice species (Westcott et al., 2008).

Bioassay results have suggested that antiparasitic treatments can drive genetic selection towards a much more resistant salmon lice population. EC_{50} for emamectin benzoate was reported to increase in *C. rogercresseyi* between the second and seventh generation from 152 ppb to 256 ppb (Bravo et al., 2010). Espedal et al. (2013) showed that reduced sensitivity was retained in all four generations of *L. salmonis* exposed to a gradient of 0–800 ppb of emamectin benzoate. Igboeli et al. (2014) reported that laboratory-reared *L. salmonis* maintained their emamectin sensitivity status up to the F3 generation. *Caligus elongatus* showed more than twofold lower EC_{50} values compared with *L. salmonis* collected from the same site. These results are in accordance with the differences in survival between sensitive and resistant *L. salmonis* following treatment, which demonstrated the ability of medicinal treatments to drive genetic selection towards a much more resistant salmon lice population within a very short time span if there is no influx of sensitive genotypes (Jensen et al., 2017).

In a typical treatment with an antiparasitic, all sensitive parasites are eliminated while the resistant individuals remain on the fish. Unless there is a considerable influx of sensitive parasites to the farms, resistance can develop rapidly, as modelled by McEwan et al. (2015). Thus, bioassays can also be used to identify fully sensitive sea lice populations that can serve as refugia and slow down the emergence of resistance.

One of the most important lessons provided by bioassays has been the recognition of a spatial structure of sea lice sensitivity for multiple control agents for both *L. salmonis* and *C. rogercresseyi*. The identification of sensitivity profiles across geographical areas contributes to the decision-making process regarding treatment options and regulation implementation. In addition, differentiation of sensitive and resistant populations has allowed the implementation of surveillance programs on the farms. Since only one or two concentrations capable of discriminating sensitive and resistant sea lice populations are used, the time and effort required to perform the bioassays are significantly reduced. Recent results have demonstrated that *L. salmonis* have a concentration-dependent response to non-medicinal treatments such as reduced salinity or increased temperature. Salinity and thermal tolerance of sea lice show that baseline sensitivity responses need to be estimated as early as possible in all geographical distributions of the sea lice species since environmental factors vary across geographical areas and populations may display a wide range of tolerances. Reduced efficacy of standard treatments (0‰ for freshwater and 31–34°C for thermal treatments) combined with a deviation from the baseline sensitivity will be a good indication of increased tolerance (Andrews and Horsberg, 2020, 2021).

Bioassays are suitable tools to identify spatial and temporal structures of sensitivity profiles driven by the interaction of, for example, (i) the connectivity among farms (Samsing et al., 2019) as currents may mix populations of different sensitivity profiles; ii) environmental conditions as they may affect survival and length of sea lice life cycle; iii) sea lice management by farms (Denholm et al., 2002) as they expose sea lice to sublethal concentrations of chemotherapeutants; and (iv) the degree of exposure to the chemotherapeutants, for example the number of treatments applied in each farm

(Jansen et al., 2012). To better understand the evolution of sensitivity of sea lice species to a given control agent it is necessary to perform studies based on the integration of several approaches such as bioassays, molecular studies, treatment efficacy, and environmental conditions. Agusti et al. (2016) explored the sensitivity of *C. rogercresseyi* towards azamethiphos, pyrethroids, and emamectin benzoate at four farms located in Los Lagos and Aysén regions in Chile. Louse infestation and sensitivity were studied at several levels: weekly monitoring of lice abundance at the farms, treatment efficacy, re-infestation rate post-treatment, and 24 h exposure bioassays (six concentrations per bioassay). The combination of these methods and data proved to be useful for detecting resistance issues and problematic areas regarding lice infestation. In addition, correlation between treatment efficacy and bioassay results were analyzed. Bioassay results also revealed louse gender differences in chemical sensitivity. Integrated studies may also be used for anticipating problems in the future, which is crucial for improving the use of anti-louse chemicals and reducing the risk of resistance development.

For parasites, the maximum applicable amount of a given antiparasitic is determined by both the host ability to cope with the toxicity and the parasite susceptibility to the antiparasitic (Aaen et al., 2015). The therapeutic index to control sea lice is small for several chemotherapeutants, and consequently only minor reductions in sea lice susceptibility may present as clinical resistance in the field (Aaen et al., 2015). Classical bioassays may be required in the initial stages of a given product's use, to detect sensitivity changes when small changes need to be detected because, as tolerance develops within a sea lice population, it appears to reach a "tipping point," after which the tolerance spreads rapidly through the remaining population and less sensitive tools may be used (Whyte et al., 2013). When using one-dose or two-dose bioassays, occasional classical bioassays are recommended for testing the accuracy of the one-or two-dose bioassays.

16.7 Recommendations

Sea lice sensitivity-resistance management is a complex task that involves knowledge integration from different attributes of sea lice biology and ecology, as well as sanitary practice on farms. These attributes are usually measured using different tools that should provide feedback to farmers and local authorities to design, execute, and evaluate sea lice management programs. Figure 16.2 shows a conceptual representation of these interactions, in which bioassays to test sea lice sensitivity is one of these tools.

Bioassays for sea lice sensitivity monitoring as a tool for managing sensitivity/resistance to chemical and non-chemical treatment still have methodological challenges. Standardization of protocols is the solution but, as they are executed under very different contexts, the reality is that there will always be deviation from a standardized protocol. It is necessary to provide bioassay details as much as possible when communicating the results to be able to make comparisons across both geographical areas and sea lice species.

Classical six-concentration bioassays must be the starting point for evaluating sea lice sensitivity to any control agent used to treat sea lice as they allow early detection of reduced sensitivity. The baseline sensitivity should be established even before the product is launched. In addition, this type of bioassay must be used to distinguish those sensitive populations from those resistant, which would allow development of simplified bioassays and molecular markers of sensitive/resistant populations. Less information exists regarding bioassays for testing sensitivity of sea lice to CSIs. Since they affect

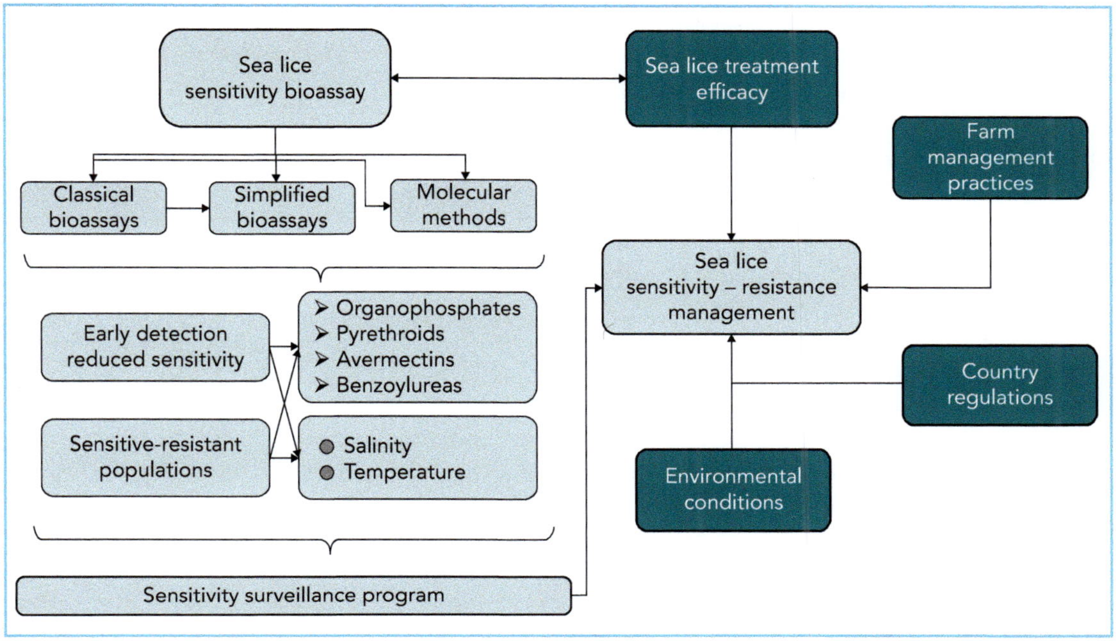

Fig. 16.2 Conceptual representation of interactions among tools that could comprise a sea lice sensitivity – resistance management program.

lice moult the experimental design of the bioassay is a challenge. However, efforts must be concentrated on generating a bioassay that can be related to sea lice infestations and development on the fish to prevent resistance development.

Sensitivity surveillance programs have been suggested as part of the integrated pest management (IPM) approach for sea lice. These programs should be linked to efficacy monitoring to associate sea lice response in a controlled environment to that observed on the farm. When the surveillance program is based on two-concentration bioassays, relationships between bioassay results and farm efficacy would permit confirmation that sensitive and resistant populations are still discriminated by the concentrations used in the bioassay.

References

Aaen, S.M., Helgesen, K.O., Bakke, M.J., Kaur, K. and Horsberg, T.E. (2015) Drug resistance in sea lice: a threat to salmonid aquaculture. *Trends in Parasitology* 31, 72–81.

Agustí, C., Bravo, S., Contreras, G., Bakke, M.J., Helgesen, K.O., Winkler, C., Silva, M.T., Mendoza, J. and Horsberg, T.E. (2016) Sensitivity assessment of *Caligus rogercresseyi* to anti-louse chemicals in relation to treatment efficacy in Chilean salmonid farms. *Aquaculture* 458, 195–205.

Agusti-Ridaura, C., Dondrup, M., Horsberg, T.E., Leong, J.S., Koop, B.F., Bravo, S., Mendoza, J. and Kaur, K. (2018) *Caligus rogercresseyi* acetylcholinesterase types and variants: a potential marker for organophosphate resistance. *Parasites & Vectors* 11, 570.

Agusti-Ridaura, C., Hamre, L., Espedal. P., Øines, Ø., Horsberg, T.E. and Kaur, K. (2019) First report on sensitivity of *Caligus elongatus* towards anti-louse chemicals and identification of mitochondrial cytochrome C oxidase I genotypes. *Aquaculture* 507, 190–195.

Agusti-Ridaura, C., Bakke, M.J., Helgesen, K.O., Sundaram, A.Y.M., Bakke, S.J., Kaur, K. and Horsberg, T.E. (2020) Candidate genes for monitoring hydrogen peroxide resistance in the salmon louse, *Lepeophtheirus salmonis*. *Parasites & Vectors* 13, 344.

Andrews, M. and Horsberg, T.E. (2020) Sensitivity towards low salinity determined by bioassay in the salmon louse, *Lepeophtheirus salmonis* (Copepoda: Caligidae). *Aquaculture* 514, 734511.

Andrews, M. and Horsberg, T.E. (2021) In vitro bioassay methods to test the efficacy of thermal treatment on the salmon louse, *Lepeophtheirus salmonis*. *Aquaculture* 532, 736013.

Arriagada, G. and Marín, S.L. (2018) Evaluating the spatial range of the effect of synchronized antiparasitic treatments on the abundance of the sea lice, *Caligus rogercresseyi* (Boxshall & Bravo 2000) in Chile. *Aquaculture Research* 49, 816–831.

Arriagada, G., Stryhn, H., Campistó, J.L, Rees, E., Sanchez, J., Ibarra, R., Medina, M. and St-Hilaire, S. (2014) Evaluation of the performance of pyrethroids on different life stages of *Caligus rogercresseyi* in southern Chile. *Aquaculture* 231–237, 426–427.

Bakke, M., Agustí, C., Bruusgaard, J., Sundaram, A.Y.M. and Horsberg, T.E. (2018) Deltamethrin resistance in the salmon louse, *Lepeophtheirus salmonis* (Krøyer): maternal inheritance and reduced apoptosis. *Scientific Reports* 8, 8450.

Bravo, S., Pozo, V. and Silva, M. (2008) The tolerance of *Caligus rogercresseyi* to salinity reduced in southern Chile. *Bulletin of the European Association of Fish Pathologists* 28, 197–204.

Bravo, S., Sevatdal, S. and Horsberg, T.E. (2008) Sensitivity assessment of *Caligus rogercresseyi* to emamectin benzoate in Chile. *Aquaculture* 282, 7–12.

Bravo, S., Sevatdal, S. and Horsberg, T.H. (2010) Sensitivity assessment in the progeny of *Caligus rogercresseyi* to emamectin benzoate. *Bulletin of the European Association of Fish Pathologists* 30(3), 99–105.

Brooks, K.M. (2009) Considerations in developing an integrated pest management programme for control of sea lice on farmed salmon in Pacific Canada. *Journal of Fish Diseases* 32, 59–73.

Carmona-Antoñanzas, G., Humble, J.L., Carmichael, S.N., Heumann, J., Christie, H.R.L., Green, D.M., Bassett, D.I., Bron, J.E. and Sturm, A. (2016) Time-to-response toxicity analysis as a method for drug susceptibility assessment in salmon lice. *Aquaculture* 464, 570–575.

Carmona-Antoñanzas, G., Bekaert, M., Humble, J.L., Boyd, S., Roy, W., Bassett, D.I., Houston, R.D., Gharbi, K., Bron, J.E. and Sturm, A. (2017) Maternal inheritance of deltamethrin resistance in the salmon louse *Lepeophtheirus salmonis* (Krøyer) is associated with unique mtDNA haplotypes. *PLoS ONE* 12, e0180625.

Chapman, P.F., Crane, M., Wiles, J.A., Noppert, F. and McIndoe, E.C. (1996) *Asking the Right Questions: Ecotoxicology and Statistics*. Society of Environmental Toxicology and Chemistry (SETAC), Europe. p. 37.

Denholm, I., Devine, G.J., Horsberg, T.E., Sevatdal, S., Fallang, A., Nolan, D.V. and Powell, R. (2002) Analysis and management of resistance to chemotherapeutants in salmon lice, *Lepeophtheirus salmonis* (Copepoda: Caligidae). *Pest Management Science* 58, 528–536.

Espedal, P.G., Glover, K.A., Horsberg T.E and Nilsen, F. (2013) Emamectin benzoate resistance and fitness in laboratory reared salmon lice (*Lepeophtheirus salmonis*). *Aquaculture* 416–417, 111–118.

Ffrench-Constant, R.H. and Roush, R.T. (1990) Resistance detection and documentation: the relative roles of pesticidal and biochemical assays. In: Roush, R.T. and Tabashnik, B.E. (eds) *Pesticide Resistance in Arthropods*. Chapman & Hall, New York. pp. 5–37.

González, L. and Carvajal, J. (2003) Life cycle of *Caligus rogercresseyi*, (Copepoda: Caligidae) parasite of Chilean reared salmonids. *Aquaculture* 220, 101–117.

González Gómez, M.P., Ovalle, L., Menanteau, M., Spinetto, C., Oyarzún, R., Rivas, M. and Oyarzo, C. (2019) Susceptibility of *Caligus rogercresseyi* collected from the native fish species *Eleginops maclovinus* (Cuvier) to antiparasitics applied by immersion. *Journal of Fish Diseases* 42, 1143–1149.

Grøntvedt, R., Jansen, P., Horsberg, T.E., Helgesen, K. and Tarpai, A. (2014–2016) The surveillance programme for resistance to chemotherapeutants in salmon lice (*Lepeophtheirus salmonis*) in Norway 2013–2015. *Surveillance Programmes for Terrestrial and Aquatic Animals in Norway. Annual Reports.* Norwegian Veterinary Institute, Oslo, Norway.

Harðardóttir, H., Male, R., Nilsen, F. and Dalvin, S. (2019) Effects of chitin synthesis inhibitor treatment on *Lepeophtheirus salmonis* (Copepoda: Caligidae) larvae. *PLoS ONE* 14(9), e0222520.

Helgesen, K.O. and Horsberg, T.E. (2013) Single-dose field bioassay for sensitivity testing in sea lice, *Lepeophtheirus salmonis*: development of a rapid diagnostic tool. *Journal of Fish Diseases* 36, 261–272.

Helgesen, K.O, Bravo, S., Sevatdal, S., Mendoza, J. and Horsberg, T.E. (2014) Deltamethrin resistance in the sea louse *Caligus rogercresseyi* (Boxhall and Bravo) in Chile: bioassay results and usage data for antiparasitic agents with references to Norwegian conditions. *Journal of Fish Diseases* 37, 877–890.

Helgesen, K.O., Romstad, H., Aaena, S.M. and Horsberg, T.E. (2015) First report of reduced sensitivity towards hydrogen peroxide found in the salmon louse *Lepeophtheirus salmonis* in Norway. *Aquaculture Reports* 1, 37–42.

Helgesen, K.O., Bakke, M.J., Kaur, K. and Horsberg, T.E. (2017) Increased catalase activity – a possible resistance mechanism in hydrogen peroxide resistant salmon lice (*Lepeophtheirus salmonis*). *Aquaculture* 468, 135–140.

Helgesen, K.O., Jansen, P., Horsberg, T.E. and Tarpai, A. (2017–2018) The surveillance programme for resistance to chemotherapeutants in salmon lice (*Lepeophtheirus salmonis*) in Norway 2016–2017. *Surveillance Programmes for Terrestrial and Aquatic Animals in Norway. Annual Reports.* Norwegian Veterinary Institute, Oslo, Norway.

Helgesen, K.O., Horsberg, T.E. and Tarpai, A. (2019–2020) The surveillance programme for resistance to chemotherapeutants in salmon lice (*Lepeophtheirus salmonis*) in Norway 2018–2019. *Surveillance Programmes for Terrestrial and Aquatic Animals in Norway. Annual Reports.* Norwegian Veterinary Institute, Oslo, Norway.

Hemmingsen, W., MacKenzie, K., Sagerup, K., Remen, M., Bloch-Hansen, K. and Dagbjartarson Imsland, A. (2020) *Caligus elongatus* and other sea lice of the genus Caligus as parasites of farmed salmonids: a review. *Aquaculture* 522, 735160.

Igboeli, O.O., Fast. M.D., Heumann, J. and Burka, J.F. (2012) Role of P-glycoprotein in emamectin benzoate (SLICE®) resistance in sea lice, *Lepeophtheirus salmonis*. *Aquaculture* 344, 40–47.

Igboeli, O.O., Purcell, S.L., Wottom, H., Poley, J., Burka, J.F. and Fast, M.D. (2013) Immunostimulation of *Salmo salar* L., and its effect on *Lepeophtheirus salmonis* (Krøyer). P-glycoprotein mRNA expression following subsequent emamectin benzoate exposure. *Journal of Fish Diseases* 36, 339–351.

Igboeli, O., Burka, J. and Fast, M. (2014a) Sea lice population and sex differences in P-glycoprotein expression and emamectin benzoate resistance on salmon farms in the Bay of Fundy, New Brunswick, Canada. *Pest Management Science* 70, 905–914.

Igboeli, O., Burka, J. and Fast, M. (2014b) *Lepeophtheirus salmonis*: a persisting challenge for salmon aquaculture. *Animal Frontier* 1, 22–32.

Jansen, P., Kristoffersen, A., Viljugrein, H., Jimenez, D., Aldrin, M. and Stien, A. (2012) Sea lice as a density-dependent constraint to salmonid farming. *Proceedings of the Royal Society B* 279, 2330–2338.

Jansen, P.A., Grøntvedt, R.N., Tarpai, A., Helgesen, K.O. and Horsberg, T.E. (2016) Surveillance of the sensitivity towards antiparasitic bath-treatments in the salmon louse (*Lepeophtheirus salmonis*). *PLoS ONE* 11(2), e0149006.

Jensen, M.E., Sevatdal, S, Bakke, M.J, Kaur, K. and Horsberg, T.E. (2017) A selection study on a laboratory-designed population of salmon lice (*Lepeophtheirus salmonis*) using organophosphate and pyrethroid pesticides. *PLoS ONE* 12(5), e0178068.

Jensen, E.M., Horsberg, T.E., Sevatdal, S. and Helgesen, K.O. (2020) Trends in de-lousing of Norwegian farmed salmon from 2000–2019 – consumption of medicines, salmon louse resistance and non-medicinal control methods. *PLoS ONE* 15(10), e0240894.

Kaur, K., Helgesen, K.O., Bakke, M.J. and Horsberg, T.E. (2015) Mechanism behind resistance against the organophosphate azamethiphos in salmon lice (*Lepeophtheirus salmonis*). *PLoS ONE* 10, e0124220.

Kaur, K., Jansen, P.A., Aspehaug, V.T. and Horsberg., T.E. (2016) Phe362Tyr in AChE: a major factor responsible for azamethiphos resistance in *Lepeophtheirus salmonis* in Norway. *PLoS ONE* 11, e0149264.

Lees, F., Gettinby, G. and Revie, C.W. (2008) Changes in epidemiological patterns of sea lice infestation on farmed Atlantic salmon, *Salmo salar* L., in Scotland between 1996 and 2006. *Journal of Fish Diseases* 31, 259–268.

Marín, S.L., Ibarra, R., Medina, M.H. and Jansen, P.A. (2015) Sensitivity of *Caligus rogercresseyi* (Boxshall & Bravo, 2000) to pyrethroids and azamethiphos measured using bioassay tests – a large scale spatial study. *Preventive Veterinary Medicine* 122, 33–41.

Marín, S.L., González, M.P., Madariaga, S.T., Mancilla, M., Mancilla, J. and Woywood, D. (2017) Response of *Caligus rogercresseyi* (Boxshall & Bravo 2000) to hydrogen peroxide: recovery of parasites, fish infestation and egg viability under experimental conditions. *Journal of Fish Diseases* 41, 861–873.

Marín, S.L., Mancilla, J., Hausdorf, M.A., Bouchard, D., Tudor, M.S. and Kane, F (2018) Sensitivity assessment of sea lice to chemotherapeutants: current bioassays and best practices. *Journal of Fish Diseases* 41, 995–1003.

McEwan, G.F., Groner, M.L., Fast, M.D., Gettinby, G. and Revie, C.W. (2015) Using agent-based modelling to predict the role of wild refugia in the evolution of resistance of sea lice to chemotherapeutants. *PLoS ONE* 10(10), e0139128.

Medina, M., Barata, C., Telfer, T. and Baird, D. (2002) Age-and sex related variation in sensitivity to the pyrethroid cypermethrin in the marine copepod *Acartia tonsa* Dana. *Archives of Environmental Contamination and Toxicology* 42(1), 17–22.

Robertson, J.L and Preisler, H.K. (1992) *Pesticide Bioassays with Arthropods*, 1st edn. CRC-Press, Boca Raton, USA.

Robertson, P., Black, K., Adams, M., Willis, K., Buchan, F., Orr, H., Lawton, L. and McCullagh, C. (2009) A new generation of biocides for control of crustacea in fish farms. *Journal of Photochemistry and Photobiology B: Biology* 95, 58–63.

Saksida, S.M., Morrison, D., McKenzie, P., Milligan, B., Downey, E., Boyce, B. and Eaves, A. (2013) Use of Atlantic salmon, *Salmo salar* L., farm treatment data and bioassays to assess for resistance of sea lice, *Lepeophtheirus salmonis*, to emamectin benzoate (SLICE®) in British Columbia. *Canadian Journal of Fish Diseases* 36(5), 515–520.

Samsing, F., Johnsen, I., Treml, E.A. and Dempster, T. (2019) Identifying 'firebreaks' to fragment dispersal networks of a marine parasite. *International Journal for Parasitology* 49, 277–286.

Scott, J.G. (1995) Effects of temperature on insecticide toxicity. In: Roe, R.M. and Kuhr, R.J. (eds) *Reviews in Pesticide Toxicology Vol 3*. Toxicology Communications Inc., Raleigh, NC. pp. 111–135.

SEARCH (2006) Sea lice resistance to chemotherapeutants: a handbook in resistance management. Second edition. 52 p. Available at: htpp://www.aquamedicine.no/uploaded_fagfiles.aspe?fag=7&meny=17 (accessed 5 April 2010).

Sevatdal, S. and Horsberg, E.T. (2003) Determination of reduced sensitivity in sea lice (*Lepeophtheirus salmonis* Krøyer) against the pyrethroid deltamethrin using bioassays and probit modelling. *Aquaculture* 218, 21–31.

Sevatdal, S., Copley, L., Wallace, C., Jackson, D. and Horsberg, T.E. (2005) Monitoring of the sensitivity of sea lice (*Lepeophtheirus salmonis*) to pyrethroids in Norway, Ireland and Scotland using bioassays and probit modelling. *Aquaculture* 244, 19–27.

Valenzuela-Muñoz, V., Nuñez-Acuña, G. and Gallardo-Escárate, C. (2014) Molecular characterization and transcription analysis of P-glycoprotein gene from the salmon louse *Caligus rogercresseyi*. *Aquaculture Research and Development* 5(4), 1–7.

Westcott, J.D., Stryhn, H., Burka, J.F. and Hammell, K.L. (2008) Optimization and field use of a bioassay to monitor sea lice *Lepeophtheirus salmonis* sensitivity to emamectin benzoate. *Diseases of Aquatic Organisms* 79, 119–131.

Whyte, S.K., Westcott, J.D., Elmoslemany, A., Hammell, K.L. and Revie, C.W. (2013) A fixed-dose approach to conducting emamectin benzoate tolerance assessments on field-collected sea lice, *Lepeophtheirus salmonis*. *Journal of Fish Diseases* 36, 283–292.

Whyte, S.K., Westcott, J.D., Jimenez, D., Revie, C.W. and Hammell, K.L. (2014) Assessment of sea lice (*Lepeophtheirus salmonis*) management in New Brunswick, Canada using deltamethrin (AlphaMax®) through clinical field treatment and laboratory bioassay responses. *Aquaculture* 422–423, 54–62.

Whyte, J.D., Westcott, C.W., Revie, K.L. and Hammell, L. (2016) Sensitivity of salmon lice (*Lepeophtheirus salmonis*) in New Brunswick, Canada, to the organophosphate Salmosan® (w/w 50% azamethiphos) using bioassays. *Aquaculture* 464, 593–600.

CHAPTER 17

Selective breeding for improved resistance to sea lice in farmed salmonids

Smaragda Tsairidou, Diego Robledo and Ross D. Houston

The Roslin Institute and Royal (Dick) School of Veterinary Studies, University of Edinburgh

17.1 Introduction

17.1.1 Genetic improvement of salmonid species through selective breeding

Selective breeding is a widely used method for improving the performance and health of farmed animal populations. The process involves choosing parents with high genetic merit for target traits, thereby producing offspring with improved performance for those traits. Selective breeding in livestock and aquaculture species has resulted in cumulative and permanent improvement of a wide range of important production, fitness, health and welfare traits, such as growth and disease resistance. While selective breeding programmes have been successfully implemented for many decades for most terrestrial livestock and crop species, their application in aquaculture is more recent. Atlantic salmon *Salmo salar* is the aquaculture species with the most advanced breeding programmes, and the technology and methods used in salmon tend to be transferred to other aquaculture sectors. Selective breeding targets improvement of traits which are desirable for production of the species of interest, and requires the existence of genetic variation in those target traits. Aquaculture species have been domesticated only recently compared to terrestrial livestock species, and are closely related to their wild counterparts (Mignon-Grasteau et al., 2005). Hence, they typically retain high levels of genetic variation in most traits of economic importance. This genetic variation can be used to achieve genetic improvement via selective breeding with impressive rates of genetic gain, despite the relatively long generation intervals (for example, three to four years in Atlantic salmon). For example, the estimated average genetic improvement of growth rate in Atlantic salmon is 13% per generation (Gjedrem and Rye, 2018). The phenotypic variation observed in a particular target trait in a population (for example, growth or resistance to a disease) is the result of both genetic and environmental variation. The genetic variance can be further categorised as additive, dominance or epistatic variance. A key parameter for selective breeding is the proportion of total phenotypic variance which is due to the additive genetic component for a target trait, termed the narrow-sense heritability. The additive genetic value of individuals in a breeding programme for this trait is referred to as a breeding value or the genetic merit of the individual. Selective breeding aims to choose the individuals with the highest breeding values for a trait, or traits, of interest to use them as the parents of the next generation.

The importance of heritability and its relationship to genetic gain is captured by the breeders' equation which shows that response to selection (or the genetic gain) for a given trait depends on the additive genetic variance, the selection intensity (which depends on the proportion of the population included in the selected group), the prediction accuracy (how accurately breeding values can be predicted) and the generation interval of the species (Falconer and Mackay, 1996). This prediction accuracy (or its square, known as the reliability) is a measure of the quality of the predicted breeding values, theoretically calculated as the correlation between the predicted and the true breeding values. However, in settings where prediction accuracy is being tested using empirical data, the true breeding value is unknown, therefore the accuracy can be estimated as the correlation between the predicted breeding values and the trait phenotype, corrected by the square root of the trait heritability.

Most salmonid species breeding programmes, including those for Atlantic salmon, tended to focus initially on growth. In more recent years, the breeding goals have broadened to include multiple traits, some of which are not measurable on the selection candidates themselves, such as disease resistance, rate of sexual maturation and fillet traits (Gjedrem and Rye, 2018). These traits are possible to include in the programme by taking advantage of features of salmon reproductive biology, including high levels of fecundity. As a result, a typical salmonid breeding programme structure may include a breeding nucleus of several hundred families where each full sibling family is split into groups, with some full siblings remaining in the breeding nucleus (selection candidates) while others are used for production and performance testing (Fig. 17.1). This process is known as sib testing, and allows genetic progress to be made for the traits that are difficult or impossible to measure on the candidates themselves. This breeding programme structure is particularly beneficial for the improvement of disease resistance traits as, although the breeding stock (selection candidates) cannot typically be exposed to diseases for biosecurity reasons, disease challenge experiments and performance testing can be conducted on their siblings. The tracking of individuals to establish family relationships and enable construction of pedigrees is key, and this is usually achieved by rearing individual full sibling families separately until the stage where they are large enough to be tagged. Alternatively, genetic markers have been applied for the same purpose where families are mixed at an early life stage, which can avoid common environmental effects associated with family separation. In early breeding programmes, pedigree data was used to predict breeding values at the family level (for example, Gjedrem et al., 1991a, 2012), where all the full siblings are considered genetically identical and predicted to have the same breeding values for the target trait (for example, disease resistance). This method captures between-family genetic variation, but does not capitalise on individual variation, that is, within-family variation due to the Mendelian sampling variance. Mendelian sampling refers to the inheritance of a random sample of a parent's alleles to offspring in meiosis. While pedigree methods assume that full siblings share 50% of their alleles, in reality this can vary widely between any two full siblings due to Mendelian sampling. With the advent of genomic information, it became possible to capture this variation, and therefore include both between- and within-family information into breeding decisions for traits measured on siblings of selection candidates, as discussed in section 17.1.2.

In summary, selective breeding utilises additive genetic variation to achieve cumulative and permanent improvement in traits of economic importance, by selecting as parents of the next generation the individuals with the best breeding values for the trait. Family-based selective breeding programmes have been successfully applied in more recent years for Atlantic salmon, and have allowed

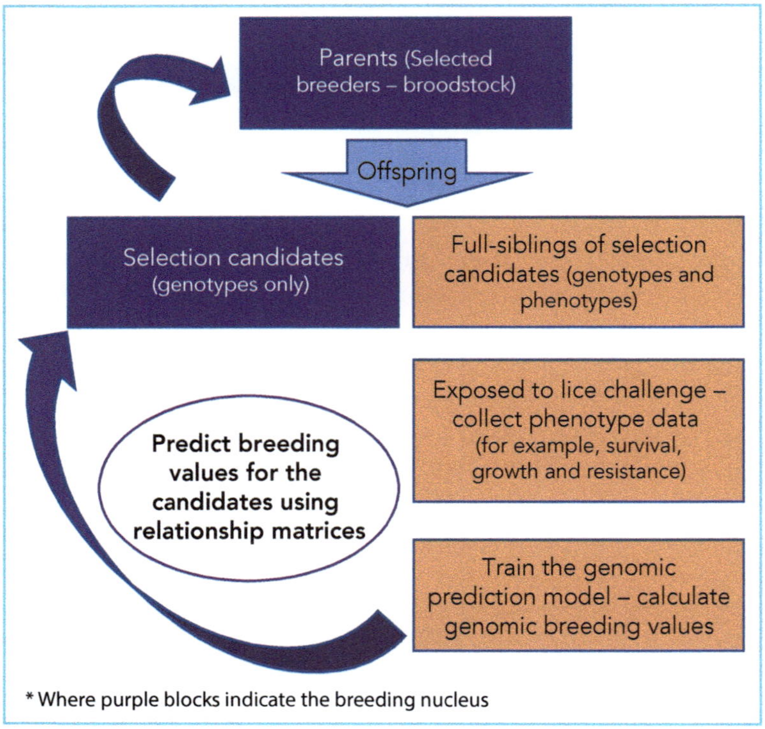

Fig. 17.1 Selection programme flowchart. Credit: Smaragda Tsairidou, Diego Robledo, Ross Houston.

the inclusion of traits which are not possible to measure on the selection candidates. These programmes have also begun to incorporate genomic tools, and the full potential of genomic selection will become more apparent in the following sections, where the specific case of selection for improved resistance to sea lice will be discussed in detail.

17.1.2 Breeding for disease resistance

Infectious diseases present a major challenge for sustainable aquaculture production of salmonids worldwide. Fully effective prevention or treatment methods are currently lacking for many important diseases, while several studies have reported an increasing number of pathogens that can develop resistance to common treatments (Tully and McFadden, 2000; Sevatdal and Horsberg, 2003). In addition, due to the continuous contact of farmed marine species in the open ocean with the wild environment, there is potential for disease transmission from farmed to wild hosts and vice versa (Lafferty et al., 2015). Breeding for disease resistance can be a complementary approach to vaccination and chemical treatments, to support long-term disease control strategies. There is significant genetic variation in resistance to most major viral, bacterial and parasitic infections (Yáñez et al., 2014a). For example, in Atlantic salmon previous studies have reported genetic variation in resistance to the infectious pancreatic necrosis (IPN) virus (Houston et al., 2008, 2014a), to the bacterium *Aeromonas salmonicida* subspecies *salmonicida*, the causative agent of furunculosis (Gjedrem et al., 1991b), and as will be discussed in detail in this chapter, to the ectoparasite sea lice (Gjerde et al., 2011). Hence, selective breeding can be used to improve resistance to a wide range of infectious diseases in aquaculture species. However, disease resistance cannot be measured directly in the selection candidates (broodstock) for biosecurity reasons. Instead, disease resistance of the candidates can be predicted using traditional family-based selective breeding, via the use of sib testing as described in the previous section. Sib testing is facilitated by the high fecundity of salmonids (and other aquatic species), and therefore the availability of large full sibling families. In these studies, disease resistance is commonly measured via challenge experiments where the fish are exposed to specific pathogens (via cohabitation with infected animals or direct injection of the pathogen). Such tests take place in controlled environments where survival can be systematically recorded, either as a binary dead/alive trait at a certain time-point (for example, at 50% cumulative mortality or at the end of the experiment), or as continuous time-to-death (Ødegård et al., 2011). In addition, it is possible to accurately measure resistance by recording mortalities during natural outbreaks (for example, Houston et al., 2008); however, a practical difficulty with this approach is that it relies on opportunistic data collection from naturally occurring outbreaks. When using field disease data, incomplete exposure to infection and the imperfect sensitivity and specificity of the diagnostic methods used to define and classify phenotypes can lead to underestimating heritability; however, it has been shown that it is still possible to demonstrate genetic differences in resistance from such data, and the true extent of genetic variation can be predicted (Bishop and Woolliams, 2010). Resistance can also be measured indirectly, for example through physiological and immune parameters that are genetically correlated with disease resistance; however, those have been reported to explain only a very small proportion of the observed variation in survival (Yáñez et al., 2014a). The trait definition is also important, and while resistance is used as an all-encompassing term in practice, 'resilience' and 'tolerance' are also used. 'Resilience' refers to the ability of a host to maintain high production levels when challenged with infectious pathogens (Lough et al., 2018), and this comprises both resistance and tolerance. Tolerance is the ability to reduce the impact of pathogens on host performance (Doeschl-Wilson et al., 2012). Regardless of the trait measurement used, it is important to assess the relationship between the trait measured in an experimental setting (for example, a disease challenge experiment) and the commercial setting (for example, mortality or parasite burden in the field). In cases where this relationship has been assessed, the genetic correlation has generally been found to be high (Yáñez et al., 2014), which bodes well for the use of experimental challenges. The use of genomic information is now routinely allowing use of both within- and between-family genetic variation in breeding value estimation, meaning that selection can be performed at the individual level rather than just the family level, even for sib-tested traits such as disease resistance. The method used to incorporate genomic information depends on the genetic architecture of the trait in question. The genetic architecture refers to the location, number and effect size of the genomic regions affecting a quantitative trait, known as quantitative trait loci (QTL). Traits can be controlled by a single QTL of large effect, or they can be polygenic controlled by many different QTLs each with small effects. When the trait is controlled by a QTL of major effect, marker-assisted selection (MAS) can be used to select individuals that carry the beneficial QTL allele. An early example of successful implementation of MAS for disease resistance is breeding for resistance to IPN in Atlantic salmon (Houston et al., 2008, 2010; Moen et al., 2009). IPN is caused by a birnavirus which is responsible for high mortality in both fry and post smolts. A single major genomic locus which explained almost all genetic variation in resistance was identified using survival data from 'natural' IPN virus outbreaks in seawater. This has been successfully implemented into commercial breeding programmes by using small numbers of QTL-linked genetic markers (single nucleotide polymorphisms (SNPs), or microsatellites) to predict and improve resistance to the IPN virus (Houston et al., 2020). This resulted in the reduction of the number of salmon mortalities due to IPN from tens of millions down to near zero over the course of approximately five years (Norris, 2017). Other examples of major QTLs affecting disease resistance include the cases of pancreas disease caused by the salmonid alphavirus (Gonen et al., 2015), resistance to the infectious salmon anaemia virus (Moen et al., 2007) and resistance to *Flavobacterium psychrophilum* in rainbow trout (Vallejo et al., 2014), some of which are also implemented commercially using MAS. Nonetheless, the majority of economically important traits, including disease resistance, have a polygenic genetic architecture, and therefore require approaches that utilise multiple genomic regions in predicting breeding values. The use of genome-wide SNP marker data to predict breeding values is typically known as genomic selection, and has been widely shown to be an effective approach for improving polygenic traits including disease resistance (Meuwissen et al., 2001; Tsairidou et al., 2014), even when the QTLs involved are not known. This approach exploits both between- and within-family variation and many studies have shown the increased prediction accuracy of using genomic data versus using pedigree information alone (Habier et al., 2007; Ødegård et al., 2014). While the genomic selection concept was developed around the turn of the century (Haley and Visscher, 1998; Meuwissen et al., 2001), it is the capability to routinely genotype large numbers of individuals for large numbers of SNPs across the entire genome that has made it a practical, albeit still expensive, reality. The development of high-density SNP arrays for several aquaculture species (Zenger et al., 2019) has been crucial in achieving this, including in Atlantic salmon (Houston et al., 2014b) and rainbow trout (Palti et al., 2015).

17.2 Genetic variation in resistance to sea lice in Atlantic salmon

17.2.1 Measuring resistance to sea lice

Sea lice (*L. salmonis* and *C. elongatus* in Europe, and *C. rogercresseyii* in Chile) present the largest disease problem for salmon farming, and therefore improving host resistance to this parasite is one of the main objectives of current salmon breeding programmes. As discussed in detail in previous chapters, sea lice cause significant financial losses to the salmon industry due to treatment costs, loss of performance, morbidity and mortality due to infestation, and secondary bacterial and fungal infections. The total cost of sea lice control was estimated to be €305 million in 2006 (Costello, 2009), and is now estimated to be almost €1 billion per annum. Host resistance to lice is commonly measured using experimental challenge, and the most commonly used phenotype to measure resistance is the number of parasites attached to the fish after challenge (sea lice count). In such trials, pre or post smolt fish are typically given a moderate dose of copepodid larvae and monitored until the lice have moulted into the early chalimus stages. Sampling of the fish is typically performed one to two weeks following the challenge, with number of lice attached showing little change following the first few days. Typically, at the stage of sampling, fish are humanely euthanised, and the number of parasites attached to the fish is assessed using a stereo-microscope or magnifying glass. While experimental challenges are typically performed on relatively small post smolt fish, counting individual lice is still a very laborious process. To increase the throughput of the process, it can be performed on one section of the fish only (for example, left or right side), which has a high correlation with the total lice count. Fish with higher numbers of sea lice attached are considered more susceptible and vice versa. The sea lice count of the fish is frequently corrected by the size of the fish since larger fish tend to have more parasites; sea lice density can be calculated by correcting for an approximate measure of the surface skin area of the fish (Ødegård et al., 2014).

17.2.2 Heritability of resistance to lice

There is substantial genetic variation in resistance to sea lice in farmed Atlantic salmon populations, and therefore, there is potential for improving resistance via selective breeding. Reported heritabilities for sea lice resistance measured using sea lice counts range between 10% and 35%, and are relatively consistent across the three most important lice species (Table 17.1). Sea lice density has been found to have slightly lower heritability than sea lice count (Table 17.1), although these two measurements of lice resistance show a relatively high genetic correlation of 0.89 (0.03) (Gjerde et al., 2011).

The studies described above have measured heritability of resistance to sea lice using data collected from challenge experiments in controlled tank environments (Gjerde et al., 2011; Lhorente et al., 2012; Yáñez et al., 2014b; Gharbi et al., 2015; Tsai et al., 2016), natural lice infections (Mustafa and McKinnon, 1999), challenge experiments in sea cages (Ødegård et al., 2014) or a combination of natural infection and experimental challenge data (Kolstad et al., 2005) (Table 17.1). The heritability for sea lice count has been found to be lower when estimated from natural lice infection compared to controlled challenge tests (Table 17.1). This can be attributed to the reduced sea lice prevalence and intensity of infection in the sea cage environment compared to an experimental tank challenge, meaning that many individuals in a natural infection can be uninformative due to lack of variation in the trait. In addition, there is temporal variation in exposure to lice in natural conditions, which is likely to correspond to a higher environmental variance, and a resulting lower heritability (Kolstad et al., 2005). However, the high genetic correlation between experimental and field conditions (0.88) suggests that sea lice count recorded from tank challenge experiments can be used as a proxy for resistance both in controlled tank and sea cage environments (Kolstad et al., 2005). This is very encouraging for the use of experimental challenges to improve genetic resistance of salmon to lice in the commercial sea cage environment. The timing of when the trait is recorded is also likely to be important, including the age of the fish and whether they have previously been exposed to sea lice. For example, the heritability of sea lice count has been shown to differ depending on whether the fish are naïve to sea lice infestation or have been previously exposed, with lower estimated heritabilities after previous exposures and chemical treatments (Kolstad et al., 2005). Interestingly, the genetic correlation for sea lice count between naïve fish and later sea lice infection has been found to be very low, suggesting that those may partly be genetically different traits (Kolstad et al., 2005).

17.2.3 Genetic architecture of resistance to sea lice

The genetic architecture of a trait (as described in section 17.1.2) is usually studied using genome-wide association studies (GWAS). GWAS approaches typically involve the collection of large quantities of genotype and trait data on a population of individuals, and the testing of the statistical association between SNPs and the trait phenotype. With a few notable exceptions (for example, the IPN resistance QTL described in section 17.1.2), most traits of importance to aquaculture breeding have a polygenic architecture. This is important for the optimal use of genetic markers to improve disease resistance – that is, either via MAS or genomic selection. GWAS approaches typically test the association between each SNP and the trait independently. As such, it can be challenging to detect SNPs with significant effects when the trait in question is polygenic, due to the large numbers of separate statistical tests performed and the need for multiple testing correction to avoid false positives. Alternatively, groups of SNPs in a particular genomic region or window can be analysed together, for

Table 17.1 Heritability of resistance to sea lice infection in salmonids.

Species	h^2 (s.e.) of sea lice count	h^2 (s.e.) of sea lice density	Genetic model	Study design	Sample size and genotype data	Citation
Atlantic salmon (*Salmo salar*)						
Lepeophtheirus salmonis	0.33 (0.08)	–	Linear animal mixed model using genomic relationship matrix	Experimental challenge in tanks	531 post smolts genotyped with Affymetrix Axiom SNP array that included ~132 K SNPs	(Tsai et al., 2016)
	0.27 (0.08)	–	Linear animal mixed model using pedigree-based relationship matrix	Experimental challenge in tanks	531 post smolts genotyped with Affymetrix Axiom SNP	(Tsai et al., 2016)
	0.30 (0.06) (log-transformed)	–	Generalised linear animal mixed model using pedigree information	Experimental challenge in tank	1400 smolts one year post-hatching	(Gharbi et al., 2015)
	–	0.14 (0.03)*	Linear animal mixed model using pedigree-based numerator relationship matrix	Experimental challenge in sea net-cages	1963 individuals genotyped with a 220 k Affymetrix genome-wide SNP-chip	(Ødegård et al., 2014)
	0.33 (0.05)	0.26 (0.05)*1	Multivariate linear animal mixed model using pedigree information	Experimental challenge in controlled tanks	2206 post smolts	(Gjerde et al., 2011)
	0.07 (0.02) (where the genetic variance included the additive genetic, dominance and common environment effect)	–	Linear animal model using pedigree relationships	Marine cage in sea water	1198 smolts	(Glover et al., 2005)
	0.14 (0.02) (log-transformed)	–	Multi-trait linear animal mixed model using pedigree relationships	First-pulse natural infection in net pens in the sea	8193 fish across three year-classes	(Kolstad et al., 2005)
	0.26 (0.07) (log-transformed)	–	Multi-trait linear animal mixed model using pedigree relationships	Experimental challenge in tank	848 fish from one year class	(Kolstad et al., 2005)

Caligus rogercresseyii	0.29 (0.04)	0.28 (0.04)*2	Linear animal mixed model using genomic relationship matrix	Experimental challenge in tanks	2345 post smolts genotyped with a panel of 968 SNPs	(Robledo et al., 2019)
	0.12 (p value < 0.00001)	–	Polygenic mixed model using genomic relationship matrix	Experimental challenge in tanks	2628 smolts genotyped with a 50 K SNP Affymetrix® Axiom® myDesignTM Genotyping Array	(Correa et al., 2017b)
	0.10 (0.03)	–	Multi-trait linear animal model using pedigree information	Experimental challenge in tanks	2628 fish	(Yáñez et al., 2014b)
	0.34 (0.07)	0.22 (0.06)*3	Linear mixed animal model using pedigree information	Experimental challenge in tanks	1511 smolts	(Lhorente et al., 2012)
Caligus elongatus	0.22	–	Unknown	Sea pen exposed to natural infection from nearby fish	419	(Mustafa and MacKinnon, 1999)
Rainbow trout (*Oncorhynchus mykiss*)						
Caligus rogercresseyi	0.08 (0.02)	–	Linear multi-trait mixed animal model using pedigree information	Experimental challenge in tanks	2580 fish	(Bassini et al., 2019)

* Lice density is typically calculated as the log-transformed density: $log_e \frac{LC+1}{\sqrt[3]{BW^2}}$ (Ødegård et al., 2014).

*1 Lice density was calculated as $LC/_{BW^{2/3}}$, where LC is lice count and BW is body weight (Gjerde et al., 2011).

*2 Lice density was estimated as the log-transformed density: $log_e\left(LC/_{BW^{2/3}}\right)$ (Robledo et al., 2019).

*3 Lice density calculated as total numbers of lice per fish per unit of body weight: $TSL/_{BW}$, (Lhorente et al., 2012).

example using regional heritability mapping (Nagamine et al., 2012), and can be more powerful for detecting associations when the effects of individual SNPs are small.

For resistance to sea lice, all evidence to date points to a polygenic architecture. No major QTLs have been detected for resistance to *L. salmonis* (Tsai et al., 2016) or *C. rogercresseyi* (Correa et al., 2017a; Robledo et al., 2019). Tsai et al. (2016) used two commercial populations (n_1 = 531 and n_2 = 588) of salmon post smolts, genotyped at high density (with the Affymetrix Axiom ~132 K SNP array, and a custom Affymetrix Axiom ~35 K array, respectively), to explore the genetic architecture of resistance; the GWAS did not reveal any genome-wide significant associations, suggesting that resistance is a polygenic trait. Correa et al. (2017b) tested n = 2628 Atlantic salmon smolts genotyped with a 50 K SNP array, and identified, via GWAS, one SNP located on chromosome 21 which was only significant at the chromosome-wide level. Robledo et al. (2019) used data from 2345 post smolts from a commercial breeding programme population, and identified, via regional heritability mapping, a putative QTL of moderate effect explaining 13.4% of the genetic variation in sea lice density. In conjunction with gene expression analyses, this QTL helped to identify some potentially interesting candidate genes and candidate functional mutations.

As the power to detect QTL is related to sample size, future studies involving larger numbers of phenotyped, and densely genotyped, individuals may identify significant QTLs. Discovering the causative mutations and genes controlling resistance to sea lice will improve our understanding of the biological pathways underlying disease resistance, and enable testing of genome editing approaches that could facilitate the introgression of resistance alleles into populations where they are not present.

17.3 Genomic selection for resistance to sea lice

17.3.1 Genomic prediction of resistance to sea lice

The substantial genetic variation in resistance to sea lice means that this trait can be improved every generation through selective breeding from the animals with the best breeding values. The polygenic architecture of the trait (that is, many QTLs affecting the trait, but each having a small effect) means that the best methods for selecting for sea lice resistance are those that consider the whole genome of each animal (in contrast to marker-assisted selection, which only assesses a few genomic positions) – that is, genomic selection. Genomic selection requires a training or reference population with both phenotypes and genotypes, in which to train the genomic prediction model and estimate the marker effects. In the typical sib-testing salmonid breeding programme as described in section 17.1.1, this reference population comprises the siblings of the selection candidates, and the model is used to predict breeding values for those candidates who only have genotypes.

SNP arrays are the most common method used to obtain genotypes for both selection candidates and the challenged fish in the reference population. These genotypes are used to calculate the genomic relationship matrix, which contains the pairwise relationships between all individuals in the dataset (Meuwissen et al., 2001; VanRaden, 2008). This genomic relationship matrix, along with the collected phenotypes, is used to calculate the genomic estimated breeding value (GEBV) for each individual, using genomic best linear unbiased prediction methods (GBLUP; VanRaden, 2008), or more sophisticated weighted single-step GBLUP (Tiezzi and Maltecca, 2015; Liu et al., 2020), and Bayesian methods (for example, BayesRC (MacLeod et al., 2016) that assign different weight to SNPs or genomic regions, depending on their variance and size of effects. The discovery of

QTLs could contribute to improving sea lice resistance more efficiently through using such methods to apply larger weight on putative QTLs affecting the trait, or through using genomic selection in conjunction with MAS.

The performance of genomic selection is assessed through the genomic prediction accuracy, which can be estimated by measuring the performance of the progeny of parents with GEBVs, or via cross-validation statistical methods. The latter method is most practical in tests of genomic prediction accuracy in aquaculture, and therefore most commonly applied. In cross-validation, the phenotypes of a proportion of the animals (full-siblings of selection candidates) are masked (typically one fifth of the data), their EBVs are predicted, and then the correlation between the EBVs and the phenotypes assessed. Due to the fact that this approach deals with phenotypes rather than true breeding values, the correlation between the phenotypes and EBVs is adjusted by the square root of the heritability. The prediction accuracy is a useful metric for assessing the performance of genomic selection, and comparison to other methods for predicting breeding values (that is, pedigree-based prediction).

A number of studies have shown that resistance to sea lice can be predicted using genome-wide SNP genotypes with moderate to high prediction accuracies in the range of 0.5–0.6 (Tsai et al., 2016; Correa et al., 2017a; Tsairidou et al., 2020). Similarly to what has been observed for other economically important traits, it has been found that genomic prediction for sea lice resistance performs better than prediction based on pedigree data (Ødegård et al., 2014; Tsai et al., 2016). These results are summarised in Table 17.2. Breeding companies have developed advanced breeding programmes that include genomic selection for sea lice resistance in salmon eggs and juvenile fish.

A feature of modern breeding programmes is the simultaneous selection for multiple traits, so that disease resistance traits can be improved while still selecting for other economically important traits (for example, growth). This can be achieved by developing selection indices that allow to optimally combine and weight a number of different traits of interest. For this purpose, it is important to know the relationships between the different traits, as selection for one trait can generate favourable or unfavourable changes in another trait if they are genetically correlated. Genetic correlations of sea lice count with growth traits (body weight and length) have been found to be slightly negative but very close to zero (Tsai et al., 2016), or moderately positive (with body weight) (Kolstad et al., 2005). Therefore, genomic selection for sea lice resistance would not be expected to generate adverse correlated responses in commercially important growth traits.

17.3.2 Optimising marker density and training population design

As discussed previously, genomic selection requires large numbers of densely genotyped individuals for SNPs across the entire genome, and that is still expensive. However, genomic selection using medium density genotypes can provide prediction accuracies for sea lice resistance similar to those obtained from full high density genome-wide genotype data (Tsairidou et al., 2020). This is in agreement with similar observations in other aquaculture species and traits (Kriaridou et al., 2020). This likely reflects the typical population structure in aquaculture breeding programmes, where selection candidates and training populations are highly related and hence share long genomic segments, which allows to achieve relatively high genomic prediction accuracy with relatively low density SNP genotypes.

Further, genotype imputation methods have been proposed as an effective tool to optimise cost-efficiency for Atlantic salmon breeding programmes (Tsai et al., 2017; Tsairidou et al., 2020).

Table 17.2 Pedigree-based and genomic prediction accuracy for resistance to sea lice in Atlantic salmon.

Trait	Pedigree-based accuracy (P-BLUP)	Genomic prediction accuracy (G-BLUP)	Sample size	Genotypes	Citation
Logarithmically transformed sea lice count	–	0.54 (for selected optimal 5 K SNP panel)	520 post smolts	Genotyped with the Affymetrix Axiom 132 K Atlantic salmon SNP chip	(Tsairidou et al., 2020)
Sea lice count	0.41	0.50	2404 Atlantic salmon smolts	Genotyped using a 50 K SNP Affymetrix® Axiom® myDesignTM Genotyping Array	(Correa et al., 2017a)
Log-transformed sea lice density	~ 0.49 (for 33 K SNPs)	~ 0.60 (for 33 K SNPs)	Population 1: 531 post smolts Population 2: 588	Population 1 was genotyped with Affymetrix Axiom SNP array that included ~132 K SNPs Population 2 was genotyped with a custom Affymetrix Axiom ~35 K array	(Tsai et al., 2016)
Logarithmically transformed lice density	0.34 (reliability)	Relative increase in reliability using GS was 52% for GBLUP with 220 K, compared to the pedigree-based model	1963 individuals	Genotyped with a 220 K Affymetrix genome-wide SNP-chip	(Ødegård et al., 2014)

Genotype imputation is a technique that allows inference of high density genotypes from low density SNP panels by using information from haplotypes that are shared between related individuals (Li et al., 2009). Hence it requires dense genotyping only for a chosen subset of the population, thereby substantially reducing the costs of genotyping. Imputation has been tested in several studies using simulated (Li et al., 2010; Dufflocq et al., 2019) and empirical data (Sanna et al., 2008; Willer et al., 2008; Tsai et al., 2017; Yoshida et al., 2018), and is one promising avenue for reducing the costs of genomic selection while maintaining or enhancing the prediction accuracy. For sea lice resistance in Atlantic salmon, it has been shown that genomic selection using very low density genotypes (for example, 200 SNPs) imputed to medium density (for example, imputing from 200 to 5000 SNPs in offspring) can provide genomic prediction accuracies for resistance similar to those obtained from high density directly genotyped data (Tsairidou et al., 2020).

The relationship between the training and test data has an impact on the achieved genomic prediction accuracy (Meuwissen et al., 2001); higher accuracies are typically observed when predictions are being made between highly related animals, while accuracies are lower across more distantly related populations. This may be due to differences in the linkage disequilibrium patterns between SNPs and QTL. Indeed, for sea lice resistance, Tsai et al. (2016) observed that prediction accuracies were highest when training and test populations were closely related.

17.3.3 Functional genomics to understand genetic resistance

Functional studies allow us to understand the biological mechanisms underlying disease resistance, which can improve selective breeding via targeted selection, and also contribute towards solving the sea lice problem via informing treatments and genome engineering approaches. Most functional studies have focused on understanding the response of the salmon to the lice, which is conditioned by the lice-induced immunosuppression (reviewed in Braden et al., 2020). The host responses are characterised by an initial inflammatory response, which is then repressed during most of the infection and development of the lice, until a second inflammatory spike occurs (Fast et al., 2006; Skugor et al., 2008). However, differences in this immune response against the lice have been observed depending on the infestation level, with T-helper (Th) related genes in particular showing a significant negative correlation with lice counts (Holm et al., 2015). Interestingly, differences in this Th immune response have been detected between genetically resistant and genetically susceptible families, suggesting that more resistant fish could show enhanced Th1 immunity (Holm et al., 2015). In fact, the transducer of ERBB2-1 (*tob1*) gene, which inhibits T-cell proliferation, is located in a sea lice resistance QTL on the third chromosome of the Atlantic salmon genome (Robledo et al., 2019). *Tob1* is downregulated in T-cells upon antigen stimulation, which leads to T-cell activation and expansion (Baranzini, 2014). The functional relevance and colocalisation with a QTL suggest *tob1* may be involved in the mechanisms of resistance to sea lice, and is therefore a good candidate for further functional validation and prioritisation in selection models. Several other immune-related genes showed significant correlations with lice loads, both positive (such as the immune receptors macrophage mannose receptor or C-type lectin receptor A (*cleca*)) and negative (such as the immune receptors CD97 and suppressor of cytokine signalling 5) (Robledo et al., 2018). In the same study, three immune receptors (including CLECA), two haem oxygenase genes and several genes related to muscle fibres showed differential expression between family-matched genetically resistant and susceptible fish (Robledo et al., 2018), suggesting the potential involvement of these genes and mechanisms in a successful response to sea lice. One limitation of these studies is that it is not possible to disentangle cause and consequence – for example, whether the animals with low expression of these genes have higher lice counts, or if having higher lice counts decreases the expression of these genes. Further functional studies directly manipulating the expression or activity of these genes are necessary to clarify their role in resistance to sea lice.

Another source of information to understand the causes of resistance to sea lice are the differences between species. There is substantial variation in resistance to sea lice across different salmonid species. For example, certain Pacific salmon species, such as coho salmon (*Oncorhynchus kisutch*) and pink salmon (*Oncorhynchus gorbuscha*), are largely resistant to sea lice, while Atlantic salmon and rainbow trout are highly susceptible. The mechanisms underlying these inter-species differences are not fully known, but it has been established that the resistant species are able to mount a successful immune response against the parasite in the very early stages of attachment, ultimately resulting in rejection of the parasite (Jones et al., 2008; Sutherland et al., 2011). This rejection has been connected with fast and consistent innate immunity and inflammation at the site of attachment in resistant species (Braden et al., 2012), and differences in Th2 response, wound repair and iron metabolism have also been found between resistant and susceptible species (Braden et al., 2020). These cross-species comparisons are an outstanding resource to understand the difference between resistant and susceptible immune responses to lice, and to better understand the dynamics of the host-parasite interaction.

Whether it comes from intra-species or cross-species comparisons, knowledge of the causative genes and mechanisms underlying genetic resistance to lice can be used to prioritise genetic markers on these genes or affecting these genes in selection models, which can increase the selection accuracy for resistance to sea lice, and therefore lead to faster genetic gains and more healthy salmon stocks. Furthermore, with the advent of genome editing, this knowledge may, eventually, allow the generation of resistance mechanisms from the resistant to the susceptible animals/species, as discussed in section 17.4.

17.4 Future outlook

17.4.1 Research priorities for breeding for improved resistance to lice

While the vast majority of farmed Atlantic salmon, and certain other salmonid species, are now selected for improved resistance to sea lice, there are various avenues by which the process can be improved.

As described in section 17.2.2, the use of sea lice counts following challenge has been shown to be a practical and effective trait for application in breeding programmes. However, it is unlikely to be the optimum trait, and further consideration should be given to alternative and complementary methods of measuring sea lice resistance. For example, automation of trait recording could allow more accurate and detailed records of louse attachment and rejection, potentially using the latest imaging and computational technology on farms or experimental sea cages. Analyses of these data could be considered in conjunction with larger scale, and potentially real time, recording of environmental parameters (temperature, water quality and other factors) to improve models of the genetic and environmental components of variation in the traits. Furthermore, consideration should be given to the holistic impact of resistance to sea lice in terms of the epidemiology of disease outbreaks, with the ultimate target phenotype being suggested to be the basic reproductive number R_0, that is, the average number of cases generated by one infectious individual; if the R_0 can be brought below one, the infection outbreak would die on its own. For sea lice in particular, the economic value of R_0 has been estimated to be €0.065/unit R_0/kg production, depending on whether management is optimised and whether expenditures with constant losses are reduced via genetic improvement (Janssen et al., 2018). Genomic selection for more resistant individuals would indirectly reduce the number of secondary infections, thus reducing R_0. Further, the number of secondary infections can be reduced directly through targeting reduced host infectivity in selective breeding programmes. It has been shown through simulations that breeding strategies which incorporate genetic selection for both increased disease resistance and reduced infectivity (where infectivity is defined as the host's ability to transmit infection) can reduce the number and severity of epidemics more efficiently and after fewer generations of selection (Tsairidou et al., 2019). Although increasing evidence suggests that infectivity is genetically controlled and super-shedders have been detected in several major epidemics, estimating infectivity genetic effects from field data remains a challenge. Recently developed statistical methods that allow simultaneous estimation of resistance and infectivity are a promising avenue for overcoming this challenge (Biemans et al., 2017; Anacleto et al., 2019).

Mapping and understanding the causative or functional variants that have an impact on resistance to sea lice has potential additional benefits for increasing rates of genetic gain in breeding either through improved selection accuracy or as targets for genome editing (as discussed below). As genomic technologies improve and become cheaper, it is likely that whole genome sequence data will be increasingly applied in salmonid selective breeding programmes. Use of sequence data can facilitate the use of models which put increased weighting on putative functional genomic variants (for example, BayesRC) to increase accuracy, albeit data from the terrestrial livestock world suggest these improvements may be relatively minor. In theory, the use of functional variants affecting resistance to sea lice should improve the accuracy of prediction of breeding values across populations, and across year-groups of a salmon breeding programme. This would then reduce the requirement for yearly challenge testing, with positive implications for animal welfare and cost. However, given the aforementioned effectiveness of using low density panels in the sib-testing scheme, the increase in selection accuracy would need to be quite substantial to justify routine use of sequence data. It is also plausible that applying some of the aforementioned functional genomic data (gene expression, epigenomic) may also improve prediction accuracy, and may also provide early-life markers to predict sea lice resistance in later life (Clark et al., 2020).

It is also essential when considering the future of breeding for resistance to sea lice to consider the interaction between the host and the parasite, and whether selection of the host puts selection pressure on the lice to evolve to overcome resistance. While this is certainly a possibility, and lice are well known to have evolved resistance to some of the common delousing drugs (Tully and McFadden, 2000; Sevatdal and Horsberg, 2003), the risk for selective breeding approaches is likely to be minimal. This is because, unlike chemical treatments which typically target one pathway in the louse, host resistance is polygenic and selection will likely act on many genes and pathways. Therefore, it is unlikely that the lice would evolve mechanisms to simultaneously overcome each of the incremental mechanisms by which the host is better able to resist the parasite. Nonetheless, it is important to be aware of the impact host resistance may have on the parasite particularly when it comes to the possibility of genome editing (discussed in the following section), since this may target one or two genes and the induced change may be of major effect. Therefore, while mechanisms of improving host resistance through breeding (or editing) are a critical component of preventing and mitigating lice, they do need to be complemented by other ongoing and potential control measures such as those discussed throughout this book.

17.4.2 Genome editing and cross-species variation in lice resistance

Genome editing and specifically CRISPR/Cas9 systems have huge potential for future applications to advance our understanding of salmon biology and improve important production and welfare traits such as resistance to sea lice. CRISPR has already been applied in salmonid species for gene knockout, for example to generate sterile salmon (Wargelius et al., 2016). The CRISPR/Cas9 technique works via the enzyme Cas9 making a double-strand cut in the DNA at a site in the genome corresponding to a guide RNA, which results in either small insertions or deletions that can lead to loss-of-function mutations (non-homologous end joining) or user-defined edits to the genome based on a provided DNA template (homology-directed repair). The CRISPR approach can also be used to insert gene activation or suppression domains to modulate gene expression of target genes, without altering the gene sequence.

Genome editing approaches are useful research tools, for example, for assessing gene function. This may include assessing the function of putative genes involved in host resistance to sea lice from GWAS and gene expression approaches, as described in section 17.3.3. In addition to the research possibilities, genome editing also has potential for applications in salmonid breeding to develop strains with improved disease resistance, including to sea lice.

These applications fall into three main categories (Gratacap et al., 2019). The first is increasing the frequency or fixing favourable alleles for production traits, which has been shown in simulation studies to increase genetic gain substantially faster than genomic selection alone (Jenko et al., 2015). The second is introgression-by-editing whereby favourable alleles from different strains or species are transferred into the target species via genome editing, without the negative side-effects associated with traditional introgression such as linkage drag of unfavourable alleles. The third is the creation of completely *de novo* alleles which have favourable effects on the trait, such as the case of resistance to PRRSV by targeted editing of the CD163 receptor in pigs (Burkhard et al., 2018).

The second of these approaches holds particular promise for sea lice, given the demonstrated inter-species differences in resistance. This possibility of transferring resistance mechanisms from resistant species to Atlantic salmon and rainbow trout using genome editing has arisen and is the focus of substantial research efforts. The key is to identify the factors underlying the relative differences in host resistance mechanisms, such that gene structure or expression could be modified in susceptible species to mimic the response to sea lice exhibited by resistant species. New genomic technologies such as single-cell RNA-sequencing, genome-scale proteomics and genome editing can help provide improved knowledge of the specific differences in early response to sea lice between the resistant and susceptible species, which can then lead to targets to modulate using genome editing in the susceptible species. These edited fish can then be compared to unedited controls for sea lice resistance using a standard experimental sea lice challenge, which may reveal the mechanisms underlying resistance to lice. Resistance mechanisms may then be engineered in susceptible species via targeted editing of the coding sequence and/or modulation of the regulatory sequence to enhance or suppress the expression of key host response genes. The transfer of resistance mechanisms from the resistant to the susceptible species would have major benefits for the aquaculture industry, producing animals immune to sea lice and significantly improving the health of Atlantic salmon stocks.

As with all potential applications of genome editing in animal breeding, it is important to consider practical and regulatory aspects of the application to production. If an edit could be identified that drastically increased resistance of Atlantic salmon to lice, then it may prompt a strong desire among the salmon industry to apply genome editing commercially. From a practical point of view, it would be necessary to develop means to introduce edits into breeding programmes, for example via routine editing of broodstock or multipliers. This would need to be coupled with stringent testing to ensure that there were no off-target or pleiotropic effects of the edit on the health, welfare or performance of the salmon. Furthermore, since escapes do occur, it may be necessary to introduce sterility to the farmed salmon stocks to ensure they cannot interbreed with their wild counterparts. This connects to the regulatory environment for genome editing, which is currently very restrictive in Europe (Schmidt et al., 2020). Nonetheless, the genetically modified AquAdvantage salmon (AquaBounty) has recently been approved for human consumption in North America, and it is likely that genome edited fish will become a practical reality in certain countries in the coming years. Therefore, an appropriate dialogue with all relevant stakeholders is necessary to guide legislation to ensure safe and appropriate use of the technology, while also avoiding undue restriction leading to significant delay in the application of innovations in this field.

References

Anacleto, O., Cabaleiro, S., Villanueva, B., Saura, M., Houston, R.D., Woolliams, J.A. and Doeschl-Wilson, A.B. (2019) Genetic differences in host infectivity affect disease spread and survival in epidemics. *Scientific Reports* 9, 4924.

Baranzini, S.E. (2014) The role of antiproliferative gene *Tob1* in the immune system. *Clinical and Experimental Neuroimmunology* 5, 132–136.

Bassini, L.N., Lhorente, J.P., Oyarzún, M., Bangera, R., Yánez, J.M. and Neira, R. (2019) Genetic parameters for *Piscirickettsia salmonis* resistance, sea lice (*Caligus rogercresseyi*) susceptibility and harvest weight in rainbow trout (*Oncorhynchus mykiss*). *Aquaculture* 510, 276–282.

Biemans, F., De Jong, M.C.M. and Bijma, P. (2017) A model to estimate effects of SNPs on host susceptibility and infectivity for an endemic infectious disease. *Genetics Selection Evolution* 49, 53.

Bishop, S.C. and Woolliams, J.A. (2010) On the genetic interpretation of disease data. *PLoS ONE*, 5, 0008940.

Braden, L.M., Barker, D.E., Koop, B.F. and Jones, S.R.M. (2012) Comparative defense-associated responses in salmon skin elicited by the ectoparasite *Lepeophtheirus salmonis*. *Comparative Biochemistry and Physiology – Part D Genomics and Proteomics* 7, 100–109.

Braden, L.M., Monaghan, S.J. and Fast, M.D. (2020) Salmon immunological defence and interplay with the modulatory capabilities of its ectoparasite *Lepeophtheirus salmonis*. *Parasite Immunology* 42, e12731.

Burkard, C., Opriessnig, T., Mileham, A.J., Stadejek, T., Ait-Ali, T., Lillico, S.G., et al. (2018) Pigs lacking the scavenger receptor cysteine-rich domain 5 of CD163 are resistant to porcine reproductive and respiratory syndrome virus 1 infection. *Journal of Virology* 92, e00415–8.

Clark, E.L., Archibald, A.L., Daetwyler, H.D., Groenen, M.a.M., Harrison, P.W., et al. (2020) From FAANG to fork: application of highly annotated genomes to improve farmed animal production. *Genome Biology* 21, 285.

Correa, K., Bangera, R., Figueroa, R., Lhorente, J.P. and Yáñez, J.M. (2017a) The use of genomic information increases the accuracy of breeding value predictions for sea louse (*Caligus rogercresseyi*) resistance in Atlantic salmon (*Salmo salar*). *Genetics Selection Evolution* 49, 15.

Correa, K., Lhorente, J.P., Bassini, L., López, M.E., Di Genova, A., et al. (2017b) Genome wide association study for resistance to *Caligus rogercresseyi* in Atlantic salmon (*Salmo salar* L.) using a 50K SNP genotyping array. *Aquaculture* 472, 61–65.

Costello, M.J. (2009) The global economic cost of sea lice to the salmonid farming industry. *Journal of Fish Diseases* 32, 115–118.

Doeschl-Wilson, A.B., Villanueva, B. and Kyriazakis, I. (2012) The first step toward genetic selection for host tolerance to infectious pathogens: obtaining the tolerance phenotype through group estimates. *Frontiers in Genetics* 3, 265.

Dufflocq, P., Pérez-Enciso, M., Lhorente, J.P. and Yáñez, J.M. (2019) Accuracy of genomic predictions using different imputation error rates in aquaculture breeding programs: a simulation study. *Aquaculture* 503, 225–230.

Falconer, D.S. and Mackay, T.F. (1996) *Introduction to Quantitative Genetics*. London, Longman.

Fast, M.D., Muise, D.M., Easy, R.E., Ross, N.W. and Johnson, S.C. (2006) The effects of *Lepeophtheirus salmonis* infections on the stress response and immunological status of Atlantic salmon (*Salmo salar*). *Fish and Shellfish Immunology* 21, 228–241.

Gharbi, K., Matthews, L., Bron, J., Roberts, R., Tinch, A., et al. (2015) The control of sea lice in Atlantic salmon by selective breeding. *Journal of the Royal Society Interface* 12(110).

Gjedrem, T. and Rye, M. (2018) Selection response in fish and shellfish: a review. *Reviews in Aquaculture* 10, 168–179.

Gjedrem, T., Gjøen, H.M. and Gjerde, B. (1991a) Genetic origin of Norwegian farmed Atlantic salmon. *Aquaculture* 98, 41–50.

Gjedrem, T., Salte, R. and Gjøen, H.M. (1991b) Genetic variation in susceptibility of Atlantic salmon to furunculosis. *Aquaculture* 97, 1–6.

Gjedrem, T., Robinson, N. and Rye, M. (2012) The importance of selective breeding in aquaculture to meet future demands for animal protein: a review. *Aquaculture* 350–353, 117–129.

Gjerde, B., Ødegård, J. and Thorland, I. (2011) Estimates of genetic variation in the susceptibility of Atlantic salmon (*Salmo salar*) to the salmon louse *Lepeophtheirus salmonis*. *Aquaculture* 314, 66–72.

Glover, K.A., Aasmundstad, T., Nilsen, F. and Storset, A. (2005) Variation of Atlantic salmon families (*Salmo salar* L.) in susceptibility to the sea lice *Lepeophtheirus salmonis* and *Caligus elongatus*. *Aquaculture* 245, 19–30.

Gonen, S., Baranski, M., Thorland, I., Norris, A., Grove, H., et al. (2015) Mapping and validation of a major QTL affecting resistance to pancreas disease (salmonid alphavirus) in Atlantic salmon (*Salmo salar*). *Heredity* 115, 405–414.

Gratacap, R.L., Wargelius, A., Edvardsen, R.B. and Houston, R.D. (2019) Potential of genome editing to improve aquaculture breeding and production. *Trends in Genetics* 35, 672–684.

Habier, D., Fernando, R.L. and Dekkers, J.C. (2007) The impact of genetic relationship information on genome-assisted breeding values. *Genetics* 177, 2389–2397.

Haley, C.S. and Visscher, P.M. (1998) Strategies to utilize marker-quantitative trait loci associations. *Journal of Dairy Science* 81, 85–97.

Holm, H., Santi, N., Kjøglum, S., Perisic, N., Skugor, S., et al. (2015) Difference in skin immune responses to infection with salmon louse (*Lepeophtheirus salmonis*) in Atlantic salmon (*Salmo salar* L.) of families selected for resistance and susceptibility. *Fish and Shellfish Immunology* 42, 384–394.

Houston, R.D., Haley, C.S., Hamilton, A., Guy, D.R., Tinch, A.E., et al. (2008) Major quantitative trait loci affect resistance to infectious pancreatic necrosis in Atlantic salmon (*Salmo salar*). *Genetics* 178, 1109–1115.

Houston, R.D., Haley, C.S., Hamilton, A., Guy, D.R., Mota-Velasco, J.C., et al. (2010) The susceptibility of Atlantic salmon fry to freshwater infectious pancreatic necrosis is largely explained by a major QTL. *Heredity* 105, 318–327.

Houston, R., Bishop, S.C., Woolliams, J. and Haley, C. (2014a) Marker-assisted selection to breed for resistance to infectious pancreatic necrosis in salmon. Research Excellence Framework (REF), impact case studies. Available at: http://impact.ref.ac.uk/CaseStudies/CaseStudy.aspx?Id=23913 (accessed 24 January 2022).

Houston, R.D., Taggart, J.B., Cézard, T., Bekaert, M., Lowe, N.R., et al. (2014b) Development and validation of a high density SNP genotyping array for Atlantic salmon (*Salmo salar*). *BMC Genomics*, 15, 90.

Houston, R.D., Bean, T.P., Macqueen, D.J., Gundappa, M.K., Jin, Y.H., et al. (2020) Harnessing genomics to fast-track genetic improvement in aquaculture. *Nature Reviews Genetics* 21, 389–409.

Janssen, K., Komen, H., Saatkamp, H.W., De Jong, M.C.M. and Bijma, P. (2018) Derivation of the economic value of R(0) for macroparasitic diseases and application to sea lice in salmon. *Genetics Selection Evolution* 50, 47–47.

Jenko, J., Gorjanc, G., Cleveland, M.A., Varshney, R.K., Whitelaw, C.B., et al. (2015) Potential of promotion of alleles by genome editing to improve quantitative traits in livestock breeding programs. *Genetics Selection Evolution* 47, 55.

Jones, S., Kim, E. and Bennett, W. (2008) Early development of resistance to the salmon louse, *Lepeophtheirus salmonis* (Krøyer), in juvenile pink salmon, *Oncorhynchus gorbuscha* (Walbaum). *Journal of Fish Diseases* 31, 591–600.

Kolstad, K., Heuch, P.A., Gjerde, B., Gjedrem, T. and Salte, R. (2005) Genetic variation in resistance of Atlantic salmon (*Salmo salar*) to the salmon louse *Lepeophtheirus salmonis*. *Aquaculture* 247, 145–151.

Kriaridou, C., Tsairidou, S., Houston, R.D. and Robledo, D. (2020) Genomic prediction using low density marker panels in aquaculture: performance across species, traits, and genotyping platforms. *Frontiers in Genetics* 11, 124.

Lafferty, K.D., Harvell, C.D., Conrad, J.M., Friedman, C.S., Kent, M.L., et al. (2015) Infectious diseases affect marine fisheries and aquaculture economics. *Annual Review of Marine Science* 7, 471–496.

Lhorente, J.P., Gallardo, J.A., Villanueva, B., Araya, A.M., Torrealba, D.A., et al. (2012) Quantitative genetic basis for resistance to *Caligus rogercresseyi* sea lice in a breeding population of Atlantic salmon (*Salmo salar*). *Aquaculture* 324–325, 55–59.

Li, Y., Willer, C., Sanna, S. and Abecasis, G. (2009) Genotype imputation. *Annual Review Genomics Human Genetics* 10, 387–406.

Li, Y., Willer, C.J., Ding, J., Scheet, P. and Abecasis, G.R. (2010) MaCH: using sequence and genotype data to estimate haplotypes and unobserved genotypes. *Genetic Epidemiology* 34, 816–834.

Liu, A., Lund, M.S., Boichard, D., Karaman, E., Guldbrandtsen, B., et al. (2020) Weighted single-step genomic best linear unbiased prediction integrating variants selected from sequencing data by association and bioinformatics analyses. *Genetics Selection Evolution* 52, 48.

Lough, G., Hess, A., Hess, M., et al. (2018) Harnessing longitudinal information to identify genetic variation in tolerance of pigs to porcine reproductive and respiratory syndrome virus infection. *Genetics Selection Evolution* 50, 50.

MacLeod, I.M., Bowman, P.J., Vander Jagt, C.J., Haile-Mariam, M., Kemper, K.E., et al. (2016) Exploiting biological priors and sequence variants enhances QTL discovery and genomic prediction of complex traits. *BMC Genomics* 17, 144.

Meuwissen, T.H.E., Hayes, B.J. and Goddard, M.E. (2001) Prediction of total genetic value using genome-wide dense marker maps. *Genetics* 157, 1819–1829.

Mignon-Grasteau, S., Boissy, A., Bouix, J., Faure, J.-M., Fisher, A.D., et al. (2005) Genetics of adaptation and domestication in livestock. *Livestock Production Science* 93, 3–14.

Moen, T., Sonesson, A.K., Hayes, B., Lien, S., Munck, H., et al. (2007) Mapping of a quantitative trait locus for resistance against infectious salmon anaemia in Atlantic salmon (*Salmo salar*): comparing survival analysis with analysis on affected/resistant data. *BMC Genetics*. 8, 53.

Moen, T., Baranski, M., Sonesson, A.K. and Kjøglum, S. (2009) Confirmation and fine-mapping of a major QTL for resistance to infectious pancreatic necrosis in Atlantic salmon (*Salmo salar*): population-level associations between markers and trait. *BMC Genomics* 10, 368.

Mustafa, A. and MacKinnon, B.M. (1999) Genetic variation in susceptibility of Atlantic salmon to the sea louse *Caligus elongatus* Nordmann, 1832. *Canadian Journal of Zoology* 77, 1332–1335.

Nagamine, Y., Pong-Wong, R., Navarro, P., Vitart, V., Hayward, C., et al. (2012) Localising loci underlying complex trait variation using regional genomic relationship mapping. *PLoS ONE*, 7, e46501.

Norris, A. (2017) Application of genomics in salmon aquaculture breeding programs by Ashie Norris: Who knows where the genomic revolution will lead us? *Marine Genomics* 36, 13–15.

Ødegård, J., Baranski, M., Gjerde, B. and Gjedrem, T. (2011) Methodology for genetic evaluation of disease resistance in aquaculture species: challenges and future prospects. *Aquaculture Research* 42, 103–114.

Ødegård, J., Moen, T., Santi, N., Korsvoll, S. A., Kjøglum, S., et al. (2014) Genomic prediction in an admixed population of Atlantic salmon (Salmo salar). *Frontiers in Genetics* 5.

Palti, Y., Gao, G., Liu, S., Kent, M.P., Lien, S., et al. (2015) The development and characterization of a 57K single nucleotide polymorphism array for rainbow trout. *Molecular Ecology Resources* 15, 662–672.

Robledo, D., Gutiérrez, A.P., Barría, A., Yáñez, J.M. and Houston, R.D. (2018) Gene expression response to sea lice in Atlantic salmon skin: RNA sequencing comparison between resistant and susceptible animals. *Frontiers in Genetics* 9.

Robledo, D., Gutiérrez, A.P., Barría, A., Lhorente, J.P., Houston, R.D., et al. (2019) Discovery and functional annotation of quantitative trait loci affecting resistance to sea lice in Atlantic salmon. *Frontiers in Genetics* 10.

Sanna, S., Jackson, A.U., Nagaraja, R., Willer, C.J., Chen, W.M., et al. (2008) Common variants in the GDF5-UQCC region are associated with variation in human height. *Nature Genetics* 40, 198–203.

Schmidt, S.M., Belisle, M. and Frommer, W.B. (2020) The evolving landscape around genome editing in agriculture. *EMBO Reports* 21, e50680.

Sevatdal, S. and Horsberg, T. (2003) Determination of reduced sensitivity in sea lice (*Lepeophtheirus salmonis* Krøyer) against the pyrethroid deltamethrin using bioassays and probit modelling. *Aquaculture* 218(1–4) 21–31.

Skugor, S., Glover, K.A., Nilsen, F. and Krasnov, A. (2008) Local and systemic gene expression responses of Atlantic salmon (*Salmo salar* L.) to infection with the salmon louse (*Lepeophtheirus salmonis*). *BMC Genomics* 9, 498.

Sutherland, B.J., Jantzen, S.G., Sanderson, D.S., Koop, B.F. and Jones, S.R. (2011) Differentiating size-dependent responses of juvenile pink salmon (*Oncorhynchus gorbuscha*) to sea lice (*Lepeophtheirus salmonis*) infections. *Comparative Biochemistry and Physiology Part D Genomics Proteomics* 6, 213–223.

Tiezzi, F. and Maltecca, C. (2015) Accounting for trait architecture in genomic predictions of US Holstein cattle using a weighted realized relationship matrix. *Genetics Selection Evolution* 47, 24.

Tsai, H.-Y., Hamilton, A., Tinch, A.E., Guy, D.R., Bron, J.E., et al. (2016) Genomic prediction of host resistance to sea lice in farmed Atlantic salmon populations. *Genetics Selection Evolution* 48, 47.

Tsai, H.-Y., Matika, O., Edwards, S.M., Antolín–Sánchez, R., Hamilton, A., et al. (2017) Genotype imputation to improve the cost-efficiency of genomic selection in farmed Atlantic salmon. *G3: Genes|Genomes|Genetics* 7, 1377–1383.

Tsairidou, S., Woolliams, J.A., Allen, A.R., Skuce, R.A., McBride, S.H., et al. (2014) Genomic prediction for tuberculosis resistance in dairy cattle. *PLoS ONE* 9, e96728.

Tsairidou, S., Anacleto, O., Woolliams, J.A. and Doeschl-Wilson, A. (2019) Enhancing genetic disease control by selecting for lower host infectivity and susceptibility. *Heredity* 122, 742–758.

Tsairidou, S., Hamilton, A., Robledo, D., Bron, J.E. and Houston, R.D. (2020) Optimizing low-cost genotyping and imputation strategies for genomic selection in Atlantic salmon. *G3: Genes|Genomes|Genetics* 10, 581–590.

Tully, O. and McFadden, Y. (2000) Variation in sensitivity of sea lice [*Lepeophtheirus salmonis* (Krøyer)] to dichlorvos on Irish salmon farms in 1991–1992. *Aquaculture Research* 31, 849–854.

Vallejo, R.L., Palti, Y., Liu, S., Evenhuis, J.P., Gao, G., et al. (2014) Detection of QTL in rainbow trout affecting survival when challenged with *Flavobacterium psychrophilum*. *Marine Biotechnology* 16, 349–360.

VanRaden, P.M. (2008) Efficient methods to compute genomic predictions. *Journal Dairy Science* 91, 4414–4423.

Wargelius, A., Leininger, S., Skaftnesmo, K., et al. (2016) Dnd knockout ablates germ cells and demonstrates germ cell independent sex differentiation in Atlantic salmon. *Scientific Reports* 6, 21284.

Willer, C.J., Sanna, S., Jackson, A.U., Scuteri, A., Bonnycastle, L.L., et al. (2008) Newly identified loci that influence lipid concentrations and risk of coronary artery disease. *Nature Genetics* 40, 161–169.

Yáñez, J.M., Houston, R.D. and Newman, S. (2014a) Genetics and genomics of disease resistance in salmonid species. *Frontiers in Genetics* 5.

Yáñez, J.M., Lhorente, J.P., Bassini, L.N., Oyarzún, M., Neira, R., et al. (2014b) Genetic co-variation between resistance against both *Caligus rogercresseyi* and *Piscirickettsia salmonis*, and body weight in Atlantic salmon (*Salmo salar*). *Aquaculture* 433, 295–298.

Yoshida, G.M., Carvalheiro, R., Lhorente, J.P., Correa, K., Figueroa, R., et al. (2018) Accuracy of genotype imputation and genomic predictions in a two-generation farmed Atlantic salmon population using high-density and low-density SNP panels. *Aquaculture* 491, 147–154.

Zenger, K.R., Khatkar, M.S., Jones, D.B., Khalilisamani, N., Jerry, D.R., et al. (2019) Genomic selection in aquaculture: application, limitations and opportunities with special reference to marine shrimp and pearl oysters. *Frontiers in Genetics* 9, 693.

CHAPTER 18

Sea lice – active monitoring methods for sea lice in the marine environment

Campbell C. Pert[1], Anna Harte[2], Alexander J. Kent[3,4], Toni Dwyer[5] and Martin Llewellyn[5]

[1]Marine Scotland Science, Aberdeen, UK
[2]London School of Hygiene and Tropical Medicine, London, UK
[3]Nottingham Trent University, Nottingham, UK
[4]Animal and Plant Health Agency, Surrey, UK
[5]University of Glasgow, Glasgow, UK

18.1 Introduction

The parasitic copepod *Lepeophtheirus salmonis* (Krøyer, 1837) is an ectoparasite that belongs to the suborder Siphonostomatoida, family Caligidae, and is one of at least 122 species of the *Lepeophtheirus* genus found worldwide (Walter and Boxshall, 2021).

The larval stages of *L. salmonis* are planktonic (Pike and Wadsworth, 1999). Development is temperature dependent, with Wootten et al. (1982) describing a development time of 46 h at 12°C from hatching to the copepodid stage. A tank-based study by Pert et al. (2012) indicated that the survival rate from egg to nauplius I was 97.1%, from egg to nauplius II 83.9%, and 57.3% to the infective copepodid stage. The settlement of *L. salmonis* larvae, and subsequent development, are affected by environmental conditions and the availability of a suitable substrate like many marine species – in this case a suitable salmonid host (Pike and Wadsworth, 1999). As *L. salmonis* transmission to the host is largely dependent on the copepodid stage which, like most free-living stages of parasitic copepods, is time limited by its stored energy supplies (Cook et al., 2010), it is important to determine the infection dynamics of *L. salmonis* copepodids (Pike and Wadsworth, 1999).

Sea lice infestations occur regularly on farmed fish (Murray, 2016; Abolofia et al., 2017; Hemmingsen et al., 2020). Several studies have linked the presence of fish farms to sea lice in the environment (Ford and Myers, 2008; Penston et al., 2008a; Torrissen et al., 2013; Serra-Llinares et al., 2014; Harte et al., 2017), subsequent infestations of wild salmonids (Middlemas et al., 2010; Vollset et al., 2016a) and detrimental population dynamics in marine aquaculture areas (Vøllestad et al., 2009; Vollset et al., 2016a). While there have been numerous studies that have shown correlations between fish farms and declines in adjacent wild salmonid populations (Bjørn et al., 2001; Butler, 2002; Krkošek et al., 2006; Shepard and Gargan, 2017), some studies have reported results that suggest otherwise (Brooks, 2005; Marty et al., 2010; Jackson et al., 2013). This creates an element of uncertainty with regard to the direct influence that salmon aquaculture may have on wild salmonids. The management and control of sea lice has therefore become an important issue in the Atlantic salmon farming industry, not only to control numbers of lice on farmed stock, but also to control the number of infective larvae being released into the wider environment and their potential to cause detrimental effects on wild salmonid populations (Butler, 2002; McKibben and Hay, 2004; Middlemas et al., 2010).

While there is extensive use of predictive models to estimate the spread of sea lice by epidemiologists to determine areas at risk and possible points of infestation, a number of authors have identified significant gaps in the biological knowledge of life cycle stage specific mortality events during their planktonic stages and during moulting for incorporation into population dynamic models (Stien et al., 2005; Salama et al., 2013). Therefore, these issues need to be urgently addressed to ensure the development of effective control strategies (Stien et al., 2005). Consequently, the use of models should be based and tested extensively using good-quality field-based biological data to ensure their robustness and accuracy prior to their use for effective pest management strategies as well as policy decisions by regulatory bodies (Stien et al., 2005; Murray and Gillibrand, 2006; Amundrud and Murray, 2009; Adams et al., 2012; Jackson et al., 2012; Salama et al., 2013, 2018; Asplin et al., 2014).

At present, monitoring of sea lice presence and abundance within salmonid pens is carried out using several methods. Direct counting of attached life cycle stages is the most widely deployed monitoring approach, and threshold levels often set out by regulators or veterinarians commonly refer to these abundances, or those of ovigerous females (Heuch et al., 2011). Direct counting can be invasive, and a poor reflection of the actual infestation pressure (Sandvik et al., 2020). Furthermore, model-based assessments of louse transfer between farms are based on planktonic louse abundances (for example, used by Sandvik et al., 2020), and verifying these models relies on assessing the abundance of these life cycle stages. Counts of planktonic lice, while more epidemiologically informative, are arguably more challenging and time consuming than direct counts on hosts. Direct counts from the zooplankton can be made via plankton tows, automated plankton identification tools (for example, techniques used by Benfield et al., 2007) and traditional microscopy. Automated fluorescence microscopy can augment the accuracy of these counts by exploiting the natural autofluorescence of the cuticle of the nauplii and copepodid stages. On this basis, progress has now been made towards the automation of nauplii detection from several louse species from the plankton using distinct emission spectra (for example, used by Nielsen et al., 2019). Sentinel fish have also previously been deployed (McVicar et al., 1993; Costelloe et al., 1995; Finstad et al., 2007; Bjørn et al., 2011; Pert et al., 2014a, 2014b), where small cages of susceptible naïve fish are deployed at defined geographic sites and louse counts made on small numbers of post smolt salmonids placed inside them.

The key to managing sea lice parasitism in Atlantic salmon producing areas is to understand the dispersal and subsequent distribution of viable lice within the relative confines of inshore waters which contain both wild and farmed salmonids. The collection of high-quality data utilising field-based environmental sampling techniques is proving increasingly important. This chapter considers three approaches and methodologies that have been developed and, in many ways, are still developing to monitor the presence of sea lice and possible estimates of infestation pressure in the marine environment, specifically: zooplankton sampling, sentinel fish held in cages and eDNA screening.

18.2 Zooplankton sampling methodologies for sea lice

Plankton sampling can provide a quantitative measure of the numbers of infective planktonic sea lice stages present in the environment, providing useful information on the actual abundance and the likelihood of infection of wild and farmed salmon (Penston et al., 2002; McKibben and Hay, 2004). Numerous environmental factors can influence the abundance and distribution of planktonic sea lice in a given area (Brooks, 2005; Krkošek et al., 2005), and different plankton sampling routines can lead to varying results depending on the method employed.

Environmental variables shown to influence sea lice fecundity and aggregation include factors such as temperature, wind direction and salinity (Boxaspen and Næss, 2000; Brooks, 2005; Bricknell et al., 2006; Amundrud and Murray, 2009). Warmer sea temperatures reduce developmental times of copepodid sea lice, resulting in more generations of lice in warmer climates during the spring, summer and autumn (Costello, 2006), whereas *L. salmonis* copepodids experienced significantly reduced survival and lower rates of successful settlement when salinity was below 29 ppt (Bricknell et al., 2006). Environmental variables that may affect the physical aggregation of larvae, such as local wind conditions, have also been shown to be important in determining larval abundance, with wind direction influencing surface currents and thus the direction and speed of larval transport (Amundrud and Murray, 2009). Other abiotic factors that have been demonstrated to influence planktonic sea lice abundance include topography and light levels (Amundrud and Murray, 2009; á Norði et al., 2015; Saksida et al., 2015).

Alongside abiotic factors that could influence sea lice abundance, other biotic factors such as human activities utilising the marine environment should be considered, including industries such as Atlantic salmon farming in inshore waters. Penston et al. (2002) investigated planktonic sea lice in offshore and sub-littoral samples taken from Loch Shieldaig, part of the Loch Torridon sea loch system on Scotland's west coast. The number of gravid sea lice on farmed fish correlated closely with the number of larval sea lice in plankton samples in offshore and sub-littoral locations in Loch Shieldaig (Penston et al., 2002). Subsequent studies carried out in the same location provided further evidence that in Loch Torridon, planktonic sea lice densities are highly dependent on the number of farmed fish that are present (Penston et al., 2004, 2008a, 2008b; Penston and Davies, 2009). However, there were also high levels of temporal variability in sea lice abundance (Penston et al., 2008a), and larvae were transported several kilometres from the point source (Penston et al., 2008b). This indicates that even if fish farms were the sole source of sea lice larvae, environmental variables may also influence their spatial and temporal distribution.

Therefore physical sampling of plankton is important to gain insight into the relative abundances of sea lice. The influence of these environmental variables suggests that accurately recording sea lice levels within the plankton may be confounded by factors such as wind speed, direction and benthic topography, and the method used to collect the samples should be carefully considered.

18.2.1 Methods of plankton sampling

Many methods and techniques can be employed to collect plankton samples, with each having advantages and trade-offs. The most common methods for collecting plankton samples for sea lice studies are horizontal or vertical hauls (Morton et al., 2011; Harte et al., 2017; Byrne et al., 2018; Moon and Park, 2019). These involve towing a fine-mesh net horizontally behind a vessel (Fig. 18.1), at a consistent depth across a set transect, or dragging it vertically upwards from a set depth to the surface. Oblique tows, where nets are lowered to a certain depth and then towed at an angle of around 45° upwards to the surface (Omori et al., 1965), are also used to provide a wider representation of the planktonic abundance in the full water column in a defined area (Gjøsæter et al., 2000). Another method is the use of an automated pump system, where an automated plankton sampler is submerged

Fig. 18.1 The deployment of plankton nets to detect sea lice, particularly the nauplius and copepodid stages, can give some indication of the possible infestation pressure in the sample area. In the image above (left) a 'Bongo net' is deployed from a vessel on Scotland's west coast. On the right (centre of image) is an *L. salmonis* copepodid found during screening of plankton samples.

and attached to a buoy or other such static marine object, and seawater is actively pumped through a net containing a range of mesh sizes (Waite and O'Grady, 1980). Automated pumps were developed specifically for collecting the two main parasitic sea lice species found in UK waters (Kilburn et al., 2010), and successfully detected the presence of *L. salmonis* in a substantial proportion of samples.

The sampling efficiencies of four different techniques to collect *L. salmonis* from the water column were investigated in Norway, named as 'go-flo', 'pump', 'vertical haul' and 'horizontal haul', with compared estimates being produced by existing hydrodynamic and particle tracking numerical models (Skarðhamar et al., 2019). Go-flo is a specialised water bottle with a close-open-close operation, which enables samples of a set volume of water to be collected at a specific depth, without contamination from the microlayer on the surface or shallow water contaminants (Flanders Marine Institute, VLIZ). The study demonstrated high variation in lice counts between all four techniques, with modelling estimates consistently lower than observed. However, models had high accuracy with regard to the fluctuations in abundance and concentrations over time (Skarðhamar et al., 2019). Depth was shown to be an important factor, with higher lice counts in water samples taken from the 0–3 m depth range, compared to the 5 m and 10 m samples.

Many of these relatively simple sampling methods provide accurate and spatially extensive datasets when monitoring is performed regularly over a long period of time. In Scotland, for example, Marine Scotland Science (MSS) collects plankton samples on a weekly basis at a set distance offshore from Stonehaven on the east coast of Aberdeenshire, using a fine-mesh plankton net towed horizontally and parallel to the coast, at a depth of 1 m (Harte et al., 2017). Additionally, a sampling programme focused primarily on sea lice has been running for over 15 years in Loch Shieldaig, west Scotland (Harte et al., 2017). The samples collected using towed plankton nets provided a consistent measure of planktonic sea lice levels across a relatively wide area. However, as depth and distance from shore is kept within a strict range for both sites, differences between sea lice abundance at different depths, and different benthic topographies, were not measured.

Less commonly used plankton sampling methods are also available, such as sampling sea lice specifically with the use of an LED/white light trap (Novales Flamarique et al., 2009). Sea lice have been shown to respond to the presence of light and, in the case of copepodids, to move towards it, a biological response termed phototaxis (Novales Flamarique et al., 2000). Recent technological developments include automated identification devices such as the plankton image analysis (PIA) system, which continuously pumps water, takes images of the passing particles and identifies the zooplankton organisms present (Pitois et al., 2018). A video plankton recorder (VPR) uses a high-resolution digital camera to capture and automatically identify plankton between 100 μm to 10 mm in size to major taxa (Davis et al., 2005). A semi-automatic towing sampler method, the continuous automatic litter and plankton sampler (CALPS), has also been developed. CALPS uses a pump system to sample at a set depth, using up to six nets of different mesh sizes along a predetermined transect as the ship sails, enabling the collection of a wide range of size fractions of plankton (Pitois et al., 2016).

An alternative approach to the labour-intensive and time-consuming manual screening of samples by microscopy has been developed which utilises metabarcoding of whole samples (Abad et al., 2016). DNA metabarcoding is a process involving the sequencing of a short section of a specific gene. The variations within that gene enable the identification of an organism to species level. A study which compared the number of species in samples from the continuous plankton recorder (CPR), identified using the traditional microscopy techniques, with those indicated by DNA metabarcoding found that overall DNA metabarcoding increased the number of species found (Deagle et al., 2018). This result aligns with the outcome of a double-blinded trial comparing microscopy with molecular identification, which demonstrated that microscopy overestimated the abundance of *L. salmonis* while underestimating *Caligus elongatus* (McBeath et al., 2006). One possible reason for this is the difficulty associated with identifying the species in nauplii stage. The best means of differentiating between species at this stage uses pigmentation characteristics, but fixative solutions can often bleach samples and remove natural pigmentation. Microscopy also requires a highly skilled morphologist, the time required for screening a diverse environmental sample, and potential unintentional bias introduced by pre-existing knowledge. For example, one of the four plates classified as negative for *L. salmonis* using microscopy was found to be positive using PCR (McBeath et al., 2006). This particular sample came from an aquaculture facility culturing Atlantic cod (*Gadus morhua* L.) and, as cod are not known hosts to *L. salmonis*, these samples would not be expected to contain *L. salmonis*. Assuming that most environmental sampling studies would not be performed using the double-blind trial method utilised by McBeath et al. (2006), it is possible that prior knowledge of the likelihood of the presence of a sea lice species could subconsciously bias results. Equally, as particular sea lice species have known host preferences (Boxaspen, 2006), subjective identification could result in rare instances of *C. elongatus* being identified as *L. salmonis* and vice versa, particularly for the nauplii stages as it can be difficult to differentiate between the two species. However, molecular techniques have their own issues. Deagle et al. (2018) showed that the presence of DNA from large species such as krill tended to mask the DNA of much smaller species. This could be problematic when looking for species with low abundance.

Another interesting issue presented by metabarcoding is highlighted in *Caligus elongatus*. A study investigated the use of DNA barcodes for the identification of crustaceans in the North Sea (Raupach et al., 2015) and found that this particular species is made up of two sibling species, with intraspecific distances of 14.87% between the two monophyletic clusters. This highlights the information that can be gained using new technologies, as well as creating potential problems with regard to selecting the correct region of the genome to use for identification to species level. This trade-off between the sensitivity and specificity of molecular versus microscopy techniques has led to a suggestion of a hybrid approach, where automated microscopy is used to select and separate out caligid copepods, and PCR is used to confirm the species (McBeath et al., 2006). This combination of methods would be considered best practice and would limit the overestimation and misidentification of important lice species.

18.2.2 Issues and limitations utilising plankton sampling

All of these techniques provide useful measures of sea lice abundance within the plankton. However, there is always a trade-off between methods with regard to specificity versus the volume of water sampled. One of the main limitations of field research is that measuring plankton abundance across all depths and areas of interest is not realistically feasible due to time and financial constraints. Decisions must be made regarding the most important aspect of any study and the best measurement to answer the research question. Depth of sampling, for example, is an important factor to be considered, as sea lice are generally found in the surface layers (Pike and Wadsworth, 1999). Issues can occur such as weather conditions affecting both mixing of the water column and loss of sample volume, as high waves would mean a deeper tow would be necessary to ensure the net never breaches the surface and thus a consistent volume of water is sampled. The choice of net-mouth diameter is related to this as a larger mouth will mean a deeper sample in a surface trawl relative to a smaller mouth surface trawl.

18.2.3 Future development around the methodology and improvements

Sampling consistency is key in producing good quality and comparable data sets. Therefore, sampling efforts should ideally focus on one method that can be used in a wide range of environments and weather conditions (Brooker et al., 2018). Skarðhamar et al. (2019) demonstrated that while a horizontal tow was the most used method, it was not appropriate for turbulent sea conditions. The authors suggested that vertical net hauls and pumping water from a specific depth to be the most feasible methods in all conditions, and with lower variability between repeated samplings.

An alternative method suggested for monitoring sea lice abundance is the use of chemical language, or 'infochemical communication' to attract parasites in the local area, and thus measure their relative abundance in a set location (Saha et al., 2019). Sea lice are attracted to chemicals produced by salmon and move towards this chemical (Bailey et al., 2006). Studies on *L. salmonis* showed activation and attraction to salmon conditioned water (SCW) and SCW extracts, where SCW water was purified using solid-phase extraction techniques to identify low molecular weight components of fish odours (Ingvarsdóttir et al., 2002). One isolated component, isophorone, was tested alone in a slow-release system and induced significant movement of sea louse towards vials containing isophorone over control chemicals in tank trials (Ingvarsdóttir et al., 2002). The use of this chemical as a parasite attractant would, however, require synthesising the chemical in large quantities and runs the risk of attracting many non-target parasitic copepods that also utilise generic attractants such as isophorone in locating hosts. Morphological identification and hence accurate determination of *L. salmonis* larvae is potentially more difficult. While environmental factors such as current and wind direction would likely still heavily influence the presence of planktonic sea lice, in a similar fashion to that of the light trap method, a chemical attractant could have the potential to aggregate those sea lice within a set range into a smaller area, and thus reduce the risk of data loss.

Existing advancements in technology beyond the simple plankton net hauls as described here, such as automated image analysis, have the potential to improve the amount of data available and the area of sea that can be covered. However, regardless of the efficiency of these methods, the same issues apply to these new technologies as to the traditional haul methodologies, with considerations such as physical placement within the water column. Alternative methods of sea lice quantification that can surmount these issues, such as the use of sentinel fish in small cages deployed at strategic locations, may be more suitable for future studies, depending on abiotic and biotic variables.

18.3 Sentinel cage sampling for sea lice

As identified in the previous section, one of the main challenges when estimating infestation pressure that both wild and farmed salmonids may experience in inshore waters from sea lice utilising point sampling methodologies, such as plankton nets and pumps, is that the data tends to be indicative, but are typically variable in nature. One of the main issues of these methodologies is that they fail to take account of high spatio-temporal variability in abundance, unpredictable currents and the viability of larvae and how these might vary for larvae in different physical environments (Johnson and Albright, 1991; Gaines and Bertness, 1993; Penston et al., 2004, 2008a, 2008b; Bricknell et al., 2006; Penston and Davies, 2009; Kilburn et al., 2010).

A more efficient method for measuring numbers of viable planktonic larvae is the provision of a suitable habitat for settlement (Moksnes and Wennhage, 2001). The deployment of cages containing sentinel fish has previously been utilised to determine the presence of parasites such as the monogenean *Benedenia seriolae* (Yamaguti 1934) and the myxosporean *Myxobolus cerebralis* (Hofer 1903) within the aquatic environment (Sollid et al., 2004; Chambers and Ernst, 2005). As *L. salmonis* is largely specific to salmonid hosts (Pike and Wadsworth, 1999), the use of Atlantic salmon in field trials should provide a natural settlement opportunity for the parasite within the water column, with observed settlement levels allowing estimation of sea lice infestation pressure in the locality (Pert et al., 2014a).

18.3.1 Methods of sampling sentinel cages

Sentinel cages have been deployed at specific locations in Ireland, Scotland, Canada, USA and Norway to investigate infestation pressure from sea lice, in particular *L. salmonis* (Tully, 1989; McVicar et al., 1993; Costelloe et al., 1995; Finstad et al., 2007; Bjørn et al., 2011; Pert et al., 2014a, 2014b; Frederick, 2018).

Sentinel cages (Fig. 18.2) are typically small mesh fish cages which are deployed at predetermined locations and anchored to the sea bed. Those deployed in Maine, USA, (Frederick, 2018) were relatively large (2 m by 2 m) square metal cages compared with those deployed extensively in Norway's Hardanger fjord (Bjørn et al., 2011). These were smaller circular cages measuring 0.8 m in diameter and 0.9 m in length. The 'Scottish design' which has been used in Scotland by MSS since 2006, and also in trials in Ireland and Canada, is a cylindrical cage measuring 1.5 m in diameter and 2 m in depth with the cage structure being supported by four rings made of polyethylene pipe (Pert et al., 2014a; Salama et al., 2018). An important addition to all the cage designs described here is knotless mesh netting as this ensures there is no abrasion which could remove sea lice from the epidermis of sentinel salmon.

Parasitic infection intensity, as well as the succession in generations of *C. elongatus* and *L. salmonis*, were examined on sentinel fish (Tully, 1989). Costelloe et al. (1995) also investigated the levels of

Fig. 18.2 Sentinel cages are typically small mesh fish cages which are deployed at predetermined locations anchored to the seabed. The cage featured in the image was of a design deployed at a number of locations in sea lochs located on the west coast of Scotland.

infestation in four cages stocked with Atlantic salmon located at varying distances between a salmon farm and a river system in Killary Harbour, on Ireland's west coast. The highest levels of settlement occurred on sentinel fish located closest to the river. It was proposed that the initial infections may have originated from wild fish. The farms subsequently were self-infected and, considering topography and hydrographical data for the region, led to the authors concluding that the sea lice originated from more than one source.

The first use of sentinel cages to investigate sea lice infestation pressure in a Scottish sea loch was carried out on the west coast by McVicar et al. (1993). The study involved deploying sentinel cages at three locations with each of the three cages containing sea trout (*Salmo trutta* L.) smolts at distances ranging from 0.2 km to 19.31 km from marine Atlantic salmon fish farms. The sea trout in cages closest to a fish farm in Little Loch Broom had significantly higher burdens of sea lice than two locations in Holy Loch and Loch Eribol (McVicar et al., 1993).

A second study in Scotland utilised sentinel cages containing Atlantic salmon smolts to estimate the monthly infestation pressure of sea lice at three locations along a salmonid migration route (Pert et al., 2014). During the 30-month study, covering a production cycle and fallow period at nearby fish farms, over 5000 sea lice were sampled and over 99% of these were *L. salmonis*. Infestation pressure was higher in cages nearer the head of the loch and the River Shieldaig estuary and *L. salmonis* copepodids were the main infective life cycle stage recovered, although a few (0.5% of total lice sampled) were mobile pre-adult and adult stages, which were also found on sentinel fish. Infestation pressure on the sentinel fish was correlated with numbers of gravid *L. salmonis* on nearby farmed salmon.

An extensive study carried out in Hardanger fjord, Norway, (Finstad et al., 2007) involved placing 15 sentinel cages containing Atlantic salmon at strategic locations covering almost the entire length of the fjord to determine the infection pressure on wild fish at specific points within the system. The settlement by *L. salmonis* was low on sentinel fish and comparable to levels found on migratory post smolts with the exception of the Etne/Ølen area of the fjord.

In another Norwegian study sentinel caged fish had significantly higher infestation levels in intensively farmed areas compared to a protected zone (Bjørn et al., 2011), although there were no significant differences in gravid lice numbers on farmed fish inside and outside of the protected zone. This suggests that sentinel caged fish may provide a more informative measure of infestation pressure within an inshore marine system than individual farm counts. Therefore, the deployment of sentinel cages is a useful tool for gathering data on the infestation pressure on wild salmonids, at fixed points, within the aquatic environment and in relation to environmental conditions (Costelloe et al., 1995; Finstad et al., 2007; Bjørn et al., 2011; Pert et al., 2014a, 2014b).

18.3.2 Issues and limitations of the sentinel cage methodology

The infestation pressure at a given location depends on the sea lice sources in the area, dispersal from these sources and the survival and viability of sea lice delivered to that location (Pert et al., 2014a). It is difficult to quantify whether the infestation pressure at a specific fixed sentinel cage site is representative of the wider infestation pressure in the locality given the spatial and temporal variability in sea lice dispersal and concentration (Penston et al., 2008a; Amundrud and Murray, 2009; Pert et al., 2014a). Therefore, this still requires some evaluation, and taking account of wild fish movements, often migrating rapidly, through inshore areas (Thorstad et al., 2004; Middlemas et al., 2017). Even with a few monitoring locations, the resources required to routinely deploy sentinel cages can be considerable. As such, there is a need for novel approaches to simulate the actual infestation pressure experienced by wild migrating salmon through inshore areas containing active salmon farms.

18.3.3 Future development of the sentinel cage methodology

It is often asked how representative is the fixed sentinel cage methodology with regard to the infestation pressure on migrating smolts in the immediate and wider area of the cage's deployment? Therefore, an important development of the sentinel cage technique would be to try to replicate under experimental field conditions, as far as reasonably practicable, the infestation pressure on wild smolts as they migrate through inshore waters containing marine aquaculture by towing them in adapted cages. Towing cages of salmon could potentially be used to determine specific points of infestation within the marine environment and provide important data for future particle tracking models by identifying specific locations along wild salmonid migratory routes where salmonids, particularly smolts, are possibly at risk as a result of high levels of infestation pressure from sea lice.

Atlantic salmon smolts have previously been towed from the estuary of the river Vosso, Norway, to points along a long fjordic system towards the open sea to avoid mortalities from predation (Vollset et al., 2014, 2016b). There have been studies looking at the effect of migration behaviour on mortality (for example, Vollset et al., 2016b). However, no study has been carried out to design and deploy equipment to attempt to better measure the specific infestation pressure Atlantic salmon smolts may experience from sea lice as they enter the marine environment and actively migrate down a fjordic-like system towards the open sea in a region containing salmonid aquaculture facilities.

Several field-based pilot studies were carried out in Loch Linnhe (Fig. 18.3), west Scotland, to assess the representativeness of fixed cages (Pert et al., 2017). The loch is approximately 60 km long, and had 11 active salmon farms with a consented biomass of 17,352 tonnes during the study.

Between 2015 and 2018 a range of field-based pilot trials investigated adapted fixed sentinel cage methodology with a towed sentinel cage array (Fig. 18.4) stocked with farmed Atlantic salmon smolts (Pert et al., 2017). Another aim was to determine if sea lice within the water column would infest

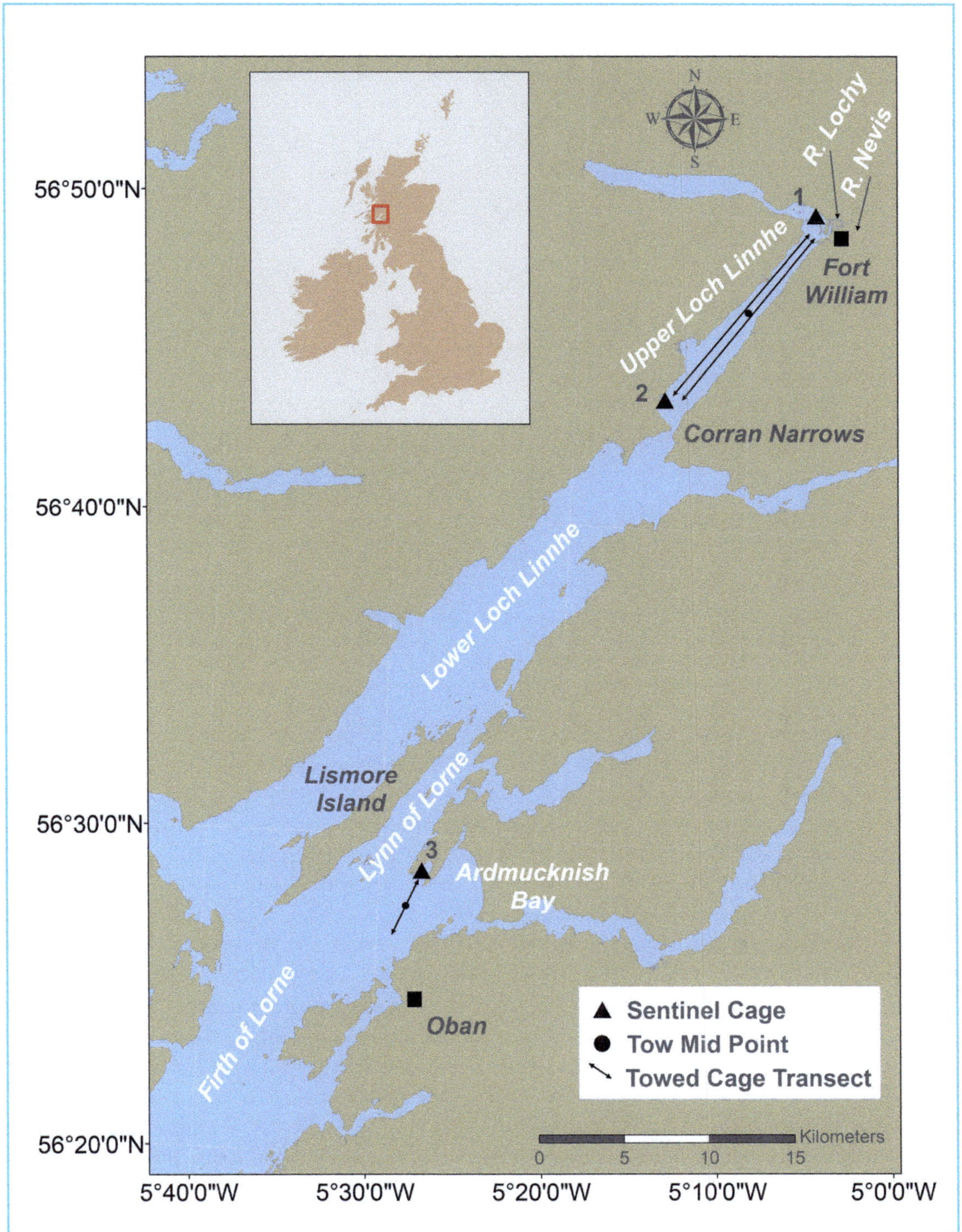

Fig. 18.3 Map of Loch Linnhe, west coast of Scotland, indicating the location of towed sentinel cage transects, sentinel cage sites and important geographic place names.

Fig. 18.4 The towed sentinel cage comprised a rigid steel ring (left) onto which a three point bridal was attached to prevent deformation during towing (right) with the other rings comprising polyethylene pipe (38 mm inside diameter × 51 mm outside diameter) with 10 mm holes drilled at regular intervals to ensure the cage sank to a target depth of 2.5 m with the aid of flotation buoys when deployed.

sentinel salmon inside the towed cage and also to compare towed sentinel cage data with that of static sentinel cages that had been extensively used since 2006 (Pert et al., 2014).

A pilot study compared sea lice settlement on fish in fixed sentinel cages (FSC), towed sentinel cages (TSC) and plankton net trawls (Pert et al., 2017). As sampling trials were undertaken for different durations, settlement was calculated on numbers of sea lice collected per hour by each methodology. This demonstrated that salmon in FSCs had the highest level of settlement at 1.28 sea lice/h, TSCs 0.86 sea lice/h and plankton trawls 0.96 sea lice/h. However, issues were identified with the towing speed which, due to the size of the vessel, could not be reduced to speeds equivalent to smolt swimming speeds (Thorstad et al., 2004; Samsing et al., 2015).

Further pilot trials in Loch Linnhe between 2016 and 2018 (Pert et al., unpublished) utilised smaller, twin engine rigid inflatable boats (RIBs) to reduce the towing speed as much as practicable. Similar to the previous trials sentinel salmon held in FSCs had a higher rate of settlement (0.23 sea lice/h) compared to TSCs (0.19 sea lice/h). The higher numbers of settled sea lice on FSC fish could be related to the fact that the towed cage salmon were constantly moving forward, potentially reducing the opportunity for successful settlement by sea lice copepodids. This hypothesis is supported by Revie et al. (2003), in which current speed was identified as a significant variable in settlement numbers of *L. salmonis* copepodids. Samsing et al. (2015) also found that current velocities strongly influenced *L. salmonis* attachment with infestation levels twice as high in low-speed compared to high velocity currents. It is therefore likely that the average towing speed achieved during the second pilot trial of 39 cm s^{-1} was still considerably faster than the 'burst' speed of 9 cm s^{-1} of an *L. salmonis* copepodid (Heuch and Karlsen, 1997), and this could explain a lower settlement rate in the TSCs compared to FSCs. If these sea lice settlement rates are reflective of migratory salmonid exposure to sea lice, stationary salmon smolts are potentially exposed to a ~25% greater rate of settlement than during periods of active migratory swimming. This may be the case during periods of localised residency, for example during the acclimatisation period when smolts first enter the marine environment (Middlemas et al., 2017).

Therefore, the towed cage methodology, if developed further, could improve estimation of the accumulative infestation pressure experienced by salmonid smolts as they migrate through fjordic-like inshore systems towards the open sea. The study demonstrated that the TSC methodology is capable of detecting infestation pressure at different locations on migrating, rather than static, salmon, and represents a potentially versatile, quickly deployable tool to increase robustness of data when determining sea lice infestation pressure. However, newer techniques, such as environmental DNA (eDNA) analysis, are being developed and may provide even more rapid indications of sea lice densities at lower cost in the future.

18.4 eDNA-based monitoring for sea lice

There have been significant advances in the development and application of DNA techniques within environmental monitoring and research over the last 15 years. The use of eDNA analysis is a tool that can be applied to the monitoring of sea lice presence and infestation pressure which does not rely on direct sea lice larval counts or microscopy associated with traditional zooplankton sampling and sentinel fish. eDNA analysis relies on genetic material, which is ubiquitous in both modern (seawater, freshwater, soil and air) and ancient (sediment, ice or permafrost) natural environments (Thomsen and Willerslev, 2015). This can be present in the form of whole cells, extracellular DNA and potentially whole organisms (Ruppert et al., 2019). eDNA is the direct retrieval of DNA from environmental material and implicitly assumes that the whole organism is not present in the sampled medium (Taberlet et al., 2012). Therefore eDNA identifies trace DNA from sources such as faeces, mucus, skin cells, organelles, gametes or even extracellular DNA (Deiner et al., 2017). This differs from community DNA, whereby targeted groups are collected in bulk (for example, soil, malaise trap or net) and individuals are removed from other sample debris and pooled prior to bulk DNA extraction (Creer et al., 2016). Many sampling methodologies collect both types of DNA (community and true 'eDNA') in parallel.

The use of eDNA dates from 1987, when the first extraction protocol for eDNA in sediments was documented (Ogram et al., 1987). The first DNA metabarcoding study was published three years later (Giovannoni et al., 1990). The development and commercialisation of high-throughput next-generation sequencing (NGS) technologies since 2004 have multiplied sequencing capabilities in comparison to first-generation DNA sequencing techniques (for example, chain termination and fragmentation techniques; Sanger et al., 1977; Maxam and Gilbert, 1977), and stimulated significant activity using metabarcoding and metagenomics. DNA metabarcoding had expanded to macro-organisms by the late 2000s, using DNA within faeces for dietary analyses (Deagle et al., 2010). More recently, a series of studies on freshwater macro-organisms has involved both single-species detections (barcoding) (Dejean et al., 2012), and the identification of multiple taxa through metabarcoding (Thomsen et al., 2012; Valentini et al., 2016). Environmental DNA analysis is now regularly deployed (Fig. 18.5) to detect and monitor the presence of eukaryotic micro- and macro-microbial communities and populations (Ficetola et al., 2008; Zamor et al., 2012). eDNA metabarcoding of sediment communities as a means of monitoring environmental quality and environmental impact of farm sites is also an emerging field in aquaculture (Turner et al., 2015). Detection of pathogens and planktonic threats in freshwater (Gomes et al., 2017) and marine aquaculture settings (Peters et al., 2018) has also been suggested as a possible application of eDNA-based barcoding and metabarcoding.

Sea lice species are also a potential target for eDNA-based assessment of abundance and diversity for biomonitoring. Planktonic life cycle stages (nauplii, copepodids) are small enough to be filtered directly from the water column by conventional eDNA filters, while eDNA is likely to be regularly sloughed off host-associated life cycle stages during moulting. Molecular approaches have been used for louse detection, principally qPCR-based enumeration of louse DNA in the water column (McBeath et al., 2006; Ingebrigtsen, 2017). These approaches are reviewed here, along with the practical and technical challenges associated with them, and the prospects for future research.

18.4.1 Methods of eDNA sampling of sea lice

The use of eDNA in biomonitoring has been widely promoted in different ecosystems as a low cost, non-invasive tool (Thomsen et al., 2012; Goldberg et al., 2016; Deiner et al., 2017; Peters et al., 2018; Harper et al., 2019; Ruppert et al., 2019; Beng and Corlett, 2020). However, multiple practical and theoretical challenges exist in successfully applying the technique. Many of these technical hurdles apply to sea lice monitoring. First, the volume of water filtered must be standardised and optimised to achieve the necessary sensitivity. Large volumes of water may improve sensitivity, but standard 0.22 μm filters often clog rapidly with large particulates (Hunter et al., 2019). Secondary 0.45 μm pore size filters can be deployed to screen larger debris but can also filter planktonic organisms of interest and can significantly increase running costs. As such, DNA analysis of various filter grades may be necessary (Majaneva et al., 2018). Practically, larger volumes take longer to filter, and time could be limiting for farm staff. There is no consensus on volumes used in other systems (Moyer et al., 2014; Schultz and Lance, 2015; Mächler et al., 2016). eDNA yields can be improved using multiple filters in parallel (Hunter et al., 2019), and filtration of large water volumes compared to the standard 500–1000 ml range (Schabacker et al., 2020). DNA of sea lice is currently detected in the field from plankton tow filtrates (Boxaspen, 2006; Jevne et al., 2020). Although mesocosm experiments suggest that sea lice are potentially detectable in less than one litre of filtered sea water, it is unclear how relevant such volumes are to louse eDNA concentrations in the field (Peters et al., 2018). Once eDNA has been filtered from the water column, DNA must be purified from the sample. To achieve standardisation, many groups deploy silica column-based kits to allow standardisation and these allow DNA to be purified and PCR inhibitors are removed by washing steps (Sanches et al., 2020). This process usually takes several hours and is thus typically time-limited by the protein lysis step (Natarajan et al., 2016). Once DNA has been obtained, overall eDNA concentrations can be low and sensitive to degradation and samples must be rapidly processed or stored at low temperatures (Thomsen and Willerslev, 2015). PCR-based amplification and characterisation of target sequences occurs after DNA extraction. Quantitative or 'real-time' PCR (qPCR) involves oligonucleotide primers that target unique sequences characteristic of a given

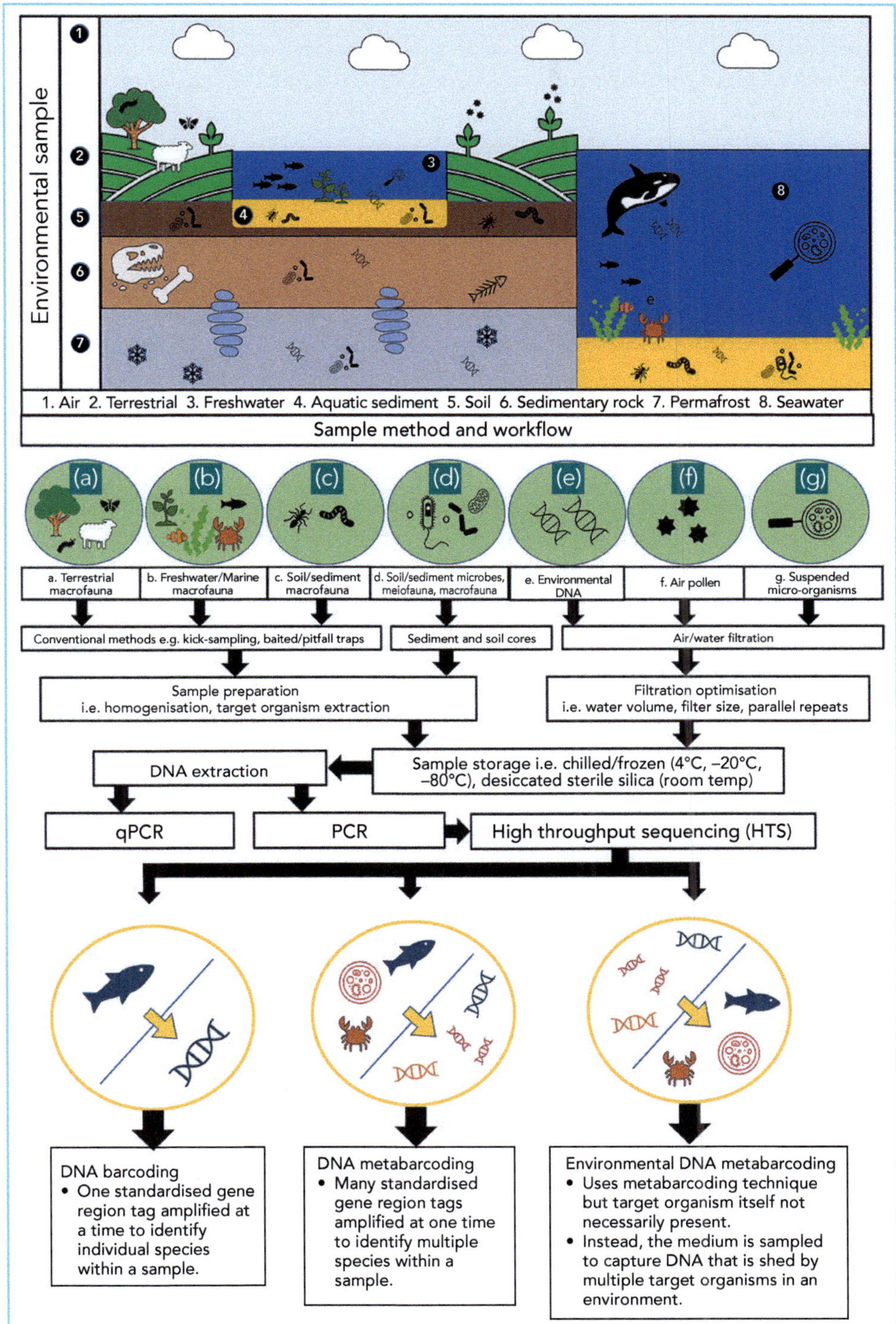

Fig. 18.5 Schematic representing the decisions and workflow involved from eDNA sampling to molecular analysis. Organisms or samples are collected by different means from different environments and then processed in distinctive ways in order to answer a specific biological question. Metabarcoding can be used to give an idea of the diversity and abundance of species across whole communities, while barcoding (and single-plex qPCR) targets one organism within a sample.

organism. The rate of accumulation of PCR product can be tracked during the reaction and used to estimate the abundance of the starting material. qPCR has been successfully deployed to estimate *L. salmonis* eDNA from the water around farms (Shea et al., 2020). As an alternative to qPCR, eDNA metabarcoding can also be used to characterise target sequences and has been applied to lice in mesocosms only (Peters et al., 2018). In this case a barcoding marker (several are in regular use, see later for further discussion) is amplified prior to massively parallel DNA sequencing. Sequences belonging to the target species are enumerated *in silico* with reference to a reference database. Both metabarcoding and qPCR approaches can provide some level of relative quantitation between samples, but the former performs poorly in terms of absolute quantification of starting material (Lamb et al., 2019). Whether deploying qPCR or metabarcoding approaches, eDNA studies must take care to standardise sample volumes and downstream protocols to ensure samples are comparable, as well as replicate measurements.

18.4.2 Challenges of using eDNA analysis to monitor sea lice

Sampling of sea lice eDNA offers several potential advantages over traditional monitoring approaches. There is ease and cost of sampling as water samples are relatively simple to collect and filter. Multiple samples can therefore be rapidly taken systematically across a water body (Zaiko et al., 2018). Invasive fish counts and microscopy for plankton tows are costly by comparison in time, money and expertise (Heuch et al., 2011; Berry et al., 2019).

Monitoring sea lice using eDNA, however, is currently in the early stages of development and faces several challenges. The stability of eDNA in the water column must be first considered and is dependent on multiple abiotic and biotic characteristics of the marine environment, for example temperature, turbidity, UV penetration, pH, salinity (Collins et al., 2018). Spatial-temporal variation in these factors can therefore influence species diversity and abundance estimates. Sample storage may also be difficult. Traditionally, eDNA samples have required a rigorous cold chain to maintain sample integrity (Williams et al., 2016). Recently, storage in silica gel desiccant has proven effective in the medium term as this simplifies sample storage at remote sites (Majaneva et al., 2018).

As previously mentioned, DNA extraction from filters is time consuming, costly and requires specialist equipment – especially where commercial DNA extraction kits are deployed. As such, onsite processing of eDNA samples, especially with a high degree of standardisation, is almost impossible. Marker-based detection of sea lice using qPCR assays can be both sensitive and highly specific, if marker loci are identified and suitably validated (Takeuchi et al., 2019). Similarly, careful choice of metabarcoding loci can identify short (150–300 bp) DNA fragments that can successfully provide species-level taxonomic resolution for sea lice (Makiola et al., 2020) and potentially identify several louse species from a sample simultaneously. However, while both sensitive and specific, PCR-based diagnostics are very sensitive to target DNA contamination and resultant false positives (Makiola et al., 2020), and specialised laboratories are required for sample processing (for example clean rooms, and decontamination protocols). Furthermore, DNA sequencers, qPCR thermocyclers and other equipment required for sequence characterisation may be beyond the scope of routine analysis. As such, the expense and expertise required to implement eDNA analyses using current technologies likely prohibit the adoption of this approach as a routine diagnostic. Finally, the inability of this approach to distinguish between different life cycle stages may also impede adoption as a monitoring tool without thorough prior benchmarking against traditional monitoring approaches.

18.4.3 Future development around routine eDNA monitoring of sea lice

Routine sea louse monitoring using current eDNA methods may not be possible. There are recent innovations in portable DNA detection methodologies that may support development of routine molecular diagnostics. Several manufacturers offer portable real-time PCR instruments, some with bespoke rapid DNA extraction methodologies (for example, https://biomeme.com/), although these are expensive. Sequencing platforms are increasingly being miniaturised to achieve portability (for example, that found at: https://nanoporetech.com). However, while the sequencers are compact, sequence library preparation remains time consuming and expensive. Paper-based portable DNA diagnostics, deployed in the developing countries to detect infectious disease agents in humans (for example, Reboud et al., 2019), offer some promise, especially as the loop-based isothermal DNA amplification (LAMP) deployed by such devices negates the necessity for PCR thermocyclers. Accurate DNA quantitation using paper-based devices remains a distant goal.

DNA sensor technologies – autonomous eDNA sampling systems – also hold significant promise (Yamahara et al., 2019). In such cases, louse abundances might be monitored in real time alongside physical features of the marine environment around farms such as temperature, pH and salinity. The development of scalable, quantitative and multi-species DNA detection sensors is likely to be supported by recent innovations in CRISPR-CAS methodologies (Ackerman et al., 2020), and these have recently been deployed to detect salmon eDNA (Williams et al., 2019).

Biomonitoring using eDNA has enjoyed an explosion of interest in recent years. The technique has potential for monitoring sea lice populations. However, given the status of the technology in terms of cost and portability, as well as the lack of the extensive benchmarking required against traditional counting methodologies, routine deployment of eDNA techniques to monitor lice is still some way off. Fortunately, technological advances, both in DNA detection and sequencing, now pave the way for the development of tools for routine deployment. As such, research investment in eDNA benchmarking trials for sea lice is now needed to deliver a potentially disruptive technology to enhance the sustainable growth of the aquaculture sector.

18.5 Conclusions

The global salmon farming industry continues to grow and, despite a range of chemical treatments and novel equipment, sea lice infestation of wild and farmed stocks of salmonids remains a significant problem (Middlemas et al., 2010; Vollset et al., 2016a; Shepard and Gargan, 2017). Monitoring the effects of sea lice life stages on farmed and wild salmon is therefore important for both industry and researchers. The methods described in this chapter cover an array of techniques available for investigating sea lice abundance and infestation pressure in the marine environment ranging from well-established plankton towing and sentinel cage techniques that contribute to long-term and geographically broad datasets, to the peak of modern technology with the application of NGS sequencing in eDNA metabarcoding. These techniques all come with their own benefits, as well as significant drawbacks, both relating to accuracy, cost and feasibility. Consequently, there are often trade-offs to be made with regard to the size of the area or volume of water to be tested, so any study will likely have to compromise between coverage and accuracy.

If resources are not a limiting factor, the utilisation of all three methodologies covered in this chapter would be the optimal strategy thereby providing data on numbers of attached lice through sentinel cages, direct counts of larval stages from plankton tows and estimates of overall abundance through eDNA sequencing. However, due to the logistical and considerable resource implications this would entail, this would unlikely be feasible. A more realistic approach would be to identify the factors most important to the study and select a suitable method. Therefore, a realistic compromise would be to utilise either the sentinel cage or plankton methodology in combination with the developing eDNA technology to generate good, and reliable, data to identify levels of infestation pressure from sea lice in the marine environment.

The generation of reproducible data from field studies allows the further development, and accuracy, of predictive hydrodynamic particle transport models to determine the interactions between sea lice and salmonid stocks both wild and farmed. Output from these models can then be utilised by science, industry and governmental bodies for the development and optimisation of effective pest management strategies and policies to ensure the sustainability of both wild and farmed salmonid stocks for future generations.

References

á Norði, G., Simonsen, K., Danielsen, E., Eliasen, K., Mols-Mortensen, A., Christiansen, D. H., Steingrund, P., Galbraith, M. and Patursson, Ø. (2015) Abundance and distribution of planktonic *Lepeophtheirus salmonis* and *Caligus elongatus* in a fish farming region in the Faroe Islands. *Aquaculture Environment Interactions* 7, 15–27.

Abad, D., Albaina, A., Aguirre, M., Laza-Martínez, A., Uriarte, I., Iriarte, A. and Estonba, A. (2016) Is metabarcoding suitable for estuarine plankton monitoring? A comparative study with microscopy. *Marine Biology* 163(7) 1–13.

Abolofia, J., Asche, F. and Wilen, J.E. (2017) The cost of lice: quantifying the impacts of parasitic sea lice on farmed salmon. *Marine Resource Economics* 32, 329–349.

Ackerman, C.M., Myhrvold, C., Thakku, S.G., Freije, C.A., Metsky, H.C., Yang, D.K., Simon, H.Y., Boehm, C.K., Kosoko-Thoroddsen, T.S.F., Kehe, J. and Nguyen, T.G. (2020) Massively multiplexed nucleic acid detection with Cas13. *Nature* (582) 277–282.

Adams, T.P., Black, K., MacIntyre, C., MacIntyre, I. and Dean, R. (2012) Connectivity modelling and network analysis of sea lice infection in Loch Fyne, west coast of Scotland. *Aquaculture and Environmental Interactions* 3, 51–63.

Amundrud, T.L. and Murray, A.G. (2009) Modelling sea lice dispersion under varying environmental forcing in a Scottish sea loch. *Journal of Fish Diseases* 32, 27–44.

Asplin, L., Johnsen, I.A., Sandvik, A.D., Albretsen, J., Sundfjord, V., Aure, J. and Boxaspen, K.K. (2014) Dispersion of salmon lice in the Hardangerfjord. *Marine Biology Research* 10, 216–225.

Bailey, R.J., Birkett, M.A., Ingvarsdóttir, A., Mordue, A.J., Mordue, W., O'Shea, B., Pickett, J.A. and Wadhams, L.J. (2006) The role of semiochemicals in host location and non-host avoidance by salmon louse (*Lepeophtheirus salmonis*) copepodids. *Canadian Journal of Fisheries and Aquatic Sciences* 63, 448–456.

Benfield, M.C., Grosjean, P., Culverhouse, P.F., Irigoien, X., Sieracki, M.E., Lopez-Urrutia, A., Dam, H.G., Hu, Q., Davis, C.S., Hansen, A., Pilskaln, C.H., Riseman, E.M., Schultz, H., Utgoff, P.E. and Gorsky, G. (2007) RAPID: Research on automated plankton identification. *Oceanography* 20(2) 172–187.

Beng, K.C. and Corlett, R.T. (2020) Applications of environmental DNA (eDNA) in ecology and conservation: opportunities, challenges and prospects. *Biodiversity and Conservation* 29(7) 2089–2121.

Berry, T.E., Saunders, B.J., Coghlan, M.L., Stat, M., Jarman, S., Richardson, A.J., Davies, C.H., Berry, O., Harvey, E.S. and Bunce, M. (2019) Marine environmental DNA biomonitoring reveals seasonal patterns in biodiversity and identifies ecosystem responses to anomalous climatic events. *PLoS Genetics* 15(2), e1007943.

Bjørn, P.A., Finstad, B. and Kristoffersen, R. (2001) Salmon lice infection of wild sea trout and Arctic char in marine and freshwaters: the effects of salmon farms. *Aquaculture Research* 32, 947–962.

Bjørn, P.A., Sivertsgård, R., Finstad, B., Nilsen, R., Serra-Llinares, R.M. and Kristoffersen, R. (2011) Area protection may reduce salmon louse infection risk to wild salmonids. *Aquaculture Environmental Interactions* 1, 233–244.

Boxaspen, K. (2006) A review of the biology and genetics of sea lice. *ICES Journal of Marine Science* 63(7), 1304–1316.

Boxaspen, K. and Næss T. (2000) Development of eggs and the planktonic stages of salmon lice (*Lepeophtheirus salmonis*) at low temperatures. *Contributions to Zoology* 69, 51–55.

Bricknell, I.R., Dalesman, S.J., O'Shea, B., Pert, C.C. and Luntz, A.J.M. (2006) Effect of environmental salinity on sea lice *Lepeophtheirus salmonis* settlement success. *Diseases of Aquatic Organisms* 71, 201–212.

Brooker, A.J., Skern-Mauritzen, R., Bron, J.E. and Fields, D. (2018) Production, mortality, and infectivity of planktonic larval sea lice, *Lepeophtheirus salmonis* (Krøyer, 1837): current knowledge and implications for epidemiological modelling. *ICES Journal of Marine Science* 75(4), 1214–1234.

Brooks, K.M. (2005) The effects of water temperature, salinity, and currents on the survival and distribution of the infective copepodid stage of sea lice (*Lepeophtheirus salmonis*) originating on Atlantic salmon farms in the Broughton Archipelago of British Columbia, Canada. *Reviews in Fisheries Science* 13 177–204.

Butler, J.R.A. (2002) Wild salmonids and sea louse infestations on the west coast of Scotland: sources of infection and implications for the management of marine salmon farms. *Pest Management Science* 58, 595–608.

Byrne, A.A., Pearce, C.M., Cross, S.F., Jones, S.R., Robinson, S.M., Hutchinson, M.J. and Johnson, D.L. (2018) Planktonic and parasitic stages of sea lice (*Lepeophtheirus salmonis* and *Caligus clemensi*) at a commercial Atlantic salmon (*Salmo salar*) farm in British Columbia, Canada. *Aquaculture* 486 130–138.

Chambers, C.B. and Ernst, I. (2005) Dispersal of the skin fluke *Benedenia seriolae* (Monogenea: Capsalidae) by tidal currents and implications for sea-cage farming of *Seriola* spp. *Aquaculture* 250, 60–69.

Collins, R.A., Wangensteen, O.S., O'Gorman, E.J., Mariani, S., Sims, D.W. and Genner, M.J. (2018) Persistence of environmental DNA in marine systems. *Communications Biology* 1, 1–11.

Cook, P.F., McBeath, S.J., Bricknell, I.R. and Bron, J.E. (2010) Determining the age of individual *Lepeophtheirus salmonis* (Krøyer, 1837) copepodids by measuring stored lipid volume, proof of principle. *Journal of Microscopy* 240, 83–86.

Costelloe, J., Costelloe, M. and Roche, N. (1995) Variation in sea lice infestation on Atlantic salmon smolts in Killary Harbour, West Coast of Ireland. *Aquaculture International* 3, 379–393.

Costello, M.J. (2006) Ecology of sea lice parasitic on farmed and wild fish. *Trends in Parasitology* 22, 475–483.

Creer, S., Deiner, K., Frey, S., Porazinska, D., Taberlet, P., Thomas, W.K., Potter, C. and Bik, H.M. (2016) The ecologist's field guide to sequence-based identification of biodiversity. *Methods in Ecology and Evolution* 7, 1008–1018.

Davis, C.S., Thwaites, F.T., Gallager, S.M. and Hu, Q. (2005) A three-axis fast-tow digital video plankton recorder for rapid surveys of plankton taxa and hydrography. *Limnology and Oceanography: Methods* 3, 59–74.

Deagle, B.E., Chiaradia, A., McInnes, J. and Jarman, S.N. (2010) Pyrosequencing faecal DNA to determine diet of little penguins: is what goes in what comes out? *Conservation Genetics* 11, 2039–2048.

Deagle, B.E., Clarke, L.J., Kitchener, J.A., Polanowski, A.M. and Davidson, A.T. (2018) Genetic monitoring of open ocean biodiversity: an evaluation of DNA metabarcoding for processing continuous plankton recorder samples. *Molecular Ecology Resources* 18, 391–406.

Deiner, K., Bik, H.M., Mächler, E., Seymour, M., Lacoursière-Roussel, A., Altermatt, F., Creer, S., Bista, I., Lodge, D.M., De Vere, N. and Pfrender, M.E. (2017) Environmental DNA metabarcoding: transforming how we survey animal and plant communities. *Molecular Ecology* 26, 5872–5895.

Dejean, T., Valentini, A., Miquel, C., Taberlet, P., Bellemain, E. and Miaud, C. (2012) Improved detection of an alien invasive species through environmental DNA barcoding, the example of the American bullfrog *Lithobates catesbeianus*. *Journal of Applied Ecology* 49, 953–959.

Ficetola, G.F., Miaud, C., Pompanon, F. and Taberlet, P. (2008) Species detection using environmental DNA from water samples. *Biology Letters* 4, 423–425.

Finstad, B., Økland, F., Uglem, I., Boxaspen, K., Skaala, Ø., Skilbrei, O., Aspelin, L., Bjorn, P.A., Butterworth, K., McKinley, R.S., Olsen, R.S., Malkenes, R., Ritchie, G., Heuch, P.A. and Kvenseth, P.G. (2007). Final report for NFR-project no. 163869: The Hardangerfjord salmon lice project – 2004–2007. 1–36 s.

Flanders Marine Institute, VLIZ. Available at: http://www.vliz.be/en/go-flobottle (accessed 11 June 2021).

Ford, J.S. and Myers, R.A. (2008) A global assessment of salmon aquaculture impacts on wild salmonids. *PLoS Biology* 6(2), e33.

Frederick, C.A. (2018) The infection dynamics of sea lice on Atlantic salmon (*Salmo salar*) in Cobscook Bay, Maine, USA. Electronic Theses and Dissertations, 3018. Available at: https://digitalcommons.library.umaine.edu/etd/3018 (accessed 11 June 2021).

Gaines, S.D. and Bertness, M. (1993) The dynamics of juvenile dispersal: why field ecologists must integrate. *Ecology* 74, 2430–2435.

Giovannoni, S.J., Britschgi, T.B., Moyer, C.L. and Field, K.G. (1990) Genetic diversity in Sargasso Sea bacterioplankton. *Nature* 345, 60–63.

Gjøsæter, H., Dalpadado, P., Hassel, A. and Skjoldal, H.R. (2000) A comparison of performance of WP2 and MOCNESS. *Journal of Plankton Research* 22 1901–1908.

Goldberg, C.S., Turner, C.R., Deiner, K., Klymus, K.E., Thomsen, P.F., Murphy, M.A., Spear, S.F., McKee, A., Oyler-McCance, S.J., Cornman, R.S. and Laramie, M.B. (2016) Critical considerations for the application of environmental DNA methods to detect aquatic species. *Methods in Ecology and Evolution* 7, 1299–1307.

Gomes, G.B., Hutson, K.S., Domingos, J.A., Chung, C., Hayward, S., Miller, T.L. and Jerry, D.R. (2017) Use of environmental DNA (eDNA) and water quality data to predict protozoan parasites outbreaks in fish farms. *Aquaculture* 479, 467–473.

Harper, L.R., Buxton, A.S., Rees, H.C., Bruce, K., Brys, R., Halfmaerten, D., Read, D.S., Watson, H.V., Sayer, C.D., Jones, E.P. and Priestley, V. (2019) Prospects and challenges of environmental DNA (eDNA) monitoring in freshwater ponds. *Hydrobiologia* 826, 25–41.

Harte, A.J., Bowman, A.S., Salama, N.K. and Pert, C.C. (2017) Factors influencing the long-term dynamics of larval sea lice density at east and west coast locations in Scotland. *Diseases of Aquatic Organisms* 123, 181–192.

Hemmingsen, W., MacKenzie, K., Sagerup, K., Remen, M., Bloch-Hansen, K. and Imsland, A.K.D. (2020) *Caligus elongatus* and other sea lice of the genus *Caligus* as parasites of farmed salmonids: a review. *Aquaculture* (522), 735160.

Heuch, P.A. and Karlsen, E. (1997) Detection of infrasonic water oscillations by copepodids of *Lepeophtheirus salmonis* (Copepoda: Caligida). *Journal of Plankton Research* 19, 735–747.

Heuch, P.A., Gettinby, G. and Revie, C.W. (2011) Counting sea lice on Atlantic salmon farms – empirical and theoretical observations. *Aquaculture* 320, 49–153.

Hunter, M.E., Ferrante, J.A., Meigs-Friend, G. and Ulmer, A. (2019) Improving eDNA yield and inhibitor reduction through increased water volumes and multi-filter isolation techniques. *Scientific Reports* 9, 1–9.

Ingebrigtsen, H. (2017) Molecular quantification of sea lice in and around sea cages – a study comparing the molecular quantification method qPCR against a conventional method. Master's thesis, Norwegian University of Science and Technology.

Ingvarsdóttir, A., Birkett, M.A., Duce, I., Genna, R.L., Mordue, W., Pickett, J.A., Wadhams, L.J. and Mordue, A.J. (2002) Semiochemical strategies for sea louse control: host location cues. *Pest Management Science* 58, 537–545.

Jackson, D., O'Donohoe, P., Kane, F., Kelly, S., McDermott, T., Drumm, A., Lyons K. and Nolan, G. (2012) Result of an epidemiological study of sea lice infestation in South Connemara, West of Ireland. *Aquaculture* 364–365, 118–123.

Jackson, D., Cotter, D., Newell, J., McEvoy, S., O'Donohoe, P., Kane, F., McDermott, T., Kelly, S. and Drumm, A. (2013) Impact of *Lepeophtheirus salmonis* infestations on migrating Atlantic salmon, *Salmo salar* L., smolts at eight locations in Ireland with an analysis of lice-induced marine mortality. *Journal of Fish Diseases* 36(3), 273–281.

Jevne, L.S., Øvrelid, M.S., Hagemann, A., Bloecher, N., Steinhovden, K.B., Båtnes, A.S., Olsen, Y. and Reitan, K.I. (2020) Biofouling on salmon pen nets and cleaner fish shelters does not harbor planktonic stages of sea lice. *Frontiers in Marine Science* 7, 727.

Johnson, S.C. and Albright, L.J. (1991) Development, growth, and survival of *Lepeophtheirus salmonis* (Copepoda: Caligidae) under laboratory conditions. *Journal of Marine Biological Association of the United Kingdom* 71, 425–436.

Kilburn, R., Bricknell, I.R., Cook, P., Pert, C.C., Dunn, J. and Matejusova, I. (2010) Design and application of a portable, automated plankton sampler for the capture of the parasitic copepods *Lepeophtheirus salmonis* (Krøyer 1837) and *Caligus elongatus* (von Nordmann 1832). *Journal of Plankton Research* 32, 967–970.

Krkošek, M., Lewis, M.A. and Volpe, J.P. (2005) Transmission dynamics of parasitic sea lice from farm to wild salmon. *Proceedings of the Royal Society B: Biological Sciences* 272, 689–696.

Krkošek, M., Lewis, M.A., Morton, M., Frazer, L.N. and Volpe, J.P. (2006) Epizootics of wild fish induced by farm fish. *Proceedings of the National Academy of Sciences* 103(42), 15506–15510.

Lamb, P.D., Hunter, E., Pinnegar, J.K., Creer, S., Davies, R.G. and Taylor, M.I. (2019) How quantitative is metabarcoding: a meta-analytical approach. *Molecular Ecology* 28, 420–430.

Mächler, E., Deiner, K., Spahn, F. and Altermatt, F. (2016) Fishing in the water: effect of sampled water volume on environmental DNA-based detection of macroinvertebrates. *Environmental Science & Technology* 50, 305–312.

Makiola, A., Compson, Z.G., Baird, D.J., Barnes, M.A., Boerlijst, S.P., Bouchez, A., Brennan, G., Bush, A., Canard, E., Cordier, T. and Creer, S. (2020) Key questions for next-generation biomonitoring. *Frontiers in Environmental Science* 7, 197.

Majaneva, M., Diserud, O.H., Eagle, S.H., Boström, E., Hajibabaei, M. and Ekrem, T. (2018) Environmental DNA filtration techniques affect recovered biodiversity. *Scientific Reports* 8, 1–11.

Marty, G.D., Saksida, S.M. and Quinn, T.J.II (2010) Relationship of farm salmon, sea lice, and wild salmon populations. *Proceedings of the National Academy of Sciences of the United States of America* 107(52), 22599–22604.

Maxam, A.M. and Gilbert, W. (1977) A new method for sequencing DNA. *Proceedings of the National Academy of Sciences* 74, 560–564.

McBeath, A.J., Penston, M.J., Snow, M., Cook, P.F., Bricknell, I.R. and Cunningham, C.O. (2006) Development and application of real-time PCR for specific detection of *Lepeophtheirus salmonis* and *Caligus elongatus* larvae in Scottish plankton samples. *Diseases of Aquatic Organisms* 73, 141–150.

McKibben, M.A. and Hay D.W. (2004) Distributions of planktonic sea lice larvae *Lepeophtheirus salmonis* in the inter-tidal zone in Loch Torridon, Western Scotland in relation to salmon farm production cycles. *Aquaculture Research* 35, 742–750.

McVicar, A.H., Sharp, L.A. and Pike, A.W. (1993) Infectious diseases of Scottish sea trout and salmon. In: *Problems with Sea Trout and Salmon in the Western Highlands.* Atlantic Salmon Trust, Perthshire, UK. pp. 48–60.

Middlemas, S.J., Raffell, J.A., Hay, D.W., Hatton-Ellis, M. and Armstrong, J.D. (2010) Temporal and spatial patterns of sea lice levels on sea trout in western Scotland in relation to fish farm production cycles. *Biological Letters* 6, 548–551.

Middlemas, S., Stewart, D., Henry, J., Wyndham, M., Ballantyne, L. and Baum, D. (2017) Dispersal of post-smolt Atlantic salmon and sea trout within a Scottish sea loch system. In: G. Harris (ed.) *Sea Trout: Science and Management. Proceedings of the 2nd International Sea Trout Symposium*. Troubador, Dundalk, Ireland. pp. 339–353.

Moksnes, P.O. and Wennhage, H. (2001) Methods for estimating decapod larval supply and settlement: importance of larval behavior and development stage. *Marine Ecology Progress Series* 209, 257–273.

Moon, S.Y. and Park, J.S. (2019) Occurrence of sea lice, *Caligus undulatus* Shen and Li, 1959 (Copepoda: Siphonostomatoida: Caligidae) in plankton samples collected from Korea. *Journal of Species Research* 8, 365–372.

Morton, A., Routledge, R., McConnell, A. and Krkošek, M. (2011) Sea lice dispersion and salmon survival in relation to salmon farm activity in the Broughton Archipelago. *ICES Journal of Marine Science* 68, 144–156.

Moyer, G.R., Diaz-Ferguson, E., Hill, J.E. and Shea, C. (2014) Assessing environmental DNA detection in controlled lentic systems. *PLoS ONE* 9(7), e103767.

Murray, A.G. (2016) Increased frequency and changed methods in the treatment of sea lice (*Lepeophtheirus salmonis*) in Scottish salmon farms 2005–2011. *Pest Management Science* 72(2), 322–326.

Murray A.G. and Gillibrand P.A. (2006) Modelling salmon lice dispersal in Loch Torridon, Scotland. *Marine Pollution Bulletin* 53, 128–135.

Natarajan, V.P., Zhang, X., Morono, Y., Inagaki, F. and Wang, F. (2016) A modified SDS-based DNA extraction method for high quality environmental DNA from seafloor environments. *Frontiers in Microbiology* (7), 986.

Nielsen, J.H., Pedersen, C., Kiørboe, T., Nikolajsen, T., Brydegaard, M. and Rodrigo, P.J. (2019) Investigation of autofluorescence in zooplankton for use in classification of larval salmon lice. *Applied Optics* 58, 7022–7027.

Novales Flamarique, I., Browman, H.I., Belanger, M.A. and Boxaspen, K.A. (2000) Ontogenetic changes in visual sensitivity of the parasitic salmon louse *Lepeophtheirus salmonis*. *Journal of Experimental Biology* 203, 1649–1657.

Novales Flamarique, I., Gulbransen, C., Galbraith, M. and Stucchi, D. (2009) Monitoring and potential control of sea lice using an LED-based light trap. *Canadian Journal of Fisheries and Aquatic Sciences* 66, 1371–1382.

Ogram, A., Sayler, G.S. and Barkay, T. (1987) The extraction and purification of microbial DNA from sediments. *Journal of Microbiological Methods* 7, 57–66.

Omori, M., Marumo, R. and Aizawa, Y. (1965) A 160 cm opening-closing plankton net. *Journal of the Oceanographical Society of Japan* 21, 245–252.

Penston, M.J. and Davies, I.M. (2009) An assessment of salmon farms and wild salmonids as sources of *Lepeophtheirus salmonis* (Krøyer) copepodids in the water column in Loch Torridon, Scotland. *Journal of Fish Diseases* 32, 75–88.

Penston, M.J., McKibben, M.A., Hay, D.W. and Gillibrand, P.A. (2002) Observations of sea lice larvae distributions in Loch Shieldaig, Western Scotland. International Council for the Exploration of the Sea. ICES CM2002/T, 9.

Penston, M.J., McKibben, M.A., Hay, D.W. and Gillibrand, P.A. (2004) Observations on open-water densities of sea lice larvae in Loch Shieldaig, Western Scotland. *Aquaculture Research* 35, 793–805.

Penston, M.J., Millar, C.P. and Davies, I.M. (2008a) Spatial and temporal distribution of *Lepeophtheirus salmonis* (Krøyer) larvae in a sea loch containing Atlantic salmon, *Salmo salar* L., farms on the north-west coast of Scotland. *Journal of Fish Diseases* 31, 361–371.

Penston, M.J., Millar, C.P. and Davies, I.M. (2008b). Reduced *Lepeophtheirus salmonis* larval abundance in a sea loch on the west coast of Scotland between 2002 and 2006. *Diseases of Aquatic Organisms* 81, 109–117.

Pert, C.C., Mordue, A.J., O'Shea, B. and Bricknell, I.R. (2012) The settlement and reproductive success of *Lepeophtheirus salmonis* (Krøyer, 1837; Copepoda: Caligidae) on atypical hosts. *Aquaculture* 43, 799–805.

Pert, C.C., Fryer, R.J., Cook, P., Kilburn, R. McBeath, S., McBeath, A., Matejusova, I., Urquhart, K., Weir, S.J., McCarthy, U., Collins, C., Amundrud, T. and Bricknell, I.R. (2014a). Using sentinel cages to estimate infestation pressure on salmonids from sea lice in Loch Shieldaig, Scotland. *Aquaculture Environmental Interactions* 5, 49–59.

Pert, C.C., Middlemas, S.J., Collins, C.M., Baum, D. and Salama, N.K.G. (2014b) Identifying variations in the potential infestation pressure from sea lice on wild salmonids in a Scottish salmonid aquaculture region. Oral presentation at Sea Lice 2014 conference, Portland, Maine, USA.

Pert C.C., Rabe, B., Dunn, J., Salama, N.K.G. and Collins, C.M. (2017) Developing a method to investigate infestation pressure from sea lice on migratory salmonids utilising towed & static sentinel cages. *Oceans 2017 – Aberdeen.* Institute of Electrical and Electronic Engineers (IEEE). pp. 1–8.

Peters, L., Spatharis, S., Dario, M.A., Dwyer, T., Roca, I.J., Kintner, A., Kanstad-Hanssen, Ø., Llewellyn, M.S. and Praebel, K. (2018) Environmental DNA: a new low-cost monitoring tool for pathogens in salmonid aquaculture. *Frontiers in Microbiology* (9) 3009.

Pike, A.W. and Wadsworth, S. (1999) Sea lice on salmonids: their biology and control. *Advances in Parasitology* 44, 233–337.

Pitois, S.G., Bouch, P., Creach, V. and Van Der Kooij, J. (2016) Comparison of zooplankton data collected by a continuous semi-automatic sampler (CALPS) and a traditional vertical ring net. *Journal of Plankton Research* 38, 931–943.

Pitois, S.G., Tilbury, J., Bouch, P., Close, H., Barnett, S. and Culverhouse, P.F. (2018) Comparison of a cost-effective integrated plankton sampling and imaging instrument with traditional systems for mesozooplankton sampling in the Celtic Sea. *Frontiers in Marine Science* (5) 5.

Raupach, M.J., Barco, A., Steinke, D., Beermann, J., Laakmann, S., Mohrbeck, I., Neumann, H., Kihara, T.C., Pointner, K., Radulovici, A. and Segelken-Voigt, A. (2015) The application of DNA barcodes for the identification of marine crustaceans from the North Sea and adjacent regions. *PLoS ONE* 10(9), e0139421.

Reboud, J., Xu, G., Garrett, A., Adriko, M., Yang, Z., Tukahebwa, E.M., Rowell, C. and Cooper, J.M. (2019) Paper-based microfluidics for DNA diagnostics of malaria in low resource underserved rural communities. *Proceedings of the National Academy of Sciences* 116, 4834–4842.

Revie, C.W., Gettinby, G. and Ritchie, G. (2003) The use of a general linear model to identify epidemiological factors affecting the abundance of chalimus stages of the sea louse (*Lepeophtheirus salmonis*) on Scottish salmon farms. In: *Proceedings of the 10th Symposium of the International Society for Veterinary Epidemiology and Economics*. Available at: http://www.sciquest.org.nz/elibrary/edition/5016 (accessed 16 June 2021).

Ruppert, K.M., Kline, R.J. and Rahman, M.S. (2019) Past, present, and future perspectives of environmental DNA (eDNA) metabarcoding: a systematic review in methods, monitoring, and applications of global eDNA. *Global Ecology and Conservation* 17, e00547.

Saha, M., Berdalet, E., Carotenuto, Y., Fink, P., Harder, T., John, U., Not, F., Pohnert, G., Potin, P., Selander, E. and Vyverman, W. (2019) Using chemical language to shape future marine health. *Frontiers in Ecology and the Environment* 17, 530–537.

Saksida, S.M., Bricknell, I., Robinson, S. and Jones, S. (2015) Population ecology and epidemiology of sea lice in Canadian waters. Research Document 2015/004. Canadian Science Advisory Secretariat. Ottawa, ON, Canada. Available at: https://digitalcommons.library.umaine.edu/seagrant_pub/75/ (accessed 2 February 2021).

Salama, N.K.G., Collins, C.M., Fraser, J.G., Dunn, J., Pert, C.C., Murray, A.G. and Rabe, B. (2013) Development and assessment of a biophysical dispersal model for sea lice. *Journal of Fish Diseases* 36, 323–337.

Salama, N.K.G., Dale, A.C., Ivanov, V.V., Cook, P.F., Pert, C.C., Collins, C.M. and Rabe, B. (2018) Using biological-physical modelling for informing sea lice dispersal in Loch Linnhe, Scotland. *Journal of Fish Diseases* 41, 901–919.

Samsing, F., Solstorm, D., Oppedal, F., Solstorm, F. and Dempster, T. (2015) Gone with the flow: current velocities mediate parasitic infestation of an aquatic host. *International Journal of Parasitology* 45(8), 559–565.

Sanches, T.M. and Schreier, A.D. (2020) Optimizing an eDNA protocol for estuarine environments: balancing sensitivity, cost and time. *PLoS ONE* 15(5), e0233522.

Sandvik, A.D., Johnsen, I.A., Myksvoll, M.S., Sævik, P.N. and Skogen, M.D. (2020) Prediction of the salmon lice infestation pressure in a Norwegian fjord. *ICES Journal of Marine Science* 77, 746–756.

Sanger, F., Nicklen, S. and Coulson, A.R. (1977) DNA sequencing with chain-terminating inhibitors. *Proceedings of the National Academy of Sciences* 74, 5463–5467.

Schabacker, J.C., Amish, S.J., Ellis, B.K., Gardner, B., Miller, D.L., Rutledge, E.A., Sepulveda, A.J. and Luikart, G. (2020) Increased eDNA detection sensitivity using a novel high-volume water sampling method. *Environmental DNA* 2, 244–251.

Schultz, M.T. and Lance, R.F. (2015) Modelling the sensitivity of field surveys for detection of environmental DNA (eDNA). *PLoS ONE* 10(10), e0141503.

Serra-Llinares R.M., Bjørn, P.A., Finstad, B., Nilsen, R., Harbitz, A., Berg, M. and Asplin, L. (2014) Salmon lice infection on wild salmonids in marine protected areas: an evaluation of the Norwegian 'National Salmon Fjords'. *Aquaculture and Environmental Interactions* 5, 1–16.

Shea, D., Bateman, A., Li, S., Tabata, A., Schulze, A., Mordecai, G., Ogston, L., Volpe, J.P., Neil Frazer, L., Connors, B. and Miller, K.M. (2020) Environmental DNA from multiple pathogens is elevated near active Atlantic salmon farms. *Proceedings of the Royal Society B* 287(1937), 20202010.

Shephard, S. and Gargan, P. (2017) Quantifying the contribution of sea lice from aquaculture to declining annual returns in a wild Atlantic salmon population. *Aquaculture Environment Interactions* 9, 181–192.

Skarðhamar, J., Fagerli, M.N., Reigstad, M., Sandvik, A.D. and Bjørn, P.A. (2019) Sampling planktonic salmon lice in Norwegian fjords. *Aquaculture Environment Interactions* 11, 701–715.

Sollid, S.A., Lorz, H.V., Stevens, D.G., Reno, P.W. and Bartholomew, J. L. (2004) Prevalence of *Myxobolus cerebralis* at juvenile salmonid acclimation sites in northeastern Oregon. *North American Journal of Fisheries Management* 24(1), 146–153.

Stien, A., Bjorn, P.A., Heuch, P.A. and Elston, D.A. (2005) Population dynamics of salmon lice *Lepeophtheirus salmonis* on Atlantic salmon and sea trout. *Marine Ecology Progress Series* 290, 263–275.

Taberlet, P., Coissac, E., Hajibabaei, M. and Rieseberg, L.H. (2012) Environmental DNA. *Molecular Ecology* 21, 1789–1793.

Takeuchi, A., Sado, T., Gotoh, R.O., Watanabe, S., Tsukamoto, K. and Miya, M. (2019) New PCR primers for metabarcoding environmental DNA from freshwater eels, genus *Anguilla*. *Scientific Reports* 9, 1–11.

Thomsen, P.F. and Willerslev, E. (2015) Environmental DNA – an emerging tool in conservation for monitoring past and present biodiversity. *Biological Conservation* 183, 4–18.

Thomsen, P.F., Kielgast, J.O.S., Iversen, L.L., Wiuf, C., Rasmussen, M., Gilbert, M.T.P., Orlando, L. and Willerslev, E. (2012) Monitoring endangered freshwater biodiversity using environmental DNA. *Molecular Ecology* 21, 2565–2573.

Thorstad, E.B., Økland, F., Finstadt, B., Sivertsgard, R., Bjørn, P.A. and McKinley, R.S. (2004) Migration speeds and orientation of Atlantic salmon and sea trout post-smolts in a Norwegian fjord system. *Environmental Biology of Fishes* 71(3), 305–311.

Torrissen, O., Jones, S., Asche, F., Guttormsen, A., Skilbrei, O.T., Nilsen, F., Horsberg, T.E. and Jackson, D. (2013) Salmon lice – impact on wild salmonids and salmon aquaculture. *Journal of Fish Diseases* 36(3), 171–194.

Tully, O. (1989) The succession of generations and growth of the caligid copepods *Caligus elongatus* and *Lepeophtheirus salmonis* parasitising farmed Atlantic salmon smolts (*Salmo salar* L.). *Journal of the Marine Biological Association of the United Kingdom* 69, 279–287.

Turner, C.R., Uy, K.L. and Everhart, R.C. (2015) Fish environmental DNA is more concentrated in aquatic sediments than surface water. *Biological Conservation* 183, 93–102.

Valentini, A., Taberlet, P., Miaud, C., Civade, R., Herder, J., Thomsen, P.F., Bellemain, E., Besnard, A., Coissac, E., Boyer, F. and Gaboriaud, C. (2016) Next-generation monitoring of aquatic biodiversity using environmental DNA metabarcoding. *Molecular Ecology* 25, 929–942.

Vollset, K.W., Barlaup, B.T., Mahlum, S., Skår, B., Skoglund, H., Normann, E.S., Holst, J.C., Wennevik, V., Skaet, G. and Davidsen, J.G. (2014) Migration and predation of Atlantic salmon smolts from Vosso. Final report FHF project #900778, LFI Uni Miljø Thormøhlensgt. 48B 5006 Bergen, Norway.

Vollset, K.W., Barlaup, B.T., Mahlum, S., Bjørn, P.A. and Skilbrei, O.T. (2016a) Estimating the temporal overlap between post-smolt migration of Atlantic salmon and salmon lice infestation pressure from fish farms. *Aquaculture Environment Interactions* 8, 511–525.

Vollset, K.W., Krontveit, R.I., Jansen, P.A., Finstad, B., Barlaup, B.T., Skilbrei, O.T., Krkošek, M., Romunstad, P., Aunsmo, A., Jensen, A.J. and Dohoo, I. (2016b) Impacts of parasites on marine survival of Atlantic salmon: a meta-analysis. *Fish and Fisheries* 17, 714–73.

Vøllestad, L.A., Hirst, D., L'Abe´e-Lund, J.H., Armstrong, J.D., MacLean, J.C., Youngson, A.F. and Stenseth, N.C. (2009) Divergent trends in anadromous salmonid populations in Norwegian and Scottish rivers. *Proceedings of the Royal Society B: Biological Sciences* 276, 1021–1027.

Waite, S.W. and O'Grady, S.M. (1980) Description of a new submersible filter-pump apparatus for sampling plankton. *Hydrobiologia* 74, 187–191.

Walter, T.C. and Boxshall, G. (2021) World of Copepods Database. Available at: http://www.marinespecies.org/copepoda (accessed 11 June 2021).

Williams, K.E., Huyvaert, K.P. and Piaggio, A.J. (2016) No filters, no fridges: a method for preservation of water samples for eDNA analysis. *BMC Research Notes* 9, 1–5.

Williams, M.A., O'Grady, J., Ball, B., Carlsson, J., de Eyto, E., McGinnity, P., Jennings, E., Regan, F. and Parle-McDermott, A. (2019). The application of CRISPR–Cas for single species identification from environmental DNA. *Molecular Ecology Resources* 19, 1106–1114.

Wootten, R., Smith, J.W. and Needham, E.A. (1982) Aspects of the biology of the parasitic copepods *Lepeophtheirus salmonis* and *Caligus elongatus* on farmed salmonids and their treatment. *Proceedings of the Royal Society of Edinburgh, Section B: Biological Sciences* 81, 185–197.

Yamahara, K.M., Preston, C.M., Birch, J.M., Walz, K.R., Marin III, R., Jensen, S., Pargett, D., Roman, B., Zhang, Y., Ryan, J. and Ussler, B. (2019) In-situ autonomous acquisition and preservation of marine environmental DNA using an autonomous underwater vehicle. *Frontiers in Marine Science* (6), 373.

Zaiko, A., Pochon, X., Garcia-Vazquez, E., Olenin, S. and Wood, S.A. (2018) Advantages and limitations of environmental DNA/RNA tools for marine biosecurity: management and surveillance of non-indigenous species. *Frontiers in Marine Science* (5), 322.

Zamor, R.M., Glenn, K.L. and Hambright, K.D. (2012) Incorporating molecular tools into routine HAB monitoring programs: using qPCR to track invasive *Prymnesium*. *Harmful Algae* 15, 1–7.

CHAPTER 19

Biological control of sea lice

Jim Treasurer[1], Albert Imsland[2], Pat Reynolds[3] and Daniel Carcajona[4]

[1]FAI Farms, Fort William, Scotland
[2]AkvaPlan Niva, Iceland, and University of Bergen, Norway
[3]GIFAS, Norway
[4]Sustainable Aquaculture Innovation Centre, Stirling

19.1 Introduction

Biological control in fishes has been documented frequently (Potts, 1973, 1974) and is a common form of symbiotic relationship between cleaners and fishes and shellfish universally (Grutter, 1999; Vaughan et al., 2017). Several reviews have examined the use of cleaner fish in aquaculture (for example Powell et al., 2017; Brooker et al., 2018; Haugland et al., 2020) and also epibionts, ciliates and bacteria as biological controls of sea lice (Treasurer, 2002) (Table 19.1). The use of cleaner fish to date has been the only large-scale biological control measure that has been applied successfully in aquaculture (Treasurer, 2018). The advantages of cleaner fish are environmental with a reduction in medicine use (Treasurer, 2002); they are a natural form of control (Bjordal, 1988) and provide continuous cleaning of lice (Brooker et al., 2018). However, there are difficulties in measuring efficacy in large pens (Overton et al., 2020), the sustainability of wild wrasse stocks has to be monitored (Halvorsen et al., 2017), the welfare of cleaner fish should be maintained (Brooker et al., 2018) and losses of fish can be high (Mo and Poppe, 2018).

Cleaner fish have been demonstrated to work effectively in small-scale tank and cage trials (Bjordal, 1988, 1991; Treasurer, 1993; Imsland et al., 2014) and there is increasing evidence of efficacy in larger commercial pens (Imsland et al., 2018a, 2020a) and also as an integral part of integrated pest management (IPM) measures for sea lice (Haugland et al., 2020). Issues remain in the use and efficacy of cleaner fish, in the health and welfare, vaccination, farm application and rearing of the numbers of cleaner fish required by salmon farmers (Haugland et al., 2020). Recently there have been several studies critical of cleaner fish use, or at least questioning welfare, survival (Mo and Poppe, 2018; Stien et al., 2020) and efficacy issues (Bui et al., 2020; Gentry et al., 2020). This review examines the current status and context of biological control in the treatment of sea lice, and the magnitude of cleaner fish stocked, and identifies production bottlenecks and the need for improvements in the management of cleaner fish including health measures, welfare and vaccination. We also examine alternative biological control methods that can be applied in the future, together with other potential cleaner fish species in addition to wrasse and lumpfish, *Cyclopterus lumpus*.

Table 19.1 Literature reviews of cleaner fish use.

Review authors and date	Species covered	Subjects covered	Conclusions	Comments
Darwall et al., 1992	Wrasse	Trials and efficacy, stocking density	Variable efficacy; small pens	Early study; mainly small cages, no ballan wrasse
Powell et al., 2018	Lumpfish	Biology, diseases, life cycle, prospects	Positive case for lumpfish farming	Very extensive review of biology, health and economics
Scholz et al., 2018	Wrasse and lumpfish	Description of diseases and treatment	Need for veterinary health plan	Original work and review by veterinarians
Brooker et al., 2018	Wrasse and lumpfish	Farm application, life cycle, diseases Regulatory; fishery and farming statistics	Highlights issues & research priorities Regulations for rearing and stocking wrasse	
Treasurer, 2018	Wrasse and lumpfish	Regulatory; fishery and farming statistics	Regulations for rearing and stocking wrasse	Regulations in all countries summarised
Overton et al., 2020	Wrasse and lumpfish	Review of evidence base	Need for more trials in large commercial pens	
Barrett et al., 2020	Wrasse and lumpfish	Review of farm stocking and efficacy	Suboptimal use	
Haugland et al., 2020	Wrasse and lumpfish	Efficacy, rearing systems, health	Requirements for successful rearing and stocking	Novel summary of diseases
Erkinharju et al., 2021	Wrasse and lumpfish	Diseases and vaccination	Possibilities for disease transfer to salmon	Detailed review of cleaner fish diseases
Imsland et al., 2020a	Lumpfish	Review of cleaning evidence base	Positive outcomes Evidence for lumpfish cleaning *Caligus*	Review of literature and asks expert panel

19.2 Filtration feeders as a biological control

Earlier studies have found that the blue mussel *Mytilus edulis* filtered the planktonic stages of sea lice such as copepodids (Molloy et al., 2011). Copepodids were identified in the buccal cavity of *M. edulis* and also in the stomach contents confirmed by molecular analysis. In a similar study blue mussel and the Atlantic sea scallop *Placopecten magellanicus* filtered and consumed the copepodid stage of *L. salmonis* in laboratory experiments in differing flows and light conditions, and at different copepodid densities (Bartsch et al., 2013). Both of these shellfish species consumed equivalent numbers of copepodids and between 18% and 38% of copepodids presented per hour. This principle was tested on salmon farms by comparing sea lice larval densities inside pens with and without Pacific oysters *Crassostrea gigas* (Byrne, 2012). Byrne concluded that nauplii stages of sea lice were quickly dispersed from the salmon pens. In Scotland mussels are commonly found on netting and structures but sea lice may still be abundant. In Region X (Puerto Montt, Chiloe) 400,000 tonnes of mussels are grown per year in around 619 salmon farms but sea lice numbers can be high (S. Bravo, August 2021, personal communication).

19.3 Pathogens of sea lice

A wide range of pathogens and epibionts of sea lice was summarised (Treasurer, 2002), including fungi, dinoflagellates, microspora, ciliates, monogeneans, turbellaria, cestodes, nematodes, polychaetes and opportunist bacteria (Table 19.2). None of these has been found to adversely affect sea lice viability and often the presence of external commensals provides colouring and camouflage to sea lice. The most promising candidate appeared to be the microsporidian *Desmozoon lepeophtherii* (Freeman and Sommerville, 2011).

Ephelota sp. is a suctorian which is commonly present on both *L. salmonis* and *C. elongatus* (Stone and Bruno, 1989) and is concentrated on the genital segment and egg strings (Fig. 19.1), but is a commensal and does not inhibit movement and feeding in sea lice. A monogenean *Udonella caligorum* is commonly attached to sea lice by a rounded anterior sucker (Fig. 19.1), and moves to the genital segment before oviposition (Stone and Bruno, 1989). *Udonella* is a commensal that grazes on the epidermal tissue removed from sea lice (Kabata, 1973). Baculoviruses can be diseases of Crustacea and occlusion bodies can damage the digestive system and lead to morbidity (Stuck and Overstreet, 1994). No such virus has been found in sea lice but this is an area for further study. *Bacillus thuringiensis* has been regarded as an insect pathogen (Schnepf et al., 1998). It initiates a process termed colloidal osmotic lysis where there is an influx of water to host cells along with ions, resulting in cell swelling and then lysis. The prospects for application to sea lice are unlikely.

Table 19.2 Examples of commensals and pathogens recorded on Crustacea by taxonomic group.

	Commensal/pathogen	Host species	Reference
Algae	*Ulva* species	*L. salmonis*	Minchin and Jackson, 1993
Fungus	*Lagenidium giganteum*	Copepods	Nestrud and Anderson, 1994
Dinoflagellates	*Gyrodinium undulans*	Copepod and rotifer eggs	Drebes and Schnepf, 1998
Microspora	Microsporidians	*Artemia* species	Ascunsion-Matinez et al., 1992
Ciliates	*Ephelota* sp.	*L. salmonis*	Stone and Bruno, 1989
	Epistylis sp.	*L. salmonis*	Gresty and Warren, 1993
Foraminifera	Sponges	Malacostraca	Becker, 1996
Monogenea	*Udonella*	*C. elongatus, L. salmonis*	Minchin and Jackson, 1993
Turbellaria	Planktonic turbellaria	Zooplankton	Blaustein et al., 1990
Cestodes	*E. crassum*	Copepods	Kennedy, 1978
Nematodes	Nematoda	Gammarids	Voigt, 1991
Polychaetes		Copepods	Capaccioni et al., 1993
Bacteria	*A. salmonicida*	*L. salmonis*	Nese and Enger, 1993

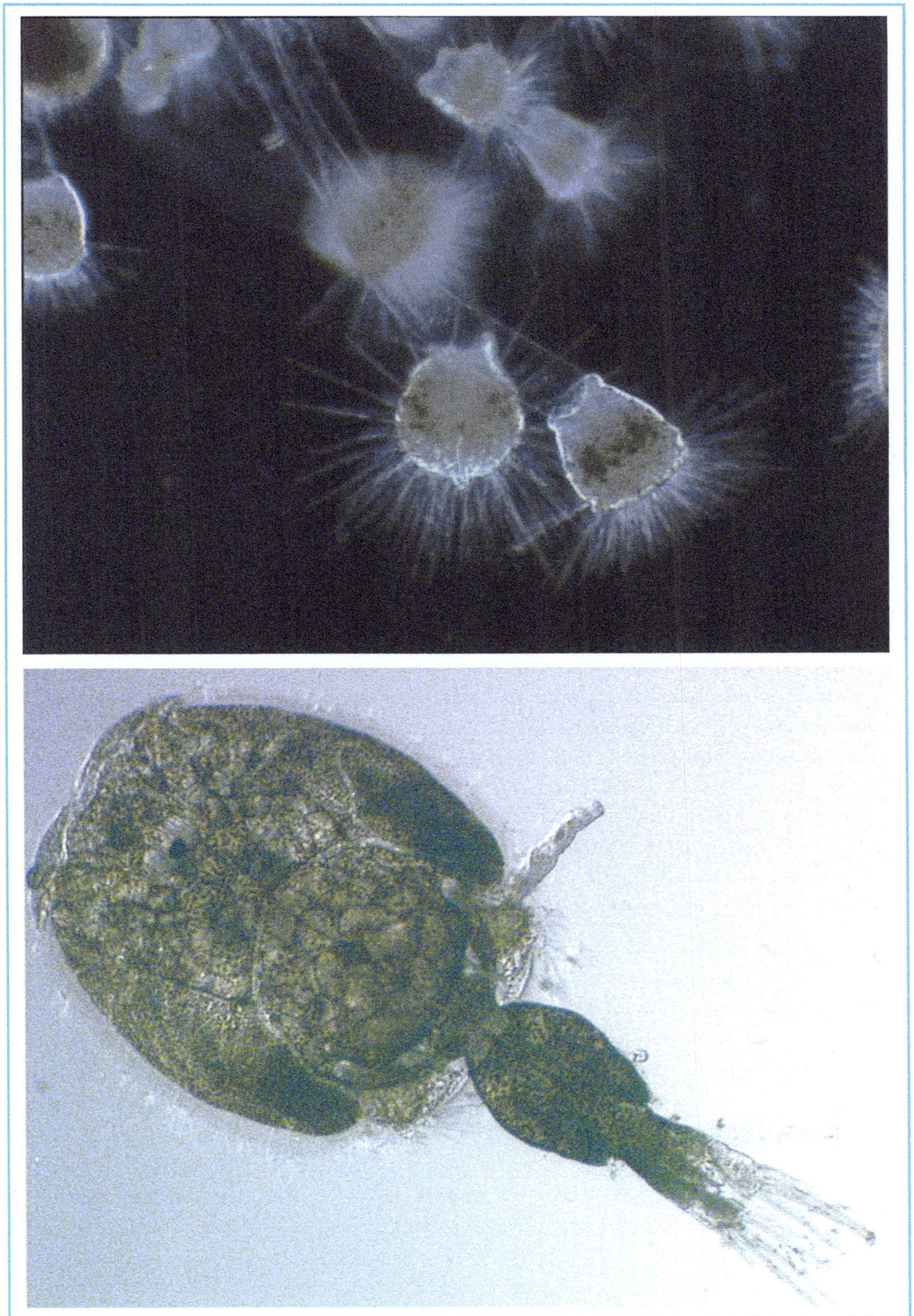

Fig. 19.1 Commensals of sea lice. (a) *Ephelota* (10 mm = 55 μm) (b) *Udonella caligorum* (10 mm = 500 μm). (Credit: David Bruno)

19.4 Development of cleaner fish use in aquaculture

The cleaning activity of wrasse species in Europe was observed in studies on the corkwing wrasse *Crenilabrus melops* and goldsinny *Ctenolabrus rupestris* in the Plymouth Aquarium and also offshore in the UK (Potts, 1973, 1974). The Institute of Marine Research in Norway recognised the potential for this in cleaning sea lice on salmon farms and Bjordal (1988) and Bjordal and Costello (1990) carried out a series of small-scale trials in 5 m^2 and 12 m^2 pens with three wrasse species and at different stocking densities (1:43 in large pens). The advantages of using cleaner fish in sea lice control included environmental benefits and a reduction in medicine use. The technique was widely adopted by salmon farmers in Norway, Scotland and Ireland in the early 1990s (Bjordal, 1988, 1991; Darwall et al., 1992; Treasurer, 1993) and was reported at an international wrasse conference held in 1994 (Sayer et al., 1996). The technique had declined in use by the late 1990s with the advent of new bath treatments such as cypermethrin and the in-feed treatment emamectin benzoate (Treasurer et al., 2002) and, in Scotland, an outbreak of infectious salmon anaemia (ISA) which also discouraged the capture and stocking of wild caught fish for biosecurity reasons. The use of cleaner fish was adopted on a larger scale later from 2008 when wrasse were first farmed on a commercial basis. Lumpfish were also utilised as cleaner fish from around 2014 in Norway and later in several other countries (Imsland et al., 2020a). Wrasse proved technically difficult to rear in large numbers (Lekva and Grøtan, 2018) and most of the wrasse stocked are fished using traps and fyke nets (Halvorsen et al., 2017; Kousoulaki et al., 2018a). These amount to around 22 million fish per year in Norway and up to two million wrasse in Scotland (Fig. 19.2). In contrast, lumpfish have proven to be easier to rear with much higher rates of survival in the hatchery from hatching to stocking in sea pens, being as high as 50% in many cases (Powell et al., 2018), and up to 40 million farmed lumpfish were stocked each year in Norway from 2017 to 2020 (Fig. 19.2).

19.5 How cleaner fish use contributes to integrated pest management

Cleaner fish are now used in integrated pest management (IPM) strategies by several fish farm companies (for example, Mowi, Scottish Sea Farms, Cermaq). IPM involves the use of several approaches to pest management, including medicinal, preventative and biological control to contain pest infestations (see Chapter 15). Although the use of cleaner fish has been adopted by most salmon farming companies, this is only one part of an IPM strategy along with preventative methods and medicinal treatments, and also increasingly mechanical methods such as thermolicers and hydrolicers where heated water and high power jets are used respectively (Barrett et al., 2020). The use of cleaner fish reduces lice loading and the frequency of treatments in many farms, but there may still be a need for other treatments as one species of cleaner fish is rarely completely effective through an entire salmon farming cycle. Wrasse are less effective in the winter as activity is low at lower water temperatures (Sayer et al., 1996) and lumpfish feed on other prey such as jellyfish, plankton and other invertebrates in the summer if these are available (Johanessen et al., 2018).

As part of this IPM approach, wrasse of small size are generally stocked from the first summer and medicinal treatments may be required at the end of the first year of salmon production with the onset of winter. Lumpfish of around 30 g are stocked as the water temperature drops. In the second

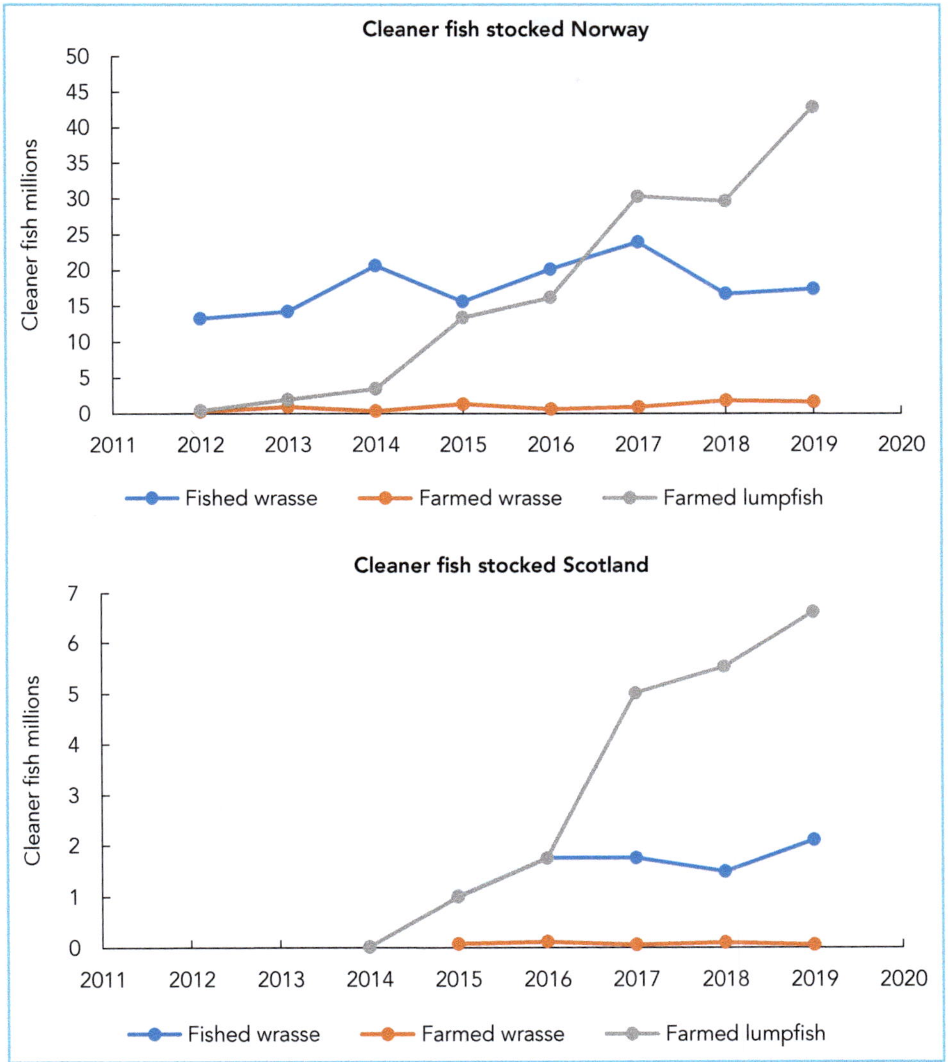

Fig. 19.2 Numbers of cleaner fish stocked in Norway and Scotland. Note that there has been little change in these numbers up to 2022.

year more wrasse may be stocked and these are larger, from 70 g or above, and larger lumpfish may also be stocked. The cleaner fish are harvested at the same time as the salmon and either stocked in a second salmon farming production cycle or harvested as fish meal or for a limited human market. Further uses are being sought to deploy the cleaner fish more sustainably when the salmon have been harvested (Nøstvold et al., 2016).

19.6 Does the biological control technique work?

Wrasse cleaned salmon of sea lice in tank trials (Table 19.3) and the same has been shown in land-based tank trials with lumpfish (Imsland et al., 2014) and in large-scale trials (Imsland et al.,

Table 19.3 A selection of trials on cleaner fish efficacy in sea lice control.

Location	Trial and stocking ratio cleaner fish: salmon	Findings	Reference
Norway	1:10–1:25 mixed wrasse species, 5 and 12 m² cages	reduced lice numbers compared with controls	Bjordal, 1988; Bjordal and Costello, 1990
Norway	1:20 wrasse:salmon, 5 m² cages	efficient delousing of salmon	Skiftesvik et al., 2013
Norway	lumpfish	95–97% removal if mature female lice compared with controls	Imsland et al., 2014
Norway	5% and 10% lumpfish 5 m² cages	sea lice found and counted in stomachs	Imsland et al., 2014
Norway	commercial size pens, 4–8% lumpfish October–May	significantly lower numbers of *L. salmonis* and *Caligus* Feb–April	Imsland et al., 2014
Norway	review of *C. elongatus* grazing in lumpfish	sig. lower numbers in numerous small and large scale trials	Imsland et al., 2020
Faroe Islands	review of cleaning efficacy of lumpfish	clear grazing but seasonal effect; higher lice grazing Sept–April, lower in summer	Eliasen et al., 2018
Scotland	various ratios wrasse to salmon, normally 1:10	lice control, wrasse moved between cages where required	Treasurer, 2013
Scotland	full farm trial 1:10 wrasse	no medicinal treatments	Hempleman, 2017
Scotland	wrasse and lumpfish	zero mechanical treatments; lice numbers <1 all stages	Nielsen, 2021
Scotland	1:25 goldsinny and rock cook, 28 × 5 m² cages	no medicinal treatments with wrasse	Treasurer, 1993
Ireland	commercial trials goldsinny and corkwing <1:250	maintained low levels of lice	Deady et al., 1995
Ireland	commercial trials 3–6% wrasse, 6–15% lumpfish	reduced lice numbers, less effective in larger cages	Bolton-Warberg et al., 2018
New Brunswick	5–9% ratio, 5 pens with cunner, 6 with lumpfish	36% lice reduction with cunner and 20% with lumpfish	Boyce et al., 2018
Chile	two cages, 10% and 20% Patagonian blenny	reduction in sea lice numbers by 47% and 27%	Sanchez et al., 2018

2018a, 2020a). Early work indicated cleaning activity in smaller pens of 5 m² and also in pens of 12 m² and 16 m² (Darwall et al., 1992; Treasurer, 1993). It has been more difficult to demonstrate cleaning activity of cleaner fish in commercial pens due to the large size of the pens and the logistical difficulty of precise replication in commercial farms (Treasurer, 2013; Barrett et al., 2020). Day-to-day management of the farm makes trials in large pens difficult. The need to stock all pens with wrasse to ensure adequate coverage and control of sea lice numbers often precludes the use of control pens with no cleaner fish. Examples of cleaner fish stocking include stocking of ballan wrasse (Treasurer, 2013) and this was based on two farms. Sea lice were micromanaged in the first in Loch Spelve in individual pens by moving the wrasse between pens to target those with salmon with the highest numbers of sea lice. As no replicated trials were possible, comparison was made with the number of medicinal treatments that were required when wrasse were stocked and in a year class before wrasse were used; or comparisons were made of lice numbers on a cage group with wrasse (eight pens) compared with a control group with no wrasse. A second option was to compare sea lice numbers on nearby farms, one that stocked wrasse and the other with no form of biological control.

Wrasse were shown to clean *L. salmonis* from salmon in smaller pens with a considerable degree of replication in the case of pens with goldsinny wrasse (n = 25) compared with n = 5 control pens (Treasurer, 1993), and also by Bjordal for five wrasse species in this specific study (Bjordal, 1988), by Darwall et al. (1992), and with corkwing wrasse *Symphodus melops* (Tully et al., 1996). Cleaning activity was also demonstrated with a mixture of wrasse species in commercial circular pens (Treasurer, 2013). Lumpfish cleaning of sea lice from salmon has been demonstrated in 12 m^2 pens and also in circular pens in the Faroe Islands (Table 19.3). It is recognised that this type of trial is more difficult to carry out in large circular pens (Barrett et al., 2020).

There has also been a range of commercial trials with cleaner fish stocked with in an entire farm, for example Marine Harvest Ireland with 18 trials on multiple pens (Bolton-Warberg et al., 2018); 17 pens in the Faroe Islands (Johannesen et al., 2018); and in several pens in Newfoundland (Boyce et al., 2018). Contemporary use of lumpfish in large commercial pens has shown 95–97% removal of mature female lice compared with controls (Imsland et al., 2014), and there were significantly lower numbers of *L. salmonis* and *C. elongatus* from February to April with a 4–8% stocking density of lumpfish (Imsland et al., 2018a) (Table 19.3). In small- and large-scale trials with lumpfish in Norway there were significantly lower numbers of *C. elongatus* (Imsland et al., 2020a). Cleaning efficacy of lumpfish was reviewed in the Faroe Islands, with a seasonal effect with higher lice grazing from September to April and a reduced efficacy in the summer being noted (Eliasen et al., 2018). While commercial trials utilising large pens have rarely been reported in the scientific literature, the industry has regularly reported efficacy of cleaner fish at industry, technical and conference events, for example an international conference on wrasse stocking in 1995 (Sayer et al., 1996), a cleaner fish workshop at the European Aquaculture Society conference in Edinburgh in 2016 and an international cleaner fish conference in Glasgow in 2017 (Sustainable Aquaculture Innovation Centre (SAIC)/ Forskningsfinansiering (FHF)). Imsland et al. (2020a) also recently reviewed data on the cleaning ability of lumpfish towards *C. elongatus* in large-scale pens and found that with stocked lumpfish there were significantly lower numbers of *C. elongatus* during a large part of the salmon production cycle.

19.7 Numbers of cleaner fish stocked worldwide

Most of the wrasse being stocked in northern Europe are from fisheries as hatchery production has been limited to date and hatcheries still find weaning of wrasse difficult. Also, there are chronic trickling losses through a prolonged juvenile period as wrasse growth is slow. It may take 18–24 months to rear ballan wrasse to a minimum size for stocking of 60–70 g. However, some salmon farmers have stocked wrasse in the first summer at smaller sizes of around 50 g which will make production more rapid and the fish may be transferred to the pens within 12 months. Production was based mainly on ballan wrasse in the period 2010–2020 and has consistently been around the figure of two million wrasse in Norway and from 100,000–200,000 fish per annum in Scotland (Fig. 19.2). Due to the ease in farming lumpfish relatively few farmers have focused on ballan wrasse with the main producers being Mowi in Norway, and Machrihanish Marine Farms and Otter Ferry Seafish in Scotland, and Ocean Matters, Anglesey, Wales.

Wild caught wrasse, mainly goldsinny, rock cook *Centrolabrus exoletus*, corkwing and ballan, are the prime source of wrasse for stocking (Fig. 19.3). The annual catch in Norway has increased sharply

Fig. 19.3 Cleaner fish species: (a) lumpfish, and the wrasses (b) ballan, (c) goldsinny, (d) corkwing, (e) rock cook, and (f) cuckoo wrasse. (Credits: Claire Treasurer (a); Mark Woombs, WA Marine & Environment (b), (c), (e), (f); Felix Sproll (d).)

since 2012 and is around 22 million wrasse (Fig. 19.2), although it did peak at 27 million fish in 2017. Total catch quotas are allocated for northern, mid and southern Norway (Kousoulaki et al., 2018). Strict rules regulating size limits are applied to ensure conservation of much of the breeding stock of smaller wrasse species and also of broodstock ballan over 24 cm in length. Recording of catches was not mandatory in the wrasse fishery in Scotland until 2014. There is also a highly visible fishery along the English Channel to supply farms in north west Scotland with both fisheries initially contributing equally, giving a total retained annual catch from one to two million wrasse (Treasurer et al., 2018; Kousoulaki et al., 2018a) (Fig. 19.2). However, from 2018 the wrasse fishery in Scotland has been more important. The size limits for wrasse retained in Scotland are over 12 cm for small wrasse species and over 17 cm for corkwing and ballan, and ballan over 25 cm are also not retained; this is now a regulatory requirement. Marine Scotland has introduced regulations relating to the fishery to ensure that only experienced and qualified fishermen who adhere to welfare guidelines are certificated to fish for wrasse (Marine Scotland, 2020). The numbers of wrasse caught in Ireland, mainly corkwing wrasse, are small and undisclosed (Bolton-Warberg et al., 2018), and the same is the case for cunner *Tautogolabrus adspersus* in Atlantic Canada (Boyce et al., 2018). There is also a small fishery for broodstock lumpfish in Iceland and Norway (Steinarsson and Arnason, 2018). Most of the lumpfish that are stocked are from farmed sources with some hatcheries using wild caught broodstock and others, having closed the life cycle in the hatchery, using hatchery bred fish (Wittwer and Treasurer, 2018). In several hatcheries reared lumpfish have matured at one year old, unlike wrasse, in which males are at least six years old on sex reversal (Grant et al., 2016). Lumpfish production figures have increased rapidly in all countries with the annual figure now up to 40 million in Norway (Fig. 19.2), six million in the UK, two million in Iceland and smaller numbers in the Faroes, Ireland and Canada (Table 19.4). Farming of cleaner fish species such as the Patagonian blenny *Eleginops maclovinus* and also *Malapterus reticulatus* in Chile is still at an early stage.

It appears that farmed lumpfish are meeting demand but further improvement in survival and overcoming production bottlenecks is required for ballan wrasse and cunner. A main issue in lumpfish production is in the quality of the fish and ensuring that sucker disc abnormalities are minimised

Table 19.4 Cleaner fish production, fishery capture and stocking by country (in 2020).

Country	Cleaner fish species	Fished million/year	Hatchery reared million/year	Number hatcheries in 2020	References/Source
Norway	B, CW, GS, RC LP	Fishery 20–22	WR: 2 LP: 30–40	43	Norwegian Directorate of Fisheries
Iceland	LP	LP broodstock only	2	2	Benchmark Ltd, pers. comm.
Faroe Islands	LP	LP broodstock Small numbers	<0.5	1	Fiskaaling, pers. comm.
UK	LP, B, CW, GS	Wrasse: 0.5	WR: 0.2	6	Marine Scotland, 2021 a,b
	RC, CK	England, 1 Scotland	LP: 5.5		
Ireland	LP, WR	small numbers	none currently	1	Bolton-Warberg et al., 2018
Canada	LP, Cunner	LP broodstock		1	D. Boyce, pers. comm
Maine, USA	LP	nil	experimental	1	I. Bricknell, pers. comm.
Chile	*Eleginops maclovinus* *Malapterus reticulatus*	small numbers	experimental	1	Sanchez et al., 2018

LP = lumpfish, WR = wrasse, B = ballan, GS = goldsinny, CW = corkwing, RC = rock cook

(Rabadan et al., 2021). The requirement for these species may be reduced by re-utilising wrasse in subsequent salmon production cycles and also in improving survival of cleaner fish in the pens through improved supplementary feeding and development of more effective vaccines, and also improved care in handling and welfare of cleaner fish.

19.8 Evidence of fishing effects

Marine Scotland has issued mandatory conditions on the wrasse fishery in Scotland such as size retention limits, gear used and numbers of traps, and record keeping, and also the requirement for fishermen to be licensed by Marine Scotland and adhere to welfare and capture good practice (Marine Scotland, 2020). Only fishermen with two years' experience are now permitted to fish for wrasse. The recording of species and numbers of fish captured will inform fishery management measures (Marine Scotland, 2020). These include a close season for capture from November to May and size restrictions to exclude fish of less than 12 cm length and greater than 25 cm length. There are also limits on the numbers of traps fished of 250 per boat. Given the high price received for these fish the salmon farmers are also closely monitoring the welfare and condition of wrasse.

The capture methods, fisheries, catch data and conservation measures and methods of assessing stock were reviewed (Kousoulaki et al., 2018a). These indicated that assessment of stock size was difficult and could be done in four ways: by diver surveys over marked transects (Sayer et al., 1996), by analysis of catch per unit effort (CPUE) data over the same fishing grounds, by comparison of population age and length structure over time (Varian et al., 1996) and by comparison of abundance and population biology between fished and non-fished areas (such as marine protected areas (MPAs)) with similar seabed, current flow and algal characteristics (Halvorsen et al., 2017). These conserved areas are ones that have been designated as MPAs where there is a fishing ban or restrictions. There have been several objections about the effect of wrasse fisheries on stocks and local ecology, but adequate fish population censuses have not been carried out to adequately demonstrate this and there was no systematic reporting of catches in Scotland until 2014 (Riley et al., 2017). A greater priority is the capture and storage arrangements for wrasse and these have been improved, and also to ensure that by-catch is safely returned to the sea.

19.9 Production cycles and systems for cleaner fish

19.9.1 Wrasse

Several former cod *Gadus morhua* hatcheries have been converted for cleaner fish rearing in Norway and the UK. An area in hatcheries is allocated for broodstock tanks and these vary from 4 to 10 m diameter and up to 2 m deep (Fig. 19.5b). Broodstocks are established on photoperiod regimes. Hatcheries normally use egg collectors attached to the exit flow from broodstock tanks, as in the case of cod and other gadoids. This system works well with goldsinny wrasse as they also spawn pelagic eggs, but the other wrasse species such as ballan lay sticky eggs that are attached to rocks, seaweeds and other substrates. In the hatchery setting artificial substrates such as mats and tiles are used to allow wrasse to spawn on these and they are used to collect and remove ballan wrasse eggs from the tanks (Fig. 19.4) (Grant et al., 2016; Helland et al., 2014; Wittwer, 2017). Mats are placed in the bottom of tanks at the

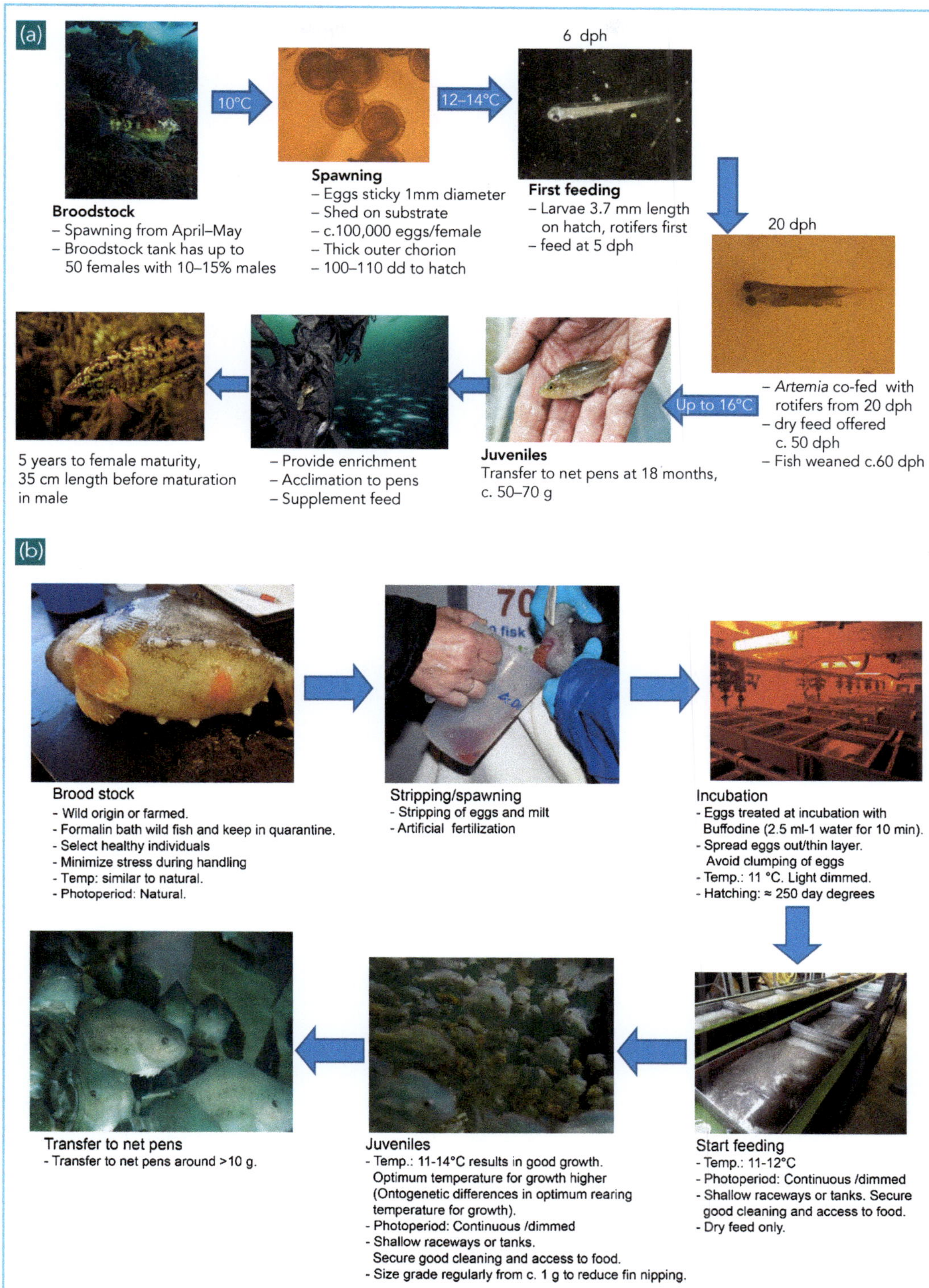

Fig. 19.4 The rearing cycles of wrasse (a) and lumpfish (b). (Credits (from top left clockwise): Macduff Marine Aquarium; J. Treasurer; Alan Dykes; Atle N. Kristofferson, OK Marine; Felix Sproll; Ingrid Lein; Claudia Wittwer; Stephen Murphy O'Sullivan; Claire Treasurer.)

Fig. 19.5 Rearing facilities for wrasse and lumpfish, from top left clockwise (a) cleaner fish hatchery (Credit: Otter Ferry Seafish); (b) broodstock unit (Credit: Øygarden, Mowi); (c) lumpfish egg incubation (Credit: Dorset Cleanerfish); (d) egg incubation (Credit: raceways at MRI Aquaculture Research Station, Iceland); (e) larval rearing (Credit: Dorset Cleanerfish); (f) juvenile tanks (Credit: NAFC hatchery).

onset of the spawning season and these are often coarse carpet tiles or astroturf mesh (Fig. 19.5). These are checked daily for eggs and are cleaned. Eggs are retained on the mats and incubated in a tank that holds several mats hung vertically, or eggs are stocked directly in larval rearing tanks and larvae hatch directly from the matting. The tanks are circular and from 2 to 28 m^3 volume. The nursery area often doubles for juvenile on growing as in the case of lumpfish or there can be a separate nursery area for wrasse. It may take two years for ballan wrasse to grow to a size suitable for stocking in salmon pens. Other areas of the hatchery are allocated to live feed production of rotifers in the case of ballan wrasse and also *Artemia*. Some hatcheries also use greenwater techniques with inhouse microalgal production units, or alternatively algal pastes, for example from Reed Aquaculture,.or clay suspension. Ballan wrasse larvae are first fed five to seven days post-hatching on rotifers. This is continued for around 20 days, and then enriched *Artemia* is offered for a further four to six weeks, before weaning ballan to dry feed (Table 19.5; Fig. 19.4). Many hatcheries now use *Artemia* replacement diets such as those produced by Skretting (Gemma diet). Fish are graded and reared in circular tanks from 2 to 28 m^3 volume or in raceways. Both flow through and recirculation systems are used for wrasse (Sveier and Breck, 2018), but rearing in heated water in recirculation systems improves growth rate and reduces stocking time from 3 years to 18 months. Cultivation systems for the cunner *Tautogolabrus adspersus*, found in Canada, are similar to ballan wrasse (Boyce et al., 2018).

19.9.2 Lumpfish

Lumpfish eggs are shed in clumps in the wild, but in the hatchery eggs are normally collected by hand stripping the fish (Fig. 19.4b) (Jonassen et al., 2018). Incubators may be long shallow hatchery troughs with water flow through. The eggs are stripped from the fish and fertilised dry and are moulded into the shape of the troughs. The egg layer is shallow, around 2 cm, to permit good waterflow and oxygenation through the network of joined eggs. There may be several other incubator designs with some being bucket-shaped, and others mounted alongside the main rearing tank with larvae hatching and flowing in the waterflow to the main rearing tank (Fig. 19.4). The lumpfish hatch at around 270 degree days (dd) and larvae are larger than in wrasse, 4–5 mm, and able to consume *Artemia* within five days of hatching (Table 19.5). The larvae are well developed on hatching and can attach to any available surface from day one with a ventral suction disc. Sufficient substrate should be available for larvae to prevent aggression and fin nipping. Fish are fed initially on *Artemia* to give a rapid growth rate and wean quickly to dry feed from *c.* 15 days. Many hatcheries in Norway wean directly to dry feed with no live feed required, first feeding with dry feed particles of 150–250 μm diameter. Frequent grading is required after larvae are *c.* 2 g to prevent tail nipping and cannibalism. Vaccination takes place at around 10 g weight and fish are transported to sea sites after five to seven months at around 20 g, although some sea sites stock lumpfish as small as 10 g. The immune induction period is 500 dd (Haugland et al., 2018). Growth in the salmon pens is rapid and there is a need to selectively breed slower-growing fish as lumpfish over 100 g are less likely to eat sea lice. Overall survival from hatch to stocking in sea pens is high and around 50% in many hatcheries.

19.10 Bottlenecks with hatchery production of cleaner fish

The production methods for cultivation of cleaner fish (Helland et al., 2014) have been established and are based on the technologies for rearing marine finfish such as sea bass *Dicentrarchus labrax* and

Table 19.5 Rearing protocols for ballan wrasse and lumpfish.

Period	Phase/protocol	Ballan wrasse	Lumpfish
Broodstock	Source	wild and farmed	wild mainly but some hatchery
	Size	female up to 28 cm, male >28 cm	c. 400 g male, up to 4 kg female
	Age	females mature 3–6 yrs, m >6 yrs	some 1st year fish mature
	Stocking density	20 to 30 fish, 2 to 4 males	20 to 30, 50:50 sex ratio
	Substrate	provide matting at spawning	provide rock nest area
	Hierarchy		
	Preparation	set up stocks before spawning	
	Temperature	12°C constant if possible	<12°C
	System	recirculaton if possible	
	Diet	enhance with fresh shellfish	
Spawning		May–June, but can be advanced/delayed with PP and temp. several egg batches 8 week spawning	March–April 1 batch if hand stripped over 8 weeks
Egg collection	Method	remove mats disinfect with formalin and/or bronopol hang in separate tanks	remove egg strand disinfect incubation units in trays or tanks
Incubation	temperature °C	12–14	8–10
	days dd	100	280 usually in upwelling tanks or raceways
Stocking tanks		mats transferred once fertilisation rate is in desired range and worth retaining	larvae transferred into rearing tank
Larval rearing	duration (dd)	300	250
	temperature (°C)	12	10
	lighting (lux)	300 to 600	300 to 600
	duration (days)	60	30
	stocking density ($kg\ m^{-3}$)	15	<25
	tank diameter (m)	2 clay	4 none
	green water technique	rotifers, *Artemia* after 30 d	most hatcheries dry feed only, Norway
	live feed	Multigain, or Origreen	
	enrichments (example)	around 60 but variable, Sparos/Otohime	Multigain
	first dry feed (days)		around 35 but variable, Sparos/Otohime
Juvenile period	duration (months)	18	8
Stocking in pens	weight (g)	from 50 but up to 70	regularly at 20, but up to 40
	age (months)	18	8
Salmon farm requirements	stocking density	1–5% wild wrasse, 4–10% for farmed	10% ± 2%
	timing	in July–Oct of 1st year, May 2nd yr	Nov 1st year salmon production
	shelter	heavy duty artificial seaweed	artificial seaweeds
	additional feeding	feed blocks, or WorldFeeds	pellets, e.g. Biomar or PT Aqua; trials with feedblocks

cod *Gadus morhua*. Production of lumpfish has increased rapidly and survival from hatching to stocking is high for a marine finfish species under hatchery conditions. There is the bonus that lumpfish grow quickly and can reach stocking size of 30 g within six to eight months. However, there remain production, disease, vaccination and quality issues. In many hatcheries the sucker of the fish can be deformed and does not permit attachment to a substrate. These fish are unlikely to survive for long in sea pens and, even in the hatchery, have to rest on the bottom of the tank often on their side. These fish may be culled at an early stage. There may be genetic or nutritional causes for these abnormalities, but the condition needs to be fully investigated. Lumpfish are susceptible to atypical furunculosis and vibriosis, although this may not be apparent until stocking in sea pens. Amoebic gill disease (AGD) has been a major pathogen of lumpfish both in sea pens and in the hatchery. Lumpfish are vaccinated at a small size, around 10 g, and 500 dd before seawater transfer to give a good immune response. However, there may be a requirement for autogenous vaccines to give better protection against local strains of *Vibrio*. Survival has been improved in the hatchery by treatment of AGD with freshwater and lumpfish have good tolerance of freshwater (Treasurer and Turnbull, 2018). Powell et al. (2017) summarised production issues and key areas for research in lumpfish rearing, and these included genetic selection for individuals that were demonstrable cleaners of sea lice, slower growth and also genetic selection for disease resistance.

19.11 Improving cleaner fish welfare and measurement of welfare

Fish welfare measurement has been of increasing importance in fish rearing in recent years. Fish welfare has been described by Huntingford et al. (2006) and had initially been explored in detail in salmonids (Turnbull et al., 2005; North et al., 2006; Adams et al., 2007). Fish veterinarians and health managers have utilised external fish condition as a marker of fish health and diagnosis, including bacterial and viral disease, eyes and physical abrasion from farm operations. Examination of eyes, gills, fins, operculum and skin have therefore been key indicators in health diagnosis. Welfare measures such as operational welfare indices (OWIs) are used for monitoring the ongoing condition of fish. Condition factor, for example Fulton's K as Wt/length^3 × 100, is rarely assessed on live fish that are being sampled for inspection on a group of pens. A length relative to weight graph has been proposed by Rabadan et al. (2021), but it can be difficult to measure weight of fish at sea in salmon pens, unless fish are sacrificed and measured onshore. Assessment of visual signs on graded scales such as fin erosion and fin splitting, skin erosion, eye condition and abnormalities such as snout and operculum damage and lumpsucker disc anomalies is therefore more practical for routine welfare assessment. This can also reflect a quality issue with fish rearing and lumpfish with sucker disc abnormalities should be graded to prevent welfare issues in salmon pens and also higher levels of mortality. Late and poor timing of grading in the hatchery and unsuitable feed inputs can give rise to aggression and severe issues with fin erosion and especially the tail although the pectoral and other fins may also be affected (Geitung et al., 2020; Rabadan et al., 2021).

The widely used welfare assessment scheme for fish is the RENSVEL classification developed by NOFIMA for salmon and adapted for application with cleaner fish (Table 19.6). This assesses damage on a two- or three-point scale (Noble et al., 2019a, 2019b). The three-point scale one, two and three is used to assess condition of the fins, skin and eyes, where one means no damage, two means minor damage and three means a fin split greater than 60%. A two-point grading is applied to abnormalities such as deformed jaw and sucker disc, and snout and opercula damage. These four indices are graded

Table 19.6 Published operational welfare index scoring systems for wrasse and lumpfish.

Species	Scoring system	Main OWIs	'2ndry'/other OWIs	Comments
Wrasse	Treasurer and Feledi, 2014	Fin splitting and erosion, K	Snout, skin, eyes	5 point scale
	Arechavala-Lopez et al., 2013	Fin splitting and erosion	None	5 point scale
	Gentry et al., 2020	Fins, skin, eyes, sucker disc	Jaw, snout, opercula	
	Noble et al., 2019a	Fins, skin, eyes, sucker disc	Jaw, snout, opercula	3 point scale
Lumpfish	Treasurer et al., 2018	Fin erosion, especially tail		5 point scale
	Rabadan et al., 2021	Skin, caudal fin, eye, suction disc	Relative weight	Use 5 indices=LOWSI
	Noble et al., 2019b	Fins, skin, eyes, sucker disc	Jaw, snout, opercula	3 point scales
	Imsland et al., 2020b	'Body condition' wounds/sores, fin damage/erosion, other fins damage/erosion	Sucker and spine deformities, cataracts, eye ulceration and condition factor	Mainly 4 point scales
Both WR and LP	Geitung et al., 2020	SGR, fins, skin, eyes, jaw & sucker disc deformity, snout, opercula		Used floy tags

with zero meaning no damage and one being damage or wound present. In lumpfish, Rabadan et al. (2021) use an overall index termed the lumpsucker operational welfare score index (LOWSI), which incorporates five indices. These are also on a three-point scale, but rather than one, two and three as in RENSVAL, zero means no damage, one means moderate damage and two means severe damage. Scores for skin damage, caudal fin damage, eye condition, suction disc abnormality and relative weight give a total LOWSI score. This rapid scoring system gives a visual table that assists scoring and takes *c.* 2 min for each fish. This score index would allow changes in condition on a farm to be tracked on time, and each score band level initiates a sequence of actions in mitigation of changes in welfare status. A 'health assessment score' was used for lumpfish during a study of the effects of using feeding blocks (Imsland et al., 2020b). This is equivalent to an individual OWI for lumpfish. Seven indices were used based mainly on a four-point scale, although a five-point scale was used for cataracts and eye ulceration. The indices assigned were: 'body condition' wounds/sores, fin damage/erosion, other fins damage/erosion, sucker and spine deformities, cataracts, eye ulceration and condition factor, with overall scores calculated for all indices and grouped in three score categories. A more detailed five-point scale of fin erosion and splitting index on all fins based on Arechavala-Lopez et al. (2013) was applied to five species of wrasse stocked as cleaner fish (Treasurer and Feledi, 2014) (Table 19.6) and this was used to compare the relative damage between the five species. The fin splitting index (FSI) for the tail was significantly higher than for the other fins and the fin erosion index (FEI) was relatively low. The study demonstrated that the FEI and FSI did not decline in ballan wrasse in commercial pens over an 18-month period. Other indices used were eye damage, skin erosion and condition factor, jaw damage and opercula damage, and the scores for these were low and overall incidence negligible. The FSI and FEI scales were also applied successfully to lumpfish in the hatchery. Unlike wrasse, fin splitting in lumpfish was negligible but the tail was subject to erosion and largely due to aggressive interaction and observed tail biting.

Welfare of cleaner fish is maintained through regular supplementary feeding and also cover provided by artificial seaweeds and hides.

19.12 Wrasse versus lumpfish efficacy

As the production of wrasse is more difficult the price of farmed fish is considerably higher than for lumpfish. It has been recognised that both wrasse and lumpfish can clean sea lice from salmon. Salmon farmers therefore see both species as having a role in controlling sea lice with small wrasse stocked in the first salmon production year when water temperatures are greater than 10°C, and lumpfish stocked at the onset of winter as they are active at lower temperatures. Geitung et al. (2020) reported that ballan did not grow over winter and condition factor declined indicating that ballan were inactive at colder temperatures. Some companies will supplement wrasse numbers with larger wrasse in the second salmon production year. Farmers would like more wrasse to be supplied as farmed rather than fished wrasse, but this is not currently possible.

It has been suggested that larger lumpfish are aggressive to wrasse (Imsland et al., 2016a) and force them away from certain areas of the pens, and this is exacerbated as lumpfish grow rapidly compared with slow-growing wrasse. However, the different species of cleaner fish occupy different depth strata and may avoid competition, for example lumpfish are commonly found near the water surface, ballan wrasse at the bottom of the pen and goldsinny are mainly found at depths of 3–4 m (Geitung et al., 2020).

19.13 Comparison between uptake and practice between countries

The rearing and application of cleaner fish for sea lice control has been implemented in several countries (Table 19.4). While Norway is the leading nation with 22 million fished wrasse, two million farmed wrasse and over 40 million reared lumpfish, cleaner fish have been widely used in the Scottish industry with over one million wild caught wrasse, 150,000 farmed wrasse and over six million reared lumpfish (Fig. 19.2). Ireland was a leader in the use of wrasse to clean farmed salmon in the 1990s and has also trialled stocking of lumpfish and wild caught wrasse *c.* 2014–2018. There has been lumpfish research and also production and selective breeding in Iceland, with over two million juvenile lumpfish produced and also large volumes of eggs have been supplied to other customers. Many of the lumpfish were shipped to the Faroe Islands but these now have a lumpfish hatchery at Fiskaaling. The cleaner fish cunner and lumpfish have been reared and stocked in Atlantic Canada in New Brunswick (Boyce et al., 2018). Work is at an exploratory stage in rearing Patagonian blenny *Eleginops maclovinus* and *Malapterus reticulatus* in Chile (Sanchez et al., 2018). Some trials have been carried out with stocking on salmon farms. It has still to be determined if these species will be suitable cleaner fish in a farm setting and whether they will feed on *Caligus rogercresseyi*.

19.14 Advances in wrasse rearing and deployment: latest developments and research needs

The expansion of farmed wrasse production has led to new production challenges but also opportunities. The sector needs to be constantly innovative in the use of cleaner fish to adapt to the new challenges and make the best of the opportunities, while exploring new ways of ensuring wrasse are provided with the best environment and tailored husbandry practices.

19.14.1 Recent scientific and development advances

The farming of cleaner fish is a growing sector, and keen in embracing innovation. This is best reflected by the most recent advances, briefly outlined here.

Wrasse broodstock

Broodstock management is essential to have a stable and reliable supply of good quality eggs and robust larvae. Ballan wrasse are protogynous hermaphrodites (Dipper, 1987). The number of males is estimated to be only 10% in natural populations (Dipper et al. 1977; Lein and Synnøve, 2014). This can restrict production and genetic makeup of the next generations. Sex induction, to achieve extra males, has been carried out in commercial hatchery conditions by using hormones such as methyl testosterone and fadrozole, applied shortly after spawning to induce fish to become male after spawning season (SAIC project). Wrasse hatched in captivity at Otter Ferry Seafish, Scotland, spawned for the first time, and this represented the closure of the life cycle in ballan wrasse. All broodstock were chip tagged and genotyped which, accompanied with parentage assignment, allowed targeted selection to begin and the species to be improved for commercially important production traits. This is a significant step forward in the scaled farming of the species.

Nutrition in wrasse

First feeding in marine fish species can be a critical bottleneck in production. Ballan wrasse larvae require live feed with specific nutrient requirements such as rotifers and *Artemia* nauplii. These require culturing and 'enrichment' to increase the nutritional value. Marine fish larvae are vulnerable during early development and require strict biotic and abiotic conditions to survive, develop and grow (Hamre et al., 2013). The lack of these could cause deformities in their skeleton, especially jaw and head. There are several commercial live feed enrichment products but little is known of the micronutrient requirements for wrasse. Commercial enrichment products lack phospholipids and essential long-chain polyunsaturated fatty acids. Marine lecithin is a good candidate to enrich *Artemia* nauplii in both phospholipid and long-chain polyunsaturated fatty acids. Cavrois-Rogacki et al. (2019a) tested soya lecithin through a process referred to as 'boosting'. This alternative did not significantly increase the long-chain polyunsaturated fatty acids content of the phospholipid fraction in comparison to standard marine lecithin. However, boosting with marine lecithin resulted in nauplii presenting similar essential fatty acid and/or phospholipid profiles to that of wild copepods (Cavrois-Rogacki et al., 2019a).

Deformities have been observed at post-weaning. Their aetiology appears to be primarily during early larval development (Cavrois-Rogacki et al., 2021). It is thought that by supplying exactly the right minerals at the right time, deformities can be avoided or substantially minimised, and the welfare of the larvae/juvenile improved, which will translate into more robust fish available for deployment and higher survival rates with better performance once deployed (D. Carcajona, June 2021, personal observation). During weaning, ballan wrasse larvae are fed crustacean meal based formulated diets (Skiftesvik and Bjelland, 2003). Dietary inclusion of full fishmeal from the onset of weaning to artificial diets significantly reduces feed acceptance and weaning survival in this species (Kousoulaki et al., 2014). Prolonged feeding periods with hot extruded pellets during ballan wrasse larvae weaning could result in low body mineralisation and the development of severe skeletal deformities (Kousoulaki et al., 2021).

Little is known of the distinct digestive system of wrasse. The anterior segment of the ballan wrasse intestine does not appear to function as a food storage region, but it is key in macronutrient digestion (Le et al., 2019a). They suggest that macronutrients were digested and absorbed along the whole intestinal tract but mainly in the first segment, where nutrient absorption was highest, and that feed moisture content does also affect protein and carbohydrate digestibility. On the other hand, lipid digestibility did not seem to be affected by feed moisture levels. Cholecystokinin (CCK) may modulate intestinal motility differently in ballan wrasse than that known in gastric vertebrates (Le et al., 2019b). CCK inhibits the propagating contractions in the foregut and reduces the velocity and amplitude of these. The standing contractions mixing the gut contents are maintained, improving the digestion and absorption of nutrients, and consequently improving the efficiency of the short digestive tract. Growth of ballan wrasse is slow and this constrains production, as at least one year is required to reach deployment size. Cavrois-Rogacki et al. (2019b) compared growth at 10°C, 13°C and 16°C over a period of three months and using two different diets. Wrasse juveniles showed enhanced growth performance (that is, weight gain, specific growth rate (SGR) and thermal growth coefficient (TGC)) when reared at a constant temperature of 16°C. Neither TGC or SGR showed any differences due to diets and thus the pooled dataset demonstrated a constant increase suggesting that the temperature at which maximum growth is obtained for the species lies beyond 16°C.

Health management in wrasse

The high mortalities during larval stages are one of the main bottlenecks in wrasse production, which is common in many other marine species (May, 1974). The greatest health challenge restricting productivity of ballan wrasse are losses due to atypical furunculosis caused by the bacterial pathogen atypical *Aeromonas salmonicida*. This is the predominant bacterial pathogen isolated during disease outbreaks in salmon pens and hatchery sites (Brooker et al., 2018). No commercial vaccine is currently available. Two types of *A. salmonicida* (predominantly V, but also VI) have been reported in diseased ballan wrasse in Scotland and Norway (Brooker et al., 2018; Papadopoulou et al., 2021). Atypical *A. salmonicida* (aAs) vapA type V is the predominant bacterial species in sea sites in Scotland (Papadopoulou et al., 2020). The available multivalent autogenous vaccines have been shown to be highly protective against homologous and heterologous strains of atypical *A. salmonicida* when administrated through injection. Injection vaccination offered good protection in pre-deployment ballan wrasse (greater than 25 g) and the vaccine elicited specific antibody responses (IgM) in the fish (Ramirez-Paredes et al., 2020). However, current autogenous vaccines did not seem to offer protection for ballan wrasse in early development stages when used as an immersion vaccine. It is unknown whether the immune system is fully competent in young fish and further studies should investigate immersion vaccination in order to offer protection from early development stages.

Wrasse husbandry and welfare

The pen environment is very different to the controlled conditions in the hatchery and nursery (such as constant temperature/light, reduced pathogen load). It is essential that farmed ballan wrasse produced in such environments are conditioned to cope with such a radical, stressful and challenging change as salmon pen deployment. Pre-deployment acclimation involves using the same hides as the receiving pens, introducing natural photoperiod (light/dark cycles) and providing the same type of feed blocks in the hatchery as on farms. Post-deployment acclimation happens at sea in temporary keep nets for maintenance of the wrasse and they are separated from the salmon. Sufficient artificial kelp should be provided to accommodate all the fish and feed blocks should also be used. These should be checked daily and regularly replenished for a period of time, while behaviour and mortality should be closely monitored (D. Carcajona, June 2021, personal observation).

Passive-acoustic telemetry (PAT) has been used to track cleaner fish movements and behaviour studied in four experiments carried out in salmon pens (SAIC, 2021). PAT is a useful technique for observing the fine-scale behaviour of individual cleaner fish in salmon sea cages (Leclercq et al., 2018).

Brooker et al. (2020) described the behaviour of pre- and post-deployment acclimated ballan wrasse. Acclimated fish displayed behaviour more similar to wild counterparts; they were active at shallower depths, showed diurnal rhythms and established larger home ranges more rapidly than non-acclimated fish. This behaviour is similar to wild ballan wrasse that also have larger territories and higher activity than non-acclimated farmed wrasse. These findings indicate wider use of the pen volume by wild and acclimated wrasse than non-acclimated fish, thus increasing the opportunities for interaction with salmon. Extending the acclimation period may further shift their overall behaviour towards that of wild wrasse, although further investigation is required (Brooker et al., 2020). Likewise, exposing wrasse to sea lice infested salmon during pre-deployment acclimation could improve de-licing performance at sea. Ballan wrasse consumed more lice when pre-exposed to lice-infested salmon (0.79 ± 0.4 SD lice found/stomach) compared with wrasse that had not been pre-exposed to salmon (0.15 ± 0.1 lice, $p = 0.003$) or wrasse pre-exposed to salmon without lice (0.01 ± 0.09 lice, $p = 0.03$) (Gentry et al., 2021). Despite this, higher lice numbers found in the wrasse stomachs did not translate to lower lice infection on salmon or improved welfare of ballan wrasse during the experiment. However, sea lice numbers do not depend only on cleaner fish performance, but are also influenced by exterior sea lice infection pressure (D. Carcajona, June 2021, personal observation).

19.14.2 Future research and development needs for wrasse

Although research and innovation are being made at a good pace and there is increased knowledge, cleaner fish farming is still in its infancy. Knowledge gains have still to be made in biology, environment, nutritional requirements, social interactions and delousing behaviour that, in turn, should translate in to better husbandry and handling practices.

Wrasse genetic selection

Enhancing the robustness of the farmed stocks and increasing hatchery outputs, to meet the demands from the salmon sector and to protect wild stocks from overfishing, is the main goal for the farmed wrasse sector in the coming years. The scaled production of a new marine aquaculture species is a significant challenge that demands sustained effort. Closing the species life cycle has therefore been a major first step forward. However, as not all farmed wrasse batches seem to be equal in their ability to delouse and cope with stress and change of environment, identifying genomic markers for relevant and desirable traits should be a focus.

Ballan wrasse display structured spawning behaviour and cohabiting females make unequal contributions within the spawning group. Spawning events occur in cycles of approximately 12 days

(Grant et al., 2016). This could further limit the performance of a hatchery, hampering the progress towards genetic selection. Tools for parental assignment and contribution are essential. Microsatellite markers have been recently developed for various North Atlantic wrasse species (Quintela et al., 2014; Jansson et al., 2016). These could help enhance parentage assignment and, together with genotyping, allow farmers best management to maintain genetic diversity and improve selection.

Wrasse malformations and nephrocalcinosis

High prevalence of vertebral malformations (up to 33%), jaw/operculum malformations (up to 13.5%) and nephrocalcinosis (up to 25%), and with high severity levels in some cases (Cavrois-Rogacki et al., 2021), are issues in early development. This may lead to downgrading of part of the production causing subsequent economic losses for hatcheries and reducing the numbers of wrasse available for the salmon sector. In Norway the figures are higher, with reports suggesting the presence of severe vertebral malformations ranging between 17% and 53% of farmed wrasse (Fjelldal et al., 2020). The feeding apparatus of marine fish larvae ossifies first (Koumoundouros et al., 1999; Boglione et al., 2013). It could be expected that nutritional deficiencies during early development would first become evident in the mouthparts and related head structures together with fins and back bone.

The mechanisms behind the development of nephrocalcinosis and skeletal deformities are not yet fully understood, and a knowledge gap that needs to be addressed. For example, the SAIC project *Boosting ballan wrasse hatchery production (Ballan Plus)*, current in 2022, is characterising the pathogenesis of malformations identified in farmed ballan wrasse, comparing prevalence and severity between hatcheries and identifying potential nutritional impacts of different live feed/dietary regimes on fish malformation development, prevalence and severity, together with immunostimulant assessment on the fish immune system and resilience.

Wrasse nutrition

Suitable diets for all development stages, from larvae to sea phase, with inclusion of specific raw materials that do not compromise intestinal health or digestive functions, does have a positive impact on their health and welfare and therefore survival and performance. It is crucial to develop the best enrichment strategies for live prey such as *Artemia* to mimic the macro- and micronutrient profile of natural prey. This has been proven difficult due to the limited efficiency of the enrichment process, along with the occurrence of undesired metabolic conversions that live prey exerts on enrichment products (Cavrois-Rogacki et al., 2019a).

Wrasse health management

Health and welfare management should be a priority research area. Excellent husbandry protocols and vaccines are key to improve and safeguard health and welfare, increase survival and hatchery output and further reduce the use of antimicrobials. *A. salmonicida* is the main pathogen restricting cleaner fish farming in Scotland and for which no fully licensed vaccines are currently available. Due to the lack of commercial vaccines available for ballan wrasse, the use of autogenous vaccines has increased in Scotland (Brooker et al., 2018). However, atypical *A. salmonicida* persists in causing losses during all development stages. The SAIC project *Optimal* Aeromonas salmonicida *ballan wrasse vaccination through informed immunological assessment* will aim to determine the optimal timing, regime and route of administration of vaccines, together with formulating improvements by refining antigen selection and identification of new immunogenic antigens. Likewise, understanding wrasse immune responsiveness and immune competence is essential to develop effective dip vaccines and vaccination regimes to protect highly susceptible young fish especially, and with a relative ease of delivery to minimise handling and stress.

Wrasse behaviour and cohabitation

Health and welfare and delousing behaviour post-deployment can be improved by more knowledge of wrasse behaviour in captivity, both in hatchery tanks and in sea pens. This includes cohabitation with other cleaner fish species and interactions with salmon stocks. This should be combined with the continuous refinement of current acclimation protocols while exploring new ones. The likely mechanism behind greater lice consumption by lice-acclimated wrasse is learning. Lice-acclimated wrasse had more opportunities to practise lice-feeding behaviours and optimise prey handling (Gentry et al., 2021). Further work investigating lice consumption performance under different acclimation protocols and in different environments is therefore required.

Once deployed, excellence in husbandry, such as the provision of adequate hides (material, configuration, depth according to the season), clean nets and supplementary food always available (especially important before winter), and routine health monitoring are essential. It has become clear that wrasse acclimation prior to transfer and on farm arrival significantly improves overall wrasse behaviour (used as a proxy for delousing) (Brooker et al., 2020). New knowledge will help to improve this during the hatchery phase and post-deployment to safeguard cleaner fish health and welfare and further enhance delousing efficacy. The SAIC project *Aquawrasse*, current in 2022, is examining this through an experiment comparing wild and farmed wrasse. Swimming behaviour in association with hatchery acclimation, fish personality, interaction with salmon (mutual behaviour) and coping with a routinely applied stressor in a sea pen are being studied.

Although research on cleaner fish behaviour to date has focused on the cleaner fish, the behaviour of salmon cohabiting with cleaner fish remains undocumented. Studying wrasse behaviour and interactions with salmon could also help understand and improve hide configurations or feeding regimes that promote cleaner fish health and welfare and improve salmon-cleaner fish interactions.

A summary of key bottlenecks, health and welfare issues, and research needs in wrasse and lumpfish rearing and deployment is given in Table 19.7.

19.15 Lumpfish rearing and deployment

As lumpfish tolerate lower temperatures than wrasse, stocking as cleaner fish was boosted principally in the northern parts of Norway. In the beginning, the production of juveniles was solely based on wild caught broodfish, where eggs were stripped, incubated, hatched and reared to a suitable size for transfer to commercial cages. With the demand for lumpfish steadily increasing, this has resulted in gradually moving from wild caught broodfish towards intensive cultivation. Lumpfish production is currently utilising technology and techniques used in halibut *H. hippoglossus*, wrasse and cod aquaculture, and a breeding programme for lumpfish was established in 2017. Several R&D projects are currently in progress in Norway on lumpfish broodstock (for example, RCN269043 – STAMINA, FHF901418 – CYCLOBREED). Two commercial breeding programmes have also been initiated focusing on lice grazing, moderate growth, robustness/survival and disease resistance (Imsland et al., 2016b, 2021). Although wild caught broodfish still are predominant, the cultivation of lumpfish is gradually moving towards intensive production. In 2019 more than 22 million lumpfish were transferred to cages at commercial marine growth sites for salmonids, which constituted 46% of all

Table 19.7 Main bottlenecks in production of cleaner fish, health issues and research priorities.

Item	Wrasse	Lumpfish
Bottlenecks rearing	Genetic selection Larval survival and first feeding Stable and predictable egg production and fertilisation rates Nutritional requirements	Genetic selection Create broodstock. Move away from wild and imported Milt and egg quality Stable and predictable egg production and fertilisation rates
Bottlenecks pen deployment	Early availability Size, larger fish, improved growth rate enhance survival Refine acclimatisation pre and post deployment Right nutrition, feeding regime and shelter	Transport and handling Size, larger fish enhances survival Acclimatisation pre and post deployment Right nutrition, feeding regime and shelter
Health issues hatchery	Furunculosis Aggression and fin nipping Nephrocalcinosis Jaw and opercula deformities	Aggression and fin nipping AGD Bacterial infections
Health issues pen deployment	Furunculosis Gill challenges Damage during day to day work and interventions Survival Right nutrition, feeding regime and shelter	Bacterial and fungal infections Gill challenges Damage during day to day work and interventions Survival Right nutrition, feeding regime and shelter
Welfare issues wrasse and lumpfish	Furunculosis and other infections Gill challenges Damage during day to day work and interventions Appropiate hides, amount and location Appropiate feed, location and feeding regime	

cleaner fish used (Sommerset et al., 2020). Lumpfish is therefore the single most important cleaner fish species to date.

As lumpfish is a novel species to aquaculture there are knowledge gaps that often endanger the health and welfare of the fish (Powell et al., 2018; Treasurer et al., 2018; Imsland et al., 2019a, 2020b). Three duplicate groups of individually tagged lumpfish (mean initial weight: 22.3 ± 2.5 g) were established. Although lumpfish seem to be prime candidates for biological control of sea lice, current farming conditions and procedures often compromise their health, and consequently their performance, rendering the fish ineffective and even raising welfare concerns.

A study conducted by the Norwegian Food Safety Authorities (Mattilsynet, 2020) revealed over 40% mortality of lumpfish deployed in Atlantic salmon net pens in Norway. Results from the survey showed that farmers associate lumpfish mortalities with the occurrence of disease, but also handling and mechanical procedures such as mechanical delousing. Lumpfish vaccination without anaesthesia is common practice, despite the lack of scientific studies showing any benefit of avoiding the use of anaesthesia during vaccination. This methodology can potentially contribute to events of extreme acute stress responses that can often trigger other secondary issues (Martos-Sitcha, 2020). Despite the common use of vaccination, the consensus among farmers was that these vaccines were not effective and required further development. Findings based on additional assessments from the lumpfish and other cleaner fish species use in Norwegian aquaculture accord with the Mattilsynet study. Reports from the Havforskningsinstituttet (Grefsrud et al., 2019) and the NTNU Samfunnsforskning (Størkersen and Amundsen, 2019) depict a grim scenario of lumpfish aquaculture, from hatcheries to Atlantic salmon farmers, where lumpfish welfare is questioned. The reports found lack of standardised practices, inadequate feeding, lack of dedicated personnel to follow up and constantly monitor health, and even lack of fundamental knowledge regarding species-specific needs (Stien et al., 2020).

Reduction of lumpfish welfare status has also been reported after transportation and deployment in sea pens, which can pose significant stress. Conditions such as rapid onset of cataract occurrence have also been demonstrated in deployed lumpfish (Jonassen et al., 2017; Imsland et al., 2020b), which is often associated with suboptimal feeds (Imsland et al., 2018b). Further understanding of lumpfish biological requirements and stress physiology is necessary to develop better methods that safeguard lumpfish wellbeing and meet their needs.

Although the surveys performed on lumpfish aquaculture stakeholders highlighted the current grim situation, it was also pointed out that a significant part of the data was not included in the studies, as the reported data lacked coherency and method. This emphasises the urgent need of standardisation of methodology such as in health monitoring, and the great upside potential for improving lumpfish welfare by adopting strategies that safeguard lumpfish wellbeing. Continuous health and welfare monitoring are essential to help identify when and what procedures and operations are detrimental, and thus allow adaptation and improvement of practices (Rimstad et al., 2017; Noble et al., 2019a, 2019b). Practical and user-friendly approaches are necessary to encourage the adoption of standardised monitoring in a more comparable way.

19.15.1 Hatcheries

There is an urgent need to determine potential health issues from first feeding after hatching and up to pre-transfer to sea cages. In addition, many hatcheries have their own production management strategies which vary widely between them. For example, feeding strategies vary widely between hatcheries with some employing constant feeding regimes while others use more restrictive regimes based on natural daylight strategies. In addition, there is no clear standardised approach to specific photoperiod lengths during the tank rearing phase. Some farmers utilise a 24:0 LD regime, and others may utilise an 18:6 LD regime, with others preferring a 12:12 LD regime (Jonassen et al., 2018).

There is little information on mortality in hatcheries. Further, it has become evident that lumpfish reared in hatcheries have varying incidences of cataracts. Some hatcheries have suggested that up to 45% of their fish have well developed cataracts just prior to transfer to sea cages. These fish will have limited visual acuity which may be the reason why mortalities are observed a few weeks after transfer. If 40% of the lumpfish transferred to sea cages are visually impaired then they are unlikely to graze on attached sea lice and the remaining fish may be unable to maintain a low sea lice burden in the cage. The future reporting of lumpfish mortalities and assessment of health status within hatcheries would lead to improvements in their current management and identify favourable and unfavourable husbandry conditions.

19.15.2 Sea cages

Recent reports have suggested low survival rates of lumpfish in sea cages with Atlantic salmon (Klakegg et al., 2020). Most mortalities in hatcheries and sea cages have been attributed to bacterial infection. However, infection may not always be the primary cause and other causal agents may be responsible. In a survey in five Atlantic salmon net pen sites with 79,000 lumpfish, conducted in 2013 in Norway, an average of 48% mortality was reported, varying from 39% to 100%. Of the 13,864 dead lumpfish categorised regarding cause of death in the same survey, 75% were assigned as bacterial infections (Nilsen et al., 2014). In contrast, data derived from one commercial farm where lumpfish were deployed at two different sites with a stocking density between 7% and 8% (Fig. 19.6) showed weekly mortality never exceeded 1% (Patrick Reynolds, unpublished data). The levels of mortality recorded vary widely between commercial sites and management strategies vary between users. This may explain some of the differences observed. Research is urgently required to elucidate good and poor management and husbandry strategies to facilitate further use of lumpfish as cleaner fish and seek all ways to improve the welfare and survival of the fish.

Bacterial infections are considered to be the main cause of the high mortality rates observed in lumpfish (Hjeltnes et al., 2017; Klakegg et al., 2020). However, a previous study has shown that the prevalence of cataracts can vary between 20% and 100% in lumpfish populations (Jonassen et al., 2017). Further, fish with severe cataracts have reduced ability to locate food items and as a result this affects feed intake leading to reduced growth and weakened immunity and robustness of the fish (reduced stress tolerance) which can result in high mortality rates. This has been observed in a recent study with over 80% of lumpfish which developed severe cataracts and, of these, 50% were found to have severe weight loss (Patrick Reynolds, unpublished data). There is a lack of knowledge as to the primary cause of death of lumpfish, particularly in sea cages. The cause of death may be multifactorial, for example poor nutritional input and/or inadequate feeding

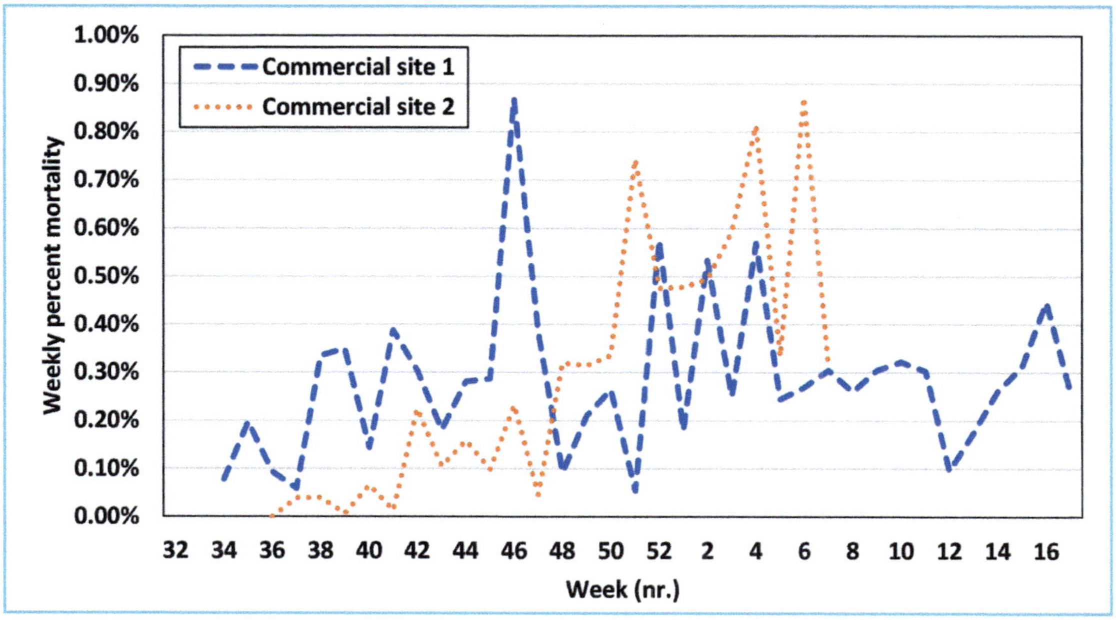

Fig. 19.6 Mean weekly percentage mortality rates of lumpfish from two commercial sites.

strategies. Environmental perturbations may also be a factor, as studies have shown that both low temperatures (less than 4°C) (Imsland et al., 2018a) and high temperatures (18°C) (Hvas et al., 2018) can also increase mortalities. Further, the lack of vaccine development and evaluation may also be involved.

High mortality can also occur after transfer to commercial cages (Fig. 19.6) due to poor handling and has been linked to chemical and mechanical delousing practices (for example, Kousoulaki et al., 2018b). Further, mortalities have been attributed to net cleaning, bath treatments or other farming operations (Rimstad et al., 2017; MarinHelse, 2018).

19.15.3 Applied research to improve lumpfish welfare and sea lice grazing

In a previous study, lumpfish exposed to live prey and frozen sea lice prior to transfer to sea pens with salmon were shown to have the ability to learn and recognise this prey after transfer to sea pens (Imsland et al., 2019b). This skill was evident throughout the study period as the cages stocked with lumpfish exposed to live prey had lower infestation levels compared to lumpfish reared on pelleted feed. The effect of the conditioning treatment was also manifested in the behaviour as they spent more time resting, swimming along the net sides and eating at biofouling. The study concluded that conditioning to live feed and frozen sea lice prior to deployment of the lumpfish into open net pens can alter the fish's efficiency in removing sea lice from salmon.

Breeding programmes have shown clear differences in sea lice grazing efficacy, growth and cataract prevalence between different lumpfish families (Imsland et al., 2016b, 2021). The series of studies has shown differences in sea lice grazing preference between families and results to date indicated that the smallest size classes of lumpfish (40–140 g) exhibited higher sea lice grazing potential compared to the larger size classes within families. Further, differences in health status and cataract prevalence show that health and grazing potential can be enhanced using targeted family production and, if this behaviour has a genetic basis, it may be further enhanced through selection and targeted breeding programmes.

The potential heritability of grazing preference traits was investigated in two separate studies where a total of 19 lumpfish families (nine for study one (Imsland et al., 2016b) and ten for study two (Imsland et al., 2021)) were deployed at sea. Results indicated a paternal effect on grazing behaviour, where specific lumpfish families were found to have higher predisposed preference for natural food sources including *C. elongatus* (Imsland et al., 2014, 2019c). Both studies showed variation in lice grazing between families. In particular, the second study revealed that two families consuming the highest numbers of *C. elongatus* shared the same father but different mothers. This suggests that grazing preference may be influenced by passed paternal genetic information.

Lumpfish feed and nutrition

Previous studies have shown that utilising feed blocks as the main source of maintenance nutrition results in controlled growth, reduced prevalence of cataracts and better health compared to lumpfish fed with pelleted feed in land-based tanks (Imsland et al., 2018c, 2019c). In a recent study in commercial salmon cages, lumpfish fed with pelleted feed were more sensitive towards mechanical delousing treatments compared to lumpfish fed with feed blocks. This sensitivity was manifested in higher mortality during treatments (Imsland et al., 2020b). Operational welfare indicators (OWI) were used to assess health and condition for both groups. The OWIs showed a lower score in the group offered feed blocks indicating better health status. Overall, the findings of the study show that feed blocks can give more controlled growth, increased survival and better overall health status of lumpfish in sea pens (Imsland et al., 2020b).

19.15.4 Research needs for lumpfish rearing and deployment

Improvement of existing vaccines and development of new vaccines is a priority and a *Pasteurella* spp. vaccine is urgently needed. Optimal feeding strategies for lumpfish should be developed during the hatchery phase and also further research in sea cages is required. Controlling growth of lumpfish is essential to maintain health and lice grazing efficacy. Research to determine optimal husbandry strategies in sea cages includes, for example, the size (and age) of lumpfish at transfer, habitat enrichment (identifying the most successful shelters/substrates to use) and at what time of year lumpfish should be transferred to sea cages (A. Imsland, 2022, personal communication).

19.16 Future developments and further cleaner fish species that may be suitable for aquaculture application

The stocking of wrasse and lumpfish is well developed on salmon farms in northern Europe. Cleaner fish may be assessed and used in farming sea bass and sea bream, although opportunities to assess goldsinny *Ctenolabrus rupestis* as a cleaner of farmed fish and the use of other potential wrasse species in the Mediterranean have not been pursued. However, cunner and lumpfish are being deployed with salmon in New Brunswick and Maine and there are promising trials with two species of cleaner fish in Chile. Preventative methods generally are seen as an area that can be developed further in IPM of sea lice, and cleaner fish have a central role in reducing sea lice infection and as part of this IPM approach. Large-scale trials in a replicated way on large pens are difficult due to scaling of pens and farm operations (Overton et al., 2020) and representative sampling of cleaner fish in large pens is difficult. Ongoing work in establishing cleaning efficiency in large pens with replication would be welcomed.

Fish health and welfare considerations in rearing and stocking cleaner fish are priority considerations for salmon farmers to meet welfare standards. Ongoing work to improve vaccines for cleaner fish (Haugland et al., 2018) and welfare guidelines for capture, stocking and use of cleaner fish (RSPCA, 2021) are now available for farmers. These set good practice for rearing cleaner fish to improve welfare. Protocols for safe transport of fish and also preparations prior to stocking on farms to receive cleaner fish will also improve welfare. A range of welfare indices is now available to regularly monitor the condition factor and physical condition of cleaner fish (for example Treasurer and Feledi, 2014; Rabadan et al., 2021) to ensure that threats to cleaner fish welfare can be identified early and prompt action taken to manage or treat fish or provide additional environmental enrichment and supplementary feed. This requires farm protocols for cleaner fish use, following welfare standards for capture and stocking (RSPCA, 2021), a farm staff with responsibility for overseeing the use of cleaner fish on farms, and also staff training in cleaner fish use. Guidance is available on cleaner fish use and protocols are available in the Norwegian cleaner fish manual (Norwegian Seafood Research Fund (FHF), 2017), and in *Cleaner Fish Biology and Aquaculture Applications* (Treasurer, 2018).

The supply of sufficient numbers of lumpfish has been achieved through the opening of hatcheries in Norway, Iceland and the UK. However, despite the high survival from egg to stocking of up to 50%, further improvements in selective breeding can give greater resistance to disease, selection of families with higher cleaning ability and a reduction of attachment sucker abnormalities. The annual shortfall in farmed wrasse is in the region of 25 million fish in Norway and two million in Scotland, and greater efforts are required to improve survival in the larval period and throughout the prolonged weaning and juvenile hatchery stages.

References

Adams, C.E., Turnbull, J.F., Bron, J.E. and Huntingford, F. (2007) Multiple determinants of welfare in farmed fish: stocking density, disturbance, and aggression in Atlantic salmon (*Salmo salar*). *Canadian Journal of Fisheries and Aquaculture Science* 64, 336–344.

Arechavala-Lopez, P., Sanchez-Jerez, P., Izquierdo-Gomez, D., Toledo-Guedes, K. and Bayle-Sempere, J.T. (2013) Does fin damage allow discrimination among wild, escaped and farmed *Sparus aurata* (L.) and *Dicentrarchus labrax* (L.)? *Journal of Applied Ichthyology* 29, 352–357.

Ascunsion-Matinez, I., Vivares, C.P., Rocha, de M., Fonseca, A.C., Andral, B., Bouix, G. and De Meideiros, R. (1992) Microsporidiosis on *Artemia* (Crustacea, Anostraca): light and electron microscopy of *Vavraia anostraca* sp nov (Microsporidia, Pleistophoridae) in the Brazilian solar salterns. *Aquaculture* 107, 229–237.

Barrett, L.T., Overton, K., Stien, L.H., Oppedal, F. and Dempster, T. (2020) Effect of cleaner fish on sea lice in Norwegian salmon aquaculture: a national scale data analysis. *International Journal for Parasitology* 50, 787–796.

Bartsch, A., Robinson, S.M.C., Liutkus, M., Ang, P.K., Webb, J. and Pearce, C.M. (2013) Filtration of sea louse, *Lepeophtheirus salmonis*, copepodids by the blue mussel, *Mytilus edulis*, and the Atlantic sea scallop, *Placopecten magellanicus*, under different flow, light and copepodid-density regimes. *Journal of Fish Diseases* 36, 361–370.

Becker, K. (1996) Epibionts on carapaces of some malacostracans from the Gulf of Thailand. *Journal of Crustacean Research* 16, 92–104.

Bjordal, Å. (1988) Cleaning symbiosis between wrasse (Labridae) and lice infested salmon (*Salmo salar*) in mariculture. *ICES Mariculture Committee* 1988/F17 8 pp.

Bjordal, Å. (1991) Wrasse as cleaner fish for farmed salmon. *Progress in Underwater Science* 16, 17–28.

Bjordal, Å. and Costello, M. (1990) How good is this natural control of sea-lice? *Fish Farmer* 13(3), 44–46.

Blaustein, L., Dumont, H.J. and Roche, K. (1990) Typhoplanid flatworms (*Mesostoma* and related genera): mechanisms of predation and evidence that they structure aquatic invertebrate communities. *Hydrobiologia* 198, 61–77.

Boglione, C., Gavaia, P., Koumoundouros, G., Gisbert, E., Moren, M., Fontagné, S. and Witten, P.E. (2013) Skeletal anomalies in reared European fish larvae and juveniles. Part 1: normal and anomalous skeletogenic processes. *Reviews in Aquaculture* 5, S99–S120

Bolton-Warberg, M., O'Sullivan, S.M., Power, A.M., Moore, A.I., Wilson, L., Sproll, F., Schlittenhardt, S., Yusof, F., D'Arcy, J., Jackson, D. and FitzGerald, R.D. (2018) Cleaner fish use in Ireland. In: Treasurer, J.W. (ed.) *Cleaner Fish Biology and Aquaculture Applications*. 5M Publications, Sheffield, UK. pp. 386–413.

Boyce, D., Ang, K.P. and Prickett, R. (2018) Cunner and lumpfish as cleaner fish species in Canada. In: Treasurer, J.W. (ed.) *Cleaner Fish Biology and Aquaculture Applications*. 5M Publications, Sheffield, UK. pp. 444–467.

Brooker, A.J., Papadopoulou, A., Gutierrez, C., Rey, S., Davie, A. and Migaud, H. (2018) Sustainable production and use of cleaner fish for the biological control of sea lice: recent advances and current challenges. *Veterinary Record* 183, 383.

Brooker, A.J., Davie, A., Leclercq, E., Zerafa, B. and Migaud, H. (2020) Pre-deployment acclimatisation of farmed Ballan wrasse (*Labrus bergylta*) to sea-cage conditions promotes behaviour analogous to wild conspecifics when used as cleaner fish in Atlantic salmon (*Salmo salar*) farms. *Aquaculture* 520, 734771.

Bui, S., Stien, L.H., Nilsson, J., Trangereid, H. and Oppedal, F. (2020) Efficiency and welfare impact of long-term simultaneous in situ management strategies for salmon louse reduction in commercial sea cages. *Aquaculture* 520, 734934.

Byrne, A. (2012) Pacific oyster (*Crassostrea gigas*) and Atlantic salmon (*Salmo salar*) integrated multi-trophic aquaculture in British Columbia: investigation of bivalve growth and natural sea lice mitigation. MSc thesis, University of Victoria.

Capaccioni, R., Gras, D. and Carbonell, E. (1993) A copepod of the genus *Rhodinicola* Levinson, 1878 (Poecilostomatoida Clausiidae) parasitic on *Clymenura clypeata* (Saint-Joseph, 1894) (Polychaeta, Maldanidae) from the Alfaques Inlet (Ebro River delta, Spain, western Mediterranean). *Crustaceana* 64(2), 129–136.

Cavrois-Rogacki, T., Davie, A., Migaud, H. and Monroig, O. (2019a) Short term lecithin enrichments can enhance the phospholipid and DHA contents of the polar lipid fraction of *Artemia* nauplii. *Aquaculture* 510, 122–130.

Cavrois-Rogacki, T., Davie, A., Monroig, O. and Migaud, H. (2019b) Elevated temperature promotes growth and feed efficiency of farmed ballan wrasse juveniles (*Labrus bergylta*). *Aquaculture* 511, 734237.

Cavrois-Rogacki, T., Drabikova, L., Migaud, H. and Davie, A. (2021) Deformities prevalence in farmed ballan wrasse (*Labrus bergylta*) in relation to hatchery origin and life stage. *Aquaculture* 533, 736212.

Darwall, W., Costello, M.J. and Lysaght, S. (1992) Wrasse: how well do they work? *Aquaculture Ireland* 50, 26–29.

Deady, S., Varian, S.J.A. and Fives, J.M. (1995) The use of cleaner-fish to control sea lice on two Irish salmon (*Salmo salar*) farms with particular reference to wrasse behaviour in cages. *Aquaculture* 131, 73–90.

Dipper, F. (1987) *British Sea Fishes*. Underwater World Publication Ltd, London.

Dipper, F.A., Bridges, C.R. and Menz, A. (1977) Age, growth and feeding in the ballan wrasse *Labrus bergylta* Ascanius 1767. *Journal of Fish Biology* 11, 105–120

Drebes, G. and Schnepf, E. (1998) *Gyrodinium undulans* Hulburt, a marine dinoflagellate feeding on the bloom-forming diatom *Odontella aurita*, and on copepod and rotifer eggs. *Helgol-Meeresuntuts* 52, 1–14.

Eliasen, K., Danielsen, E., Johannesen, A., Joensen, L. and Patursson, E. (2018) The cleaning efficacy of lumpfish (*Cyclopterus lumpus* L.) in Faroese salmon (*Salmo salar* L.) farm pens in relation to lumpfish size and seasonality. *Aquaculture* 488, 61–65.

Erkinharju, T., Dalmo, R.A., Hansen, M. and Seternes, T. (2021) Cleaner fish in aquaculture: review on diseases and vaccination. *Reviews in Aquaculture* 13, 189–237.

Fjelldal, P.G., Madaro, A., Hvas, M., Stien, L.H., Oppedal, F. and Fraser, T.W. (2020) Skeletal deformities in wild and farmed cleaner fish species used in Atlantic salmon *Salmo salar* aquaculture. *Journal of Fish Biology* 98, 1049–1058.

Freeman, M. and Sommerville, C. (2011) Original observations of *Desmozoon lepeophtherii*, a microsporidian hyperparasite infecting the salmon louse *Lepeophtheirus salmonis* and its subsequent detection by other researchers. *Parasites & Vectors* 4, 231.

Geitung, L., Wright, D.W., Oppedal, F., Stien, L.H., Vågseth,T. and Madaro, A. (2020) Cleaner fish growth, welfare and survival in Atlantic salmon sea cages during an autumn-winter production. *Aquaculture* 528, 735623.

Gentry, K., Bui, S., Oppedal, F. and Dempster, T. (2020) Sea lice prevention strategies affect cleaner fish delousing efficacy in commercial Atlantic salmon sea cages. *Aquaculture Environment Interactions* 12, 67–80.

Gentry, K, Bui, S, Oppedal, F, Bjelland R., Nola, V. and Dempster, T (2021) Acclimatisation with lice-infested salmon improves cleaner fish lice consumption. *Aquaculture Environment Interactions* 13, 41–49.

Grant, B., Davie, A., Taggart, J.B., Selly, S.L., Picchi, N., Bradley, C., Prodohl, P., Leclerq, E. and Migaud, H. (2016) Seasonal changes in broodstock spawning performance and egg quality in ballan wrasse (*Labrus bergylta*). *Aquaculture* 464, 505–514.

Grefsrud, E.S., Svåsand, T., Glover, K., Husa, V., Hansen, P.K., Samuelsen, O., Nandlund, N. and Stien, L.H. (2019) Risikorapport Norsk Fiskeoppdrett 2019 – Miljøeffekter av lakseoppdrett. (Risk report Norwegian fish farming 2019 – Environmental effects of salmon farming.) In: *Fisken og Havet (Fish and the Sea)* 2019–5. Institute of Marine Research, Bergen, Norway.

Gresty, K.A. and Warren, A (1993) Incidence of ciliate epibionts on *Lepeophtheirus* study. In: Boxshall, G. and Defaye, D. (eds) *Pathogens of Wild and Farmed Fish: Sea Lice*. Ellis Horwood, Chichester, UK. pp. 356–362.

Grutter, A.S. (1999) Cleaner fish really do clean. *Nature* 398, 672–673.

Halvorsen, K.T., Larsen, T., Sørdalen, T.K., Vøllestad, L.A., Knutsen, H. and Olsen, E.M. (2017) Impact of harvesting cleaner fish for salmonid aquaculture assessed by replicated coastal marine protection areas. *Marine Biology Research* 13, 359–369.

Hamre, K., Yúfera, M., Rønnestad, I., Boglione, C., Conceição, L.E.C. and Izquierdo, M. (2013) Fish larval nutrition and feed formulation: knowledge gaps and bottlenecks for advances in larval rearing. *Reviews in Aquaculture* 5, S26–S58.

Haugland, G.T., Rønneseth, A. and Wergeland, H.I. (2018) Immunology and vaccinology of lumpfish and wrasse. In: Treasurer, J.W. (ed.) *Cleaner Fish Biology and Aquaculture Applications*. 5M Publications, Sheffield, UK. pp. 179–196.

Haugland, G.T., Imsland, A.K.D., Reynolds, P. and Treasurer, J. (2020) Application of biological control: use of cleaner fish. In: Kibenge, F.S.B. and Powell, M.D. (eds) *Aquaculture Health Management*. Academic Press, London, UK. pp. 319–369.

Helland, S., Wiborg, D.S,. Hamre, K,. Skiftesvik, A.B., Lein, I., Kousoulaki, K., Øie, G., Browman, H., Evjemo, J.O., Kjørsvik, E. and Grøtan, E. (2014) Best practices for farming of ballan wrasse. In: Helland, S., Wiborg Dahle, S., Hough, C. and Borthen, J. (eds) *Production of Ballan Wrasse. Science and Practice.* The Norwegian Seafood Research Fund, Norway. pp. 123–128.

Hempleman, C. (2017) International perspectives: Scotland. International Cleaner Fish Summit, Glasgow, SAIC/FHF.

Hjeltnes, B., Borno, G., Jansen, M.D., Haukaas, A. and Walde, C. (2017) Fiskehelserapporten 2016 Veterinærinstituttet. Ufullstendig ref.

Huntingford, A., Adams, C.V.A., Braithwaite,V., Kadri, S., Pottinger, T.G., Sandøe, P. and Turnbull, J.F. (2006) Current issues in fish welfare. *Journal of Fish Biology* 68, 332–372.

Hvas, M, Folkedal, O., Imsland., A. and Oppedal, F. (2018) Metabolic rates, swimming capabilities, thermal niche and stress response of the lumpfish, *Cyclopterus lumpus*. *Biology Open*, 7(9), bio036079.

Imsland, A.K., Reynolds, P., Eliassen, G., Hangstad, T.A., Foss, A., Vikingstad, E. and Elvegård, T.A. (2014) The use of lumpfish (*Cyclopterus lumpus* L.) to control sea lice (*Lepeophtheirus salmonis* Krøyer) infestations in intensively farmed Atlantic salmon (*Salmo salar* L.). *Aquaculture* 424–425, 18–23.

Imsland, A.K., Reynolds, P., Eliassen, G., Hangstad, T.A., Jónsdóttir, Ó.D.B., Elvegård, T.A., Lemmens, S.C.A., Rydland, R. and Nytrø, A.V. (2016a) Investigation and quantification of behavioural interactions between lumpfish (*Cyclopterus lumpus*) and goldsinny (*Ctenolabrus rupestris*) under controlled conditions. *Aquaculture International* 24, 1509–1521.

Imsland, A.K., Reynolds, P., Eliassen, G., Mortensen, A., Hansen, Ø.J., Puvanendran, V., Hangstad, T.A., Jónsdóttir, Ó.D.B., Emaus, P.A., Elvegård, T.A., Lemmens, S.C.A., Rydland, R., Nytrø, A.V. and Jonassen, T.M. (2016b) Is cleaning behavior in lumpfish (*Cycloptherus lumpus*) parentally controlled? *Aquaculture* 459, 156–165.

Imsland, A.K.D., Hanssen, A., Nytrø, A.V., Reynolds, P., Jonassen, T.M., Hangstad, T.A., Elvegård, T.A., Urskog, T.C. and Mikalsen, B. (2018a) It works! Lumpfish can significantly lower sea lice infestation in large-scale salmon farming. *Biology Open*, 7(9), bio036301.

Imsland, A.K.D., Reynolds, P., Jonassen, T.M., Hangstad, T.A., Elvegård, T.A., Urskog, T.C. and Mikalsen, B. (2018b) Effects of three commercial diets on growth, cataract development and histopathology of lumpfish (*Cyclopterus lumpus* L.). *Aquaculture Research* 49(9), 3131–3141.

Imsland, A.K.D., Reynolds, P., Hangstad, T.A., Jónsdóttir, Ó.D.B., Noble, T., Wilson, M., Mackie, A., Elvegård, T.A., Ursog, T.C. and Mikalsen, B. (2018c) Feeding behaviour and growth of lumpfish (*Cyclopterus lumpus* L.) fed with feed blocks. *Aquaculture Research* 49(5), 2006–2012.

Imsland, A.K.D., Reynolds, P., Jonassen, T.M., Hangstad, T.A., Elvegård, T.A., Urskog, T.C. and Mikalsen, B. (2019a) Effects of different feeding frequencies on growth, cataract development and histopathology of lumpfish (*Cyclopterus lumpus* L.). *Aquaculture* 501, 161–168.

Imsland, A.K., Frogg, N., Stefansson, S.O. and Reynolds, P. (2019b) Improving sea lice grazing of lumpfish (*Cyclopterus lumpus* L.) by feeding live feeds prior to transfer to Atlantic salmon (*Salmo salar* L.) net-pens. *Aquaculture* 511 1–9.

Imsland, A.K.D., Reynolds, P, Jonassen, T.M., Hangstad, T.A., Adrons, J., Elvegård, T.A., Ursog, T.C., Hanssen, A. and Mikalsen, B. (2019c) Comparison of diet composition, feeding, growth and health of lumpfish (*Cyclopterus lumpus* L.) fed either feed blocks or pelleted commercial feed. *Aquaculture Research* 1–12.

Imsland, A.K.D., Reynolds, P., Remen, M., Bloch-Hansen, K., Sagerup, K., Hemmingsen, W., Mathisen, R. and Myklebust, E.A. (2020a) The possible use of lumpfish against *Caligus elongatus*: a mini review. *Journal of Oceans University of China* 19, 1133–1139.

Imsland, A.K.D., Reynolds, P., Lorentzen, M., Eilertsen, R.A., Micallef, G. and Tvenning, R. (2020b) Improving survival and health of lumpfish (*Cyclopterus lumpus* L.) by the use of feed blocks and operational welfare indicators (OWIs) in commercial Atlantic salmon cages. *Aquaculture* 527 1–9.

Imsland, A.K.D., Reynolds, P., Hangstad, T.A., Madura, S., Hagen, S., Jónsdóttir, Ó.D.B., Spetland, F. and Lindberg, K.S. (2021) Quantification of grazing efficacy, growth and health score of different lumpfish (*Cyclopterus lumpus* L.) families: possible size and gender effects. *Aquaculture* 530, 735925.

Jansson, E., Taggart, J.B., Wehner, S., et al. (2016) Development of SNP and microsatellite markers for goldsinny wrasse (*Ctenolabrus rupestris*) from ddRAD sequencing data. *Conservation Genetic Resources* 8, 201–206.

Johannesen, A., Arge, R. and Eliasen, K. (2018) Rearing, farm application, and behaviour of lumpfish (*Cyclopterus lumpus*) in the Faroes. In: Treasurer, J.W. (ed.) *Cleaner Fish Biology and Aquaculture Applications*. 5M Publications, Sheffield, UK. pp. 414–428.

Jonassen, T.M., Hamadi, M., Remø, S.C. and Waagbø, R (2017) An epidemiological study of cataracts in wild and farmed lumpfish (*Cyclopterus lumpus* L.) and the relation to nutrition. *Journal of Fish Diseases* 40, 1903–1914.

Jonassen, T.M., Lein, I. and Nytrø, A.V. (2018) Hatchery management of lumpfish. In: Treasurer, J.W. (ed.) *Cleaner Fish Biology and Aquaculture Applications*. 5M Publications, Sheffield, UK. pp. 114–136.

Kabata, Z. (1973) The species of *Lepeophtheirus* (Copepoda: Caligidae) from fishes of British Columbia. *Journal of the Fisheries Research Board of Canada* 30, 729–759.

Kennedy, C.R. (1978) Studies on the biology of *Eubothrium salvelini* and *E. crassum* in resident and migratory *Salvelinus alpinus* and *Salmo trutta* and in *S. salar* in north Norway and the islands of Spitsbergen and Jan Mayen. *Journal of Fish Biology* 12, 147–162.

Klakegg, O., Myhren, S., Juell, R.A., Aase, M., Salonius, K. and Sørum, H. (2020) Improved health and better survival of farmed lumpfish (*Cyclopterus lumpus*) after a probiotic bath with two probiotic strains of *Aliivibrio*. *Aquaculture* 518, 734810

Koumoundouros, G., Divanach, P. and Kentouri, M. (1999) Osteological development of the vertebral column and of the caudal complex in *Dentex dentex*. *Journal of Fish Biology* 54, 424–436.

Kousoulaki, K., Grøtan, E., Kvalheim, K., Høstmark, Ø., Klinge, M. and Bogevik, A.S. (2014) Protein quality, commodity options and B. wrasse weaning performance. In: Helland, S., Wiborg Dahle, S., Hough, C.

and Borthen, J. (eds) *Production of Ballan Wrasse. Science and Practice.* The Norwegian Seafood Research Fund, Norway. pp. 96–100.

Kousoulaki, K., Treasurer, J. and FitzGerald, R (2018a) Fisheries for cleaner fish species in Europe. In: Treasurer, J.W. (ed.) *Cleaner Fish Biology and Aquaculture Applications.* 5M Publications, Sheffield, UK. pp. 330–358.

Kousoulaki, K., Migaud, H. and Davie, A. (2018b) Cleaner fish species nutrition and feeding practices. In: Treasurer, J.W. (ed.) *Cleaner Fish Biology and Aquaculture Applications.* 5M Publications, Sheffield, UK. pp. 179–196.

Kousoulaki, K., Grøtan, E., Kortner, T.M., Berge, G.M., Haustveit, G., Krogdahl, Å., Nygaard, H., Sæle, Ø., Chikwati, E.M. and Lein. I. (2021) Technical feed quality influences health, digestion patterns, body mineralization and bone development in farming of the stomachless cleaner fish ballan wrasse (*Labrus bergylta*). *Animal Feed Science and Technology* 274, 114830.

Leclercq, E., Zerafa, B., Brooker, A.J., Davie, A. and Migaud, H. (2018) Application of passive-acoustic telemetry to explore the behaviour of ballan wrasse (*Labrus bergylta*) and lumpfish (*Cyclopterus lumpus*) in commercial Scottish salmon sea-pens. *Aquaculture* 495, 1–12.

Le, H.T.M.D., Shao X., Krogdahl Å., Kortner T.M., Lein I., Kousoulaki K., Lie K.K. and Sæle Ø. (2019a) Intestinal function of the stomachless fish, ballan wrasse (*Labrus bergylta*). *Frontiers in Marine Science* 6, 140.

Le, H.T.M.D., Lie K.K., Giroud-Argoud J., Rønnestad I. and Sæle Ø. (2019b) Effects of cholecystokinin (CCK) on gut motility in the stomachless fish ballan wrasse (*Labrus bergylta*). *Frontiers in Neuroscience* 13, 553.

Lein, I. and Synnøve, H. (2014) Establishment of ballan wrasse brood stock. In: Helland, S., Wiborg Dahle, S., Hough, C. and Borthen, J. (eds) *Production of Ballan Wrasse. Science and Practice.* The Norwegian Seafood Research Fund, Norway. pp. 10–13.

Lekva, A. and Grøtan, E. (2018) Rearing of ballan wrasse. In: Treasurer J.W. (ed.) *Cleaner Fish Biology and Aquaculture Applications.* 5M Publications, Sheffield, UK. pp. 37–53.

Losey, G.S. (1987) Cleaning symbiosis. *Symposium on Marine Symbioses* 4, 229–256.

Marine Scotland (2020) Wild wrasse harvesting consultation. Available at: https://consult.gov.scot/marine-scotland/wild-wrasse-harvesting (accessed 18 September 2021).

Marine Scotland (2021a) Scottish Sea Fisheries Statistics. Available at: https://data.marine.gov.scot/dataset/2020-scottish-sea-fisheries-statistics-fishing-effort-and-quantity-and-value-landing-ices (accessed 3 February 2022).

Marine Scotland (2021b) Scottish Fish Farming Production Survey 2020. Available at: www.gov.scot/publications/scottish-fish-farm-production-survey-2020/ (accessed 3 February 2022).

MarinHelse (2018) Veileder i Dødfiskkategorisering. 101pp. Developed by MarinHelse AS, Tromsø. Available at: https://www.marinhelse.no/produkter/veileder-dodfiskkategorisering/ (accessed 23 June 2021).

Martos-Sitcha, J.A., Mancera, J.M., Prunet, P. and Magnoni, L.J. (2020) Editorial: Welfare and stressors in fish: challenges facing aquaculture. *Frontiers in Physiology* 11, 162.

Mattilsynet (2020) Nasjonal tilsynskampanje 2018/2019 Velferd Hos Rensefisk. Available at: https://www.mattilsynet.no/fisk_og_akvakultur/akvakultur/rensefisk/mattilsynet_sluttrapport_rensefiskkampanje_2018__2019.37769/binary/Mattilsynet%20sluttrapport%20rensefiskkampanje%202018%20-%202019 (accessed 23 June 2021).

May, R.C. (1974) Larval mortality in marine fishes and the critical period concept. In: Blaxter, J.H.S. (ed.) *The Early Life History of Fish.* Springer, Berlin, Heidelberg. pp. 3–19.

Minchin, D. and Jackson, D. (1993) *Udonella caligorum* Johnston, 1835 (Platyhelminthes: Udonellidae) associated with caligid copepods on farmed salmon. In: Boxshall, G.A. and Defaye, D. (eds) *Pathogens of Wild and Farmed Fish: Sea Lice.* Ellis Horwood, Chichester, UK. pp. 346–355.

Mo, T.A. and Poppe, T.T. (2018) Risk of using cleaner fish in fish farming (In Norwegian). Norsk Veterinaertidsskrift nr. 2-2018. Available at: www.vetnett.no/norsk-veterinartidsskrift-nr-2-2018 (accessed 23 June 2021).

Molloy, S., Pietrak, M., Bouchard, D. and Bricknell, I. (2011) Ingestion of *Lepeophtheirus salmonis* by the blue mussel *Mytilus edulis*. *Aquaculture* 311, 61–64.

Nese, L. and Enger, Ø. (1993) Isolation of *Aeromonas salmonicida* from salmon lice *Lepeophtheirus salmonis* and marine plankton. *Diseases of Aquatic Organisms* 16, 79–81.

Nestrud, L.B. and Anderson, R.L. (1994) Aquatic safety of *Lagenidium giganteum*: effects of freshwater fish and invertebrates. *Journal of Invertebrate Pathology* 64, 228–233.

Nielsen, A. (2021) Wrasse and lumpfish duality for an all-year round cover. International Ballan Wrasse Conference 2021, Oral presentation, SAIC and Otter Ferry Seafish.

Nilsen, A., Viljugrein, H., Rosag, M.V. and Colquhon, D. (2014) Rensefiskhelse – kartlegging av dodelighets-sarsaker. *Veterinarinstituttets Rapportserie* 12. (In Norwegian.)

Noble, C., et al. (2019a) RENSVEL OWI Fact Sheet Series: An introduction to operational and laboratory-based welfare indicators for lumpfish (*Cyclopterus lumpus* L.). NTNU open. Available at: https://ntnuopen.ntnu.no/ntnu-xmlui/handle/11250/2599803 (accessed 18 August 2021).

Noble, C., et al. (2019b) RENSVEL OWI Fact Sheet Series: An introduction to operational and laboratory-based welfare indicators for ballan wrasse (*Labrus bergylta* L.). Available at: https://nofima.no/wp-content/uploads/2019/10/Rensvel-fact-sheet-series-Ballan-Wrasse.pdf (accessed 18 August 2021).

North, B.P., Turnbull, J., Ellis, M.J. Porter, H., Migaud, J., Bron, J. and Bromage, N.R. (2006) The impact of stocking density on the welfare of rainbow trout (*Oncorhynchus mykiss*). *Aquaculture* 255, 466–479.

Norwegian Seafood Research Fund (FHF) (2017) Cleanerfish – best practice guidelines from the Norwegian seafood research fund (FHF). Available at: www.sustainableaquaculture.com/media/1470/fhf-cleaner-fish-guidelines.pdf (accessed 26 January 2022).

Nøstvold, B.H., Kvalvik, I., Voldnes, G. and Jentoft, A.R. (2016) Nofima rapportserie (43/2016). Available at: https://nofima.no/pub/1397108 (accessed 3 June 2021).

Overton, K., Barrett, L.T., Oppedal, F., Kristiansen, T.S. and Dempster, T. (2020) Sea lice removal by cleaner fish in salmon aquaculture: a review of the evidence base. *Aquaculture Environment Interactions* 12, 31–44.

Papadopoulou A., Wallis T., Ramirez-Paredes J.G., Monaghan, S.J., Davie, A., Migaud, H. and Adams, A. (2020) Atypical *Aeromonas salmonicida* vapA type V and *Vibrio* spp. are predominant bacteria recovered from ballan wrasse *Labrus bergylta* in Scotland. *Diseases of Aquatic Organisms* 140, 47–54.

Papadopoulou, A, Davie A., Monaghan, S.J., Migaud, H. and Adams, A. (2021) Development of diagnostic assays for differentiation of atypical *Aeromonas salmonicida* vapA type V and type VI in ballan wrasse (*Labrus bergylta*, Ascanius). *Journal of Fish Diseases* 40, 1–9.

Potts, G.W. (1973) Cleaning symbiosis among British fish with special reference to *Crenilabrus melops* (Labridae). *Journal of the Marine Biological Association of the UK* 53, 1–10.

Potts, G.W. (1974) The colouration and its behavioural significance in the corkwing wrasse, *Crenilabrus melops*. *Journal of the Marine Biological Association of the UK* 54, 925–938.

Powell, A., Treasurer, J.W., Pooley, C.L., Keay, A.J., Lloyd, R., Imsland, A.K. and Garcia de Leaniz, C. (2017) Use of lumpfish for sea-lice control in salmon farming: challenges and opportunities. *Reviews in Aquaculture* 10, 683–702.

Powell, A., Pooley, C., Scolamacchia, M. and Garcia de Leaniz, C. (2018) Review of lumpfish biology. In: Treasurer, J.W. (ed.) *Cleaner Fish Biology and Aquaculture Applications.* 5M Publishing, Sheffield, UK. pp. 90–113.

Quintela, M., Danielsen, E.A. and Svåsand, T. (2014) Isolation and characterization of twenty microsatellite loci for the ballan wrasse, *Labrus bergylta*. *Conservation Genetic Resources* 6, 425–428.

Rabadan, C.G. (2018) Improving the use of wrasse (Labridae) stocked in commercial sea cages with Atlantic salmon (*Salmo salar*) for sea lice control: key features for a successful deployment. In: Treasurer, J.W. (ed.) *Cleaner Fish Biology and Aquaculture Applications.* 5M Publishing, Sheffield, UK. pp. 54–68.

Rabadan, C.G., Spreadbury, C., Consuegra, S. and de Leaniz, C.G. (2021) Development, evaluation and testing of an operational welfare score index for farmed lumpfish *Cyclopterus lumpus* L. *Aquaculture* 531, 735777.

Ramirez-Paredes, J.G., Verner-Jeffreys, D.W., Papadopoulou, A., Monaghan, S.J., Smith, L., Haydon, D., Wallis, T.S., Davie, A., Adams, A. and Migaud, H. (2020) A commercial autogenous injection vaccine protects ballan wrasse (*Labrus bergylta*, Ascanius) against *Aeromonas salmonicida* vapA type V. *Fish and Shellfish Immunology*. 107, Part A, 43–53.

Riley, A., Jeffrey, K., Cochrane-Dyer, T., White, P. and Ellis, J. (2017) Northern European wrasse – summary of commercial use, fisheries and implications for management. Available at: https://www.gov.scot/binaries/content/documents/govscot/publications/foi-eir-release/2018/03/foi-18-00461/documents/foi-18-00461-documents-release-2-pdf/foi-18-00461-documents-release-2-pdf/govscot:document/FoI-18-00461%20%20Documents%20for%20release%202.pdf (accessed 7 February 2022).

Rimstad, E., Basic, D., Gulla, S., Hjeltnes, B. and Mortensen, S. (2017) Report from the Norwegian Scientific Committee for Food and Environment (VKM) 2017:32 Risk assessment of fish health associated with the use of cleaner fish in aquaculture. Opinion of the Panel on Animal Health and Welfare of the Norwegian Scientific Committee for Food and Environment. Available at: www.vkm.no (accessed 4 May 2021).

RSPCA (2021) Welfare Standards for Farmed Atlantic Salmon. RSPCA, Horsham, UK. Available at: www.rspca.org.uk/sciencegroup/famanimals/standards (accessed 26 January 2022).

SAIC (2021) Scaling up the use of cleaner fish. Available at: www.sustainableaquaculture.com/projects/health-and-welfare/details/scaling-up-the-use-of-cleaner-fish (accessed 3 June 2021).

Sánchez, J.C., Mancilla, J., Hevia, M. and Saez, P.J. (2018) The Patagonian blenny (*Elginops maclovinus*): a Chilean native fish with potential to control sea lice (*Caligus rogercresseyi*) infestations in salmonids. In: Treasurer, J.W. (ed.) *Cleaner Fish Biology and Aquaculture Applications*. 5M Publications, Sheffield, UK. pp. 179–196.

Sayer, M.D.J. and Davenport, J. (1996) Hypometabolism in torpid goldsinny wrasse subjected to rapid fluctuations in seawater temperature. *Journal of Fish Biology* 49, 64–75.

Sayer, M.D.J., Reader, J. and Davenport, J. (1996) Survival, osmoregulation and oxygen consumption of wrasse at low salinity and/or low temperature. In: Sayer, M.D.J., Treasurer, J. and Costello, M. (eds) *Wrasse Biology and Use in Aquaculture*. Fishing News Books, Oxford, UK. pp. 119–135.

Scholz, F., Glosvik, H. and Marcos-López, M. (2018) Cleaner fish health. In: Treasurer, J.W. (ed.) *Cleaner Fish Biology and Aquaculture Applications*. 5M Publications, Sheffield, UK. pp. 221–257.

Schnepf, E., Crickmore, N., Ven Rie, J., Lereclus, D., Baum, J., Feitelson, J., Zeigler, D.R. and Dean, D.H. (1998) *Bacillus thuringiensis* and its pesticidal crystal proteins. *Microbiology and Molecular Biology Reviews* 62, 775–806.

Skiftesvik, A.B. and Bjelland, RM. (2003) Breeding of ballan wrasse (*Labrus bergylta*). *Norsk Fiskeoppdrett* 28, 5.

Skiftesvik, A.B., Bjelland, R.M., Durif, C.M.F., Johansen, I.S. and Browman, H.I. (2013) Delousing of Atlantic salmon (*Salmo salar*) by cultured vs. wild ballan wrassse (*Labrus bergylta*). *Aquaculture* 402–403, 113–118.

Sommerset, I., Walde, C. S., Bang Jensen, B., Bornø, B. and Haukaas, A. (2020) Fiskehelserapporten 2019 (Norwegian Fish Health Report 2019). In: Brun, E. (ed.): Veterinærinstituttet rapportserie nr 5a/2020. Norwegian Veterinary Institute, Ås, Norway. (In Norwegian.)

Steinarsson, A. and Arnason, T. (2018) Rearing of cleaner fish in Iceland. In: Treasurer, J.W. (ed.) *Cleaner Fish Biology and Aquaculture Applications*. 5M Publications, Sheffield, UK. pp. 414–428.

Stien, L.H., Størkersen, K.V. and Gåsnes, S.K. (2020) *Analyse Av Dødelighetsdata Fra Spørreundersøkelse Om Velferd Hos Rensefisk*. Available at: https://www.mattilsynet.no/fisk_og_akvakultur/akvakultur/rensefisk/analyse_av_dodelighetsdata_fra_sporreundersokelse_om_velferd_hos_rensefisk.37771/binary/Analyse av dødelighetsdata fra spørreundersøkelse om velferd hos rensefisk (accessed 4 May 2021).

Stone, J. and Bruno, D.W. (1989) A report of *Ephelota* sp., a suctorian found on the sea lice *Lepeophtheirus salmonis* and *Caligus elongatus*. *Bulletin of the European Association of Fish Pathologists* 9, 113–115.

Størkersen, K. And Amundsen, V. (2019) *Resultat av spørreundersøkelse til matfisklokaliteter med rensefisk Rensefiskens ve og vel i merdene Rapport*. Available at: www.samforsk.no (date accessed 5 May 2021).

Stuck, K.C. and Overstreet, R.M. (1994) Effect of *Baculovirus penaei* on growth and survival of experimentally infected postlarvae of the Pacific white shrimp, *Penaeus vannamei*. *Journal of Invertebrate Pathology* 64, 18–25.

Sveier, H. and Breck, O. (2018) Cleaner fish application in Norway. In: Treasurer J.W. (ed.) *Cleaner Fish Biology and Aquaculture Applications*. 5M Publications, Sheffield, UK. pp. 359–369.

Treasurer, J.W. (1993) Management of sea lice (Caligidae) with wrasse (Labridae) on Atlantic salmon (*Salmo salar* L.) farms. In: Boxshall, G.S. and Defaye, D. (eds) *Pathogens of Wild and Farmed Fish: Sea Lice*. Ellis Horwood, Chichester, UK. pp. 335–345.

Treasurer, J.W. (2002) A review of potential pathogens of sea lice and the application of cleaner fish in biological control. *Integrated Pest Management in Salmon Farming* 58(6), 546–558.

Treasurer, J.W. (2013) Use of wrasse in sea lice control (SARF068). Scottish Aquaculture Research Forum, Birnam. Previously available at: www.sarf.org.uk/cms-assets/documents/124859-172398.sarf068.pdf (accessed 15 June 2020).

Treasurer, J.W. (2018) An introduction to sea lice and the rise of cleaner fish. In: Treasurer, J.W. (ed.) *Cleaner Fish Biology and Aquaculture Applications*. 5M Publications, Sheffield, UK. pp. 179–196.

Treasurer, J. and Feledi, T. (2014) The physical condition and welfare of five species of wild-caught wrasse stocked under aquaculture conditions and when stocked in Atlantic salmon, *Salmo salar*, production cages. *Journal of the World Aquaculture Society* 45, 213–219.

Treasurer, J.W. and Turnbull, T. (2018) Tolerance of lumpfish, *Cyclopterus lumpus*, to freshwater bath treatment for amoebic gill disease, *Neoparamoeba perurans*, infection and efficacy of different treatment regimens. *Journal of the World Aquaculture Society* 50, 42–53.

Treasurer, J.W., Wallace, C. and Dear, G. (2002) Control of sea lice on farmed Atlantic salmon *S. salar* with the oral treatment emamectin benzoate (SLICE). *Bulletin of the European Association of Fish Pathologists* 22, 375–380.

Treasurer, J., Noble, C., Puvanendran, V., Planellas, S.R. and Iversen, M. (2018) Cleaner fish welfare. In: Treasurer, J. (ed.) *Cleaner Fish Biology and Aquaculture Applications*. 5M Publications, Sheffield, UK. pp. 281–312.

Tully, O., Daly, P., Lysaght, S., Deady, S. and Varian, S.J.A. (1996) Use of cleaner-fish (*Centrolabrus exoletus* (L.) and *Ctenolabrus rupestris* (L.)) to control infestations of *Caligus elongatus* Nordmann on farmed Atlantic salmon. *Aquaculture* 142, 11–24.

Turnbull, J., Bell, A., Adams, C., Bron, J. and Huntingford, F. (2005) Stocking density and welfare of cage farmed Atlantic salmon: application of multivariate analysis. *Aquaculture* 243, 121– 132.

Varian, S.J.A., Deady, S. and Fives, J.M. (1996) The effect of intensive fishing of wild wrasse populations in Lettercallow Bay, Connemara, Ireland: implications for the future management of the fishery. In: Sayer, M.D.J., Treasurer, J.W. and Costello, M.J. (eds) *Wrasse Biology and Use in Aquaculture*. Fishing News Books, Oxford, UK. pp. 100–118.

Vaughan, D.B., Grutter, A.S., Costello, M.J. and Hutson, K.S. (2017) Cleaner fishes and shrimp diversity and re-evaluation of cleaning symbioses. *Fish and Fisheries* 18, 698–716.

Voigt, M.O.C. (1991) Community structure of the helminth parasite fauna of gammarids (Crustacea: Amphiopoda) in Kiel Bay, western Baltic Sea. *Meeresfirschung Rep Marine Research* 3, 66–274.

Wittwer, C. and Treasurer, J. (2014) Desticking eggs of ballan wrasse to improve incubation. *Fish Farmer* 37, 35–37.

Wittwer, C. and Treasurer, J. (2018) Egg collection by hand-stripping and natural spawning in lumpfish *Cyclopterus lumpus* broodstock. In: Treasurer J.W. (ed.) *Cleaner Fish Biology and Aquaculture Applications*. 5M Publications, Sheffield, UK. pp. 137–150.

CHAPTER 20

Vaccination

Jessica Piesz and Ian Bricknell

Aquaculture Research Institute and School of Marine Sciences, University of Maine, USA

20.1 Introduction

Vaccines against sea lice have long been an ambition of vaccinologists, especially following the success of fish vaccines in the early to mid-1990s against *Aeromonas salmonicida* (Evelyn, 1997; Tafalla et al., 2014), assorted vibrios (Gudding and Goodrich, 2014; Kashulin et al., 2017; Press and Lillehaug, 1995), and viral hemorrhagic septicemia virus (VHSV) (Lorenzen and Olesen, 1997; Biering et al., 2005; Pereiro et al., 2016). Not all bacterial and viral fish diseases can currently be controlled by vaccines, as is the case for bacterial kidney disease (BKD) caused by *Renibacterium salmoninarum* and infectious salmon anemia caused by the infectious salmon anemia virus (ISAV), for which a vaccine remains elusive.

Development of an effective vaccine against the salmon louse *Lepeophtheirus salmonis* (Krøyer, 1837) is arguably one of the most urgent vaccine challenges in finfish aquaculture. Not because of the number of fish impacted, which is relatively small when compared to a disease such as *Saprolegnia parasitica*, but due to the economic impact the parasite has on salmon farming, costing over $1 billion annually (Abolofia et al., 2017). As a proportion, sea lice control has been estimated to cost up to 25% of the total production costs of Atlantic salmon farming (Abolofia et al., 2017). This figure represents not only the loss of fish to sea lice directly due to mortality or downgrading of carcass quality at harvest due to lesions, but also the capital investment in mechanical delousing systems such as a thermolicer and hydrolicer, the engineering costs of building lice deterring snorkel cages, the cost of medications, and the additional cost of culturing and deploying cleaner fish. For these reasons, a vaccine is seen as a very desirable goal.

20.2 The history of sea lice vaccines

Historically, the first sea lice vaccine studies appeared in the research literature in the mid-1990s (Roper et al., 1995; Grayson et al., 1995; Woo, 1997). These early vaccines were little more than crude whole lice preparations or fractionated louse proteins (Roper et al., 1995; Grayson et al., 1995). Although Woo (1997) suggested there were 20 candidate vaccines developed through recombinant protein technology being evaluated against *L. salmonis* and *Caligus elongatus*, these vaccine candidates were not successful (Raynard et al., 2002; Bricknell et al., 2003). The whole louse gut-based approach of these early vaccines was a result of some earlier work that suggested that naturally exposed Atlantic salmon and marine rainbow trout develop a suite of antibodies against lice following exposure (Ruane et al., 1995; Roper et al., 1995; Grayson et al., 1995).

The second phase of sea lice vaccines concentrated on identifying functional proteins in the digestive tract of the sea louse. The success of vaccines against the Australian cattle tick (*Rhipicephalus (Boophilus) microplus* (Canestrini, 1888)) was a major driver of sea lice vaccines targeting aspects of the louse digestive tract. Ticks take large blood meals, often three to four times their own body weight, and then leave the host to digest their meal. In the tropical habitat favored by *R. microplus*, the environmental temperature is typically in the range of 20–35°C, very similar to that of the host cattle which is 38.6°C. It is hypothesized that there is a long period following tick feeding where the conditions in the digestive tract are compatible with the functioning of immune defenses, such as antibody binding to the antigens found along the gut of the tick. (For further details of tick vaccines see the reviews by Almazan et al. (2018); De La Fuente and Contreras (2015); De La Fuente et al. (2016); Klouwens et al. (2016).) The first antigen to be identified as protective in ticks was BM86 (Willadsen et al., 1995, 1996; Willadsen, 1997; De Rose et al., 1999). BM86 is a mucin-like membrane glycoprotein (McKenna et al., 1998) associated with the endophagocytic cells of the digestive tract of *R. microplus*. IgG antibodies raised against this protein by vaccination do not damage the parasite's digestive tract directly but interfere with the absorption of nutrients and, in turn, lower fecundity in female ticks. For such a vaccine to be effective, it has to be used for multiple years to reduce the infectious pressure within a vaccinated herd of cattle. While not ideal, this was the first commercially available vaccine to be offered for an ectoparasite. TickGARD (PLUS) made by Intervet Australia and Gavac® (Heber Biotec) were shown to be 46.9% and 49.2% effective, respectively, against *R. microplus* (Patarroyo et al., 2002). These are currently considered the most effective ectoparasite vaccines currently developed.

20.3 Vaccines against adult sea lice

Vaccines against adult sea lice took a similar approach for many years. Numerous groups investigated the potential of antigens (including BM86-like molecules) in the digestive tract of adult *L. salmonis* (Woo, 1997; Carpio et al., 2011; Casuso et al., 2019; Raynard et al., 2002). There was certainly some success in identifying proteins (Fig. 20.1); however, none have given reproducible results in sea lice challenge models. The reason for this is unclear, but the conditions found in the digestive tract of sea lice are very different from that of ticks.

Sea lice are not truly hematophagous as are ticks, but instead eat a mixture of mucus, skin, muscle, necrotic tissues, and other material which they scrape off the surface of the host with their mouthparts. So how much blood does a sea louse ingest during feeding? Bricknell and Pietrak (2022, in review) used hemoglobin as a proxy to measure the amount of blood in the digestive tract of sea lice. In fed lice that did not have obvious blood in the digestive tract, the hemoglobin level was 21.4 ± SD 3.8 mg l^{-1}. In fed lice where blood was visible in the digestive tract, a mean hemoglobin level of 317 ± SD 20.4 mg l^{-1} was found. The hemoglobin level in the salmon blood the lice were feeding on was 76,465 ± SD 312 mg l^{-1}. This strongly indicates that there is a considerable dilution of blood with sea water and other material, such as mucus, during feeding. Converting these to immunological

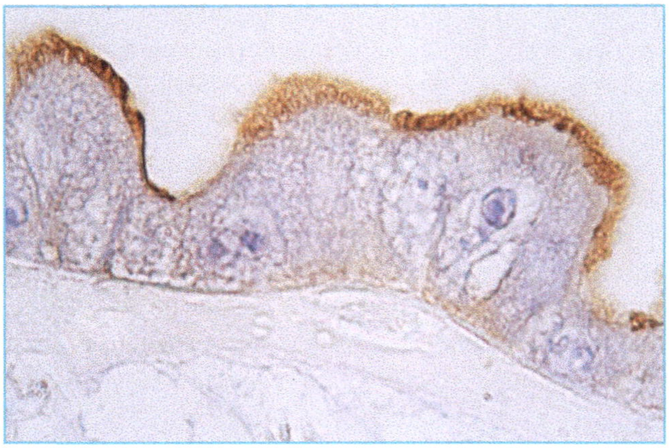

Fig. 20.1 Sea lice antigens stained brown (IHC HRP stain), were identified along the epithelial surface of the digestive tract of adult *L. salmonis*.

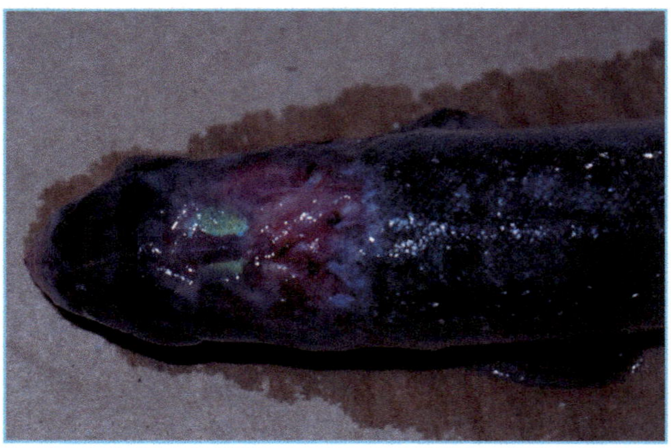

Fig. 20.2 An experimental Atlantic salmon with a severe skin lesion due to sea lice grazing (note: this is a rare occurrence on fish farms today as therapeutic intervention is triggered by agreed lice threshold levels).

titers, it is estimated you would need a titer of 1:245 to have any immunoglobulin at the highest hemoglobin level seen in the fed louse digestive tract and a titer of 1:3700 to have any antibody in the pre-lesion lice. Fig. 20.2 shows the typical degree of lesion that is required before a significant number of feeding adult lice will show the presence of blood in the digestive system. As Bricknell and Pietrak (2022, in review) showed, higher hemoglobin levels correlate with higher levels of host antibody. Unfortunately, allowing these serious lesions to develop on fish before the salmon louse encounters the vaccine is unacceptable because of welfare issues.

It is assumed that the digestive tract is used for osmoregulation in sea lice. Measurements by Bricknell and Pietrak (2022, in review) suggest that the digestive tract of fed sea lice has an osmolarity of about 789 milliosmols/l which is very close to that of sea water at 1000 milliosmols/l. It is reasonable to suggest that the slight dilution observed is due to mixing of sea water with tissue fluids absorbed from the host during feeding. If that is the case, then this would suggest a dilution factor of about 1:4 of food to seawater. Given that high salt solutions are commonly used to precipitate proteins from solution (Shih et al., 1992) it is feasible that any IgM generated by a vaccine and ingested during feeding may precipitate or at the very least undergo conformational changes due to the high salinity and may lose functionality due to stereochemical changes.

The pH of the digestive tract is less well documented, the main problem being the accurate measurement of internal gut pH. To date, there is only one study which is currently in review (Bricknell and Pietrak, 2022, in review). One difficulty encountered was the injection of pH sensitive dye into the digestive tract. First, if the louse had fed and blood was visible, the red from the hemoglobin obscured the true color of the dye. When injected into a starved gut the color was visible but did not reflect any pH change caused by acid secreting cells for example. There was also the issue that the dye itself would dilute the contents of the digestive tract and hence may skew the proportion of H^+ and OH^- ions present. There are some clues that support the observation that the pH in the gut is higher than the physiological pH of fish blood (pH 7.4) (Mota et al., 2020). The high osmolarity observed by Bricknell and Pietrak (in review) suggests that a considerable amount of sea water is ingested during feeding or for normal osmoregulation. Seawater is a very well-buffered mix of ions typically with a pH range of 8.2–8.4. Even with a dilution effect of 20% or so, as shown in the Bricknell and Pietrak paper, it is still likely that the digestive tract pH is close to its natural range. This is supported by the observation of Bricknell who recorded a pH of about 8.0 using pH sensitive dyes, suggesting that much of the contents of the digestive tract is composed of ingested seawater. Granted, there could be an equivalent of the vertebrate parietal cells (hydrochloric acid secreting cells) in the digestive tract, but these have not been reported in anatomical studies of sea lice digestive tracts (Bron et al., 1993).

It has been well documented that salmon lice, being largely obligate salmonid parasites, have evolved a series of enzymes that will digest salmon proteins very successfully (Muhlia-Almazan et al., 2008; Nylund et al., 1992; Skern-Mauritzen et al., 2009; Valenzuela-Miranda and Gallardo-Escárate, 2016). Small-scale experiments by one of the authors (Bricknell, 1997, unpublished) showed complete breakdown of Atlantic salmon IgM within four hours of exposure to adult sea lice digestive tract contents at 14°C by western blot analysis of native gels.

Given the nature of the internal environment of the digestive tract of sea lice it is clear that it is a hostile environment for the functioning and long-term survival of salmon IgM and therefore not a great target for sea louse vaccine development. To recap, first, sea lice ordinarily do not take in a significant blood meal when feeding and therefore gut proteins are less than optimal targets. Second, the salinity of the digestive tract is close to the salinity of sea water, usually in the range 800–1000 milliosmol/l and will cause a decrease in antibody binding or at worst precipitation of the antibodies once in the gut (Bricknell et al., 2002). Furthermore, the pH of the gut is around 8.0, which is much closer to seawater than the physiological pH of salmon blood (pH 7.4) (Mota et al., 2020) and will more likely impact antibody function. When combined with the digestive processes that have evolved to digest host proteins, it is fair to assume that any antibodies targeted against the adult louse's digestive tract will not be able to function for long.

As for many candidate vaccines it is quite possible to vaccinate Atlantic salmon against adult sea lice and get a good serum antibody response (Leal et al., 2019; Raynard et al., 2002; Woo, 1997) but to hope that any fish antibodies will survive for a sufficiently long enough time to attach, neutralize, or fix complement and damage the parasite is a long shot. These vaccines are directed against an animal that has evolved to parasitize and digest Atlantic salmon. Such a situation might be compared to one where, as noted by Bricknell (Anon., 2021), "You can vaccinate antelope against lions, but by the time the antibodies get into the lion the damage is done." Even if a localized immune response develops against the pre-adults or adults, the high mobility of these life stages complicates the matter as the parasite can move around on the host or jump to a new host to avoid a localized immune response.

20.4 The attached stages of sea lice and the immune response

The chalimus I and II stages of the parasite life cycle do not have the ability to avoid a localized immune response as they are physically attached to the host. They attach by means of a frontal filament that represents an extension of the anterior cuticle of the parasite and is secured by a terminal "basal plate" that adheres to host tissue (Bron et al., 1991) (Fig. 20.3).

It was observed by Johnson et al. in the 1990s (Johnson and Albright, 1992a, 1992b; Johnson et al., 1996) that there was a significant difference in immune response following copepodid settlement and the moult to chalimus between different salmonid species. In Atlantic salmon there

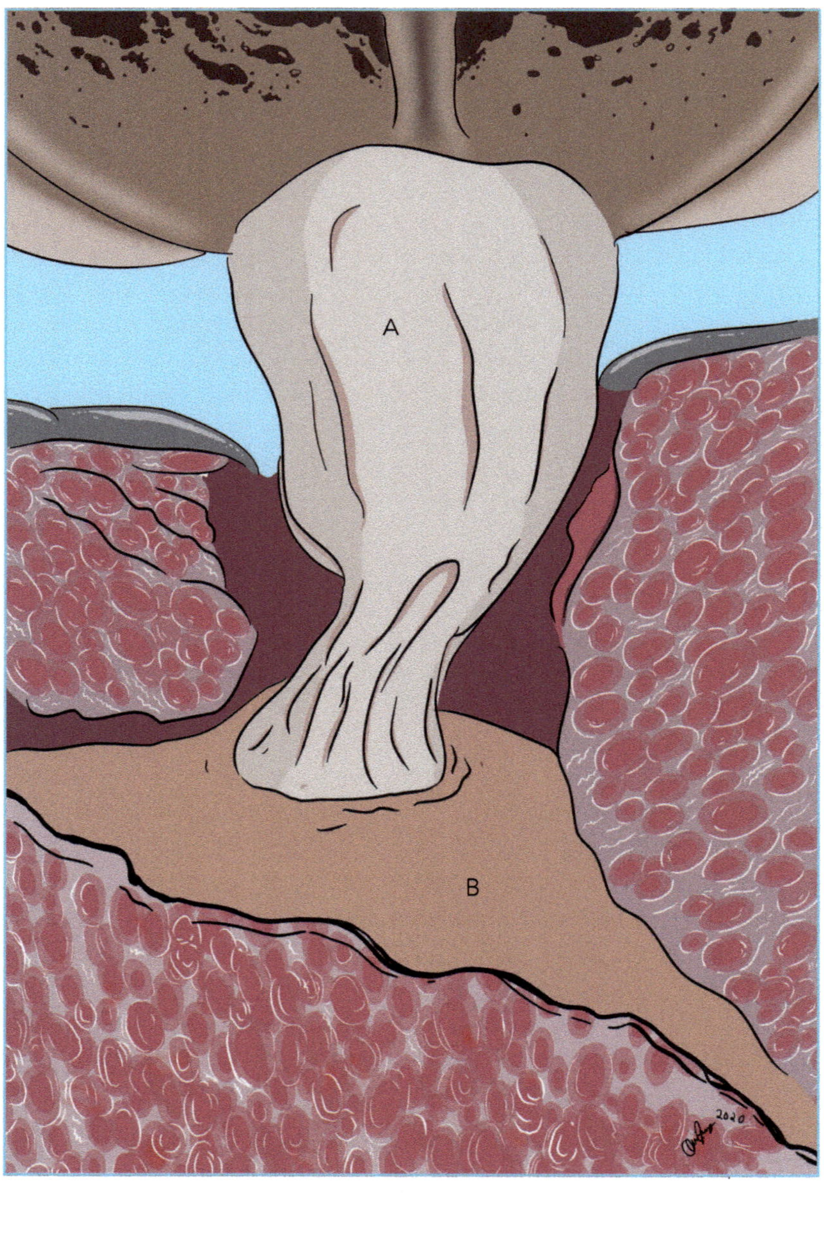

Fig. 20.3 Schematic of frontal filament (A) and basal plate (B) (from Bron et al., 1991; image by Olivia Joyce used with permission).

was virtually no inflammatory response, while chinook salmon mounted a very strong response, characterized by hyperplasia that often served to physically enclose the chalimus in host tissue and kill the sea louse before it could moult to the mobile stages. There seemed to be a clear evolutionary relationship to this response in the oncorynchiform salmonids with pink salmon having a very mild hyperplasic response, and coho a moderate response that was intermediate between pink and chinook salmon (Johnson and Albright, 1992a). This moderate inflammatory response to the chalimus stages was also observed in rainbow trout (Ruane et al., 2000; Fast et al., 2002; Skov et al., 2014; Dalvin et al., 2020), while both the sea-going subpopulation of brown trout (*Salmo trutta* L.) and migratory Arctic charr (*Salvelinus alpinus* (L.)) showed no observable inflammatory response to the chalimus stages (Dawson et al., 1997; Fjelldal et al., 2019; Subramanian et al., 2007). This poses interesting questions about the evolution of *L. salmonis*, suggesting it evolved in the Atlantic Ocean (Skern-Mauritzen et al., 2014) and became superbly host adapted by downregulating the immune responses of the two anadromous *Salmo* species and one *Salvelinus* species that it would regularly encounter. It is worth noting that no studies seem to have been conducted on sea lice and brook trout (*Salvelinus fontinalis* (Mitchill, 1814)) in eastern North America so there is an interesting evolutionary study waiting to be carried out.

The hyperplasia response observed was clearly induced by an immune reaction, as immunosuppressing the salmon with cortisol completely removed the immune activity (Johnson and Albright, 1992a, 1992b), allowing the chalimus stages to moult successfully into the pre-adult stages. This observation went mostly unnoticed by vaccine researchers but was the initial observation that suggested the skin's immune responses, including inflammation and hyperplasia, played an important role in the natural resistance of salmonids to sea lice. By suppressing these responses with cortisol, the sea lice were able to settle and develop in much greater numbers on naturally resistant fish species such as coho salmon.

20.5 Secreted proteins

A more recent approach to sea lice vaccines has been to target individual proteins in the saliva or salivary glands of the louse. This proteomics approach has been used in ticks since the late 1990s after discovering the role of saliva in pathogen transmission (Bell et al., 1979). From that initial discovery, a multitude of tick salivary compounds have been identified and characterized that modulate both innate and adaptive host immunity (Anguita et al., 2002; Hannier et al., 2004; Hajnická et al., 2005; Daix et al., 2007; Hovius et al., 2008; Guo et al., 2009). An example that has shown promise as a recombinant protein vaccine is the salivary gland metalloprotease, BrRm-MP4, from the cattle tick *R. microplus*. It was shown to provide 60% protection from infection by decreasing tick number, egg laying capacity, and egg hatching (Ali et al., 2015). In sea lice, early work by Firth et al. (2000) and Fast et al. (2003) identified trypsin-like low molecular weight proteases in the mucus of *L. salmonis* infected Atlantic salmon. These data were the first to suggest that *L. salmonis* secretes compounds in their saliva while feeding. To explore this further, Fast et al. (2004) isolated the secretory/excretory products (SEPs) by incubating the lice with dopamine. Dopamine is a neurotransmitter that has been shown to induce salivation in ticks when applied to the arthropods' cuticle (Tatchell, 1967). Using LC-MS analysis, prostaglandin E_2 (PGE_2) was identified in sea lice secretions (Fast et al., 2004). Prostaglandins are vasodilatory lipid compounds that have several important functions; most notably, they modulate cellular inflammation and host defense (Kalinski, 2012). Fast et al. (2005)

were able to show that PGE_2 inhibits interleukin-1β (IL-1β), cyclooxygenase-2 (COX-2), and major histocompatibility (MH) class I and II immune related genes in an Atlantic salmon macrophage-like cell line, providing support that the PGE_2 from the SEPs of *L. salmonis* may be partially responsible for the limited inflammatory response to the parasite in Atlantic salmon (Fast et al., 2005).

Since this initial discovery, Hamilton et al. (2018) have identified 187 proteins from the excretory/secretory products (E/S) of adult and pre-adult lice. Some of the suspected roles of these proteins were in proteolysis, lipid transport, and metabolism. In pre-adult E/S products, it was determined that 11% of proteins were serine type endopeptidases, specifically trypsins, which supports the earlier findings of Firth et al. (2000). Hamilton et al. (2018) suggested that these proteins may be important for host immune evasion, considering that they were identified in high quantities in pre-adult E/S products, and they have been identified in other models to have immunomodulatory potential. The protein profile of the adult E/S products was found to be much more diverse than the pre-adult E/S products leading the authors to speculate that these differentially expressed proteins may be important for feeding, as the adult stages are more prone to feed on blood and three cysteine proteases including cathepsin L were identified in adult E/S products. As support for their hypothesis, the authors reference a cathepsin L protein from the hard tick, *I. ricinus*, which has been shown to be important in blood meal digestion (Franta et al., 2011). McCarthy et al. (2012) previously observed cathepsin L activity in all parasitic life stages and in the SEPs of *L. salmonis* adults.

McCarthy et al. (2012), like Hamilton et al. (2018), proposed that *L. salmonis* cathepsin L1 and L2 may be involved in the digestion of host mucus, skin, and blood during feeding, considering that activity was observed in all life stages of the parasite, and the pH optimum for cathepsin L activity is similar to the physiological pH of salmon mucus and skin. Further support for its role in blood digestion is provided by Eichner et al. (2008), who identified an *L. salmonis* cysteine peptidase that shares sequence similarity with a cysteine proteinase from the malaria parasite *Plasmodium falciparum*, Welch, 1897. Other E/S products worth noting from Hamilton et al. (2018) were the antioxidants thioredoxin 2 and thiol peroxiredoxin isoform X3. Antioxidants play important roles in immune modulation by protecting parasites from the oxidative effects of the host's immune response, and they have been shown to have potential as vaccine candidates (Sangpairoj et al., 2018). The antioxidants identified by Hamilton et al. (2018) are hypothesized to have a similar immunomodulatory role in sea lice, though their function has not been characterized. Functional characterization studies in helminths have led to the identification of vaccine targets that reduce egg laying capacity, maturation, and survival in helminth parasites (Elridi et al., 1997; Galvani, 2003; Hillyer, 2005). Sea lice vaccine development would benefit from using a similar approach for screening vaccine candidates. Targeting molecules essential for parasite growth and survival has proven to be an effective strategy in other arthropod models and has led to the development of effective vaccines.

High throughput next generation sequencing (NGS) is the latest tool for identifying potential sea louse vaccine targets. NGS has the advantages of being less time-consuming than the proteomics approach of identifying candidates from fractionated sea lice extracts, as it can screen tens of thousands of genes and hundreds of pathways simultaneously. Techniques such as microarray (Yasuike et al., 2012) and RNA sequencing (Gallardo-Escárate et al., 2014) are used for identifying genes and measuring the differential expression of genes on a temporal scale due to external factors such as infection or stress, or to examine gene expression changes during development. These techniques can reveal a great deal about the regulation of genes during the course of an infection and are useful for identifying gene targets for sea lice vaccine development. In addition, they are useful for monitoring the response of genes to treatments with drugs (Valenzuela-Miranda and Gallardo-Escárate, 2016) or functional feeds (Sutherland et al., 2017). Eichner et al. (2018) used RNA-seq to investigate changes in gene expression between chalimus and pre-adult 1 stages of *L. salmonis* to identify genes involved in moulting and suggested using key moulting genes for vaccine or drug development. Umasuthan et al. (2020) used microarrays to measure the early transcriptomic responses in pelvic fins of Atlantic salmon parasitized with stage I chalimus. They were able to show a downregulation in immune defense-related genes at chalimus attachment sites and at sites adjacent to chalimus attached sites, concluding that chalimus can immunomodulate the host's defense pathways (Umasuthan et al., 2020). In addition, the authors noted key physiological processes impacted by lice attachment. They included pathways involved in extracellular matrix degradation, wound healing, defense, and chemotaxis (Umasuthan et al., 2020). Studies like the ones conducted by Umasuthan et al. (2020), along with the work of numerous others (Skugor et al., 2008; Dalvin et al., 2009; Tadiso et al., 2011; Sutherland et al., 2011, 2014; Gallardo-Escárate et al., 2014; Gonçalves et al., 2014; Robledo et al., 2018), are paving the way to a deeper understanding of the genes and pathways involved in a variety of physiological processes such as establishing an infection on the host, sea lice development and metabolism, and salmon immunity (see Chapter 12).

Our limited knowledge of the complex interactions between pathogen and host has thus far restricted the development of more effective treatment strategies for managing sea lice infection; however, next-generation technologies may prove to be game changers for the future of sea lice vaccine and drug development. Post-sequencing screening of sea lice candidates by RNA interference (RNAi) is a valuable method for assessing the function of target genes *in vivo* and for identifying essential molecules for vaccine development (Campbell et al., 2009; Dalvin et al., 2009; Marr et al., 2014; Eichner et al., 2014; Tröße et al., 2014). Carpio et al. (2011) used RNAi to evaluate the function of a novel gene in *C. rogercresseyi*, my32, that shares sequence homology with akirin-2 from *L. salmonis*. The authors were able to confirm with RNAi gene knock down and RT-qPCR that the my32 transcript level in *C. rogercresseyi* was 70% less in the my32-dsRNA group versus the control and this correlated with a significant reduction in the number of parasites in this group. Consequently, when evaluated as a recombinant protein vaccine, my32 significantly delayed parasite development and reduced the number of parasites from the second generation by 57% at 24 days post-challenge (Carpio et al., 2011). It is important to note that RNAi loss of function phenotypes do not necessarily always correlate with vaccine efficacy as seen in Carpio et al. (2011). For example, immunization with the protective antigen BM86 reduced egg laying capacity of engorged *R. microplus* female cattle ticks. However, gene knock down of BM86 in female ticks failed to reduce tick fecundity (Willadsen et al., 1989). As sea lice genomic and transcriptomic data become more available, functional characterization methods like RNAi will be important for screening candidates and identifying targets likely to be effective *in vivo* (that is, protective antigens; targets that will adversely impact the growth and survival of the parasite). In addition, RNAi provides the added benefit of reducing the time and cost of screening ineffective targets if left to vaccine efficacy testing alone.

20.6 Antigen identification

One of the big challenges facing sea lice vaccine development is identifying targets that not only elicit a high specific antibody titer in salmon but also provide adequate protection against sea lice infection. While numerous attempts have been made to identify effective sea lice targets, most are met with limited or no success. Chile approved the first and only sea lice vaccine in 2014 for the management of *C. rogercresseyi* infections (Flores-Kossack et al., 2020). The subunit vaccine, Providean Aquatec Sea Lice®, is reported to reduce juvenile and adult *C. rogercresseyi* up to 73% (Villegas, 2015). The vaccine reduces infection by stimulating the proliferation of mucus secreting cells and by inducing specific antibodies in mucus and serum for up to 300 days post-vaccination. Currently this vaccine is not approved for use in *L. salmonis* infections.

A more recent vaccine that shows promise is a recombinant vaccine that was developed using a reverse vaccinology approach. When evaluated *in vivo* the vaccine was found to be up to 90% effective at 25 days post-infection (Gallardo-Escárate et al., 2019; Flores-Kossack et al., 2020). The protective response elicited by the vaccine is not fully understood but Flores-Kossack et al. (2020) hypothesize that the immune mechanism could be related to a non-specific mucosal neutralizing response acting on the parasite.

A study by Contreras et al. (2020) revisits the previous work of Raynard et al. (2002) by evaluating *L. salmonis* gut plasma membrane proteins suspected to function in louse feeding and development as vaccine candidates. The recombinants, P33 (amino acid transporter) and P30 (putative toll-like receptor 6), were shown to provide moderate protection from adult female lice (35%) and chalimus stage lice (31%), respectively, post-challenge (Contreras et al., 2020). The authors speculate that vaccine efficacy is a result of salmon IgM specific antibodies binding to target proteins on the apical membrane of intestinal cells of the louse.

A ribosomal peptide identified from the ribosomal P0 protein of *L. salmonis* was shown to have good efficacy (56%) against *L. salmonis* infection when administered intraperitoneally (Swain et al., 2020). The vaccine reduced adult female lice counts and fecundity and appeared to have a greater impact on the F1 lice generation rather than the lice that initially settled on the host. To increase immunogenicity of this vaccine in salmon, the peptide was constructed to contain T cell epitopes from tetanus toxin and measles virus at its N-terminal. The incorporation of T cell epitopes was shown to increase IgM specific responses in salmon (Swain et al., 2020). While the data are very promising, the authors recognize that additional efficacy testing on a larger scale is warranted.

While recombinant or purified protein vaccines have shown some promise as effective vaccine delivery systems, we should not limit ourselves from investigating other key technologies in vaccinology. For example, DNA vaccines have been shown to be effective vaccine delivery systems in protozoal and helminth parasites (Stutzer et al., 2018) and more recent evidence suggests they might be useful for ticks (Noon and Aroian, 2017; Tebeje et al., 2016; Wei et al., 2016). A subolesin-based DNA vaccine combined with a ribosomal acidic protein o (Po) peptide boost provided cross protection against *Haemaphysalis longicornis* infection, resulting in significant mortalities, reduced fecundity, and vaccine efficacies of 86.6% (Hassan et al., 2020). A recent study by Faber et al. (2019) suggests DNA vaccines may be an effective vaccine delivery system against myxozoan parasites. In this study, the authors evaluated three DNA vaccines against *Tetracapsuloides bryosalmonae*, the causative agent of proliferative kidney disease (PKD) in trout, and found some formulations decreased the number of parasites while reducing disease-associated kidney pathology (Abos et al., 2018). This delivery approach may be a useful tool for sea lice control as DNA vaccines have the advantages of being more stable and can stimulate both B cell and T cell responses that are longer lived than in a recombinant delivery system (Nascimento and Leite, 2012).

With the recent success of mRNA vaccines (that is, the malaria vaccine Mosquirix) mRNA vaccine delivery systems are beginning to be evaluated for use against other parasitic diseases (Garcia et al., 2018; Versteeg et al., 2019). Some vaccines in the pipeline target *Leishmania donovani* and *Toxoplasma gondii*. However, there are currently no published studies of mRNA vaccines in sea lice.

The complex life cycle and advanced immune evasion strategies of sea lice makes developing effective vaccines challenging, so when traditional vaccines fall short (that is, recombinant protein vaccines and subunit vaccines), the consideration of novel vaccine platforms such as DNA or mRNA as alternative delivery systems will be important.

20.7 Conclusion

As we draw this chapter to a close, the reader may think that we are no further forward in developing effective vaccines for sea lice than we were at the start of salmon aquaculture back in the 1960s. This, however, is not the case, as a recent patent has suggested that targeting the copepodid stage may be a more successful strategy for sea lice vaccines. Indeed, targeting the skin of fish for vaccination may be one of the key areas that we have overlooked in fish vaccinology. The discovery of IgT in the late 1990s (Litman et al., 1999) suggests that the immune system of fish is compartmentalized between systemic and mucosal responses and that antigens need to be presented differently to the fish to optimize the response in the skin. Intraperitoneal (IP) vaccination has been the standard way to vaccinate fish since the 1970s as it is easily reproducible and ensures the target amounts of antigen and adjuvant (an immune enhancing compound) are administered to each fish appropriately. This has some advantages, such as giving an accurate and repeatable dose of antigen and adjuvant (an immunity enhancing compound) to each fish as well as being confident that vaccination was correctly performed.

It is well known that IP vaccinations induce strong IgM responses in Atlantic salmon (Bricknell et al., 1997), but there does not seem to have been any systematic study to evaluate IgT responses in salmon after IP vaccination. There are a few studies that look at the agglutination potential of fish mucus after IP vaccination and ELISAs to determine IgM titers (Hiney et al., 1994; Swan et al., 2008; Palaksha et al., 2008; Valdenegro-Vega et al., 2013; Makesh et al., 2015; Tartor et al., 2020). However, none of these studies saw either agglutination activity or ELISA-based titers that were close to the much higher titers seen in the serum.

Could it be we have not been successful in developing an effective vaccine against sea lice because we have simply made vaccines that develop strong IgM titers and low IgT titers? It is possible that future sea lice vaccines will need to be formulated to develop strong immune responses in the mucosal surfaces of salmon, especially the skin. How we accomplish this remains unclear. Will future fish vaccinologists have to think about bath vaccines or intradermal injection systems to ensure that good responses in the skin are induced? After all, this is where the sea lice spend most of their lives and, in resistant salmonid species, it seems to be an acute immune response in the skin that prevents settlement and development of the copepodid/chalimus stages. If this is the case, then considerable research is needed to identify optimal strategies for vaccination of the skin. Not just in how to deliver the vaccine to the skin, or the antigen dose required, or if monovalent or polyvalent vaccines are needed, but in developing novel adjuvants to optimize the immune responses in these tissues as well as the optimum tissues to target when vaccinating the fish. A recent discovery is that the nares of zebrafish play an important role in the presentation of antigens to the immune system, especially mucosal surfaces. What role the nares play in salmon immunology remains unknown. However, it is very clear that the search for a sea lice vaccine is not over and there are still major obstacles that need to be overcome before we have an effective vaccine as part of the integrated pest management toolbox to manage the sea lice problem. As Bricknell et al. (2003) posed in their review of sea lice vaccines in 2003, “Are sea lice vaccines just around the corner or a holy grail?” From current evidence, it would seem that sea lice vaccines are closer to being a holy grail, given that the first major sea lice vaccine projects were generated under the EU FAIR IV program in the late 1980s, and there is still no effective vaccine to date.

References

Abolofia, J., Wilen, J.E. and Asche, F. (2017) The Cost of Lice: Quantifying the Impacts of Parasitic Sea Lice on Farmed Salmon. *Marine Resource Economics* 32, 329–349.

Abos, B., Estensoro, I., Perdiguero, P., Faber, M., Hu, Y.F., Rosales, P.D., Granja, A.G., Secombes, C.J., Holland, J.W. and Tafalla, C. (2018) Dysregulation of B Cell Activity During Proliferative Kidney Disease in Rainbow Trout. *Frontiers in Immunology* 9, 1203.

Ali, A., Fernando Parizi, L., Garcia Guizzo, M., Tirloni, L., Seixas, A., Silva Vaz Jr, I.D. and Termignoni, C. (2015) Immunoprotective Potential of a *Rhipicephalus* (*Boophilus*) *microplus* Metalloprotease. *Veterinary Parasitology* 207, 107–114.

Almazan, C., Tipacamu, G.A., Rodriguez, S., Mosqueda, J. and De Leon, A.P. (2018) Immunological Control of Ticks and Tick-Borne Diseases That Impact Cattle Health and Production. *Frontiers in Bioscience-Landmark* 23, 1535–1551.

Anguita, J., Ramamoorthi, N., Hovius, J.W.R., Das, S., Thomas, V., Persinski, R., Conze, D., Askenase, P.W., Rincon, M., Kantor, F.S. and Fikrig, E. (2002) Salp15, an *Ixodes scapularis* Salivary Protein, Inhibits Cd4(+) T Cell Activation. *Immunity* 16, 849–859.

Anonymous (2021) *Sea Lice 2021 Online* Faroe Islands: Sea Lice 2021 online. Available at: https://www.youtube.com/watch?v=p3H0eStazQ8&t=44s (accessed 8 October 2021).

Bell, J.F., Stewart, S.J. and Wikel, S.K. (1979) Resistance to Tick-Borne *Francisella tularensis* by Tick-Sensitized Rabbits: Allergic Klendusity. *The American Journal of Tropical Medicine and Hygiene* 28, 876–880.

Biering, E., Villoing, S., Sommerset, I. and Christie, K.E. (2005) Update on Viral Vaccines for Fish. In: Midtlyng, P.J. (ed.) *Progress in Fish Vaccinology*. Karger, Basel. pp. 97–113.

Bricknell, I.R., Bowden, T.J., Lomax, J. and Ellis, A.E. (1997) Antibody Response and Protection of Atlantic Salmon (*Salmo salar*) Immunised with an Extracellular Polysaccharide of *Aeromonas salmonicida*. *Fish and Shellfish Immunology* 7, 1–16.

Bricknell, I.R., Bisset, P.F. and Bowden, T.J. (2002) The Function of Atlantic Salmon (*Salmo Salar* L.) Antibodies under Extremes of pH and Osmolarity. *Fish and Shellfish Immunology* 13, 215–219.

Bricknell, I.R., Hayes, H.E., Billingsley, P.F., Vigneau, A.J., Sommerville, C. and Raynard, R.S. (2003) Vaccines against the Salmon Lice (*Lepeophtheirus salmonis*) – a Holy Grail, or Just around the Corner? *Journal of Parasitology* 89, 149–157.

Bron, J.E., Sommerville, C., Jones, M. and Rae, G.H. (1991) The Settlement and Attachment of Early Stages of the Salmon Louse, *Lepeophtheirus salmonis* (Copepoda: Caligidae) on the Salmon Host, *Salmo salar*. *Journal of Zoology* 224, 201–212.

Bron, J.E., Sommerville, C. and Rae, G.H. (1993) The Functional-Morphology of the Alimentary Canal of Larval Stages of the Parasitic Copepod *Lepeophtheirus salmonis*. *Journal of Zoology* 230, 207–220.

Campbell, E.M., Pert, C.C. and Bowman, A.S. (2009) RNA-Interference Methods for Gene-Knockdown in the Sea Louse, *Lepeophtheirus salmonis*: Studies on a Putative Prostaglandin E Synthase. *Parasitology* 136, 867–874.

Carpio, Y., Basabe, L., Acosta, J., Rodriguez, A., Mendoza, A., Lisperger, A., Zamorano, E., Gonzalez, M., Rivas, M., Contreras, S., Haussmann, D., Figueroa, J., Osorio, V.N., Asencio, G., Mancilla, J., Ritchie, G., Borroto, C. and Estrada, M.P. (2011) Novel Gene Isolated from *Caligus rogercresseyi*: A Promising Target for Vaccine Development against Sea Lice. *Vaccine* 29, 2810–2820.

Casuso, A., Nunez-Acuna, G., Valenzuela-Munoz, V., Leal, Y. and Gallardo-Escarate, C. (2019) Identification of Peritrophins as Potential Vaccine Candidates against Sea Lice: A Reverse Vaccinology Approach. *Fish and Shellfish Immunology* 91, 393–393.

Contreras, M., Karlsen, M., Villar, M., Olsen, R.H., Leknes, L.M., Furevik, A., Yttredal, K.L., Tartor, H., Grove, S., Alberdi, P., Brudeseth, B. and De La Fuente, J. (2020) Vaccination with Ectoparasite Proteins Involved in Midgut Function and Blood Digestion Reduces Salmon Louse Infestations. *Vaccines* 8, 32.

Dalvin, S., Frost, P., Biering, E., Hamre, L.A., Eichner, C., Krossøy, B. and Nilsen, F. (2009) Functional Characterisation of the Maternal Yolk-Associated Protein (Lsyap) Utilising Systemic RNA Interference in the Salmon Louse (*Lepeophtheirus salmonis*) (Crustacea: Copepoda). *International Journal for Parasitology* 39, 1407–1415.

Dalvin, S., Jorgensen, L.V., Kania, P.W., Grotmol, S., Buchmann, K. and Overgard, A.C. (2020) Rainbow Trout *Oncorhynchus mykiss* Skin Responses to Salmon Louse *Lepeophtheirus salmonis*: From Copepodid to Adult Stage. *Fish and Shellfish Immunology* 103, 200–210.

Dawson, L.H.J., Pike, A.W., Houlihan, D.F. and Mcvicar, A.H. (1997) Comparison of the Susceptibility of Sea Trout (*Salmo trutta* L.) and Atlantic Salmon (*Salmo salar* L.) to Sea Lice (*Lepeophtheirus salmonis* (Krøyer, 1837)) Infections. *ICES Journal of Marine Science* 54, 1129–1139.

Daix, V., Schroeder, H., Praet, N., Georgin, J.-P., Chiappino, I., Gillet, L., de Fays, K., Decrem, Y., Leboulle, G., Godfroid, E., Bollen, A., Pastoret, P.-P., Gern, L., Sharp, P.M. and Vanderplasschen, A. (2007) Ixodes Ticks Belonging to the *Ixodes ricinus* Complex Encode a Family of Anticomplement Proteins. *Insect Molecular Biology* 16(2), 155–166.

De La Fuente, J. and Contreras, M. (2015) Tick Vaccines: Current Status and Future Directions. *Expert Review of Vaccines* 14, 1367–1376.

De La Fuente, J., Kopacek, P., Lew-Tabor, A. and Maritz-Olivier, C. (2016) Strategies for New and Improved Vaccines against Ticks and Tick-Borne Diseases. *Parasite Immunology* 38, 754–769.

De Rose, R., McKenna, R.V., Cobon, G., Tennent, J., Zakrzewski, H., Gale, K., Wood, P.R., Scheerlinck, J.P.Y. and Willadsen, P. (1999) Bm86 Antigen Induces a Protective Immune Response against *Boophilus microplus* Following DNA and Protein Vaccination in Sheep. *Veterinary Immunology and Immunopathology* 71, 151–160.

Eichner, C., Frost, P., Dysvik, B., Jonassen, I., Kristiansen, B. and Nilsen, F. (2008) Salmon Louse (*Lepeophtheirus salmonis*) Transcriptomes During Post Molting Maturation and Egg Production, Revealed Using Est-Sequencing and Microarray Analysis. *BMC Genomics* 9, 126.

Eichner, C., Nilsen, F., Grotmol, S. and Dalvin, S. (2014) A Method for Stable Gene Knock-Down by RNA Interference in Larvae of the Salmon Louse (*Lepeophtheirus salmonis*). *Experimental Parasitology* 140, 44–51.

Eichner, C., Dondrup, M. and Nilsen, F. (2018) RNA Sequencing Reveals Distinct Gene Expression Patterns During the Development of Parasitic Larval Stages of the Salmon Louse (*Lepeophtheirus salmonis*). *Journal of Fish Diseases* 41, 1005–1029.

Elridi, R., Ozaki, T., Sato, H., Inaba, T. and Kamiya, H. (1997) Immunization of Mice with Ultraviolet-Attenuated Cercariae of *Schistosoma mansoni* Transiently Reduces the Fecundity of Challenge Worms. *International Journal for Parasitology* 27, 581–586.

Evelyn, T.P.T. (1997) A Historical Review of Fish Vaccinology. In: Gudding, R., Lillehaug, A., Midtlyng, P. and Brown, F. (eds) *Fish Vaccinology*. Karger, Basel. pp. 3–12.

Faber, M.N., Holland, J.W. and Secombes, C.J. (2019) Vaccination Strategies and IgM Responses against PKD in Rainbow Trout. *Fish and Shellfish Immunology* 91, 423.

Fast, M.D., Ross, N.W., Mustafa, A., Sims, D.E., Johnson, S.C., Conboy, G.A., Speare, D.J., Johnson, G. and Burka, J.F. (2002) Susceptibility of Rainbow Trout *Oncorhynchus mykiss*, Atlantic Salmon *Salmo salar* and Coho Salmon *Oncorhynchus kisutch* to Experimental Infection with Sea Lice *Lepeophtheirus salmonis*. *Diseases of Aquatic Organisms* 52, 57–68.

Fast, M.D., Burka, J.F., Johnson, S.C. and Ross, N.W. (2003) Enzymes Released from *Lepeophtheirus salmonis* in Response to Mucus from Different Salmonids. *The Journal of Parasitology* 89, 7–13.

Fast, M.D., Ross, N.W., Craft, C.A., Locke, S.J., Mackinnon, S.L. and Johnson, S.C. (2004) *Lepeophtheirus salmonis*: Characterization of Prostaglandin E2 in Secretory Products of the Salmon Louse by Rp-Hplc and Mass Spectrometry. *Experimental Parasitology* 107, 5–13.

Fast, M.D., Ross, N.W. and Johnson, S.C. (2005) Prostaglandin E(2) Modulation of Gene Expression in an Atlantic Salmon (*Salmo salar*) Macrophage-Like Cell Line (Shk-1). *Developmental and Comparative Immunology* 29, 951–963.

Firth, K.J., Johnson, S.C. and Ross, N.W. (2000) Characterization of Proteases in the Skin Mucus of Atlantic Salmon (*Salmo salar*) Infected with the Salmon Louse (*Lepeophtheirus salmonis*) and in Whole-Body Louse Homogenate. *Journal of Parasitology* 86, 1197, 1199–1205.

Fjelldal, P.G., Hansen, T.J., Karlsen, O. and Wright, D.W. (2019) Effects of Laboratory Salmon Louse Infection on Arctic Char Osmoregulation, Growth and Survival. *Conservation Physiology* 7, 11.

Flores-Kossack, C., Montero, R., Kollner, B. and Maisey, K. (2020) Chilean Aquaculture and the New Challenges: Pathogens, Immune Response, Vaccination and Fish Diversification. *Fish and Shellfish Immunology* 98, 52–67.

Franta, Z., Sojka, D., Frantova, H., Dvorak, J., Horn, M., Srba, J., Talacko, P., Mares, M., Schneider, E., Craik, C.S., Mckerrow, J.H., Caffrey, C.R. and Kopacek, P. (2011) Ircl1 – the Haemoglobinolytic Cathepsin L of the Hard Tick, *Ixodes ricinus*. *International Journal of Parasitolo*gy 41, 1253–1262.

Gallardo-Escárate, C., Valenzuela-Muñoz, V. and Nuñez-Acuña, G. (2014) RNA-Seq Analysis Using *de novo* Transcriptome Assembly as a Reference for the Salmon Louse *Caligus rogercresseyi*. *PLoS ONE* 9, e92239.

Gallardo-Escárate, C., Arriagada, G., Carrera, C., Gonçalves, A.T., Nuñez-Acuña, G, Valenzuela-Miranda, D. and Valenzuela-Muñoz, V. (2019) The Race between Host and Sea Lice in the Chilean Salmon Farming: A Genomic Approach. *Reviews in Aquaculture* 11, 325–339.

Galvani, A.P. (2003) Immunity, Antigenic Heterogeneity, and Aggregation of Helminth Parasites. *Journal of Parasitology* 89, 232–241.

Garcia, A.B., Siu, E., Sun, T., Exler, V., Brito, L., Hekele, A., Otten, G., Augustijn, K., Janse, C.J., Ulmer, J.B., Bernhagen, J., Fikrig, E., Geall, A. and Bucala, R. (2018) Neutralization of the Plasmodium-Encoded Mif Ortholog Confers Protective Immunity against Malaria Infection. *Nature Communications* 9, 2714.

Gonçalves, A.T., Farlora, R. and Gallardo-Escárate, C. (2014) Transcriptome Survey of the Lipid Metabolic Pathways Involved in Energy Production and Ecdysteroid Synthesis in the Salmon Louse *Caligus rogercresseyi* (Crustacea: Copepoda). *Comparative Biochemistry and Physiology B Biochemistry and Molecular Biology* 176, 9–17.

Grayson, T.H., John, R.J., Wadsworth, S., Greaves, K., Cox, D., Roper, J., Wrathmell, A.B., Gilpin, M.L. and Harris, J.E. (1995) Immunization of Atlantic Salmon against the Salmon Louse: Identification of Antigens and Effects on Louse Fecundity. *Journal of Fish Biology* 47, 85–94.

Gudding, R. and Goodrich, T. (2014) *The History of Fish Vaccination*. Blackwell Science Publications, Oxford, UK.

Guo, X., Booth, C.J., Paley, M.A., Wang, X., DePonte, K., Fikrig, E., Narasimhan, S. and Montgomery, R.R. (2009) Inhibition of neutrophil function by two tick salivary proteins. *Infection and Immunity* 77, 2320–2329.

Hamilton, S., Mclean, K., Monaghan, S.J., Mcnair, C., Inglis, N.F., Mcdonald, H., Adams, S., Richards, R., Roy, W., Smith, P., Bron, J., Nisbet, A.J. and Knox, D. (2018) Characterisation of Proteins in Excretory/Secretory Products Collected from Salmon Lice, *Lepeophtheirus salmonis*. *Parasites & Vectors* 11, 294.

Hannier, S., Liversidge, J., Sternberg, J.M. and Bowman, A.S. (2004) Characterization of the B-cell Inhibitory Protein Factor in *Ixodes ricinus* Tick Saliva: A Potential Role in Enhanced *Borrelia burgdoferi* Transmiss ion. *Immunology* 113, 401–408.

Hajnická, V., Vančová, I., Kocáková, P., Slovák, M., Gašperík, J., Slávikovà, M, Hails, R.S., Labuda, M. and Nuttall, P.A. (2005) Manipulation of Host Cytokine Network by Ticks: A Potential Gateway for Pathogen Transmission. *Parasitology* 130, 333–342.

Hassan, I.A., Wang, Y.N., Zhou, Y.Z., Cao, J., Zhang, H.S. and Zhou, J.L. (2020) Cross Protection Induced by Combined Subolesin-Based DNA and Protein Immunizations against Adult *Haemaphysalis longicornis*. *Vaccine 38*(4), 907–915.

Hillyer, G.V. (2005) Fasciola Antigens as Vaccines against Fascioliasis and Schistosomiasis. *Journal of Helminthology* 79, 241–247.

Hiney, M.P., Kilmartin, J.J. and Smith, P.R. (1994) Detection of *Aeromonas-salmonicida* in Atlantic Salmon with Asymptomatic Furunculosis Infections. *Diseases of Aquatic Organisms* 19, 161–167.

Hovius, J.W.R., de Jong, M.A.W.P., den Dunnen, J., Litjens, M., Fikrig, E., van der Poll, T., Gringhuis, S.I. and Geijtenbeek, T.B.H. (2008) Salp15 Binding to DC-SIGN Inhibits Cytokine Expression by Impairing Both Nucleosome Remodeling and mRNA Stabilization. *PLoS Pathogens* 4(2), e31.

Johnson, S.C. and Albright, L.J. (1992a). Comparative Susceptibility and Histopathology of the Response of Naive Atlantic, Chinook and Coho Salmon to Experimental-Infection with *Lepeophtheirus salmonis* (Copepoda: Caligidae). *Diseases of Aquatic Organisms* 14, 179–193.

Johnson, S.C. and Albright, L.J. (1992b). Effects of Cortisol Implants on the Susceptibility and the Histopathology of the Responses of Naive Coho Salmon *Oncorhynchus kisutch* to Experimental-Infection with *Lepeophtheirus salmonis* (Copepoda: Caligidae). *Diseases of Aquatic Organisms* 14, 195–205.

Johnson, S.C., Blaylock, R.B., Elphick, J. and Hyatt, K.D. (1996) Disease Induced by the Sea Louse (*Lepeophtheirus salmonis*) (Copepoda: Caligidae) in Wild Sockeye Salmon (*Oncorhynchus nerka*) Stocks of Alberni Inlet, British Columbia. *Canadian Journal of Fisheries and Aquatic Sciences* 53, 2888–2897.

Kalinski, P. (2012) Regulation of Immune Responses by Prostaglandin E2. *Journal of Immunology* 188, 21–28.

Kashulin, A., Seredkina, N. and Sorum, H. (2017) Cold-Water Vibriosis. The Current Status of Knowledge. *Journal of Fish Diseases* 40, 119–126.

Klouwens, M.J., Trentelman, J.J. and Hovius, J.W.R. (2016) Anti-Tick Vaccines to Prevent Tick-Borne Diseases: An Overview and a Glance at the Future. *Ecology and Control of Vector-Borne Diseases* 4, 295–316.

Leal, Y., Velazquez, J., Hernandez, L., Swain, J.K., Rodriguez, A.R., Martinez, R., Garcia, C., Ramos, Y., Estrada, M.P. and Carpio, Y. (2019) Promiscuous T Cell Epitopes Boosts Specific Igm Immune Response against a Po Peptide Antigen from Sea Lice in Different Teleost Species. *Fish and Shellfish Immunology* 92, 322–330.

Litman, G.F., Anderson, M.K. and Rast, J. (1999) Evolution of Antigen Binding Receptors. *Annual Review of Immunology* 17, 109–147.

Lorenzen, N. and Olesen, N.J. (1997) Immunization with Viral Antigens: Viral Haemorrhagic Septicaemia. In: Gudding, R., Lillehaug, A., Midtlyng, P. and Brown, F. (eds) *Fish Vaccinology*. Karger, Basel. pp. 201–209.

Makesh, M., Sudheesh, P.S. and Cain, K.D. (2015) Systemic and Mucosal Immune Response of Rainbow Trout to Immunization with an Attenuated *Flavobacterium psychrophilum* Vaccine Strain by Different Routes. *Fish and Shellfish Immunology* 44, 156–163.

Marr, E.J., Sargison, N.D., Nisbet, A.J. and Burgess, S.T.G. (2014) RNA Interference for the Identification of Ectoparasite Vaccine Candidates. *Parasite Immunology* 36, 614–624.

McCarthy, E., Cunningham, E., Copley, L., Jackson, D., Johnston, D., Dalton, J.P. and Mulcahy, G. (2012) Cathepsin L Proteases of the Parasitic Copepod, *Lepeophtheirus salmonis*. *Aquaculture* 264–271, 356–357.

McKenna, R.V., Riding, G.A., Jarmey, J.M., Pearson, R.D. and Willadsen, P. (1998) Vaccination of cattle against the *Boophilus microplus* using a mucin-like membrane glycoprotein. *Parasite Immunology* 20, 325–336.

Mota, V.C., Nilsen, T.O., Gerwins, J., Gallo, M., Kolarevic, J., Krasnov, A. and Terjesen, B.F. (2020) Molecular and Physiological Responses to Long-Term Carbon Dioxide Exposure in Atlantic Salmon (*Salmo salar*). *Aquaculture* 519.

Muhlia-Almazan, A., Sanchez-Paz, A. and Garcia-Carreno, F.L. (2008) Invertebrate Trypsins: A Review. *Journal of Comparative Physiology B-Biochemical Systems and Environmental Physiology* 178, 655–672.

Nascimento, I.P. and Leite, L.C.C. (2012) Recombinant Vaccines and the Development of New Vaccine Strategies. *Brazilian Journal of Medical and Biological Research* 45, 1102–1111.

Noon, J.B. and Aroian, R.V. (2017) Recombinant Subunit Vaccines for Soil-Transmitted Helminths. *Parasitology* 144, 1845–1870.

Nylund, A., Okland, S. and Bjorknes, B. (1992) Anatomy and Ultrastructure of the Alimentary Canal in *Lepeophtheirus salmonis* (Copepoda: Siphonostomatoida). *Journal of Crustacean Biology* 12, 423–437.

Palaksha, K.J., Shin, G.W., Kim, Y.R. and Jung, T.S. (2008) Evaluation of Non-Specific Immune Components from the Skin Mucus of Olive Flounder (*Paralichthys olivaceus*). *Fish and Shellfish Immunology* 24, 479–488.

Patarroyo, J., Portela, R., Castro, R., Pimentel, J., Guzmán, F., Patarroyo, M., Vargas Viloria, M., Prates, A.A. and Mendes, M. (2002) Immunization of Cattle with Synthetic Peptides Derived from the *Boophilus microplus* Gut Protein (Bm86). *Veterinary Immunology and Immunopathology* 88, 163–172.

Pereiro, P., Figueras, A. and Novoa, B. (2016) Turbot (*Scophthalmus maximus*) Vs. VHSV (Viral Hemorrhagic Septicemia Virus): A Review. *Frontiers in Physiology* 7, 10.

Press, C.M. and Lillehaug, A. (1995) Vaccination in European Salmonid Aquaculture – a Review of Practices and Prospects. *British Veterinary Journal* 151, 45–69.

Raynard, R.S., Bricknell, I.R., Billingsley, P.F., Nisbet, A.J., Vigneau, A. and Sommerville, C. (2002) Development of Vaccines against Sea Lice. *Pest Management Science* 58, 569–575.

Robledo, D., Gutiérrez, A.P., Barría, A., Yáñez, J.M. and Houston, R.D. (2018) Gene Expression Response to Sea Lice in Atlantic Salmon Skin: RNA Sequencing Comparison between Resistant and Susceptible Animals. *Frontiers in Genetics* 9, 287.

Roper, J., Grayson, T.H., Jenkins, P.G., Hone, J.V., Wrathmell, A.B., Russell, P.M. and Harris, J.E. (1995) The Immunocytochemical Localization of Potential Candidate Vaccine Antigens from the Salmon Louse *Lepeophtheirus salmonis* (Krøyer 1837). *Aquaculture* 132, 221–232.

Ruane, N., Mccarthy, T.K. and Reilly, P. (1995) Antibody Response to Crustacean Ectoparasites in Rainbow Trout, *Oncorhynchus mykiss* (Walbaum), Immunized with *Argulus foliaceus* L Antigen Extract. *Journal of Fish Diseases* 18, 529–537.

Ruane, N.M., Nolan, D.T., Rotllant, J., Costelloe, J. and Bonga, S.E.W. (2000) Experimental Exposure of Rainbow Trout *Oncorhynchus mykiss* (Walbaum) to the Infective Stages of the Sea Louse *Lepeophtheirus salmonis* (Krøyer) Influences the Physiological Response to an Acute Stressor. *Fish and Shellfish Immunology* 10, 451–463.

Sangpairoj, K., Apisawetakan, S., Changklungmoa, N., Kueakhai, P., Chaichanasak, P., Sobhon, P. and Chaithirayanon, K. (2018) Potential of Recombinant 2-Cys Peroxiredoxin Protein as a Vaccine for *Fasciola gigantica* Infection. *Experimental Parasitology* 194, 16–23.

Shih, Y.C., Prausnitz, J.M. and Blanch, H.W. (1992) Some Characteristics of Protein Precipitation by Salts. *Biotechnology and Bioengineering* 40, 1155–1164.

Skern-Mauritzen, R., Frost, P., Dalvin, S., Kvamme, B.O., Sommerset, I. and Nilsen, F. (2009) A Trypsin-Like Protease with Apparent Dual Function in Early *Lepeophtheirus salmonis* (Krøyer) Development. *BMC Molecular Biology* 10, 11.

Skern-Mauritzen, R., Torrissen, O. and Glover, K.A. (2014) Pacific and Atlantic *Lepeophtheirus salmonis* (Krøyer, 1838) Are Allopatric Subspecies: *Lepeophtheirus salmonis salmonis* and *L. salmonis oncorhynchi* Subspecies novo. *BMC Genetics* 15, 9.

Skov, J., Mehrdana, F., Marana, M.H., Bahlool, Q.Z.M., Jaafar, R.M., Sindberg, D., Jensen, H. M., Kania, P.W. and Buchmann, K. (2014) Parasite Infections of Rainbow Trout (*Oncorhynchus mykiss*) from Danish Mariculture. *Aquaculture* 434, 486–492.

Skugor, S., Glover, K.A., Nilsen, F. and Krasnov, A. (2008) Local and Systemic Gene Expression Responses of Atlantic Salmon (*Salmo salar* L.) to Infection with the Salmon Louse (*Lepeophtheirus salmonis*). *BMC Genomics* 9, 498.

Stutzer, C., Richards, S.A., Ferreira, M., Baron, S. and Maritz-Olivier, C. (2018) Metazoan Parasite Vaccines: Present Status and Future Prospects. *Frontiers in Cellular and Infection Microbiology* 8, 67.

Subramanian, S., Mackinnon, S.L. and Ross, N.W. (2007) A Comparative Study on Innate Immune Parameters in the Epidermal Mucus of Various Fish Species. *Comparative Biochemistry and Physiology B – Biochemistry and Molecular Biology* 148, 256–263.

Sutherland, B.J., Jantzen, S.G., Sanderson, D.S., Koop, B.F. and Jones, S.R. (2011) Differentiating Size-Dependent Responses of Juvenile Pink Salmon (*Oncorhynchus gorbuscha*) to Sea Lice (*Lepeophtheirus salmonis*) Infections. *Comparative Biochemistry and Physiology – Part D Genomics Proteomics* 6, 213–223.

Sutherland, B.J., Koczka, K.W., Yasuike, M., Jantzen, S.G., Yazawa, R., Koop, B.F. and Jones, S.R. (2014) Comparative Transcriptomics of Atlantic Salmon *Salmo salar*, Chum *Oncorhynchus keta* and Pink Salmon *O. gorbuscha* During Infections with Salmon Lice *Lepeophtheirus salmonis*. *BMC Genomics* 15, 200.

Sutherland, B.J.G., Covello, J.M., Friend, S.E., Poley, J.D., Koczka, K.W., Purcell, S.L., Macleod, T.L., Donovan, B.R., Pino, J., González-Vecino, J.L., Gonzalez, J., Troncoso, J., Koop, B.F., Wadsworth, S.L., Fast, M.D. and Trudeau, V.L. (2017) Host–Parasite Transcriptomics During Immunostimulant-Enhanced Rejection of Salmon Lice (*Lepeophtheirus salmonis*) by Atlantic Salmon (*Salmo salar*). *FACETS* 2, 477–495.

Swain, J.K., Carpio, Y., Johansen, L.H., Velazquez, J., Hernandez, L., Leal, Y., Kumar, A. and Estrada, M.P. (2020) Impact of a Candidate Vaccine on the Dynamics of Salmon Lice (*Lepeophtheirus salmonis*) Infestation and Immune Response in Atlantic Salmon (*Salmo salar* L.). *PLoS ONE* 15, e0239827.

Swan, C.M., Lindstrom, N.M. and Cain, K.D. (2008) Identification of a Localized Mucosal Immune Response in Rainbow Trout, *Oncorhynchus mykiss* (Walbaum), Following Immunization with a Protein-Hapten Antigen. *Journal of Fish Diseases* 31, 383–393.

Tadiso, T.M., Krasnov, A., Skugor, S., Afanasyev, S., Hordvik, I. and Nilsen, F. (2011) Gene Expression Analyses of Immune Responses in Atlantic Salmon During Early Stages of Infection by Salmon Louse (*Lepeophtheirus salmonis*) Revealed Bi-Phasic Responses Coinciding with the Copepod-Chalimus Transition. *BMC Genomics* 12, 141–141.

Tafalla, C., Bogwald, J., Dalmo, R.A., Munang'andu, H.M. and Evensen, O. (2014) *Adjuvants in Fish Vaccines*. Blackwell Science Publications, Oxford.

Tartor, H., Monjane, A.L. and Grove, S. (2020) Quantification of Defensive Proteins in Skin Mucus Of Atlantic Salmon using Minimally Invasive Sampling and High-Sensitivity ELISA. *Animals* 10, 17.

Tatchell, R.J. (1967) A Modified Method for Obtaining Tick Oral Secretion. *The Journal of Parasitology* 53, 1106–1107.

Tebeje, B.M., Harvie, M., You, H., Loukas, A. and Mcmanus, D.P. (2016) Schistosomiasis Vaccines: Where Do We Stand? *Parasites & Vectors* 9, 528.

Trößе, C., Nilsen, F. and Dalvin, S. (2014) RNA Interference Mediated Knockdown of the Kdel Receptor and Copb2 Inhibits Digestion and Reproduction in the Parasitic Copepod *Lepeophtheirus salmonis*. *Comparative Biochemistry and Physiology B Biochemistry and Molecular Biology* 170, 1–9.

Umasuthan, N., Xue, X., Caballero-Solares, A., Kumar, S., Westcott, J.D., Chen, Z. Y., Fast, M.D., Skugor, S., Nowak, B.F., Taylor, R.G. and Rise, M.L. (2020) Transcriptomic Profiling in Fins of Atlantic Salmon Parasitized with Sea Lice: Evidence for an Early Imbalance between Chalimus-Induced Immunomodulation and the Host's Defense Response. *International Journal of Molecular Sciences* 21(7), 1–45.

Valdenegro-Vega, V.A., Crosbie, P., Vincent, B., Cain, K.D. and Nowak, B.F. (2013) Effect of Immunization Route on Mucosal and Systemic Immune Response in Atlantic Salmon (*Salmo salar*). *Veterinary Immunology and Immunopathology* 151, 113–123.

Valenzuela-Miranda, D. and Gallardo-Escárate, C. (2016) *Caligus rogercresseyi* Serine Proteases: Transcriptomic Analysis in Response to Delousing Drugs Treatments. *Aquaculture* 465, 65–77.

Versteeg, L., Almutairi, M.M., Hotez, P.J. and Pollet, J. (2019) Enlisting the mRNAVaccine Platform to Combat Parasitic Infections. *Vaccines* 7(4), 122.

Villegas, A. (2015) First Ever Sea Lice Vaccine Launched in Chile. *Undercurrent News*. Available at: https://www.undercurrentnews.com/2015/11/24/first-ever-sea-lice-vaccine-launched-in-chile/ (accessed 23 Jan 2021).

Wei, F., Song, M.X., Liu, H.H., Wang, B., Wang, S.C., Wang, Z.D., Ma, H.Y., Li, Z.Y., Zeng, Z., Qian, J. and Liu, Q. (2016) Molecular Detection and Characterization of Zoonotic and Veterinary Pathogens in Ticks from Northeastern China. *Frontiers in Microbiology* 7, 1913.

Willadsen, P. (1997) Novel Vaccines for Ectoparasites. *Veterinary Parasitology* 71, 209–222.

Willadsen, P., Riding, G.A., Mckenna, R.V., Kemp, D.H., Tellam, R.L., Nielsen, J.N., Lahnstein, J., Cobon, G.S. and Gough, J.M. (1989) Immunologic Control of a Parasitic Arthropod. Identification of a Protective Antigen from *Boophilus microplus*. *Journal of Immunology* 143, 1346–1351.

Willadsen, P., Bird, P., Cobon, G.S. and Hungerford, J. (1995) Commercialization of a Recombinant Vaccine against *Boophilus-microplus*. *Parasitology* 110, S43–S50.

Willadsen, P., Cobon, G. and Mckenna, R.V. (1996) Comparative Vaccination of Cattle against *Boophilus microplus* with Recombinant Antigen BM86 Alone or in Combination with Recombinant BM91. *Parasite Immunology* 18, 241–246.

Woo, P.T.K. (1997) Immunization against Parasitic Diseases of Fish. In: Gudding, R., Lillehaug, A., Midtlyng, P. and Brown, F. (eds) *Fish Vaccinology*. Karger, Basel. pp. 231–241.

Yasuike, M., Leong, J., Jantzen, S.G., Von Schalburg, K.R., Nilsen, F., Jones, S.R. and Koop, B.F. (2012) Genomic Resources for Sea Lice: Analysis of ESTs and Mitochondrial Genomes. *Marine Biotechnology (NY)* 14, 155–166.

CHAPTER 21

Functional feeds and sea lice control

Stanko Skugor[1] *and Simon L. Wadsworth*[2]

[1]Cargill Aqua Nutrition, Cargill, and Sea Lice Research Centre (SLRC), Oslo, Norway
[2]Nord University, Bodø, Norway

21.1 Sea lice control

The cost of salmon production in Norway has increased by 60% over a six-year period (2012–2018), mainly due to the increasing impact of sea lice control measures. The production price has risen from 23.40 NOK per kg in 2012, to 38.4 NOK per kg by 2018, representing a total additional cost of US$1,900,000,000 to the industry (Ytreberg, 2020). During the same period, the overall production volume of salmon in Norway did not increase, mainly due to restrictions imposed by sea lice control (Iversen et al., 2020). Similar issues with sea lice control and an associated increase in production cost have been seen in other salmon producing regions, especially Scotland, Chile and Eastern Canada. In these regions the major lice issues are *Lepeophtheirus salmonis* (Krøyer 1837) in Europe, and *Caligus rogercresseyi* (Boxshall & Bravo 2000) in Chile.

The Norwegian lice surveillance and control programme is rigorous and requires each farm to develop a general plan for the prevention, control and treatment of salmon lice. Lice control plans include regular counting and recording, delousing operations and assessment of treatment efficacy as well as routine farm fallowing. The farm's lice management plans are updated annually, and the details provided to Mattilsynet, the Norwegian Food Safety Authority. Sea lice levels are strictly regulated by law and cannot exceed an average of 0.5 adult female lice per fish during the year. The lice burden on farmed salmon is reduced to 0.2 adult lice per fish during the seawater migration of salmon smolts in late spring (Torrissen et al., 2013; Overton et al., 2018). Further regulatory controls (the 'traffic light system') have been imposed on sites in areas that present a higher sea lice challenge on the local wild salmon populations, resulting in significant reduction in allowable biomass for those production sites affected (Myksvoll et al., 2018; Svendsen, 2019).

For a farm to remain compliant, many salmon farming companies have implemented a range of control methods which has seen a dramatic move away from the use of anti-sea-lice medicines (Overton et al., 2018). Currently, sea lice are being controlled via mechanical treatments; here fish are crowded in the net pen before being pumped past a low-pressure water spray system to physically remove the lice (see Chapter 22). Variations on this theme include an inverse water turbulence system, where lice are vacuumed off the fish. In addition to mechanical removal, the use of thermal treatments, often combined with low-pressure sprays, is also being deployed. In these systems fish are crowded in the net pen, before being pumped into a treatment chamber containing seawater at temperatures up to 34°C and held for 20 to 30 seconds (Holan et al., 2017). The sea lice become inactive following the temperature shock and detach from the fish (Grøntvedt et al., 2015; Roth, 2016). Efficacy of the physical treatment methods has been recorded within the range of 80–100% for mobile lice, although effects on attached lice remain unclear (Gismervik et al., 2017).

The physical treatments have rapidly become an essential tool in salmon farming operations but are considered to have a negative impact on fish welfare. The treatments significantly reduce feed intake due to the pre-treatment food withholding, as well as loss of appetite from stress and physical damage that has been observed post-treatment (Overton et al., 2018). To address the compromised welfare of fish during these operations and help with wound healing, thermal challenge and wound healing models were established resulting in the development of functional feeds that can accelerate wound healing by 40% (Skugor, 2018).

Several species of cleaner fish have been widely used by the industry for over 30 years and their use has increased four-fold since 2012. In Norway, 50 million cleaner fish are deployed at 75% of all salmon farms (Norwegian Directorate of Fisheries, 2021). Further focus on cleaner fish for managing sea lice entails improved understanding of species-specific nutritional needs, and development of feeds and feeding practices with the potential to boost the delousing efficiency of cleaner fish (Skiftesvik et al., 2013; Leclercq et al., 2015; Johannesen et al., 2018). Lice are a scarce food item for cleaner fish as their numbers must be kept at low levels, and supplemental feeding is necessary to stop them from starving due to a lack of prey items. Additionally, lice may not be the preferred prey item of cleaner fish if more appetising members of the zooplankton community (Eliasen et al., 2018) or crustaceans (Gentry et al., 2020) are present. In some studies, only around 10% of lumpfish used in the Faroe Islands (Eliasen et al., 2018) and 11% of corkwing wrasse on farms in Western Norway (Gentry et al., 2020) were found to be actively consuming lice. Furthermore, a study by Gentry et al. (2020) revealed interactions of corkwing wrasse and other in-cage lice control methods resulting in reduced delousing efficacy of cleaner fish.

Sea lice are known to concentrate in surface waters and a variety of methods have been developed to maintain fish at depth, away from the infective stages of sea lice. These systems include subsurface feeding, submerged lights and snorkel nets that restrict the areas of immediate access of the fish to the surface. Planktonic nets are also deployed around the pens, from the surface down to 8 m depth to reduce the influx of sea lice from the surrounding surface waters (Nilsson et al., 2017; Wright et al., 2017; Stien et al., 2016, 2018; Grøntvedt et al., 2018).

21.2 Development of functional feeds against sea lice

Functional feeds have been used as an additional tool for sea lice control and have been successfully implemented along with the wider range of control measures (Jensen et al., 2015; Holm et al., 2016a; Bui et al., 2020; Gentry et al., 2020). In-feed applications hold even more promise in the future, as in comparison to other control measures, they are easy to administer, handling stress is avoided, they can be effective against multiple stages of lice and they are less likely to result in the development of resistance (Burka et al., 1997; Sutherland et al., 2017).

The prerequisites for the development of functional feeds against lice include understanding of the complex salmon–louse, host–pathogen interactions, knowledge of physicochemical and bioactive properties of candidate functional ingredients, a dedicated laboratory with established protocols for *in vitro* screening, a facility for nutritional and lice challenge studies in salmonid fish and a stable supply of infective copepodids. The discovery rate of effective functional ingredients is low in general and sufficient laboratory capacity is required to screen many potential candidate ingredients per year. The discovery pipeline typically involves pre-screening by several different *in vitro* tools, such as the simple LD50 assay that measures the number of dead and live copepodids exposed to the increasing concentration of the test product.

The lower the concentration that kills copepodids the more effective the product is expected to be when included in an in-feed application. An important caveat is that such *in vitro* tools do not take into account that tested functional ingredients may not have a direct toxic effect against lice but may nevertheless possess anti-parasitic properties exerted by the activation of host protection mechanisms. The research of lice repellents of botanical origin (O'Shea et al., 2017) and semiochemicals involved in host-finding and non-host avoidance (Bailey et al., 2006) was greatly advanced by the development of the Y-tube behavioural bioassay (Ingvarsdóttir et al., 2002a, 2002b; Pino-Marambio et al., 2007). The Y-tube assay relies on the use of adult males or copepodids which are given a choice between two olfactory stimuli. The high activation and directional response of lice towards or away from the olfactory stimuli originating from the two arms of the glass Y-shaped tube arena are recorded with a camera and used to estimate the attractant or repellent nature of the tested product. The *Caligus* frontal filament extrusion assay uses copepodids and a simple petri dish set-up. Copepodids are incubated for 48 hours in seawater that overlays agar which is mixed either with salmon mucus only or with mucus and the test product (Núñez-Acuña et al., 2016). Infective copepodids of *C. rogercresseyi* can quickly extrude the preformed frontal filament, a structure that helps them attach to the fish skin once exposed to the appropriate Atlantic salmon host signals. In the absence of Atlantic salmon mucus, or in the presence of mucus combined with the functional ingredient that has anti-lice properties, extrusion of the frontal filament is inhibited. The *ex vivo* fin settlement assay is more time and resource demanding as it requires that fish be fed the test feed (for example, containing a phytogenic compound with putative anti-lice properties) and control feed for about two weeks prior to the *ex vivo* fin experiment. Pairs of pectoral fins are removed from fish and placed in glass containers filled with seawater and incubated with *L. salmonis* copepodids which readily attach to the fins under these conditions. Lower numbers of lice settling on fins originating from fish exposed to the test feed suggests that the tested phytogenic product changes olfactory (or nutritional) properties of skin and mucus thus negatively affecting the early phases of lice attachment (Skugor, 2017).

Finally, *in vivo* efficacy trials executed in the controlled laboratory environment are needed to validate *in vitro* findings. The reasons for a product not being effective *in vivo* while promising results are observed in the LD50, Y-tube, frontal filament or *ex vivo* fin assays could be many. Bioactive compounds contained in the tested functional ingredient may be lost during the feed production process or degraded in the fish gastrointestinal tract, the compounds may not pass the gut barrier, or they may be absorbed but degraded in the blood or metabolised by the liver before reaching peripheral blood circulation, and in particular skin and mucus as the most important target tissues. Further, negative interactions may occur with other substances present in the feed, including other functional ingredients. The length of dietary exposure and the dose needed to achieve effective levels in target tissues are also of importance. Typically, a series of trials is required to determine the optimal dose and duration of treatment. To reliably detect low to moderate effects of feeds against lice, which typically do not exceed a 30% reduction in the total number of lice, a sufficient number of tank replicates and experimental animals must be ensured. Anti-parasitic effects may involve more pronounced effects against one gender as well as delayed development. Therefore skilled personnel that can classify lice by gender and developmental stage can contribute to deeper insights into the effects of the functional ingredient compared to observing only the reduction in total lice numbers. Final validation of successful *in vivo* laboratory trials is done in sea cages, under the real-world farming conditions.

The sector of functional foods and food supplements has greatly expanded during the last decades as dietary approaches in human and veterinary (predominantly mammalian) medicine have shown potential to reduce obesity, dental diseases, osteoporosis, gastrointestinal tract and cardiovascular problems, cancer risk, diabetes and allergies. The need to prevent diseases and improve health and welfare of farmed animals has ushered in the application of functional feeds fortified with special dietary ingredients (for example, nucleotides, β-glucans, plant secondary metabolites) that promote beneficial immune and physiological responses thus resulting in superior performance. A wide range of functional feeds have been developed for aquatic species and many have been in commercial use for over 20 years. Diets rich in vitamin E were used for the recovery of acute infectious pancreatic necrosis and pancreas disease in Europe in the early to mid-1990s (Raynard et al., 1991; McCoy et al., 1994). Anti-lice functional feeds that have been used to date were mainly based on the inclusion of special dietary ingredients of microbial and herbal origin.

Another approach that has emerged within the field of functional or health feeds shares similarities with recently popularised concepts in human nutrition, which rely on changes in dietary formulations and inherent properties of macronutrients to convey health benefits (Sears, 2015). Such examples include a very low in carbohydrate and high in fat ketogenic diet for the long-term management of obesity (Bueno et al., 2013) or the omega-3 fatty acid rich Mediterranean diet that reduces cardiovascular risk and lowers chronic low-grade inflammation (Estruch, 2010). Examples of this approach in aqua health feeds are the anti-viral diets, based on reduction in dietary lipid along with increased levels of the omega-3 fatty acid EPA, that exert anti-inflammatory effects leading to lowered heart inflammation and heart lesions caused by Atlantic salmon reovirus infection (Martinez-Rubio et al., 2012, 2013), and a more recent work by Dessen et al. (2019) who showed that diets with an increased dietary protein/lipid ratio improved survival of Atlantic salmon naturally affected by pancreas disease. Findings that susceptible Atlantic salmon and resistant coho salmon exhibit differential metabolic changes in response to *C. rogercresseyi* infestation suggests that diets differing in their composition of macronutrients could be a dietary intervention leading to increased protection (Vargas-Chacoff et al., 2016). Also worth mentioning here is the groundwork laid by Katan et al. (2020) and Xue et al. (2020) who showed that effective stimulation of the pro-inflammatory state can be achieved in tissues (head kidney and liver) involved in shaping the immune response to lice in Atlantic salmon by increasing omega-6/omega-3 fatty acid ratio.

21.3 Host protection mechanisms against lice

The work which addressed what constitutes protective host responses against lice has paved much of the way for the development of functional feeds against lice. Different approaches were used to improve the understanding of endogenous protection mechanisms: transcriptional profiling of skin and immune tissues of lice-infected Atlantic salmon (Skugor et al., 2008; Tadiso et al., 2011; Valenzuela-Muñoz et al., 2016, 2017; Holm et al., 2016b; Umasuthan et al., 2020), including comparative analysis of Atlantic salmon infested with *C. rogercresseyi* and *L. salmonis* (Valenzuela-Muñoz and Gallardo-Escárate, 2017); lice and virus co-infected Atlantic salmon (Carvalho et al., 2020);

interspecific comparison between susceptible salmonids and resistant coho salmon (Fast et al., 2002; Braden et al., 2015; Valenzuela-Muñoz et al., 2017) and pink salmon (Sutherland et al., 2014; Braden et al., 2012); within-species comparison of susceptible and resistant families of Atlantic salmon (Holm et al., 2015; Robledo et al., 2018); studies of hormonal interventions that promote resistance (Krasnov et al., 2015); and work exploring differences in skin olfactory properties of salmonids differing by their level of resistance (see Chapter 14). Collectively, research has shown that resistant phenotypes involve one or more of the following characteristics: 1) unappealing olfactory properties of skin and mucus, 2) the ability of the host to avoid immunosuppression by lice, and the capacity to mount an early pro-inflammatory Th1 immune response, followed by the Th2-type response with a strong regulatory component, and finally, 3) regulation of iron metabolism so that the acquisition of iron and haem becomes difficult for the parasite.

Diet-mediated changes in the composition of skin and mucus could potentiate endogenous protective mechanisms related to olfaction. It is hypothesised that lice in the early phases of the attachment process can be confused by diet-derived compounds that mask the specific smell cues of the preferred Atlantic salmon host. In addition, bioactives contained in the functional ingredient may taste bad or interfere with digestive processes in lice, which will consequently feed less aggressively, resulting in lowered immunosuppressive potential, allowing the host to mount an appropriate immune response (see Chapter 20). More than one protection mechanism can be modulated by nutritional interventions to the benefit of the host; for example, feeding glucosinolate-containing feeds to Atlantic salmon challenged by *L. salmonis* was shown to modulate both skin immunity and iron metabolism in multiple tissues (Holm et al., 2016a; Skugor et al., 2016). It is expected that functional feeds that can stimulate more than one protection mechanism will produce a stronger final anti-parasitic effect. For that reason, combinations of different functional ingredients that stimulate distinct protective mechanisms to produce a more resistant phenotype are often investigated in nutritional challenge trials. The study of molecular underpinnings of diet-mediated protection in salmonids has been made easier by mid- and high-throughput technologies that can measure a great number of transcripts and proteins involved in a biological process. Findings from dietary studies confirmed the importance of Th1 pro-inflammatory mucosal responses (Holm et al., 2016a; Núñez-Acuña et al., 2015), and outlined nutritional immunity based on the regulation of iron (Covello et al., 2012a; Sutherland et al., 2014; Skugor et al., 2016). Haem is an essential metalloprotein that cannot be synthesised by sea lice (as is the case with most blood-consuming parasites), and instead must be acquired from the host. The positive outcome in the resistant pink salmon and other resistant salmonids undergoing *L. salmonis* infestation can be attributed to the induction of iron sequestration and homeostasis mechanisms (Braden et al., 2012, 2015, 2017; Sutherland et al., 2014). In addition to the importance of iron and haem regulation, a possible role of cholesterol metabolism and detoxification pathways in the outcome of lice infestations has been shown in studies of the protective role of glucosinolates against *L. salmonis* (Skugor et al., 2016).

Salmon skin mucus comprises a complex matrix that provides mechanical protection from the environment. It contains enzymes, antibacterial agents, lysozyme and other immune-related compounds (Firth et al., 2000; Fast et al., 2002; Easy and Ross, 2009). Mucus is also a promising candidate for biomarker identification, given its ease of sampling. Sea lice infections have been shown to reduce the number of epidermal mucous cells (Nolan et al., 1999), while it has been shown that diet can influence epidermal thickness (Kato et al., 2014) and mucous cell density (Pittman et al., 2013). The protective effect of increased mucus production against lice infections is questionable. The immune suppression of Atlantic salmon by *L. salmonis* and the diversion towards a less effective

Th2-like response has been reported in several studies (Fast et al., 2007; Skugor et al., 2008; Tadiso et al., 2011; Umasuthan et al., 2020). One well-described aspect of the mammalian Th2 response is activation of goblet cell hyperplasia and an increase in mucin production and secretion to aid expulsion of parasites from mucosal surfaces (Díaz and Allen, 2007). Studies of Atlantic salmon and salmon louse have provided somewhat conflicting results in this area. Families of Atlantic salmon that show significantly higher susceptibility to lice infection showed reduction in the expression of Th1 pro-inflammatory genes and higher expression of genes likely associated with a humoral antibody response, suggesting diversion towards Th2 immune response (Holm et al., 2015). It was noted in the same study that fish with lower lice resistance had thicker skin, and the highest number and density of mucous cells. This study highlighted the importance of an effective immune response over physical characteristics such as thickness of the skin and mucus production. However, as the mucus layer provides the initial interface between the attaching copepodid stage and the host fish, changing physicochemical properties of mucus, its nutritional value, associated microbiota or expression of anti-parasitic molecules in mucus by means of dietary interventions represent promising venues for future research.

21.4 The biology of lice attachment, host recognition and immunomodulation

Technological advances in the functional genomics of two louse species (*L. salmonis* and *C. rogercresseyi*) have also greatly expanded the knowledge of salmon–louse host–pathogen interactions. In addition to weak immunological and physiological protective responses of the Atlantic salmon host, the high preference for *Salmo* species by *L. salmonis* (Braden et al., 2017) contributes to a large degree to the outcome of an infestation. Immunomodulation of the host response appears to be dependent on the yet undefined physicochemical properties of the skin and mucus which define host specificity and are involved in host recognition. The mode of action of several commercially available functional feeds presumably involves a reduced attraction of lice towards Atlantic salmon and/or deterrence (for example, Shield (Skretting) and Robust (Cargill)). Changes in olfactory and possibly nutritional properties of skin and mucus mediated by the inclusion of compounds of plant origin with repellent or masking properties is thus another promising strategy for the development of functional feeds (O'Shea et al., 2017).

The *Lepeophtheirus* genus of sea lice are highly specialised parasites that are usually adapted to a narrow range of hosts. For example, *Lepeophtheirus hippoglossi* (Krøyer 1837) is a parasite restricted to Atlantic halibut *Hippoglossus hippoglossus* (L.), *Lepeophtheirus thompsoni*, Baird, 1850, to turbot (*Scopthalamus maximus* (L.)), and so on; currently there are over 120 valid *Lepeophtheirus* species within the genus (Boxshall, 1974; Schram and Haug, 1988). All the *Lepeophtheirus* sp. chalimus stages attach via a hold-fast that they drive into the tissue of their chosen host (Kabata, 1979). The sea lice must be able to suppress (and divert) the immune reaction of the host against this invasive structure to defend the hold-fast and be able to complete their life cycle. This is achieved through the release of several secretory/excretory products (SEP) from the salivary glands of the sea lice. The secretory products include prostaglandins and a range of proteases (McCarthy et al., 2012), as well as peroxinectin, an important modulator of immunity (Fast et al., 2007; Wotton et al., 2013).

Accurate host identification plays a key role in the immune suppression of the fish hosts, and there is an intimate relationship between host recognition and the effectiveness of immune suppression.

When exposed to mucus from Atlantic salmon or rainbow trout (*Oncorhynchus mykiss* (Walbaum, 1792)), *L. salmonis* released a cluster of secreted proteases. However, only a small percentage of the dose was released when *L. salmonis* encountered mucus from other species, such as winter flounder, *Pseudopleuronectes americanus* (Walbaum, 1792). The lowest response was observed in *L. salmonis* exposed to mucus from coho salmon (*Oncorhynchus kisutch* (Walbaum, 1792)) (Fast et al., 2003). A host receiving a lower dose of immune-suppressive factors from the parasite potentially enables the host to mount a much more effective immune response (Braden et al., 2017). This may, in part, explain the improved defences mounted by fish fed anti-attachment diets where lice recognition becomes compromised and the lice are likely to release fewer immune-suppressive compounds (Núñez-Acuña et al., 2014; Holm et al., 2016a).

L. salmonis can attach to a non-host species to which they are not adapted; however, they do not remain attached for very long (Pert et al., 2009). Correct host identification is a life and death decision that needs to be taken before the louse fully attaches. Consequently, sea lice have invested heavily in the biological resources required for accurate host identification. Several advanced ionotropic glutamate receptors are contained within the antennae positioned on each side of the head of *L. salmonis* (Mordue and Birkett, 2009; Komisarczuk et al., 2017). These antennae can deploy into the water column when the lice are free-swimming and pick up fish molecules 'off-host'. Once the lice have landed on a fish, the antennae can be pushed into the mucus to detect specific compounds. The receptors in the antennae can detect a range of host molecules including isophorone and 6-methyl-5-hepten-2-one, which can trigger the attachment process (Devine et al., 2000; Ingvarsdóttir et al., 2002a, 2002b). Alternatively, the receptors can identify non-host molecules signalling a high-risk environment, encouraging the lice to leave and search for a more appropriate species. Both 4-methylquinazoline and 2-aminoacetophenone from turbot *S. maximus* have been shown to disrupt the settlement and attachment process of *L. salmonis* and act as a repellent (Bailey et al., 2006; Mordue and Birkett, 2009; Hastie and Bowman, 2014).

Sea lice belonging to the *Caligus* genus are much less host-specific than *Lepeophtheirus* and are known to parasitise a much wider range of species. *Caligus elongatus*, Nordmann 1860, is known to parasitise over 80 species including squid (*Loligo*) and minke whales, *Balaenoptera acutorostrata* (Lacépède, 1804) (Kabata, 1979; Hemmingsen et al., 2020). *Caligus* remove the epidermal tissue from scales before applying a basal plate and deploying a long hold-fast to avoid a strong host reaction. This surface attachment avoids an intimate association with the host tissue and the long hold-fast allows *Caligus* to remain at some distance from any immunological response induced by the host (Pike et al., 1993).

Caligus also secrete a greater range and quantity of serine proteases than *L. salmonis*, which may assist in overcoming a wide range of host defences (Fast, 2014). Despite these adaptions *Caligus* are still vulnerable to protective responses of resistant fish, such as coho salmon (Vargas-Chacoff et al., 2016). Coho salmon were able to remove 100% of *C. rogercresseyi* by 14 days post-infection, whereas Atlantic salmon showed no reduction in parasite load over the same infection period (Núñez-Acuña et al., 2016). Thirty years after first being found on farmed salmon in Southern Chile, *C. rogercresseyi* has developed a preference for Atlantic salmon and now will preferentially settle on Atlantic salmon over their original hosts such as the blenny *Hypsoblennius sordidus* (E.T. Bennett, 1828) (Pino-Marambio et al., 2007). *Caligus* can deploy an advanced sensory array for host identification and have ionotropic glutamate receptors that can accurately detect host molecules (Núñez-Acuña et al., 2014).

21.5 Phytogenic feed additives

Host location and attachment of sea lice can be disrupted by a wide range of plant-based ingredients. A significant reduction in host location of *L. salmonis* was achieved in a series of *in vitro* assessments using oils from rosemary *Rosmarinus officinalis*, garlic *Allium sativum*, lavender *Lavandula angustifolia* and bog myrtle *Myrica gale*. Purified plant compounds were also assessed including diallyl-disulfide from garlic as well as allyl, propyl, butyl, 4-pentenyl and 2-phenylethyl isothiocyanates, which are derived from glucosinolate precursors, found in the species of the *Brassica* genus. Glucosinolates are recognised as plant defence products that protect against grazing by herbivores. Once plant cells suffer mechanical damage, a myrosinase enzyme encounters the glucosinolate and converts it to isothiocyanate, via hydrolysis. These are volatile, pungent compounds that are able to mask host odours and may also be toxic to invertebrates upon ingestion. Both the garlic and glucosinolate-derived compounds maintained efficacy down to very low concentrations (parts per trillion) (Wadsworth et al., 2011; O'Shea et al., 2017).

Similar effects were also observed using a selection of these plant-based compounds in disrupting host-location behaviour of *C. rogercresseyi* (Pino-Marambio et al., 2017). Glucosinolates fed to Atlantic salmon also significantly reduced infection of *C. rogercresseyi* by 28%. In this study there was no reduction in feed intake or fish growth, effects that had been encountered at higher doses with glucosinolates in studies with *L. salmonis* (Holm et al., 2016a; Skugor et al., 2016). Lice exposed to glucosinolates showed an over-expression in ionotropic receptors responsible for host recognition and attachment. There was increasing expression of ionotropic glutamate receptors (*IR-25*, *IR-kainate-2*) in the lice associated with increasing dose. Highest expression of the receptors was achieved with the high level of dietary glucosinolates when combined with an immune stimulant (Núñez-Acuña et al., 2014). It is likely that the anti-attachment compounds were binding to the receptors, with increasing effects at the higher doses, as well as when using the combination of products (Núñez-Acuña et al., 2015, 2016).

Incorporating *Brassica* ingredients with a high content of glucosinolates in an Atlantic salmon diet reduced infection of *L. salmonis* by 25%. In addition to anti-parasitic sensorial properties of isothiocyanates, positive effects on several host protection mechanisms were also observed. Pathways associated with upregulated detoxification pathways in the liver, cholesterol metabolism in skin and iron regulation in multiple tissues have also been observed in the functional-feed fed groups. In the control group there was a skewed immune response towards a general type-2 humoral response that has previously been associated with increased sea lice levels (Holm et al., 2016a; Skugor et al., 2016). Collectively, these findings suggest that in addition to the induction of host protective responses, isothiocyanates possess repellent properties and may also be directly toxic to parasites upon ingestion when consumed together with salmon tissues.

A large set of type-1 immune genes was upregulated in the group exposed to dietary glucosinolates, including 50 genes associated with anti-viral protection. This is of particular interest as it is becoming increasingly clear that immunosuppression by lice predisposes salmon to several microbial pathogens. Reduced anti-viral defences associated with sea lice infection have important implications to overall disease control and go some way to explaining the greater incidence of viral infections associated with sea lice infections on salmon production sites. During the period 2007–2008 the industry in Chile suffered a devastating infectious salmon anaemia (ISA) outbreak that reduced farmed salmon volumes by over 60% (Asche et al., 2018). This was closely associated with rapid reduction in the efficacy of emamectin benzoate against *C. rogercresseyi* and widespread problems encountered with sea lice control (Bravo et al., 2008). Although not through vectors, sea lice can transmit ISA to healthy salmon (Oelckers et al., 2014). Moreover, Atlantic salmon pre-exposed to sea lice are at a four-fold higher risk to subsequent ISA infection, compared to salmon with no prior sea lice exposure (Barker et al., 2012, 2019).

Essential oils form a diverse group of volatile plant extracts, often containing a strong odour and taste. Examples include eucalyptus, oregano, clove, peppermint and tea tree among many others, and these can be extracted from various parts of the plant including leaves, bark, roots and seeds. The oils have been used extensively to protect both crops and animals against insect pests and are known to restrict feeding and disrupt digestion and often have direct toxicity against invertebrates (Khater et al., 2009). A few essential oils have been tested against *L. salmonis* and *C. rogercresseyi* as well as other ectoparasites affecting aquaculture species (Steverding et al., 2005; Yiagnisis et al., 2009; Skugor, 2018). Protection may be achieved by a specific oil or compound, but there is often a synergistic effect observed when using different oils, where a combination produces a higher bioactivity and efficacy (Nerio et al., 2010).

The addition of carvacrol and thymol in diets of Atlantic salmon resulted in a 24% reduction of *C. rogercresseyi*. Similar effects were also observed in rainbow trout against *C. rogercresseyi* (Troncoso et al., 2011). Further studies have shown that a blend of essential oils fed to Atlantic salmon for 30 days reduced infection of *L. salmonis* by 20%. No differences were observed in the thickness of the skin or the level of mucus produced on fish between the dietary groups, although there was a trend for lower mucus production on the fish fed the essential oils with lower lice levels. The fish with lower lice infections also showed a different proteomic profile in the mucus with reduced levels of heat shock proteins produced (Jensen et al., 2015). Essential and botanical oils are thought to have host-masking properties but also have direct effects against the lice (Brasil et al., 2019).

A range of additional plant-based compounds have been reported to have a direct effect upon sea lice. The soap bark tree, *Quillaja saponaria*, is native to Chile and the inner bark has been used for numerous pharmaceutical applications, including vaccine adjuvants. Compounds include a high concentration of glycoside saponin with broad spectrum activity against a range of pathogens. Feeding Atlantic salmon *Q. saponaria* extract at 0.1% diet^{-1} for 50 days reduced *C. rogercresseyi* numbers by 44%. Saponins from soy are known to act as anti-nutritional factors in salmon; however, there were no differences in feed intake or weight gain observed between the dietary groups (Zelaya and Eggers, 2015).

Neem-tree extracts from *Azadirachta indica* have been widely used throughout the Indian subcontinent as a pest control for crops. They have become an important tool within integrated pest management systems as an alternative to synthetic pesticides. The active ingredient azadirachtin disrupts feeding in arthropods, reduces growth and interferes with hormone-mediated regulation of development. Extracted neem oil has shown high efficacy in topical treatments against freshwater lice *Argulus* sp. infecting carp *Carassius auratus* L. (Kumar et al., 2012), as well as *Caligus* sp. on Asian seabass, *Lates calcarifer* (Khoa et al., 2019). Finally, concentrated neem oil incorporated into diets for Atlantic salmon at 25 mg kg feed^{-1} and fed for seven days reduced 92% of chalimus stages and 74% of mobile stages of *L. salmonis* (Mason and Strang, 2014).

21.6 Functional ingredients of microbial origin

It is expected that algae, filamentous fungi and yeast, and bacteria will provide increasingly larger quantities of protein for livestock and human consumption in the future (Ritala et al., 2017). Safety of such ingredients is one of the key concerns, as products of microbial origin may contain microbial toxins or other harmful substances derived from the substrate (for example, waste from other industries that may contain heavy metals). The rapidly proliferating microbial cells have high nucleic acid content, which can lead to the pathological deposition of uric acid in mammals and birds. This is not a problem for fish which degrade uric acid by the enzyme uricase to allantoin and further down the uricolytic pathway to urea (Andersen et al., 2006). However, greater functional benefits are achieved with fractionation and purification of the microbial biomass. As by-products of single cell protein production, specific functional ingredients of microbial origin, such as components of the cellular wall and fractions enriched for nucleic acid content, hold great promise for aquaculture. Numerous products from different species and strains grown on diverse substrates have been assessed as functional ingredients under specific challenging conditions, and quite a few have become indispensable components of functional feeds against lice. The overall effect of most bioactives of microbial origin (β-glucans, peptidoglycan and CpG-oligodeoxynucleotides) tested in aqua feeds is considered to be immunostimulatory although more research is needed to fully understand the scope of their effects. In all likelihood, at least some of these preparations exert immunomodulatory properties in fish, as has been shown for the shiitake mushroom β-glucan (Djordjevic et al., 2009). The study revealed that a β-glucan-based functional diet enhanced beneficial immune responses while at the same time it reduced the potentially detrimental immune responses, which would be the most desirable property of an immunomodulator. Non-selective immunostimulation warrants attention as farmed salmon often experience co-infections of lice and microbial pathogens where overly activated immune responses may aggravate co-infection outcomes. Stimulation of immunity against co-infections of lice and bacterial pathogens that cause skin lesions would also most likely need to be avoided. Recent research has addressed responses in Atlantic salmon simultaneously exposed to lice and microbial pathogens (Cai et al., 2022; Caballero-Solares et al., 2022, in press) in an attempt to better understand which mechanisms constitute protection in such situations so that appropriate functional feeds can be designed for each specific case – co-infection of lice and virus vs lice and bacterial pathogen. Functional feeds that do not stimulate responses detrimental for the viral infection while at the same time being effective against lice may be based on functional ingredients that interfere with olfaction and early attachment of parasites while diets targeting both bacteria and lice co-infecting the salmon host may be based on utilising ingredients with iron chelating properties, as these have been shown to be not only anti-parasitic but also effective in reducing the bacterial load of *P. salmonis*, the causative agent of SRS disease, by making iron less available to bacteria (Díaz et al., 2021). Novel functional feeds targeting co-infections would, however, need to be comprehensively tested for each specific co-infection of lice and a microbial pathogen before they can be applied in real-life farm settings.

21.7 Nucleotides

Nucleotides provide the building blocks for RNA, DNA and ATP molecules and are required for all cell division, protein synthesis, and tissue growth and repair. In addition, they have a range of other essential functions, including roles in signalling as well as co-enzyme components. Nucleotides may be classed as semi-essential nutrients and can become limiting during periods of rapid growth, stress or disease challenge. Dietary nucleotides provided to terrestrial animals during periods of rapid growth or during infections with pathogens may have a protein sparing effect through moderation of *de novo* synthesis from amino acid precursors (Carver, 1999). They may also have a beneficial role in moderating the costly metabolic process of nucleotide salvaging (Sandersen and He, 1994). Additionally, haematopoietic cells and cells of the immune system have limited capacity for *de novo* synthesis and depend on the supply by the salvage pathway (Zöllner, 1982). It has been proposed that nucleotides do not stimulate the immune system directly, but rather support the cell division of the immune cell type that is required (Low et al., 2003). Provision of exogenous dietary nucleotides have shown a range of beneficial effects in aquatic animals as well, including boosting both the specific and non-specific immune pathways, as well as improving protection against pathogens (Burrells et al., 2001b; Li and Gatlin, 2006; Ringø et al., 2012).

Early documentation of anti-parasitic functional feeds included the assessment of nucleotides in reducing infection of *L. salmonis* in Atlantic salmon. Provision of dietary nucleotides (Optimun, Chemoforma) reduced sea lice levels by 38% compared to control diets (Burrells et al., 2001a). It may be beneficial to include dietary nucleotides when functional feeds contain ingredients with immunostimulatory properties, due to the very high demand for nucleotides during the immune response when they may become limiting (Ringø et al., 2012).

Composition of nucleotide mixture preparations in feeds may differ depending on microbial species, strains, or their combinations, substrates used for the production of the microbial biomass, and downstream processing. More research should be done to evaluate how differences of nucleotide products available on the market translate to differences in functionality under various challenging conditions.

21.8 β-glucans

Several dietary immune stimulants have been assessed against sea lice. Beta glucans (β-1,3/1,6-glucans) are high molecular weight polysaccharides and are one of the most common immune stimulants used in aquaculture diets (Bricknell and Dalmo, 2005; Ringø et al., 2012). They can be obtained from several sources including the cell walls of yeast, bacteria, algae, fungi and cereal grains. The specific structure of β-glucans affects their ability to bind to the target immune cells via toll-like and other receptors (Bricknell and Dalmo, 2005) and this largely determines the effectiveness of the subsequent immune response (Sakai, 1999; Mello et al., 2019). β-glucans from fungi exhibit a variety of physiological properties including anti-tumour and immune-modulating properties (Murphy et al., 2020). The use of β-glucans in Atlantic salmon diets has been shown to contribute to the substantial reduction in lice numbers (Refstie et al., 2010).

The physiological effects of β-glucans vary widely depending on the structural characteristics, including molecular weight, chemical composition and chain conformation (Wasser, 2002). Efforts to characterise the structural and biological properties of β-glucans and also to modify their structure by incorporating new functional groups into the polymer chains are ongoing. The branching structure of β-glucans plays a crucial role in immunomodulatory effects and these effects could be enhanced by modulating the critical branching structure of β-glucan.

A study by Boutros et al. (2022) revealed that each of four tested commercial products of *Saccharomyces cerevisiae* β-glucans boast a unique purity profile, branching and linkage patterns. The strain and manufacturing process were shown to strongly impact the final β-glucan structure, and likely the health effects associated with their consumption.

21.9 Peptidoglycan

Peptidoglycan is a structural component of bacterial cell walls and can constitute up to 90% dry weight of Gram-positive bacteria. It is formed from two alternating amino sugars that produce a strong lattice structure responsible for cell strength. Aquatic animals have protein receptors capable of reacting specifically to peptidoglycan and these receptors are distributed in a range of tissues including the gills, intestine, liver, skin and immune cells such as phagocytes. Due to their presence in the cell walls of many pathogenic bacteria (*Streptococcus*, *Flavobacteria*, *Vibrio*), peptidoglycans can produce a strong immune response when exposed to the host immune system (Chang and Nie, 2008; Casadei et al., 2009). Significant protection has been achieved using dietary peptidoglycan against a range of pathogens in aquatic species using the immune stimulant, both alone and when combined with nucleotides (Itami et al., 1998; Vecino and Wadsworth, 2011).

Several antimicrobial peptides (AMP), including hepcidin, were rapidly upregulated following oral administration of peptidoglycan to rainbow trout, at doses between 5 mg and 100 mg per kg diet (Casadei et al., 2013). AMPs can kill pathogens by interacting directly with their cell membrane and hepcidin is known to be involved in iron metabolism and can bind with ferroportin, thus blocking the release of iron by macrophages and intestinal intake. Lower iron availability is a limiting factor for many pathogens, therefore the up regulation of hepcidin can indirectly interfere with pathogen survival (Ong et al., 2006; Casadei et al., 2015). Regulating iron is an important defence mechanism for salmon species that are highly resistant to *L. salmonis*, such as pink salmon *Oncorhynchus gorbuscha* (Walbaum, 1792) (Sutherland et al., 2014).

A combination of nucleotides and peptidoglycan fed to Atlantic salmon resulted in a 50% reduction in the number of *L. salmonis* post-challenge, compared to control diets (Sutherland et al., 2017). Fish fed the nucleotide and peptidoglycan diet showed an upregulation of a number of pro-inflammatory immune genes in the skin, including interleukin-1β. This cytokine is an early responder within the innate immune system and drives a cascade of increasing responses. The final response can generate a profound inflammation that engulfs and kills sea lice attached to resistant salmon species (Johnson and Albright, 1992; Braden et al., 2012; Sutherland et al., 2014). In fish fed the control diet, over 550 genes responsible for a range of immune and physiological responses were downregulated and this correlated to the higher lice infection (Covello et al., 2012a). During attachment to Atlantic salmon, *L. salmonis* can suppress the immune system (Fast et al., 2007; Lewis et al., 2014) and prevent a protective inflammatory response (Skugor et al., 2008; Tadiso et al., 2011; Umasuthan et al., 2020). The functional diet was able to overcome the suppression by the parasite and enable the upregulation of a pro-inflammatory response, and eventually reduced the lice population by 50%.

Interestingly, lice sampled from fish fed the functional diets showed a downregulation in numerous genes associated with host suppression including papilin, an extracellular glycoprotein (Tirloni et al., 2014), as well as several proteases. The lice population also showed much lower expression of factors normally associated with *L. salmonis* feeding on a susceptible host such as Atlantic salmon. There was a marked under-expression of metalloproteases and chymotrypsin as well as trypsin-like proteases (Sutherland et al., 2017). The expression of factors was similar to lice sampled from resistant salmon species such as coho salmon, which are able to significantly disrupt feeding of the lice (Braden et al., 2017).

21.10 CpG-oligodeoxynucleotides (CpG-ODN)

CpG is a synthetic, single-stranded DNA molecule that contains cytosine and guanine deoxynucleotides. These simple molecular sequences are common in pathogen genomes but rare in vertebrate systems. Because they mimic the presence of a pathogen, they are quickly recognised by several receptors in the host species and stimulate a strong pro-inflammatory response. CpG is an effective immune stimulant; however, due to its high cost and product registration issues it is not in use within commercial aquaculture operations. CpG has still proven to be a valuable research tool and has highlighted effective immune pathways that are protective for salmon. A study by Xue et al. (2019) revealed that dietary CpG modulation of immunity in Atlantic salmon may be regulated through modulation of the expression of microRNAs.

A series of studies have shown that feeding CpG to Atlantic salmon at 20 mg kg diet^{-1} can reduce *L. salmonis* infections up to 48%. Protection was correlated to an increased expression of pro-inflammatory genes such as interleukin-1β and interleukin-12b. In the control population, pro-inflammatory responses were suppressed, and the immune system had been diverted towards a humoral (antibody) type response which is less effective against attached ectoparasites such as *L. salmonis* (Covello et al., 2012b; Purcell et al., 2012; Poley et al., 2013). The control fish also showed a marked reduction in the level of anti-viral defences such as MHC-1b, Mx and galectin (Covello et al., 2012b).

21.11 Mannan oligosaccharides

As lice control methods have shifted towards mechanical treatments, the incidence of physical damage to the fish has increased dramatically. New nutritional solutions that can aid wound healing and help fish regain homeostasis after thermal/mechanical delousing treatments could become an important tool in the integrated pest management toolbox. This will be more effective if they also contribute to fish becoming more resistant to new lice infestations. Dietary supplementation with the indigestible mannan oligosaccharide (MOS) products derived from the outer cell wall of yeast is a well-recognised nutritional intervention that supports intestinal homeostasis in numerous farmed species.

A recent study has shown that MOS also exerts a similar positive effect on the mucosal skin layer in fish. Moreover, the increased skin mucus and goblet cell density in Atlantic salmon mediated by MOS were associated with a moderate reduction in susceptibility to the copepodid *L. salmonis* (Leclercq et al., 2020). A study by Fernández-Montero et al. (2019) also showed a protective effect of MOS against a marine monogenean worm ectoparasite *Neobenedenia girellae*. Functional feed supplemented with the concentrated MOS product reduced the load and size of the ectoparasite *N. girellae* in greater amberjack juveniles which was accompanied by the upregulation of several immune genes in mucosal tissues (skin, gills and proximal gut), and in the head kidney and spleen. It should be noted that gene encoding an antimicrobial peptide and a key iron regulator, hepcidin, potentially involved in the sequestration of iron thus effectively making it less available to parasites, was consistently upregulated (except in spleen) in fish that received MOS (Fernández-Montero et al., 2019). Feeding diets supplemented with MOS appears to reduce the sea lice burdens on salmon

(Sweetman et al., 2010). However, studies by Refstie et al. (2010) and Jensen et al. (2015) have not shown protective effects of dietary MOS in an *L. salmonis* Atlantic salmon challenge. The discrepancies could be caused by structural and compositional differences of tested MOS products, different supplementation regimes (duration of exposure and dietary concentration) and other factors.

Proprietary manufacturing processes determine the chemical structure and composition of MOS. Furthermore, chemical modification of MOS to improve specific characteristics are of great interest in the future.

21.12 Future perspectives

The deciphering of salmon–louse interactions in recent decades has greatly facilitated the development of anti-parasitic functional feeds; however, there is still great room for innovation in this space. The design of new effective dietary interventions against the two lice species parasitising Atlantic salmon will require continued and concerted efforts to further understand the cellular and molecular underpinnings of protective host responses and how they can be stimulated by dietary interventions. In parallel, virulence mechanisms deployed by lice and their physiology need to be further studied. In particular, the completion of proteomes and genomes of two lice species is expected to aid in our improved understanding of the mechanisms involved in the infectious process. Together, approaches addressing responses in both the salmonid host and the parasite will shed more light on the nature of protection mediated by a functional ingredient, which can be mediated indirectly via stimulation of host protective responses or directly upon olfactory detection and ingestion. Such knowledge will also be helpful in assessing the most effective combinations of functional ingredients that can stimulate or inhibit multiple targets – various evidence suggests that modulating one single mechanism, either in the salmonid host or lice, is going to be less effective in comparison to the simultaneous modulation of several mechanisms. Refinement of *in vitro* tools to study direct anti-parasitic (semiochemical and toxic) properties in a high-throughput manner will be crucial for the initial screening of candidate dietary additives. The list of novel functional ingredients is continually growing, including yet unexplored phytogenics and novel bioactives of microbial origin. Substantial research will be necessary to warrant their safe and effective usage in the future.

References

Andersen, Ø., Aas, T.S., Skugor, S., Takle, H., van Nes, S., Grisdale-Helland, B., Helland, S.J. and Terjesen, B.F. (2006) Purine-induced expression of urate oxidase and enzyme activity in Atlantic salmon (*Salmo salar*). Cloning of urate oxidase liver cDNA from three teleost species and the African lungfish *Protopterus annectens*. *The FEBS Journal* 273(13), 2839–2850.

Asche, F., Cojocaru, A. and Sikveland, M. (2018) Market shocks in salmon aquaculture. The impact of the Chilean disease crisis. *Journal of Agricultural and Applied Economics* 50(2), 255–269.

Bailey, R., Birkett, M., Ingvarsdóttir, A., Mordue (Luntz), A., O'Shea, B., Pickett, J. and Wadhams, L. (2006) The role of semiochemicals in host location and non-host avoidance by salmon louse (*Lepeophtheirus salmonis*) copepodids. *Canadian Journal of Fisheries and Aquatic Sciences* 63, 448–456.

Barker, S.E., Bouchard, D., Covello, J., Fast, M.D., Wolters, W. and Bricknell, I.R. (2012) Are 'lousy' fish more susceptible to ISAV? And what is the role of lice in transmitting ISAV. *Proceedings of the ACFFA Annual Meeting*. The Huntsman Marine Science Centre, St. Andrews, New Brunswick, Canada.

Barker, S.E., Bricknell, I.R., Covello, J., Purcell, S., Fast, M.D., Wolters, W. and Bouchard, D.A. (2019) Sea lice, *Lepeophtheirus salmonis* (Krøyer 1837), infected Atlantic salmon (*Salmo salar* L.) are more susceptible to infectious salmon anemia virus. *PLoS ONE* 14(1), e0209178.

Boutros, J.A., Magee, A.S. and Cox, D. (2022) Comparison of structural differences between yeast β-glucan sourced from different strains of *Saccharomyces cerevisiae* and processed using proprietary manufacturing processes. *Food Chemistry* 367, 130708.

Boxshall, G. (1974) The population dynamics of *Lepeophtheirus pectoralis* Müller. Seasonal variation in abundance and age structure. *Parasitology* 69, 361–371.

Braden, L.M., Barker, D.E., Koop, B.F. and Jones, S.R.M. (2012) Comparative defense associated responses in salmon skin elicited by the ectoparasite *Lepeophtheirus salmonis*. *Comparative Biochemistry and Physiology* D7, 100–109.

Braden, L.M., Koop, B.F. and Jones, S.R.M. (2015) Signatures of resistance to *Lepeophtheirus salmonis* include a TH2-type response at the louse-salmon interface. *Developmental and Comparative Immunology* 48, 178–191.

Braden, L.M., Sutherland, B.J.G., Koop, B.F. and Jones S.R.M. (2017) Enhanced transcriptomic responses in the Pacific salmon louse *Lepeophtheirus salmonis oncorhynchi* to the non-native Atlantic salmon *Salmo salar* suggests increased parasite fitness. *BMC Genomics* 18, 110.

Brasil, E., Figueredo, A., Cardoso, L. and Santos, M. (2019) *In vitro* and *in vivo* antiparasitic action of essential oils of *Lippia* spp. in koi carp (*Cyprinus carpio*) fed supplemented diets. *Brazilian Journal of Veterinary Pathology* 12(3), 88–100.

Bravo, S., Sevatdal, S. and Horsberg, T.E. (2008) Sensitivity assessment of *Caligus rogercresseyi* to emamectin benzoate in Chile. *Aquaculture* 282, 7–12.

Bricknell, I. and Dalmo, R.A. (2005) The use of immunostimulants in fish larval aquaculture. *Fish and Shellfish Immunology* 19, 457–472.

Bueno, N.B., de Melo, I.S., de Oliveira, S.L. and da Rocha Ataide, T. (2013) Very-low-carbohydrate ketogenic diet v. low-fat diet for long-term weight loss: a meta-analysis of randomised controlled trials. *British Journal of Nutrition* 110(7), 1178–1187.

Bui, S., Stien, L.H., Nilsson, J., Trengereid, H. and Oppedal, F. (2020) Efficiency and welfare impact of long-term simultaneous *in situ* management strategies for salmon louse reduction in commercial sea cages. *Aquaculture* 520, 734934.

Burka, J.F., Hammell, K.L., Horsberg, T.E., Johnson, G.R., Rainnie, D.J. and Speare, D.J. (1997) Drugs in salmonid aquaculture, a review. *Journal of Veterinary Pharmacology and Therapeutics* 20, 333–349.

Burrells, C., Williams, P. and Forno, P. (2001a) Dietary nucleotides: a novel supplement in fish feeds. 1. Effects on resistance to disease in salmonids. *Aquaculture* 199, 159–169.

Burrells, C. Williams, P., Southgate P. and Wadsworth S. (2001b) Dietary nucleotides: a novel supplement in fish feeds. 2. Effects on stress tolerance, physiology and growth of Atlantic salmon (*Salmo salar* L). *Aquaculture* 199, 171–184.

Caballero-Solares, A., Umasuthan, N., Xue, X., Katan, T., Kumar, S., Westcott, J.D., et al. (2022) Interacting effects of sea louse (*Lepeophtheirus salmonis*) infection and formalin-killed *Aeromonas salmonicida* on Atlantic salmon skin transcriptome. *Frontiers in Immunology* (in press).

Cai, W., Kumar, S., Navaneethaiyer, U., Caballero-Solares, A., Carvalho, L.A., Whyte, S.K., et al. (2022) Transcriptome analysis of Atlantic salmon (*Salmo salar*) skin in response to sea lice and infectious salmon anemia virus co-infection under different experimental functional diets. *Frontiers in Immunology* 3(12), 787033.

Carvalho, L.A., Whyte, S.K., Braden, L.M., Purcell, S.L., Manning, A.J., Muckle, A. and Fast, M.D. (2020) Impact of co-infection with *Lepeophtheirus salmonis* and *Moritella viscosa* on inflammatory and immune responses of Atlantic salmon (*Salmo salar*). *Journal of Fish Diseases* 43(4), 459–473.

Carver, J.D. (1999) Dietary nucleotides: effects on the immune and gastrointestinal systems. *Acta Paediatrica Suppl.* 88(430), 83–88.

Casadei, E., Wang, T., Zou, J., González Vecino, J., Wadsworth, S.L. and Secombes, C.J. (2009) Characterization of three novel defensin antimicrobial peptides in rainbow trout (*Oncorhynchus mykiss*). *Molecular Immunology* 46, 3358–3366.

Casadei, E., Bird, S., Vecino, J.G., Wadsworth, S. and Secombes, C.J. (2013) The effect of peptidoglycan enriched diets on antimicrobial peptide gene expression in rainbow trout (*Oncorhynchus mykiss*). *Fish and Shellfish Immunology* 34, 529–537.

Casadei, E., Bird, S., Wadsworth, S.L. and Secombes, C.J. (2015) The longevity of the antimicrobial response in rainbow trout (*Oncorhynchus mykiss*) fed a peptidoglycan (PG) supplemented diet. *Fish and Shellfish Immunology* 44, 316–320.

Chang, M.X. and Nie, P. (2008) RNAi suppression of zebrafish peptidoglycan recognition protein mediated differentially expressed genes involved in toll-like receptor signalling pathway and caused increased susceptibility to *Flavobacterium columnare*. *Veterinary Immunology and Immunopathology* 124, 295–301.

Covello, J.M., Purcell, S.L., Pino, J., Gonzalez, J., Vecino, J.L., Troncoso, J., Fast, M.D. and Wadsworth, S.L. (2012a) Effects of orally administered immune stimulants on Atlantic salmon (*Salmo salar*) transcriptional responses and subsequent sea lice (*Lepeophtheirus salmonis*) infection. *Proceeding of the Sea Lice 2012 Meeting*. Bergen, Norway.

Covello, J., Friend, S.E., Purcell, S.L., Burka, J.F., Markham, R.J.F., Donkin, A., Groman, D.B. and Fast, M.D. (2012b) Orally administered immunostimulants affect inflammatory gene expression and sea lice (*Lepeophtheirus salmonis*) burdens on Atlantic salmon (*Salmo salar*). *Aquaculture* 366–367, 9–16.

Díaz, A. and Allen, J.E. (2007) Mapping immune response profiles: the emerging scenario from helminth immunology. *European Journal of Immunology* 37(12), 3319–3326.

Díaz, R., Troncoso, J., Jakob, E. and Skugor, S. (2021). Limiting access to iron decreases infection of Atlantic salmon SHK-1 cells with bacterium *Piscirickettsia salmonis*. *BMC Veterinary Research* 17, 155.

Dessen, J.E., et al. (2019) Increased dietary protein-to-lipid ratio improves survival during naturally occurring pancreas disease in Atlantic salmon, *Salmo salar* L. *Journal of Fish Diseases* 42, 21–34.

Devine, G.J., Ingvarsdóttir, A., Mordue, W., Pike, A.W., Pickett, J., Duce, I. and Mordue (Luntz), A.J. (2000) Salmon lice, *Lepeophtheirus salmonis*, exhibit specific chemotactic responses to semiochemicals originating from the salmonid, *Salmo salar*. *Journal of Chemical Ecology* 26, 1833–1847.

Djordjevic, B., Skugor, S., Jørgensen, S.M., Øverland, M., Mydland, L.T., Krasnov, A. (2009) Modulation of splenic immune responses to bacterial lipopolysaccharide in rainbow trout (*Oncorhynchus mykiss*) fed lentinan, a beta-glucan from mushroom *Lentinula edodes*. *Fish and Shellfish Immunology* 26(2), 201–209.

Easy, R.H. and Ross, N.W. (2009) Changes in Atlantic salmon (*Salmo salar*) epidermal mucus protein composition profiles following infection with sea lice (*Lepeophtheirus salmonis*). *Comparative Biochemistry and Physiology Part D: Genomics and Proteomics* 4, 159–167.

Eliasen, K., Danielsen, E., Johannesen, Á., Joensen, L.L. and Patursson, E.J. (2018) The cleaning efficacy of lumpfish (*Cyclopterus lumpus* L.) in Faroese salmon (*Salmo salar* L.) farming pens in relation to lumpfish size and seasonality. *Aquaculture* 488, 61–65.

Estruch, R. (2010) Anti-inflammatory effects of the Mediterranean diet: the experience of the PREDIMED study. *Proceedings of the Nutrition Society* Aug 69(3), 333–340.

Fast, M. (2014) Fish immune responses to parasitic copepod (namely sea lice) infection. *Developmental and Comparative Immunology* 43, 300–312.

Fast, M.D., Ross, N.W., Mustafa, A., Sims, D.E., Johnson, S.C., Conboy, G.A., et al. (2002) Susceptibility of rainbow trout *Oncorhynchus mykiss*, Atlantic salmon *Salmo salar* and coho salmon *Oncorhynchus kisutch* to experimental infection with sea lice *Lepeophtheirus salmonis*. *Diseases of Aquatic Organisms* 52(1), 57–68.

Fast, M.D., Burka, J.F., Johnson, S.C. and Ross, N.W. (2003) Enzymes released from *Lepeophtheirus salmonis* in response to mucus from different salmonids. *Journal of Parasitology* 89, 7–13.

Fast, M.D., Johnson, S.C., Eddy, T.D., Pinto, D. and Ross, N.W. (2007) *Lepeophtheirus salmonis* secretory/excretory products and their effects on Atlantic salmon immune gene regulation. *Parasite Immunology* 29, 179–189.

Fernández-Montero, Á., Torrecillas, S., Izquierdo, M., Caballero, M.J., Milne, D.J., Secombes, C.J., et al. (2019) Increased parasite resistance of greater amberjack (*Seriola dumerili* Risso 1810) juveniles fed a cMOS supplemented diet is associated with upregulation of a discrete set of immune genes in mucosal tissues. *Fish and Shellfish Immunology* 86, 35–45.

Firth, K.J., Johnson, S.C. and Ross, N.W. (2000) Characterization of proteases in the skin mucus of Atlantic salmon (*Salmo salar*) infected with the salmon louse (*Lepeophtheirus salmonis*) and in whole-body louse homogenate. *Journal of Parasitology* 86, 1199–1205.

Gentry, K., Bui, S., Oppedal, F. and Dempster T. (2020) Sea lice prevention strategies affect cleaner fish delousing efficacy in commercial Atlantic salmon sea cages. *Aquaculture Environment Interactions* 12, 67–80.

Gismervik, K., Nielsen, K.V., Lind, M.B. and Vijugrein, H. (2017) Norwegian Veterinary Institute's Report Series: 6–2017. Mekanisk avulsing med FLS-avlusersystem, dokumentasjon av fiskevelferd og effekt mot lus. Norwegian Veterinary Institute, Norway.

Grøntvedt, R.N., Nerbøvik, I.K.G., Viljugrein, H., Lillehaug, A., Nilsen, H. and Gjevre A.G. (2015) Norwegian Veterinary Institutes Report Series: 13. Thermal de-licing of salmonid fish; documentation of fish welfare and effect. Norwegian Veterinary Institute, Norway.

Grøntvedt, R.N., Kristoffersen, A.B. and Jansen, P.A. (2018) Reduced exposure of farmed salmon to salmon louse (*Lepeophtheirus salmonis* L.) infestation by use of plankton nets: estimating the shielding effect. *Aquaculture* 495, 865–872.

Hastie, L. and Bowman, A. (2014) Methods and compositions for inhibiting sea lice. Patent. WO 2014/020339 Al.

Hemmingsen, W., MacKenzie, K., Sagerup, K., Remen, M., Bloch-Hansen, K. and Imsland, A. (2020) *Caligus elongatus* and other sea lice of the genus *Caligus* as parasites of farmed salmonids: a review. *Aquaculture* 522, 735160.

Holan, A.B., Roth, B., Breiland, M.S.W., Jelena, K., Hansen, Ø.J. and Iversen, A. (2017) Beste praksis for medikamentfrie metoder for lakselus kontroll (MEDFRI). Rapport 10/2017, NOFIMA.

Holm, H.J., Santi, N., Kjøglum, S., Perisic, N., Skugor, S. and Evensen, Ø. (2015) Difference in skin immune responses to infection with salmon louse (*Lepeophtheirus salmonis*) in Atlantic salmon (*Salmo salar* L.) of families selected for resistance and susceptibility. *Fish and Shellfish Immunology* 42, 384–394.

Holm, H.J., Wadsworth, S., Bjelland, A.K., Krasnov, A., Evensen, Ø. and Skugor, S. (2016a) Dietary phytochemicals modulate skin gene expression profiles and result in reduced lice counts after experimental infection in Atlantic salmon. *Parasites & Vectors* 9, 271.

Holm, H.J., Skugor, S., Bjelland, A.K., Radunovic, S., Wadsworth, S., Koppang, E.O. and Evensen, Ø. (2016b) Contrasting expression of immune genes in scaled and scaleless skin of Atlantic salmon infected with young stages of *Lepeophtheirus salmonis*. *Developmental and Comparative Immunology* 67, 153–165.

Itami, T., Asano, M., Tokushige, K., Kubono, K., Nakagawa, A., Takeno, N., Nishimura, H., Maeda, M., Kondo, M. and Takahashi, Y. (1998) Enhancement of disease resistance of kuruma shrimp, *Penaeus japonicus*, after oral administration of peptidoglycan derived from *Bifidobacterium thermophilum*. *Aquaculture* 164, 277–288.

Ingvarsdóttir, A., Birkett, M.A., Duce, I., Genna, R.L., Mordue, W., Pickett, J.A., et al. (2002a) Semiochemical strategies for sea louse control: host location cues. *Pest Management Science* 58, 537–545.

Ingvarsdóttir, A., Birkett, M.A., Duce, I., Mordue, W., Pickett, J.A., Wadhams, L.J. and Mordue (Luntz), A.J. (2002b) Role of semiochemicals in mate location by the parasitic sea louse, *Lepeophtheirus salmonis*. *Journal of Chemical Ecology* 28, 2107–2017.

Iversen, A., Asche, F., Hermansen, Ø. and Nystøyl, R. (2020) Production cost and competitiveness in major salmon farming countries 2003–2018. *Aquaculture* 522, 735089.

Jensen, L.B., Proyan, F., Larssen, E., Bron, J.E. and Obach, A. (2015) Reducing sea lice (*Lepeophtheirus salmonis*) infestation of farmed Atlantic salmon (*Salmo salar* L.) through functional feeds. *Aquaculture Nutrition* 21, 983–993.

Johannesen, A., Joensen, N. E. and Magnussen, E. (2018) Shelters can negatively affect growth and welfare in lumpfish if feed is delivered continuously. *PeerJ* 6, e4837.

Johnson, S.C. and Albright, L.J. (1992) Comparative susceptibility and histopathology of the response of naive Atlantic, chinook and coho salmon to experimental infection with *Lepeophtheirus salmonis* (Copepoda: Caligidae). *Diseases of Aquatic Organisms* 14, 179–193.

Kabata, Z. (1979) *Parasitic Copepoda of British Fishes*. Ray Society, London, UK.

Katan, T., Xue, X., Caballero-Solares, A., Taylor, R.G., Rise, M.L. and Parrish, C.C. (2020) Influence of dietary long-chain polyunsaturated fatty acids and ω6 to ω3 ratios on head kidney lipid composition and expression of fatty acid and eicosanoid metabolism genes in Atlantic salmon (*Salmo salar*). *Frontiers in Molecular Biosciences* 7, 602587.

Kato, K., Yamamoto, M., Peerapon, K., Fukada, H., Biswas, A., Yamamoto, S., et al. (2014) Effects of dietary taurine levels on epidermal thickness and scale loss in red sea bream, *Pagrus major*. *Aquaculture Research* 45(11), 1818–1824.

Khater, H.F., Ramadan, M.Y. and El-Madawy, R.S. (2009) Lousicidal, ovicidal and repellent efficacy of some essential oils against lice and flies infesting water buffaloes. *Egyptian Veterinary Parasitology* 164, 257–266.

Khoa, T.N.D., Mazelan, S., Muda, S. and Shaharom-Harrison, F. (2019) Use of neem oil (*Azadirachta indica*) to control caligid copepod infestation on Asian seabass (*Lates calcarifer*). *Aquaculture Research* 50, 1885–1892.

Komisarczuk, A.Z., Grotmol, S. and Nilsen, F. (2017) Ionotropic receptors signal host recognition in the salmon louse (*Lepeophtheirus salmonis*, Copepoda). *PLoS ONE* 12(6), e0178812.

Krasnov, A., Wesmajervi Breiland, M.S., Hatlen, B., Afanasyev, S. and Skugor, S. (2015) Sexual maturation and administration of 17β-estradiol and testosterone induce complex gene expression changes in skin and increase resistance of Atlantic salmon to ectoparasite salmon louse. *General Comparative Endocrinology* 212, 34–43.

Kumar, S., Raman, P., Kumar, K., Pandey, P., Kumar, N., Mohanty, S. and Kumar, A. (2012) *In vitro* and *in vivo* antiparasitic activity of azadirachtin against *Argulus* spp. in *Carassius auratus* (Linn. 1758). *Parasitology Research* 110, 1795–1800.

Leclercq, E., Graham, P. and Migaud, H. (2015) Development of a water-stable agar-based diet for the supplementary feeding of cleaner fish ballan wrasse (*Labrus bergylta*) deployed within commercial Atlantic salmon (*Salmon salar*) net pens. *Animal Feed Science and Technology* 208, 98–106.

Leclercq, E., Pontefract, N., Rawling, M., Valdenegro, V., Aasum, E., Andujar, L.V., Migaud, H., Castex, M. and Merrifield, D. (2020) Dietary supplementation with a specific mannan-rich yeast parietal fraction enhances the gut and skin mucosal barriers of Atlantic salmon (*Salmo salar*) and reduces its susceptibility to sea lice (*Lepeophtheirus salmonis*). *Aquaculture* 529, 735701.

Lewis, D., Barker, D. and McKinley, R. (2014) Modulation of cellular innate immunity by *Lepeophtheirus salmonis* secretory products. *Fish and Shellfish Immunology* 38, 175–183.

Li, P. and Gatlin, D.M. (2006) Nucleotide nutrition in fish: current knowledge and future applications. *Aquaculture* 251, 141–152.

Low, C., Wadsworth, S., Burrells, C. and Secombes, C.J. (2003) Expression of immune genes in turbot *Scophthalmus maximus* fed a nucleotide-supplemented diet. *Aquaculture* 221, 23–40.

Martinez-Rubio, L., Morais, S., Evensen, Ø., Wadsworth, S., Ruohonen, K., Vecino, J.L.G., Bell, G. and Tocher, D.R. (2012) Functional feeds reduce heart inflammation and pathology in atlantic salmon (*Salmo salar* L.) following experimental challenge with Atlantic salmon reovirus (ASRV). *PLoS ONE* 7(11), e40266.

Martinez-Rubio, L., Morais, S., Evensen, Ø., Wadsworth, S., Vecino, J.G., Ruohonen, K., et al. (2013) Effect of functional feeds on fatty acid and eicosanoid metabolism in liver and head kidney of Atlantic salmon (*Salmo salar* L.) with experimentally induced heart and skeletal muscle inflammation. *Fish and Shellfish Immunology* 34(6), 1533–1545.

Mason, M.P. and Strang, R.H. (2014) Supplemented fish feed. Patent. WO 2014/140623 Al.

McCarthy, E., Cunningham, E., Copley, L., Jackson, D., Johnston, D., Dalton, J.P. and Mulcahy, G. (2012) Cathepsin L proteases of the parasitic copepod, *Lepeophtheirus salmonis*. *Aquaculture* 356–357, 264–271.

McCoy, M.A., McLoughlin, M.F., Rice, D.A. and Kennedy, D.G. (1994) Pancreas disease in Atlantic salmon (*Salmo salar*) and vitamin E supplementation. *Comparative Biochemistry and Physiology Part A: Physiology* 109(4), 905–912.

Mello, M., Faria, C., Sabbadin, F. and Urbinatia, C. (2019) β-glucan modulates cortisol levels in stressed pacu (*Piaractus mesopotamicus*) inoculated with heat-killed *Aeromonas hydrophila*. *Fish and Shellfish Immunology* 93, 1067–1083.

Mordue, A.J. and Birkett, M.A. (2009) A review of host finding behaviour in the parasitic sea louse, *Lepeophtheirus salmonis* (Copepoda: Caligidae). *Journal of Fish Diseases* 32(1), 3–13.

Murphy, E.J., Rezoagli, E., Major, I., Rowan, N.J. and Laffey, J.G. (2020) β-glucan metabolic and immunomodulatory properties and potential for clinical application. *Journal of Fungi* 6(4), 356.

Myksvoll, M.S., Sandvik, A.D., Albretsen, J., Asplin, L., Johnsen, I.A., Karlsen, O., et al. (2018) Evaluation of a national operational salmon lice monitoring system – from physics to fish. *PLoS ONE* 13, e0201338.

Nerio, L.S., Olivero-Verbel, J. and Stashenko, E. (2010) Repellent activity of essential oils: a review. *Bioresource Technology* 101, 372–378.

Nilsson, J., Stien, L.H. and Oppedal, F. (2017) *Environment, Lice Levels, Welfare and Salmon Swim Depth at Kobbavika Site with Surface or Deep Feeding Combined with Artificial Light.* Havforskningen. Institute of Marine Research, Bergen.

Nolan, D.T., Reilly, P. and Bonga, S.E.W. (1999) Infection with low numbers of the sea louse (*Lepeophtheirus salmonis*) induces stress-related effects in postsmolt Atlantic salmon (*Salmo salar*). *Canadian Journal of Fisheries and Aquatic Sciences* 56(6), 947–959.

Norwegian Directorate of Fisheries (2021) Statistics for aquaculture. Available at: www.fiskeridir.no/English/Aquaculture/Statistics/Cleanerfish-Lumpfish-and-Wrasse (accessed 18 February 2022).

Núñez-Acuña, G., Valenzuela-Muñoz, V., Pino J., Wadsworth S. and Gallardo-Escárate, C. (2014) Insights into the olfactory system in the ectoparasite *Caligus rogercresseyi*: molecular characterization and gene transcription analysis of novel ionotropic receptors. *Experimental Parasitology* 145, 99–109.

Núñez-Acuña, G., Gonçalves, A.T., Valenzuela-Muñoz, V., Pino-Marambio, J., Wadsworth, S. and Gallardo-Escárate, C. (2015) Transcriptome immunomodulation of in-feed additives in Atlantic salmon *Salmo salar* infested with sea lice *Caligus rogercresseyi*. *Fish and Shellfish Immunology* 47, 450–460.

Núñez-Acuña, G., Vera-Bizama, F., Boltaña, S., Hawes, C., Pino-Marambio, J., Wadsworth, S. and Gallardo-Escárate, C. (2016) In-feed additives modulate ionotropic receptor genes from the sea louse *Caligus rogercresseyi*: a comparative analysis in two host salmonid species. *Aquaculture* 451, 99–105.

Oelckers, K., Vike, S., Duesund, H., Gonzalez, J., Wadsworth, S. and Nylund, A. (2014) *Caligus rogercresseyi* as a potential vector for transmission of infectious salmon anaemia (ISA) virus in Chile. *Aquaculture* 420, 126–132.

Ong, S.T., Shan Ho, J.Z., Ho, B. and Ding, J.L. (2006) Iron-withholding strategy in innate immunity. *Immunobiology* 211(4), 295–314.

O'Shea, B., Wadsworth, S., Pino, J., Birkett, M., Pickett, J. and Mordue, J. (2017) Disruption of host seeking behaviour by the salmon louse, *Lepeophtheirus salmonis*, using botanically-derived repellents. *Journal of Fish Diseases* 40(4), 495–505.

Overton, K., Dempster, T., Oppedal, F., Kristiansen, T., Gismervik, K. and Stien, L. (2018) Salmon lice treatments and salmon mortality in Norwegian aquaculture: a review. *Reviews in Aquaculture* 11(4), 1398–1417.

Pike, A.W., MacKenzie, K. and Rowand, A. (1993) Ultrastructure of the frontal filament in chalimus larvae of *Caligus elongatus* and *Lepeophtheirus salmonis* from Atlantic salmon, *Salmo salar*. In: Boxshall, G.A. and Defaye, D. (eds) *Pathogens of Wild and Farmed Fish: Sea Lice*. Ellis Horwood, Chichester, UK. pp. 99–113.

Pino-Marambio, J., Mordue, A.J., Birkett, M., Carvajal, J., Asencio, G., Mellado, A. and Quiroz, A. (2007) Behavioural studies of host, non-host and mate location by the sea louse, *Caligus rogercresseyi* (Boxshall & Bravo, 2000) (Copepoda: Caligidae). *Aquaculture* 271(1–4), 70–76.

Pino-Marambio, J., Holm, H., Skugor, S. and Wadsworth, S. (2017) Feed composition for prevention and treatment of parasite infection of fish. Patent. WO 2017/039454 Al.

Pittman, K., et al. (2013) Body size matters: an evaluation and application of a novel histological methodology on the quantification of mucous cells in the skin of Atlantic salmon, *Salmo salar* L. *Journal of Fish Diseases* 36, 115–127.

Pert, C.C., Mordue, A. J., Fryer, R.J., O'Shea, B. and Bricknell, I.R. (2009) The settlement and survival of the salmon louse, *Lepeophtheirus salmonis* (Krøyer, 1837), on atypical hosts. *Aquaculture* 288(3–4), 321–324.

Poley, J., Purcell, S.L., Igboeli, O.O., Donkin, A., Wotton, H. and Fast, M.D. (2013) Combinatorial effects of administration of immune-stimulatory compounds in feed and follow up administration of triple dose SLICE (emamectin benzoate) on Atlantic salmon (*Salmo salar*) infection with *Lepeophtheirus salmonis. Journal of Fish Diseases* 36(3), 299–309.

Purcell, S.L., Friend, S.E., Covello, J.M., Donkin, A., Groman, D.B., Poley, J. and Fast, M.D. (2012) CpG inclusion in feed reduces sea lice (*Lepeophtheirus salmonis*) numbers following re-infection. *Journal of Fish Diseases* 36(3), 229–240.

Raynard, R.S., McVicar, A.H., Bell, J.G., Youngson, A., Knox, D. and Fraser, C.O. (1991) Nutritional aspects of pancreas disease of Atlantic salmon: the effects of dietary vitamin E and polyunsaturated fatty acids. *Comparative Biochemistry and Physiology. A: Comparative Physiology* 98(1), 125–131.

Refstie, S., Baeverfjord, G., Seim, R. and Elvebø, O. (2010) Effects of dietary yeast cell wall β-glucans and MOS on performance, gut health, and salmon lice resistance in Atlantic salmon (*Salmo salar*) fed sunflower and soybean meal. *Aquaculture* 305, 109–116.

Ringø, E., Olsen, R., González-Vecino, J., Wadsworth, S. and Song, K. (2012) Use of immune-stimulants and nucleotides in aquaculture: a review. *Journal of Marine Science: Research and Development* 2, 104.

Ritala, A., Häkkinen, S.T., Toivari, M. and Wiebe, M.G. (2017) Single cell protein-state-of-the-art, industrial landscape and patents 2001–2016. *Frontiers in Microbiology* 8, 2009.

Robledo, D., Gutierrez, A.P., Berria, A., Yanex, J.M. and Houston, R.D. (2018) Gene expression response to sea lice in Atlantic salmon skin: RNA sequencing comparison between resistant and susceptible animals. *Frontiers in Genetics* 287.

Roth, B. (2016) Avlusing av laksefisk med Optilicer: Effekt på avlusing og fiskevelford. Rapport 59/2016. NOFIMA.

Sakai, M. (1999) Current research status of fish immunostimulants. *Aquaculture* 172, 63– 92.

Sanderson, I.R. and He, Y. (1994) Nucleotide uptake and metabolism by intestinal epithelial cells. *Journal of Nutrition* 124(1 Suppl), 131–137.

Schram, T. and Haug, T. (1988) Ectoparasites on the Atlantic halibut *Hippoglossus hippoglossus* from N. Norway. Potential pests in halibut aquaculture. *Sarsia* 73(3), 213–227.

Sears, B. (2015) Anti-inflammatory diets. *Journal of the American College of Nutrition* 34(1), 14–21.

Skiftesvik, A.B., Bjelland, R.M., Durif, C.M.F., Johansen, I.S. and Browman, H.I., (2013) Delousing of Atlantic salmon (*Salmo salar*) by cultured vs. wild ballan wrasse (*Labrus bergylta*). *Aquaculture* 402–403, 113–118.

Skugor, S. (2017) Work Package-2. Sea Lice Research Centre (SLRC), Annual report – 2017. SLRC-Annual-Report-2018.pdf. Available at: uib.no (accessed 28 October 2021).

Skugor, S. (2018) Work Package-2. Sea Lice Research Centre (SLRC), Annual Report – 2018. SLRC-Annual-Report-2017_web.pdf. Available at: uib.no (accessed 28 October 2021).

Skugor, S., Glover, K.A., Nilsen, F. and Krasonov, A. (2008) Local and systemic gene expression responses of Atlantic salmon (*Salmo salar* L.) to infection with the salmon louse (*Lepeophtheirus salmonis*). *BMC Genomics* 9, 498.

Skugor, S., Holm, H.J., Bjelland, A.K., Evensen, Ø., Krasnov, A. and Wadsworth, S. (2016) Nutrigenomic effects of glucosinolates on liver, muscle and distal kidney in parasite free and salmon louse infected Atlantic salmon. *Parasites & Vectors* 9, 639.

Steverding, D., Morgan, E., Tkaczynski, P., Walder, F. and Tinsley, R. (2005) Effect of Australian tea tree oil on *Gyrodactylus* spp. infection of the three-spined stickleback *Gasterosteus aculeatus*. *Diseases of Aquatic Organisms* 66, 29–32.

Stien, L.H., Dempster, T., Bui, S., Glaropoulos, A., Fosseidengen, J.E. and Wright, D.W. (2016) 'Snorkel' sea lice barrier technology reduces sea lice loads on harvest-sized Atlantic salmon with minimal welfare impacts. *Aquaculture* 458, 29–37.

Stien, L.H., Lind, M.B., Oppedal, F., Wright, D.W. and Seternes, T. (2018) Skirts on salmon production cages reduced salmon lice infestations without affecting fish welfare. *Aquaculture* 490, 281–287.

Sutherland, B.J.G., Koczka, K.W., Yasuike, M., Jantzen, S.G., Yazawa, R. and Koop, B.F. (2014) Comparative transcriptomics of Atlantic *Salmo salar*, chum *Oncorhynchus keta* and pink salmon *O. gorbuscha* during infections with salmon lice *Lepeophtheirus salmonis*. *BMC Genomics* 15, 200.

Sutherland, B., Covello, J., Friend, S., Poley, J., Koczka, J., Purcell, S., et al. (2017) Host–parasite transcriptomics during immunostimulant-enhanced rejection of salmon lice (*Lepeophtheirus salmonis*) by Atlantic salmon (*Salmo salar*). *Facets* 2(1), 477–495.

Svendsen, B.M. (2019) Implementation of the traffic light system in Norwegian salmon aquaculture – success or failure for whom? MSc Thesis. Norwegian College of Fishery Science, Norway.

Sweetman, J.W., Torrecillas, S., Dimitroglou, A., Rider, S., Davie, S.J. and Izquierdo M.S. (2010) Enhancing the natural defences and barrier protection of aquaculture species. *Aquaculture Research* 41, 345–355.

Tadiso, T.M., Krasnov, A., Skugor, S., Afanasyev, S., Hordvik, I. and Nilsen, F. (2011) Gene expression analyses of immune responses in Atlantic salmon during early stages of infection by salmon louse (*Lepeophtheirus salmonis*) revealed bi-phasic responses coinciding with the copepod-chalimus transition. *BMC Genomics* 12, 141.

Tirloni, L., Reck, J., Terra, R.M.S., Martins, J.R., Mulenga, A. and Sherman, N.E. (2014) Proteomic analysis of cattle tick *Rhipicephalus* (*Boophilus microplus)* saliva: a comparison between partially and fully engorged females. *PLoS ONE* 9, e94831.

Torrissen, O., Jones, S., Asche, F., Guttormsen, A., Skillbrei, O.T., Nilsen, F., et al. (2013) Salmon lice – impact on wild salmonids and salmon aquaculture. *Journal of Fish Diseases* 36, 171–194.

Troncoso, J., Vecino, J.G., Pino-Marambino, J. and El-Mowafi, A. (2011) Carvacrol and thymol composition for the prevention of an ectoparasitic infection in fish. Patent. WP 2011/119049.

Umasuthan, N., Xue, X., Caballero-Solares, A., Kumar, S., Westcott, J.D., Chen, Z., et al. (2020) Transcriptomic profiling in fins of Atlantic salmon parasitized with sea lice: evidence for an early imbalance between chalimus-induced immunomodulation and the host's defense response. *International Journal of Molecular Science* 31, 2417.

Valenzuela-Miranda, D. and Gallardo-Escárate, C. (2016) Novel insights into the response of Atlantic salmon (*Salmo salar*) to *Piscirickettsia salmonis*: interplay of coding genes and lncRNAs during bacterial infection. *Fish and Shellfish Immunology 59*, 427–438.

Valenzuela-Muñoz, V. and Gallardo-Escárate, C. (2017) Iron metabolism modulation in Atlantic salmon infested with the sea lice *Lepeophtheirus salmonis* and *Caligus rogercresseyi*: a matter of nutritional immunity? *Fish and Shellfish Immunology* 60, 97–102.

Valenzuela-Muñoz, V., Boltaña, S. and Gallardo-Escárate, C. (2016) Comparative immunity of *Salmo salar* and *Oncorhynchus kisutch* during infestation with the sea louse *Caligus rogercresseyi*: an enrichment transcriptome analysis. *Fish and Shellfish Immunology* 59, 276–287.

Valenzuela-Muñoz, V., Novoa, B., Figueras, A. and Gallardo-Escárate, C. (2017) Modulation of Atlantic salmon miRNome response to sea louse infestation. *Developmental Comparative Immunology* 76, 380–391.

Vargas-Chacoff, J.L.P., Muñoz, C., Hawes, R., Oyarzún, J.P., Pontigo, J., et al. (2016) Atlantic salmon (*Salmo salar*) and coho salmon (*Oncorhynchus kisutch*) display differential metabolic changes in response to infestation by the ectoparasite *Caligus rogercresseyi*. *Aquaculture* 464, 469–479.

Vecino, J.G. and Wadsworth, S. (2011) Functional feed composition. Patent. US 2011/0152177 A1.

Wadsworth, S., Vecino, J.G., Mordue, J. and Pino, J. (2011) Methods of feed compositions for masking of fish semiochemicals. Patent. WO2011/068415 A3.

Wasser, S.P. (2002) Medicinal mushrooms as a source of antitumor and immunomodulating polysaccharides. *Applied Microbiology and Biotechnology* 60(3), 258–274.

Wotton, H.J., Covello, J.M., Purcell, S.L. and Fast, M.D. (2013) *Lepeophtheirus salmonis* responses during a cohabitation of infected and uninfected Atlantic salmon. *Proceedings of the Canadian Society Zoological Meeting*. Guelph, ON.

Wright, D.W., Stien, L.H., Dempster, T., Vågseth, T., Nola, V. and Fosseidengen, J. (2017) 'Snorkel' lice barrier technology reduced two co-occurring parasites, the salmon louse (*Lepeophtheirus salmonis*) and the amoebic gill diseases causing agent (*Neoparamoeba perurans*), in commercial salmon sea-cages. *Preventative Veterinary Medicine* 140, 97–105.

Yiagnisis, M., Alexix, M.N., Bitcgava, K., Govaris, A. and Athanassopoulou, F. (2009) Effect of dietary oregano essential oil supplementation on combined infections by pathogenic bacteria and parasites (sea lice and copepods) in European sea bass *Dicentrarchus labrax* L. P139. In: *Proceedings of the 14th EAFP International Conference*, Prague, Czech Republic. p. 169.

Ytreberg, R. (2020) Miljøtrøbbel koster laksdsenæringen milliarder: Kostnadene Løper Løpsk. *Dagens Næringsliv*. 7–12.

Xue, X., Woldemariam, N.T, Caballero-Solares, A., Umasuthan, N., Fast, M.D., Taylor, R.G., Rise, M.L. and Andreassen, R. (2019) Dietary immunostimulant CpG modulates microRNA biomarkers associated with immune responses in Atlantic salmon (*Salmo salar*). *Cells* 8(12), 1592.

Xue, X., Hall, J.R., Caballero-Solares, A., Eslamloo, K., Taylor, R.G., Parrish, C.C. and Rise, M.L. (2020) Liver transcriptome profiling reveals that dietary DHA and EPA levels influence suites of genes involved in metabolism, redox homeostasis, and immune function in Atlantic salmon (*Salmo salar*). *Marine Biotechnology (NY)* 22(2), 263–284.

Zelaya, A.X.C. and Eggers, M.A. (2015) Compound for use in the prevention and treatment of an ectoparasitic copepod infestation in fish. Patent. WO 2015/155293 Al.

Zöllner, N. (1982) Purine and pyrimidine metabolism. *Proceedings of the Nutrition Society* 41(3), 329–342.

CHAPTER 22

Physical prevention and control of sea lice

Frode Oppedal[1], Lars Helge Stien[1], Samantha Bui[1], Tina Oldham[1] and Luke T. Barrett[2]

[1]Institute of Marine Research, Norway
[2]University of Melbourne, Australia

22.1 Historical background

Sea lice have impacted the salmon farming industry since its inception in Norway in the late 1960s. The expansion of the industry benefitted sea lice, and farmers were initially powerless to control infestations. The introduction of chemotherapeutants during the 1970s, and to a lesser extent cleaner fish, together with new fallowing regimes and single year classes in some areas, enabled the industry to continue expanding through the 1980s to 2000s while having some control over louse densities (see Chapter 15). From 2009 onwards, there was a dramatic increase in the volume of chemotherapeutants used in Norway, likely linked to declining sensitivity of lice to the active compounds used in Norway and the British Isles (Aaen et al., 2015; Hannisdal et al., 2020; Myhre Jensen et al., 2020). This, coupled with public concerns about food contamination and environmental impacts (Burridge et al., 2010), motivated a transition to non-chemical methods of control in many farming areas, led by Norway during the 2010s. Chemotherapeutants remained relatively effective outside Europe (Kreitzman et al., 2017). At the same time, cleaner fish use surged in Norway and the British Isles (see Chapter 19) and was followed by the introduction of mechanical and thermal immediate delousing methods that, by 2017, had overtaken chemotherapeutant use in Norway (Overton et al., 2019). However, these immediate methods have several drawbacks, most notably that: (i) the salmon are subjected to the same stressful procedure as the lice, which necessitates interruptions to feeding and risks salmon welfare and survival (Nilsson et al., 2019; Overton et al., 2019), and (ii) they are primarily reactive, being applied only once lice densities approach mandated or voluntary 'trigger levels'. This means that rather than being maintained at constant low levels, louse densities follow a sawtooth pattern of exponential increase, followed by an immediate drop once delousing is conducted. Cleaner fish can slow louse population growth between delousing treatments but are also most effective when lice are numerous (Barrett et al., 2020c). In Norway, these factors, coupled with legislation that makes industry expansion contingent on maintaining very low louse densities, have resulted in a slowing of industry growth since 2012. Similarly, the Scottish industry grew only 3% per annum on average from 1999–2019 (Munro, 2020). In all the major production countries, the industry faces pressure from the public and regulators relating to fish welfare, impacts on wild salmonids and chemical use.

High treatment rates in the latter half of the 2010s prompted an increasing focus on preventative methods that aim to reduce exposure and susceptibility to infective sea lice (Barrett et al., 2020b). A range of concepts have been developed to achieve this, including installing physical barriers, influencing salmon behaviour or immunity, and removing or incapacitating infective lice while they drift in the water column.

22.2 The principle of physical prevention

The principles of physical prevention stemmed from an improved understanding of the biology and behaviour of the infective stage of sea lice. Current physical prevention methods were developed in Norway, where salmon lice *Lepeophtheirus salmonis* (Krøyer, 1837) is the most problematic species and the focus of most research. Many of the principles and strategies outlined in this chapter will also be applicable to other sea lice species, but it should be noted that strategies developed to exploit the behavioural traits of salmon lice may not have the same efficacy against other sea lice.

The life cycle of the salmon louse is relatively simple for a parasite, with three planktonic larval stages and only a single host type occupied during their juvenile and adult stages. After hatching from eggs, the larvae have a finite energy source which must suffice while they progress through two nauplii stages into the infective copepodid stage, when they begin host-searching behaviours. Fuelled by this finite energy source, survival time of copepodids is dictated by their age and temperature (Samsing et al., 2016a); similarly, the behaviour of this infective stage is correlated to their age and the quantity of their remaining lipid reserves (Skern-Mauritzen et al., 2020). The larvae remain pelagic until a host is found and successfully infected.

22.2.1 Salmon lice dispersal

As salmon lice hatch directly into the water, they are at the mercy of hydrodynamic forces and can be transported long distances during their larval phase (Asplin et al., 2011). The direction and speed of currents also interacts with the behaviour of the larvae, along with environmental conditions such as salinity and temperature, to influence their dispersal characteristics (Johnsen et al., 2014). Numerous studies on the behaviour of nauplii and copepodids were published prior to the year 2000, but the topic has again come into focus to verify and describe these responses at a finer scale. The core interest is in the vertical movement of larvae within the water column, together with the environmental factors that influence this behaviour.

22.2.2 Larval behaviour

Many marine planktonic species exhibit some form of diel vertical migration, whereby the pattern of movement is from shallow depths at night to deeper distribution during the day (Hays, 2003). The ecological significance of this behaviour is typically related to optimal feeding and growth while reducing predation risk. However, as non-feeding planktonic larvae, the parasitic salmon louse has alternative drivers of vertical behaviour. Nauplii are the dispersal stage during which lice must maximise survival, while copepodids must actively seek a host, and have only limited time to do

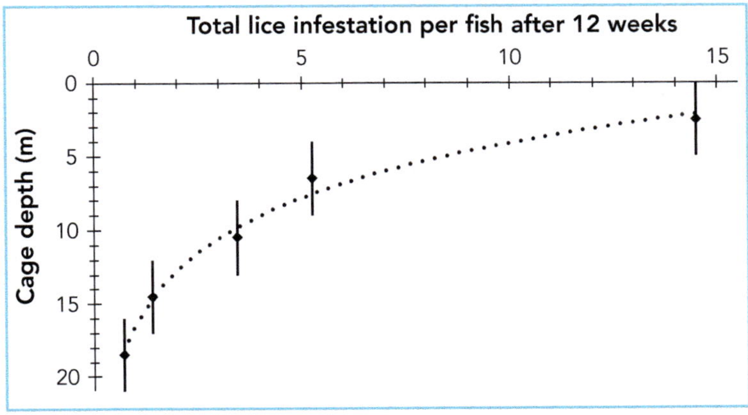

Fig. 22.1 Positive phototaxis of infective salmon lice copepodid larvae results in higher abundance near the surface, often named 'lice belt'. This pattern has been indirectly measured on groups of salmon kept at different depths (Oppedal et al., 2017). Points represent the middle depth within a 5 m cage depth interval (whiskers). A logarithmic trendline is fitted.

so. Intuitively, copepodids position themselves to maximise the chance of encountering a potential host, which requires them to respond to environmental cues such as salinity, light, temperature and possibly pressure (Mordue (Luntz) and Birkett, 2009). This typically manifests as a shallower daytime depth distribution, with infestation risk higher for salmon in the surface waters (Fig. 22.1) (Oppedal et al., 2017).

22.2.3 Response to light

Salmon lice larvae have optic photoreceptors developed enough to detect light stimuli in the marine environment, in particular polarisation cues (Bron and Sommerville, 1998). They are phototactic, reacting positively and directionally towards light (Novales Flamarique et al., 2000). A peak response is observed at light wavelengths of 500–550 nm (Gravil, 1996), which corresponds to the transmitted light quality at twilight. Although this morphological trait may be linked to host detection (that is, discerning the silvery underbelly of a salmonid), copepodids do not require light to successfully infect a host, suggesting that mechano- or chemosensory stimuli are more important (Browman et al., 2004). The phototactic behaviour of copepodids is likely the reason for the reverse diurnal migration behaviour that has been recorded in the sea, whereby larvae distribute closer to the surface during the day and sink to deeper depths at night (Heuch et al., 1995). However, copepodids have higher sensitivity to light intensity than nauplii, and no such reverse diurnal migration pattern or strong attraction to light has been observed in nauplii stages, suggesting there is a separate driver for their behaviour (Novales Flamarique et al., 2000). In a laboratory scale study in 2019, however, the proportion of nauplii that aggregated at the surface did increase with light intensity (Szetey et al., 2021). The study also found that copepodid behaviour differed to nauplii as they swam upwards in both light and fully dark conditions, and surface aggregations increased with dispersal duration. Further field or mesocosm studies are needed to document the true vertical distribution during hourly and daily time scales to carve out the exact relevance of the larval responses to light in view of physical prevention.

22.2.4 Response to salinity

Both nauplii and copepodids display clear preferences for full-salinity seawater, with avoidance of lower salinities when confronted with a step-gradient change in water quality (Heuch, 1995; Crosbie

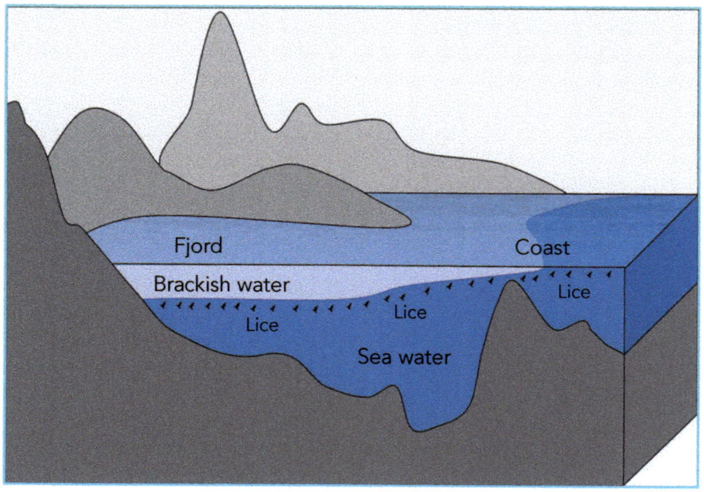

Fig. 22.2 Salmon lice larvae (black copepodites in figure) typically move towards the light and aggregate towards the surface, but simultaneously avoid brackish water. Drawing by Lars Helge Stien.

et al., 2019). More recent work investigated the sensitivity of both stages to smaller step-gradients in a controlled laboratory study and found that behaviours differed between the larval stages: nauplii exhibited strong avoidance of even a two parts per thousand (ppt) drop in salinity, while there was a more gradual decline in preference for more saline waters in copepodids (Crosbie et al., 2019). Both nauplii and copepodids aggregated just below the step-gradient (that is, the halocline) whenever one was present (sketched in Fig. 22.2). Copepodids occasionally situated themselves within the brackish layer down to 16–20 ppt, indicating that perhaps short-term tolerance is possible (Gravil, 1996). Being able to cope with brief exposure to low-salinity waters would increase chances of infection success; however, low salinities can reduce copepodid activity (Gravil, 1996), survival and fecundity (Sievers et al., 2019; Tucker et al., 2000; Bricknell et al., 2006), and infection success (Genna et al., 2005).

Based on the behavioural response of pelagic lice to light and salinity independently, in a homogenous water column it can be assumed that copepodids will orient themselves towards the surface when light is present (Heuch, 1995; Crosbie et al., 2019). However, in the presence of a brackish layer, lice would aggregate just below the halocline in an equilibrium between attraction to surface light and aversion to low-salinity waters. It is thought that copepodids can hold themselves in this position below the halocline by actively swimming up, coupled with slower swimming once low-salinity water is detected, as they do in response to light or chemosensory stimuli (Gravil, 1996; Mordue (Luntz) and Birkett, 2009). This could occur irrespective of the actual halocline depth.

22.2.5 Response to temperature

While both salinity and light have been well studied as environmental cues for salmon lice larvae, response to temperature was not investigated thoroughly until 2019. Empirical testing in laboratory experiments exposed both nauplii and copepodids to thermoclines with a base of 12°C and surface layer of between 6°C and 18°C. Crosbie et al. (2020) showed that nauplii displayed preference for cooler waters whether they were in the upper or lower section, whereas copepodids preferred the upper section regardless of temperature. This divergence in behaviour between stages indicates that cues other than temperature impart a stronger influence on copepodids. However, nauplii respond strongly to temperature compared to, for example, light.

22.2.6 Response to pressure

Although likely to be innately coupled with other cues, there is some indication that copepodids can detect and respond to changes in pressure, therefore exhibiting depth preferences (Bron et al., 1993). A study tested this response in a small-scale controlled environment that simulated hydrostatic pressures experienced at depths of 0 m, 5 m and 10 m, and found a strong upward migration in response to pressure change (Coates et al., 2020). Further, the authors also tested for the possibility that this could be a heritable trait and found evidence for inter-family differences in this behaviour (Coates et al., 2020).

22.2.7 Observations from plankton studies

Through the knowledge gleaned from laboratory and a few small-scale field tests, the widely accepted behavioural profile of salmon lice larvae is that their positive phototaxis and avoidance of low-salinity waters are the primary drivers. In reality, it is likely multiple cues orient larvae, and it would be difficult to disentangle the effects of individual environmental factors. Salmon lice also have strong responses to semiochemical and mechanosensory cues (Ingvarsdóttir et al., 2002; Heuch et al., 2007), and perhaps other stimuli that have not yet been discovered. Nevertheless, field studies which sample the plankton and describe the natural distribution of larvae reveal some of the combined outcomes of these various behavioural responses. Data for planktonic distribution of copepodids is limited because most studies with regimented sampling schemes find nauplii but very few copepodids. Nevertheless, most larvae are found in the upper 5 m of the water column (Penston et al., 2004), but can also be found in samples taken from 30 m depth (Nelson et al., 2017). However, some field collections have found results contrary to laboratory studies: á Norði et al. (2015) found distributions of nauplii that suggested they seek the highest possible temperature in the water column, as opposed to the cold-seeking behaviour observed by Crosbie et al. (2020). Similarly, samples taken from nearby salmon farms have found a weak diel vertical migration in nauplii with higher densities in the surface during the night, not the reverse as would be expected for copepodids (Nelson et al., 2017).

22.2.8 Vertical distribution of salmonid hosts

The overarching theory is that the sensory capabilities and migration patterns of salmon lice copepodids have evolved to maximise encounters with a potential host; wild salmonids mostly swim in the upper 3 m during their time in coastal waters or fjords (Rikardsen et al., 2007; Thorstad et al., 2012; Eldøy et al., 2017), which can also be in the brackish layer close to rivers and estuaries (Davidsen et al., 2008). In the ocean, Atlantic salmon likewise mainly occupy surface waters, typically at depths less than 10 m (exemplified by Hedger et al., 2017; Strøm et al., 2018; Einarsson et al., 2018). Comparably, salmon production cages are typically 10 m to 50 m deep. The farmed salmon are often close to the surface for feeding or refilling their swim bladders (Furevik et al., 1993; Juell et al., 1994), and will remain near the surface if light and temperature are optimal (Johansson et al., 2009). However, in general, satiated salmon in production cages display the opposite migration pattern to copepodids whereby they swim closer to the surface at night, and deeper in the water column during the day (reviewed by Oppedal et al., 2011 and exemplified in Fig. 22.3). Thus, farmed salmon would be exposed to a population of sinking and rising

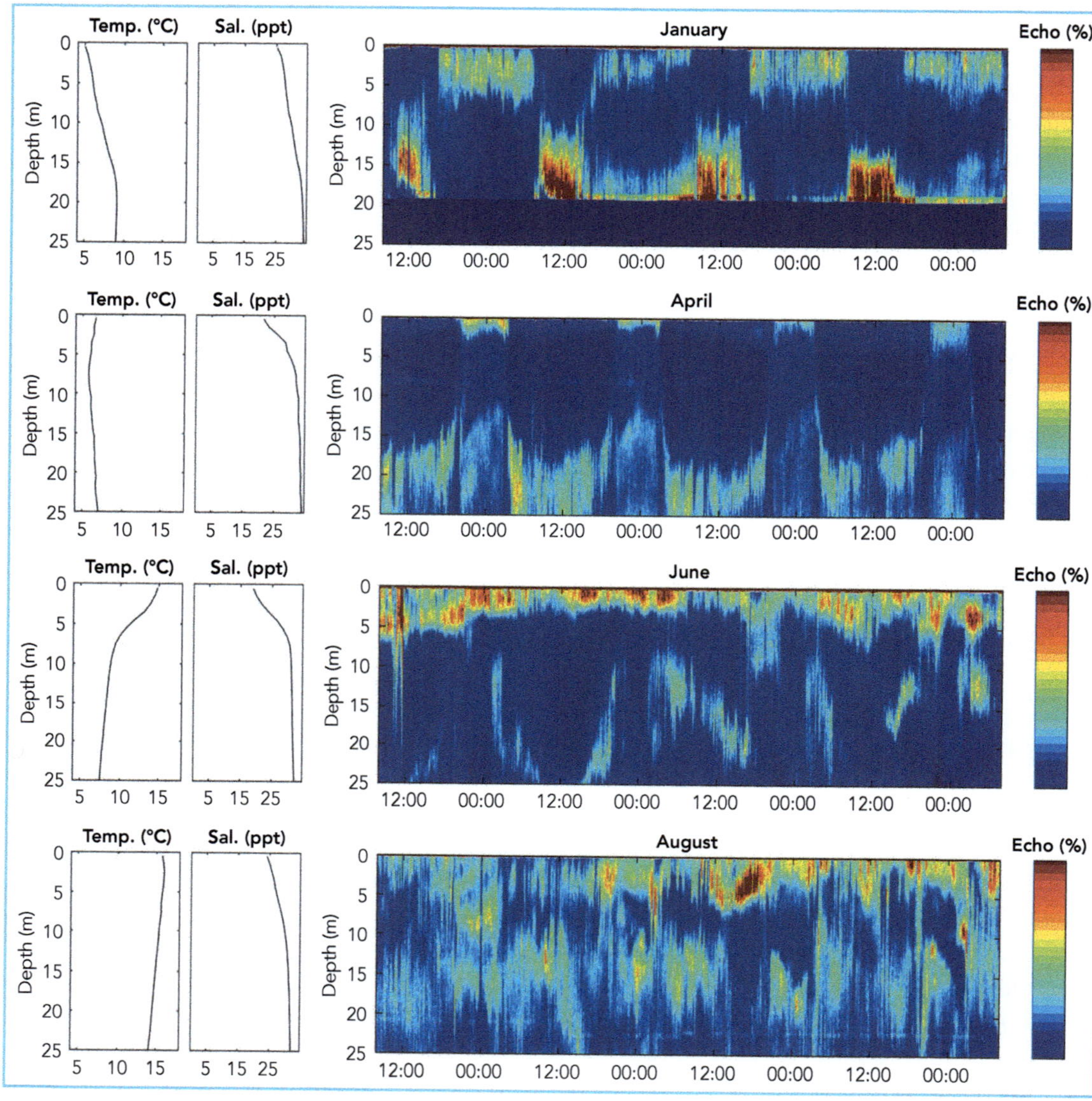

Fig. 22.3 Temperature (Temp.) and salinity (Sal.) profiles, averaged over four days, plotted with vertical behaviour in standard cages at a salmon farm in January, April, June and August (unpublished data from the same trial described in Bui et al., 2020a). The echo sounder data are displayed as percentage echo with depth per time unit. In January, the low surface temperature is avoided during the day but occupied during the long night, when swimming speed is also reduced. In April, warmer layers both at the surface and at depth were preferred at night, while all fish swam in the warmer deep water during the day. In June, the much (5–7°C) warmer surface water is preferred by most fish both day and night. In August, salmon generally prefer the warmest surface water with some fish in the slightly cooler, deeper waters below the halocline.

copepodids every day (Heuch et al., 1995). In addition to this, the farmed fish display distinct temperature preferences and avoid cold surface waters in winter, prefer warmest water all year and avoid surface waters when temperature exceed 16°C with cooler waters deeper down (Oppedal et al., 2011).

22.2.9 Prevention methods to minimise contact between host and parasite

From the evidence of shallow depth preference during the infective copepodid stage, a preventative approach emerged which aims to create a mismatch in the vertical distributions of salmon and lice to reduce encounters. This is loosely termed 'depth-based prevention' and functions by harnessing the spatial and temporal behaviours of both host and parasite, leading the species to occupy different depth distributions through physical barriers or attraction to specific cues. For the salmon aquaculture industry, it is much easier to control the fish in their production cages than influence the dispersal of planktonic larvae. As such, new developments seek to reduce encounter rates pre-emptively by changing the cage structure to 'shield' the fish from inflowing surface water, or by manipulating salmon behaviour to draw them away from the surface and/or the halocline. Barriers made of semi- or fully impermeable materials can prevent particles from entering the sea cage near the surface and have been adapted to 'skirts', 'snorkels' and fully enclosed sea cages. Changing the depth distribution of the school has been explored using submerged attraction lights and feeding zones, or submergence of the whole sea cage to below the parasite-risky shallow depths. These technologies and cage structures are described in sections 22.3.1–22.3.6, and sketched out in Fig. 22.4.

However, some innovations have also been aimed at the lice themselves, either at the planktonic stage or while attached to the host (see section 22.3.7). These have had varying success as preventative methods to date but do explore interesting concepts that attempt to harness behavioural traits of the lice.

22.3 Focused examples

Preferred lice management strategies are generally specific to sites, regions and farming companies, influenced by local conditions and availability of technologies. At the time of writing (2022), mechanical or thermal immediate delousing, cleaner fish and chemotherapeutants remain the predominant lice control strategies (often used in combination at a site), but a range of novel physical prevention and continuous control methods are either in development or already on the market (Table 22.1).

Commercially available lice prevention products include lice barriers, methods to draw up deep lice-free water, continuous delousing by automated lasers and methods to actively filter lice larvae before they encounter salmon (Table 22.1). Artificial lights are also widely used, often submerged in a manner that (together with deep feeding) allows the farmer to strategically manipulate salmon swimming depth with little financial investment (F. Oppedal, personal communication, February 2022). Various other physical prevention strategies remain in research and development, including cages that submerge salmon below the most lice-dense surface waters (Oppedal et al., 2020; Warren-Myers et al., 2022), lice traps and filters (Blue Lice and Vard Aqua), ultrasonic cavitation (Skjelvareid et al., 2018; Svendsen et al., 2018), low frequency sounds (Solé et al., 2021) and electric pulses (Harbor). Of the methods that are currently on the market, barrier technologies and closed containment have the most scientific support for efficacy (Fig. 22.5).

22.3.1 Skirt barriers

One of the earliest lice prevention strategies used was the 'skirt' – a barrier which completely encircles the uppermost portion of a cage. Skirts, and later closed tarpaulins, were originally developed

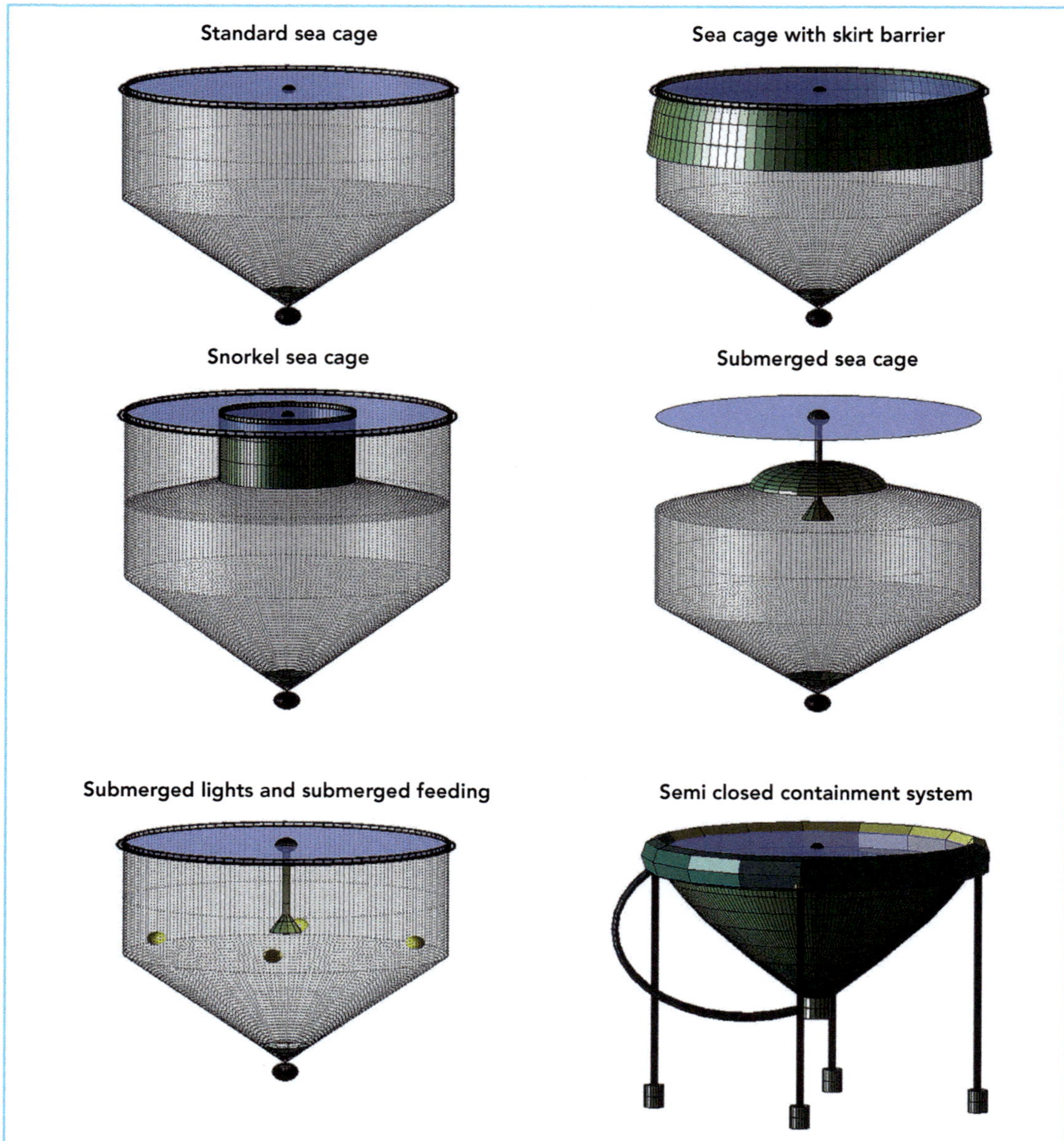

Fig. 22.4 Sketches of the main prevention technologies to spatially mismatch salmon lice larvae and the salmon host. A standard sea cage is open to surface-dwelling infective louse copepodids. Skirts create a surface barrier around the entire cage. Snorkel cages create a surface barrier and encourage the host deeper while allowing surface access for swim bladder refilling. Submerged light and feeding attract the salmon to voluntary swim deeper at night and day, respectively. The submerged cage forces the host to swim deep, but an air-filled pocket is required to allow for swim bladder refilling, and underwater feed delivery is necessary. A semi-closed or closed cage entirely blocks water flow and uses deeper (or shallow filtrated), parasite-free water to circulate within the production volume. Drawings by Lars Helge Stien.

Table 22.1 Assessment of the independent scientific evidence (at commercial scale) for prevention and physical continuous control methods. We also suggest some priorities for future research, and note the level of commercial development (developers and market availability).

	Scientifically validated?	Research priorities	Commercial developer	On market?
Prevention				
Closed containment	Y	Production efficiency, large commercial scale	Various	Y
Skirt/shield	Y	Production efficiency, oxygenation, interactions with other lice control methods	Various	Y
Snorkel cage	Y	Ensuring welfare with deep snorkels, interactions with other lice control methods, water quality	AKVA group/ Egersund Net	Y
Submergence	Y	Ensuring welfare under long-term submergence, with large fish, management and optimal environment with depth	AKVA group/ AOF*	Y*
Filter/trap	N	Quantification of effect	Vard Aqua, Fjord Miljø, Blue Lice	Y
Deep lights/feeding	Y	Integrating with data on lice depth for optimal timing and positioning in variable environments	Various	Y
Incapacitation (for example, electricity, ultrasonic cavitation)	N	Quantification of effect at commercial scale	Harbor	Y
Continuous control				
Laser	N	Quantification of effect in standard sea cages	Stingray	Y

* In 2021, the Atlantis cage (atlantisfarming.no) was tested at a commercial site in replicates for a production cycle and can be purchased from AKVA group. Norwegian Royal Salmon are building two commercial cages called Arctic Offshore Farming (arcticoffshorefarming.no) that will be used from 2022 onwards.

to reduce the flushing of water through cages during chemical delousing treatments (for example, used by Vigen (2008) and exemplified in Fig. 22.6a). Skirts were later adapted as a protective barrier to reduce introduction of infective louse larvae to cages. The materials utilised as the barrier can be impermeable (using canvas- or tarpaulin-like material) or semi-permeable to fluid inflow while filtering out copepodids (using plankton-mesh or similar material that has microscopic mesh sizes). Salmon lice copepodids are of a minimum width of 250 µm (Schram, 2004), so any mesh sizes smaller than the louse should potentially block the larvae from entering through the skirt. Skirts have demonstrated efficacy as a lice prevention tool and are now widely used (estimated at more than 300 cages per year), exemplified in Fig. 22.6b. Even so, more research is needed to optimise their use, as skirt effectiveness has varied widely in scientific trials (Barrett et al., 2020b). In one trial when surface waters were warmest and likely the cage area most preferred by the salmon, cages equipped with 10 m-deep fluid-permeable skirt barriers had 80% fewer lice than control cages after 12 weeks, with no impact on fish welfare or growth (Stien et al., 2018). However, in a larger, multi-year trial spanning five commercial sites, the average reduction in louse infestation was only 30% in cages

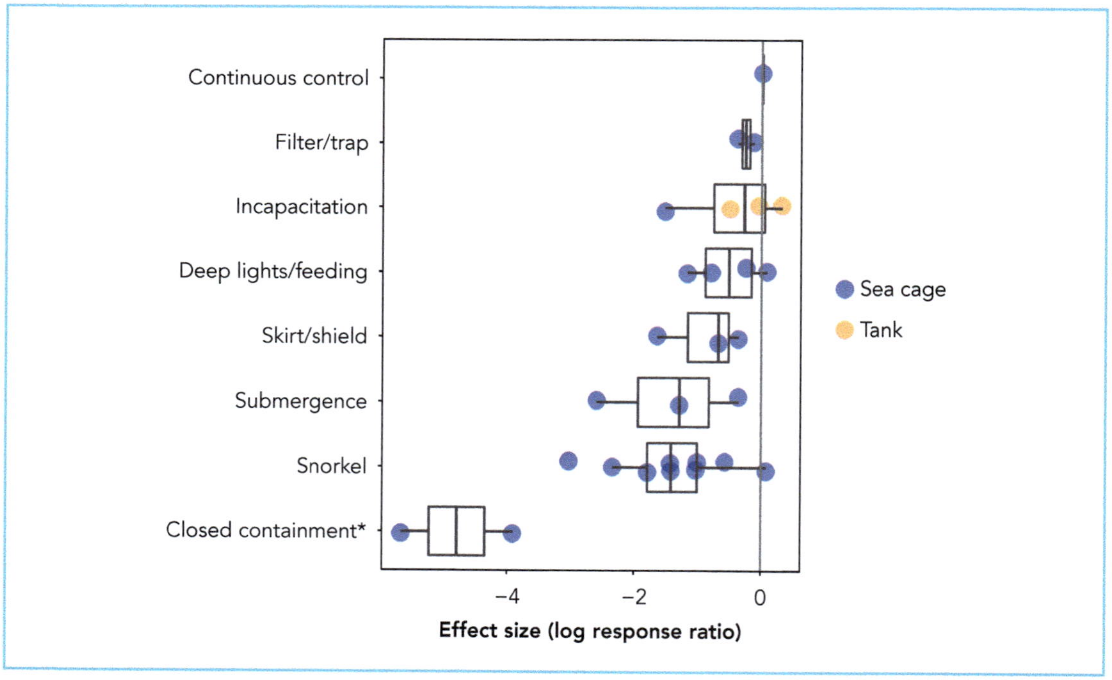

Fig. 22.5 Comparison of reported efficacy for physical prevention and physical control methods. Effect sizes are taken from published studies that reported relative infestation rates between sea cages or tanks with or without the intervention in question, presented as a log response ratio: RR = ln(I/R), where I is the group with the intervention applied and R is the reference group. Negative effect sizes indicate reduced infestation associated with the intervention. *Not shown is a closed containment study (Nilsen et al., 2020) that prevented 100% of infestations across multiple sites (RR = undefined). Adapted and updated from Barrett et al. (2020b).

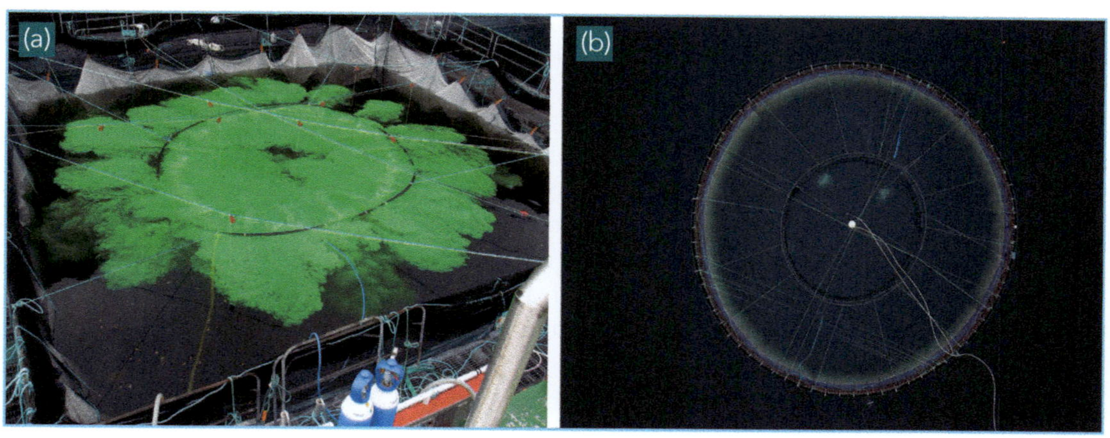

Fig. 22.6 a) Closed tarpaulin skirt used in a 12 m × 12 m cage during topical delousing in 2011 with green dye, fluorescein, displaying the distribution of the added pyrethroid drug. Photo by Frode Oppedal. b) Aerial view of combined use of a 6 m deep skirt (white collar inside the floater), central deep feeding and two bubble aeration systems (at 10 and 2 o'clock within the central bird net support floater) in a 200 m diameter commercial salmon cage in 2020. Credit: Tina Oldham.

equipped with either 6 m or 10 m fluid-permeable skirts relative to control cages (Grøntvedt et al., 2018). It is likely that the variability in observed effectiveness of skirts is related to both the depth and type of skirt, as well as with the environmental conditions at the site (Samsing et al., 2016b; Bui et al., 2020a). Skirts can also be lifted by several metres during periods of strong current or if there are significant differences in water density between the inside and outside.

One of the major challenges when using skirts is maintenance of suitable water quality within the protected water volume. The functional principle of skirts, to intentionally reduce water flow into the cage (Frank et al., 2015; Klebert and Su, 2020), can lead to rapid, dramatic drops in dissolved oxygen of up to 30% saturation within one hour of deployment (Oldham et al., 2017; Solstorm et al., 2018). If water flow through the cage is stopped, so is oxygen replenishment. For example, when a 3 m-deep skirt was left on the cage for several days at a commercial farm, median dissolved oxygen plummeted within the tarpaulin where high densities of fish were congregating and remained at or below 55% saturation despite reference dissolved oxygen consistently being above 85% saturation (Stien et al., 2012). Technology is ever-improving, however, and there are now numerous producers and types of skirts, with the most common being semi-permeable plankton nets which are intended to block particles but allow water flow, helping to mitigate the oxygen problem (Stien et al., 2018). Another potential optimisation is supplemental oxygenation or aeration to maintain water quality, either by directly adding oxygen or increasing the refreshment rate via circulation of deeper water (Bergheim et al., 2006; Endo et al., 2008; Berillis et al., 2016). Nevertheless, even when using mitigation measures to maintain dissolved oxygen levels, the use of any skirt necessitates close monitoring to avoid poor conditions which can reduce salmon growth, and in extreme cases, cause mortalities. Most importantly, aeration or oxygenation must be deployed at critical times according to dissolved oxygen variability at the site (Solstorm et al., 2018). Another concern when using skirts are the increased forces acting on the cages and moorings because of the larger surface area. As current speeds increase, deformations will occur to a larger extent than in standard cages (Lader et al., 2008), and more surface water will enter the skirted volume reducing the lice barrier effect (Frank et al., 2015). It has also been suggested that louse eggs which hatch within the skirt may not disperse as normal and develop to the infective stage within the skirt volume (F. Oppedal, personal communication, February 2022). An unrepeated/unreplicated study found that planktonic nauplii were more abundant inside cages with skirts than directly outside the cage, whereas the same pattern was not observed between the inside and outside of standard cages (Øvrelid, 2017). This hypothesis, however, remains to be fully tested.

Future developments to improve the use of skirts may include (i) active water quality management via exchange, such as pumps, propellers or mixing via aeration, (ii) strategic deployment based on current environmental conditions, (iii) deployment in combination with salmon behavioural modification or biological control measures, and (iv) adjustable depth skirts which can be responsively positioned to match the predicted depths of highest copepodid density.

22.3.2 Snorkel cages

The snorkel cage is a version of skirt barrier technology combined with submergence. A submerged net roof opening into a central barrier tube (snorkel) is added to standard cages to move salmon deeper while retaining surface access (Fig. 22.4, 22.7 and 22.8). The impermeable material enclosing the snorkel prevents particles in the surface layers from entering the cage, while still allowing surface access for the salmon to refill swim bladders. Salmon often choose to reside below the level of the snorkel, meaning that the barrier simultaneously encourages salmon to swim below the depths of highest infestation risk, while also protecting any individuals that use the surface layers, for example while feeding or refilling the swim bladder. Typically, snorkel circumferences are 30–90 m within 160–240 m commercial cages, and snorkel depths range from 10 to 20 m within 30–50 m-deep cages. The snorkel has a much smaller volume compared to conventional skirt barriers that surround

Fig. 22.7 Three generations of research snorkels, from 2006 (a and b), 2012/2013 (c and d) to 2014/2017 (e and f). Credit: Frode Oppedal.

Fig. 22.8 A commercial snorkel cage in 2014. Credit: Frode Oppedal.

the whole cage and is therefore easier to manage in terms of both water quality and cage deformation. Feed delivery can be at the surface but is commonly deeper, both to improve water quality and to avoid unnecessary attraction of fish to the snorkel volume.

The snorkel cage concept was hatched within the Centre for Research-based Innovation in Aquaculture (CREATE) funded by the Research Council of Norway and industry partners, with the Institute of Marine Research, University of Melbourne and Egersund Net/AKVA group as main players. The principle was then developed in medium-scale trials (Stien et al., 2016; Oppedal et al., 2017, 2019; Wright et al., 2018b, 2019) and commercial-scale trials (Wright et al., 2017; Geitung et al., 2019; Bui et al., 2020a), together with a range of synergistic research efforts to better understand how snorkel cage use interacts with production practices, and salmon behaviour and welfare (Samsing et al., 2016b; Bui et al., 2013, 2016, 2020b; Geitung et al., 2020; Wright et al., 2018a). As of 2020, more than 120 commercial snorkel cages have been used during partial or full commercial production cycles.

Of the physical prevention methods tested at commercial scale thus far, snorkel barriers are the most consistently effective. The prevention effect increases exponentially with depth, with infections lowered by 63%, 75%, 90% and 95% in cages with 4 m, 8 m, 12 m or 16 m snorkel depth, respectively, during mid-winter (Oppedal et al., 2017) (Fig. 22.1). In numbers, the final infestation level after 12 weeks within the surface cage was 14.5 lice per fish, which reduced to 0.7 lice per fish in the 16 m deep snorkel cage. Other trials reported an average of 47% reduced infection during autumn (Stien et al., 2016) and 76% during summer (Oppedal et al., 2019), using 4 m snorkels. Other than snorkel depth, another cause of variability in effectiveness relates to environmental conditions, particularly salinity. Brackish surface water pushes copepodids deeper, at or below the halocline (Crosbie et al., 2019), and thus snorkel efficacy is reduced in stratified fjord environments (Oppedal et al., 2019). Nevertheless, when 10 m-deep snorkels were used in a commercial test an 84% reduction in louse infection was observed compared to control cages (Wright et al., 2017). Similarly, in the most comprehensive commercial-scale snorkel trial yet performed, there was an overall 75% reduction in newly attaching lice over a full grow-out cycle, with variability in parasite reduction depending upon environmental variables (Geitung et al., 2019).

Snorkel cages also come with challenges. Management and handling require more labour and equipment. Surface area for net washing increases and sampling of fish can be challenging, while delousing baths or mechanical treatments often require that snorkels be dismantled and reassembled.

Water quality may degrade and requires continuous monitoring but can be maintained with proper water exchange (Oppedal et al., 2017; Wright et al., 2017). At smolt transfer, it is critical that in-mixing of well-boat water does not deteriorate the snorkel volume water quality. A solution may be to transfer smolt into the cage below the base of the snorkel, in the larger cage volume section. A further problem can occur when pumping or mixing cool, high-salinity deep water into the snorkel volume (to improve water quality), causing large water density differences to occur between the snorkel volume and outside. Similarly, this difference can occur when a surface brackish layer exists, leading to the snorkel deforming into an hourglass shape, reducing the usable space available for fish and depth of the protective barrier. This effect can be reduced by adding lower density water to the snorkel, such as brackish water either filtered from the surface or generated through reverse osmosis.

Overall, fish welfare scores from snorkel cage trials are within the normal ranges observed in standard cages, and fluctuate with season, environment, lice treatments, parasitic infections and disease outbreaks (as seen in Wright et al., 2017; Bui et al., 2020a). Snorkels may also interfere with the effect of other lice control measures. For example, there is evidence that cleaner fish may be less effective in skirt cages (Gentry et al., 2020), while the only independent test of a vision-based laser system (www.stingray.no) found no effect in snorkel cages (Bui et al., 2020b). In addition, some parasites increase in abundance within snorkel cages, most notably amoebic gill disease (AGD); however, this disease is treatable by adding freshwater to the snorkel volume (Wright et al., 2017, 2018a, 2018b). Other parasites, such as tapeworms, may be reduced in salmon held in snorkel cages (Geitung et al., 2021), while a range of others remain unstudied.

Future work to fine-tune the use of snorkel cages may include trials of continuous addition of freshwater into the snorkel to impair the infective ability of copepodids, while alleviating AGD outbreaks; oxygenation, aeration or improved water exchange within the snorkel; improved use of cleaner fish within snorkel cages with hides and feeding; and monitoring and control of other pathogens and parasites.

22.3.3 Behavioural modification using submerged lighting

After the discovery that exposure to continuous lighting between the winter and summer solstices could suppress incidence of sexual maturation in pre-harvest salmon, the use of artificial lights in sea cages became widespread (Hansen et al., 1992, 2017; Taranger et al., 1998, 2010; Oppedal et al., 1997, 2006; Porter et al., 1999; Endal et al., 2000; Bromage et al., 2001; Leclercq et al., 2011). Thousands of submerged lights have been sold to the salmon industry since 1992. In practice, anti-maturation lights are installed either above cages, or more commonly now, submerged just under the water's surface (Bromage et al., 2001; Migaud et al., 2007), illustrated in Fig. 22.9. But the impacts of long-term light use extend beyond suppressing maturation.

In Atlantic salmon, light is a critical modulator of not only reproduction and growth, but also behaviour (Oppedal et al., 2001, 2007, 2011). Continuously illuminated artificial lights are highly attractive to salmon in sea cages, both when surface mounted and submerged (Juell et al., 2003; Juell and Fosseidengen, 2004; Dempster et al., 2009; Wright et al., 2015). Early observations of behaviour at night in response to artificial lighting found that when lights were surface mounted on commercial cages, salmon crowded in the upper 3 m, sometimes at up to 20 times the calculated fish density. Conversely, when lights were submerged, the fish were more dispersed and swam deeper on average (Juell et al., 2003; Oppedal et al., 2007). Thus, artificial light attraction presents both a challenge and an opportunity for louse management. If not deployed with consideration of environmental

Fig. 22.9 Submerged lights used in the trial conducted by Oppedal et al. (2007), in which different light deployment depths were tested for two weeks at a time. Credit: Frode Oppedal.

conditions, artificial lights can attract salmon into the cage areas where infective stages of *L. salmonis* are most abundant (Hevrøy et al., 2003). On the other hand, when used as a lice prevention strategy, submerged artificial lights can be a minimally invasive tool to attract salmon away from the surface and into the areas of the cage with the most favourable conditions (Bui et al., 2019). Since the mid-2010s, farmers have become more aware of the potential to reduce lice infestations by deploying lights deeper (F. Oppedal, personal communication, February 2022), and light-emitting diode (LED) sources have replaced metal halide lamps as they are more efficient and equally effective (Hansen et al., 2017).

Few studies have investigated the impacts of artificial lighting, in isolation, on louse infestation. In the trials which have, the results are reasonably consistent: the fish which swim shallower experience more infestation. If the lights are placed above the cage, fish in the lit cages experience the same or greater infestation than natural light cages (Hevrøy et al., 2003); if the lights are submerged deeply, fish in the lit cages have similar or lower infestation (Frenzl et al., 2014; Bui et al., 2020a). In the only full-scale commercial study which attempted to use submerged artificial lights alone as a lice prevention tool, fish in cages exposed to submerged lights at 10 m depth had 40–50% fewer lice than those in cages with lights positioned at 1.5 m, in two out of three monthly sampling events (Frenzl et al., 2014). Using deep lights and deep feeding, in combination with skirts, throughout a production cycle, infestation was reduced by 51.3–63.3% compared to two groups without cage alterations (Bui et al., 2020a).

Despite the clear behavioural response of salmon to artificial lights, the attractive effect of lights is less clear when it competes with the numerous other behavioural drivers present in sea cages. Salmon can respond rapidly to lighting changes, and even follow lights as they are moved vertically through the cage at up to 4 m min^{-1} (Juell and Fosseidengen, 2004; Wright et al., 2015). But salmon do not necessarily respond when intensity is too dim and less when the colour band is within the upper or lower extremes of visible light (Stien et al., 2014). Temperature preferences also overrule the light preference increasingly with the temperature gradient (Oppedal et al., 2007; Stien et al., 2014). Individual salmon must balance preferred temperature, salinity, dissolved oxygen, light, turbidity, conspecific behaviour and feed availability simultaneously and, while light is an important driver of salmon behaviour, it can be over ridden by other factors (Oppedal et al., 2011; Føre et al., 2013).

Challenges for submerged lighting as lice prevention include use of artificial lights outside the anti-maturation window, especially at decreasing day lengths, which may recruit individuals to commence sexual maturation and thereby reduce slaughter quality and welfare (mature fish are less osmotically adjusted for seawater and males can begin fighting). Thus, year-round usage of artificial lighting still needs to be investigated in detail to determine the appropriate light quality to attract the fish without inducing maturation at decreasing day length. A proposed hypothesis to test is using violet or low-intensity blue-green light year-round in combination with full light during increasing day lengths (T. Oldham and F. Oppedal, personal communications, February 2022). The copepodid lice larvae may also be attracted downwards to the submerged light, but with normal water flow through the cages, limited time is available for the larvae to swim several metres in response to lighting. Typical continuous swimming speeds for lice larvae are 0.2–0.3 cm s^{-1}, equivalent to 12–18 cm min^{-1} (Fields et al., 2018). Environmental light pollution has also been raised in terms of unfavourable aesthetics for humans and attraction of other species to the farm area, including plankton, fish and birds. Animals attracted to the lights may be affected via hormonal changes or predation risk. Another concern is that continuous lighting does not allow the fish to rest. However, salmon naturally experience continuously lit conditions in the higher latitudes of their range and may therefore be adapted to such conditions. Finally, overruling the environmental preferences of the fish (for example, temperature or feeding versus light) may limit the beneficial effects, for example by inhibiting behavioural thermoregulation.

22.3.4 Behavioural modification using feeding

After temperature, and to some extent light, a critical determinant of vertical behaviour in farmed salmon is hunger. Salmon in sea cages generally migrate downwards during daylight hours in the absence of husbandry activities (Juell et al., 1994; Fernö et al., 1995; Oppedal et al., 2011); however, behavioural changes occur as feed is scattered across the cage surface throughout the day. Thus, the vertical distribution of salmon in sea cages can usually be explained by trade-offs between hunger, day-time surface avoidance and temperature (Huse and Holm, 1993; Fernö et al., 1995). Food acquisition often trumps all other considerations for caged salmon, at least in the short term (Oppedal et al., 2011). Despite their natural inclination to swim deeper during the day, salmon will crowd the surface during feeding, and only gradually return to their pre-feeding preferred depth as satiation is reached (Huse and Holm, 1993; Juell et al., 1994). Further, when food is withheld, such as prior to planned handling events, hungry fish will crowd the site of feed distribution in anticipation (Oppedal et al., 2011). In a detailed study which tracked environmental conditions and vertical fish distribution in cages at four commercial farming sites with different hydrodynamic conditions, Johansson et al. (2007) found that higher fish densities were consistently predicted in the surface during feeding periods. While the attraction of salmon to the surface to feed can be problematic with regard to louse infestation, like anti-maturation lights, it is also an opportunity. The amount of time that salmon spend in the risky surface waters and halocline can be dramatically reduced simply by submerging feed distribution (Fig. 22.10) – a shift which would decouple both the anticipatory and active feeding behaviours from the cage areas where lice are most abundant.

Although no trials have investigated the efficacy of submerged feeding in isolation, a few studies have tested it in combination with other cage adjustments and provide useful insights. In one study where entire cages were submerged to a depth of 3 m for 22 days, including feeding, no significant differences were detected in specific growth rate nor appetite between the control and submerged cages (Dempster et al., 2009). Another study by Frenzl et al. (2014) compared three treatments – a) surface feeding and lights, b) surface feeding and submerged lights and c) submerged feeding and lights – throughout the final three months of production at a commercial site. In this trial, although

Fig. 22.10 Submerged feeder developed by AKVA group. Credit: Henning Holm.

fish in the cages with submerged feeding were observed to move into the feeding corridor below 5 m, and no significant differences were detected in feed conversion ratio between treatments, fish in the submerged feeding group were smaller than those in the other two treatments at the end of the trial (Frenzl et al., 2014). Further, while louse prevalence was significantly lower in the submerged feeding group than the surface-fed controls, it did not differ significantly from the surface-fed and submerged light treatment group (Frenzl et al., 2014). Critically, this trial spanned only a single summer season. Although little information was provided regarding the environmental conditions during the trial, the fish were observed to promptly return to the surface after satiation, similar to the behaviour observed in other trials where the warmest waters were at the surface during summer (Oppedal et al., 2001, 2007). It is therefore logical that louse infestation was not affected, as the intended behavioural shift to deeper areas of the cage was not achieved.

Clearly, more data are needed to understand how best to implement submerged feeding to achieve maximum efficacy throughout the year. One important consideration is feeding strategy (Juell, 1995). Juell et al. (1994) found that while growth did not differ between fish fed continuously throughout the day to satiation and those fed in a restricted, high-intensity regime, vertical behaviour was considerably altered. Further, Ang and Petrell (1998) suggest that when fish are continuously fed throughout the day, they form a ring-like structure, characterised by organised foraging. In contrast, when fed in batches, salmon initially form a spiral-like structure followed by foraging in a disorganised style until feed becomes unavailable.

Summarised, fish fed continuously respond less and spend more time in the feeding corridor, while fish fed at high intensity move quickly in and out of the feeding corridor. Choosing the appropriate depth for feed dispersal is important and should consider the depth at which louse copepodids are most abundant as well as the seasonal depth preference of the fish. At coastal sites without a brackish surface layer, continuous deep feeding could considerably reduce louse infections. However, at more inland sites, such as within fjords, determining optimal feeding depth can be more complicated. One possibility is to place stationary feeders several metres below the maximum halocline depth of the site, to attract the fish below the cage area with highest louse abundance as much as possible. Perhaps a more promising alternative, especially at sites with frequent, strong brackish layers, is variable depth feeding. Using this strategy, the feeder could be positioned either above or below the halocline, according to the current preferred fish depth, so that the fish have minimal motivation to cross the halocline (where copepodids are most abundant). The addition of a visual or auditory cue at the time of feeding could be used to help the fish adjust to changes in feed position (Bratland et al., 2010; Macaulay et al., 2020c). If submerged feed dispersal is not possible, then a high-intensity surface-feeding regime to satiation, rather than continuous feeding, can at least minimise the time which the fish spend crowding in the surface and allow them maximum opportunity to disperse according to their preference.

By 2020, at least three companies in Norway have sold more than 200 submerged feeding units (H. Holm, Nærøysund Aquaservice/AKVA group, Norway, personal communication, December 2020). Some challenges may be considered in their operation. Providing feed underwater could be less efficient than surface feeding if fish do not descend below the feed entrance depth, the space below the feed entrance point is too limited or if there is insufficient horizontal spreading. In addition, attracting fish by feed delivery will mainly be useful during the feeding time, and for the individuals participating in the meal. As observed in Frenzl et al. (2014), fish may return to the depth they prefer when satiated and thus the mismatch to lice larvae is likely to be negligible. Finally, although behavioural adaptation of the fish is sometimes very fast, it can also be a protracted process for reasons yet unclear. Future trials which could improve the implementation and effectiveness of submerged feeding include testing adaptable feeding based on current environmental conditions and positioning of the fish, testing the use of 'training' techniques to help the fish recognise when and where feeding is occurring, and further study of how salmon behaviour modification affects the efficacy of other prevention strategies with which it could be used in combination, such as cleaner fish and/or skirt barriers. While it is unlikely that any behavioural modification strategy will perform optimally for all individuals in all conditions, working with rather than against the innate preferences and behaviour of farmed salmon can certainly reduce louse infestation pressure during production (Bui et al., 2019). Optimal implementation of any behavioural modification strategy will require knowledge of the site-specific environmental conditions (such as the strength and depth of the halocline), and real-time fish behaviour.

22.3.5 Submerged cages

Submerging salmon cages is not a recent concept, having been considered as early as the 1970s at the beginning of commercial salmon farming. However, unlike other species which are commonly farmed in submerged cages, salmon have a physostomous swim bladder and must gulp air into their swim bladder to regulate buoyancy (Fänge, 1953). Salmon need to refill their swim bladders regularly to be neutrally buoyant down to their maximum neutral buoyance depth (Macauley et al., 2020a). As a result, the idea lay dormant for decades until the lice problem drove farmers and researchers back to the drawing board. In theory, submerged cages can completely avoid the surface waters where infective louse copepodids reside. In practice, the challenge of farming physostomous fish in submerged cages is considerable.

To overcome the need for air access, two different adapted submergence schemes have been tested: (i) short-term submergence with occasional resurfacing to allow refilling, and (ii) continuous submergence with provision of air-pockets within the cage volume (Fig. 22.11). Providing compressed air deeper down will also give the salmon a deeper maximum neutral buoyancy depth and open possibilities for utilisation of extensive volumes of water. An early short-term trial found that fish in submerged cages had a lower specific growth rate than their counterparts in standard cages, possibly because temperatures were lower deeper in the water column than in the surface (Dempster et al., 2008), while submergence for longer than 40 days led to reduced growth and welfare, particularly fin and snout damage (Korsøen et al., 2009; Sievers et al., 2018). Trials have found that short-term submergence of 7–22 days without access to air, followed by brief periods of surface access, does not negatively impact fish growth or welfare (Dempster et al., 2009; Glaropoulos et al., 2019). However, periodically resurfacing submerged cages is logistically difficult, requires exposing the fish to exactly the water layer that submergence is aiming to avoid and is risky for fish production performance and welfare. Interest has therefore shifted away from periodic resurfacing as a solution.

The most exciting developments in submerged salmon farming are cages where air pockets are provided, via a large, submerged dome centrally positioned in the net ceiling (Fig. 22.4, 22.11 and 22.12) or multiple pockets integrated into the cage structure (Fig. 22.13). At medium scale, the provision of a central air dome in cages continuously submerged to 15 m for 35–49 days resulted in no negative impacts on fish growth or welfare (Oppedal et al., 2020). When submerged for a full production cycle, however, the fish in cages with air domes had reduced growth and welfare, likely due to poorer temperature and oxygen conditions deeper down (Warren-Myers et al., 2022).

Fig. 22.11 Submerged cage trial at Smørdal site of Institute of Marine Research (Oppedal et al., 2020), involving use of an air dome. A 10 cm layer of air is held under the white dome (in photo, held up by the boat-crane) and held at depth by a weight (red concrete object in photo) connected with poles. Credit: Frode Oppedal.

Fig. 22.12 Renders of the 'Atlantis' submerged cage with air dome. Left: Details of the air dome. Right: Atlantis cage in submerged position. Credit: Trude Olafsen, AKVA group.

Fig. 22.13 Renders of a planned Arctic offshore farming sea cage, where salmon are to be produced from 2021. Left: In submerged position. Right: In floating service position. Four rectangular air domes can be seen fitted around the perimeter. Credit: Klaus Hatlebrekke, Norway Royal Salmon.

Further, during 2019–2020, commercial testing of a single submerged cage equipped with an air dome was conducted for a full production cycle with 25,000 and 100,000 fish for 112 and 61 days respectively, without any issues reported (Atlantis cage, Fig. 22.12 and 22.14). Research is ongoing to test whether air dome fitted cages and flexible submergence depths (following ideal environmental conditions through seasons) can solve the growth and welfare issues created by forcing fish into cooler and hypoxic areas of the water column (F. Oppedal, personal communication, February 2022). In Oppedal et al. (2020) an unpublished trial concluded that a 3 m-diameter air dome was sufficiently large for medium-scale commercial cages (12 m × 12 m, 10,000 fish; or 20 m × 20 m, 30,000 fish), but to date this has only been tested at a shallow roof depth of less than 0.5 m from the surface with a maximum submergence period of 18 days, with no measurement of animal-based fish welfare variables. Increased use of air domes was seen after acclimating the fish as juveniles (Macauley et al., 2020b). Other trials are testing whether air bubbles can be used by the fish for swim bladder refilling underwater, with or without air domes (O. Folkedal, personal communication, February 2022). By 2022, two companies will be testing commercial-scale submerged cages with air pockets throughout

Fig. 22.14 The Atlantis cage coming to the surface after four months of submergence. Half of the collar is floating at surface, while the other half is still ascending. Credit: Frode Oppedal.

full production cycles (www.atlantisfarming.no, Fig. 22.12 and 22.14; www.arcticoffshorefarming.no, Figure 22.13).

The prevention efficiency of submergence should follow the exponential increase with depth as described for snorkel cages (Fig. 22.1, section 22.3.2). A 72% reduction in infestation was observed during submergence to 10 m in winter (Sievers et al., 2018). In a later trial when cages were submerged to 15 m depth during winter, up to a 97% reduction in newly attached lice was recorded in winter, with an overall reduction of 65% until summer (Warren-Myers et al., 2022). In contrast, when fish were subjected to weekly, repetitive submergence and resurfacing, the fish in test cages had on average 1.4 times as many lice as the fish in control cages (Glaropoulos et al., 2019). There are two key differences between the two continuous submersion trials which successfully reduced louse loads, and the resurfacing trial which did not: the presence of a brackish surface layer pushing copepodids deeper while the cage position prevented salmon from positioning themselves in the brackish layer, and fish being raised through the halocline into the surface waters once per week, during which they were susceptible to louse infection. Thus, as with other depth-based prevention strategies, efficiency of submergence is reduced in stratified fjord environments compared to coastal. However, unlike other depth-based prevention strategies, when armed with knowledge of the prevailing site conditions submerged cages can relatively easily be positioned below the deepest halocline depth, even at sites with frequent and strong brackish layers. The operation of a fully submerged sea cage brings logistical challenges. Even more so than snorkel cages, management and handling of submerged cages requires increased labour and equipment. Net washing is required less frequently but is more difficult to carry out. Sampling of fish can be logistically difficult; lice counts or delousing requires that cages be surfaced and partly opened. In addition, monitoring the behaviour, welfare and growth of fish in submerged cages requires more effort than in standard cages, and is more critical, because the ability to quickly respond to problems is reduced.

Future work to develop the use of submerged cages should focus on:

- long-term production trials in variable environments at different depths
- how different cleaner fish species with supporting hides and feeding may cope in submerged cages
- establishing baseline behavioural expectations for salmon in submerged cages and identifying deviant behaviour as a welfare monitoring tool
- determining the need for and optimal deployment of artificial lighting in submerged cages, where natural light is reduced
- examining the impact of submergence on the prevalence of other salmon parasites and pathogens.

22.3.6 Closed/semi-closed floating cages

Closed floating cages, or closed containment systems, are in principle completely isolated floating rearing units with waterproof barriers between outside water and the water inside the unit. An example of an active farm with several cages is provided in Fig. 22.15. The objective is that water pumped into the unit and subsequently released again is free of pathogens, and faeces, feed particles, and so on. Fully achieving this is difficult, and in recognition that there typically is some risk of negative outside influence to and from the unit, it is more accurate to use the term semi-closed containment system (S-CCS). In practice, some systems are more closed than others.

The S-CCS typically pumps water into the rearing unit from the deep to avoid infestation by salmon lice larvae, while also collecting faeces and other particles that settle in the cage bottom to minimise the effect of the unit on the local environment. This isolation means that the S-CCS can be placed at shallower, more sheltered locations with the minimum risk of damage from waves and stormy weather. This is especially important since high waves outside can lead to violent 'sloshing' water movements within an enclosed cage, potentially damaging the cage structure or harming the fish (Kristiansen et al., 2018).

Although in theory it should be almost impossible to have infestations in the S-CCS, some studies using the S-CCS still report salmon lice on the fish (for example, Nilsen et al., 2017; Øvrebø, 2020). On the other hand, it has also been reported that, even though present, lice are so rare that they are not able to reproduce effectively within the S-CCS (Nilsen et al., 2017). In five S-CCS cages stocked with post smolts, the first lice were found on fish 15–66 days after stocking, and infestation persisted for 253 days. However, louse abundance gradually reduced without intervention, and no delousing treatments were required (Nilsen et al., 2017). A subsequent study in a similar system reported no lice (Nilsen et al., 2020).

Aside from the avoidance of salmon lice, the S-CCS also makes it possible to exert more control over cage environmental characteristics, including temperature, salinity, oxygenation and current

Fig. 22.15 A sea cage farm consisting of the semi-closed containment system (S-CCS) in Hamnsundet, Vevelstad commune, Norway. This type of system was used in the work of Nilsen et al. (2017, 2019, 2020). Credit: Visual 360, Brønnøysund, for Akvafuture AS.

flow, in order to maintain suitable and stable conditions for fish welfare and growth (Nilsen et al., 2019, 2020). Deep water is often much colder than surface waters from spring to the start of winter. Selecting the depth of the deep-water inlet therefore becomes a trade-off between increasing growth performance of the fish with a relatively shallow inlet, or reducing risk of salmon louse infestation by going further down. In winter, the opposite occurs, with the warmest water deeper down. Important improvements to the S-CCS will therefore be the development of pumping systems that filter the intake water and treat for harmful pathogens, allowing the intake of water from the most suitable depth while minimising risk of infection or infestation (or even intake of large quantities of jellyfish leading to mass mortality (Hosteland, 2016)).

One inescapable disadvantage of the S-CCS is that due to high operational costs, fish density must be much higher than in standard sea cages for it to be profitable for the farmer under current regulatory conditions. High stocking density can stress the fish and lead to decreased growth, fin damage and cataracts (Calabrese et al., 2017). A stocking density of 75 kg m^{-3} has been suggested as a compromise between economy and fish welfare, but reduced growth rate was observed at densities as high as 50 kg m^{-3} (Calabrese et al., 2017). High stocking density also brings added risk of accumulation of metabolites and CO_2, and depletion of O_2. It is therefore vital that the S-CCS replenishes with new water at sufficient rates pending internal water flow, fish behaviour and 'water consumption' by the fish (Nilsen et al., 2020). In addition, the S-CCS generally also needs to have systems for oxygenating the water (Nilsen et al., 2020).

This technology is still in its juvenile phase, with many different systems being tested and improved and, to date, there are relatively few published studies on fish welfare in the S-CCS. Given the many potential risks and the complexity of the S-CCS, it is important to proceed with caution. Future directions and challenges for the S-CCS will likely involve the development of larger units that can be constructed and operated more cost-effectively, and advanced systems for filtered and depth-adjustable water intake.

22.3.7 Active methods

With the paradigm shift away from chemotherapeutants, control methods for continuous or acute delousing have also developed further, with innovations aiming to provide control options that sit outside of the standard treatment approach. These generally focus on removing, immobilising or deterring salmon lice of some or all stages. The following technologies are in various stages of development, from being still in the concept phase (for example, UVC sterilisation, bubbling, ultrasonic cavitation, electric pulses) to commercially tested or available (laser delousing and lice traps). However, some innovations have been tried without substantial success (large filtration, ultrasound), but are frequently revisited when new knowledge becomes available.

22.3.7.1 Filtration, traps or deterrence

One strategy for prevention is to remove infective copepodids from inside or around sea cages before they encounter a host. Lice traps work on the concept of attracting copepodids into a trap, using light or olfactory stimuli, which then removes them from the water. Field tests using light traps have found little efficacy in prevention when deployed adjacent to a salmon farm (Novales Flamarique et al., 2009). However, recently this technology has been further developed into several commercial products (for example, BlueLice, 'Strømmen-rør' by Fjord Miljø or 'NS Collector' by Vard Aqua). The literature on host-searching behaviours in copepodids is growing and may present new opportunities to adapt these traps by using chemosensory attractants (Ingvarsdóttir et al., 2002; Fields et al., 2018) and optimal lighting (Nordtug et al., 2020).

Alternatively, copepodids could be removed from the water column using biological filtration. Filter-feeding organisms such as mussels or oysters have been deployed on lines or racks around sea cages to consume lice larvae before they enter the cage. This approach appears to be effective if a large number of filter-feeding animals are used to spatially protect the cage or site (Byrne et al., 2018; Montory et al., 2020). It is not well known how such arrays will affect dissolved oxygen levels (for example, via oxygen consumption by filter feeders or via reduced current velocity). Filtration devices have also been explored for the collection of lice and egg strings during delousing or harvesting procedures (O'Donohoe and McDermott, 2014).

Finally, a concept that has been discussed frequently but not yet scientifically investigated or developed is the use of bubbling to deter copepodids from entering the cage, air-lift lice-free water or impair the larval ability to attach to a host. 'Bubble curtains' have been anecdotally trialled, whereby air bubbles are continuously released at the edge of the sea cage to form a barrier. Even so, in observations during a laboratory study, no bubbles successfully adhered to lice or lice shells and any direct effects of the bubbles themselves could not be observed (Solvang-Garten et al., 2015). An indirect effect of the vertical water flow created by the ascending bubbles is more likely. The use of bubbling has been adopted in the industry to introduce oxygen and improve water quality within cages, in particular in skirt and snorkel cages – this has been considered a possible lice prevention measure if copepodids cannot detect a host because of mechanosensory disturbance.

22.3.7.2 Incapacitation or removal

Several methods focus on the lice themselves and attempt to remove, immobilise or sterilise them within the sea cage. The use of optical delousing is currently being used by a number of farming sites and companies, and comes in the form of nodes within the cage that emit laser pulses targeted at lice attached to the salmon. Although a relatively successful innovation commercially, only one independent study has aimed to validate its use in a commercial setting (deployed for three months within snorkel cages), without success (Bui et al., 2020b). Sterilisation of egg strings is an alternative method explored recently in controlled tank trials, with UVC irradiation shown to reduce the hatching success of eggs (Barrett et al., 2019). This was carried to the next stage by testing the effective exposure on egg strings still carried by female lice attached to salmon in tanks; although the technology was successful at reducing development success of larvae, effective doses were also harmful to the salmon (Barrett et al., 2020a). The use of UVC is better suited to killing or sterilising lice either before attachment or after removal by delousing.

Electrical fences around cages which incapacitate lice larvae have been tested in pilot scale at a commercial fish farm, with promising results reported by the manufacturer (Harbor Fence, Harbor.no). Efficacy remains to be tested independently. Ultrasonic cavitation has been tested to immobilise or kill pelagic copepodids around or in the sea cage, but with little success in showing lethal thresholds or preventing attachment (Skjelvareid et al., 2018).

Methods that trap, filter or kill free-swimming copepodids also risk killing non-target planktonic species, depending on their specificity. A range of organisms from other copepods to fish larvae could be vulnerable. This should be considered before deploying such technologies at scale that could lead to unintended consequences for local ecosystems. It is currently unclear whether any of these methods will eventually be effective enough at commercial scale to be adopted as industry common practice. However, the exploration of concepts is the basis for innovation which may one day provide a viable application. Although no single method is likely to be the complete solution, these are possible lice-focused strategies that complement the salmon-focused strategies discussed in sections 22.3.1–22.3.6, and summarised in Table 22.1.

22.4 Economics of physical prevention strategies

Management of sea louse infestations is often costly, but necessary to avoid the even larger direct and indirect costs associated with inaction.

Infestations cost farmers via at least three mechanisms:

1. Infestations stress fish (Øverli et al., 2014), reducing growth rates and in severe cases, leading to death by osmotic stress or secondary complications (Fjelldal et al., 2020). All can result in lost production (Abolofia et al., 2017).
2. The monitoring, prevention and treatment of infestations is costly and requires labour, and depending on the method, may interrupt feeding and itself cause stress or even mortality among farmed salmon (Overton et al., 2019).
3. To maintain animal welfare and reduce the threat to wild salmonids from lice larvae produced at salmon farms (Krkošek et al., 2013; Thorstad et al., 2015; Vollset et al., 2018), responsible

regulators will typically set limits on louse infestation abundances and regulate the growth of the industry. This leads to opportunity costs for individual sites and the entire industry whenever reduced biomass or early harvesting is mandated.

The total cost of sea lice for the salmon farming industry is difficult to estimate but is potentially over US$1,000,000,000 annually worldwide (Costello, 2009; Liu and Bjelland, 2014; Iversen et al., 2015; Abolofia et al., 2017). Iversen et al. (2020) estimated production costs for all major salmon farming countries, where delousing costs were grouped with net cleaning and other services, together totalling US$0.38–1.2 per kg. This is relative to combined costs of US$2.92–3.76 per kg for feed, smolts and labour, such that miscellaneous costs, primarily delousing, are 10–24% of on-water production costs.

There is currently no broad consensus on the most cost-effective lice management strategy, both because optimal lice management tends to reflect local conditions and availability of methods, and because production costs – including those associated with louse management – are generally held as commercial secrets. A complicating factor is also that for a single farmer it may be economically smart, at least short term, to delay delousing and implementation of costly preventive technology. A recent study found that isolated individual farms would benefit economically from having high treatment thresholds, while a network of farms, infecting each other, maximise profit by having low treatment thresholds (Kragesteen et al., 2019). That some can benefit from acting contrary to the common good is part of the reason it is so difficult to reach consensus on the best lice management strategy for the industry as a whole (Stien et al., 2020b). Such 'tragedies of the commons' may be avoided by agreements between companies or spatial arrangement of farms according to ownership, such that each farming company bears primary responsibility for, and consequences of, louse levels within a specific hydrographic zone or zones.

In Norway, most sites rely on a combination of methods, including cleaner fish, immediate delousing when needed and, increasingly, preventative methods (Barrett et al., 2020b), while industries in other countries rely on a smaller range of measures (especially in North America and Chile, where chemotherapeutants still dominate). Prices for cleaner fish in Norway have reportedly varied between US$1.7–4 per fish in recent years, with 60 million cleaner fish stocked in 2019 at a cost of US$144,000,000 (mean US$2.4 per fish) (Directorate of Fisheries, 2020). The cost of hired mechanical or thermal delousing services is not publicised, but the recent acquisition of a single thermal delousing vessel in Scotland reportedly cost the farming company more than US$5,000,000 not including operating costs. A 2015 report estimated up to US$17,000 per hour for bath treatments (Iversen et al., 2015), which can span several days and likely total upwards of US$340,000 per treatment per site.

There are little cost data available for preventative methods such as skirts and snorkel cages, but a Scottish company reported purchasing skirts for 11 sites for approximately US$1,000,000, while the largest salmon farming company, MOWI, invested about US$11,000,000 to install snorkel cages at seven farming locations in 2020. Snorkel cages are associated with increased labour during setup but have relatively low ongoing costs during the production cycle while reducing lice levels by around 75% at suitable locations (Geitung et al., 2019; Barrett et al., 2020b), and can be reused for multiple production cycles. The pay-off time for snorkel cages relative to post-infestation control methods is not yet known, but may be in the order of years if the use of cleaner fish and immediate delousing can be largely avoided (assuming similar production efficiency and US$1–2,000,000 in avoided lice control costs per site per year). Costs of a submerged feeder, as exemplified by the AKVA group version, lay within the region of US$28,000 (H. Holm, Nærøysund Aquaservice/AKVA group, personal communication, Norway, 2020).

Fully enclosed methods are also effective (Nilsen et al., 2017), and commercial trials are proliferating worldwide with some Norwegian examples: Akvafuture system (akvafuture.com), the Ecocage (ecomerden.com), Aquatraz cage (aquatraz.com) and the extended smolt farm Preline (preline.no). The competitiveness of these methods is highly dependent on production costs – closed containment is almost guaranteed to prevent infestations, but maintaining sufficient oxygenation requires stocking much lower densities of fish or energy-intensive pumping of oxygen or oxygenated seawater. In the extreme, lice can be entirely avoided by farming in land-based facilities, but with dramatically higher energy and land-use requirements.

22.5 Future development and research needs

The future of lice management may be best centred around a combination of physical and non-physical prevention and continuous control methods, ideally with multiple methods selected according to site-specific conditions and deployed in combination. The main argument for a focus on prevention and continuous control is that the industry benefits when louse densities are maintained at very low levels (that is, below mandated trigger levels), thus reducing infestation pressure for all sites and reducing the required expenditure on louse control (Kragesteen et al., 2019; Sandvik et al., 2021). This broad aim is currently hindered by a reliance on immediate delousing methods, in that farmers must balance the wish to keep lice levels low against the considerable financial and animal welfare costs of frequent delousing operations.

By contrast, physical prevention methods such as lice barriers are less stressful and can be left in place for an entire production cycle (Geitung et al., 2019), while non-invasive continuous delousing methods may be able to remove lice that do succeed in attaching to the salmon, allowing sites to constantly maintain low lice densities. The use of multiple methods will also be important in reducing directional selection for resistant lice traits. Lice have already developed resistance to chemical treatments (Aaen et al., 2015), are likely developing resistance to thermal and freshwater delousing (Ljungfeldt et al., 2017), and may have the capacity to develop resistance to depth-based preventions such as skirts, snorkels or submerged cages (Coates et al., 2020). Thus, a management approach that utilises multiple tools could reduce the rate of potential adaption by the parasites.

A key research pillar will be to properly document and mitigate animal welfare risks associated with current and future lice management strategies. Novel lice control methods are eagerly adopted by the industry, often without rigorous testing prior to widespread deployment (Brakstad et al., 2019), and welfare concerns may not be documented until much later (Nilsson et al., 2019; Stien et al., 2020a). A central issue will be mitigating the adverse environmental conditions that may occur in the salmon farms of the future, including (i) loss of buoyancy over time when surface access is restricted, as in submerged cages (Oppedal et al., 2020), (ii) exposure to unsuitable temperatures when the depth range is restricted and salmon cannot choose a preferred position in the thermocline (Oppedal et al., 2011), and (iii) low oxygen conditions within skirt cages or fully enclosed cages (Stien et al., 2012, 2018). Additionally, it will be necessary to understand and improve the effect of various management strategies when used in combination, especially with physical preventions such as barriers and deep cages that alter the working environment for continuous delousing methods.

For example, snorkel cages likely reduce the effect of cleaner fish (Gentry et al., 2020), and perhaps interfere with optical methods (Bui et al., 2020b). In the future, it may be that all farms are at least partially enclosed by lice barriers. Fully enclosed cages are ~100% effective at excluding sea lice while maintaining fish welfare at acceptable levels (Nilsen et al., 2017, 2020) but, because of the need to pump water, can be at a disadvantage in a marketplace where regulations allow salmon to be produced cheaply in open sea cages. Regulation, together with technological innovation, could play a significant role in directing future developments in lice management (for example, Myksvoll et al., 2018; Stien et al., 2020b).

References

á Norði, G., Simonsen, K., Danielsen, E., Eliasen, K., Mols-Mortensen, A., Christiansen, D.H., Stingrund, P., Galbraith, M. and Patursson, Ø. (2015) Abundance and distribution of planktonic *Lepeophtheirus salmonis* and *Caligus elongatus* in a fish farming region in the Faroe Islands. *Aquaculture Environment Interactions* 7, 15–27.

Aaen, S.M., Helgesen, K.O., Bakke, M.J., Kaur, K. and Horsberg, T.E. (2015) Drug resistance in sea lice: a threat to salmonid aquaculture. *Trends in Parasitology* 31, 72–81.

Abolofia, J., Asche, F. and Wilen, J.E. (2017) The cost of lice: quantifying the impacts of parasitic sea lice on farmed salmon. *Marine Resource Economics* 32, 329–349.

Ang, K.P. and Petrell, R.J. (1998) Pellet wastage, and subsurface and surface feeding behaviours associated with different feeding systems in sea cage farming of salmonids. *Aquacultural Engineering* 18, 95–115.

Asplin, L., Boxaspen, K. and Sandvik, A.D. (2011) Modelling the distribution and abundance of planktonic larval stages of *Lepeophtheirus salmonis* in Norway. In: Jones, S. and Beamish, R. (eds) *Salmon Lice: An Integrated Approach to Understanding Parasite Abundance and Distribution.* John Wiley & Sons, Chichester, UK. pp. 29–50.

Arctic Offshore Farming Available at: www.arcticoffshorefarming.no (accessed 26 January 2022).

Atlantis Subsea Farming (2021) Submersible fish-farming facilities for salmon. Available at: www.atlantisfarming.no (accessed 26 January 2022).

Barrett, L.T., Pert, C.G., Bui, S., Oppedal, F. and Dempster, T. (2019) Sterilization of sea lice eggs with ultraviolet C light: towards a new preventative technique for aquaculture. *Pest Management Science* 76, 901–906.

Barrett, L.T., Bui, S., Oppedal, F., Bardal, T., Olsen, R.E. and Dempster, T. (2020a) Ultraviolet-C light suppresses reproduction of sea lice but has adverse effects on host salmon. *Aquaculture* 520, 734954.

Barrett, L.T., Oppedal, F., Robinson, N. and Dempster, T. (2020b) Prevention not cure: a review of methods to avoid sea lice infestations in salmon aquaculture. *Reviews in Aquaculture* 12, 2527–2543.

Barrett, L., Overton, K., Stien, L.H., Oppedal, F. and Dempster, T. (2020c) Effect of cleaner fish on sea lice in Norwegian salmon aquaculture: a national-scale data analysis. *International Journal for Parasitology* 50, 787–796.

Bergheim, A., Gausen, M., Næss, A., Hølland, P.M., Krogedal, P. and Crampton, V. (2006) A newly developed oxygen injection system for cage farms. *Aquacultural Engineering* 34, 40–46.

Berillis, P., Mente, E., Nikouli, E., Makridis, P., Grundvig, H., Bergheim, A. and Gausen, M. (2016) Improving aeration for efficient oxygenation in sea bass sea cages. Blood, brain and gill histology. *Open Life Sciences* 11, 270–279.

Brakstad, O.M., Hagspiel, V., Lavrutich M.N. and Matanovic, D. (2019) Optimal investment decisions in lice-fighting technologies: a case study in Norway. *Aquaculture* 504, 300–313.

Bratland, S., Stien. L.H., Braithwaite, V., Juell, J-E., Folkedal, O., Nilsson, J., Oppedal, F. and Kristiansen, T.S. (2010) From fright to anticipation: using aversive light stimuli to investigate reward conditioning in large groups of Atlantic salmon (*Salmo salar* L.). *Aquaculture International* 18, 991–1001.

Bricknell, I.R., Dalesman, S.J., O'Shea, B., Pert, C.C. and Mordue Luntz, A.J. (2006) Effect of environmental salinity on sea lice *Lepeophtheirus salmonis* settlement success. *Diseases of Aquatic Organisms* 71, 201–212.

Bromage, N., Porter, M. and Randall, C. (2001) The environmental regulation of maturation in farmed finfish with special reference to the role of photoperiod and melatonin. *Aquaculture* 197, 63–98.

Bron, J. and Sommerville, C. (1998) The functional and comparative morphology of the photoreceptors of the copepodid larva of the salmon louse *Lepeophtheirus salmonis* (Krøyer, 1837) (Crustacea: Copepoda: Caligidae). *Zoologischer Anzeiger* 237, 113–126.

Bron J.E., Sommerville C. and Rae G.H. (1993) Aspects of the behaviour of copepodid larvae of the salmon louse *Lepeophtheirus salmonis* (Krøyer, 1837). In: Boxshall, G.A. and Defaye, D. (eds) *Pathogens of Wild and Farmed Fish: Sea Lice.* Ellis Horwood Ltd, Chichester, UK. pp. 125–142.

Browman, H.I., Boxaspen, K. and Kuhn, P. (2004) The effect of light on the settlement of the salmon louse, *Lepeophtheirus salmonis*, on Atlantic salmon, *Salmo salar* L. *Journal of Fish Diseases* 27, 701–708.

Bui, S., Oppedal, F., Korsøen, Ø. and Dempster, T. (2013) Modifying Atlantic salmon behaviour with light or feed stimuli may improve parasite control techniques. *Aquaculture Environment Interactions* 3, 125–133.

Bui, S., Oppedal, F., Stien, L.H. and Dempster, T. (2016) Sea lice infestation level alters salmon swimming depth in sea-cages. *Aquaculture Environment Interactions* 8, 429–435.

Bui, S., Oppedal, F., Sievers, M. and Dempster, T. (2019) Behaviour in the toolbox to outsmart parasites and improve fish welfare in aquaculture. *Reviews in Aquaculture* 11, 168–186.

Bui, S., Stien, L.H., Nilsson, J., Trengereid, H. and Oppedal, F. (2020a) Efficiency and welfare impact of long-term simultaneous in situ management strategies for salmon louse reduction in commercial sea cages. *Aquaculture* 520, 734934.

Bui, S., Geitung, L., Oppedal, F. and Barrett, L.T. (2020b) Salmon lice survive the straight shooter: a commercial scale sea cage trial of laser delousing. *Preventive Veterinary Medicine* 181, 105063.

Burridge, L., Weis, J.S., Cabello, F., Pizarro, J. and Bostick, K. (2010) Chemical use in salmon aquaculture: a review of current practices and possible environmental effects. *Aquaculture* 306, 7–23.

Byrne, A.A., Pearce, C.M., Cross, S.F., Jones, S.R., Robinson, S.M., Hutchinson, M.J., Miller, M.R., Haddad, C.A. and Johnson, D.L. (2018) Field assessment of Pacific oyster (*Crassostrea gigas*) growth and ingestion of planktonic salmon louse (*Lepeophtheirus salmonis*) larvae at an Atlantic salmon (*Salmo salar*) farm in British Columbia, Canada. *Aquaculture* 490, 53–63.

Calabrese, S., Nilsen, T.O., Jolarevic, J., Ebbesson, L.O.E., Pedrosa, C., Fivelstad, S., Hosfeld, C., Stefansson, S.O., Terjesen, B.F., Takle, H., Martins, C.I.M., Sveier, H., Mathisen, F., Imsland, A.K. and Handeland, S.O. (2017) Stocking density limits for post-smolt Atlantic salmon (*Salmo salar* L.) with emphasis on production performance and welfare. *Aquaculture* 468, 363–370.

Coates, A., Phillips, B.L., Oppedal, F., Bui, S., Overton, K. and Dempster, T. (2020) Parasites under pressure: salmon lice have the capacity to adapt to depth-based preventions in aquaculture. *International Journal for Parasitology* 50, 865–872.

Costello, M.J. (2009) The global economic cost of sea lice to the salmonid farming industry. *Journal of Fish Diseases* 32, 115–118.

Crosbie T., Wright, D.W., Oppedal, F., Johnsen, I., Samsing, F. and Dempster, T. (2019) Effects of step salinity gradients on salmon lice larvae behaviour and dispersal. *Aquaculture Environment Interactions* 11, 181–190.

Crosbie, T., Wright, D.W., Oppedal, F., Dalvin, S., Myksvoll, M.S. and Dempster, T. (2020) Impact of thermoclines on the vertical distribution of salmon lice larvae. *Aquaculture Environment Interactions* 12, 1–10.

Davidsen, J.G., Plantalech Manel-la, N., Økland, F., Diserud, O.H., Thorstad, E.B., Finstad, B., Sivertsgård, R., McKinley, R.S. and Rikardsen, A.H. (2008) Changes in swimming depths of Atlantic salmon *Salmo salar* post-smolts relative to light intensity. *Journal of Fish Biology* 73, 1065–1074.

Dempster, T., Juell, J.-E., Fosseidengen, J.E., Fredheim, A. and Lader, P. (2008) Behaviour and growth of Atlantic salmon (*Salmo salar* L.) subjected to short-term submergence in commercial scale sea-cages. *Aquaculture* 276, 103–111.

Dempster, T., Korsøen, O., Folkedal, O., Juell, J.-E. and Oppedal, F. (2009) Submergence of Atlantic salmon (*Salmo salar* L.) in commercial scale sea-cages: a potential short-term solution to poor surface conditions. *Aquaculture* 288, 254–263.

Directorate of Fisheries (2020) Statistics for aquaculture: cleaner fish (lumpfish and wrasse). Available at: https://www.fiskeridir.no/English/Aquaculture/Statistics/Cleanerfish-Lumpfish-and-Wrasse (accessed 15 May 2019).

Einarsson, S.M., Guðjónsson, S., Jónsson, I.R. and Guðbrandsson, J. (2018) Deep-diving of Atlantic salmon (*Salmo salar*) during their marine feeding migrations. *Environmental Biology of Fishes* 101, 1707–1715.

Eldøy, S.H., Davidsen, J.G., Thorstad, E.B., Whoriskey, F.G., Aarestrup, K., Næsje, T.F., Rønning, L., Rikardsen, A.H. and Arnekleiv, J.V. (2017) Marine depth use of sea trout *Salmo trutta* in fjord areas of central Norway. *Journal of Fish Biology* 91, 1268–1283.

Endal, H.P., Taranger, G.L., Stefansson, S.O. and Hansen, T. (2000) Effects of continuous additional light on growth and sexual maturity in Atlantic salmon, *Salmo salar*, reared in sea cages. *Aquaculture* 191, 337–349.

Endo, A., Srithongouthai, S., Nashiki, H., Teshiba, I., Iwasaki, T., Hama, D. and Tsutsumi, H. (2008) DO-increasing effects of a microscopic bubble generating system in a fish farm. *Marine Pollution Bulletin* 57, 78–85.

Fänge, R. (1953) The mechanism of gas transport in the euphysoclist swimbladder. *Acta Physiologica Scandinavia* 30, 1–133.

Fernö, A., Huse, I., Juell, J.-E. and Bjordal, Å. (1995) Vertical distribution of Atlantic salmon (*Salmo salar* L.) in net pens: trade-off between surface light avoidance and food attraction. *Aquaculture* 132, 285–296.

Fields, D.M., Skiftesvik, A.B. and Browman, H.I. (2018) Behavioural responses of infective-stage copepodids of the salmon louse (*Lepeophtheirus salmonis*, Copepoda: Caligidae) to host-related sensory cues. *Journal of Fish Diseases* 41, 875–884.

Fjelldal, P.G., Hansen, T.J. and Karlsen, Ø. (2020) Effects of laboratory salmon louse infection on osmoregulation, growth, and survival in Atlantic salmon. *Conservation Physiology* 8, coaa023.

Føre, M., Dempster, T., Alfredsen, J.A. and Oppedal, F. (2013) Modelling of Atlantic salmon (*Salmo salar* L.) behaviour in sea-cages: using artificial light to control swimming depth. *Aquaculture* 137–146, 388–391.

Frank, K., Gansel, L., Lien, A. and Birkevold, J. (2015) Effects of a shielding skirt for prevention of sea lice on the flow past stocked salmon fish cages. *Journal of Offshore Mechanics and Arctic Engineering* 137, 011201.

Frenzl, B., Stien, L.H., Cockerill, D., Oppedal F., Richards, R.H., Shinn, A.P., Bron, J.E. and Migaud, H. (2014) Manipulation of farmed Atlantic salmon swimming behaviour through the adjustment of lighting and feeding regimes as a tool for salmon lice control. *Aquaculture* 183–188, 424–425.

Furevik, D.M., Bjordal, Å., Huse, I. and Fernö, A. (1993) Surface activity of Atlantic salmon (*Salmo salar* L.) in net pens. *Aquaculture* 110, 119–128.

Geitung, L., Oppedal, F., Stien, L.H., Dempster, T., Karlsbakk, E., Nola, V. and Wright, D.W. (2019) Snorkel sea-cage technology decreases salmon lice infestation by 75% in a full-cycle commercial test. *International Journal for Parasitology* 49, 843–846.

Geitung, L., Wright, D.W., Oppedal, F., Stien, L.H., Vågseth, T. and Madaro, A. (2020) Cleaner fish behaviour and welfare in Atlantic salmon sea-cages during an autumn-winter stratified scenario. *Aquaculture* 528, 735623.

Geitung, L., Wright, D.W., Stien, L.H., Oppedal, F. and Karlsbakk, E. (2021) Tapeworm (*Eubothrium* sp.) infection decreased in lice barrier snorkel sea cages during commercial-scale study. *Aquaculture* 541, 736774.

Genna, R.L., Mordue, W., Pike, A.W. and Mordue, A.J. (2005) Light intensity, salinity, and host velocity influence presettlement intensity and distribution on hosts by copepodids of sea lice, *Lepeophtheirus salmonis*. *Canadian Journal of Fisheries and Aquatic Sciences* 62, 2675–2682.

Gentry, K., Bui, S., Oppedal, F. and Dempster, T. (2020) Sea lice prevention strategies affect cleaner fish delousing efficacy in commercial Atlantic salmon sea cages. *Aquaculture Environment Interactions* 12, 67–80.

Glaropoulos, A., Stien, L.H., Folkedal, O., Dempster, T. and Oppedal, F. (2019) Welfare, behaviour and feasibility of farming Atlantic salmon in submerged cages with weekly surface access to refill their swim bladders. *Aquaculture* 502, 332–337.

Gravil, H.R. (1996) Studies on the biology and ecology of the free swimming larval stages of *Lepeophtheirus salmonis* (Krøyer, 1838) and *Caligus elongatus* Nordmann, 1832 (*Copepoda: Caligidae*). PhD thesis, University of Stirling, UK.

Grøntvedt, R.N., Kristoffersen, A.B. and Jansen, P.A (2018) Reduced exposure of farmed salmon to salmon louse (*Lepeophtheirus salmonis* L.) infestation by use of plankton nets: estimating the shielding effect. *Aquaculture* 495, 865–872.

Hannisdal, R., Nøstbakken, O.J., Hove, H., Madsen, L., Horsberg, T.E. and Lunestad, B.T. (2020) Anti-sea lice agents in Norwegian aquaculture; surveillance, treatment trends and possible implications for food safety. *Aquaculture* 521, 735044.

Hansen, T., Stefansson, S. and Taranger, G.L. (1992) Growth and sexual maturation in Atlantic salmon, *Salmo salar* L., reared in sea cages at two different light regimes. *Aquaculture Research* 23, 275–280.

Hansen, T.J., Fjelldal, P.G., Folkedal, O., Vågseth, T. and Oppedal, F. (2017) Effects of light source and intensity on sexual maturation, growth and swimming behaviour of Atlantic salmon in sea cages. *Aquaculture Environment Interactions* 9, 193–204.

Harbor Fence. Electrical fence. Available at: www.harbor.no (accessed 31 January 2022).

Hays, G.C. (2003) A review of the adaptive significance and ecosystem consequences of zooplankton diel vertical migrations. *Hydrobiologia* 503, 163–170.

Hedger, R.D., Rikardsen, A.H., Strøm, J.F., Righton, D.A., Thorstad, E.B. and Næsje, T.F. (2017) Diving behaviour of Atlantic salmon at sea: effects of light regimes and temperature stratification. *Marine Ecology Progress Series* 574, 127–140.

Heuch, P.A. (1995) Experimental evidence for aggregation of salmon louse copepodids (*Lepeophtheirus salmonis*) in step salinity gradients. *Journal of the Marine Biological Association of the UK* 75, 927–939.

Heuch, P.A., Parsons, A. and Boxaspen, K. (1995) Diel vertical migration: a possible host-finding mechanism in salmon louse (*Lepeophtheirus salmonis*) copepodids? *Canadian Journal of Fisheries and Aquatic Sciences* 52, 681–689.

Heuch, P.A., Doall, M.H. and Yen, J. (2007) Water flow around a fish mimic attracts a parasitic and deters a planktonic copepod. *Journal of Plankton Research* 29, i3–i16.

Hevrøy, E., Boxaspen, K., Oppedal, F., Taranger G.L. and Holm, J.C. (2003) The effect of artificial light treatment and depth on the infestation of the sea louse *Lepeophtheirus salmonis* on Atlantic salmon (*Salmo salar* L.) culture. *Aquaculture* 220, 1–14.

Hosteland, L.T.S. (2016) Had to slaughter out the Ecomerd (In Norwegian: Måtte slakte ut Ecomerden). Available at: https://www.kyst.no/article/maatte-slakte-ut-ecomerden/ (accessed January 2022).

Huse, I. and Holm, J.C. (1993) Vertical distribution of Atlantic salmon (*Salmo salar*) as a function of illumination. *Journal of Fish Biology* 43, 147–156.

Ingvarsdóttir, A., Birkett, M.A., Duce, I., Genna, R.L., Mordue, W., Pickett, J.A., Wadhams, L.J. and Mordue, A.J. (2002) Semiochemical strategies for sea louse control: host location cues. *Pest Management Science* 58, 537–545.

Iversen, A., Hermansen, Ø., Andreassen, O., Brandvik, R.K., Marthinussen, A. and Nystøyl, R. (2015) Cost drivers in salmon farming. NOFIMA report 41/2015, Tromsø, Norway.

Iversen, A. Asche, F., Hermansen, Ø. and Nystøyl, R. (2020) Production cost and competitiveness in major salmon farming countries 2003–2018. *Aquaculture* 522, 735089.

Johansson, D., Juell, J.-E., Oppedal, F., Stiansen, J.-E. and Ruohonen, K. (2007) The influence of the pycnocline and cage resistance on current flow, oxygen flux and swimming behaviour of Atlantic salmon (*Salmo salar* L.) in production cages. *Aquaculture* 265, 271–287.

Johansson, D., Ruohonen, K., Juell, J.-E. and Oppedal, F. (2009) Swimming depth and thermal history of individual Atlantic salmon (*Salmo salar* L.) in production cages under different ambient temperature conditions. *Aquaculture* 290, 296–303.

Johnsen, I.A., Fiksen, Ø., Sandvik, A.D. and Asplin, L. (2014) Vertical salmon lice behaviour as a response to environmental conditions and its influence on regional dispersion in a fjord system. *Aquaculture Environment Interactions* 5, 127–141.

Juell, J.-E. (1995) The behaviour of Atlantic salmon in relation to efficient cage-rearing. *Reviews in Fish Biology and Fisheries* 5, 320–335.

Juell, J.-E. and Fosseidengen, J.E. (2004) Use of artificial light to control swimming depth and fish density of Atlantic salmon (*Salmo salar*) in production cages. *Aquaculture* 233, 269–282.

Juell, J.-E., Fernö, A., Furevik, D. and Huse, I. (1994) Influence of hunger level and food availability on the spatial distribution of Atlantic salmon, *Salmo salar* L., in sea cages. *Aquaculture Research* 25, 439–451.

Juell, J.-E., Oppedal, F., Boxaspen, K. and Taranger, G.L. (2003) Submerged light increases swimming depth and reduces fish density of Atlantic salmon *Salmo salar* L. in production cages. *Aquaculture Research* 34, 469–478.

Klebert, P. and Su, B. (2020) Turbulence and flow field alterations inside a fish sea cage and its wake. *Applied Ocean Research* 98, 102113.

Korsøen Ø., Dempster, T., Fjelldal P.G., Oppedal, F. and Kristiansen, T. (2009) Long-term culture of Atlantic salmon (*Salmo salar* L.) in submerged cages during winter affects behaviour, growth and condition. *Aquaculture* 296, 373–381.

Kragesteen, T.J., Simonsen, K., Visser, A.W. and Andersen, K.H. (2019) Optimal salmon lice treatment threshold and tragedy of the commons in salmon farm networks. *Aquaculture* 512, 734329.

Kreitzman, M., Ashander, J., Driscoll, J., Bateman, A.W., Chan, K.M.A., Lewis, M.A. and Krkošek, M. (2017) Wild salmon sustain the effectiveness of parasite control on salmon farms: conservation implications from an evolutionary ecosystem service. *Conservation Letters* 11, 1–13.

Kristiansen, D., Endresen, P.C., Lader, P., Su, B., Volent, Z. and Aksnes, V. (2018) Sjøegenskaper og forankring til flytende lukkede oppdrettsanlegg. Sluttrapport: SJØFLO, Rapportnr. 2018, 00191. 15 February 2018. SINTEF Ocean, Trondheim, Norway.

Krkošek, M., Revie, C.W., Gargan, P.G., Skilbrei, O.T., Finstad, B. and Todd, C.D. (2013) Impact of parasites on salmon recruitment in the Northeast Atlantic Ocean. *Proceedings of the Royal Society B* 280, 20122359.

Lader, P., Dempster, T., Fredheim, A. and Jensen, Ø. (2008) Current induced net deformation in full-scale sea-cages for Atlantic salmon (*Salmo salar*). *Aquacultural Engineering* 38, 52–65.

Leclercq, E., Taylor, J.F., Sprague, M. and Migaud, H. (2011) The potential of alternative lighting-systems to suppress preharvest sexual maturation of 1+ Atlantic salmon (*Salmo salar*) post-smolts reared in commercial sea-cages. *Aquacultural Engineering* 44, 35–47.

Liu, Y. and Bjelland, H.V. (2014) Estimating costs of sea lice control strategy in Norway. *Preventive Veterinary Medicine* 117, 469–477.

Ljungfeldt, L.E.R., Quintela, M., Besnier, F., Nilsen, F. and Glover, K.A. (2017) A pedigree-based experiment reveals variation in salinity and thermal tolerance in the salmon louse, *Lepeophtheirus salmonis*. *Evolutionary Applications* 10, 1007–1019.

Macaulay, G., Wright, D.W., Oppedal, F. and Dempster, T. (2020a) Buoyancy matters: establishing the maximum neutral buoyancy depth of Atlantic salmon. *Aquaculture* 519, 734925.

Macaulay, G., Bui, S., Oppedal, F. and Dempster, T. (2020b) Acclimating salmon as juveniles prepares them for a farmed life in sea-cages. *Aquaculture* 523, 735227.

Macauley, G., Bui, S., Oppedal, F. and Dempster, T. (2020c) Challenges and benefits of applying fish behaviour to improve production and welfare in industrial aquaculture. *Reviews in Aquaculture* 1–15.

Migaud, H., Cowan, M., Taylor, J. and Ferguson, H.W. (2007) The effect of spectral composition and light intensity on melatonin, stress and retinal damage in post-smolt Atlantic salmon, *Salmo salar*. *Aquaculture* 270, 390–404.

Montory, J.A., Chaparro, O.R., Averbuj, A., Salas-Yanquin, L.P., Buchner-Miranda, J.A., Gebauer, P., Cumillaf, J.P. and Cruces, E. (2020) The filter-feeding bivalve *Mytilus chilensis* capture pelagic stages of *Caligus rogercresseyi*: a potential controller of the sea lice fish parasites. *Journal of Fish Diseases* 43, 475–484.

Mordue (Luntz), A.J. and Birkett, M.A. (2009) A review of host finding behaviour in the parasitic sea louse, *Lepeophtheirus salmonis* (Caligidae: Copepoda). *Journal of Fish Diseases* 32, 3–13.

Munro, L.A. (2020) *Scottish Fish Farm Production Survey 2019*. Scottish Government statistics, Scotland.

Myhre Jensen, E., Horsberg, T.E., Sevatdal, S. and Helgesen, K.O. (2020) Trends in de-lousing of Norwegian farmed salmon from 2000–2019 – consumption of medicines, salmon louse resistance and non-medicinal control methods. *PLoS ONE* 15, e0240894.

Myksvoll, M.S., Sandvik, A.D., Albretsen, J., Asplin, L., Johnsen, I.A., Karlsen, Ø., Kristensen, N.M., Melsom, A., Skardhamar, J. and Ådlandsvik, B. (2018) Evaluation of a national operational salmon lice monitoring system – from physics to fish. *PLoS ONE* 13, e0201338.

Nelson, E.J., Robinson, S.M.C., Feindel, N., Sterling, A., Byrne, A. and Pee Ang, K. (2017) Horizontal and vertical distribution of sea lice larvae (*Lepeophtheirus salmonis*) in and around salmon farms in the Bay of Fundy, Canada. *Journal of Fish Diseases* 41, 885–899.

Nilsen, A., Nielsen, K.V., Biering, E. and Bergheim, A. (2017) Effective protection against sea lice during the production of Atlantic salmon in floating enclosures. *Aquaculture* 466, 41–50.

Nilsen, A., Hagen, Ø., Johnsen, C.A., Prytz, H., Zhou, B., Nielsen, K.V. and Bjørnevik, M. (2019) The importance of exercise: increased water velocity improves growth of Atlantic salmon in closed cages. *Aquaculture* 501, 537–546.

Nilsen, A., Nielsen, K.V. and Bergheim, A. (2020) A closer look at closed cages: growth and mortality rates during production of post-smolt Atlantic salmon in marine closed confinement systems. *Aquacultural Engineering* 91, 102124.

Nilsson, J., Moltumyr, L., Madaro, A., Kristiansen, T.S., Gåsnes, S.K., Mejdell, C.M., Gismervik, K. and Stien, L.H. (2019) Sudden exposure to warm water causes instant behavioural responses indicative of nociception or pain in Atlantic. *Veterinary and Animal Science* 8, 100076.

Nordtug, T., Kvæstad, B. and Hagemann, A. (2020) Responses and preferences of salmon louse (*Lepeophtheirus salmonis* Krøyer 1836) copepodids to underwater artificial light sources. *Aquaculture* 532, 736036.

Novales Flamarique, I., Browman, H.I., Belanger, M. and Boxaspen, K. (2000) Ontogenetic changes in visual sensitivity of the parasitic salmon louse *Lepeophtheirus salmonis*. *Journal of Experimental Biology* 203, 1649–1657.

Novales Flamarique, I., Vinebrooke, R., Gulbransen, C., Galbraith, M. and Stucchi, D. (2009) Monitoring and potential control of sea lice using an LED-based light trap. *Canadian Journal of Fisheries and Aquatic Sciences* 66, 1371–1382.

O'Donohoe, P. and McDermott, T. (2014) Reducing sea lice re-infestation risk from harvest water at a salmon farm site in Ireland using a bespoke sieving and filtration system. *Aquacultural Engineering* 60, 73–76.

Oldham, T., Dempster, T., Fosse, J.O. and Oppedal, F. (2017) Oxygen gradients affect behavior of caged Atlantic salmon *Salmo salar*. *Aquaculture Environment Interactions* 9, 145–153.

Oppedal, F., Taranger, G.L., Juell, J.-E. Fosseidengen, J.E. and Hansen, T. (1997) Light intensity affects growth and sexual maturation of Atlantic salmon (*Salmo salar*) postsmolts in sea cages. *Aquatic Living Resources* 10, 351–357.

Oppedal, F., Juell, J.-E., Taranger, G.L. and Hansen, T. (2001) Artificial light and season affects vertical distribution and swimming behaviour of post-smolt Atlantic salmon in sea cages. *Journal of Fish Biology* 58, 1570–1584.

Oppedal, F., Berg, A., Olsen, R.E., Taranger, G.L. and Hansen, T. (2006) Photoperiod in seawater influence seasonal growth and chemical composition in autumn sea-transferred Atlantic salmon (*Salmo salar* L.) given two vaccines. *Aquaculture* 254, 396–410.

Oppedal, F., Juell, J.-E. and Johansson, D. (2007) Thermo-and photoregulatory swimming behaviour of caged Atlantic salmon: implications for photoperiod management and fish welfare. *Aquaculture* 265, 70–81.

Oppedal, F., Dempster, T. and Stien, L.H. (2011) Environmental drivers of Atlantic salmon behaviour in sea-cages: a review. *Aquaculture* 311, 1–18.

Oppedal, F., Samsing, F., Dempster, T., Wright, D.W., Bui, S. and Stien, L.H. (2017) Sea lice infestation levels decrease with deeper "snorke" barriers in Atlantic salmon sea-cages. *Pest Management Science* 73, 1935–1943.

Oppedal, F., Bui, S., Stien, L.H., Overton, K. and Dempster, T. (2019) Snorkel technology to reduce sea lice infestations: efficacy depends on salinity at the farm site, but snorkels have minimal effects on salmon production and welfare. *Aquaculture Environment Interactions* 11, 445–457.

Oppedal, F., Folkedal, O., Stien, L.H., Vågseth, T., Fosse, J.O., Dempster, T. and Warren-Myers, F. (2020) Atlantic salmon cope in submerged cages when given access to an air dome that enables fish to maintain neutral buoyancy. *Aquaculture* 525, 735286.

Øverli, Ø., Nordgreen, J., Mejdell, C.M., Janczak, A.M., Kittilsen, S., Johansen, I.B. and Horsberg, T.E. (2014) Ectoparasitic sea lice (*Lepeophtheirus salmonis*) affect behavior and brain serotonergic activity in Atlantic salmon (*Salmo salar* L.): perspectives on animal welfare. *Physiology & Behavior* 132, 44–50.

Overton, K., Dempster, T., Oppedal, F., Kristiansen, T.S., Gismervik, K. and Stien, L.H. (2019) Salmon lice treatments and salmon mortality in Norwegian aquaculture: a review. *Reviews in Aquaculture* 11, 1398–1417.

Øvrebø, T.K. (2020) Growth performance and welfare of post-smolt (*Salmo salar* L.) reared in semi closed containment systems (S-CCS) – a comparative study. Master's thesis, University of Bergen, Norway.

Øvrelid, M.S. (2017) Characterization of planktonic sea lice distribution and association to fish farm installations. Master's thesis, Norwegian University of Science and Technology, Norway.

Porter, M.J.R., Duncan, N.J., Mitchell, D. and Bromage, N.R. (1999) The use of cage lighting to reduce plasma melatonin in Atlantic salmon (*Salmo salar*) and its effects on the inhibition of grilsing. *Aquaculture* 176, 237–244.

Penston, M.J., McKibben, M.A., Hay, D.W. and Gillibrand, P.A. (2004) Observations on open-water densities of sea lice larvae in Loch Shieldaig, Western Scotland. *Aquaculture Research* 35, 793–805.

Rikardsen, A.H., Diserud, O.H., Elliott, J.M., Dempson, J.B., Sturlaugsson, J. and Jensen, A.J. (2007) The marine temperature and depth preferences of Arctic charr (*Salvelinus alpinus*) and sea trout (*Salmo trutta*), as recorded by data storage tags. *Fisheries Oceanography* 16, 436–447.

Schram, T.A. (2004) Practical identification of pelagic sea lice larvae. *Journal of the Marine Biological Association of the United Kingdom* 84, 103–110.

Sandvik, A.D., Bui, S., Huserbråten, M., Karlsen, Ø., Myksvoll, M., Ådlandsvik, B. and Johnsen, I. (2021) The development of a sustainability assessment indicator and its response to management changes as derived from salmon lice dispersal modelling. *ICES Journal of Marine Science* 78, 1781–1792.

Samsing, F., Oppedal, F., Dalvin, S., Johnsen, I., Vågseth, T. and Dempster, T. (2016a) Salmon lice (*Lepeophtheirus salmonis*) development times, body size, and reproductive outputs follow universal models of temperature dependence. *Canadian Journal of Fisheries and Aquatic Sciences* 73, 1841–1851.

Samsing, F., Johnsen, I., Stien, L.H., Oppedal, F., Albretsen, J., Asplin, L. and Dempster, T. (2016b). Predicting the effectiveness of depth-based technologies to prevent salmon lice infection using a dispersal model. *Preventive Veterinary Medicine* 129, 48–57.

Sievers, M., Korsøen, Ø., Dempster, T., Fjelldal, P.G., Kristiansen, T., Folkedal, O. and Oppedal, F. (2018) Growth and welfare of submerged Atlantic salmon under continuous lighting. *Aquaculture Environment Interactions* 10, 501–510.

Sievers, M., Oppedal, F., Ditria, E. and Wright, D.W. (2019) The effectiveness of hyposaline treatments against host-attached salmon lice. *Scientific Reports* 9, 6976.

Skern-Mauritzen, R., Sissener, N.H., Sandvik, A.D., Meier, S., Sævik, P.N., Skogen, M.D., Vågseth, T., Dalvin, S., Skern-Mauritzen, M. and Bui, S. (2020) Parasite development affect dispersal dynamics; infectivity, activity and energetic status in cohorts of salmon louse copepodids. *Journal of Experimental Marine Biology and Ecology* 530–531, 151429.

Skjelvareid, M.H., Breiland, M.S.W. and Mortensen, A. (2018) Ultrasound as potential inhibitor of salmon louse infestation – a small-scale study. *Aquaculture Research* 49, 2684–2692.

Solé, M., Lenoir, M., Fortuño, J.-M., De Vreese, S., van der Schaar, M. and André, M. (2021) Sea lice are sensitive to low frequency sounds. *Journal of Marine Science and Engineering* 9, 765.

Solstorm, D., Oldham, T., Solstorm, F., Klebert, P., Stien, L.H., Vågseth, T. and Oppedal, F. (2018) Dissolved oxygen variability in a commercial sea-cage exposes farmed Atlantic salmon to growth limiting conditions. *Aquaculture* 486, 122–129.

Solvang-Garten, T., Steen Bondø, M. and Gjøsund, S.H. (2015) Flotation of salmon lice. SINTEF report.

Stien, L.H., Nilsson, J., Oppedal, F., Kristiansen, T.S., Lien, A.M. and Folkedal, O. (2012) Skirt around a salmon sea cage to reduce infestation of salmon lice resulted in low oxygen levels. *Aquacultural Engineering* 51, 21–25.

Stien, L., Fosseidengen, J.E., Malm, M., Sveier, H., Torgersen, T., Wright, D.W. and Oppedal, F. (2014) Low intensity light of different colours modifies Atlantic salmon depth use. *Aquacultural Engineering* 62, 42–48.

Stien, L.H., Dempster, T., Bui, S., Glaropoulos, A., Fosseidengen, J.E., Wright, D. and Oppedal, F. (2016) 'Snorkel' sea lice barrier technology reduces sea lice loads on harvest-sized Atlantic salmon with minimal welfare impacts. *Aquaculture* 458, 29–37.

Stien, L.H., Lind, M.B., Oppedal, F., Wright, D.W. and Seternes, T. (2018) Skirts on salmon production cages reduced salmon lice infestations without affecting fish welfare. *Aquaculture* 490, 281–287.

Stien, L.H., Størkersen, K.V. and Gåsnes, S.K. (2020a) Analysis of mortality data from survey on cleaner fish welfare, *Rapport fra Havforskningen*. Bergen, 2020–6. Available at: https://www.hi.no/hi/nettrapporter/rapport-fra-havforskningen-2020-6. (accessed 31 January 2022).

Stien, L.H., Tørud, B., Gismervik G., Lien, M.E., Medaas, C., Osmundsen, T., Kristiansen, T.S. and Størkersen, K.V. (2020b) Governing the welfare of Norwegian farmed salmon: three conflict cases. *Marine Policy* 117, 103969.

Strøm, J.F., Thorstad, E.B., Hedger, R.D. and Rikardsen, A.H. (2018) Revealing the full ocean migration of individual Atlantic salmon. *Animal Biotelemetry* 6, 2.

Svendsen, E., Dahle, S.W., Hagemann, A., Birkevold, J., Delacroix, S. and Andersen, A.B. (2018) Effect of ultrasonic cavitation on small and large organisms for water disinfection during fish transport. *Aquaculture Research* 49, 1166–1175.

Szetey, A., Wright, D., Oppedal, F. and Dempster, T. (2021) Salmon lice nauplii and copepodids display different vertical migration patterns in response to light. *Aquaculture Environment Interactions* 13, 121–131.

Taranger, G.L., Haux, C., Stefansson, S.O., Björnsson, B.T., Walther, B.T. and Hansen, T. (1998) Abrupt changes in photoperiod affect age at maturity, timing of ovulation and plasma testosterone and oestradiol-17β profiles in Atlantic salmon, *Salmo salar*. *Aquaculture* 162, 85–98.

Taranger, G.L., Carrillo, M., Schulz, R.W., Fontaine, P., Zanuy, S., Felip, A., Weltzien, F.-A., Dufour, S, Karlsen, Ø., Norberg, B., Andersson, E. and Hansen, T. (2010) Control of puberty in farmed fish. *General and Comparative Endocrinology* 165, 483–515.

Thorstad, E.B., Whoriskey, F., Uglem, I., Moore, A., Rikardsen, A.H. and Finstad, B. (2012) A critical life stage of the Atlantic salmon *Salmo salar*: behaviour and survival during the smolt and initial post-smolt migration. *Journal of Fish Biology* 81, 500–542.

Thorstad, E.B., Todd, C.D., Uglem, I., Bjørn, P.A., Gargan, P.G., Vollset, K.W., Halttunen, E., Kålås, S., Berg, M. and Finstad, B. (2015) Effects of salmon lice *Lepeophtheirus salmonis* on wild sea trout *Salmo trutta* – a literature review. *Aquaculture Environment Interactions* 7, 91–113.

Tucker, C.S., Sommerville, C. and Wootten, R. (2000) The effect of temperature and salinity on the settlement and survival of copepodids of *Lepeophtheirus salmonis* (Krøyer, 1837) on Atlantic salmon, *Salmo salar* L. *Journal of Fish Diseases* 23, 309–320.

Vigen, J. (2008) Oxygen variation within a sea cage. Master's thesis, University of Bergen, Norway.

Vollset, K.W., Dohoo, I., Karlsen, Ø., Halttunen, E., Kvamme, B.O., Finstad, B., Wennevik, V., Diserud, O.H., Bateman, A., Friedland, K.D., Mahlum, S., Jørgensen, C., Qviller, L., Krkošek, M., Åtland, Å. and Barlaup, B.T. (2018) Disentangling the role of sea lice on the marine survival of Atlantic salmon. *ICES Journal of Marine Science* 75, 50–60.

Warren-Myers, F., Vågseth, T., Folkedal, O., Stien L.H., Fosse, J.O., Dempster, T. and Oppedal, F. (2022) Full production cycle, commercial scale culture of salmon in submerged sea-cages with air domes reduces lice infestation, but creates production and welfare challenges. *Aquaculture* 548, 737570.

Wright, D.W., Glaropoulos, A., Solstorm, D., Stien, L.H. and Oppedal, F. (2015) Atlantic salmon *Salmo salar* instantaneously follow vertical light movements in sea cages. *Aquaculture Environment Interactions* 7, 61–65.

Wright, D.W., Stien, L.H., Dempster, T., Vågseth, T., Nola, V. and Oppedal, F. (2017) "Snorkel" lice barrier technology reduced two co-occurring parasites, the salmon louse (*Lepeophtheirus salmonis*) and the amoebic gill disease causing agent (*Neoparamoeba perurans*), in commercial salmon sea-cages. *Preventive Veterinary Medicine* 140, 97–105.

Wright, D.W., Nowak, B., Oppedal, F., Crosbie, P., Stien, L.H. and Dempster, T. (2018a) Repeated sublethal freshwater exposures reduce the amoebic gill disease parasite, *Neoparamoeba perurans*, on Atlantic salmon. *Journal of Fish Diseases* 41, 1403–1410.

Wright, D.W., Geitung, L., Karlsbakk, E., Stien, L.H., Dempster, T., Oldham, T., Nola, V. and Oppedal, F. (2018b). Surface environment modification in Atlantic salmon sea-cages: effects on amoebic gill disease, salmon lice, growth and welfare. *Aquaculture Environment Interactions* 10, 255–265.

Wright, D., Stien, L., Dempster, T. and Oppedal, F. (2019) Differential effects of internal tagging depending on depth treatment in Atlantic salmon: a cautionary tale for aquatic animal tag use. *Current Zoology* 65, 665–673.

PART III

Global impacts of sea lice

CHAPTER 23

Impacts of sea lice

Karin Kroon Boxaspen, Ørjan Karlsen, Terje Svåsand and Lars Asplin

Institute of Marine Research, Bergen, Norway

23.1 Introduction

Aquaculture of salmonids, and especially Atlantic salmon, has developed to be a successful industry in several countries. However, environmental challenges have also increased, and have limited further increase of open cage marine aquaculture in many countries (ICES, 2020a). Impacts of salmon lice from aquaculture are seen as one of the major environmental risks for production of salmonids in areas overlapping with the natural habitat of wild salmonids (Taranger et al., 2015; Forseth et al., 2017; ICES 2020a). As an example, in Norway, the largest producing country, the production capacity for Atlantic salmon and rainbow trout is regulated according to the impact from salmon lice on wild post smolt of Atlantic salmon (traffic light system).

Stock sizes and changes for many wild salmonids are poorly described over time, except for Atlantic salmon (*Salmo salar* L.) where a major decrease in the stocks has been seen throughout the North Atlantic (ICES, 2020b) since the 1950s. Open cage aquaculture is the usual production system for Atlantic salmon, and several environmental impacts have been identified, with the spread of sea lice from fish farming being considered as one of the most serious threats to wild salmonids (ICES, 2020a). Since the start of the farming industry in the early 1970s, the parasitic salmon lice (*Lepeophtheirus salmonis* (Krøyer, 1837)) has posed a problem (Brandal et al., 1976; Brandal and Egidius, 1977; Torrissen et al., 2013). Sea lice shed from fish farms may infest wild salmonids. In addition the treatments of lice may potentially have a suite of effects on non-target species, including wild fishes and invertebrates feeding on waste feed or faeces, both directly and indirectly. The effect of sea lice is not restricted to negative impacts on wild salmonids, but also includes effects on the environment from the use of different chemical treatments and the use of chemotherapeutants (see Chapter 14).

23.2 A brief introduction to farming

Marine fish farming is increasing globally. In 2019, production was nearly 5,700,000 metric tonnes, and Atlantic salmon accounted for 45.5% of this production, based on farm gate values. Small

Table 23.1 Production (tonnes) of Atlantic salmon in marine environments 2014–2019 (FAO, 2021).

Country	2014	2015	2016	2017	2018	2019
Norway	1,258,356	1,303,346	1,233,619	1,236,353	1,282,003	1,364,042
Chile	622,992	608,352	531,897	613,611	660,645	701,331
United Kingdom	179,391	172,137	163,132	189,707	156,025	190,500
Canada	86,347	121,926	123,522	120,553	123,184	118,630
Faroe Islands	86,454	80,600	83,300	86,800	78,900	95,000
Australia	41,591	48,331	56,115	52,580	61,227	56,989
Russian Federation	18,675	10,834	12,857	13,016	20,566	32,343
United States of America	18,719	18,719	16,185	14,685	16,107	16,491
Ireland	9368	13,116	16,300	18,342	11,984	11,333
Other	32	28	26	74	60	72
Total	2,321,924	2,377,388	2,236,953	2,345,721	2,410,701	2,586,731

Table 23.2 Production (tonnes) of rainbow trout in marine environments 2014–2019 (FAO, 2021).

Country	2014	2015	2016	2017	2018	2019
Norway	68,910	72,921	87,775	66,902	68,216	83,290
Chile	131,315	94,508	84,213	74,585	78,255	81,714
Finland	10,732	11,740	11,367	11,752	11,119	11,781
Turkey	4812	6187	4643	4972	9235	9411
Denmark	10,733	12,143	11,087	13,036	9737	9188
United Kingdom	1909	4678	3759	3482	3363	3400
Sweden	2485	1919	2424	2856	2870	1518
Other	14	25	285	1236	266	499
Total	230,910	204,121	205,553	178,822	183,060	200,801

quantities of Atlantic salmon (less than 1% of the total production) are also produced in brackish and fresh water (FAO, 2021) and in onshore recirculating aquaculture systems. Atlantic salmon aquaculture has developed to be important for the sea food production in several countries such as Norway, Chile, the UK and Canada with annual marine productions between 119,000 and 1,364,000 metric tonnes in these countries in 2019 (Table 23.1).

Rainbow trout are produced both in fresh water and in marine waters; the marine production has remained at about 200,000 tonnes between 2014 and 2019, where Chile and Norway are the largest producers (Table 23.2).

Farming of salmonid species occurs worldwide and at present produces over 2,500,000 tonnes.

23.3 Sea lice and salmon lice biology

Sea lice are parasites affecting salmonid fish (Costello, 2009b; Hamre et al., 2013). Several sea lice species cause problems for farmed salmonids, and this includes the salmon louse (*L. salmonis*) and various *Caligus* spp., especially *Caligus rogercresseyi* (Boxshall & Bravo 2000) (Carvajal et al., 1998; Bravo, 2003) and *Caligus clemensi* (Parker & Margolis 1964). However, as *Caligus* spp. presently do not seem to be a major problem to farmed and wild salmonids in the Northern Hemisphere, we will focus on *L. salmonis*.

L. salmonis attaches to the body of the salmonids either via the frontal filament in the chalimus stages or as free-living pre-adults and adults. The salmon louse feeds on skin, blood and mucus (Pike and Wadsworth, 1999; Boxaspen, 2006). The life cycle of the salmon louse has eight stages separated by moults (Johnson and Albright, 1991; Hamre et al., 2013) (see Chapter 3). The first three stages of the larval louse are planktonic (the two nauplii stages and the infectious copepod). Over the next five stages the successful louse is attached to a fish (see Chapter 3). In the final stage on the fish, the ovigerous female louse repeatedly produces hundreds of eggs which hatch into the first nauplius stage at intervals, depending on the water temperature, hence closing the life cycle (Brooker et al., 2018).

The first two nauplius stages are non-infective (Johnson and Albright, 1991). The duration of these first two stages is temperature dependent but typically lasts between three and five days, while the duration of the infective copepodite stage is another one to three weeks (Samsing et al., 2016; Hamre et al., 2019). The lice have a massive reproduction ability with hundreds of offspring potentially being produced by each female during her lifespan. The larval lice live for two to three weeks in the planktonic stages and, since the nauplius is not infective, these organisms may spread a long way from their origin. This requires a regional mindset when trying to assess the interaction of salmon lice with the environment. High variability of fjord and coastal currents in both time and space leads to a high geographical variability of the distribution of the planktonic lice (reviewed by Brooker et al., 2018).

23.4 Environmental impact of sea lice on salmonids

Salmon lice infections can negatively affect the welfare of the host fish as the parasite feeds on their skin and blood, leading to general stress, wounds, osmotic stress, impaired immune capacity and higher susceptibility to secondary infections (Grimnes and Jakobsen, 1996; Bjørn and Finstad, 1997, 1998; Finstad et al., 2000; Wagner et al., 2003, 2004; Wells et al., 2006, 2007; Tveiten et al., 2010; Barker et al., 2019; Gallardi et al., 2019). Salmon lice may affect behaviour and swimming capacity of salmonid fish (Øverli et al., 2014; Bui et al., 2016) and increase the risk of being predated (Godwin et al., 2015; Peacock et al., 2015). The pathology caused by salmon lice is mainly associated with the mobile pre-adult and adult stages (Jones et al., 1990; Jónsdóttir et al., 1992; Grimnes and Jakobsen, 1996).

Even though the salmon louse is a natural parasite of salmonids in the Northern Hemisphere, and infects both farmed and wild salmonids, the number of hosts for salmon lice has increased as the farming industry has grown (Heuch and Mo, 2001; Dempster et al., in press). Wild fish populations have always represented a reservoir for salmon lice, while the presence of farmed fish in large numbers is necessary for epidemic situations (Heuch et al., 2005). Even if the number of lice on a farmed fish is limited by various treatments, the total amount of farmed fish together with the high reproduction ability and spreading of offspring by the lice can cause epidemics in wild fish (Skaala et al., 2014; Skarðhamar et al., 2018).

Water currents, temperature, light and salinity affect the dispersion, growth and behaviour of the pelagic salmon louse. The level of information necessary to assess regional distribution of salmon lice copepodites is only achieved from numerical model results. In recent years, several coupled dispersion modelling systems utilising results from hydrodynamic current models have been developed (Asplin et al., 2011, 2014; Foreman et al., 2012; Adams et al., 2016; Samsing et al., 2018; Cantrell et al., 2018, 2020; Kragesteen et al., 2018; Rabe et al., 2020). These models are becoming increasingly sophisticated with more realistic biological behaviour included (Johnsen et al., 2014, 2016) (see Chapter 9). Dispersal modelling is gradually becoming a valuable tool for estimating regional infestation pressure of salmon lice on the wild fish populations, and a modelling system used in Norway (Asplin et al., 2020; Sandvik et al., 2020) is sufficiently accurate to discriminate between regions with high and low infestation pressures (Sandvik et al., 2016; Myksvoll et al., 2018). Thus, numerical model results can be utilised to estimate regional infestation pressure and eventually the carrying capacity for lice on out-migrating salmon smolts (Johnsen et al., 2020) or regionally residing sea trout (Sandvik et al., 2020). In areas with several salmon farms, the dispersion of lice between them can lead to an increased lice population and a non-sustainable infestation pressure on the wild fish. Using results from salmon lice dispersion models, it is possible to identify the connectivity between salmon farms and find those acting as central key nodes that hold the dispersion network together over a large area (Adams et al., 2015; Samsing et al., 2019; Kragesteen et al., 2021). Actions can be taken to break such dispersion networks, hence reducing the evolution of the total salmon lice population in the region.

Typically, farmed fish will outnumber wild salmonids, except for some regions of Canada (ICES, 2020b). In Norway, the wild salmonid populations consist of about 500,000 returning mature Atlantic salmon that return to more than 400 rivers spread along the coast, in addition to a million sea trout (*Salmo trutta* L.) and Arctic char (*Salvelinus alpinus* L.). Escaped farmed salmonids may carry lice but are also a source of lice that is often overlooked. Since the number of wild salmonids is far lower than the number of farmed fish, the source of lice on wild fish in farm-intensive areas is probably driven, to some extent, by released lice from active salmonid fish farms (Serra-Llinares et al., 2014, 2016; Fjørtoft et al., 2017, 2019). In Norway, lice in fish farms are counted weekly and, by calculating numbers of hatched lice from all farms, a yearly trend in lice density is calculated. Usually, due to farming practices and regulations, the number of released lice increases during spring and early summer, so fish that leave the rivers early in the season are less affected than fish leaving late, or fish that have a long distance from the rivers to the ocean as these will spend more time exposed, and the lice levels will increase during the season.

Atlantic salmon may become infested during their migration from the rivers to the feeding areas in the ocean, while at the feeding area or when returning to the coast as mature individuals. Since sea lice will not survive prolonged periods in fresh or brackish water, the adult lice will die one to two weeks after the mature fish has entered low-salinity waters (Hahnenkamp and Fyhn, 1985).

Wild Atlantic salmon smolts are more susceptible to sea lice infections due to their small size (15–25 g), as the surface area to volume ratio will be large in these small fish and hence they will be more susceptible to osmotic disturbance from lice lesions (Grimnes and Jakobsen, 1996). Additionally they are coping with the transition from fresh water to full-salinity sea water, which is physiologically and energetically very demanding and considered a period of high disease and mortality risk (Grimnes and Jakobsen, 1996). The coastal zone is also considered the area with the highest densities of salmon lice (although little work has looked at levels of lice on the feeding grounds or migratory routes) as the majority of salmon farms are found within 15 km of the shore. However, in an analysis of returning salmon in Scotland and England, a model was created that considered that the condition of returning salmon was correlated to the number of lice the fish was carrying (Susdorf et al., 2018b), and model simulations showed that this reduced condition could affect stock development (Susdorf et al., 2018a). However, this was not the case for *S. truttae* in a study by Urquhart et al. (2010) who found little difference in the burden of lice on returning fish across several sites in Scotland.

Assessing the effect of sea lice on the stock levels of Atlantic salmon is difficult, as the number of out-migrating fish is usually based on estimated smolt production, and the survival of salmon during their feeding migration in the sea varies between years, probably due to natural variations in the ecosystem. What is clear is that the number of returning Atlantic salmon has declined year on year since the 1950s and possibly earlier throughout the natural range of Atlantic salmon, probably due to anthropogenic factors such as dams, at-sea fishing, acid rain, ocean acidification, warming temperature and the recovery of piscivorous marine mammal species (reviewed in Bricknell et al., 2021). Early sea water mortality and smolt performance have been estimated by treating smolts with anti-lice drugs prior to their release and comparing their recapture rates as adults with untreated control groups (Gargan et al., 2012; Jackson et al., 2013; Vollset et al., 2016; Bøhn et al., 2020). The duration of the protection is, typically, limited to a few weeks or months (Stone et al., 1999; Skilbrei et al., 2008). It is assumed that the survival during their seaward migration is affected by lice infestations during the first period of the seaward migration, particularly in the estuarine and coastal environment. It is also assumed that the salmon smolts are most vulnerable shortly after they have left the rivers due to the physiological challenges as they transition from fresh to salt water (Strand and Finstad, 2007; Strand et al., 2011; Thorstad et al., 2012) and predation (Hvidsten and Lund, 1988; Handeland et al., 1996). Norwegian data from experiments comparing fish protected from salmon lice infestations with untreated control fish has shown that the effect of the protection varies with the general survival of the groups (Vollset et al., 2016). In years where the general survival was high, lice protection had no statistical effect, while in years with a general low survival, the survival in the protected groups increased. Using data from releases in Daleelva, western Norway, it was shown that lice infestations caused about 15% smolt mortality between 1997 and 2009 (Skilbrei et al., 2013). In a larger meta-analysis, including more Norwegian data, it was estimated that an overall mortality of 18% was attributed to sea lice (Vollset et al., 2016). In most years untreated smolts were slightly smaller (~0.1 kg) when returning after one year in the sea, which may indicate that lice caused a growth depression even when the mortality seems unaffected by lice.

In a 26-year-long time-series study, in the river Erriff in western Ireland, analysis indicated that after years in which there were higher lice numbers in nearby farms there were fewer returning 1-seawinter salmon (grilse) in the river (Shephard and Gargan, 2017). There are obviously several differences between cultivated and wild fish, but the results still show that salmon lice may be an important cause of mortality for wild salmon populations.

Sea trout and Arctic char feed along the coastal margin of fjords or along the coast and are exposed to lice infestations for a much longer period than out-migrating salmon. The at-sea feeding period usually occurs over the summer, lasting several months, before the fish return to fresh water to overwinter. Similarly to salmon, the first sea trout smolts, in early May, are considered the most vulnerable to sea lice infections, though sea trout post smolts are typically larger than salmon post smolts. However, sea trout may mitigate lice infestations behaviourally, by seeking refuge in brackish or fresh water (Birkeland and Jakobsen, 1997). Sea lice larvae do not tolerate sea water below 29 parts per thousand (ppt) very well (Bricknell et al., 2006; Crosbie et al., 2019). Also, the lice larval density is probably lower in brackish estuaries and, in addition, the sea trout may end their feeding period and return to the rivers if the infestations become severe. This phenomenon has been termed 'premature return'.

Several experiments have shown that lice infestations may affect the behaviour of sea trout. Sea trout smoltify and leave the rivers in late spring and early summer, much the same as salmon. The growth is usually far better in the sea than in fresh water, and increased growth reduces risk of predation and results in increased reproductive success (Stearns, 1992). Consequently, reduced growth and increased mortality due to salmon lice may reduce the advantages of the anadromous life history.

A study of sea trout in Ireland and Scotland found that sea trout caught near fish farms carried more lice, particularly in warm years, and had a lower condition index (Shephard et al., 2016). Salmon lice will avoid water with low salinities (Heuch, 1995), and the lice survival will be lower with lower water salinity (Gravil, 1996; Bricknell et al., 2006; Wright et al., 2016), and sea trout may therefore seek water with lower salinities to rid themselves of the lice infestation (Birkeland and Jakobsen, 1997; Serra-Llinares et al., 2018, 2020). An extensive literature review of the effects of salmon lice on sea trout has recently been published (Thorstad et al., 2015, 2016). Several behavioural studies using acoustically tagged sea trout have shown the behavioural effects of salmon lice. The studies indicate that in years with low infestation pressure the trout used a larger feeding area and spent a longer time in the sea than in years with high infestation pressure (Halttunen et al., 2018), but no effect on mortality was observed for the fish with mean weights of 190–302 g. In a pilot experiment in an area with naturally low infestation pressure in southern Norway, artificially infested sea trout showed higher mortalities than the non-infested control group, and behavioural differences were observed (Serra-Llinares et al., 2020). Mortality was seen in nearly 20% of heavily infected mature sea trout that prematurely returned to the river (Birkeland, 1996).

23.5 Economic impacts of sea lice

Overall analysis of production costs in salmon farming lists feed as the most important cost at over 50% (Iversen et al., 2020) but the prevention and treatment of sea lice has become a major cost component as well (Abolofia et al., 2017; Iversen et al., 2020). In Norway, binding government regulations due to environmental concerns and salmon lice issues have stagnated biomass growth since 2012. Despite this, the costs of salmon lice control have continued to rise (Bjørndal and Tusvik, 2019).

Several authors have calculated the cost of sea lice from the early days in the 1990s and upwards. Most of the analyses are made on a regional or national scale and with varying degree of details in the analysis. The first global analysis (Costello, 2009a) defines sea lice as the most damaging parasite to the salmonid farming industry in both Europe and the Americas. Tabulated in this publication are some references stretching from 1994 to 2002 including analysis from Chile (Carvajal et al., 1998; Bravo, 2003), Canada (Pike and Wadsworth, 1999; Mustafa et al., 2001), Ireland (Mordue (Luntz) and Pike, 2002), Scotland (Sinnott, 1998; Pike and Wadsworth, 1999) and Norway (Pike and Wadsworth, 1999). The cost in these early estimates range from €0.003 to €0.298/kg salmon (in April 2008) but most estimated values were between €0.1 and €0.2/kg. Coupled with the total production per country as reported by FAO (2006), the global cost of sea lice was reported as approximately €305,000,000, with the main producers Chile and Norway making up 83% of the total global cost around 2006.

Abolofia et al. (2017) conclude that there have been few attempts beyond surveys and laboratory trials to assess the impact of the cost of sea lice. Economic considerations carried out over the last years have mainly been applied in Norway. We postulate that this could be due to the regulatory approach in Norway, with an open data policy that makes direct empirical analysis and calculations feasible. The overall sea lice levels in individual farms are registered every week and reported to the Norwegian Food Safety Authority and are available for download on www.barentswatch.no. Likewise, the biomass development is available from the Directorate of Fisheries homepage at www.fiskdir.no or www.fiskeridir.no/Akvakultur/Tall-og-analyse/Akvakulturstatistikk-tidsserier/Laks-regnbueoerret-og-oerret/Matfiskproduksjon.

The economic impact of sea lice in aquaculture is a tangled issue and can be related to economic cost on fish farms, including prevention, control and reduction measures typically divided into direct and indirect cost. Likewise, effects ascribed to other industries and socio-economic issues complicates the picture.

Costello (2009a), for example, divides the cost of sea lice into treatment costs, reduced fish growth and reduced food conversion efficiency. Direct cost to the salmon farmer includes the cost of chemotherapeutants (in-feed being the most expensive), manual labour and use and lease of technical equipment such as well boats. Biological losses due to reduced growth are included in several studies. Increased mortality in the fish population as a direct effect of the treatment is also a rising problem. This effect is described in a review of Norwegian information and summarises data from 2012 to 2017 (Overton et al., 2019a) and includes the increasing use of mechanical treatment regimes, such as warm water treatment (Overton et al., 2019b; Moltumyr et al., 2021), mechanical flushing and brushing, which are used while the fish are being pumped between farm pens (Stien et al., 2020; Moltumyr et al., 2021). In two studies by Overton et al. (2019a) and Bui et al. (2020) it was acknowledged that keeping salmon lice regulations may diminish animal welfare. The same issue is addressed by Liu and Bjelland (2014) who give a concise overview of different treatment methods ranging from medical in-feed treatment, chemical baths and cleaner fish for farm-specific Norwegian salmon farming. Their analysis shows that treatment costs were dependent on the treatment types used and timing of treatments conducted. The cost varied from 2.28 NOK/kg to 5.03 NOK/kg (2021: €0.24 to €0.6) reducing the profit margin by about 50% for the most expensive treatment regimes. Comparably the corresponding cost was estimated to be 0.79 NOK/kg in 2011. Combining this with the overall biomass production for 2011 (Directorate of Fisheries) gives an estimate of the total cost for salmon farming in Norway of 841 million NOK in 2011.

Abolofia et al. (2017) are the first to publish a detailed empirical analysis of biomass reduction due to salmon lice infections consisting of data from all Norwegian salmon farms over an 84-month cycle. They included a bio-econometric model of fish biomass growth that incorporates productive and biophysical inputs (for example, feed, stocking density, fish size and water temperature), harmful inputs (for example, parasites) and damage abatement inputs (for example, parasiticide applications). In contrast, they state the existing literature up to this point uses farm questionnaires and accounting techniques to estimate the costs of lice infestations on specific farms in particular regions of the world. By using their approach for 2006 they thus find the value stipulated in Costello (2009a) of US$165,000,000 likely to be an underestimation of the estimated total economic impacts for Norway at US$301,000,000 (8.81% of total production value for 2006). By using the same method for 2011 the numbers for Norway are US$436,000,000 and 8.70% correspondingly.

Although there are few publications addressing direct cost of salmon lice over the last years on a global scale, some conclusions can be drawn. Iversen et al. (2020) analysed the total production cost for the major salmon producing countries Canada, Chile, Faroe Islands, Norway and Scotland. They broke down the cost for 2003 and 2018 respectively and noted that what was grouped as 'miscellaneous costs' have increased significantly between the two years. In Norway this is often considered to represent the increased lice treatment costs.

Anecdotal reports (in 2018) estimate the total cost of salmon lice in Norway to be 5.2 billion NOK/year. If the cost in all salmon producing countries consists of the same biological and technical challenges described here it is possible to propose a total global cost of sea lice. A simple calculation uses the facts that 1) Norway produces 53% of the total global production, 2) average cost in Norway is approximately 20% higher than in the EU (https://ec.europa.eu/eurostat, 2018) and 3) the cost in Chile, which produces 27% of the global production, is assumed to be 30% lower than Norway (authors' estimation). The total global cost without adjusting for national price difference would be 9.8 billion NOK or €0.96 billion. Adjusting for estimated price differences the total global cost would be slightly lower at 8.1 billion NOK, or €790 million, per year.

23.6 Socio-economic impacts of salmon lice in and from fish farms

Atlantic salmon aquaculture is a large and economically important industry with an export value of more than 67 billion NOK/year in Norway alone. This, of course, creates jobs and economic activities. Olavsen et al. (2011) estimate that the number of people employed in aquaculture-related jobs is twice that of people directly involved in the aquaculture core operation. The need for sea lice treatments on a much larger scale than before has also generated a growing supply and research industry. Development of new and innovative production technology and delousing practices creates new companies and jobs. Reflecting the practical needs for delousing today, this requires well boats, veterinarians, specialised treatment teams and new pharmaceuticals.

The effects of salmon lice from fish farms on an iconic species like wild salmon makes the issue more problematic. Returning wild salmon is a much sought-after commodity for a large group of dedicated anglers. Negative impacts on wild salmon stocks from farming activities are perceived to reduce wild catches for anglers and decrease the demand for fishing licences (Liu et al., 2011). All rivers in Norway are owned privately or in the stewardship of the state and, as such, salmon fishing is strongly regulated. The rights to angling are therefore sold for substantial sums. The industry around this comprises not only the right to fishing, but whole local economies centred around the rivers. Olaussen et al. (2015) present an interesting study to quantify the economic effects of sea lice from farmed fish on harvest of wild fish. The loss is analysed both by potentially reduced harvesting value due to sea lice and by the 'non-consumptive' value of the wild stocks. They find that the sea lice-induced reduction of wild salmon stock value varies from 64% to 66%, and the reduced harvest from 65% to 70% reduction in total monetary value. The range depended on whether they use optimal fishing mortality or optimal uniform fishing mortality combined with managing for harvest value or also includes non-consumptive values (summarised in table 7, Olaussen et al. (2015)). They also approximate the number of recreational anglers in Norway alone to be over 100,000 individuals (1.8% of the population).

The overall economic effects due to strict regulations can result in the loss of production capacity and subsequent revenue that has been proposed by Norwegian fish farmers. Torrissen et al. (2013) reviewed the legislation for the salmon producing countries up to approximately 2009. Legislative and regulatory management of sea lice was recently reviewed in a working group report from ICES, together with other environmental impacts from marine aquaculture in Europe and Northern

America (ICES, 2020a). Sea lice left untreated on the fish farms can result in severe damage to the salmon and eventually lead to welfare issues and mortality. All salmon producing countries in Europe (except Denmark) and North America have regulations on how many sea lice a farmed fish can have, as well as rules for mitigation (ICES, 2020a). Due to the low salinity of the inner Danish waters, treatments here are considered unnecessary nor are they legally permitted.

23.7 Regulation of sea lice

In recent years, negative effects of sea lice originating from farms on the wild salmon populations have likewise gained increased attention, and legislation aiming at protecting wild ecosystems is in place in most of the salmon producing countries (ICES, 2020a). Due to various levels of production, differences in wild ecosystems and other naturally occurring differences, the legislation and detailed regulations vary between countries, although there are some similarities (ICES, 2020a). A common way is to monitor the sea lice level on the farms, and mitigating measures must be taken if the sea-lice levels exceed pre-set values (for example, maximum numbers of mature sea lice per fish).

Such regulations are not an effective way to regulate the impact from sea lice on wild salmonids, since numbers of mature sea lice per fish do not take into account the number of fish in the farms or in the region, dispersal and survival of salmon lice hatched in the farms, nor interaction and infestations of wild salmonids (see Chapter 9).

Such considerations were implemented in the regulations of commercial production of Atlantic salmon, trout and rainbow trout in Norway in 2017, called the 'traffic-light system' (Greaker et al., 2020). The coast of Norway is divided into 13 production zones and, within each zone, the impact on the wild salmon is assessed through a combination of field surveillance and modelling of sea lice dispersion. Based on predefined levels of lice-induced mortality of the populations (%), the production areas are defined as being one of three zones, green (<10%), yellow (10–30%) or red (> 30%), which, in turn, gives the farmers an opportunity to increase production volume (biomass) by 6% if green but they might have to reduce the biomass by 6% in red areas. The same production volume can be kept in yellow areas.

Today, there are still gaps in our knowledge and understanding of how sea lice from fish farms impact wild salmonids. The legislation and regulatory management of sea lice need to be revised as new knowledge is discovered or when new productions and lice control methods are implemented (ICES, 2020a).

23.8 Public perception

Agriculture is approximately 10,000 years old and, globally, many people still have knowledge and closeness to this system of food production. Agriculture also underwent a radical transformation during the agricultural revolution between the 1700s and 1900s leading to fundamental changes in the approach to issues including parasite control and animal health (Jackson et al., 2018). Although aquaculture has been practised for millennia, it is often not widely accepted in the western world, and aquaculture is perceived as a new way of producing food. Only farmed salmon has achieved commodity status so far and this is considered a global scale sector and ranked as number eight by tonnage for globally cultured fish (Glover et al., 2017).

Osmundsen et al. (2017), addressing the complicated issue of managing and regulating aquaculture, refers to what political scientists will call a 'wicked problem' (Rittel and Webber, 1973). Uncertainty and lack of knowledge and, maybe more importantly, a lack of consensus when it comes to societal and scientific interpretations, solutions and unresolved problems can give rise to strange (social *sic*) media stories. Interestingly, the conclusion is that relative to its added economic value, aquaculture sectors are the subject of a much larger and more critical debate than most other sectors in many countries including Europe and North America (Osmundsen et al., 2017). Analysing media coverage and framing of aquaculture in Norwegian newspapers, Olsen and Osmundsen (2017) found that environmental issues (including sea lice) were presented in a more sensational manner compared to the topics of politics and salmon as industry. They also conclude from their study that the media give risks more focus than benefits which will also affect the overall public perception. The public will source much of their knowledge about aquaculture from media and social media. General knowledge on aquaculture has not yet found its way into many educational systems at a lower grade level. One anecdotal example of how difficult it can be to change the public perception comes from Norway. Antibiotics were widely used in the salmon industry in the 1980s due to bacterial diseases such as vibriosis and furunculosis, both of which have now been completely managed by vaccines since the early 1990s. Thirty years hence it is still a 'known fact' that aquaculture uses a lot of antibiotics although Atlantic salmon aquaculture has the lowest use of antibiotics of any large mass food production. Data from the Norwegian Medicines Agency state an average use of 8.0 mg/kg for land-based protein production and only 0.16 mg/kg for salmon (numbers for 2016 used from Anon. (2017)).

Criticism of the way the salmon farming industry has handled the salmon lice problem results in strong negative publicity on the issue, and also a general negative effect on the public perception of aquaculture (Jackson et al., 2018). The fish farms' 'licence to operate' in 'the commons' is currently unresolved and, for several countries, still hinges on the perception of sea lice related issues. The core of the conflicts has been disagreement on the scale of the impact of salmon lice or their therapeutants on the decline of wild salmon populations or non-targeted species (Torrissen et al., 2013).

References

Abolofia, J., Asche, F. and Wilen, J.E. (2017) The cost of lice: quantifying the impacts of parasitic sea lice on farmed salmon. *Marine Resource Economics* 32, 329–349.

Adams, T.P., Proud, R. and Black, K.D. (2015) Connected networks of sea lice populations: dynamics and implications for control. *Aquaculture Environment Interactions* 6, 273–284.

Adams, T.P., Aleynik, D. and Black, K.D. (2016) Temporal variability in sea lice population connectivity and implications for regional management protocols. *Aquaculture Environment Interactions* 8, 585–596.

Anonymous (2017) NORM/NORM-VET 2017. Usage of Antimicrobial Agents and Occurrence of Antimicrobial Resistance in Norway. Available at: https://www.vetinst.no/overvaking/antibiotikaresistens-norm-vet/_/attachment/download/3b568168-1005-4925-8b83-535a384275d7:591d6da99e48213389ab94e986aab3f23707b13f/NORM%20NORM-VET%202017.pdf (accessed January 2022).

Asplin, L., Boxaspen, K.K. and Sandvik, A.D. (2011) Modeling the distribution and abundance of planktonic larval stages of *Lepeophtheirus salmonis* in Norway. In: Jones, S. and Beamish, R. (eds) *Salmon Lice: An Integrated Approach to Understanding Parasite Abundance and Distribution*. Wiley-Blackwell, Oxford, UK. pp. 29–50.

Asplin, L., Johnsen, I.A., Sandvik, A.D., Albretsen, J., Sundfjord, V., Aure, J. and Boxaspen, K.K. (2014) Dispersion of salmon lice in the Hardangerfjord. *Marine Biology Research* 10, 216–225.

Asplin, L., Albretsen, J., Johnsen, I.A. and Sandvik, A.D. (2020) The hydrodynamic foundation for salmon lice dispersion modeling along the Norwegian coast. *Ocean Dynamics* 70, 1151–1167.

Barker, S.E., Bricknell, I.R., Covello, J., Purcell, S., Fast, M.D., Wolters, W. and Bouchard, D.A. (2019) Sea lice, *Lepeophtheirus salmonis* (Krøyer, 1837), infected Atlantic salmon (*Salmo salar* L.) are more susceptible to infectious salmon anemia virus. *PLoS ONE* 14, e0209178.

Birkeland, K. (1996) Consequences of premature return by sea trout (*Salmo trutta*) infested with the salmon louse (*Lepeophtheirus salmonis* Krøyer): migration, growth, and mortality. *Canadian Journal of Fisheries and Aquatic Sciences* 53, 2808–2813.

Birkeland, K. and Jakobsen, P.J. (1997) Salmon lice, *Lepeophtheirus salmonis*, infestation as a causal agent of premature return to rivers and estuaries by sea trout, *Salmo trutta*, juveniles. *Environmental Biology of Fishes* 49, 129–137.

Bjørn, P.A. and Finstad, B. (1997) The physiological effects of salmon lice infection on sea trout postsmolts. *Nordic Journal of Freshwater Research* 73, 60–72.

Bjørn, P.A. and Finstad, B. (1998) The development of salmon lice (*Lepeophtheirus salmonis*) on artificially infected post smolts of sea trout (*Salmo trutta*). *Canadian Journal of Zoology* 76, 970–977.

Bjørndal, T. and Tusvik, A. (2019) Economic analysis of land based farming of salmon. *Aquaculture Economics & Management* 23, 449–475.

Bøhn, T., Gjelland, K.Ø., Serra-Llinares, R.M., Finstad, B., Primicerio, R., Nilsen, R., Karlsen, Ø., Sandvik, A.D., Skilbrei, O.T., Elvik, K.M.S., Skaala, Ø. and Bjørn, P.A. (2020) Timing is everything: survival of Atlantic salmon *Salmo salar* postsmolts during events of high salmon lice densities. *Journal of Applied Ecology* 57, 1149–1160.

Boxaspen, K. (2006) A review of the biology and genetics of sea lice. *ICES Journal of Marine Science* 63, 1304–1316.

Brandal, P.O. and Egidius, E. (1977) Preliminary report on oral treatment against salmon lice, *Lepeophtheirus salmonis*, with Neguvon. *Aquaculture* 10, 177–178.

Brandal, P.O., Egidius, E. and Romslo, I. (1976) Host blood – major food component for parasitic copepod *Lepeophtheirus salmonis* (Krøyer, 1838) (Crustacea: Caligidae). *Norwegian Journal of Zoology* 24, 341–343.

Bravo, S. (2003) Sea lice in Chilean salmon farms. *Bulletin of the European Association of Fish Pathologists* 23, 197–200.

Bricknell, I.R., Dalesman, S.J., O'Shea, B., Pert, C.C. and Mordue (Luntz), A.J. (2006) Effect of salinity on sea lice *Lepeophtheirus salmonis* settlement success. *Diseases of Aquatic Organisms* 71, 201–212.

Bricknell, I.R., Birkel, S.D., Brawley, S.H., Van Kirk, T., Hamlin, H., Capistrant-Fossa, K., Huguenard, K., Van Walsum, G.P., Liu, Z.L., Zhu, L.H., Grebe, G., Taccardi, E., Miller, M., Preziosi, B.M., Duffy, K., Byron, C.J., Quigley, C.T.C., Bowden, T.J., Brady, D., Beal, B.F., Sappati, P.K., Johnson, T.R. and Moeykens, S. (2021) Resilience of cold water aquaculture: a review of likely scenarios as climate changes in the Gulf of Maine. *Reviews in Aquaculture* 13(1), 460–503.

Brooker, A.J., Skern-Mauritzen, R. and Bron, J.E. (2018) Production, mortality, and infectivity of planktonic larval sea lice, *Lepeophtheirus salmonis* (Krøyer, 1837): current knowledge and implications for epidemiological modelling. *ICES Journal of Marine Science* 75, 1214–1234.

Bui, S., Dempster, T., Remen, M. and Oppedal, F. (2016) Effect of ectoparasite infestation density and life-history stages on the swimming performance of Atlantic salmon *Salmo salar*. *Aquaculture Environment Interactions* 8, 387–395.

Bui, S., Stien, L.H., Nilsson, J., Trengereid, H. and Oppedal, F. (2020) Efficiency and welfare impact of long-term simultaneous in situ management strategies for salmon louse reduction in commercial sea cages. *Aquaculture* 520, 734934.

Cantrell, D.L., Rees, E.E., Vanderstichel, R., Grant, J., Filgueira, R. and Revie, C.W. (2018) The use of kernel density estimation with a bio-physical model provides a method to quantify connectivity among salmon farms: spatial planning and management with epidemiological relevance. *Frontiers in Veterinary Science* 5, 269.

Cantrell, D.L., Groner, M.L., Ben-Horin, T., Grant, J. and Revie, C.W. (2020) Modeling pathogen dispersal in marine fish and shellfish. *Trends in Parasitology* 36, 239–249.

Carvajal, J., González, L. and George-Nascimento, M. (1998) Native sea lice (Copepoda: Caligidae) infestation of salmonids reared in netpen systems in southern Chile. *Aquaculture* 166, 241–246.

Costello, M.J. (2009a) The global economic cost of sea lice to the salmonid farming industry. *Journal of Fish Diseases* 32, 115–118.

Costello, M.J. (2009b) How sea lice from salmon farms may cause wild salmonid declines in Europe and North America and be a threat to fishes elsewhere. *Proceedings Royal Society B Biological Sciences* 276, 3385–3394.

Crosbie, T., Wright, D.W., Oppedal, F., Johnsen, I.A., Samsing, F. and Dempster, T. (2019) Effects of step salinity gradients on salmon lice behaviour and dispersal. *Aquaculture Environment Interactions* 11, 181–190.

Dempster, T., Overton, K., Bui, S., Stien, L.H., Oppedal, F., Karlsen, Ø., Coates, A., Philips, B. and Barrett, L.T. (in press) Farmed salmonids drive the abundance, ecology and evolution of parasitic salmon lice in Norway. *Aquaculture Environment Interactions*.

Directorate of Fisheries (no date) Aquaculture statistics: food fish production of salmon, rainbow trout and trout. Available at: htttps://www.fiskdir.no, https://www.fiskeridir.no/Akvakultur/Tall-og-analyse/Akvakulturstatistikk-tidsserier/Laks-regnbueoerret-og-oerret/Matfiskproduksjon) (accessed May 2019).

FAO (2006) FAO statistics 2006. Available at: http://www.fao.org/statistics/2006 and https://www.fao.org/fishery/statistics-query/en/aquaculture/aquaculture_quantity. Input variables year: 2006, continents: Europe, Asia, Americas, species: Atlantic salmon, environment: marine (accessed May 2019).

FAO (2021) FishStatJ, a tool for fishery statistics analysis. Release: 4.01.7. Available at: fao.org/fishery/statistics/software/fishstatj/ (accessed January 2022).

Finstad, B., Bjørn, P.A., Grimnes, A. and Hvidsten, N.A. (2000) Laboratory and field investigations of salmon lice [*Lepeophtheirus salmonis* (Krøyer, 1837)] infestation on Atlantic salmon (*Salmo salar* L.) postsmolts. *Aquaculture Research* 31, 795–803.

Fjørtoft, H.B., Besnier, F., Stene, A., Nilsen, F., Bjørn, P.A., Tveten, A.-K., Finstad, B., Aspehaug, V. and Glover, K.A. (2017) The Phe362Tyr mutation conveying resistance to organophosphates occurs in high frequencies in salmon lice collected from wild salmon and trout. *Scientific Reports* 7, 14258.

Fjørtoft, H.B., Nilsen, F., Besnier, F., Stene, A., Bjørn, P.A., Tveten, A.K., Aspehaug, V.T., Finstad, B. and Glover, K.A. (2019) Salmon lice sampled from wild Atlantic salmon and sea trout throughout Norway display high frequencies of the genotype associated with pyrethroid resistance. *Aquaculture Environment Interactions* 11, 459–468.

Foreman, M.G.G., Stucchi, D.J., Garver, K.A., Tuele, D., Isaac, J., Grime, T., Guo, M. and Morrison, J. (2012) A circulation model for the Discovery Islands, British Columbia. *Atmosphere-Ocean* 50, 301–316.

Forseth, T., Barlaup, B.T., Finstad, B., Fiske, P., Gjøsæter, H., Falkegård, M., Hindar, A., Mo, T.A., Rikardsen, A.H., Thorstad, E.B., Vøllestad, L.A. and Wennevik, V. (2017) The major threats to Atlantic salmon in Norway. *ICES Journal of Marine Science* 74, 1496–1513.

Gallardi, D., Xue, X., Hamoutene, D., Lush, L. and Rise, M.L. (2019) Impact of origin (wild vs. farmed) and sea lice (*Lepeophtheirus salmonis*) infestation on expression of immune-relevant genes in Atlantic salmon (*Salmo salar* L.) skin. *Aquaculture* 499, 306–315.

Gargan, P.G., Forde, G., Hazon, N., Russel, D.J.F. and Todd, C.D. (2012) Evidence for sea lice-induced marine mortality of Atlantic salmon (*Salmo salar*) in western Ireland from experimental releases of ranched smolts treated with emamectin benzoate. *Canadian Journal of Fisheries and Aquatic Sciences* 69, 343–353.

Glover, K.A., Solberg, M.F., McGinnity, P., Hindar, K., Verspoor, E., Coulson, M.W., Hansen, M.M., Araki, H., Skaala, Ø. and Svåsand, T. (2017) Half a century of genetic interaction between farmed and wild Atlantic salmon: status of knowledge and unanswered questions. *Fish and Fisheries* 18, 890–927.

Godwin, S.C., Dill, L.M., Reynolds, J.D. and Krkošek, M. (2015) Sea lice, sockeye salmon, and foraging competition: lousy fish are lousy competitors. *Canadian Journal of Fisheries and Aquatic Sciences* 72, 1113–1120.

Gravil, H.R. (1996) Studies on the biology and ecology of the free swimming larval stages of *Lepeophtheirus salmonis* (Krøyer, 1838) and *Caligus elongatus* Nordmann, 1832 (Copepoda: Caligidae). PhD thesis, University of Stirling, UK.

Greaker, M., Vormedal, I. and Rosendal, K. (2020) Environmental policy and innovation in Norwegian fish farming: resolving the sea lice problem? *Marine Policy* 117, 103942.

Grimnes, A. and Jakobsen, P.J. (1996) The physiological effects of salmon lice infection on post-smolt of Atlantic salmon (*Salmo salar*). *Journal of Fish Biology* 48, 1179–1194.

Hahnenkamp, L. and Fyhn, H.J. (1985) The osmotic response of salmon louse, *Lepeophtheirus salmonis* (Copepoda: Caligidae), during the transition from sea-water to fresh-water. *Journal of Comparative Physiology B – Biochemical Systemic and Environmental Physiology* 155(3), 357–365.

Halttunen, E., Gjelland, K.Ø., Hamel, S., Serra-Llinares, R,M., Nilsen, R., Arechavala-Lopez, P., Skarðhamar, J., Johnsen, I.A., Asplin, L., Karlsen, Ø., Bjørn Pål, A. and Finstad, B. (2018) Sea trout adapt their migratory behaviour in response to high salmon lice concentrations. *Journal of Fish Diseases* 41, 953–967.

Hamre, L.A., Eichner, C., Caipang, C.M.A., Dalvin, S.T., Bron, J.E., Nilsen, F., Boxshall, G. and Skern-Mauritzen, R. (2013) The salmon louse *Lepeophtheirus salmonis* (Copepoda: Caligidae) life cycle has only two chalimus stages. *PLoS ONE* 8 (9), e73539.

Hamre, L.A., Bui, S., Oppedal, F., Skern-Mauritzen, R. and Dalvin, S. (2019) Development of the salmon louse *Lepeophtheirus salmonis* parasitic stages in temperatures ranging from 3 to 24°C. *Aquaculture Environment Interactions* 11, 429–443.

Handeland, S.O., Järvi, T., Fernö, A. and Stefansson, S.O. (1996) Osmotic stress, antipredatory behaviour, and mortality of Atlantic salmon (*Salmo salar*) smolts. *Canadian Journal of Fisheries and Aquatic Sciences* 53, 2673–2680.

Heuch, P.A. (1995) Experimental evidence for aggregation of salmon louse copepodids (*Lepeophtheirus salmonis*) in step salinity gradients. *Journal of the Marine Biological Association of the United Kingdom* 75, 927–939.

Heuch, P.A. and Mo, T.A. (2001) A model of salmon louse production in Norway: effects of increasing salmon production and public management measures. *Diseases of Aquatic Organisms* 45, 145–152.

Heuch, P.A., Bjørn, P.A., Finstad, B., Holst, J.C., Asplin, L. and Nilsen, F. (2005) A review of the Norwegian 'National Action Plan Against Salmon Lice on Salmonids': the effect on wild salmonids. *Aquaculture* 246, 79–92.

Hvidsten, N.A. and Lund, R.A. (1988) Predation on hatchery-reared and wild smolts of Atlantic salmon, *Salmo salar* L., in the estuary of River Orkla, Norway. *Journal of Fish Biology* 33, 121–126.

ICES (2020a) Working Group on Environmental Interactions of Aquaculture (WGEIA). *ICES Scientific Reports* 2:112:187.

ICES (2020b) Working Group on North Atlantic Salmon (WGNAS). *ICES Scientific Reports* 2:21:358.

Iversen, A., Asche, F., Hermansen, Ø. and Nystøyl, R. (2020) Production cost and competitiveness in major salmon farming countries 2003–2018. *Aquaculture* 522, 735089.

Jackson, D., Cotter, D., Newell, J., McEvoy, S., Donohoe, P., Kane, F., McDermott, T., Kelly, S. and Drumm, A. (2013) Impact of *Lepeophtheirus salmonis* infestations on migrating Atlantic salmon, *Salmo salar* L., smolts at eight locations in Ireland with an analysis of lice-induced marine mortality. *Journal of Fish Diseases* 36, 273–281.

Jackson, D., Moberg, O., Stenevik Djupevåg, E.M., Kane, F. and Hareide, H. (2018) The drivers of sea lice management policies and how best to integrate them into a risk management strategy: an ecosystem approach to sea lice management. *Journal of Fish Diseases* 41, 927–933.

Johnsen, I.A., Fiksen, Ø., Sandvik, A.D. and Asplin, L. (2014) Vertical salmon lice behaviour as a response to environmental conditions and its influence on regional dispersion in a fjord system. *Aquaculture Environment Interactions* 5, 127–141.

Johnsen, I.A., Asplin, L.C., Sandvik, A.D. and Serra-Llinares, R.M. (2016) Salmon lice dispersion in a northern Norwegian fjord system and the impact of vertical movements. *Aquaculture Environment Interactions* 8, 99–116.

Johnsen, I.A., Sandvik, A.D., Asplin, L. and Oppedal, F. (2020) Optimal estimation of lice release for aquaculture based on ambient temperatures. *Aquaculture Environment Interactions* 12, 179–191.

Johnson, S.C. and Albright, L.J. (1991) Development, growth, and survival of *Lepeophtheirus salmonis* (Copepoda: Caligidae) under laboratory conditions. *Journal of the Marine Biological Association of the United Kingdom* 71, 425–436.

Jones, M.W., Sommerville, C. and Bron, J. (1990) The histopathology associated with the juvenile stages of *Lepeophtheirus salmonis* on the Atlantic salmon, *Salmo salar* L. *Journal of Fish Diseases* 13, 303–310.

Jónsdóttir, H., Bron, J.E., Wootten, R. and Turnbull, J.F. (1992) The histopathology associated with the preadult and adult stages of *Lepeophtheirus salmonis* on the Atlantic salmon, *Salmo salar* L. *Journal of Fish Diseases* 15, 521–527.

Kragesteen, T.J., Simonsen, K., Visser, A.W. and Andersen K.H. (2018) Identifying salmon lice transmission characteristics between Faeroese salmon farms. *Aquaculture Environment Interactions* 10, 49–60.

Kragesteen, T.J., Simonsen, K., Visser, A.W. and Andersen, K.H. (2021) Estimation of external infection pressure and salmon-louse population growth rate in Faroese salmon farms. *Aquaculture Environment Interactions* 13, 21–32.

Liu, Y. and Bjelland, H.V. (2014) Estimating costs of sea lice control strategy in Norway. *Preventive Veterinary Medicine* 117, 469–477.

Liu, Y., Olaf Olaussen, J. and Skonhoft, A. (2011) Wild and farmed salmon in Norway – a review. *Marine Policy* 35, 413–418.

Moltumyr, L., Gismervik, K., Gu, J., Gåsnes, S.K., Kristiansen, T.S., Rønnestad, I., Nilsson, J. and Stien, L.H. (2021). Does the thermal component of warm water treatment inflict acute lesions on Atlantic salmon (*Salmo salar*)? *Aquaculture* 532, 736048.

Mordue, A.J. (Luntz) and Pike, A. (2002). Salmon farming: towards an integrated pest management strategy for sea lice. *Pest Management Science* 58, 513–514.

Mustafa, A., Rankaduwa, W. and Campbell, P. (2001) Estimating the cost of sea lice to salmon aquaculture in eastern Canada. *Canadian Veterinary Journal* 42, 54–56.

Myksvoll, M.S., Sandvik, A.D., Albretsen, J., Asplin, L., Johnsen, I.A., Karlsen, Ø., Kristensen, N.M., Melsom, A., Skardhamar, J. and Ådlandsvik, B., (2018) Evaluation of a national operational salmon lice monitoring system – from physics to fish. *PLoS ONE* 13, e0201338.

Olaussen, J.O., Liu, Y. and Skonhoft, A. (2015) Conservation versus harvest of wild Atlantic salmon. The cost of sea lice induced mortality. *Fisheries Research* 168, 63–71.

Olavsen, T., Sandberg, M., Henriksen, K., Bull-Berg, H., Johansen, U. and Stokka, A. (2011) The impact of the fishing and aquaculture industry in Norway in 2009 – a national and regional analysis of multiplicator effect. In: *SINTEF Report A19673*. SINTEF, Trondheim, Norway. p. 58.

Olsen, M.S. and Osmundsen, T.C. (2017) Media framing of aquaculture. *Marine Policy* 76, 19–27.

Osmundsen, T.C., Almklov, P. and Tveterås, R. (2017) Fish farmers and regulators coping with the wickedness of aquaculture. *Aquaculture Economics and Management* 21, 163–183.

Øverli, Ø., Nordgreen, J., Mejdell, C.M., Janczak, A.M., Kittilsen, S., Johansen, I.B. and Horsberg, T.E. (2014) Ectoparasitic sea lice (*Lepeophtheirus salmonis*) affect behavior and brain serotonergic activity in Atlantic salmon (*Salmo salar* L.): perspectives on animal welfare. *Physiology & Behavior* 132, 44–50.

Overton, K., Dempster, T., Oppedal, F., Kristiansen, T.S., Gismervik, K. and Stien, L.H. (2019a) Salmon lice treatments and salmon mortality in Norwegian aquaculture: a review. *Reviews in Aquaculture* 11, 1398–1417.

Overton, K., Oppedal, F., Stien, L.H., Moltumyr, L., Wright, D.W. and Dempster, T. (2019b) Thermal delousing with cold water: effects on salmon lice removal and salmon welfare. *Aquaculture* 505, 41–46.

Peacock, S.J., Krkošek, M., Bateman, A.W. and Lewis, M.A. (2015) Parasitism and food web dynamics of juvenile Pacific salmon. *Ecosphere* 6, 1–16.

Pike, A.W. and Wadsworth, S.L. (1999) Sealice on salmonids: their biology and control. *Advances in Parasitology* 44, 233–337.

Rabe, B., Gallego, A., Wolf, J., O'Hara Murray, R., Stuiver, C., Price, D. and Johnson, H. (2020) Applied connectivity modelling at local to regional scale: the potential for sea lice transmission between Scottish finfish aquaculture management areas. *Estuarine, Coastal and Shelf Science* 238, 106716.

Rittel, H.W.J. and Webber, M.M. (1973) Dilemmas in a general theory of planning. *Policy Sciences* 4, 155–169.

Samsing, F., Oppedal, F., Dalvin, S., Johnsen, I., Vagseth, T. and Dempster, T. (2016) Salmon lice (*Lepeophtheirus salmonis*) development times, body size, and reproductive outputs follow universal models of temperature dependence. *Canadian Journal of Fisheries and Aquatic Sciences* 73, 1841–1851.

Samsing, F., Johnsen, I., Dempster, T., Oppedal, F. and Treml, E.A. (2018) Network analysis reveals strong seasonality in the dispersal of a marine parasite and identifies areas for coordinated management. *Landscape Ecology* 32, 1953–1967.

Samsing, F., Johnsen, I., Treml, E.A. and Dempster, T. (2019) Identifying 'firebreaks' to fragment dispersal networks of a marine parasite. *International Journal for Parasitology* 49, 277–286.

Sandvik, A.D., Bjørn, P.A., Ådlandsvik, B., Asplin, L., Skardhamar, J., Johnsen, I.A., Myksvoll, M. and Skogen, M.D. (2016) Towards a model-based prediction system for salmon lice infestation pressure. *Aquaculture Environment Interactions* 8, 527–542.

Sandvik, A.D., Johnsen, I.A., Myksvoll, M.S., Sævik, P.N. and Skogen, M.D. (2020) Prediction of the salmon lice infestation pressure in a Norwegian fjord. *ICES Journal of Marine Research* 77, 746–756.

Serra-Llinares, R.M., Bjørn, P.A., Finstad, B., Nilsen, R., Harbitz, A., Berg, M. and Asplin, L. (2014) Salmon lice infection on wild salmonids in marine protected areas: an evaluation of the Norwegian "National Salmon Fjord". *Aquaculture Environment Interactions* 5, 1–16.

Serra-Llinares, R.M., Bjørn, P.A., Finstad, B., Nilsen, R. and Asplin, L. (2016) Nearby farms are a source of lice for wild salmonids: a reply to Jansen et al. (2016) *Aquaculture Environment Interactions* 8, 351–356.

Serra-Llinares, R.M., Freitas, C., Nilsen, R., Elvik, K.M.S., Albretsen, J., Bøhn, T., Karlsen, Ø. and Bjørn, P.A. (2018) Towards direct evidence of the effects of salmon lice (*Lepeophtheirus salmonis* Krøyer, 1837) on sea trout (*Salmo trutta* L.) in their natural habitat: proof of concept for a new combination of methods. *Environmental Biology of Fishes* 101, 1677–1692.

Serra-Llinares, R.M., Bøhn, T., Karlsen, Ø., Nilsen, R., Freitas, C., Albretsen, J., Haraldstad, T., Thorstad, E.B., Elvik, K.M.S. and Bjørn, P.A. (2020) Impacts of salmon lice on mortality, marine migration distance and premature return in sea trout. *Marine Ecology Progress Series* 635, 151–168.

Shephard, S. and Gargan, P. (2017) Quantifying the contribution of sea lice from aquaculture to declining annual returns in a wild Atlantic salmon population. *Aquaculture Environment Interactions* 9, 181–192.

Shephard, S., MacIntyre, C. and Gargan, P. (2016) Aquaculture and environmental drivers of salmon lice infestation and body condition in sea trout. *Aquaculture Environment Interactions* 8, 597–610.

Sinnott, R. (1998) Sea lice – watch out for the hidden costs. *Fish Farmer* 21(3), 45–46.

Skaala, Ø., Kålås, S. and Borgstrøm, R. (2014) Evidence of salmon lice-induced mortality of anadromous brown trout (*Salmo trutta*) in the Hardangerfjord, Norway. *Marine Biology Research* 10, 279–288.

Skarðhamar, J., Albretsen, J., Sandvik, A.D., Lien, V.S., Myksvoll, M.S., Johnsen, I.A., Asplin, L., Ådlandsvik, B., Halttunen, E. and Bjørn, P.A. (2018) Modelled salmon lice dispersion and infestation patterns in a sub-arctic fjord. *ICES Journal of Marine Science* 75, 1733–1747.

Skilbrei, O.T., Glover, K.A., Samuelsen, O.B. and Lunestad, B.T. (2008) A laboratory study to evaluate the use of emamectin benzoate in the control of sea lice in sea-ranched Atlantic salmon (*Salmo salar* L.). *Aquaculture* 285, 2–7.

Skilbrei, O.T., Finstad, B., Urdal, K., Bakke, G., Kroglund, F. and Strand, R. (2013) Impact of early salmon louse, *Lepeophtheirus salmonis*, infestation and differences in survival and marine growth of sea-ranched Atlantic salmon, *Salmo salar* L., smolts 1997–2009. *Journal of Fish Diseases* 36, 249–260.

Stearns, S.C. (1992) *The Evolution of Life Histories*. Oxford University Press, Oxford, UK.

Stien, L.H., Tørud, B., Gismervik G., Lien, M.E., Medaas, C., Osmundsen, T., Kristiansen, T.S. and Størkersen, K.V. (2020) Governing the welfare of Norwegian farmed salmon: three conflict cases. *Marine Policy* 117, 103969.

Stone, J., Sutherland, I.H., Sommerville, C.S., Richards, R.H. and Varma, K.J. (1999) The efficacy of emamectin benzoate as an oral treatment of sea lice, *Lepeophtheirus salmonis* (Krøyer, 1837), infestations in Atlantic salmon, *Salmo salar* L. *Journal of Fish Diseases* 22(4) 261–270.

Strand, J.E.T., Davidsen, J.G., Jørgensen, E.H. and Rikardsen, A.H. (2011) Seaward migrating Atlantic salmon smolts with low levels of gill Na^+, K^+-ATPase activity; is sea entry delayed? *Environmental Biology of Fishes* 90, 317–321.

Strand, R. and Finstad, B. (2007) Migratory behaviour in relation to smolt development and releasing strategies in Atlantic salmon (*Salmo salar* L.) smolts. *Aquaculture* 273, 277–283.

Susdorf, R., Salama, N.K.G. and Lusseau, D. (2018a) Influence of body condition on the population dynamics of Atlantic salmon with consideration of the potential impact of sea lice. *Journal of Fish Diseases* 41, 941–951.

Susdorf, R., Salama, N.K.G., Todd, C.D., Hillman, R.J., Elsmere, P. and Lusseau, D. (2018b) Context-dependent reduction in somatic condition of wild Atlantic salmon infested with sea lice. *Marine Ecology Progress Series* 606, 91–104.

Taranger, G.L., Karlsen, Ø., Bannister, R.J., Glover, K.A., Husa, V., Karlsbakk, E., Kvamme, B.O., Boxaspen, K.K., Bjørn, P.A., Finstad, B., Madhun, A.S., Morton, H.C. and Svåsand, T. (2015) Risk assessment of the environmental impact of Norwegian Atlantic salmon farming. *ICES Journal of Marine Research* 72, 997–1021.

Thorstad, E.B., Whoriskey, F., Uglem, I., Moore, A., Rikardsen, A.H. and Finstad, B. (2012) A critical life stage of the Atlantic salmon *Salmo salar*: behaviour and survival during the smolt and initial post-smolt migration. *Journal of Fish Biology* 81, 500–542.

Thorstad, E.B., Todd, C.D., Uglem, I., Bjørn, P.A., Gargan, P.G., Vollset, K.W., Halttunen, E., Kålås, S., Berg, M. and Finstad, B. (2015) Effects of salmon lice *Lepeophtheirus salmonis* on wild sea trout *Salmo trutta* – a literature review. *Aquaculture Environment Interactions* 7, 91–113.

Thorstad, E.B., Todd, C.D., Uglem, I., Bjørn, P.A., Gargan, P.G., Vollset, K.W., Halttunen, E., Kålås, S., Berg, M. and Finstad, B. (2016) Marine life of the sea trout. *Marine Biology* 163, 47.

Torrissen, O, Jones, S., Asche, F., Guttormsen, A., Skilbrei, O.T., Nilsen, F., Horsberg, T.E. and Jackson, D. (2013) Salmon lice – impact on wild salmonids and salmon aquaculture. *Journal of Fish Diseases* 36, 171–194.

Tveiten, H., Bjørn, P.A., Johnsen, H.K., Finstad, B. and McKinley, R.S. (2010) Effects of the sea louse *Lepeophtheirus salmonis* on temporal changes in cortisol, sex steroids, growth and reproductive investment in Arctic charr *Salvelinus alpinus*. *Journal of Fish Biology* 76, 2318–2341.

Urquhart, K., Pert, C.C., Fryer, R.J., Cook, P., Weir, S., Kilburn, R., McCarthy, U., Simons, J., McBeath, S.J., Matejusova, I. and Bricknell, I.R. (2010) A survey of pathogens and metazoan parasites on wild sea trout (*Salmo trutta*) in Scottish water. *ICES Journal of Marine Science* 67, 444–453.

Vollset, K.W., Krontveit, R.I., Jansen, P.A., Finstad, B., Barlaup, B.T., Skilbrei, O.T., Krkošek, M., Romunstad, P., Aunsmo, A., Jensen, A.J. and Dohoo, I. (2016) Impacts of parasites on marine survival of Atlantic salmon: a meta-analysis. *Fish and Fisheries* 17, 714–730.

Wagner, G.N., McKinley, R.S., Bjørn, P.A. and Finstad, B. (2003) Physiological impact of sea lice on swimming performance of Atlantic salmon. *Journal of Fish Biology* 62, 1000–1009.

Wagner, G.N., McKinley, R.S., Bjørn, P.A. and Finstad, B. (2004) Short-term freshwater exposure benefits sea lice-infected Atlantic salmon. *Journal of Fish Biology* 64, 1593–1604.

Wells, A., Grierson, C.E., MacKenzie, M., Russon, I., Middlemiss, C., Bjørn, P.A., Finstad, B., Wendelar Bonga, S.E., Todd, C.D. and Hazon, N. (2006) Physiological effects of simultaneous, abrupt seawater entry and sea lice (*Lepeophtheirus salmonis*) infestation of wild, sea-run brown trout (*Salmo trutta*) smolts. *Canadian Journal of Fisheries and Aquatic Sciences* 63, 2809–2821.

Wells, A., Grierson, C.E., MacKenzie, M., Russon, I., Reinardy, H., Sivertsgård, R., Bjørn, P.A., Finstad, B., Wendelaar Bonga, S.E., Todd, C.D. and Hazon, N. (2007) Physiological consequences of premature freshwater return; for wild sea-run brown trout (*Salmo trutta*) postsmolts infested with sea lice (*Lepeophtheirus salmonis*). *Canadian Journal of Fisheries and Aquatic Sciences* 64, 1360–1369.

Wright, D.W., Oppedal, F. and Dempster, T. (2016) Early-stage sea lice recruits on Atlantic salmon are freshwater sensitive. *Journal of Fish Diseases* 39, 1179–1186.

CHAPTER 24

Wild and farmed fish interactions with sea lice

Alan W. Pike

School of Biological Sciences, University of Aberdeen, Aberdeen

24.1 Introduction

It is paradoxical that the presence of sea lice (unless otherwise stated, this term refers to *Lepeophtheirus salmonis* (Krøyer, 1837)) on a fresh run salmon will gladden the heart of an angler regardless of the threat it poses to the long-term survival of the species. We have long known that salmon are natural hosts to sea lice, a widespread natural parasite of salmonids in both the Atlantic and Pacific oceans. Records of natural infections were collated by Pike and Wadsworth (2000) and reports continue to be added as the need for information of wild salmonid infections persists.

Although several species of copepods infect and inflict injury on salmon, most reported research deals with *Lepeophtheirus salmonis*. Separate subspecies from the Atlantic and Pacific oceans exist (Skern-Mauritzen et al., 2014) but researchers do not distinguish between them other than by inference, as they appear to exhibit allopatric distribution.

Prior to the advent of fish farming, the sole source of salmon for consumption was wild fisheries, and the life cycle of sea lice was sustained by mixing of infection-free smolts with a reservoir of infection on wild adult fish, including sea trout in the North Atlantic and Pacific salmon species in western Canada. In such natural circumstances we should assume infection is, for the most part, an incremental process that is unlikely to overwhelm post smolts as they join marine populations.

Salmon fisheries have sustained communities over past centuries but, in recent times, have been subjected to escalating anthropogenic influences including overexploitation (Lackey, 1999; Lenders et al., 2016).

Fishing for Atlantic salmon changed in the 1960s from long lines to monofilament gillnets, and catches peaked in the early 1970s with an annual harvest of 3,500,000 fish (NASCO, 2019a). From that time, catches declined in all areas of the North Atlantic, the effect of a range of anthropogenic influences (NASCO, 2019b). NASCO (2019b) reports fewer than 500,000 fish are now harvested annually. The tightly managed Pacific salmon fishery still yields large tonnages of fish, albeit from declining stock (Lackey, 1999).

Although salmon breeding has existed since at least the 19th century, largely to augment natural spawning, Atlantic salmon farming in marine net pens first began in Norway in the late 1960s

(Tilseth et al., 1991). It was established in Scotland and the Atlantic coast of Canada in the 1970s before becoming established in Ireland and the Pacific coast of Canada in the 1980s. Despite early challenges with disease, the industry has continually expanded to become a dominant supplier of fish to the consumer market (FAO, 2020).

As might have been expected, with hindsight, sea lice emerged as a health issue soon after the advent of farming, as lice from wild salmonids infected farmed fish in pens either through transfer from inward migrating adults or by infected fish seeding the water with nauplii. The problem was not initially considered serious, but infections increased in intensity, and treatment became a necessity (Rae, 1979).

An unfortunate, but inevitable, consequence of the establishment of salmon farming is that the conditions optimal for the industry exist in locations that are home to wild salmonids: the pristine cold deep waters found along remote coastlines. Frequently farms are located in marine bays downstream of rivers from which migrating salmon emerge, so smolts encounter salmon farms on their journey to the open ocean. Facilities are often situated in areas of low economic opportunity where fisheries and sport fishing traditionally offer employment, and where fish farming was initially welcomed by local populations and governments. As the scale of farming increased and the range of problems escalated, the coexistence of the fisheries and farming came under strain and salmon farming under increasing scrutiny. What began as a biological and fish health issue soon became an environmental and political issue and remains so, as an iconic, at risk, wild species fishery is dwarfed by a burgeoning farming industry that continues to expand year on year. It is reported that the Norwegian government's ambition is to see their salmon industry quintuple by 2050 (Vollset et al., 2018).

The key issue in the conflict is whether salmon farming adversely affects the survival of wild salmonids through the transmission of sea lice from farmed fish, especially during their migration to the ocean. Irrespective of the likelihood of resolving this question, are there mitigations that can be applied to prevent wild salmonid smolt exposure to sea louse copepodids of farm origin along their migratory routes?

This chapter addresses the emergence and development of the conflict, reviews relevant research, and considers mitigation processes to minimise the impact of large-scale salmon farming on Atlantic and Pacific salmon as well as on other salmonids, including sea trout.

24.2 Emergence and development of conflict

24.2.1 North Atlantic

Atlantic salmon *Salmo salar* populations in northern Europe and eastern Canada were in decline for many years before salmon farming existed, and it is accepted that this was due to a multiplicity of factors associated initially with anthropogenic activity and now with effects of climate change on sea temperature (NASCO, 2019b). The North Atlantic Salmon Conservation Organisation (NASCO) was established in 1984, in response to decline in the fisheries, to research and promote wild salmon conservation in their Atlantic distribution. Its initial focus of pressures on salmon fisheries soon expanded to include interactions with farmed salmon, especially farm escapes and potential genetic consequences. Sea lice were not mentioned as perceived threats to wild populations, although natural infections existed. Pre-2000 records of sea louse infections in the North Atlantic were reviewed by Pike and Wadsworth (2000). It was only in the following decades, with rapid expansion of salmon farming, that sea lice came to be seen as a potential, then actual, disease threat to wild populations.

The developing recognition that Atlantic salmon farming would exert detrimental effects on wild salmon led two Norwegian institutions in 1990 to convene the first of several meetings to discuss wild and farmed salmon interactions (Hansen, 1991). Sea lice were peripheral to the discussion, being seen as of 'little danger to wild salmon' (Saunders, 1991) and as an organism that 'may cause some problems in wild migratory salmon in sea water' (Håstein and Lindstad, 1991). By this time, annual production of salmon in the eastern Atlantic had risen to 180,000 tonnes (Pike and Wadsworth, 2000). At a second international meeting arranged by NASCO and the International Council for the Exploration of the Sea (ICES) held in 1997, of five papers on sea lice, only one dealt with lice on wild salmon (Jacobsen and Gaard, 1997), although McVicar (1997a) said, surely with unintentional understatement, that 'There has been concern for several years about whether diseases associated with fish farming may affect wild salmonid populations …' At this time annual production of salmon in the northeast Atlantic was approaching 500,000 tonnes (Pike and Wadsworth, 2000), and the genie was truly out of the bottle.

What proved to be a significant turn of events took place in the interval between these two conferences, first in Ireland and subsequently in Norway and Scotland, causing alarm and consternation among the angling community. This came to be known as 'premature return' of sea trout smolts, and was associated with underweight and morbid fish in Irish estuaries bearing severe infections of, mainly immature, sea lice (Tully et al., 1993). It was probably the first occasion that a link was claimed among sea lice, farmed salmon, and wild salmonids.

In 2005, a third international conference, again arranged by ICES and NASCO, convened with specific objectives including 'to ensure that aquaculture practices are sustainable and consistent with the Precautionary Approach' (Hansen and Windsor, 2006). It was further reported by Hansen and Windsor (2006) that 'In their messages, the salmon farming industry representatives acknowledged that it can no longer be claimed that salmon farming does not pose a threat to wild salmon …' which represented a departure from earlier positions held by the industry. As NASCO observed in their 2000/2001 report (NASCO, 2001), establishing a dialogue with the salmon farming industry was difficult despite the Oslo Resolution adopted in 1994 (NASCO, 1994). A liaison group was established by NASCO in 2000 to improve cooperation with the salmon farming industry, and its deliberations were incorporated into a new agreement, the Williamsburg Resolution (NASCO, 2006). This included requirements for risk assessment, minimising impact, and mitigation related to sea louse infection as well as other impacts on wild salmonids.

By 2005, the reported production of Atlantic salmon in the northeast Atlantic was in excess of 800,000 tonnes, an increase of more than 50% from the time of the previous conference (ICES data provided to NASCO (NASCO, 2020)). An additional agreement drawn up by NASCO (2010) sought to detail best management practice for sea louse control and for reporting and tracking infections on wild salmonids.

There seems to have been little more NASCO could do to cajole the salmon farming industry, or governments, to reduce the impact of sea lice on wild salmon populations, which was also in the industry's own best interests, as ineffectively managed louse infections reduce farm productivity (Silva et al., 2021).

A decade later, salmon production in the northeast Atlantic was in excess of 1,600,000 tonnes, and the scale of the industry in terms of production was approximately 1300 times that of the wild salmon fishery (NASCO, 2016) on farms commonly located in the sea lochs and bays through which Atlantic salmon and sea trout smolts migrate.

NASCO convened a special session to address what they recognised as an escalating problem and invited contributions from all salmon farming jurisdictions (NASCO, 2016). The planned meeting was not greeted with universal enthusiasm, and the invitation to the International Salmon Farmers Association (ISFA) to make a presentation was declined (NASCO, 2016). For the first time, questions were asked about the lack of progress in implementing long-standing agreements: 'Why, therefore, has there been so little change at the level of many of those jurisdictions that are party to the NASCO Convention?' (NASCO, 2016).

A recent statement has come from the NGO representation of NASCO demanding that governments of salmon farming countries honour the agreement made by the ISFA and NASCO and embodied in the Williamsburg Resolution (NASCO, 2019a).

For more than 35 years, NASCO and other organisations have sought to implement agreements to manage, *inter alia*, sea louse infections on salmon farms to avoid damage to wild populations. It is clear that this has not had the desired influence on regulation of salmon farms in the North Atlantic or restricted their production, which currently stands at almost 2,000,000 tonnes. However, current Norwegian government legislation links salmon farming management and future expansion with mitigation of effects on wild salmonids.

24.2.2 Eastern Pacific

Pacific salmon are similarly in decline and have been so for many decades, initially through over-exploitation but, more recently, from anthropogenic activity and as a result of a heating ocean. The size of the population is immensely greater than that of Atlantic salmon, but fluctuates widely with respect to both inward migrating adults and outward migrating juveniles. If we consider only pink salmon fry production 1991 through 2008, the minimum number recorded was 730,000 in 1992 and the maximum was ~409,000,000 in 2000, with an overall mean of ~37,000,000 (Jones and Beamish, 2011). Annual adult returning pink salmon numbers also vary over a wide range.

Sea lice occur naturally on Pacific salmon with similar prevalence and intensity as in wild Atlantic salmon, but some species are less susceptible to infection (Beamish et al., 2005).

The sea louse/salmon interaction in the eastern Pacific differs from that of the North Atlantic for several reasons. First, the area occupied by salmon aquaculture in British Columbia, Canada, is home to multiple species of Pacific salmon with different life histories, especially regarding time spent as young in freshwater and timing of out-migration. Second, the population size of the major species of Pacific salmon is vastly greater than that of Atlantic salmon in the North Atlantic. Nevertheless, as far as lice are concerned, apart from the Pacific louse being a separate subspecies, the overall problem is similar.

Commercial Atlantic salmon farming in the coastal waters of British Columbia began in the 1980s, although small-scale rearing of Pacific salmon species began a decade earlier (NASCO, 1994). The switch to Atlantic salmon rearing was inevitable despite its status as an alien species in the Pacific, as it is better suited to rearing in captivity. Initially the industry blossomed, with many limited-size local operations that gradually consolidated into a small number of large international companies. Production in 1990 was less than 2000 tonnes but increased tenfold by 1992, doubled by 1998, and more or less doubled again in the following decade before plateauing at almost 89,000 tonnes in 2019 (DFO, 2021). Compared to Atlantic salmon farming areas in Norway and Chile, British Columbia is a small producer.

The response to commercial Atlantic salmon farming in British Columbia has been robust and enduring, with prominent conservation organisations and other groups outspoken in their disapproval of growing Atlantic salmon in net pens, especially in locations through which migrating juvenile salmon pass. As elsewhere, salmon aquaculture in British Columbia has been a highly emotive development. A review by Bocking (2011) is recommended as a comprehensive and detailed account of the socio-economic and scientific issues involved in British Columbia aquaculture.

Morton and Williams (2003) raised the temperature of the debate with the reported discovery of frequent sea louse infections on juvenile pink salmon from within the Broughton Archipelago, a major salmon aquaculture area. The presence of infected fish was initially reported in June 2001 by a local fisherman, leading to surveys mid-June to mid-August of juvenile fish from areas upstream of fish farms and two areas near farms stocked with adult Atlantic salmon and smolts. Prevalence of infection was 68.4% upstream of farms, 90.6% adjacent to smolt farms (Yr1), and 98.4% adjacent to Yr2 farms. Infection intensity was 2.7, 5.2, and 12.5 for the respective samples. The percentage of fish with ≥1.6 lice g^{-1} taken from areas adjacent to Yr2 fish was 81.2% and for upstream fish, 15.7%. A follow-up study in the Broughton Archipelago and several other sites for ten weeks beginning mid-April 2002 reached similar conclusions: fewer sea lice in areas upstream from salmon farming with fish taken near Yr2 farms significantly more infected than those caught near smolt farms and sites not associated with farming operations (Morton et al., 2004). Most of the sea lice on juvenile fish were chalimus, indicative of recent infection and, as no adult pink salmon were present, the authors concluded that 'In the absence of any B.C. studies directed on sea lice in the salmon farms and therefore any means to relate lice on farm salmon to lice on wild salmon, there do not seem to be any direct tests possible for the concern that salmon farming is involved in this cycle of infections' (Morton et al., 2004). A survey by the Department of Fisheries and Oceans (DFO) in the area reported no serious problems. Both reports (Morton and Williams, 2003; Morton et al., 2004) have been criticised on the basis that the dipnets used to catch juvenile fish were deemed not to provide random samples. The DFO survey was also criticised for being inadequately designed and conducted. Nevertheless, the concern widely expressed when the first of these papers was published prompted the provincial government of British Columbia to establish the Broughton Archipelago Sea Lice Action Plan in spring 2003 requiring salmon farmers, for the first time, to monitor their stock for sea louse infection and to record the data for later auditing. A limit of three adult lice per fish was set for all times of year: 'The Licence Holder must follow all area-based and site-specific sea lice thresholds listed in this licence. If the licence does not list this information, the following generic threshold will apply: (a) sea lice by fish threshold: an average of 3.0 motile *Lepeophtheirus salmonis*' (DFO, 2003).

Recording sex of the lice appears not to be stipulated, but female lice are counted separately according to Saksida et al. (2007). Exceeding the limit triggers a requirement for mitigation through treatment. At the instigation of the plan, a concurrent three-year study was established to audit farm data and independently gather data for comparison (Saksida et al., 2007). Sea louse monitoring has now become a requirement for all Atlantic salmon farms under the Pacific Aquaculture Regulations.

It has been said that sea lice are 'not considered a significant health issue', as farmed salmon appear not to suffer skin damage, at least not in comparison to reported damage in European salmon farming (Saksida et al., 2011a). In an earlier paper Saksida et al. (2007) commented that despite an immigrating population of Pacific salmon six times the total population of inward migrating Atlantic salmon in Norway, assumed to be infected with sea lice, there are no recorded severe outbreaks of infection in salmon farms during this inward migration. It is assumed the authors mean no outbreaks via transfer of adult lice or females seeding the water with nauplii. Marty et al. (2010) reported, to the contrary, that 'sea lice were abundant on farm fishes in 2000' and also stated that treatment was considered in the early 1990s and used in 1996.

As the controversy continued, the BC Pacific Salmon Forum was established in 2005, with a remit to research sea lice/salmon interactions, continuing until 2009 with the publication of its final report and recommendations. The report contains a detailed analysis and review of sea lice on Pacific salmon but is no longer readily available.

In late 2009, the Canadian government established a 'Commission of Inquiry (The Cohen Inquiry) into the Decline of Sockeye Salmon in the Fraser River' (Cohen, 2012). Among 75 recommendations, it contains two that proposed suspension of Atlantic salmon farming (recommendation 18) with subsequent closure (recommendation 19) by the end of September 2020: 'On September 30, 2020, the minister of fisheries and oceans should prohibit net pen salmon farming in the Discovery Islands (fish health subzone 3–2) unless he or she is satisfied that such farms pose at most a minimal risk of serious harm to the health of migrating Fraser River sockeye salmon.'

Recommendation 20 detailed the processes involved leading up to closure. There are no public details relating to meeting the requirement for consultation with the salmon farming industry about implementing recommendations 18 and 19 but over the eight-year period it might be expected that a dialogue would have taken place before the closure, which was enacted in December 2020, in line with the Cohen recommendation, with all Atlantic salmon in the area to be removed by 2022. This was despite the Fisheries and Oceans Canada announcing in late 2020, in relation to the Cohen report recommendation, that 'Assessments concluded that the pathogens [not including sea lice – author insertion] found in Atlantic salmon farms in the Discovery Islands area pose no more than a minimal risk to migrating Fraser River Sockeye salmon' (DFO, 2020). This statement is described as their 'final update' on the department's response to the Cohen recommendations.

The Government of Canada has now published a report of their Standing Committee on Fisheries and Oceans which sets out 32 recommendations among which is: 'That the Government of Canada recognize that the Minister of Fisheries, Oceans and the Canadian Coast Guard cannot both protect wild salmon and promote the aquaculture industry, and remove the promotion of aquaculture from the mandate of the Minister of Fisheries, Oceans and the Canadian Coast Guard' (FOPO, 2021).

Given the impossibility of satisfying both interests it seems an eminently sensible recommendation.

Implementing the Cohen recommendation has raised the ire of many, including the international salmon farming companies supported by the Canadian Federation of Agriculture, and some, Mowi (2021) and Cermaq (2021) for example, have applied for a judicial review of the decision. Either this was an unanticipated decision or the farming companies had ignored 'the elephant in the room' for eight years.

Another area in which Atlantic salmon farming operates is north of the Discovery Islands in the Broughton Archipelago, a network of islands and bays into which 27 rivers flow that are major spawning grounds for pink and other Pacific salmon species. This too, unsurprisingly, is a controversial location for salmon farming, given that the farms are sited along the out-migration routes of juvenile wild salmon.

The stage is now set to test the resolve of the Government of Canada in the light of their stated intention that it 'remains committed to sustainable, environmentally conscious aquaculture, but it must be developed collaboratively and include the voices of indigenous peoples and all Canadians' (Cision Canada, 2020). It also remains to be seen what 'sustainable environmentally conscious aquaculture' means in practice, although the government have supported the concept of closed containment aquaculture systems (FOPO, 2013; DFO, 2019). Testing times lie ahead.

24.3 Evidence of interactions

Since 2005, ten published reviews (Heuch et al., 2005; Costello, 2006, 2009; Todd, 2007; Jones, 2009; Krkošek, 2010; Johansen et al., 2011; Brauner et al., 2012; Torrissen et al., 2013; Thorstad et al., 2015) and two reports (Whelan, 2010; Thorstad and Finstad, 2018) have focused specifically on the interactions between sea lice and wild and farmed salmonids. The serious reader is recommended to read these for background of the topic. For those wishing to contextualise these interactions in the wider anthropogenic activity, the review by Hodgson et al. (2020) is commended. Thorstad et al. (2012) present a comprehensive review of smolt and post smolt migration and behaviour: a useful background to the host biology at the critical migratory phase.

It is not the purpose of this chapter to present an exhaustive review of the extensive literature on sea lice/salmonid interactions, but rather to provide the reader with context to inform a view of the long-standing conflict between the aquaculture industry and the salmonid fishery interests and to engage with the urgent need for mitigation.

24.3.1 Sea trout infection

The anadromous brown trout *Salmo trutta* L., or sea trout, although of lesser value as a fished species is nevertheless an important component of sport fishery in the North Atlantic. They are naturally infected with sea lice, as confirmed by studies on the east coast of the UK where there is no salmon farming (Tingley et al., 1997; Urquhart et al., 2008) and in the Irish sea (Gargan et al., 2016). They also played a pivotal role in bringing sea louse infections of wild salmonids to our attention.

Although government scientists and others became aware of the risk that sea lice posed early in the development of salmon farming, and international symposia convened in the early 1990s to share experiences, it was reports of an epizootic of sea lice on sea trout in Ireland (Whelan, 2010) that ignited the debate about sea louse infections and salmonid survival. The epizootic coincided with an extensively documented stock collapse (1.7% survival) in 1989. Briefly, post smolts and kelts of sea trout returned to their natal rivers with substantial infections of sea lice and in an emaciated condition, although some emaciated returning adults were not infected with sea lice. A particular feature of the event was the early return of post smolts, reportedly within two to three weeks of emigration, a behaviour described as premature return. A synthesis of the data suggested that the epizootic might be attributed to (i) increased availability of copepodids from nearby salmon farms; (ii) higher water temperatures in both winter and summer, increasing the proliferation of sea lice; and (iii) changes in host susceptibility through a diminished immune response, although there was no evidence supporting such change (Tully et al., 1993). In respect to the last, McVicar (1997b) reminds us that sea lice infections may be opportunistic invasions of otherwise compromised hosts. Sea trout, as do salmon, face multiple anthropogenic perturbations that have been enumerated by Thorstad et al. (2016) including pollution, physical obstructions in freshwaters, and climate change as well as overexploitation and disease. Nevertheless, the involvement of sea lice continues to feature prominently in published work.

Studies in Scotland (Hatton-Ellis et al., 2007) and Norway (Birkeland and Jakobsen, 1997; Birkeland, 1996) confirmed the phenomenon of premature return in those areas. The interested reader is referred to Whelan (2010) for an authoritative review of the timeline of the sea trout problem in Ireland as well as instances of premature return elsewhere.

Initial research into sea louse infections of sea trout consisted of short-term studies or examination of historical records. Birkeland (1996) examined tagged sea trout post smolts from fish traps for sea lice from June to September 1992 and found 96% prevalence and mean intensity of 272, mainly copepodid and chalimus larvae. Returning fish deloused in freshwater before returning to the sea, with more older returning trout dying than post smolts. The river in which this study was conducted opens into an estuary containing salmon farms within 300 m of the river mouth. A concurrent experiment at the same location by Birkeland and Jakobsen (1997) exposed tagged out-migrating sea trout and Atlantic salmon smolts to naturally occurring sea louse copepodids in a cage experiment with unexposed fish as a control. After 12 days, infected and control fish were released into the estuary, and returning fish, entering a trap in the river, were counted. Sixty per cent of infected trout returned on the release day compared to 3% of control fish. However, control fish, now with heavy natural infections of sea lice larvae (median 175), returned within two days of release and acquired much higher infections after four days (median 450). Analysis of blood samples from infected fish on the release day showed significantly reduced chloride, total protein, and albumin, indicative of osmoregulatory stress.

Middlemas et al. (2013) completed a six-year study (2003–2009) of sea trout caught in sweep nets from 48 sites across western Scotland in which half of 6590 were infected with sea lice. Almost 75% of fish were <20 cm L_F. They found a significant negative association of infection with distance from salmon farms, and a positive association between infection and size of farmed salmon. The probability of trout suffering infections above the critical threshold of 13 mobile lice per fish was high for fish within 30 km of salmon farms. The authors conclude that the 'risk and magnitude of effect on sea trout' will be greater the closer farms are situated to natal rivers. These results could be relevant to decision-making with regard to siting of farms. In relation to the correlation between farmed salmon size and infection of sea trout, Middlemas et al. (2010) report a ten-year study of sea trout in western Scotland finding a broad relationship between sea louse infections and farms with adult fish (with a likelihood of higher parasite burdens) but caution that a range of variables may make this correlation not universal. Nonetheless, they interpret their results as supporting the concept of a link between sea trout infections and salmon farm production cycles.

Shephard et al. (2016) examined an extensive database of sea trout post smolts (n=7714) sampled May through June from 1991 through 2015 in Ireland, and May through July in Scotland from 1997 through 2015 (n=16,768). Locations were diverse and varied in proximity to salmon farms. Sea louse infections were analysed relative to water temperature, rainfall, and distance from a salmon farm, using a generalised linear mixed model (GLMM). There were significantly more sea lice on post smolts caught near salmon farms, an effect greater in years of high water temperatures. Body condition was also correlated with proximity to salmon farms, suggesting that sub-lethal effects of sea louse infection may be a risk factor for trout populations. The authors claimed that their analysis, accounting as it did for environmental variables and biogeographic variation, has implications for sea lice control measures, regulation, siting of salmon farms, and the establishment of farm-free areas.

With regard to farm-free areas, a five-year study (2008–2012) of nine large nationally protected areas at two times in the year on the west coast of Norway revealed 70% of the 857 sea trout and Arctic charr caught in the four most-protected fjords (>30 km from a salmon farm) to be louse free. Of the 42 samples taken 71% were <50% infected, and with low intensities. In the five small less-protected areas (<30 km from a salmon farm), 1268 sea trout and Arctic charr were examined. Prevalence was variable and frequently over 63% with high intensity. Linear regression analysis revealed that 41% of the variance in mean lice abundance was explained by sea lice from salmon farms (Serra-Llinares et al., 2014). The authors concluded that protected areas are effective but only when separation from salmon farming areas is sufficient to ensure low lice populations.

In Iceland, where salmon farming is still small-scale, Karbowski et al. (2019) reported louse levels, mostly observed on sea trout, to be below the critical threshold even in proximity to salmon farms. They caution that, while this is currently the case, if aquaculture expands, sea lice are likely to become problematic.

The mentioned research demonstrates the consequences to migrating sea trout of being exposed to sea lice as they transition from fresh- to seawater, a time of significant physiological challenge. Laboratory evidence amassed by Wells et al. (2006, 2007) shows that 13 mobile sea lice on fish in the weight range 19–70 g is the threshold beyond which louse infections impose alterations to a suite of physiological variables on sea trout. These variables revert to normal values when fish return to freshwater, mitigating the physiological stress, but doing little to compensate for the nutritional deficits incurred through leaving the superior feeding grounds of the ocean.

Attempts to characterise the impact of sea lice on sea trout have also focused on quantification of their effect on migrating sea trout smolt survival at sea and subsequent re-entry to natal rivers. Three studies have employed telemetry to monitor post smolts after prophylactic treatment intended to protect them during their initial migration. The data obtained by this method may lack robustness, as documented returns are low.

Skaala et al. (2014) reported sea lice counts and injuries on ascending tagged sea trout from 2001 to 2011 in the River Guddalselva, which flows into Hardangerfjord, Norway. In 2004–2005, half of the descending smolts were treated with an externally applied chitin inhibitor to protect against sea lice. For the entire 11-year period, 80–90% of returning fish exhibited damage to dorsal fins attributed to sea lice infection, with 50% of these showing moderate to severe damage. The proportion of prophylactic-treated fish that returned was 3.41%, while 1.76% of those left untreated returned. In the final year of the study, 100% of sea trout sampled in the fjord in May and June were infected, with an average intensity, in late May, of 50 lice. The authors conclude that the lower survival of untreated fish points to sea lice infections as 'an important contributor to the high mortality' of sea trout in the fjord.

Using acoustic telemetry, Gjelland et al. (2014) tracked 30 naturally infected sea trout post smolts, half of which were treated with emamectin benzoate as a prophylactic, in an area protected from salmon farming but exposed to infective larvae from adjacent areas. The prophylaxis did not affect behaviour, and most fish stayed close to the natal river from which they were released. The conclusion that this was driven by high sea louse pressure has to be tentative, as no fish were recovered to assess louse levels and the numbers of fish were small.

Halttunen et al. (2018) tagged 310 sea trout over three years, half of which were treated with one of two prophylactics before all were released for tracking. Just over a third were successfully tracked at sea or freshwater. There was no difference in survival among groups and, overall, fish spent more time closer to, or in, freshwater during high infection years than in years seeing low infection when farms were fallow. The evidence is circumstantial, but may imply that in years of high infection pressure fish remaining near, or in, freshwater would show lower growth rates with potential population consequences.

The cited studies including fish treated with prophylactics did not yield conclusive results, mainly due to low recovery numbers and small sample sizes. A different approach was adopted by Serra-Llinares et al. (2020) in a pilot study where 144 acoustically tagged sea trout post smolts (mean L_F 18.0 ± 1.4 cm), 71 uninfected and 73 sea louse copepodid pre-infected, were released into a fjord around

85 km distant from salmon farming activity, and tracked for five months. Pre-exposed trout were 100% infected with a mean intensity of 65.4 ± 30.6 lice per fish, well above the potentially lethal level. Approximately 60% of released trout migrated to sea and were not recorded again. Twelve pre-infected and six uninfected trout were tracked back to freshwater with the time to return significantly shorter for those pre-infected than for those uninfected, 18 ± 26 and 100 ± 49 days, respectively. Survival was lower in infected fish, but the small sample sizes rendered the statistical power of the results low.

These studies illustrate the challenge of obtaining reliable and robust data of the survival of exposed smolts in the ocean to complement the observational evidence of heavily infected fish returning to freshwaters in areas where salmon farming is prevalent.

There is currently substantial documentary evidence of higher louse infection on sea trout taken from areas in proximity to salmon farms compared to those in areas distant from farms. There is also correlative evidence of premature return to freshwater among sea trout exposed to higher than normal sea louse infections, in particular of chalimus stages. Investigation of premature return by field experimentation has shown that infected fish return to freshwaters prematurely, but unequivocal evidence is lacking or, in some instances, circumstantial. Application of prophylactics to determine whether uninfected fish show better survival potential has suggested this to be the case, but the evidence is again weak.

Recently, two Norwegian studies have reported on the development and use of modelling to help predict the effects of sea lice on sea trout under varying conditions. Hedger et al. (2021) created an individual based model to investigate how sea lice infections may affect sea trout abundance and premature return behaviour using a 25-year dataset from two northern rivers. The simulations indicated a decline in abundance of sea trout after week 30 due to mortality and premature return although, on the basis of the parameters used, this was not considered likely to lead to population elimination. The model, despite admitted shortcomings, will be a very useful tool to investigate the consequences of premature return for affected sea trout populations. The second study, by Finstad et al. (2021), describes the use of existing computer models to create proxies for risk assessment of exposure of sea trout to sea lice in aquaculture production areas and the consequences for their access to available marine habitats. The study evaluated two parameters: reduced marine living area and reduced marine feeding time, to predict the effect of sea lice on sea trout populations. Late migrating trout were more affected than early migrating fish, but there were serious potential negative impacts influenced by spatial and temporal variables. The authors suggest that their methodology can be used in other areas where data on sea trout and sea lice are available.

Sea trout is a valuable species for characterising interaction between farmed and wild salmonids because their seaward migrations after smolting take them to near-shore locations where they remain to potentially acquire infections from salmon farm dispersal, thus supporting a reservoir of adult sea lice to infect their own populations as well as salmon and Arctic charr. While this is a natural cycle of events in the absence of salmon farms, the question remains of how much salmon farm infections contribute to the pool of infective sea louse copepodids. Qualifying the role of sea lice, from whatever source, in long-term survival of emigrating sea trout smolts in these coastal areas remains elusive.

24.3.2 Atlantic salmon

Initially the reports of premature return seemed to have side-lined research into the interactions of wild and farmed salmon with sea lice, but McVicar (1997a) commented on the matter, posing the questions ‘does the farm disease alter the characteristics of the same disease in wild populations?’ and 'does the farm provide a focus or enhancement of disease to wild populations?' In drawing conclusions from an ICES workshop on interactions between salmon lice and salmonids (ICES, 1997), the only firm statements made referred to effects on sea trout (McVicar, 1997a). Almost 20 years after salmon farming was established, an ICES workshop did not make definitive statements although, as mentioned earlier, salmon farming was now producing 500,000 tonnes annually.

A seven-year survey (1992–1998) by Finstad et al. (2000) was among the first to record naturally acquired lice on out-migrating salmon post smolts. Only chalimi were present indicating recent infection, and prevalence varied widely among years, with a maximum of 54%. In the first five years, post smolts showed fewer than ten larvae per fish, but the intensity of infection in the final year approached the lethal level.

Skilbrei and Wennevik (2006) conducted a field experiment with wire-tagged Atlantic salmon smolts (~15 cm L_F) in which three groups of fish, including controls and prophylactic-treated groups, were released in early and late May and early June 2002 into the Dale River, an area of active salmon farming. A fish trap in the river caught returning adults (one, two, and three sea winter (SW) fish), which were weighed, and recapture rates and growth rates were calculated. Survival of 1 SW returning adults was higher in the prophylactic-treated fish released in June, but there was no difference in survival from the May releases. When the three SW categories were combined, survival was significantly higher for treated fish compared to untreated, and treated smolts were significantly heavier than untreated.

An early acoustic telemetry experiment in Romsdalsfjord involving the release of farmed post smolts of salmon and sea trout (L_T ~25 cm) with no infection, infected with copepodids, and uninfected and treated with a prophylactic revealed no difference in short-term survival. The outermost receiver in the detection array, approximately 70 km from the release site, detected salmon post smolts of all groups an average of 151 h post-release when infections would have still been chalimus. The salmon were not subsequently detected, but sea trout from all treatments were recorded in the fjord an average of 27 days post-release when louse infections would have comprised pre-adults or adults. The authors conclude that chalimi do not seriously affect migrating fish, although the fish used were larger and therefore better able to withstand the infection (Sivertsgard et al., 2007).

Prophylactic-treated and untreated tagged hatchery-reared salmon smolts were released each year from 2001 through 2008 in the Burrishoole catchment of Ireland (Jackson et al., 2011a), an area with little salmon farming. Records of returning adults showed significantly more treated fish returning in four of the ten years analysed. There was also a significant decline in survival of both treated and untreated salmon over the study period. Data from a related study (Jackson et al., 2011b) of similar design in four other geographical areas showed two locations with significantly more prophylactic-treated fish returning. The authors nevertheless concluded from these studies that sea lice are a minor factor in salmon mortality at sea in the studied areas.

A large-scale study by Gargan et al. (2012) followed prophylactically treated and untreated hatchery-reared tagged salmon smolts released at eight locations in three river systems in western Ireland, 2003–2005. Migration distance in the selected sites was short and passed active salmon farms. Fish were released in April prior to the wild salmon out-migration. The results varied, perhaps reflecting prevailing conditions in the locations, but the median ratio of treated to untreated returns was 1.8:1. Gargan et al. (2012) interpreted their results as indicating that sea lice can have a significant effect on the survival of post smolts at sea with consequences for adult returning salmon. A meta-analysis of these three studies by Jackson et al. (2013) also concluded that sea lice are not a major factor in salmon smolt mortality, a determination not shared by Gargan et al. (2012).

The data from the Irish studies have been combined with those of two studies in Norway (Skilbrei and Wennevik, 2006; Hvidsten et al., 2007) and re-analysed using three statistical protocols not employed in the original papers (Krkošek et al., 2013). The authors concluded that all collected data point to the benefits of pre-treatment with a prophylactic, although the mean odds ratio indicating that infection affects survival was small (1.29). The conclusion, the authors state, is confirmed by two further analyses revealing a 'large effect of parasites'. There has been no resolution to the conflicting conclusions reached.

Research by Skilbrei and Wennevik (2006) formed part of a longer-term study reported by Skilbrei et al. (2013) for the period 1997–2009 in which reared smolts and a mixture of wild and wild-caught reared smolts were wire tagged, and half were treated with a prophylactic and released with untreated smolts in the lower reaches of the River Dale, Norway, and in neighbouring locations. Of 35 releases, only three showed lethal effects of sea lice, but the proportion of returning adults was higher in treated fish. The authors conclude that, at an odds ratio of 1.17, somewhat smaller than the meta-analysis of Krkošek et al. (2013), treated smolts survive better than untreated smolts. The protocols were altered several times over the study period, and the authors include caveats and cautionary messages, such as noting that survival rates varied both within and among years. Later migrating smolts suffered higher marine mortalities.

Sea lice show moderate influence on salmon populations, but such impacts should be considered in conjunction with other negative pressures. Reduced growth and fecundity are also thought to be a consequence of infection, pointing to sea lice being a risk factor for small or vulnerable wild salmon populations.

A more sophisticated meta-analysis by Vollset et al. (2016) used data from Norway derived from 118 published and unpublished sources comprising over 650,000 salmon smolts in two groups: untreated or treated with a prophylactic. Although this was a large dataset, it came chiefly from a restricted area north of Bergen with fish most often released close to the ocean, meaning short migrations with little predation pressure or exposure to sea lice infection. The aim of the analyses was to (i) estimate the effect of pre-treatment on smolt survival and (ii) investigate whether differences in survival were related to infection pressure from salmon farms along migration routes. The main results were (i) treated 1 SW salmon were heavier than controls. This difference was not present in 2 or 3 SW fish, and length of returning fish did not differ; and (ii) the overall risk ratio (RR), the value derived from dividing the probability of a treated fish being recaptured by the probability of an untreated fish being recaptured, was 1.18, but showed wide heterogeneity. Location, year of release, and baseline survival (proportion of untreated fish recovered) differed significantly between the groups. Baseline survival was an important predictor, whereby for every unit increase in survival, the RR declined by 0.24 unit. The RR varied among four unequal release periods examined. Baseline survival was best predicted by distance from release to the sea, decreasing by 0.04 unit for each km distance from the ocean. Treated fish showed a significant, but slightly, higher survival rate, implying a sea louse contribution to salmon mortality with the caveat that salmon populations are also subject to other risks, and factors impacting their survival must be considered collectively.

Kristoffersen et al. (2018) collated and modelled valuable data of factors affecting smolt survival in 401 rivers in Norway and developed a risk assessment tool to evaluate sea louse-induced smolt mortality. The authors, benefitting from the extensive registration and regulatory oversight of salmon farms, modelled out-migration of salmon smolts and exposure to sea louse infection to develop an overall risk model of parasite-induced mortality. Components comprised data of infection pressure derived from farm louse infection reporting, smolt infection development from sentinel cages, salmon fecundity and migration timing in 80 selected rivers, post smolt marine out-migration, and parasite-induced mortality.

Louse egg output per farm was significantly correlated with local biomass density which, in turn, significantly increased infection pressure on fish in sentinel cages. Increased infection pressure was predicted to increase louse abundance on post smolts, depending on timing of out-migration, with later migration leading to higher infections. Production zones varied widely in predicted sea louse transmission, with western zones exhibiting higher transmission than more northerly and easterly zones, and consequently suffering higher parasite-induced mortalities. The authors consider that the risk model can effectively predict spatial and temporal sea louse risks linked to salmon farming and is intended to inform political decision-making regarding production capacity.

The most recent study of treat and release, using nearly 30,000 tagged hatchery-reared smolts divided between prophylactically treated and untreated groups, was conducted in Hardangerfjord, Norway, an area of high infection pressure (Bøhn et al., 2020). The study was repeated over two seasons with smolts released in May and June each year and returning adults recorded over the following four years.

Sentinel cage data showed infection pressure increasing from 0.06 to 0.66 lice per fish between May and June of the first season and 0.37 to 8.87 lice per fish in the same period of the second season. The annual difference in infection pressure reflects differences in salmon farming activity.

Treated and control fish usually returned after two years in the ocean, although of the first May release, the control to treated return ratio was 2:1, whereas in June of the first year and May the following year the return rates did not significantly differ. The June release in the second season showed highly significant differences, with 70 treated fish but only a single control fish returning. The overall risk ratio of 55.3:1 represents an extreme mortality rate in untreated salmon smolts, especially at the time of high infection pressure in June of the second season, albeit based on a low rate of return for both groups. The authors argue that their results provide 'compelling evidence that lice can have a devastating impact' where they are present at high densities, but may be mitigated by suppressing farm infections at migration times.

An important and intriguing modelling study is reported by Myksvoll et al. (2020) in production zone seven in Norway, covering the years 2016–2018. This zone encompasses an area that contained 50–58 active salmon farms during the study period as well as, importantly, the Namsen Fjord, protected waters from which a potential 900,000 migrating salmon can emerge from 21 rivers. Despite its importance, the production zone was designated for expansion. The study was established to model the potential effects of sea louse proliferation in the area through which migrating post smolts pass. The model incorporated data of salmon production in 2012–2018; louse counts from farm records for the period; and hydrodynamic data on currents, temperature, and salinity. Using particle tracking software, the authors modelled louse larval dispersion, which was calibrated with sentinel cage infections. Louse dispersion varied widely with season and was particularly affected by wind direction. Northerly winds drive lice offshore to disperse over a wider area, whereas south westerly winds push lice towards the shore, creating larger concentrations inshore. Neither of these conditions favour smolt survival, with modelled mortalities reaching the national sustainability threshold in two of the three years. These levels are likely to be exceeded if salmon production increases as predicted.

Numerous studies, mostly in Norway but also in Scotland and Ireland, have established that migrating salmon smolts are infected by sea lice as they enter the ocean for the first time, but, unlike sea trout, they continue their onward migration, and their fate is unknown until returning adults can be counted. Prophylactically treated smolts return in greater numbers than untreated and show better growth. Irrespective of the source of sea louse copepodids, the indications are that sea lice pose a risk for wild salmon, demanding application of a precautionary approach to aggressively manage copepodid populations along migration routes. This could be accomplished by a variety of farm management practices or, more effectively, by complete separation of farmed and wild fish.

24.3.3 Pacific salmon

Just as the publicity given to premature return of sea louse-infected sea trout catalysed research into the interaction between sea lice and wild and cultivated Atlantic salmon, so the report of sea lice on out-migrating juvenile pink salmon activated research into the putative role of salmon farms in infecting wild migrating juvenile Pacific salmon.

As was the case for Atlantic salmon, knowledge of natural infections of adult wild Pacific salmon from the Canadian Pacific coast was limited to anecdotal observations and early taxonomic surveys in British Columbia (Kabata, 1973), along with surveys conducted in the North Pacific (Nagasawa, 1985, 1987; Nagasawa et al., 1993). In response, Beamish et al. (2005) conducted a survey of all Pacific salmon species taken from two locations in August, one in a salmon farming area and one in an area free of farming. Most species of salmon were in-migrating adults and were deemed to have acquired their adult lice offshore. The source of chalimus stages, which constituted 53% of the infections, was not evaluated. Almost all fish in both areas were infected with similar infection intensity, regardless of species. The authors do not comment on this, but noted that sea louse infections on wild immigrating salmon could be a matter of concern for salmon farmers. There was no indication that infections increased on farms during the inward migration.

Since the discovery of sea lice on migrating juvenile salmon, numerous surveys have been conducted to gather more data on their occurrence. These covered the period 2003–2012 and were focused on pink and chum salmon collected by beach seine from March through July by Patanasatienkul et al. (2013) and Jones and Hargreaves (2007) in the Broughton Archipelago; Price et al. (2010) in multiple areas including Broughton and Bella Bella; Butterworth et al. (2008) in the Finlayson channel; Morton et al. (2008) in the Discovery Islands; and Saksida et al. (2011b), who sampled an area north of the Broughton Archipelago. Some programmes included sites up- and downstream, or adjacent, to salmon farms, while other reports were not specific. The surveys confirmed the presence of infections on migrating juvenile fish in all areas studied. The prevalence and intensity varied among and within studies, attributed in part to variations in sea temperature and salinity.

The reports of lice on out-migrating juvenile Pacific salmon stimulated ongoing research to determine whether Atlantic salmon farms are the source of infection and if sea lice contribute to mortality of juveniles and threaten wild Pacific salmon populations: the same question that has seen serious investment of time and financial resources for research into sea lice and wild Atlantic salmon.

Acquiring direct unequivocal evidence that salmon farms are a source of infection in wild fish is not straightforward, especially as infections occur naturally. Krkošek et al. (2005) set out to investigate the link by seine netting migrating juvenile pink and chum salmon at 1–4 km intervals along the Knight and Tribune channels in the Broughton Archipelago where fish passed near an active Atlantic salmon farm (the report withholds specific details of the site and locations, but partial information is given in Frazer (2008)). Louse infections, including developmental stages and adults, were recorded, and the accumulated data were modelled to predict the effect of a salmon farm on the migrating juveniles. The authors suggest that infection pressure on juveniles caught near a salmon rearing facility was 73 times the ambient level and persisted for 30 km downstream. Furthermore, lice infecting migrating fish were claimed to reach maturity and re-infect fish within the migrating population.

This was the first serious field investigation to address the question of origin of sea lice on migrating juvenile fish. The data were subsequently remodelled by Frazer (2008). It is unfortunate that details of the locations involved are incomplete and imprecisely presented in the two papers. However, Frazer describes the farm (second hand) as, 'At the junction of Knight Inlet and Tribune Channel there was an operating sea cage salmon farm stocked with Atlantic salmon (*Salmo salar*) carrying sea lice.' It is not certain that this is the farm in the channel; if not, no information on the salmon farm is provided. Neither paper gives information on whether the farm was stocked with Yr1 or Yr2 fish. However, additional details of the location and farming activity are given by Brooks (2005) who states: 'Doctor Islets was being harvested in the spring of 2003, and it was fallow on and after May 19, 2003. Therefore, there were no farmed fish containing gravid sea lice in this area during their final sample period when Krkošek et al. (2005) again reported peak abundance adjacent to the farm.'

The conclusions drawn by Krkošek et al. (2005) were challenged by Brooks (2005) who argued that the source of infection was not salmon farms but the multiple host species in the channels and that hydrodynamic and other abiotic variables rapidly transported louse larvae downstream of the farm. Additional exchanges (Brooks and Stucchi, 2006; Krkošek et al., 2006) have not brought resolution of this disagreement about the source of infection with sea lice. As is acknowledged, more research is required to evaluate the role of abiotic variables in louse dispersal and, thus, their availability to susceptible hosts.

The presence of planktonic sea louse larvae is another avenue for investigation taken up by Morton et al. (2011) with sampling of migrating salmon near farms differing in production phase and louse treatment. Individual farms contained broodstock, mature salmon being harvested, Yr1 and Yr2 fish, fallowed pens, and farm-free control sites. Nauplii were absent from plankton tows in control and fallow sites, and recovery of larvae from active sites showed gradation from lower to higher numbers as farmed fish age increased. Lice on juvenile fish sampled in the areas followed a similar trajectory, as might be expected. Noteworthy is the greater presence of copepodids on fish sampled near farms compared to more advanced stages on migrating fish taken further downstream of farms. Background levels of natural infection were present, including in control areas, but were low compared to farm-derived sources.

It is generally accepted that background levels of sea lice are present on juvenile fish upstream of salmon farms and in areas without salmon farms, which is unsurprising, as a natural cycle exists among wild salmon. Since inward migrating Pacific salmon carry infections of sea lice, any juvenile smolts inter-mixing with these fish might be expected to acquire new infections from this natural reservoir, albeit at a point when body size has increased to make the fish better able to tolerate infection. Morton and Williams (2006) attempted to measure this putative effect by sampling juvenile pink and chum salmon before and during the inward pink salmon migration. They were not able to detect a measurable increase in existing infection levels before the arrival of adult salmon and found no increase in copepodid numbers, which would be indicative of new infection. Gottesfeld et al. (2009) recorded an increase in all developmental stages of sea lice on juvenile pink salmon in a farm-free location from May through July that coincided with adult salmon return.

It has been frequently suggested that salmon farms can operate alongside wild fisheries and that sea lice are not a serious problem for British Columbia Atlantic salmon farmers. Beamish et al. (2006) compared infections on out-migrating juvenile pink salmon in 2003 with returning adult salmon in 2004 and collected sea louse data on salmon farms and juveniles from six rivers contributing the bulk of pink salmon in the Broughton Archipelago. Prevalence of infection on migrating juveniles was 13.1% in March rising to 30.3% in June with a mean intensity of 1.7, of which chalimi predominated in March/April and adult lice contributed 30% of the abundance in June. Farmed salmon louse prevalence was highest in February (85%) declining to 46% in August. Abundance in the study period declined from 9.0 to 1.6, with greater numbers of chalimus than adults at all times. Survival, based on immigrating adult salmon, was estimated at 34%. The authors ambitiously conclude, on the basis of this single season, that sustainability of wild and farmed salmon populations in a common ecosystem is feasible. A longer-term study that would allow a more substantive assessment of what might turn out to be an exceptional event may be of value.

Krkošek and Hilborn (2011) compared escapement data for sea louse-exposed and unexposed pink salmon from the Broughton Archipelago to evaluate productivity in the two populations. The results indicated that abundance of exposed fish declined to a greater degree relative to unexposed fish populations. They conclude by recommending a large-scale experiment to investigate the effect of manipulating sea louse abundance on salmon to address a causative link between lice and salmon productivity.

Marty et al. (2010) analysed historical data of fish farm louse infections and pink salmon escapement records to address published concerns that sea lice could cause population extinction (Krkošek et al., 2007a). They report that immigrating pink salmon numbers predict female sea louse numbers on farmed salmon in the following spring, and that farm counts predict 98% of the variation of sea louse prevalence on out-migrating juveniles. They also state that wild salmon productivity is not negatively correlated to farm lice numbers or farm production and conclude that their results do not support the contention that separating farmed salmon from wild salmon (presumably migrating juveniles) will increase marine salmon survival. Shortly after this paper was published, Krkošek et al. (2011) reported on a re-analysis of the dataset from Marty et al. (2010) using different modelling techniques and claimed that sea louse abundance on farms is negatively correlated with productivity of both pink and coho salmon in the Broughton Archipelago in direct contradiction to Marty et al. (2010).

In relation to farm management for sea lice and compulsory monitoring to determine treatment Rogers et al. (2013) modelled sea louse population data from 25 farms from 2002 to 2008, considering water temperature and salinity, host density, and SLICE treatments. Their results confirmed industry best practice that medication over winter may provide the best protection for spring migrating juvenile salmon in the Broughton Archipelago.

Peacock et al. (2020) used more advanced techniques to analyse data of dispersal of larvae from farms with respect to infection processes and their development on migrating fish, factoring in spatial and temporal variables to advance understanding of how farms affect infection dynamics of migrating fish. Their conclusion, as for Rogers et al. (2013), was that pre-emptive treatment prior to migration and siting of farms as far as possible from runs would minimise their impact.

After a government-imposed fallowing event in 2003, following a dramatic fall in pink salmon escapement Morton et al. (2005) reported significantly reduced sea louse counts during this fallow period with a subsequent return to ambient levels. They conclude that these results are further evidence of a link between farmed salmon infections and those on migrating juvenile fish and suggest that improved collaboration between researchers and salmon farmers could advance knowledge of the relationship. Werkman et al. (2011) studied fallowing in Scotland by network modelling to investigate transmission routes in scenarios including synchronised, partially synchronised, and unsynchronised management of farms. The importance of epidemiologically separate management zones and synchronised fallowing was stressed. Guarracino et al. (2018) on the other hand question the effectiveness of synchronised fallowing in management zones in which high infection pressure may return via adjacent farming zones.

The confrontational tone evident in the correlational studies documented above and the unusual degree of controversy over their interpretation, as evidenced by frequent comments, rebuttals, and re-analyses, have possibly run their course, and future research would be most effectively directed towards a more consensual and collaborative approach to investigation, both within the scientific community, including government scientists, and with the aquaculture industry.

Peacock et al. (2013) documented declines in sea louse numbers on out-migrating juvenile pink salmon, which they attributed to parasite management on Atlantic salmon farms. Their results provide significant insight into the potential of basic research to create a transformation in the perception of threats to wild Pacific salmon from farmed salmon. To quote their statement in its entirety:

> The changes in parasite management we have documented occurred during a period of intensive scientific study. During this period, many multi-stakeholder processes connected research scientists with fisheries managers, aquaculture veterinarians, policy representatives, conservation organizations, First Nations, eco-tourism operators, commercial-fishing interests, and other groups. Examples include the Broughton Archipelago Monitoring Program (available online), the British Columbia Pacific Salmon Forum, and Simon Fraser University's Speaking for the Salmon Series. These processes may have been vital in exchanging and interpreting scientific information that aided a response from management during a period of high scientific progress and uncertainty.

This seems to strike the conciliatory tone that is needed.

Bateman et al. (2016), in reporting the nature and circumstances of the failure to control sea lice on salmon farms, make sensible suggestions regarding cooperative management and suggest changes to the regulation of sea louse levels on single farms and to area management practices, as exist outside Canada.

The excellent review by Brauner et al. (2012) emphasises that empirical data are needed to remove uncertainty regarding the direct impact of sea lice infestations attributed to aquaculture on wild fish populations. The work of Grant et al. (2009), Nendick et al. (2011), Sackville et al. (2011), and Brauner et al. (2012) signal the way forward in determining the effect of sea lice on Pacific salmon juvenile fitness. Without doubt, closer ties and collaboration between the salmon farming industry and researchers will facilitate long-sought changes to mitigate the burden of sea lice infections on Pacific salmon populations during an extremely vulnerable period in their development (Brauner et al., 2012; Vollset et al., 2018). Compared to the Atlantic salmon industry elsewhere, that in British Columbia is small, but there is clear political desire for expansion for socio-economic reasons. Salmon farming in Europe vastly outpaced the progress of science and regulatory policy needed to control its growth, to the point at which it expanded beyond control. The awareness and knowledge available to Canadian political, economic, and environmental interests should be marshalled to create a partnership structure in which all stakeholders have an equal voice and their respective priorities equably treated.

24.4 Mitigation

This review of selected literature on the interactions among wild and farmed salmon and sea lice confirms that sea louse infections on farmed Atlantic salmon are transmitted to wild migrating salmon in the North Atlantic and the east Pacific oceans. Agreement on the extent of negative consequences is a work in progress, as correlative studies, however sophisticated, cannot establish an unequivocal causal link, and accumulation of empirical evidence of harm lags behind modelling studies (Brauner et al., 2012; Vollset et al., 2018). Nevertheless, if the precautionary principle is invoked, there is sufficient evidence to demand that mitigation be applied, although Vollset et al. (2018) caution against over-enthusiasm in applying controls, in order to avoid adverse economic consequences. It is not clear what these consequences might be, or to what extent they should override conservation interests.

Salmon are iconic species that have been exploited by humans for hundreds of years. Their spawning habitats have suffered serious anthropogenic interference, which, along with heating oceans, threatens their existence. In light of this multiplicity of threats, the additional burden of sea lice on vulnerable young juvenile fish migrating to the oceans has to be viewed seriously and not dismissed as inconsequential in comparison with the benefits of a burgeoning salmon aquaculture industry. Critical to any debate is recognising the importance of salmon as a keystone species in the ecosystems they inhabit.

Krkošek et al. (2007b) have drawn attention to the life history of salmon, characterised by generations that are separated spatially and temporally allowing migrating juveniles to enter the ocean with little to no sea louse infection at a time when they are physiologically and immunologically vulnerable (Brauner et al., 2012). Sea lice bridge the generation gap through their presence on coastal salmonids, but acquisition of lice by juvenile fish from this source may be expected to occur at a rate leading only to sub-lethal infections as fish have grown to a size where they are able to withstand their effects. It is clear that interposing a source of infection in the form of Atlantic salmon farms in the path of migrating salmon has the potential to tip the balance against them either through parasite-induced mortality or reduced fitness. This is the crux of the debate and the threat level demands confirmation through empirical investigation.

The risk to wild salmon arising from sea lice infection from farms has been consistently dismissed wherever Atlantic salmon farming has been established, which may explain why salmon farmers have been historically reluctant to accept a share of the responsibility. Admittedly, salmon farmers stock parasite-free smolts in sea cages that become infected from external wild sources, but the consequences of dense monocultures for parasite transmission is well known in animal husbandry. As the industry grows, the monocultures proliferate into clusters locally and regionally to create islands of infection, interconnected by wild fish. Samsing et al. (2019) used dispersion models and network analysis to determine the effect of imposing 'firebreaks' between farming areas, and concluded that such firebreaks could interrupt transmission of larvae and reduce infection pressure on farms.

Salmon aquaculture is an established and expanding enterprise that has long since overtaken sport and wild fisheries in tonnage produced and as a revenue source, so sea louse infection mitigation is the only option if wild salmon and trout are to escape further harm. This is not a one-direction process, however, as unchecked sea louse infection impacts the health and welfare of farmed Atlantic salmon. It is in the interests of all to control sea lice on farmed salmon and encouraging to see that the problem is currently widely recognised as shared and as one demanding cooperation, collaboration, and mutual solutions.

Medication has been effective until resistant lice appear, as they consistently do, so other options including fallowing, cleaner fish, and physical methods are in current use. Integrated pest management first proposed by Pike (1989), following the example of insect pest management on terrestrial crops, has been adopted but needs more effective implementation.

Terrestrial farmers primarily represent independent units in which many aspects of health and welfare can be maintained independent of neighbours (except for notifiable diseases), but this choice does not exist in aquaculture, as there are no barriers to transmission of disease through the aquatic medium. This demands closer regulation. As an example, Kragesteen et al. (2019) identified two louse treatment thresholds: one high, optimising profit of an individual farm through less frequent treatment, and a low threshold, optimising average profit of an overlapping farm cluster. Individual non-compliance to a low threshold has benefits to the individual but compromises the profit of the entire cluster. Regulation therefore demands transparency and publication of louse counts to prevent cheating. The issue of non-compliance was also raised by Misund (2019) who collated views of multiple stakeholders through interviews and media reports. His conclusion is that the industry claims to comply with government guidelines, whereas there is a wider perception that they fall short of implementing the regulations on louse thresholds. Non-compliance is clearly disadvantageous to the majority, benefitting only the miscreant. However, Larsen and Vormedal (2021) argue that slavishly following the rules to stay within the required louse thresholds could damage fish welfare and may increase mortality with little overall benefit to sea louse control. It could be argued that a one-size-fits-all approach is too blunt an instrument.

Advance warning of potential high infection pressure could be facilitated by accurate hydrodynamic modelling of larval dispersal in bays, salinity and temperature records, and detailed farm louse records. Forecasting of terrestrial parasite infections, such as for *Fasciola* and gastro-intestinal parasites of cattle, is currently in use and allows avoidance action when a high risk is forecast. The DFO requirement for delousing treatments in advance of the salmon migration season is a good example of simple mitigation.

For many, close containment of net cages or land-based tanks is the only solution. Siting farms at a distance from major fish migration routes or complete exclusion from such locations could also provide effective mitigation.

Decisions about new development and expansion of existing farms are based primarily on economic and political grounds, albeit with elements of environmental risk assessment. Salmon farming has often been embraced, as it provides employment and revenue to remote areas with otherwise limited economic resources. It is a competitive business, however, and pressure for automation is high, threatening the employment opportunities it promised.

Salmon aquaculturists may argue that rearing salmon relieves fishing pressure on wild species but, ironically in the case of the salmon, poses a threat to the fish it aims to conserve. With respect to the concept that salmon culture provides an important source of protein to the world's people, it would appear that a large proportion of the production supplies populations in which animal protein is not in short supply. In areas in which salmon farming is seeking to expand, it would be in the best interests of all stakeholders, as well as the environment, to place natural capital at the centre of decision-making processes relating to approval, siting, and management regulation (Dasgupta, 2021; Volpe and Shaw, 2008) to offer hope that wild salmon can survive the current intense anthropogenic pressure.

Acknowledgement

I am indebted to Kathleen Hills for her critical review that has greatly enhanced the readability and coherence of this chapter.

References

Bateman, A.W., Peacock, S.J., Connors, B., Polk, Z., Berg, D., Krkošek, M. and Morton, A. (2016) Recent failure to control sea louse outbreaks on salmon in the Broughton Archipelago, British Columbia. *Canadian Journal of Fisheries and Aquatic Sciences* 73, 1164–1172.

Beamish, R.J., Neville, C.M., Sweeting, R.M. and Ambers, N. (2005) Sea lice on adult Pacific salmon in the coastal waters of Central British Columbia, Canada. *Fisheries Research* 76, 198–208.

Beamish, R.J., Jones, S., Neville, C.E., Sweeting, R., Karreman, G., Saksida, S. and Gordon, E. (2006) Exceptional marine survival of pink salmon that entered the marine environment in 2003 suggests that farmed Atlantic salmon and Pacific salmon can coexist successfully in a marine ecosystem on the Pacific coast of Canada. *ICES Journal of Marine Science* 63, 1326–1337.

Birkeland, K. (1996) Consequences of premature return by sea trout (*Salmo trutta*) infested with the salmon louse (*Lepeophtheirus salmonis* Krøyer): migration, growth, and mortality. *Canadian Journal of Fisheries and Aquatic Sciences* 53, 2808–2813.

Birkeland, K. and Jakobsen, P.J. (1997) Salmon lice, *Lepeophtheirus salmonis*, infestation as a causal agent of premature return to rivers and estuaries by sea trout, *Salmo trutta*, juveniles. *Environmental Biology of Fishes* 49, 129–137.

Bocking, S. (2011) Farming salmon in a crowded wilderness: exploring the history of aquaculture science in British Columbia, Canada. In: Massard-Guilbaud, G. and Mosley, S. (eds) *Common Ground: Integrating the Social and Environmental in History*. Cambridge Scholars Publishing, Cambridge, UK. pp. 246–269.

Bøhn, T., Gjelland, K.Ø., Serra-Llinares, R.M., Finstad, B., Primicerio, R., Nilsen, R., Karlsen, Ø., Sandvik, A.D., Skilbrei, O.T., Elvik, K.M.S., Skaala, Ø. and Bjørn, P.A. (2020) Timing is everything: survival of Atlantic salmon *Salmo salar* postsmolts during events of high salmon lice densities. *Journal of Applied Ecology* 57(6), 1149–1160.

Brauner, C.J., Sackville, M., Gallagher, Z., Tang, S., Nendick, L. and Farrell, A.P. (2012) Physiological consequences of the salmon louse (*Lepeophtheirus salmonis*) on juvenile pink salmon (*Oncorhynchus gorbuscha*): implications for wild salmon ecology and management, and for salmon aquaculture. *Philosophical Transactions of the Royal Society B-Biological Sciences* 367, 1770–1779.

Brooks, K.M. (2005) The effects of water temperature, salinity, and currents on the survival and distribution of the infective copepodid stage of sea lice (*Lepeophtheirus salmonis*) originating on atlantic salmon farms in the Broughton Archipelago of British Columbia, Canada. *Reviews in Fisheries Science* 13(3), 177–204.

Brooks, K.M. and Stucchi, D.J. (2006) The effects of water temperature, salinity and currents on the survival and distribution of the infective copepodid stage of the salmon louse (*Lepeophtheirus salmonis*) originating on Atlantic salmon farms in the Broughton Archipelago of British Columbia, Canada (Brooks, 2005) – a response to the rebuttal of Krkošek et al. (2005a). *Reviews in Fisheries Science* 14, 13–23.

Butterworth, K.G., Cubitt, K.F. and McKinley, R.S. (2008) The prevalence, density and impact of *Lepeophtheirus salmonis* (Krøyer, 1837) infestation on juvenile pink salmon (*Oncorhynchus gorbuscha*) from the central coast of British Columbia, Canada. *Fisheries Research* 91, 35–41.

Cermaq (2021) Statement regarding Cermaq Canada's filing of a judicial review on the DFO decision on Cermaq Canada's licences in the Discovery Islands region. Available at: https://www.cermaq.ca/news/statement-regarding-cermaq-canadas-filing-of-a-judicial-review-on-the-dfo-decision-on-cermaq-canadas-licences-in-the-discovery-island-region (accessed 10 May 2021).

Cision Canada (2020) Government of Canada moves to phase out salmon farming licences in Discovery Islands following consultations with First Nations. Available at: https://www.newswire.ca/news-releases/government-of-canada-moves-to-phase-out-salmon-farming-licences-in-discovery-islands-following-consultations-with-first-nations-816013273.html (accessed 19 February 2022).

Cohen, B.I. (2012) *The Uncertain Future of Fraser River Sockeye.* Volume 3. Commission of Inquiry into the Decline of Sockeye Salmon in the Fraser River (Canada). Recommendations, summary, process Ottawa. Publishing and Depository Services Public Works and Government Services, Ottawa, Canada. 211pp.

Costello, M.J. (2006) Ecology of sea lice parasitic on farmed and wild fish. *Trends in Parasitology* 22, 475–483.

Costello, M.J. (2009) How sea lice from salmon farms may cause wild salmonid declines in Europe and North America and be a threat to fishes elsewhere. *Proceedings Royal Society B Biological Sciences* 276, 3385–3394.

Dasgupta, P. (2021) *The Economics of Biodiversity: The Dasgupta Review.* HM Treasury, London.

DFO (2003) Marine Finfish Aquaculture Licence under the Fisheries Act. Available at: https://www.pac.dfo-mpo.gc.ca/aquaculture/licence-permis/docs/licence-cond-permis-mar/licence-cond-permis-mar-eng.pdf (accessed 8 May 2021).

DFO (2019) State of Salmon Aquaculture Technologies, 2019. Available at: https://www.dfo-mpo.gc.ca/aquaculture/publications/ssat-ets-eng.html (accessed 10 May 2021).

DFO (2020) Work to support Cohen Commission's recommendation 19. Available at: https://www.dfo-mpo.gc.ca/cohen/recomm-19-work-travaux-eng.html (accessed 8 May 2021).

DFO (2021) Aquaculture Production Quantities and Value. Available at: https://www.dfo-mpo.gc.ca/stats/aqua/aqua-prod-eng.htm (accessed 8 May 2021).

FAO (2020) *The State of World Fisheries and Aquaculture 2020. Sustainability in Action.* FAO, Rome.

Finstad, B., Bjørn, P.A., Grimnes, A. and Hvidsten, N.A. (2000) Laboratory and field investigations of salmon lice [*Lepeophtheirus salmonis* (Krøyer, 1837)] infestation on Atlantic salmon (*Salmo salar* L.) postsmolts. *Aquaculture Research* 31, 795–803.

Finstad, B., Sandvik, A.D., Ugedal, O., Vollset, K.W., Karlsen, O., Davidsen, J.G., Saegrov, H. and Lennox, R.J. (2021) Development of a risk assessment method for sea trout in coastal areas exploited for aquaculture. *Aquaculture Environment Interactions* 13, 133–144.

FOPO (2013) Closed containment salmon aquaculture. Report of the Standing Committee on Fisheries and Oceans. Available at: https://www.ourcommons.ca/DocumentViewer/en/41-1/FOPO/report-3/ (accessed 10 May 2021).

FOPO (2021) Pacific salmon: ensuring the long-term health of wild populations and associated fisheries. Report of the Standing Committee on Fisheries and Oceans. Available at: https://www.ourcommons.ca/DocumentViewer/en/43-2/FOPO/report-5/ (accessed 10 May 2021).

Frazer, L.N. (2008) Sea-lice infection models for fishes. *Journal of Mathematical Biology* 57, 595–611.

Gargan, P.G., Forde, G., Hazon, N., Russell, D.J.F. and Todd, C.D. (2012) Evidence for sea lice-induced marine mortality of Atlantic salmon (*Salmo salar*) in western Ireland from experimental releases of ranched smolts treated with emamectin benzoate. *Canadian Journal of Fisheries and Aquatic Sciences* 69, 343–353.

Gargan, P.G., Kelly, F.L., Shephard, S. and Whelan, K.F. (2016) Temporal variation in sea trout *Salmo trutta* life history traits in the Erriff River, western Ireland. *Aquaculture Environment Interactions* 8, 675–689.

Gjelland, K.Ø., Serra-Llinares, R.M., Hedger, R.D., Arechavala-Lopez, P., Nilsen, R., Finstad, B., Uglem, I., Skilbrei, O.T. and Bjørn, P.A. (2014) Effects of salmon lice infection on the behaviour of sea trout in the marine phase. *Aquaculture Environment Interactions* 5, 221–233.

Gottesfeld, A.S., Proctor, B., Rolston, L.D. and Carr-Harris, C. (2009) Sea lice, *Lepeophtheirus salmonis*, transfer between wild sympatric adult and juvenile salmon on the north coast of British Columbia, Canada. *Journal of Fish Diseases* 32, 45–57.

Grant, A., Gardner, M., Nendick, L., Sackville, M., Farrell, A.P. and Brauner, C.J. (2009) Growth and ionoregulatory ontogeny of wild and hatchery-raised juvenile pink salmon (*Oncorhynchus gorbuscha*). *Canadian Journal of Zoology* 87, 221–228.

Guarracino, M., Qviller, L. and Lillehaug, A. (2018) Evaluation of aquaculture management zones as a control measure for salmon lice in Norway. *Diseases of Aquatic Organisms* 130, 1–9.

Halttunen, E., Gjelland, K.Ø., Hamel, S., Serra-Llinares, R,M., Nilsen, R., Arechavala-Lopez, P., Skarðhamar, J., Johnsen, I.A., Asplin, L., Karlsen, Ø., Bjørn Pål, A. and Finstad, B. (2018) Sea trout adapt their migratory behaviour in response to high salmon lice concentrations. *Journal of Fish Diseases* 41, 953–967.

Hansen, L.P. (1991) Introduction. In: *Interactions between Cultured and Wild Atlantic Salmon* (Special issue). Hansen, L.P, Hastein, T., Naevdal, G., Saunders, R.L. and Thorpe, J.E. (eds) *Aquaculture* 98, ix-x.

Hansen, L.P. and Windsor, M.L. (2006) Interactions between aquaculture and wild stocks of Atlantic salmon and other diadromous fish species: science and management, challenges and solutions: an introduction by the conveners. *ICES Journal of Marine Science* 63, 1159–1161.

Håstein, T. and Lindstad, T. (1991) Diseases in wild and cultured salmon: possible interaction. *Aquaculture* 98, 277–288.

Hatton-Ellis, M., Hay, D.W., Walker, A.F. and Northcott, S.J. (2007) Sea lice *Lepeophtheirus salmonis* infestations of post-smolt sea trout in Loch Shieldaig, Wester Ross, 1999–2003. In: Harris, G. and Milner, N. (eds) *Sea Trout: Biology, Conservation and Management.* Blackwell Publishing Ltd, Oxford, UK. pp. 372–376.

Hedger, R.D., Diserud, O.H., Finstad, B., Jensen, A.J., Hendrichsen, D.K., Ugedal, O. and Naesje, T.F. (2021) Modeling salmon lice effects on sea trout population dynamics using an individual-based approach. *Aquaculture Environment Interactions* 13, 145–163.

Heuch, P.A., Bjørn, P.A., Finstad, B., Holst, J.C., Asplin, L. and Nilsen, F. (2005) A review of the Norwegian 'National Action Plan Against Salmon Lice on Salmonids': the effect on wild salmonids. *Aquaculture* 246, 79–92.

Hodgson, E.E., Wilson, S.M. and Moore, J.W. (2020) Changing estuaries and impacts on juvenile salmon: a systematic review. *Global Change Biology* 26, 186–201.

Hvidsten, N.A., Finstad, B., Kroglund, F., Johnsen, B.O., Strand, R., Arnekleiv, J.V. and Bjørn, P.A. (2007) Does increased abundance of sea lice influence survival of wild Atlantic salmon post-smolt? *Journal of Fish Biology* 71, 1639–1648.

ICES (1997) *Report of the 1996 Workshop on the Interactions between Salmon Lice and Salmonids.* International Council for the Exploration of the Sea, Copenhagen, Denmark. ICES CM 1997, M:4.

Jackson, D., Cotter, D., ÓMaoiléidigh, N., O'Donohoe, P., White, J., Kane, F., Kelly, S., McDermott, T., McEvoy, S., Drumm, A., Cullen, A. and Rogan, G. (2011a) An evaluation of the impact of early infestation with the salmon louse *Lepeophtheirus salmonis* on the subsequent survival of outwardly migrating Atlantic salmon, *Salmo salar* L., smolts. *Aquaculture* 320, 159–163.

Jackson, D., Cotter, D., ÓMaoiléidigh, N., O'Donohoe, P., White, J., Kane, F., Kelly, S., McDermott, T., McEvoy, S., Drumm, A. and Cullen, A. (2011b) Impact of early infestation with the salmon louse *Lepeophtheirus salmonis* on the subsequent survival of outwardly migrating Atlantic salmon smolts from a number of rivers on Ireland's south and west coasts. *Aquaculture* 319, 37–40.

Jackson, D., Cotter, D., Newell, J., McEvoy, S., O'Donohoe, P., Kane, F., McDermott, T., Kelly, S. and Drumm, A. (2013) Impact of *Lepeophtheirus salmonis* infestations on migrating Atlantic salmon, *Salmo salar* L., smolts at eight locations in Ireland with an analysis of lice-induced marine mortality. *Journal of Fish Diseases* 36, 273–281.

Jacobsen, J.A. and Gaard, E. (1997) Open-ocean infestation by salmon lice (*Lepeophtheirus salmonis*): comparison of wild and escaped farmed Atlantic salmon (*Salmo salar* L.). *ICES Journal of Marine Science* 54, 1113–1119.

Johansen, L.H., Jensen, I., Mikkelsen, H., Bjørn, P.A., Jansen, P.A. and Bergh, O. (2011) Disease interaction and pathogens exchange between wild and farmed fish populations with special reference to Norway. *Aquaculture* 315, 167–186.

Jones, S. and Beamish, R. (2011) *Lepeophtheirus salmonis* on salmonids in the northeast Pacific ocean. In: Jones, S. and Beamish, R. (eds) *Salmon Lice: An Integrated Approach to Understanding Parasite Abundance and Distribution.* Wiley-Blackwell, Chichester, UK. pp. 307–329.

Jones, S.R.M. (2009) Controlling salmon lice on farmed salmon and implications for wild salmon. *CAB Reviews: Perspectives in Agriculture, Veterinary Science, Nutrition and Natural Resources* 4 (048).

Jones, S.R. and Hargreaves, N.B. (2007) The abundance and distribution of *Lepeophtheirus salmonis* (Copepoda: Caligidae) on pink (*Oncorhynchus gorbuscha*) and chum (*O. keta*) salmon in coastal British Columbia. *Journal of Parasitology* 93, 1324–1331.

Kabata, Z. (1973) The species of *Lepeophtheirus* (Copepoda: Caligidae) from fishes of British Columbia. *Journal of the Fisheries Research Board of Canada* 30, 729–759.

Karbowski, C.M., Finstad, B., Karbowski, N. and Hedger, R.D. (2019) Sea lice in Iceland: assessing the status and current implications for aquaculture and wild salmonids. *Aquaculture Environment Interactions* 11, 149–160.

Kragesteen, T.J., Simonsen, K., Visser, A.W. and Andersen, K.H. (2019) Optimal salmon lice treatment threshold and tragedy of the commons in salmon farm networks. *Aquaculture* 512, 734329.

Kristoffersen, A.B., Qviller, L., Helgesen, K.O., Vollset, K.W., Viljugrein, H. and Jansen, P.A. (2018) Quantitative risk assessment of salmon louse-induced mortality of seaward-migrating post-smolt Atlantic salmon. *Epidemics* 23, 19–33.

Krkošek, M. (2010) Sea lice and salmon in Pacific Canada: ecology and policy. *Frontiers in Ecology and the Environment* 8, 201–209.

Krkošek, M. and Hilborn, R. (2011) Sea lice (*Lepeophtheirus salmonis*) infestations and the productivity of pink salmon (*Oncorhynchus gorbuscha*) in the Broughton Archipelago, British Columbia, Canada. *Canadian Journal of Fisheries and Aquatic Sciences* 68, 17–29.

Krkošek, M., Lewis, M.A. and Volpe, J.P. (2005) Transmission dynamics of parasitic sea lice from farm to wild salmon. *Proceedings of the Royal Society B: Biological Sciences* 272(1564), 689–696.

Krkošek, M., Lewis, M.A., Volpe, J.P. and Morton, A. (2006) Fish farms and sea lice infestations of wild juvenile salmon in the Broughton Archipelago – a rebuttal to Brooks (2005). *Reviews in Fisheries Science* 14, 1–11.

Krkošek, M., Ford, J.S., Morton, A., Lele, S., Myers, R.A. and Lewis, M.A. (2007a) Declining wild salmon populations in relation to parasites from farm salmon. *Science* 318.

Krkošek, M., Gottesfeld, A., Proctor, B., Rolston, D., Carr-Harris, C. and Lewis, M.A. (2007b) Effects of host migration, diversity and aquaculture on sea lice threats to Pacific salmon populations. *Proceedings of the Royal Society B-Biological Sciences* 274, 3141–3149.

Krkošek, M., Connors, B.M., Morton, A., Lewis, M.A., Dill, L.M. and Hilborn, R. (2011) Effects of parasites from salmon farms on productivity of wild salmon. *Proceedings of the National Academy of Sciences of the United States of America* 108, 14700–14704.

Krkošek, M., Revie, C.W., Gargan,. P.G., Skilbrei, O.T., Finstad, B. and Todd, C.D. (2013) Impact of parasites on salmon recruitment in the Northeast Atlantic Ocean. *Proceedings of the Royal Society B: Biological Sciences* 280, 20122359.

Lackey, R.T. (1999) Salmon policy: science, society, restoration, and reality. *Environmental Science and Policy* 2, 369–379.

Larsen, M.L. and Vormedal, I. (2021) The environmental effectiveness of sea lice regulation: compliance and consequences for farmed and wild salmon. *Aquaculture* 532, 736000.

Lenders, H.J.R., Chamuleau, T.P.M., Hendriks, A.J., Lauwerier, R.C.G.M., Leuven, R.S.E.W. and Verberk, W.C.E.P. (2016) Historical rise of waterpower initiated the collapse of salmon stocks. *Scientific Reports* 6, 29269.

Marty, G.D., Saksida, S.M. and Quinn II, T.J. (2010) Relationship of farm salmon, sea lice, and wild salmon populations. *Proceedings National Academy of Sciences USA* 107(52), 22599–22604.

McVicar, A.H. (1997a) Disease and parasite implications of the coexistence of wild and cultured Atlantic salmon populations. ICES/NASCO Symposium on Interactions Between Salmon Culture and Wild Stocks of Atlantic Salmon – The Scientific and Management Issues. Apr 18–22 Bath, England. Academic Press Ltd, 1093–1103.

McVicar, A.H. (1997b) *Report of the Workshop on the Interactions between Salmon Lice and Salmonids, 11–15 November 1996, Edinburgh, United Kingdom*. ICES, Anadromous and Catadromous Fish Committee, CM 1997/M:4 Ref.:F.

Middlemas, S.J., Raffell, J.A., Hay, D.W., Hatton-Ellis, M. and Armstrong, J.D. (2010) Temporal and spatial patterns of sea lice levels on sea trout in western Scotland in relation to fish farm production cycles. *Biology Letters* 6, 548–551.

Middlemas, S.J., Fryer, R.J., Tulett, D. and Armstrong, J.D. (2013) Relationship between sea lice levels on sea trout and fish farm activity in western Scotland. *Fisheries Management and Ecology* 20, 68–74.

Misund, A.U. (2019) From a natural occurring parasitic organism to a management object: historical perceptions and discourses related to salmon lice in Norway. *Marine Policy* 99, 400–406.

Morton, A.B. and Williams, R. (2003) First report of a sea louse, *Lepeophtheirus salmonis*, infestation on juvenile pink salmon, *Oncorhynchus gorbuscha*, in nearshore habitat. *Canadian Field Naturalist* 117, 634–641.

Morton, A.B. and Williams, R. (2006) Response of the sea louse *Lepeophtheirus salmonis* infestation levels on juvenile wild pink, *Oncorhynchus gorbuscha*, and chum, *O. keta*, salmon to arrival of parasitized wild adult pink salmon. *Canadian Field-Naturalist* 120, 199–204.

Morton, A., Routledge, R., Peet, C. and Ladwig, A. (2004) Sea lice (*Lepeophtheirus salmonis*) infection rates on juvenile pink (*Oncorhynchus gorbuscha*) and chum (*Oncorhynchus keta*) salmon in the nearshore marine environment of British Columbia, Canada. *Canadian Journal of Fisheries and Aquatic Sciences* 61, 147–157.

Morton, A., Routledge, R.D. and Williams, R. (2005). Temporal patterns of sea louse infestation on wild Pacific salmon in relation to the fallowing of Atlantic salmon farms. *North American Journal of Fisheries Management* 25, 811–821.

Morton, A., Routledge, R. and Krkošek, M. (2008) Sea louse infestation in wild juvenile salmon and Pacific herring associated with fish farms off the east-central coast of Vancouver Island, British Columbia. *North American Journal of Fisheries Management* 28, 523–532.

Morton, A., Routledge, R., McConnell, A. and Krkošek, M. (2011) Sea lice dispersion and salmon survival in relation to salmon farm activity in the Broughton Archipelago. *ICES Journal of Marine Science* 68, 144–156.

Mowi (2021) Mowi seeks judicial review of minister's decision on Discovery Islands salmon farms. Available at: https://mowi.com/caw/blog/2021/01/18/mowi-seeks-judicial-review-of-ministers-decision-on-discovery-islands-salmon-farms/ (accessed 10 May 2021).

Myksvoll, M.S., Sandvik, A.D., Johnsen, I.A., Skardhamar, J. and Albretsen, J. (2020) Impact of variable physical conditions and future increased aquaculture production on lice infestation pressure and its sustainability in Norway. *Aquaculture Environment Interactions* 12, 193–204.

Nagasawa, K. (1985) Comparison of the infection levels of *Lepeophtheirus salmonis* (Copepoda) on chum salmon captured by two methods. *Japanese Journal of Ichthyology* 32, 368–370.

Nagasawa, K. (1987) Prevalence and abundance of *Lepeophtheirus salmonis* (Copepoda: caligidae) on high-seas salmon and trout in the north Pacific Ocean. *Nippon Suisan Gakkaishi* 53, 2151–2156.

Nagasawa, K., Ishida, Y., Ogura, M., Tadokoro, K. and Hiramatsu, K. (1993) The abundance and distribution of *Lepeophtheirus salmonis* (Copepoda: Caligidae) on six species of Pacific salmon in offshore waters of the North Pacific Ocean and Bering Sea. In: Boxshall, G.A. and Defaye, D. (eds) *Pathogens of Wild and Farmed Fish: Sea Lice*. Ellis Horwood, Chichester, UK. pp. 166–178.

NASCO (1994) Report of the Eleventh Annual Meeting of the Council. Available at: https://nasco.int/annual-meeting/eleventh-annual-meeting/ (accessed 8 May 2021).

NASCO (2001) Report of the activities of the North Atlantic Salmon Conservation Organization 2000–2001. Available at: https://nasco.int/wp-content/uploads/2020/06/nasco_report2000_2001.pdf (accessed 12 Dec 2020).

NASCO (2006) Resolution by the Parties to the Convention for the Conservation of Salmon in the North Atlantic Ocean to Minimise Impacts from Aquaculture, Introductions and Transfers, and Transgenics on the Wild Salmon Stocks. The Williamsburg Resolution. Available at: https://nasco.int/document/the-williamsburg-resolution-2006/ (accessed 1 Dec 2020).

NASCO (2010) Guidance on Best Management Practices to address impacts of sea lice and escaped farmed salmon on wild salmon stocks. Available at: https://nasco.int/?s=CNL%2820%2910&post_type=document (accessed 10 Dec 2020).

NASCO (2016) Addressing impacts of salmon farming on wild Atlantic salmon: Challenges to, and developments supporting, achievement of NASCO's international goals. Available at: https://nasco.int/wp-content/uploads/2020/02/2016ThemeBasedSession.pdf (accessed 11 Nov 2020).

NASCO (2019a) Salmon farming: NGOs demand that Governments honour the Williamsburg Resolution commitments. Available at: https://nasco.int/?s=CNL%2819%2943&post_type=document (accessed 2 Dec 2020).

NASCO (2019b) State of North Atlantic salmon. Available at: https://nasco.int/wp-content/uploads/2020/05/SoS-final-online.pdf (accessed 2 Dec 2020).

NASCO (2020) ICES Advice on fishing opportunities, catch, and effort Northeast Atlantic ecoregions. Available at: https://nasco.int/?s=CNL%2820%2910&post_type=document (accessed 2 Dec. 2020).

Nendick, L., Sackville, M., Tang, S., Brauner, C.J. and Farrell, A.P. (2011) Sea lice infection of juvenile pink salmon (*Oncorhynchus gorbuscha*): effects on swimming performance and postexercise ion balance. *Canadian Journal of Fisheries and Aquatic Sciences* 68, 241–249.

Patanasatienkul, T., Sanchez, J., Rees, E.E., Krkošek, M., Jones, S.R.M. and Revie, C.W. (2013) Sea lice infestations on juvenile chum and pink salmon in the Broughton Archipelago, Canada, from 2003 to 2012. *Diseases of Aquatic Organisms* 105, 149–161.

Peacock, S.J., Krkošek, M., Proboszcz, S., Orr, C. and Lewis, M.A. (2013) Cessation of a salmon decline with control of parasites. *Ecological Applications* 23, 606–620.

Peacock, S., Krkošek, M., Bateman, A.W. and Lewis, M.A. (2020) Estimation of spatiotemporal transmission dynamics and analysis of management scenarios for sea lice of farmed and wild salmon. *Canadian Journal of Fisheries and Aquatic Sciences* 77, 55–68.

Pike, A.W. (1989) Sea lice – major pathogens of farmed Atlantic salmon. *Parasitology Today* 5, 291–297.

Pike, A.W. and Wadsworth, S.L. (2000). Sealice on salmonids: their biology and control. In: Baker, J.R., Muller, R. and Rollinson, D. (eds) *Advances in Parasitology, Vol 44.* Elsevier Academic Press Inc, San Diego, California. pp. 233–337.

Price, M.H.H., Morton, A. and Reynolds, J.D. (2010) Evidence of farm-induced parasite infestations on wild juvenile salmon in multiple regions of coastal British Columbia, Canada. *Canadian Journal of Fisheries and Aquatic Sciences* 67, 1925–1932.

Rae, G.H. (1979) On the trail of the sea lice. *Fish Farmer* 2, 22–25.

Rogers, L.A., Peacock, S.J., Mckenzie, P., Dedominicis, S., Jones, S.R.M., Chandler, P., Foreman, M.G.G., Revie, C.W. and Krkošek, M. (2013) Modeling parasite dynamics on farmed salmon for precautionary conservation management of wild salmon. *PLoS ONE* 8, e60096.

Sackville, M., Tang, S., Nendick, L., Farrell, A.P. and Brauner, C.J. (2011) Pink salmon (*Oncorhynchus gorbuscha*) osmoregulatory development plays a key role in sea louse (*Lepeophtheirus salmonis*) tolerance. *Canadian Journal of Fisheries and Aquatic Sciences* 68, 1087–1096.

Saksida, S., Constantine, J., Karreman, G.A. and Donald, A. (2007) Evaluation of sea lice abundance levels on farmed Atlantic salmon (*Salmo salar* L.) located in the Broughton Archipelago of British Columbia from 2003 to 2005. *Aquaculture Research* 38, 219–231.

Saksida, S., Morrison, D., Sheppard, M. and Keith, I. (2011a) Sea lice management on salmon farms in British Columbia, Canada. In: Jones, S. and Beamish, R. (eds) *Salmon Lice: An Integrated Approach to Understanding Parasite Abundance and Distribution.* Wiley-Blackwell, Oxford, UK. pp. 235–278.

Saksida, S.M., Greba, L., Morrison, D. and Revie, C.W. (2011b) Sea lice on wild juvenile Pacific salmon and farmed Atlantic salmon in the northernmost salmon farming region of British Columbia. *Aquaculture* 320, 193–198.

Samsing, F., Johnsen, I., Treml, E.A. and Dempster, T. (2019) Identifying 'firebreaks' to fragment dispersal networks of a marine parasite. *International Journal for Parasitology* 49, 277–286.

Saunders, R.L. (1991) Potential interaction between cultured and wild Atlantic salmon. *Aquaculture* 98, 51–60.

Serra-Llinares, R.M., Bjørn, P.A., Finstad, B., Nilsen, R., Harbitz, A., Berg, M. and Asplin, L. (2014) Salmon lice infection on wild salmonids in marine protected areas: an evaluation of the Norwegian "National Salmon Fjord". *Aquaculture Environment Interactions* 5, 1–16.

Serra-Llinares, R.M., Bøhn, T., Karlsen, Ø., Nilsen, R., Freitas, C., Albretsen, J., Haraldstad, T., Thorstad, E.B., Elvik, K.M.S. and Bjørn, P.A. (2020) Impacts of salmon lice on mortality, marine migration distance and premature return in sea trout. *Marine Ecology Progress Series* 635, 151–168.

Shephard, S., MacIntyre, C. and Gargan, P. (2016) Aquaculture and environmental drivers of salmon lice infestation and body condition in sea trout. *Aquaculture Environment Interactions* 8, 597–610.

Silva, A.L.N., Rodrigues, R.A., Siqueira, M.S., Farias, K.N.N., Kuibida, K.V., Franco-Belussi, L. and Fernandes, C.E. (2021) Transaminase profile and hepatic histopathological traits in *Piaractus mesopotamicus* exposed to insecticide diflubenzuron. *Environmental Science and Pollution Research* 9, 22002–22010.

Sivertsgard, R., Thorstad, E.B., Okland, F., Finstad, B., Bjorn, P.A., Jepsen, N., Nordal, T. and McKinley, R.S. (2007) Effects of salmon lice infection and salmon lice protection on fjord migrating Atlantic salmon and brown trout post-smolts. *Hydrobiologia* 582, 35–42.

Skaala, Ø., Kålås, S. and Borgstrøm, R. (2014) Evidence of salmon lice-induced mortality of anadromous brown trout (*Salmo trutta*) in the Hardangerfjord, Norway. *Marine Biology Research* 10, 279–288.

Skern-Mauritzen, R., Torrissen, O. and Glover, K.A. (2014) Pacific and Atlantic *Lepeophtheirus salmonis* (Krøyer, 1837) are allopatric subspecies: *Lepeophtheirus salmonis salmonis and L. salmonis oncorhynchi* subspecies novo. *BMC Genetics* 15, 32.

Skilbrei, O.T. and Wennevik, V. (2006) Survival and growth of sea-ranched Atlantic salmon, *Salmo salar* L., treated against sea lice before release. *ICES Journal of Marine Science* 63, 1317–1325.

Skilbrei, O.T., Finstad, B., Urdal, K., Bakke, G., Kroglund, F. and Strand, R. (2013) Impact of early salmon louse, *Lepeophtheirus salmonis*, infestation and differences in survival and marine growth of sea-ranched Atlantic salmon, *Salmo salar* L., smolts 1997–2009. *Journal of Fish Diseases* 36, 249–260.

Thorstad, E.B. and Finstad, B. (2018) Impacts of salmon lice emanating from salmon farms on wild Atlantic salmon and sea trout. *NINA Report.* NINA, Trondheim, Norway.

Thorstad, E.B., Whoriskey, F., Uglem, I., Moore, A., Rikardsen, A.H. and Finstad, B. (2012) A critical life stage of the Atlantic salmon *Salmo salar*: behaviour and survival during the smolt and initial post-smolt migration. *Journal of Fish Biology* 81, 500–542.

Thorstad, E.B., Todd, C.D., Uglem, I., Bjørn, P.A., Gargan, P.G., Vollset, K.W., Halttunen, E., Kålås, S., Berg, M. and Finstad, B. (2015) Effects of salmon lice *Lepeophtheirus salmonis* on wild sea trout *Salmo trutta* – a literature review. *Aquaculture Environment Interactions* 7, 91–113.

Thorstad, E.B., Todd, C.D., Uglem, I., Bjørn, P.A., Gargan, P.G., Vollset, K.W., Halttunen, E., Kålås, S., Berg, M. and Finstad, B. (2016) Marine life of the sea trout. *Marine Biology* 163, 47.

Tilseth, S., Hansen, T. and Møller, D. (1991) Historical development of salmon culture. *Aquaculture* 98, 1–9.

Tingley, G.A., Ives, M.J. and Russell, I.C. (1997) The occurrence of lice on sea trout (*Salmo trutta* L.) captured in the sea off the East Anglian coast of England. *ICES Journal of Marine Science* 54, 1120–1128.

Todd, C.D. (2007) The copepod parasite (*Lepeophtheirus salmonis* (Krøyer), *Caligus elongatus* Nordmann) interactions between wild and farmed Atlantic salmon (*Salmo salar* L.) and wild sea trout (*Salmo trutta* L.): a mini review. *Journal of Plankton Research* 29, i61–71.

Torrissen, O, Jones, S., Asche, F., Guttormsen, A., Skilbrei, O.T., Nilsen, F., Horsberg, T.E. and Jackson, D. (2013) Salmon lice – impact on wild salmonids and salmon aquaculture. *Journal of Fish Diseases* 36, 171–194.

Tully, O., Poole, W.R., Whelan, K.F. and Merigoux, S. (1993) Parameters and possible causes of epizooties of *Lepeophtheirus salmonis* (Krøyer) infesting sea trout (*Salmo trutta* L.) off the west coast of Ireland. In: Boxshall, G.A. and Defaye, D. (eds) *Pathogens of Wild and Farmed Fish: Sea Lice.* Ellis Horwood, Chichester, UK. pp. 202–213.

Urquhart, K., Pert, C.C., Kilburn, R., Fryer, R.J. and Bricknell, I.R. (2008) Prevalence, abundance, and distribution of *Lepeoptheirus salmonis* (Krøyer, 1837) and *Caligus elongatus* (Nordmann, 1832) on wild sea trout *Salmo trutta* L. *ICES Journal of Marine Science* 65, 171–173.

Vollset, K.W., Krontveit, R.I., Jansen, P.A., Finstad, B., Barlaup, B.T., Skilbrei, O.T., Krkošek, M., Romunstad, P., Aunsmo, A., Jensen, A.J. and Dohoo, I. (2016) Impacts of parasites on marine survival of Atlantic salmon: a meta-analysis. *Fish and Fisheries* 17, 714–730.

Vollset, K.W., Dohoo, I., Karlsen, O., Halttunen, E., Kvamme, B.O., Finstad, B., Wennevik, V., Diserud, O.H., Bateman, A., Friedland, K.D., Mahlum, S., Jorgensen, C., Qviller, L., Krkošek, M., Atland, A. and Barlaup, B.T. (2018) Disentangling the role of sea lice on the marine survival of Atlantic salmon. *ICES Journal of Marine Science* 75, 50–60.

Volpe, J.P. and Shaw, K. (2008) Fish farms and neoliberalism: salmon aquaculture in British Columbia. In: Gore, C. and Stotett, P. (eds) *Environmental Challenges & Opportunities: Local-Global Perspectives on Canadian Issues.* Emond Montgomery, Toronto, Canada. pp. 131–158.

Wells, A., Grierson, C.E., MacKenzie, M., Russon, I., Middlemiss, C., Bjørn, P.A., Finstad, B., Wendelar Bonga, S.E., Todd, C.D. and Hazon, N. (2006) Physiological effects of simultaneous, abrupt seawater entry and sea lice (*Lepeophtheirus salmonis*) infestation of wild, sea-run brown trout (*Salmo trutta*) smolts. *Canadian Journal of Fisheries and Aquatic Sciences* 63, 2809–2821.

Wells, A., Grierson, C.E., MacKenzie, M., Russon, I., Reinardy, H., Sivertsgård, R., Bjørn, P.A., Finstad, B., Wendelaar Bonga, S.E., Todd, C.D. and Hazon, N. (2007) Physiological consequences of premature freshwater return; for wild sea-run brown trout (*Salmo trutta*) postsmolts infested with sea lice (*Lepeophtheirus salmonis*). *Canadian Journal of Fisheries and Aquatic Sciences* 64, 1360–1369.

Werkman, M., Green, D.M., Murray, A.G. and Turnbull, J.F. (2011) The effectiveness of fallowing strategies in disease control in salmon aquaculture assessed with an SIS model. *Preventive Veterinary Medicine* 98, 64–73.

Whelan, K. (2010) A review of the impacts of the salmon louse, *Lepeophtheirus salmonis* (Krøyer, 1837) on wild salmonids. Atlantic Salmon Trust, Perth, UK. Available at: www.atlanticsalmontrust.org (accessed 1 December 2020).

PART IV

Future developments

CHAPTER 25

Epilogue: Prospects for the future

Ian Bricknell[1], Jim Treasurer[2] and James Bron[3]

[1]Aquaculture Research Institute and School of Marine Sciences, University of Maine, USA
[2]FAI Farms, UK
[3]Institute of Aquaculture, University of Stirling, UK

In the lifetime of the last major book on sea lice, *Pathogens of Wild and Farmed Fish: Sea Lice*, several unexpected technologies were developed to help control sea lice, many of which are still employed. What, then, is the prospect for sea louse research and the development of new tools and approaches over the next decade or so? Speculation is easy to undertake, but it is much harder to provide firm predictions regarding the likely identity of the one or more magic bullets that might be used to successfully manage and control sea lice in the future.

It is probably safe to assume that we will continue to see major advances across all the research areas outlined in this book, particularly for vaccination, molecular biology approaches and non-chemical control methods, but will we see a new class of sea lice specific medications that only target sea lice and have no or minimal environmental impacts?

Perhaps we will see instead the development of more technologies where the medication for removal of lice is administered in a contained treatment system before the water is discharged back into the environment or taken offsite for detoxification and disposal elsewhere. This short chapter seeks to provide an overview of the possible treatments and control measures that could emerge to be at the forefront of lice control over the next decades. Of course, the future suggested here may never come to pass, possibly superseded by changes in the basic technologies of salmon aquaculture such as the extensive development of onshore recirculating aquaculture systems or the migration of open water aquaculture systems to offshore locations where lice are no longer a problem. However, while onshore farms may provide a nominal solution to sea lice, they require much higher energy consumption in order to grow salmon while also making it more difficult to satisfy the welfare needs of fish and may not, therefore, provide the commonly envisaged universal panacea to problems of open water aquaculture.

25.1 Current treatment status and development needs

Management of sea lice remains challenging, particularly as they represent a moving target, constantly undergoing adaptation under the selective pressure of treatments. The short generation time and exposure to frequent, and sometimes less than effective, medicinal treatments has increased resistance to medicines (see Chapter 14), and this situation is likely to continue for other forms of intervention such as use of cleaner fish and physical barriers/removal approaches (see Chapter 22). Integrated pest management (IPM) (Mordue (Luntz) and Pike, 2002) has proven to be the way forward in controlling sea lice infestations. Good management practice and cooperation between companies, health staff and other stakeholders can see the implementation of strategic approaches such as area management agreements, coordinated simultaneous treatments between farms, and national sea lice treatment strategies that often lean heavily on controlling sea lice numbers in early spring of the second year of salmon production (see Chapter 15).

Many commentators (Barrett et al., 2020) have emphasised that much of sea lice control and sea lice research has been focused on the development of new treatment methods and these authors have advocated for more funding to research preventative methods. However, the main burdens for development of new medicines have largely fallen on pharmaceutical companies and the long time course, high costs and extensive regulatory hurdles incurred comprise major barriers to entering the market. Many of the recent advances in sea lice control have come from technologists and industry, such as snorkel nets, thermolicers, hydrolicers and skirts.

25.2 Signs of optimism in management and national agreements

Sea lice control and management, although undertaken in the form of medicinal treatments from the 1960s onwards, have lagged behind the rapid expansion of salmon farming and the scaling of net pens, and have often struggled to keep up with the problems posed by the parasite (see Chapter 15, for detail). Initial failure to rotate medicines has been a major disadvantage and enabled the development of resistance (Aaen et al., 2015) (see Chapter 14). This problem was compounded by the limited range of available medicines in the 1980s and for much of the 1990s (Rae, 2002; Roth et al., 1993). Dichlorvos (DDVP) was the mainstay of treatments but was banned in 1996. The stopgap measure was a more environmentally friendly hydrogen peroxide-based treatment, and it was the late 1990s before emamectin benzoate, cypermethrin, azamethiphos and insect growth regulators (IGRs) were licensed for use. Nevertheless, innovation in the form of alternative approaches such as cleaner fish and latterly preventative methods has been rapid. Management cooperation and coordination of treatments and strategies have provided welcome additional tools, leading to joint fallowing of farms across defined hydrographic zones, adding to the arsenal available for IPM and assisting national sea lice treatment plans.

25.3 Salmonid stock decline continues

Meanwhile, the decline in salmonid stocks in many areas continues due to a wide range of complex factors, many of them being extremely difficult to quantify (NASCO, 2020). The provision of opportunities to share space and goodwill for cooperation and open discussion on saving these iconic species is therefore critical. Openness and transparency between aquaculture and wild fish interests are essential components with respect to making progress in conservation of endangered stocks. As with other environmental issues, we may not have much time remaining to save the Atlantic salmon in particular. The pace of climate change may well overtake efforts to ensure the survival and continued broad distribution of Atlantic salmon populations, may impact the farming of salmon in some areas and may also have important effects on sea louse biology, life cycles and interactions with wild and farmed fish populations.

25.4 Immunological advances

Vaccines against sea lice have been described as the 'holy grail' of sea lice control (Raynard et al., 2002; Bricknell et al., 2003) but at the time of writing this chapter there is still no effective vaccine available against *L. salmonis*. There have been many advances in understanding of both fish vaccination strategies and the host–parasite relationships since the early work of the 1990s (MacKinnon, 1993; Grayson et al., 1995; Roper et al., 1995; Ruane et al., 1995). This has led to an increased understanding of lice biology and the immunomodulation of the host by the louse (Bron et al., 1991; Bricknell et al., 2006; Fast et al., 2006; Torrealba et al., 2011; Braden et al., 2012, 2015; Lewis et al., 2014; Pert et al., 2009, 2012; Piesz et al., 2016, 2020; Hamilton et al., 2018; Barker et al., 2019) and has the potential to improve the identification of potential protective antigens from sea lice. Recent work also raises the possibility of improving lice control by using targeted immunomodulators to upregulate the inflammatory responses of the host in order to overcome the inherent suppressive activity of the attached louse stages.

Advances in vaccination techniques themselves may also play a major role in the development of a successful lice vaccine. Currently fish vaccines are typically delivered by two methods: by injection (either into the peritoneum or into muscle) or by bath vaccination where the fish are immersed into a solution of vaccine and the vaccine is taken up via the mucosal surfaces or ingested. Intraperitoneal vaccination (IP) is by far the most common method used in salmon aquaculture today and induces a very strong humoral response to the antigens, not least due to the powerful oil-based adjuvants typically used in the formulation of salmon vaccines (Ma et al., 2019; Miccoli et al., 2021). Conversely, bath vaccination tends to induce a weaker humoral response in Atlantic salmon and protection is for a shorter period. As such bath vaccination is often limited to small fish before it is possible to use IP vaccination.

When one considers the biology of sea lice the most likely stages to be impacted by vaccines are the copepodid and chalimus stages, when the parasite is physically attached to the host and cannot move away from an inflammatory response. Infecting copepodids are exposed to vaccinated fish for a short time before moulting, while mobile adult and pre-adult stages can simply move away from any localised immune response developed by the host. The close association of the copepodid and chalimus with the skin of the fish would seem to suggest that a vaccine that causes a strong immune response in the skin, an aspect of mucosal immunity, rather than the systemic immune system would be a better choice of vaccination strategy and this is clearly reflected in ongoing research.

25.5 Molecular biology and bioinformatics

The advances made in molecular biology tools and approaches over the last decade have had significant impacts on our ability to understand sea louse biology and interactions. It is likely that we will see further major advances in the control of sea lice that have their basis in the tools and knowledge provided by the use of molecular biology approaches. Genomic selection using single nucleotide polymorphisms (SNPs) is already being used extensively to breed high-performing Atlantic salmon that have a measure of inherent resistance to sea lice (Houston et al., 2014). Once the mechanisms underlying resistance to sea lice are better understood, it is also potentially possible to improve lice resistance in Atlantic salmon through genetic engineering. In particular, the use of the recently developed CRISPR/Cas9 toolbox, which allows directed modification of the salmon genome, holds considerable promise for further research. While already licensed in the USA, production of genetically engineered salmon for human food use remains a controversial area in many parts of the world and will require wider consumer and societal acceptance prior to large-scale adoption. The CRISPR/Cas9 toolbox will, however, also allow the study of critical pathways in the host–parasite interface, by allowing the introduction of tailored changes to genomic elements directing the immune system of the host and the aggression pathways of the louse, allowing the identification of key components of the interaction to be elucidated. This in turn may lead to the development of specific anti-lice immunomodulators that allow the salmon to overcome the immunosuppressive secretome of lice by the use of functional feeds. CRISPR/Cas9 and allied technologies could also assist in the identification of key anti-host molecules that, if eliminated in the lice, could reduce or eliminate louse settlement success. Similarly, the same tools could be used to assist in the identification/confirmation of potential vaccine antigens.

An approach employed with some success in other arthropod pest control is the generation of sterile males and their use to reduce the population of parasites in the wild. The biggest application of this technology has been in parasites that spread human diseases, in particular the *Anopheles* mosquito which spreads malaria. Given that male mosquitos do not require a blood meal and feed on nectar there is little opposition from the general public for them to be released into the wild. Once released these sterile males go on to mate with wild type females which undergo egg production after mating (they still require a blood meal) but the eggs do not develop (Crawford et al., 2020; Kaiser et al., 2021). Thus, several releases of the genetically engineered males across the season will reduce the population size of the mosquitos (Iboi et al., 2020; Hosack et al., 2021). Ultimately this reduces the infectious pressure of both the mosquitos and malaria in the region. Of course, this technology is not without its own issues. Not least is the environmental impact of non-natural genes becoming established in the wild and the impact that reducing the numbers of mosquitos will have on species that are dependent on them for food, such as bats. This technology could be appealing for sea lice control too. However, lice currently need to be produced using host fish so that the process of initially amplifying numbers of parasites for sterilising under laboratory conditions and the need to subsequently introduce such sterile, but still pathogenic, lice into farms remain key barriers to prospects for the use of such technologies under real-world conditions.

While there may be further breakthroughs in bioinformatic analysis in the future, we are already seeing its applications in sea lice biology from genetic profiling of host resistance, characterisation of the miRNome and understanding the microbiome associated with sea lice infections, as well as more traditional applications such as transcriptomic profiling of the immune response of Atlantic salmon (Núñez-Acuña and Gallardo-Escárate, 2019; Agusti-Ridaura et al., 2020; AmbuAli et al., 2020; Dalvin et al., 2020; Kjetsa et al., 2020; Tsairidou et al., 2020; Umasuthan et al., 2020).

25.6 Biological control of sea lice

Improving control of sea lice through the use of naturally occurring predators or pathogens continues to be seen as a potentially productive and environmentally friendly approach. Considerable progress has been made in the use of cleaner fish, particularly ballan wrasse and lumpfish, as an approach to keeping sea louse numbers low on farms (see Chapter 19). Recent efforts to farm cleaner fish rather than capturing them from wild stocks and to better address the welfare needs of the cleaner fish themselves have paid dividends in terms of their effectiveness in louse removal (Brooker et al., 2020; Haugland et al., 2020; Imsland et al., 2020a, 2020b).

Despite extensive searching, no native pathogen has been identified that appears to be suitable for sea louse control; however, very little research has been carried out in this area. The published material is restricted to looking at microsporidians, the ciliate suctorian protozoa *Ephelota* sp. and electron microscopy observations of viral plaques in sea lice tissues (Fernandez-Leborans et al., 2005; Freeman et al., 2003; Freeman and Sommerville, 2009, 2011; Sommerville, 2009, 2012; Økland et al., 2014; Øvergård et al., 2018). Because of the critical importance of free-living and parasitic copepods to aquatic ecosystems, however, any potential pathogen or parasite of sea lice would need to be properly evaluated before use. In the meantime, so little research has been carried out on the pathogens of *L. salmonis* and *C. rogercresseyi* that any prospect for this method of louse control is likely some years away.

25.7 Emerging medications

Things have changed a lot since the 1980s and sea louse medication has been through several generations of treatment, including hydrogen peroxide, pyrethroids and SLICE™ (emamectin benzoate). However, the development of drug resistance has seriously limited the use of veterinary medicines for sea lice control. Although use of chemotherapeutants will continue to play an important role in future sea lice control, it is likely that they will require to be applied in ways that minimise discharge into the environment and hence environmental impacts. One approach to minimising environmental impact is the removal or neutralisation of drug-containing effluent at sea or onshore. Such an approach may rejuvenate the use of non-specific broad action anti-parasitics, at least in the short to medium term, but sea lice resistance to broad action anti-parasitics remains. The potential for the use of nicotinamides on sea lice has been proposed in Scotland through use of a CleanTreat® system that allows for its removal prior to effluent discharge, but this has already been met with considerable resistance in some quarters and garnered attendant media interest (McVeigh, 2021). This highlights the care that is required when proposing potential new sea lice medications to avoid compounding the negative image that aquaculture has in some regions.

Further pathways for developing new medicines are offered by the fields of bioprospecting and bioinformatics. As both subspecies of *L. salmonis* have been sequenced and annotated, along with *C. rogercresseyi*, there is potential to identify compounds that specifically inhibit pathways unique to the genus or even species, an approach that has been taken to protect terrestrial crops (Li and Zhang, 2016; Sadasivam et al., 2018). The risk with this approach and indeed with the proposed use of sea louse pathogens is that sea lice are copepods and it is free-living copepods that underpin most marine ecosystems.

25.8 Other control measures

In *Pathogens of Wild and Farmed Fish: Sea Lice* there was no mention of cage design or physical barriers to limit sea lice settlement on fish. Yet it is common to see snorkel cages and similar methods being widely employed on salmon farms today (see Chapter 22). Taking these advances further it is possible to speculate that a new generation of sea lice-proof fish farms will be developed. Indeed, prototype systems such as the Hauge Aqua system were being trialled in Norway in 2020. If successful, such systems may provide a new engineering solution to the sea lice problem. Of course, these systems may bring in their own problems, not least cost and scaling and the problems of infection if a pathogen can breach the biosecurity of these systems.

Sea lice traps have often been considered as possible methods of sea lice control (see Chapter 6). However, general plankton traps are considered too broad to be effective, typically pulling out beneficial and problematic zooplankton non-specifically.

One area of semiochemicals that has yet to be thoroughly explored for control is the group of lice semiochemicals that are considered to repel lice or mask host odours (see Chapter 6). These would function by making the lice avoid the fish on the farm either by masking them from the lice's senses or by making the fish repellent to the lice. Some of these compounds have been successfully incorporated in functional feeds (see Chapter 21). To date, however, an effective large-scale push chemical has yet to be identified. Interestingly it has been observed that sea lice will abandon dead salmon shortly after the host's death. Why this happens is unknown. The processes that occur when the host dies are complex. The abandonment of the fish by mobile lice may be due to the change in the tissue biomechanical properties or the charge on the fish's surface. Conversely, it may be combined with a change in the semiochemical environment on the dead fish. Compounds like cadaverine and putrescene form rapidly during a morbidity and mortality event and are strong push semiochemicals for many organisms (Heale and Vanderwolf, 1994; Hussain et al., 2013; Li et al., 2015, 2016).

25.9 Concluding remarks

The future of sea louse control remains unknown but it is probably fair to suggest that sea lice will remain a major issue in the short to medium term. One editor, IB, can recall one of his post-doc mentors, the late fish immunologist Tony Ellis, telling him that sea lice would not be a problem by 2000 as there would be a vaccine for them by then. Researchers have expended an enormous effort on these parasites. Yet we still have a sea lice issue. One thing is certain, there will continue to be some amazing insights into the biology of these parasites and the research presented in this volume outlines the very cutting-edge of parasite control. Even if someone is writing a new edition of this book in 30 years' time they can look back on a time of some outstanding parasite research and be proud of the community's efforts in helping to control these parasites. Many of the advances in sea lice control have been made by staff of aquaculture companies, husbandry staff, innovators, technologists and engineers. The great efforts of fish farm staff in treating salmon infected with sea lice, often under very difficult physical conditions, have to be acknowledged. Farm staff have been adaptive and flexible in taking on new control methods such as surface skirts, hydrolicers, thermolicers, snorkel cages and any new ideas. Communications and conferences involving scientific, farm and technology staff are therefore important in bringing to market scientific advances in sea lice control.

References

Aaen, S.M., Helgesen, K.O., Bakke, M.J., Kaur, K. and Horsberg, T.E. (2015) Drug resistance in sea lice: a threat to salmonid aquaculture. *Trends in Parasitology* 31, 72–81.

Agusti-Ridaura, C., Bakke, M.J., Helgesen, K.O., Sundaram, A.Y.M., Bakke, S.J., Kaur, K. and Horsberg, T.E. (2020) Candidate genes for monitoring hydrogen peroxide resistance in the salmon louse, *Lepeophtheirus salmonis*. *Parasites & Vectors* 13(1), 16.

AmbuAli, A., Monaghan, S.J., McLean, K., Inglis, N.F., Bekaert, M., Wehner, S. and Bron, J.E. (2020) Identification of proteins from the secretory/excretory products (SEPs) of the branchiuran ectoparasite *Argulus foliaceus* (Linnaeus, 1758) reveals unique secreted proteins amongst haematophagous ecdysozoa. *Parasites & Vectors* 13(1), 13.

Barker, S.E., Bricknell, I.R., Covello, J., Purcell, S., Fast, M.D., Wolters, W. and Bouchard, D.A. (2019) Sea lice, *Lepeophtheirus salmonis* (Krøyer 1837), infected Atlantic salmon (*Salmo salar* L.) are more susceptible to infectious salmon anemia virus. *PLoS ONE* 14 (1), e0209178.

Barrett, L.T., Oppedal, F., Robinson, N. and Dempster, T. (2020) Prevention not cure: a review of methods to avoid sea lice infestations in salmon aquaculture. *Reviews in Aquaculture* 12, 2527–2543.

Braden, L.M., Barker, D.E., Koop, B.F. and Jones, S.R.M. (2012) Comparative defense-associated responses in salmon skin elicited by the ectoparasite *Lepeophtheirus salmonis*. *Comparative Biochemistry and Physiology Part D Genomics Proteomics* 7(2), 100–109.

Braden, L.M., Barker, D.E., Koop, B.F. and Jones, S.R.M. (2015) Differential modulation of resistance biomarkers in skin of juvenile and mature pink salmon, *Oncorhynchus gorbuscha*, by the salmon louse, *Lepeophtheirus salmonis*. *Fish and Shellfish Immunology* 47(1), 7–14.

Bricknell, I.R., Hayes, H.E., Billingsley, P.F., Vigneau, A.J., Sommerville, C. and Raynard, R.S. (2003) Vaccines against the salmon lice (*Lepeophtheirus salmonis*) – a holy grail, or just around the corner? *Journal of Parasitology* 89, 149–157.

Bricknell, I.R., Dalesman, S.J., O'Shea, B., Pert, C.C. and Luntz, A.J.M. (2006) Effect of environmental salinity on sea lice *Lepeophtheirus salmonis* settlement success. *Diseases of Aquatic Organisms* 71(3), 201–212.

Bron, J.E., Sommerville, C., Jones, M. and Rae, G.H. (1991) The settlement and attachment of early stages of the salmon louse *Lepeophtheirus salmonis* (Copepoda: Caligidae) on the salmon host, *Salmo salar*. *Journal of Zoology* 224, 201–212.

Brooker, A.J., Davie, A., Leclercq, E., Zerafa, B. and Migaud, H. (2020) Pre-deployment acclimatisation of farmed Ballan wrasse (*Labrus bergylta*) to sea-cage conditions promotes behaviour analogous to wild conspecifics when used as cleaner fish in Atlantic salmon (*Salmo salar*) farms. *Aquaculture* 520, 734771.

Crawford, J.E., Clarke, D.W., Criswell, V., Desnoyer, M., Cornel, D., Deegan, B., White, B.J., et al. (2020) Efficient production of male *Wolbachia*-infected *Aedes aegypti* mosquitoes enables large-scale suppression of wild populations. *Nature Biotechnology* 38(4), 482–492.

Dalvin, S., Skaftnesmo, K.O., Malde, K., Glover, K.A., Leong, J.S., Koop, B.F. and Sandlund, N. (2020) Microbial communities associated with the parasitic copepod *Lepeophtheirus salmonis*. *Marine Genomics* 49, 100688.

Fast, M.D., Ross N.W., Muise, D.M. and Johnson, S.C. (2006) Differential gene expression in Atlantic salmon, *Salmo salar*, infected with *Lepeophtheirus salmonis* (Copepoda: Caligidae). *Journal of Aquatic Animal Health* 18, 116–127.

Fernandez-Leborans, G., Freeman, M., Gabilondo, R. and Sommerville, C. (2005) Marine protozoan epibionts on the copepod *Lepeophtheirus salmonis*, parasite of the Atlantic salmon. *Journal of Natural History* 39(8), 587–596.

Freeman, M.A., Bell, A.S. and Sommerville, C. (2003) A hyperparasitic microsporidian infecting the salmon louse, *Lepeophtheirus salmonis*: an rDNA-based molecular phylogenetic study. *Journal of Fish Diseases* 26(11–12), 667–676.

Freeman, M.A. and Sommerville, C. (2009) *Desmozoon lepeophtherii* n. gen., n. sp., (Microsporidia: Enterocytozoonidae) infecting the salmon louse *Lepeophtheirus salmonis* (Copepoda: Caligidae). *Parasites & Vectors* 2, 15.

Freeman, M.A. and Sommerville, C. (2011) Original observations of *Desmozoon lepeophtherii*, a microsporidian hyperparasite infecting the salmon louse *Lepeophtheirus salmonis*, and its subsequent detection by other researchers. *Parasites & Vectors* 4, 231.

Grayson, T.H., John, R.J., Wadsworth, S., Greaves, K., Cox, D., Roper, J. and Harris, J.E. (1995) Immunization of Atlantic salmon against the salmon louse: identification of antigens and effects on louse fecundity. *Journal of Fish Biology* 47, 85–94.

Hamilton, S., McLean, K., Monaghan, S.J., McNair, C., Inglis, N.F., McDonald, H., Adams, S., Richards, R., Roy, W., Smith, P., Bron, J., Nisbet, A.J. and Knox, D. (2018) Characterisation of proteins in excretory/secretory products collected from salmon lice, *Lepeophtheirus salmonis*. *Parasites & Vectors* 11(1), 294.

Haugland, G.T., Imsland, A.K.D., Reynolds, P. and Treasurer, J. (2020) Application of biological control: use of cleaner fish. In: Kibenge, F.S.B. and Powell, M.D. (eds). *Aquaculture Health Management.* Academic Press, London, UK. pp. 319–369.

Heale, V.R. and Vanderwolf, C H. (1994) Toluene and weasel (2-Propylthiethane) odors suppress feeding in the rat. *Journal of Chemical Ecology* 20(11), 2953–2958.

Hosack, G.R., Ickowicz, A. and Hayes, K.R. (2021) Quantifying the risk of vector-borne disease transmission attributable to genetically modified vectors. *Royal Society Open Science* 8(3), 201525.

Houston, R.D., Bean, T.P., Macqueen, D.J., Gundappa, M.K., Jin, Y.H., et al. (2020) Harnessing genomics to fast-track genetic improvement in aquaculture. *Nature Reviews Genetics* 21, 389–409.

Hussain, A., Saraiva, L.R., Ferrero, D.M., Ahuja, G., Krishna, V.S., Liberles, S.D. and Korsching, S.I. (2013) High-affinity olfactory receptor for the death-associated odor cadaverine. *Proceedings of the National Academy of Sciences of the United States of America* 110(48), 19579–19584.

Iboi, E.A., Gumel, A.B. and Taylor, J.E. (2020) Mathematical modeling of the impact of period release of sterile male mosquitoes and seasonality on the population abundance of malaria mosquitoes. *Journal of Biological Systems* 28(2), 277–310.

Imsland, A.K.D., Reynolds, P., Remen, M., Bloch-Hansen, K., Sagerup, K., Hemmingsen, W., Mathisen, R. and Myklebust, E.A. (2020a) The possible use of lumpfish against *Caligus elongatus*: a mini review. *Journal of Oceans University of China* 19, 1133–1139.

Imsland, A.K.D., Reynolds, P., Lorentzen, M., Eilertsen, R.A., Micallef, G. and Tvenning, R. (2020b) Improving survival and health of lumpfish (*Cyclopterus lumpus* L.) by the use of feed blocks and operational welfare indicators (OWIs) in commercial Atlantic salmon cages. *Aquaculture* 527, 1–9.

Kaiser, M.L., Wood, O.R., Damiens, D., Brooke, B.D., Koekemoer, L.L. and Munhenga, G. (2021) Estimates of the population size and dispersal range of *Anopheles arabiensis* in Northern KwaZulu-Natal, South Africa: implications for a planned pilot programme to release sterile male mosquitoes. *Parasites & Vectors* 14(1), 205.

Kjetsa, M.H., Odegard, J. and Meuwissen, T.H.E. (2020) Accuracy of genomic prediction of host resistance to salmon lice in Atlantic salmon *(Salmo salar)* using imputed high-density genotypes. *Aquaculture* 526, 735415.

Lewis, D.L., Barker, D.E. and McKinley, R.S. (2014) Modulation of cellular innate immunity by *Lepeophtheirus salmonis* secretory products. *Fish and Shellfish Immunology* 38(1), 175–183.

Li, H. and Zhang, Z.D. (2016) Systems understanding of plant-pathogen interactions through genome-wide protein-protein interaction networks. *Frontiers of Agricultural Science and Engineering* 3(2), 102–112.

Li, Q., Tachie-Baffour, Y., Liu, Z.K., Baldwin, M.W., Kruse, A.C. and Liberles, S.D. (2015) Non-classical amine recognition evolved in a large clade of olfactory receptors. *eLlife* 4, e10441.

Li, Q., Tachie-Baffour, Y., Liu, Z.K., Baldwin, M.W., Kruse, A.C. and Liberles, S.D. (2016) Non-classical amine recognition evolved in a large clade of olfactory receptors. *Chemical Senses* 41(7), E31–E32.

Ma, J., Bruce, T.J., Jones, E.M. and Cain, K.D. (2019) A review of fish vaccine development strategies: conventional methods and modern biotechnological approaches. *Microorganisms* 7(11), 569.

MacKinnon, B.M. (1993) Host response of Atlantic salmon (*Salmo salar*) to infection by sea lice (*Caligus elongatus*). *Canadian Journal of Fisheries and Aquatic Sciences* 50(4), 789–792.

McVeigh, K. (2021) Banned pesticide blamed for killing bees may be approved for fish farms, a series drawing attention to the dramatic changes taking place in our oceans, and the innovations under way to tackle them. Available at: https://www.theguardian.com/environment/2021/may/27/novichok-for-insects-may-be-approved-for-scottish-fish-farms (accessed 16 June 2021).

Miccoli, A., Manni, M., Picchietti, S. and Scapigliati, G. (2021) State-of-the-art vaccine research for aquaculture use: the case of three economically relevant fish species. *Vaccines* 9(2), 140.

Mordue (Luntz), A.J. and Pike, A.W. (2002) Special Issue: Integrated pest management in salmon farming. *Pest Management Science* 58, 513–635.

NASCO (2020) The state of wild north Atlantic salmon. Available at: https://nasco.int/atlantic-salmon/state-of-salmon/ (accessed 5 November 2021).

Núñez-Acuña, G. and Gallardo-Escárate, C. (2019) Characterization of the salmon louse Lepeophtheirus salmonis miRNome: sex-biased differences related to the coding and non-coding RNA interplay. *Marine Genomics* 45, 38–47.

Økland, A.L., Nylund, A., Øvergård, A.C., Blindheim, S., Watanabe, K., Grotmol, S., Arnesen, C.E. and Plarre, H. (2014) Genomic characterization and phylogenetic position of two new species in Rhabdoviridae infecting the parasitic copepod, salmon louse (*Lepeophtheirus salmonis*). *PLoS ONE* 9(11), e112517.

Øvergård, A.C., Hamre, L.A., Grotmol, S. and Nilsen, F. (2018) Salmon louse rhabdoviruses: impact on louse development and transcription of selected Atlantic salmon immune genes. *Developmental and Comparative Immunology* 86, 86–95.

Pert, C.C., Luntz, A.J.M., Fryer, R.J., O'Shea, B. and Bricknell, I.R. (2009) The settlement and survival of the salmon louse, *Lepeophtheirus salmonis* (Krøyer, 1837), on atypical hosts. *Aquaculture* 288(3–4), 321–324.

Pert, C.C., Mordue, A.J., O'Shea, B. and Bricknell, I.R. (2012) The settlement and reproductive success of *Lepeophtheirus salmonis* (Krøyer, 1837; Copepoda: Caligidae) on atypical hosts. *Aquaculture Research* 43(6), 799–805.

Piesz, J., Barker, S. and Bricknell, I. (2016) Effects of the secretory/excretory products of *Lepeophtheirus salmonis* on Atlantic salmon leukocyte migration. *Fish and Shellfish Immunology* 53, 84.

Piesz, J., Barker, S. and Bricknell, I.R. (2020) Anti-chemotactic activity in the secretory/excretory products of *Lepeophtheirus salmonis*. *Fish and Shellfish Immunology* 98, 296–300.

Rae, G.H. (2002) Sea louse control in Scotland, past and present. *Pest Management Science* 58, 515–520.

Raynard, R.S., Bricknell, I.R., Billingsley, P.F., Nisbet, A.J., Vigneau, A. and Sommerville, C. (2002) Development of vaccines against sea lice. *Pest Management Science* 58, 569–575.

Roper, J., Grayson, T.H., Jenkins, P.G., Hone, J.V., Wrathmell, A.B., Russell, P.M. and Harris, J.E. (1995) The immunocytochemical localization of potential candidate vaccine antigens from the salmon louse *Lepeophtheirus salmonis* (Krøyer 1837). *Aquaculture* 132(3–4), 221–232.

Roth, M., Richards, R. and Sommerville, C. (1993) Current practices in the chemotherapeutic control of sea lice infestations in aquaculture: a review. *Journal of Fish Diseases* 16, 1–26.

Ruane, N., McCarthy, T.K. and Reilly, P. (1995) Antibody response to crustacean ectoparasites in rainbow trout, *Oncorhynchus mykiss* (Walbaum), immunized with *Argulus foliaceus* L antigen extract. *Journal of Fish Diseases* 18(6), 529–537.

Sadasivam, G.S., Madhesu, S., Mumthas, O.Y. and Dharani, K. (2018) Crop disease protection using parallel machine learning approaches. In: Dey, N., Ashour, A.S. and Borra, S. (eds) *Classification in Bioapps: Automation of Decision Making*. Volume 26. Springer, New York. pp. 227–259.

Sommerville, C. (2009) *Controlling Parasitic Diseases in Aquaculture: New Developments*. Woodhead Publications Ltd, Cambridge, UK.

Sommerville, C. (2012) Advances in non-chemical methods for parasite prevention and control in fish. In: B. Austin (ed.), *Infectious Disease in Aquaculture: Prevention and Control.* Woodhead Publications Ltd, Cambridge, UK. pp. 480–512.

Torrealba, D.A., Toledo, X.E. and Gallardo, J.A. (2011) Artificial settlement of sea lice, *Caligus rogercresseyi* (Boxshall & Bravo 2000) (Copepoda: Caligidae) on tissues of fish used as a substrate. *Crustaceana* 84(8), 939–948.

Tsairidou, S., Hamilton, A., Robledo, D., Bron, J.E. and Houston, R.D. (2020) Optimizing low-cost genotyping and imputation strategies for genomic selection in Atlantic salmon. *G3 Genes, Genomes, Genetics* 10, 581–590.

Umasuthan, N., Xue, X., Caballero-Solares, A., Kumar, S., Westcott, J.D., Chen, Z.Y. and Rise, M.L. (2020) Transcriptomic profiling in fins of Atlantic salmon parasitized with sea lice: evidence for an early imbalance between chalimus-induced immunomodulation and the host's defense response. *International Journal of Molecular Sciences* 21(7), 45.

Index

The letter t following a page number denotes a table; the letter f following a page number denotes a figure.